Hybrid Polymer Composite Materials

Related titles

Environmentally-friendly Polymer Nanocomposites (ISBN 978-0-85709-777-4)

Ceramic Nanocomposites (ISBN 978-0-85709-338-7)

Polymer-carbon Nanotube Composites (ISBN 978-1-84569-761-7)

Woodhead Publishing Series in Composites Science and Engineering

Hybrid Polymer Composite Materials

Properties and Characterisation

Edited by

Vijay Kumar Thakur

Manju Kumari Thakur

Asokan Pappu

ELSEVIER

WP WOODHEAD PUBLISHING

An imprint of Elsevier

Woodhead Publishing is an imprint of Elsevier
The Officers' Mess Business Centre, Royston Road, Duxford, CB22 4QH, United Kingdom
50 Hampshire Street, 5th Floor, Cambridge, MA 02139, United States
The Boulevard, Langford Lane, Kidlington, OX5 1GB, United Kingdom

Notices
Knowledge and best practice in this field are constantly changing. As new research and experience broaden our understanding, changes in research methods, professional practices, or medical treatment may become necessary.

Practitioners and researchers must always rely on their own experience and knowledge in evaluating and using any information, methods, compounds, or experiments described herein. In using such information or methods they should be mindful of their own safety and the safety of others, including parties for whom they have a professional responsibility.

To the fullest extent of the law, neither the Publisher nor the authors, contributors, or editors, assume any liability for any injury and/or damage to persons or property as a matter of products liability, negligence or otherwise, or from any use or operation of any methods, products, instructions, or ideas contained in the material herein.

British Library Cataloguing-in-Publication Data
A catalogue record for this book is available from the British Library

Library of Congress Cataloging-in-Publication Data
A catalog record for this book is available from the Library of Congress

ISBN: 978-0-08-100787-7 (print)
ISBN: 978-0-08-100788-4 (online)

For information on all Woodhead Publishing publications
visit our website at https://www.elsevier.com/books-and-journals

Publisher: Matthew Deans
Acquisition Editor: Gwen Jones
Editorial Project Manager: Charlotte Rowley
Production Project Manager: Debasish Ghosh
Cover Designer: Greg Harris

Typeset by MPS Limited, Chennai, India

Contents

List of Contributors

H.P.S. Abdul Khalil Schools of Industrial Technology, Universiti Sains Malaysia, Penang, Malaysia; Cluster for Polymer Composites, Science and Engineering Research Center, University Sains Malaysia, Penang, Malaysia

Sandro Campos Amico PPGEM, Federal University of Rio Grande do Sul (UFRGS), Porto Alegre, Brazil

Alireza Ashori Department of Chemical Technologies, Iranian Research Organization for Science and Technology (IROST), Tehran, Iran

Vajiheh Behranvand Organic Polymer Chemistry Research Laboratory, Department of Chemistry, Isfahan University of Technology, Islamic Republic of Iran

Jarosław Bieniaś Department of Materials Engineering, Faculty of Mechanical Engineering, Lublin University of Technology, Lublin, Poland

Grzegorz Bubak Robotics, Brain and Cognitive Sciences Department, Italian Institute of Technology, Genoa, Italy; Italian Institute of Technology, Graphene Labs, Genoa, Italy

Ana M. Díez-Pascual Analytical Chemistry, Physical Chemistry and Chemical Engineering Department, Faculty of Biology, Environmental Sciences and Chemistry, Alcalá University, Madrid, Spain

David Gendron Robotics, Brain and Cognitive Sciences Department, Italian Institute of Technology, Genoa, Italy; Italian Institute of Technology, Graphene Labs, Genoa, Italy

Patryk Jakubczak Department of Materials Engineering, Faculty of Mechanical Engineering, Lublin University of Technology, Lublin, Poland

Mohammad Jawaid Laboratory of Biocomposite Technology, Institute of Tropical Forestry and Forest Products (INTROP), Universiti Putra Malaysia, Serdang, Selangor, Malaysia

Ayesha Kausar Nanoscience and Technology Department, National Center For Physics, Quaid-i-Azam University, Islamabad, Pakistan

Elham Khadem Organic Polymer Chemistry Research Laboratory, Department of Chemistry, Isfahan University of Technology, Islamic Republic of Iran

Shadpour Mallakpour Organic Polymer Chemistry Research Laboratory, Department of Chemistry, Isfahan University of Technology, Islamic Republic of Iran; Nanotechnology and Advanced Materials Institute, Isfahan University of Technology, Islamic Republic of Iran; Center of Excellence in Sensors and Green Chemistry, Department of Chemistry, Isfahan University of Technology, Islamic Republic of Iran

Jose Roberto Moraes d'Almeida Materials Engineering Department, Pontifical Catholic University of Rio de Janeiro (PUC), Rio de Janeiro, Brazil

Hiroshi Mutsuyoshi Department of Civil & Environmental Engineering, Saitama University, Saitama, Japan

Hai Nguyen College of Information Technology & Engineering, Marshall University, Huntington, WV, United States

M.R. Nurul Fazita Schools of Industrial Technology, Universiti Sains Malaysia, Penang, Malaysia

Wagner Mauricio Pachekoski Federal University of Santa Catarina (UFSC), Joinville, Brazil

Sergio Henrique Pezzin Center of Technological Sciences, Santa Catarina State University (UDESC), Joinville, Brazil

Masoud Rais-Rohani University of Maine, Orono, ME,United States

Shima Rashidimoghadam Organic Polymer Chemistry Research Laboratory, Department of Chemistry, Isfahan University of Technology, Islamic Republic of Iran

E. Rosamah Faculty of Forestry, Mulawarman University, East Kalimantan, Indonesia

Mohammad Rouhi Concordia University, Montreal, QC, Canada

Naheed Saba Laboratory of Biocomposite Technology, Institute of Tropical Forestry and Forest Products (INTROP), Universiti Putra Malaysia, Serdang, Selangor, Malaysia

Fabrizio Sarasini University of Rome "La Sapienza" Department of Chemical Engineering Materials Environment, Rome, Italy

N.A. Sri Aprilia Department of Chemical Engineering, Syiah Kuala University, Banda Aceh, Indonesia

Barbara Surowska Department of Materials Engineering, Faculty of Mechanical Engineering, Lublin University of Technology, Lublin, Poland

Tham Mun Wai Schools of Industrial Technology, Universiti Sains Malaysia, Penang, Malaysia

Wael Zatar College of Information Technology & Engineering, Marshall University, Huntington, WV, United States

Functional materials from polymer derivatives: properties and characterization

1

Grzegorz Bubak[1,2] *and David Gendron*[1,2]
[1]Robotics, Brain and Cognitive Sciences Department, Italian Institute of Technology, Genoa, Italy, [2]Italian Institute of Technology, Graphene Labs, Genoa, Italy

Chapter Outline

1.1 Introduction

Nowadays, the presence of functional polymeric materials can be found in a broad range of application. They offer unique and attractive properties that can be tuned to fit our everyday technological needs. This chapter aims to summarize and discuss the physicochemical properties of a broad range of functional polymeric materials. The first section of this chapter introduces electro-photoactive polymer material for optoelectronic. The optoelectronic field has grown steadily in the past decade and can therefore be declined in many subfields such as materials for organic photovoltaics, photodetectors, light-emitting diodes and transistors. In this section, we will discuss the optical and electronic properties of novel functional materials (polymer composites) as well as their performances in polymer solar cell.

Hybrid Polymer Composite Materials: Properties and Characterisation. DOI: http://dx.doi.org/10.1016/B978-0-08-100787-7.00001-9

The second section of the chapter will be centered on hybrid polymer composites for electroactive devices. More precisely, we will glance at functional materials for storing or transducing energy as well as their corresponding devices such as supercapacitors. At last, polymeric materials used in electroactive actuators will be described and the main characterization techniques will be reviewed.

1.2 Section 1. Electro-photoactive polymer materials for optoelectronics

Electroactive polymers have attracted great attention in the past decade as active materials in optoelectronic devices such as organic photovoltaic cell (OPV), organic field-effect transistor (OFET), light-emitting diodes, and photodetector. In this section, we will discuss the optical, structural, and electronical properties of polymer composites used as active material in these devices.

Before reviewing the main classes of polymer composites, we will briefly introduce the concept of polymer solar cell and discuss the main characterization techniques commonly used to obtain information about the active layer (also referred to the *conjugated polymer*—fullerene derivative composite) such as optical, structural, and electronic properties.

1.2.1 History and basics parameters of an organic solar cell

The first organic solar cell was reported by Tang in 1986 (Tang, 1986). At the time, a power conversion efficiency (PCE) of about 1% as well as a fill factor of 65% was reached. However, it is important to note that the architecture of Tang's solar cell was different from the one usually used today. Indeed, Tang constructed a bilayer device using a *p*-type copper phthalocyanine and a *n*-type bisbenzimidazo (2,1-a:2′,1′-a′)anthra(2,1,9-def:6,5,10-d′e′f′)diisoquinoline-10,21-dione. Both were vacuum-evaporated in order to construct the device. Then the concept of bulk heterojunction (BHJ) was introduced to address the exciton diffusion length problem (the exciton is defined as a Coulombically bonded hole−electron pair) (Yu et al., 1995). Indeed, contrary to the bilayer approach, in the BHJ, the donor (conjugated polymer) and the acceptor (fullerene derivative) are mixed together thus forming an interpenetrating network (which can also be seen as a polymer composite) with large interfacial areas to enhance the exciton dissociation. This polymer−fullerene mixture is also named "active layer." We will see later in this chapter that several routes can be used to optimize the morphology of the active layer, therefore improving the exciton diffusion and enhancing the solar cell device performance.

Before entering into the details concerning the chosen materials, let us first discuss how to calculate the PCE of a solar cell. Plainly, the PCE can be defined as the product of the open circuit potential (V_{OC}), the short-circuit current density (J_{SC}) and the fill factor (FF) divided by the light intensity incident on the device (P_{IN}) (Li et al., 2010). In short, the V_{OC} is related to the energy levels of the donor and acceptor and to the nanomorphology of the active layer. The J_{SC} depends on the charge mobility in the polymer and on the number of absorbed photons. At last, the FF is determined by the number of charge reaching the electrodes when the electric field is decreased towards the V_{OC}.

It is generally accepted that the BHJ solar cell mechanism can be described according to the following steps: (1) light absorption and generation of highly localized Coulombically bonded pair hole–electron (exciton), (2) exciton diffusion to the donor–acceptor interface, (3) exciton dissociation at the interface creating a charge-transfer state which then dissociate into free charge carriers (holes and electrons), and (4) charge transport and collection (Cheng et al., 2009; Günes et al., 2007).

Few years later, conjugated polymers were introduced and used as electron donor together with a fullerene derivative ($PC_{61}BM$) as electron acceptor. In the second generation of organic solar cell, conjugated polymers were used as the electron donor. The following list represents the main classes: poly(phenylenevinylene) PPV derivatives (for example: poly(2-methoxy-5-(3,7-dimethyloctyloxy)-1,4-phenylene, also named MDMO-PPV), poly(thiophene) derivatives (for example: poly(3-hexylthiophene, also named P3HT), and poly(fluorene) derivative (for example: poly(2,7-(9-(2-ethylhexyl)-9-hexyl-fluorene-co-5,5-(4′,7′-dithienyl-2′,1′,3′-benzothiadiazole) PFDTBT).

As for the third generation of organic solar cell, recent advances in the polymer design have brought forward the concept of donor–acceptor (D-A) copolymer (also named push–pull copolymer) which led to very efficient devices. It is important to avoid confusing this donor–acceptor (D-A) copolymer concept with the electron donor (conjugated polymer) and the electron acceptor (fullerene derivative). Unlike P3HT, which is a homopolymer, donor–acceptor copolymers possess low bandgaps and their energy levels (HOMO, LUMO) as well as their molecular structures can be optimized. More specifically, D-A copolymers incorporate an electron-rich moiety (donor) and an electron-deficient moiety (acceptor). This has the effect of increasing the double bond character (quinodal effect) of the single bonds in the polymer backbone which results in a reduction of the bond length alternation and thus, a reduction of the band gap of the copolymer (Dennler et al., 2009; Zhang and Tour, 1998). Moreover, an interesting feature of the D-A copolymer is that their HOMO and LUMO energy levels are greatly determined by the HOMO energy level of the donor moiety and the LUMO energy level of the acceptor moiety (Bredas et al., 2009). Therefore, it is possible to tune the energy levels of the copolymer by choosing appropriately the donor and acceptor moieties or/and by engineering of their chemical structures. Fig. 1.1 shows a series of donor and acceptor fragment units commonly used for efficient polymer solar cells.

Concerning the acceptor component (fullerene derivative), the most commonly used derivatives are the $PC_{61}BM$ ((1-(3-methoxycarbonyl)propyl-1-phenyl[6,6] C_{61})) and the $PC_{71}BM$ ((1-(3-methoxycarbonyl)propyl-1-phenyl[6,6]C_{71}) (Fig. 1.2). PCBM has been chosen as electron acceptor as it possesses the right energy levels (LUMO located at −4.3 eV, HOMO located at −6.0 eV) compared to conjugated polymer and allows an efficient exciton dissociation to free charge carriers.

As the aim of this subchapter is to discuss the different characterization techniques available to study the polymer–fullerene composite we will not go into the details of the preparation of such polymeric materials. If the reader wishes to explore in depth the field, we refer him to the following reviews (Hoppe and Sariciftci, 2004; Lu et al., 2015; Huang et al., 2014; Thompson and Fréchet, 2008).

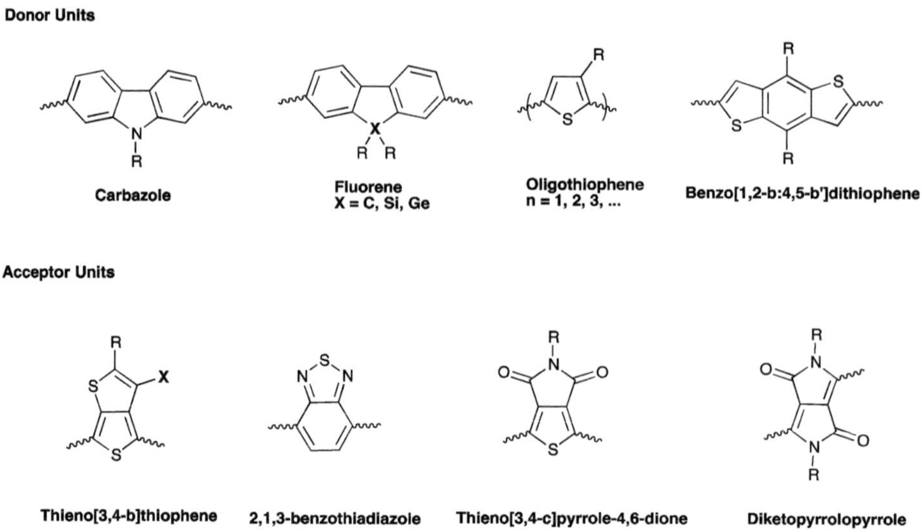

Figure 1.1 Donor units (top) and acceptor units (bottom) of D-A copolymers.

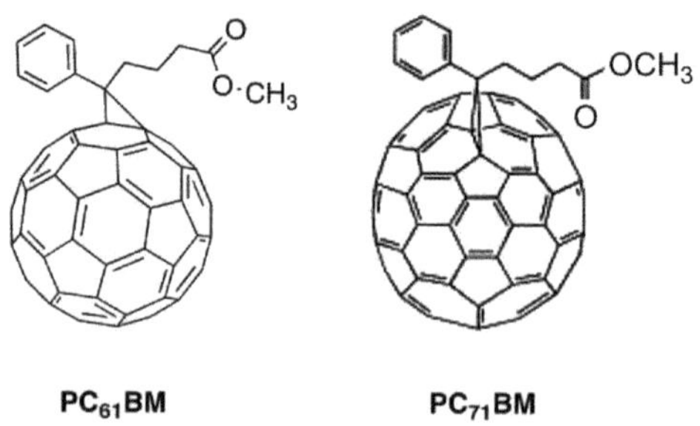

Figure 1.2 PCBM derivatives.

1.2.2 Architecture of a polymer solar cell device

A typical polymer solar cell consists of a glass substrate coated with a layer of indium tin oxide (ITO) acting as the anode (Fig. 1.3). The ITO is then covered by a hole transport layer thin film such as poly(3,4-ethylene-dioxythiophene) polystyrene sulfonate (PEDOT:PSS). Then the active layer is deposited by wet processing (spin-coating, doctor blading, screen-printing, ink-jet printing) or by evaporation (usually for the case of small molecule). At last the cathode (Al and/or Ca) is evaporated and selected to match the energy levels of the acceptor (fullerene derivative).

Let us summarize again the operating mechanism: absorption of a photon by the active layer and formation of an exciton (hole–electron bonded by Coulombian force), separation of the exciton in electron (electron polaron) and hole (hole polaron), and transport of the charges through the acceptor and donor phases of the active layer to the cathode and anode. In this regard, the bicontinuous interpenetrating network morphology need to be optimize in order to enhance the solar cell performances. Thus, the

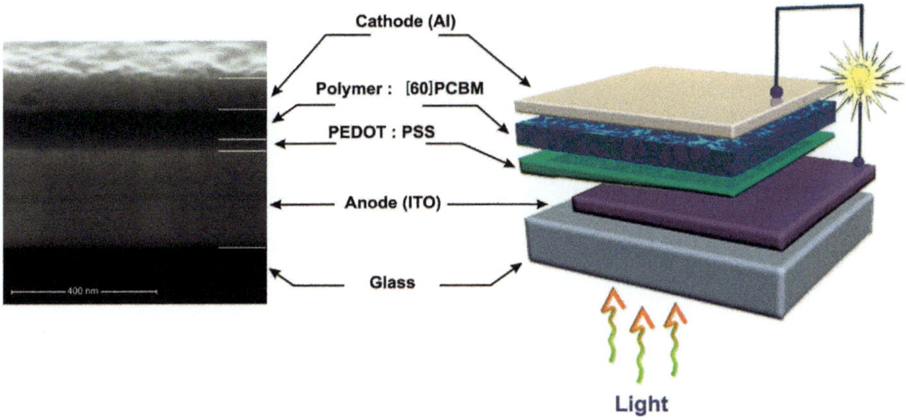

Figure 1.3 Architecture of a polymer solar cell.
Source: Reprinted from: Li, C., Liu, M., Pschirer, N.G., Baumgarten, M., Müllen, K., 2010. Chem. Rev. 110, 6817–6855.

film morphology (active layer) must be probed by several characterization techniques. In the following part, we will discuss the main characterization techniques used to analyze the active layer (conjugated polymer—fullerene derivative) composite and what kind of information we can extract from such experiments.

The grazing incidence wide-angle X-ray scattering (GIWAXS) is a scattering technique used to investigate the film morphology as well as the nanostructure of thin films. GIWAXS is particularly relevant for structural characterization of the BHJ layer (composite) due to its large sampling volume and statistical information obtained (Huang et al., 2013). GIWAXS is often used on organic thin films in order to determine both the crystalline lattice spacing from the diffraction peaks and the crystalline correlation length from the peak widths and to further determine the orientation order parameters. It is possible to perform in situ measurements of the parameters just mentioned as a function of the film-drying process and annealing. A lot of studies have been reported GIWAXS to investigate the effects of thermal annealing as well as solvent additives on the crystallization of the donor component (PCBM) of the BHJ blend (see Fig. 1.4). However, one must be careful when using GIWAXS to describe the orientation distribution and degree of crystallinity as it can be easily misinterpreted.

Similarly, as GIWAXS, GISAXS is a useful tool for studying thin film nanostructure because of the surface footprint and increased scattering volume. GISAXS can provide information about the size, shape and the interdomain correlation of the BHJ layer (Rivnay et al., 2012). That is why GISAXS is used with electron microscopy in order to investigate large-scale domain size. For instance, it is possible to observe the average lateral size of the nanoscale separation of the BHJ composite.

Transmission electron microscopy (TEM) is another useful technique based on the contrast provided by the scattering of the mass of the scattered atoms (DeLongchamp et al., 2012). For example, thicker domains in the film or regions containing heavier atoms will appear darker, while thinner domain will appear

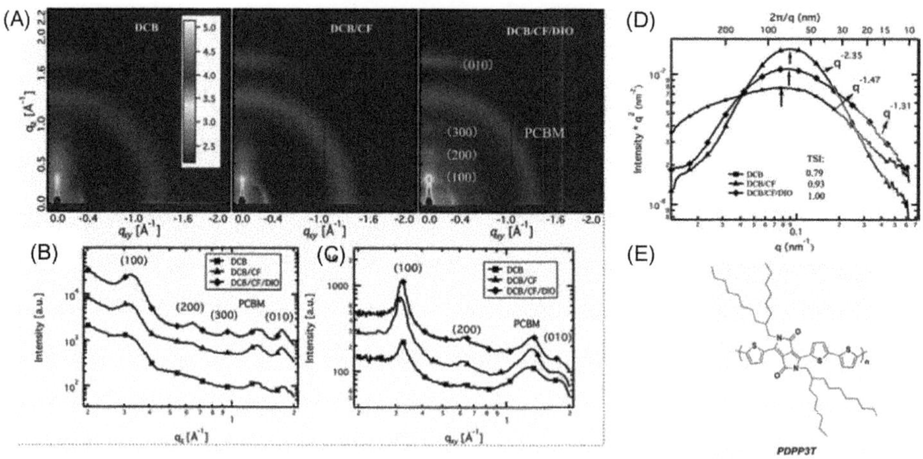

Figure 1.4 2D GIWAXS diffraction pattern (A) out-of-plane, (B) in-plane, (C) GIWAXS data of PDPP3T:PC71BM blend films prepared with 1,2-dichlorobenzene, 1,2-dichlorobenzene/chloroform (4:1, v/v), and 1,2-dichlorobenzene/chloroform/1,8-diiodooctane (76:19:5, v/v/v). (D) R-SoXS information on PDPP3T:$PC_{71}BM$ blend films prepared at various conditions: 1,2-dichlorobenzene, 1,2-dichlorobenzene/chloroform (4:1, v/v), 1,2-dichlorobenzene/chloroform/1,8-diiodooctane (76:19:5, v/v/v). (F) Chemical structure of PDPP3T.

Source: From Ye, L., Zhang, S.Q., Ma, W., Fan, B.H., Guo, X., Huang, Y., Ade, H., Hou, J. H., 2012. Adv. Mater. 24, 6335–6341.

brighter. Therefore, the contrast arises from density, mass, and thickness variation. TEM gives information about the BHJ composite crystalline behavior since different diffraction contrast can be observed. Scanning electron microscopy (SEM) provides information about the morphology of the active layer (composite material). For example, in the case of the active layer of an organic solar cell which is composed of a conjugated polymer and a C_{60} derivative, information about the blend or the presence of domains can be obtained. Cross-sectional images reveal dark and bright region of the polymer–PCBM composite, thus providing information about the donor and acceptor distribution. Fig. 1.5 shows typical TEM of a conjugated polymer (Si-PDTBT):$PC_{71}BM$ composite.

Atomic force microscopy (AFM) can be used to probe the surface topography thus revealing details about the nanoscale domain structure such as roughness on the surface of the film by height and phase images of the active layer (Huang et al., 2012a). AFM is a useful tool to acquire surface topography features and is particularly helpful when use in conjunction with other characterization techniques. A variation of AFM, the photoconductive atomic force microscopy (pcAFM), is another technique to investigate the role of the nanomorphology of the polymer–PCBM composite in terms of charge transport and device efficiency. It is able to provide information on local photocurrents and on individual microstructure (Coffey et al., 2007). Another interesting mode is the conductive atomic force

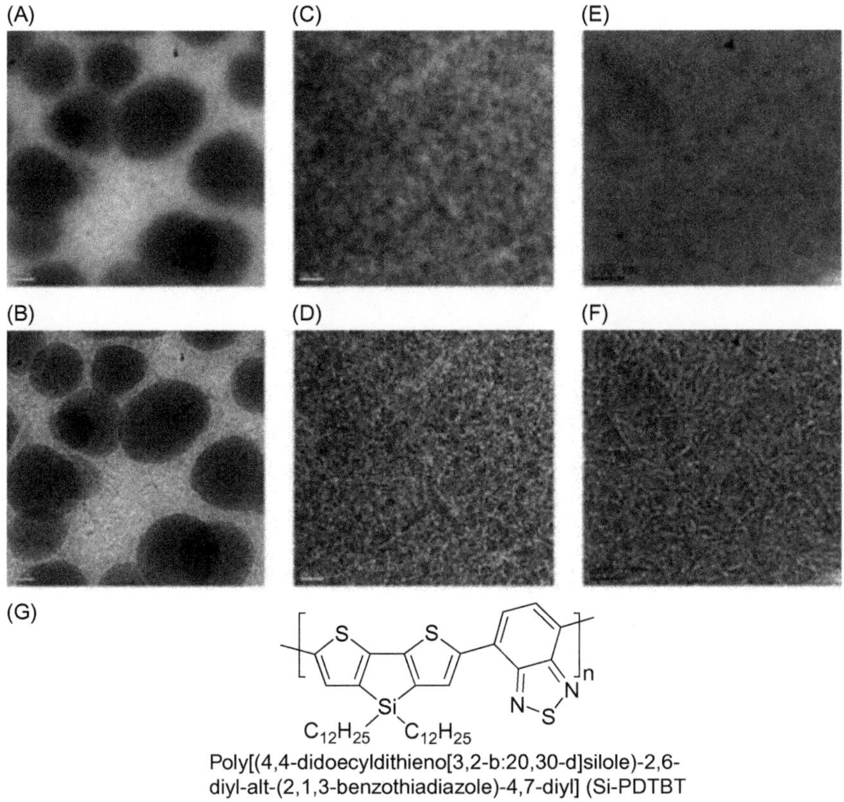

Figure 1.5 Top-down TEM images of Si-PDTBT:$PC_{71}BM$ composite: (A) in-focus without CN additive; (B) −25 μm defocus without CN additive; (C) in-focus with 1% CN additive; (D) −25 μm defocus with 1% CN additive; (E) in-focus with 4% CN additive; (F) −25 μm defocus with 4% CN additive. All scale bars represent 100 nm. (G) Chemical structure of Si-PDTBT.
Source: From Moon, J.S., Takacs, C.J., Cho, S., Coffin, R.C., Kim, H., Bazan, G.C., Heeger, A.J., 2010. Nano Lett., 10, 4005−4008.

microscopy (cAFM) that measures local conductivity and surface topography with nanometer range resolution (Pingree et al., 2009). The current flow can be monitored between the tip and the sample and thus a map of the local current density can be constructed.

UV-visible spectroscopy is one of the first characterization performed to record information about the optical properties of a material. Information that can be obtained is the absorption range, the band gap energy (particularly important for solar cell application) charge-transfer process, purity, and structural information to name only a few. Fig. 1.6 shows a typical UV-visible spectra obtained for conjugated polymer composite. Technique such as measuring the photoluminescence of chosen composite will bring complementary information about the material.

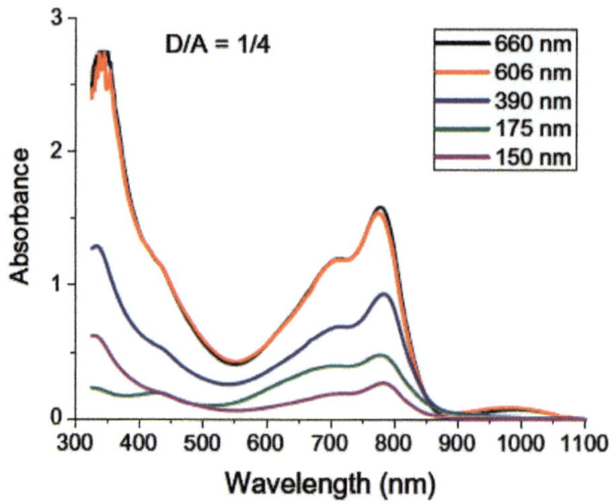

Figure 1.6 Typical UV-Vis absorption spectra of a polymer:PCBM mixture at different film thicknesses.
Source: From Murphy, L., Hong, W., Aziz, H., Li, Y., 2013. Org. Electron., 14, 3484–3492.

1.2.3 Morphology of the polymer–PCBM composite (active layer) performance relationship

Once a suitable donor (conjugated polymer) and acceptor (fullerene derivative) have been found, the following step is to optimize the morphology of the BHJ layer to maximize the exciton diffusion and charge transport to the relevant electrodes. To do so, a particular care must be put toward the morphology of the active layer such as the domain areas and size, interpenetration of the polymer phase in the fullerene phase. In general, the ideal domain size is found to be around 10–20 nm to allow an efficient exciton diffusion (Yu et al., 1995). In the upcoming section, we will discuss several elements that are useful to optimize the morphology of the active layer as well as few characterization techniques commonly used.

The first element that has a direct effect on the active layer morphology is the use of solvent additive also called processing additives (Lou et al., 2011). They are usually added to the composite (polymer-fullerene blend) during the spin-casting process to enhance the device performance. Structural studies by GIWAXS, TEM, and AFM have been carried out to investigate the morphology before and after the addition of these solvent additives. By varying the amount and the nature of the additives, it was found to significantly increase the overall efficiency of the device. One typical example of this, is the blend PCPDTBT (2,6-(4,4-bis(2-ethylhexyl)-4*H*-cyclopenta(2,1-b;3,4-b′)dithiophene-*alt*-4,7)-(2,1,3-benzothiadiazole) with $PC_{71}BM$) in which octanedithiol (ODT) was added in small amount (Fig. 1.7) (Peet et al., 2007). The PCE was increased from 2.2% to 5.5% after addition. As a result, the thin film morphology of the polymer composite was greatly improved and thus led to increased PCE. Based on analyses done by TEM and AFM, it was proposed that

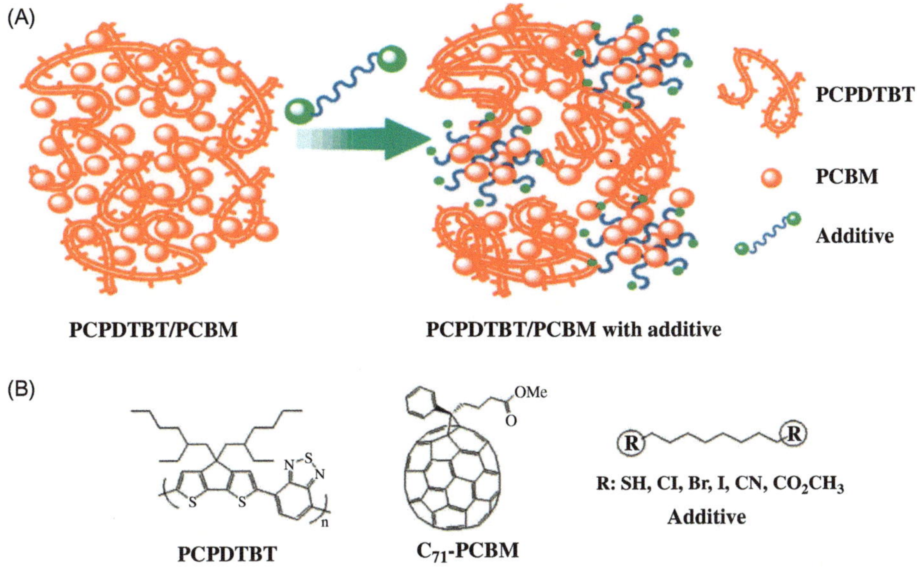

Figure 1.7 (A) Schematic of the role of the solvent additive in the self assembly of the polymer–PCBM blend composite. (B) Chemical structures of PCPDTBT, $PC_{71}BM$, and additives.
Source: From Lee, J.K., Ma, W.L., Brabec, C.J., Yuen, J., Moon, J.S., Kim, J., Lee, K., Bazan, G.C., Heeger, A.J., 2008. J. Am. Chem. Soc., 130, 3619–3623).

the octanedithiol (ODT) selectively dissolve the $PC_{71}BM$ (6,6-phenyl-C_{70}-butyric acid methyl ester) as PCPDTBT is not soluble in ODT. More precisely, $PC_{71}BM$ tends to stay dissolved in ODT longer than the polymer as octanedithiol possesses a lower vapor pressure than the other solvent used to solubilize the polymer (chlorobenzene). Thus the evaporation process affects the crystallinity of the PCPDTBT, creating phases separation which results in variation of the final composite morphology.

1.2.4 Thermal annealing and postannealing

Thermal annealing is another method to control the copolymer composite morphology and to improve the PCE of the device. A good polymer composite model to explain this method is the couple P3HT:$PC_{61}BM$. By thermal annealing of the active layer, the crystallinity of the P3HT can be enhanced. More precisely, the PCBM molecules diffuse out from the P3HT polymer chains to form larger fullerene cluster. This phenonema allows the P3HT to recrystallize into large fibrillar structures which are distributed in a matrix of PCBM clusters and amorphous P3HT (Li et al., 2005). The impact of the annealing process was investigated by TEM and cAFM in order to obtain details about the nanoscale spatial variation and the photocurrent across the matrix. Basically, they found that the total current and the spatial

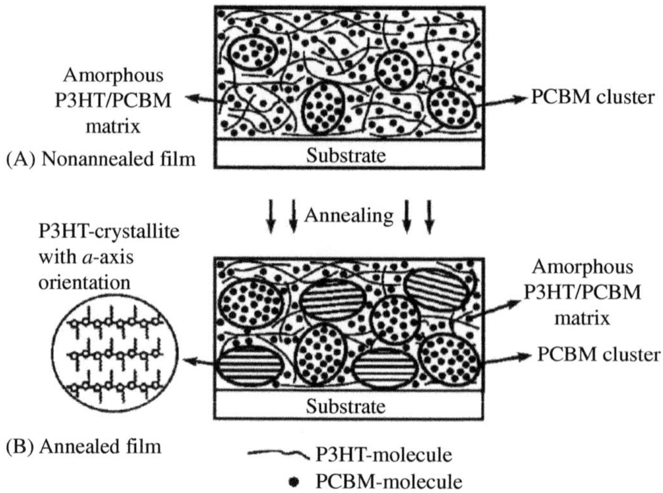

Figure 1.8 Structural changes in the blend P3HT:$PC_{61}BM$ thin films before and after thermal annealing.
Source: From Erb, T., Zhokhavets, U., Gobsch, G., Raleva, S., Stühn, B., Schilinsky, P., Waldauf, C., Brabec, C.J., 2005. Adv. Funct. Mater. 11, 1193–1196.

variation increase with the annealing process. Fig. 1.8 shows the structural changes in the blend P3HT:$PC_{61}BM$ thin films before and after thermal annealing.

The annealing step can be performed before and after the electrode evaporation (preannealing and postannealing, respectively). In the case of postannealing, several characterization techniques can be used. For instance, NEXAFS spectroscopy can determine the surface orientation of the polymer chain in the P3HT:$PC_{61}BM$ composite as well as the PCBM surface concentration near the composite–electrode interface. The post-thermal treatment induces PCBM segregation near the active layer–electrode interface and therefore increases the contribution of the face-on orientation of the P3HT polymer chains.

Solvent annealing affects the evaporation rate during the preparation of the active layer. Solvent annealing is an effective way to control the phase separation and the crystallization of the blend (polymer–PCBM) and thus modify the morphology of the active layer. Generally, solvent annealing studies are performed in a container with a chosen solvent. The type of solvent (polar or nonpolar), the drying time, the phase evolution as well as the relative crystallinity of the polymer are some parameters that are took into account. In this regard, TEM electron tomography can provide three-dimensional analysis of the nanoscale morphology of the blend (P3HT-PCBM). In other words, it is possible to observe the morphology of the active layer at different depths.

The donor:acceptor ratio (i.e., polymer–PCBM ratio) can significantly affect the solar cell performance, and thus should be carefully optimized. It has been reported that donor:acceptor ratio influences the crystallinity, the phase separation, and the morphology of the thin film. For instance, in the typical P3HT/PCBM composite blend, a 40% PCBM loading gave rise to the highest device efficiency. A dense network of P3HT nanowires was observed and the crystallinity of the matrix was investigated by electron diffraction. A crystallinity of about 45% was measured for the P3HT polymer (van Bavel et al., 2010).

1.2.5 Polymer chemical modification

The molecular architecture of the polymers materials used in conjunction with the PCBM counterpart represents a key parameter in order to optimize the performance of the device. The molecular arrangement is normally controlled by the planarity, the position and chemical nature of the solubilizing side chains, and the molecular weight. Subsequently, the increased ordering of the polymer backbone leads to enhanced charge transport and to improved performances. In this regard, a further molecular ordering can be achieved by substituting the silicon atom with a larger atom such a germanium in the cyclopentadithiophene system. Compared to the C–Si bond, the C–Ge is longer and thus, reduces the bulky side chain effect on the heterocyclic plane allowing for increased $\pi-\pi$ stacking interactions. In this scenario, a PCE up to 7.3 % was obtained (Amb et al., 2011).

Another way to relieve the steric hindrance in the acceptor unit and to increase the main chain planarity can be achieved by the introduction of a thiophene moiety in the donor unit—specifically in the case of PBDTDTTT-S-T (poly [(2-((2-hexyldecyl)-sulfonyl)-4,6-bis(thiophene-2-yl)thieno[3,4–b]thiophene-2,6-diyl)-alt-(4,8-bis((2-ethylhexyl)oxy)benzo[1,2-b:4,5-b′]- dithiophene-2,6-diyl)]) (Huang et al., 2012b). This structural modification allows to enlarge the π-overlap area between the polymer backbone and to enhance the $\pi-\pi$ stacking. Therefore, higher hole mobility is measured and higher PCE is observed. After the chemical modification, GIWAXS experiments showed a decreased *d*-spacing of the $\pi-\pi$ stacking and an increased crystallinity.

Important aspects that are often overlook when characterizing polymer solar cell are the charge transport and charge carriers' mobility. Indeed, efficient movement of charges is sought to reduce the carriers (hole and electron) recombination. In this regard, the hole mobility (μ_h) of the donor (polymer) and the electron mobility (μ_e) of the acceptor (PCBM) are important factors and will influence the overall performance of the solar cell device. It is important to mention that the electrical conductivity is proportional to the mobility, to the number of charge carrier and to the fundamental charge of the electron. This implies that the magnitude of mobility will determine the drift length of charges. If the thickness of the active layer is higher than the mean drift charges length, charge transport losses will occur. Therefore, in order to obtain an efficient system, it is important to maximize the mobility.

1.2.6 Charge transport and effect of charge carrier mobility

Measurements of the charge carriers' mobility are important in order to characterize the electronic behavior of the active layer (polymer–PCBM composite). Several methods are available to obtain information about the mobility.

The field-effect transistor (FET) mobility can be determined by first constructing an OFET in which the polymer:PCBM composite is used as the component under test. OFET measurements are usually supported by other characterization techniques due to its nonanisotropic nature. Fabricating and testing the OFET properties are relatively easy, and conclusions about the solar cell device can be drawn. A typical device is constructed as follow: a source electrode, a drain electrode, a gate electrode, a conducting channel (which in this case is the polymer:PCBM composite) and a dielectric layer separating the conducting channel and the gate electrode. Then, the mobility is measured by keeping the gate electrode voltage constant and measuring the current of the device against the drain electrode voltage. As an example, Fig. 1.9 depicts (a) the typical $I-V$ output curves at different gate voltage and (b) the transfer characteristic at the saturation regime FBT-Th_4(1,4) polymer system (Chen et al., 2014c). In this particular case, a bottom-gate top-contact was fabricated. From the curves, an average hole mobility of $1.92\ cm^2\ V^{-1}\ s^{-1}$ were obtained. It is important to point out that the devices were annealed at 100°C before testing. The authors attributed the good charge transport to the strong interchain aggregation, and to the close $\pi-\pi$ stacking distance. Solar cell made using this polymer (FBT-Th_4(1,4) mixed with $PC_{71}BM$ as the active layer led to PCE up to 7.6% showing that a high charge transport along the polymer backbone is important to obtain high performing solar cell devices.

The space charge limited current (SCLC) mobility can be used to characterize the charge carrier mobility in OPV. The SCLC device architecture consists of an active layer (of about 100 nm) sandwiched between an injection contact and an injection blocking contact. The current is then measured against the voltage and the results are fitted to obtain the mobility data (Murphy et al., 2013). For instance, SCLC has been performed on both thin films of PDBFBT and PDBFDT:$PC_{61}BM$ as shown in Fig. 1.10A and B. On the pristine PDBFBT, the authors managed to measure average hole mobility values of $2.3 \times 10^{-2}\ cm^2\ V^{-1}\ s^{-1}$. When they added an equal amount of $PC_{61}BM$ to the polymer, they observed a mobility drop to $4.5 \times 10^{-3}\ cm^2\ V^{-1}\ s^{-1}$. They attributed this decrease to the disruption of the polymer chain packing in the presence of $PC_{61}BM$. By tuning the ratios of the polymer (1:3 or 1:4) and $PC_{61}BM$, higher μ_h were reached (about $1.5 \times 10^{-2}\ cm^2\ V^{-1}\ s^{-1}$) suggesting that the PDBFBT forms a continuous phase between the electrodes. Therefore, indirect information about the morphology the polymer–PCBM composite was obtained.

The time-of-flight (TOF) mobility method consists of a thin film of material of interest sandwiched between two electrodes, one of which is transparent. Then the experiment proceeds as follow: a laser pulse excites the surface of the material through the transparent electrode. A pack of charge carriers are produced and visualized as a current drop after a certain amount of time equal to the time required for the charges to be transported between the electrodes. The mobility is then calculated. It is important to specify that the TOF technique can be used to measure the charge recombination dynamics. By using a high-intensity laser pulse, a large amount of charge will be generated and the amount of recombination can be measured to the Langevin recombination. The charge carriers will move through the device - named transit time (t_{tr})- until some later time when the extraction is

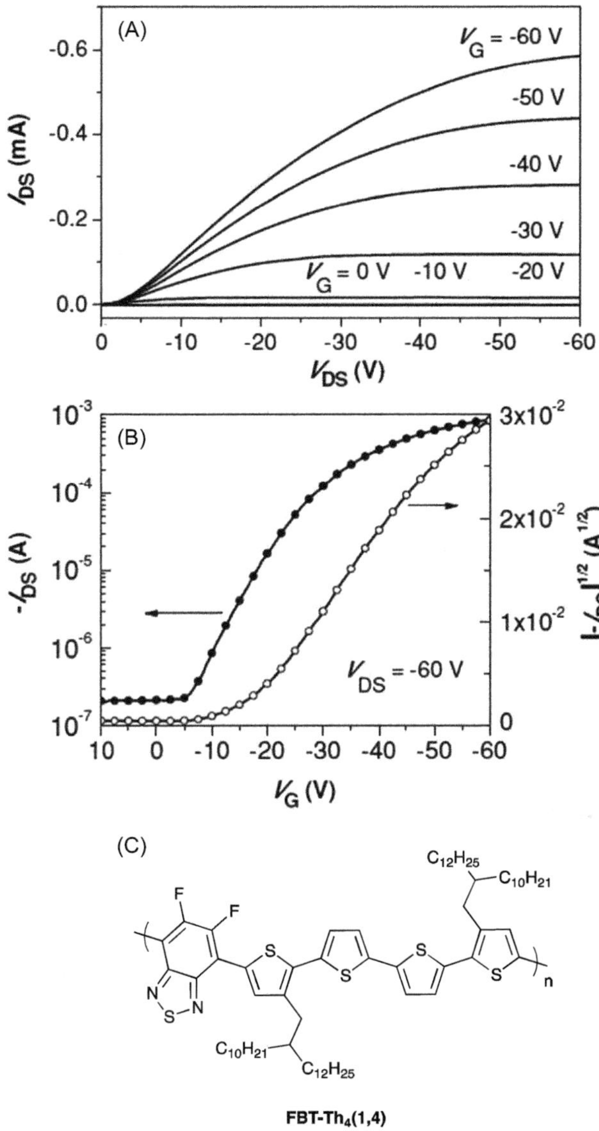

Figure 1.9 (A) $I-V$ output characteristic at different gate voltages, (B) transfer characteristics at saturation regime, (C) chemical structure of the polymer FBT-Th$_4$(1,4). *Source*: From Chen, Z., Cai, P., Chen, J., Liu, X., Zhang, L., Lan, L., et al., 2014c. Adv. Mater. 26, 2586–2591.

ended (t_e). The ratio of the times t_{tr} and t_e are related to the ratio of the recombination against the Langevin recombination. As an example, TOF measurements have been carried out on the PCDTBT:PCBM system (Fig. 1.11) (Clarke et al., 2012). It is founded that both photocurrent transients are fairly dispersive as observed by a continuous decrease in photocurrent versus time and the lack of a clear photocurrent

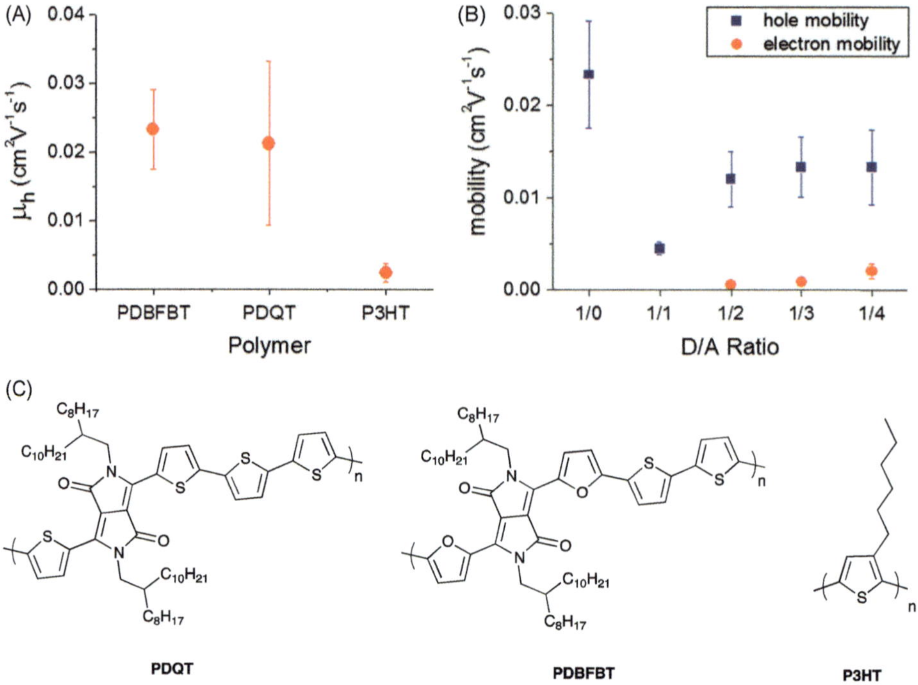

Figure 1.10 (A) Average space charge limited current (SCLC) μ_h of pristine films of PDBFBT, PDQT and P3HT, (B) Average μ_h and μ_e SCLC of blended films of BDBFBT: $PC_{61}BM$ at various donor:acceptor ratios, (C) chemical structures of polymers PDQT, PDBFBT, and P3HT (films thickness = 200–300 nm).
Source: From Murphy, L., Hong, W., Aziz, H., Li, Y., 2013. Org. Electron. 14, 3484–3492.

plateau. The hole mobility is then calculated from the transit time at each applied voltage and plotted against the square root of the electric field. At last, it is important to point out that electron and hole mobility measurements are only possible in very thick films where the extracted mobility is different compared to operational devices.

At last, the charge extraction by linearly increasing voltage (CELIV) mobility method use a similar setup as the TOF mobility method, however an increasing voltage is applied rather than a constant voltage (You et al., 2013). Thus, it is possible to use a weaker light pulse which allows some light to penetrate into the device and generate carriers inside the thin film rather than solely at the surface. The CELIV enables to calculate mobility more closely related to the actual solar cell device than the one obtained by TOF. Moreover, the CELIV approach can also determine the free carrier recombination rate using derivatives techniques such as the MIS-CELIV, photo-CELIV, i-CELIV and OTRACE (Armin et al., 2014). It is also possible to determine relatively low mobilities in organic photovoltaic systems. Fig. 1.12 shows the CELIV curves for a series polymers PBDTTFDPP C10 (C12 or EH) bearing different alkyl side chain on their polymer backbone (Gao et al., 2014). It is observed that the carrier mobility in the polymer:$PC_{71}BM$)-blend films

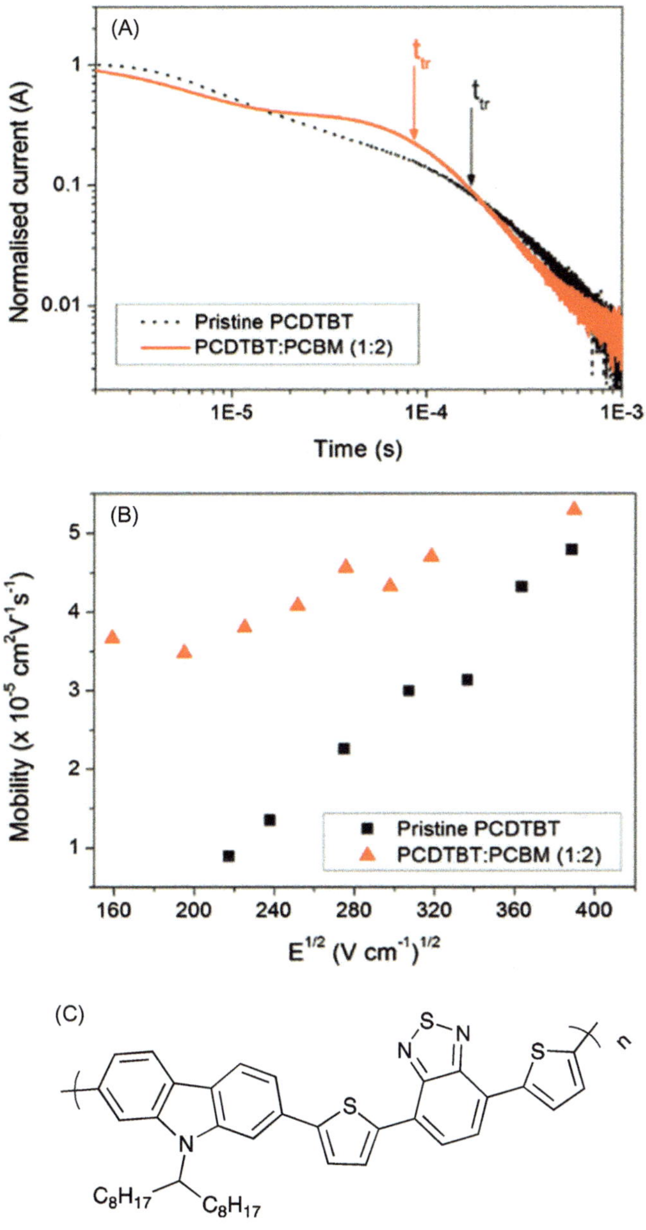

Figure 1.11 (A) Time-of-flight (TOF) performed on pristine PCDTBT and PCDTBT: PC71BM with active layer thicknesses of 5.3 and 4.0 μm, respectively. Transit time in shown by an arrow. (B) Electric field dependence of the PCDTBT and PCDTBT:PC71BM devices on the hole mobility. (C) PCDTBT polymer chemical structure.

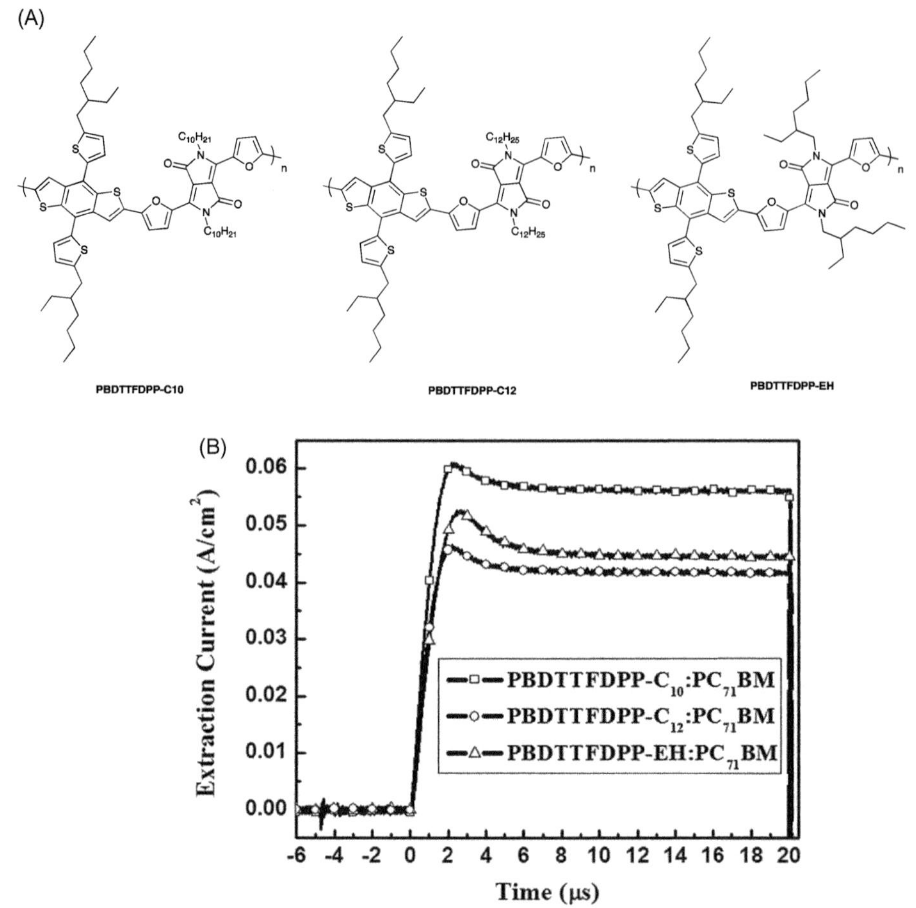

Figure 1.12 (A and B) Measured photo-CELIV curves of polymer:$PC_{71}BM$ blends thin films. *Source*: Gao, J., Dou, L., Chen, W., Chen, C.-C., Guo, X., You, J., et al., 2014. Adv. Energy Mater. 4, 1300739.

is improved by replacing the polymer branched side chains with linear chains. In a larger picture, the enhanced structural order and carrier mobility by changing the side chains was found to be an important factor contributing to improve the device performance.

The incident photon-to-current efficiency (IPCE) is a measure on how the solar cell converts the incident light into electrical energy. Two types of quantum efficiency can be defined: the EQE and the IQE.

The external quantum efficiency (EQE) is one of the most important measurement for solar cell. The EQE is defined as the measurement of the number of carriers (electrons or holes) being extracted for each photon incident on the device, and it is given in percentage (%). The experiment is carried out for each wavelength of incident light. By scanning the whole wavelength range, it is possible to obtain the short-circuit current of the solar cell and thus, can serve as a diagnostic tool.

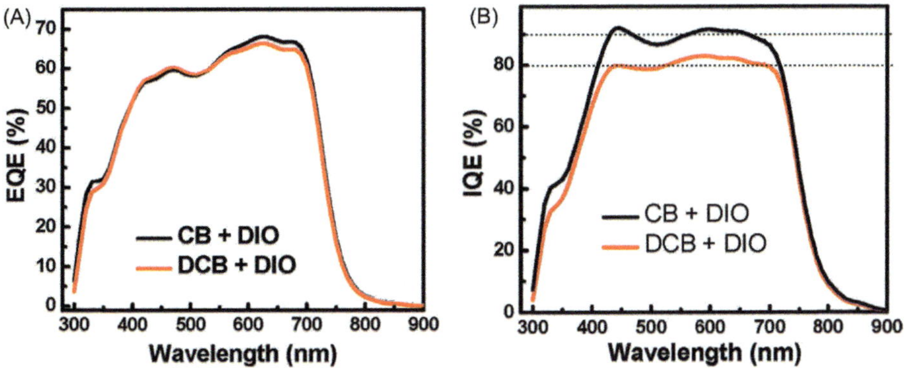

Figure 1.13 (A) EQE spectra of PTB7:PC71BM solar cell, (B) IQE spectra obtained from the EQE and absorption spectra.
Source: Liang, Y., Yu, L., 2010. Acc. Chem. Res. 43, 1227–1236.

The area under the curve represent the total number of carriers created by the device under the full spectrum range during white light illumination. Integrating the curve will give the electrical current density. At last, it is important to point out that the EQE is measured when no voltage is applied on the solar cell. Fig. 1.13A shows a typical EQE spectra of PTB7:PC_{71}BM with a certain amount of solvent additive (diiodooctane) (Liang and Yu, 2010). EQE measurement were further examined by integrating the data with the AM 1.5 G solar spectrum. The short-circuit current density (J_{SCEQE}) derived from the EQE measurement were found comparable to the short-circuit current density measured from the solar cell (J_{SCSC}) 14.4 mA cm^{-2} to 14.5 mA cm^{-2}, respectively.

Another closely related parameter is the internal quantum efficiency (IQE) which is defined as the ratio of the collected carriers to the number of photons that are actually absorbed by the solar cell. In short, the IQE is an indication of the capacity of the active layers of the solar cell to make good use of the absorbed photons. The IQE is always higher than the EQE, but should never exceed 100% with the exception of multiexciton generation.

The difference between the EQE and the IQE is important to be able to distinguish the loss mechanism between the optical absorption properties of the device and the photoconversion properties of the materials. We can note that the IPCE can be useful to investigate the aging behavior of the device.

1.3 Section 2. Polymeric materials for supercapacitors and electroactive polymer actuators

The recent advances in flexible, wearable, and mobile electronics goes in line with the development of high-performance polymer based devices for energy storage and transduction. Traditionally, polymers have been used due to their lightweight and low-cost as well as their intrinsic mechanical properties. Nowadays, those properties are being complemented by electroactive features such as electronic and ionic conductivity or high capacitance which can be obtained by creating polymer composites and hybrids when combined with other materials (nanoparticles, graphene, 2-D crystals, or carbon nanotubes). Thus applications such as all-solid state supercapacitors, polymer batteries, soft actuators, and novel biomedical devices are now feasible.

In this section, we will focus our attention on the physicochemical and electronic properties of polymer composites for supercapacitors and electroactive polymer actuators—particularly low-voltage polymer actuators based on carbonaceous materials that can operate in air. We will also discuss the main properties of the devices themselves as well as the methods to characterize them including both electrochemical and mechanical techniques.

1.3.1 Polymer derivatives in electrochemical supercapacitors

Due to the rapidly growing energy consumption and anticipated reduction of fossil fuels there is an imperative need to develop new efficient, environmentally friendly and sustainable ways of energy production, conversion, and storage. The crucial and practical components for electrochemical energy storage are batteries and supercapacitors. The latter one recently have attracted much attention due to their high power density, quick charge/discharge abilities and long cycling lifetime.

The supercapacitors are devices made of two electrodes in close proximity and an electrolyte with a proper separator in between. They are able to store electrical charges in the electric double layer (EDL) at the interface between the electrode and electrolyte (Fig. 1.14). Another names for supercapacitors are electrochemical capacitors or ultracapacitors (Wang et al., 2012a). The mechanism of a charge

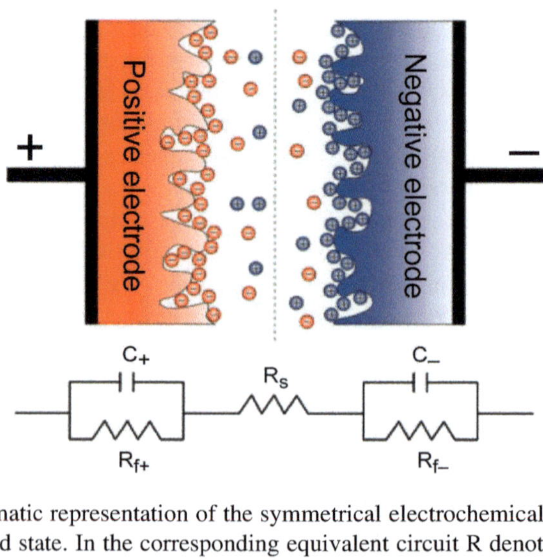

Figure 1.14 Schematic representation of the symmetrical electrochemical double layer capacitor in charged state. In the corresponding equivalent circuit R denotes resistors and C denotes capacitors.
Source: Adapted from Béguin F., Presser V., Balducci A., Frackowiak E., 2014. Adv. Mater. 26, 2219–2251.

storage can be nonfaradaic, faradaic or both; a hybrid which combines pseudocapacitance with EDL capacitance. The high specific surface area (SSA) of the electrodes allows to create double layers with a maximum number of electrolyte ions, thus with higher specific and volumetric power density as well as high capacitance of the devices. The electrochemical stability window of the electrolytes limits the operational voltage usually to around 1.0–1.3 V for aqueous electrolytes and below 3 V for organic based electrolytes. Recently, a new class of ionic liquid based electrolytes offers higher potential windows of 3.5–4.0 V (Zhong et al., 2015). Herein, we will discuss the properties, characterization methods and main advances in the supercapacitor's electrolyte and electrodes development in terms of polymeric materials.

First, to better understand the upcoming section, we will focus on the main parameters of the supercapacitors and their relation to the supercapacitor performance as well as general techniques of characterization (Table 1.1).

One of the main parameters of the supercapacitor performance is the capacitance. It can be defined as the total amount of charge stored in a fully charged system. Therefore, it is expressed in Farads (1 F = 1 C/1 V) although practically it is expressed in terms of specific capacitance in $F\,g^{-1}$ or $F\,cm^{-2}$, which is a normalization by mass or the area. Typical techniques used for the estimation of the capacitance are cyclic voltammetry, galvanostatic charge–discharge and impedance spectroscopy.

From the cyclic voltammetry results, the capacitance can be estimated as follows:

$$C = \frac{I_p/V_s}{m} \tag{1.1}$$

where I_p is the peak-to-peak current in A at open-circuit voltage, V_s is the scan rate of the voltage (in $V\,s^{-1}$) and m can be either the area of the electrode giving us the capacitance per area unit (for instance $F\,cm^{-2}$) or mass loading resulting in the specific gravimetric capacitance (F per gram of the total mass of active material or more accurately the total mass of both dry electrodes the electrode mass) (Béguin et al., 2014). During galvanostatic charge–discharge measurements the potential in function of time is recorded while a constant current density is applied. The capacitance is given by the following relation:

$$C = \frac{I\Delta t}{m\Delta V} \tag{1.2}$$

where I denotes the current, Δt is the discharge time, m states for the mass of active material ΔV is the potential range.

Besides specific capacitance, equally important parameters of the supercapacitor are the energy stored and power. When, the voltage (V) is applied to the device the charge (Q) maybe added or removed. In supercapacitors, contrary to the batteries, the voltage drops while the device is being discharged. The theoretical energy density of the supercapacitor is given by the following equation:

$$E = \int VdQ = ItV = \frac{1}{2}CV^2 = \frac{QV}{2} \tag{1.3}$$

Table 1.1 Table summarizing the main properties of the supercapacitors, their relation to the performance and characterization techniques

Property	Role	Characterization
Electrical conductivity of the electrode	Factor limiting charging/ discharging speed	Two or four probe method
Specific surface area—electrode	Influence the capacitance	Adsorption based methods
Flexibility	For flexible polymer-based supercapacitor devices	Mechanical methods coupled with electrical; the capacitance retention in function of bending cycles (Kang et al., 2016), demos like powering LED with "twisted" supercapacitor (Meng et al., 2010)
Capacitance	Total amount of stored charge. One of the most important figures of merit.	Cyclic voltammetry, galvanostatic charge–discharge, EIS, (see Equation 1.1 and 1.2)
Operating voltage/ electrochemical stability window (ESW)	Maximum voltage of the supercapacitor, device stability, main criterion for choosing the electrolyte	Cyclic voltammetry (Sylla et al., 1992)
Ionic conductivity of the electrolytes	Main criterion for choosing the electrolyte, higher ionic conductivity means lower internal device resistance (ESR).	Electrochemical impedance spectroscopy (Qian et al., 2001)
Capacitor lifetime	Life time of the device with nearly 100% of the initial capacitance, important criterion towards commercial applications.	Cyclic voltammetry, number of cycles before a significant deviation is observed

Because $C = Q/V = It/V$

The supercapacitor energy can be expressed as follows:

$$P = IV = I^2R = \frac{1}{4}\frac{V^2}{R_S} \tag{1.4}$$

From these equations, it can be seen that energy and power of the supercapacitors depends on the voltage. Hence, these values will also depend on the materials used to form the electrolyte and the electrodes. Commonly, the supercapacitors electrodes are made of carbon materials or conducting polymers such as poly(cyclopenta (2,1-b;3,4-b′-dithiophen-4-one)), poly(3-fluorophenyl)thiophene, polyaniline (PANI), polypyrrole (PPy), and poly(3,4-ethylene-dioxythiophene) (PEDOT).

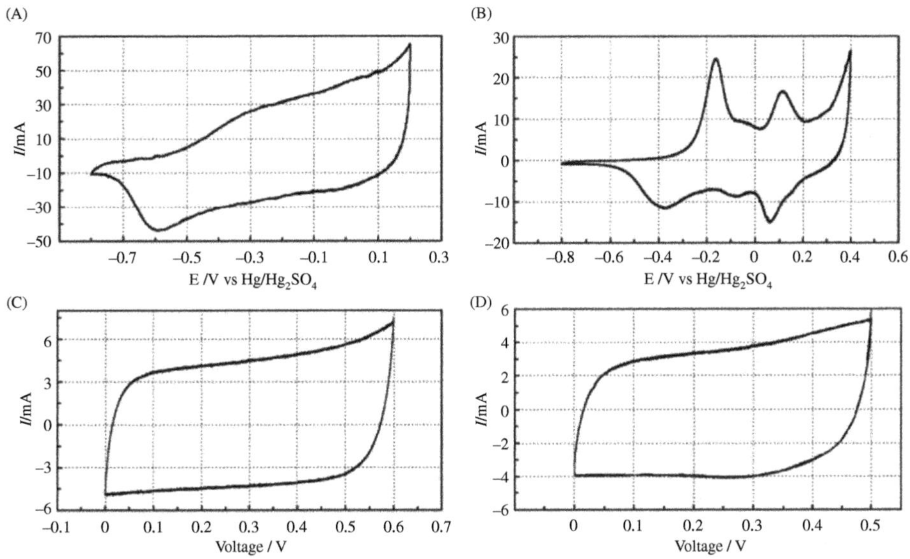

Figure 1.15 Cyclic voltammetry (A) in a three-electrode cell configuration (5 mV s^{-1}) of a PPy/MWNTs pellet electrode; mass of composite, 9.6 mg; electrolyte, 1 mol L^{-1} H2SO4, (B) in a three-electrode cell configuration (2 mV s^{-1}) of a PANI/MWNTs pellet electrode; mass of composite, 5.4 mg; electrolyte 1 mol L^{-1} H_2SO_4; scan rate, (C) in a two-electrode cell configuration (5 mV s^{-1}) of a symmetric capacitor, based on PPy/MWNTs composite electrodes, mass of each electrode, 8.6 mg, (D) in a two-electrode cell configuration (2 mV s^{-1}) of a symmetric capacitor based on PANI/MWNTs composite electrodes, mass of each electrode, 10.4 mg.
Source: Reprinted from Khomenko, V., Frackowiak, E., Béguin, F., 2005. Electrochim. Acta 50, 2499–2506.

A study by Khomenko et al. shows that the specific gravimetric capacitance can be affected by the electrode material composition, the electrochemical cell configuration and by the selection of the characterization method (Fig. 1.15 and Table 1.2) (Khomenko et al., 2005). The capacitance of the electrodes consisting of multiwalled carbon nanotubes (MWCNT) plus one of two following conducting polymers: polypyrrole (PPy) or polyaniline (PANI) has been estimated in different cell configurations as well as by galvanostatic charge–discharge. The results revealed that the values obtained using a three-electrode configuration cell cannot be simply compared with values obtained by two-electrode configuration (Khomenko et al., 2005). However, a common three-electrode configuration (a working electrode, a reference electrode and a counter electrode) is extremely useful in probing electrochemical behavior of a single electrode, for instance redox reactions. In the abovementioned report, authors have also concluded that a polymer/multiwalled carbon nanotube hybrid supercapacitor with an asymmetric configuration—positive electrode PANI/MWCNT and negative electrode PPy/MWCNT—performs better than symmetric one in terms of cyclic stability and stored energy due to the higher values of possible voltages. Moreover, the galvanostatic charge–discharge characterization in two-electrode configuration is suggested as an appropriate method for

Table 1.2 Average value of specific capacitance ($F\ g^{-1}$) of MWCNT/conducting polymers composite electrodes depending on the measurement technique.

Conducting polymer in the composite electrode	Three-electrode cell		Two-electrode cell	
	CV	Galvanostatic discharge	CV	Galvanostatic discharge
PPy	506	495	192	200
PANI	670	650	344	360

Source: Adapted from Khomenko, V., Frackowiak, E., Béguin, F., 2005. Electrochim. Acta 50, 2499–2506.

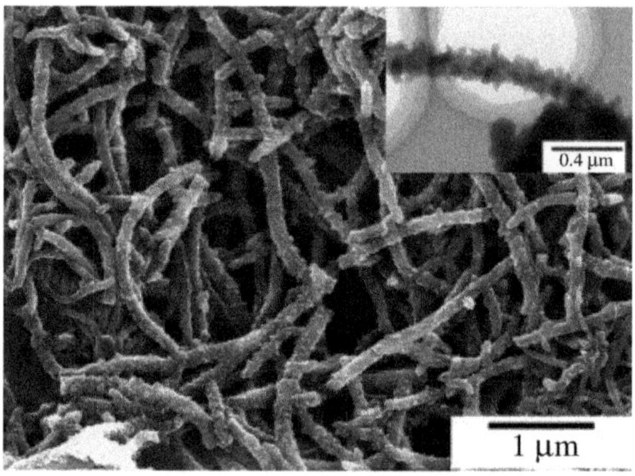

Figure 1.16 SEM images of PANI on the substrate (TEM image inset at top-right corner). Reprinted from: Li H., Wang, J., Chu, Q., Wang, Z., Zhang, F., Wang, S., 2009. J. Power Sources 190, 578–586.

comparing specific capacitance of electrodes prepared from different electronically conducting polymers.

Li et al. calculated a maximum specific capacitance of PANI electrode to be 2000 $F\ g^{-1}$ and 1000 $F\ g^{-1}$ for assembled symmetrical supercapacitor when assuming that 100% of the PANI is oxidized or reduced in the charge/discharge process. Such high capacitance is possible due to contribution of both pseudocapacitance and electrical double-layer capacitance. In practice, by electro-polymerization of 0.10 M aniline monomer in 1.0 M H_2SO_4 on stainless steel they were able to prepare electrodes resulting in around 30% of the theoretical capacitance (Li et al., 2009). The morphology of deposited PANI was investigated by SEM and TEM showing polymer nanofibers (Fig.1.16). Thanks to such a loose porous structure a high specific surface area can be developed. However, the heterogeneity of the material will lower the conductivity of PANI as the bottom part of the polymeric part of the electrode is dense due to the PANI aggregation which can then slow the diffusion of counter-ions.

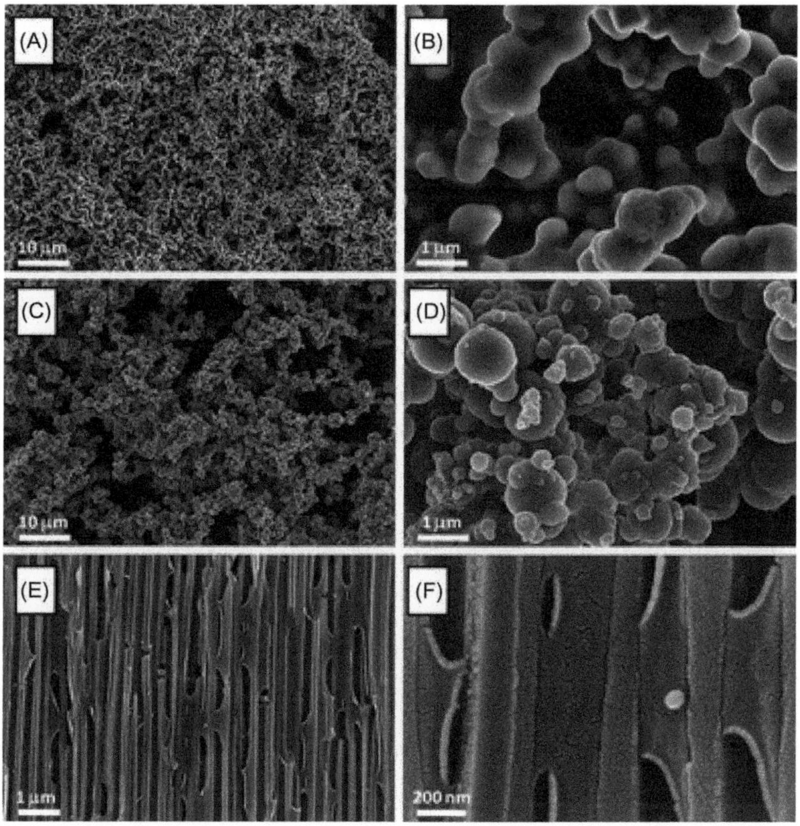

Figure 1.17 Scanning electron microscopy images of pTTPA films electrochemically deposited with cyclic voltammetry between 0 and 1 V with a rate of 100 mV s^{-1} (A) and with chronoamperometry at 0.7 V for 10 min on Au/PI (C) and Au/200 nm Anopore membranes (E). Enhanced magnification images of (A), (C), and (E) are shown in (B), (D), and (F), respectively.
Source: Reprinted from Roberts M.E., Wheeler, D.R., McKenzie, B.B., Bunker, B.C., 2009. J. Mater. Chem. 19, 6977.

By electrodeposition of another conducting amine—poly(tris(4-(thiophen-2-yl) phenyl)amine) (pTTPA)—Roberts et al. prepared conducting polymer supercapacitor electrodes (Fig. 1.17). They were able to achieve remarkably high peak capacitance of about 950 F g^{-1} (Roberts et al., 2009). The thin tubes of pTTPA leading to the specific surface area of 45 m^2 g^{-1} were obtained by using a specific current collector (working electrode) in the electrochemical deposition process—alumina membranes with 200 nm diameter pores, 25−40% porosity. Moreover, the complex secondary structure of pTTPA has a well-defined porosity which is not feasible in linear polymers such as PANI, which lead to the improved counter-ion intercalation.

Besides electrodes development, studies on the electrolytes for the applications in supercapacitors are equally important and have attracted great attention in the last decade (Zhong et al., 2015).

An ideal electrolyte should be highly ionic conductive, have a wide electrochemical stability (potential) window (ESW) together with a wide operating temperature, be chemically and electrochemically stable, possess low volatility and flammability, and preferably be environmentally friendly. Of course, as for all industrial applications its cost should be maximally low. In practice, it is difficult for a certain material composition to meet all these requirements.

Electrolytes can be divided into two main categories: liquid electrolytes (including aqueous and nonaqueous based) and solid state electrolytes (including quasisolid state). The preparation of the latter one is mostly based on polymer derivatives. As solid state electrolytes (SPE) are liquid-leakage free and simple to fabricate and to pack for storage and transport, they became highly attractive for applications in wearable, flexible and portable electronics as well as printable devices. Moreover, solid polymer electrolytes are not only ionic conductors but can also behave as the electrodes separator. Polymer-based solid electrolytes can be made in different ways; they can be formed by a polymer and a salt, in which salt ions can migrate through the polymer matrix as well as consists of polymer matrix swollen by an electrolyte (it could be a salt-solvent system or water based electrolyte). Up to date, various polymers such as thermoplastic fluoropolymers like polyvinylidene difluoride (PVDF) and its copolymer with hexafluoro-propylene (e.g. PVDF-HFP), poly (vinyl alcohol) (PVA), poly(ethylene oxide) (PEO) and poly(ethylene glycol) (PEG) derivatives, poly(methyl methacrylate) (PMMA), potassium poly(acrylate) (PAAK), and polyurethane (PU) have been studied towards solid state electrolytes (Meng et al., 2010; Gao and Lian, 2014; Yang et al., 2014).

Interestingly, PVA can be used to prepare both acidic, neutral or base electrolytes. Among others, following polymer gel electrolytes were reported: PVA/H_2SO_4, PVA/H_3PO_4, PVA/LiCl, PVA/NaCl, PVA/KCl, PVA/KOH, and PVA/NaOH (Kaempgen et al., 2009; Yuan et al., 2006; Yu et al., 2011; Wang et al., 2012b; Chen et al., 2014a). The electrolyte solution has a significant effect on the supercapacitor voltammograms and the specific capacitance (Fig. 1.18A–D). For instance, Chen et al. reported the specific capacitance of 820 $\mu F\ cm^{-2}$ for PVA/H_2SO_4 based supercapacitors while PVA/H_3PO_4 devices had a lower capacitance of 15 $\mu F\ cm^{-2}$ (Chen et al., 2014a). Interestingly, in that study supercapacitors with neutral and alkaline electrolytes had a capacitance between 0.5 and 2.5 $\mu F\ cm^{-2}$. The utilization of PVA/H_2SO_4 gel electrolyte led to the highest capacitance although it was considered to have a poor stability due to the pseudocapacitive features. To compare performance of the supercapacitors having different electrolytes as well as for the comparison of any energy-storing devices a Ragone plot can be used. Fig. 1.18E shows an example of Ragone plots for above-mentioned study (Chen et al., 2014a), revealing that capacitors prepared with PVA/H_3PO_4 outperformed neutral and alkaline electrolytes.

A severe importance of polymer-based electrolytes and electrodes can be appreciated when it comes to the realization of solid-state flexible and stretchable supercapacitors. In general, these devices take advantage from solid or quasisolid polymer electrolytes and electrodes. For instance, Meng et al. (2010) by using flexible thin film electrodes based on PANI/CNT nanocomposite and H_2SO_4-PVA gel electrolyte prepared in a simple process all-solid-state flexible supercapacitor (Fig. 1.19A and B) having a specific capacitance as high as 350 $F\ g^{-1}$—for the

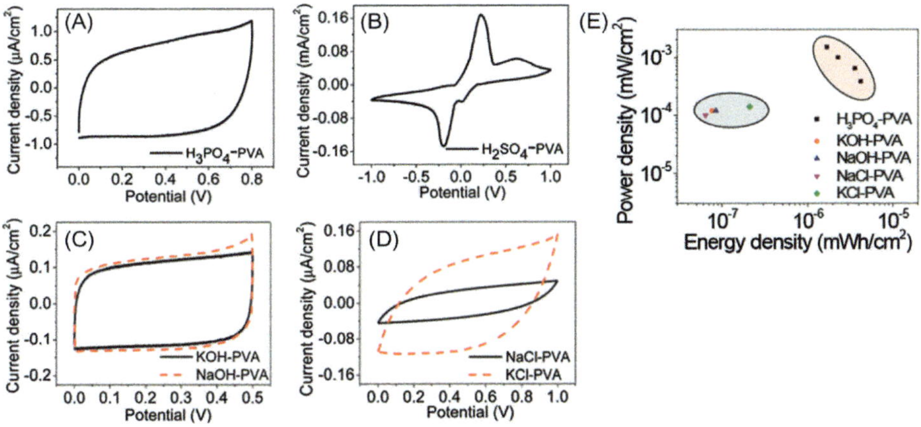

Figure 1.18 (A–D) CV curves of as-prepared solid-state supercapacitors with different electrolytes at 100 mV s^{-1} scan rate. (A) H_3PO_4–PVA and (B) H_2SO_4–PVA electrolyte. (C) Base gel electrolytes: KOH (black) and NaOH (red dashed), (D) neutral gel electrolytes: NaCl (black) and KCl (red dashed). (E) Ragone plots for the supercapacitors with the H_3PO_4–PVA, KOH-PVA, KOH-PVA, NaOH-PVA, KCl-PVA, and NaCl-PVA gel electrolytes. The plot in up-right circle region is for devices using H_3PO_4–PVA electrolyte, while the left circle is for those using base and salt electrolytes.
Source: Reprinted from Chen Q., Li X., Zang X., Cao Y., He Y., Li P., et al., 2014a. RSC Adv. 4, 36253.

electrode material. For the entire device it was 31.4 F g^{-1} and authors observed only 8.1% decay in specific capacitance after 1000 cycles. Moreover, the cyclic voltammetry curves, specific capacitance and discharge abilities were almost the same for both the aqueous electrolyte (0.5 M H_2SO_4 solution) and corresponding H_2SO_4 − PVA polymer electrolyte as shown by Fig. 1.19C and D.

Another interesting polymer selection was reported by Yang et al. (Yang et al., 2014). Specifically, potassium polyacrylate/KCl gel electrolyte was used as both supercapacitor electrode separator and solid-state electrolyte. The flexible electrodes were prepared from graphene oxide/carbon nanotube and carbon fiber paper/polypyrrole to obtain an asymmetrical supercapacitor with an energy density of 28.6 W h kg^{-1} and a power density of 15.1 kW kg^{-1} (Yang et al., 2014). This device demonstrates even better cycle stability with a capacitance decay of 7% after 2000 cycles and 82.4 F g^{-1} specific capacitance calculated on the total mass of active materials of two electrodes. Moreover, when this supercapacitor was bended the capacitive behavior and voltammograms shape become unchanged (Fig. 1.20).

We would also like to mention that thanks to the usage of polymers a proof of concept energy-storage devices as stretchable and transparent supercapacitors became possible. Yang et al. (2013a) developed a wire-like highly stretchable supercapacitors by using a subsequent coating steps of a rubber fiber with a thin layer of PVA-H_3PO_4 gel electrolyte and carbon nanotubes (Fig. 1.21). As prepared supercapacitor was stretchable due to the mechanical properties of both rubber "core" and polymer gel electrolyte.

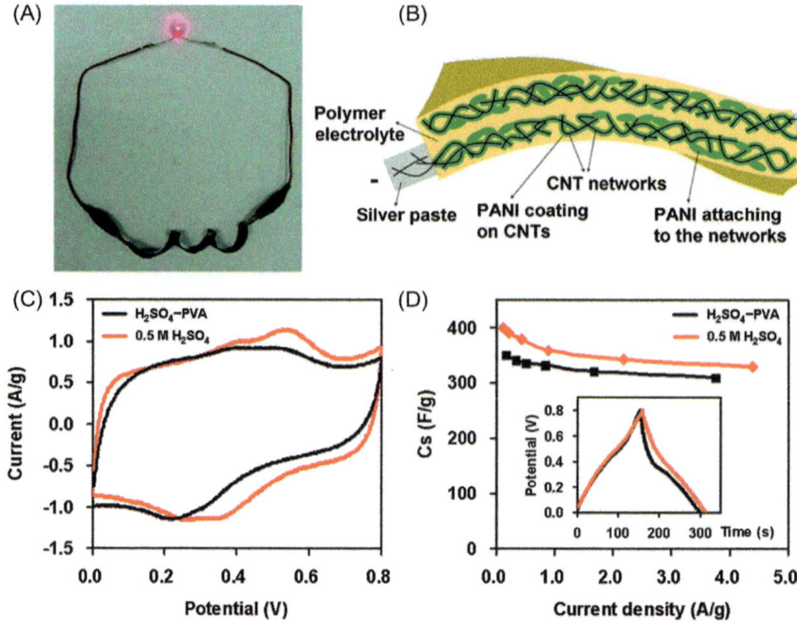

Figure 1.19 (A) Digital picture that shows the all-solid-state polymer/CNT based supercapacitor under its highly flexible state. (B) Schematic illustration of the PANI/CNT nanocomposite electrodes well solidified in the polymer gel electrolyte. (C and D) The electrochemical performance of the all-solid-state devices recorded under highly flexible conditions. (C) Comparison of cyclic voltammetry at 5 mV s^{-1} and (D) discharge abilities of the flexible PANI/CNT nanocomposite thin film electrodes in the H_2SO_4 − PVA gel electrolyte and in the 0.5 M H_2SO_4 aqueous solution. The inset in (D) shows one cycle of galvanostatic charge − discharge curves at 1 A g^{-1}.
Source: Meng C., Liu, C., Chen, L., Hu, C., Fan, S., Nano Lett. 10, 4025−4031.

Chen et al. (2014b) also employed carbon nanotubes and PVA-H_3PO_4 gel electrolyte but this time with (polydimethylsiloxane) PDMS as a substrate enabling transparency and stretchability of the supercapacitor. The supercapacitors were prepared as a multilayer film of PDMS-supported CNT electrode on the electrode coated with an electrolyte (PVA-H_3PO_4/CNT/PDMS) as shown in Fig. 1.22. The authors showed by UV/Vis spectroscopy that values of supercapacitor transmittance can reach almost 70−80%, which can find an application in the integration of supercapacitors with solar cells (Yang et al., 2013b; Chen et al., 2013).

1.3.2 Polymeric hybrid materials for electroactive carbon-based actuators

1.3.2.1 Overview

An actuator is an energy transducer, which function is to convert a certain energy source (for instance electric current/voltage, the potential of chemical

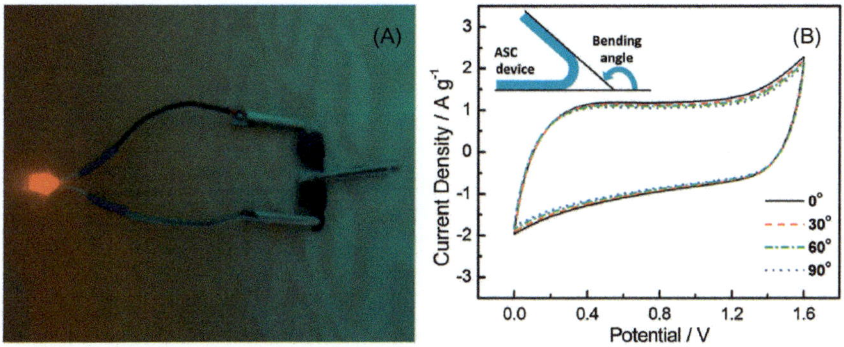

Figure 1.20 (A) Optical photograph shows a red LED lighted powered by two all-solid-state asymmetric supercapacitor (ASC) of CFP/PPy//RGO/cMWCNT connected in series.
(B) Cyclic voltammetry curves of CFP/PPy//RGO/cMWCNT ASC at different curvatures of 0°, 30°, 60°, and 90°.
Source: Reprinted from Sylla S., Sanchez, J.-Y., Armand, M., 1992. Electrochim. Acta 37, 1699–1701.

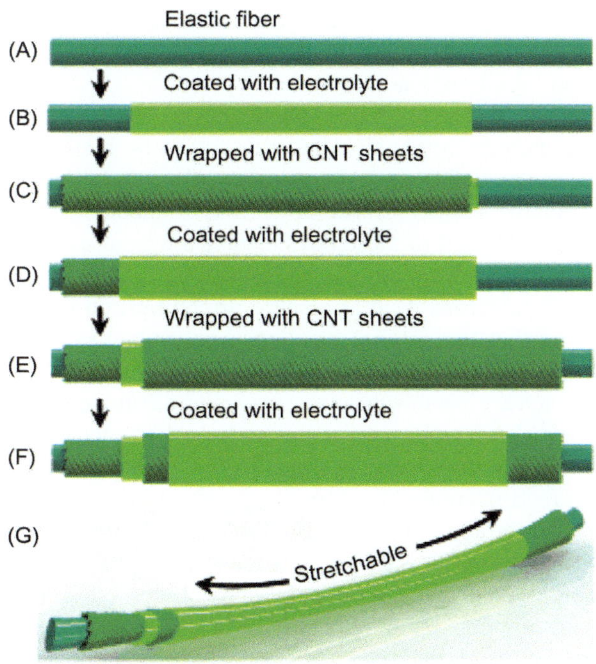

Figure 1.21 Illustrative fabrication of a highly stretchable, fiber-shaped supercapacitor with a coaxial structure.
Source: Reprinted from Yang Z., Deng, J., Chen, X., Ren, J., Peng, H., 2013a. Angew. Chem. Int. Ed. 52, 13453–13457.

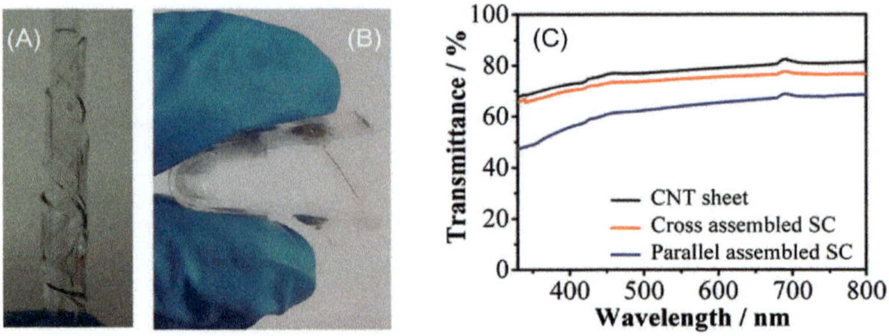

Figure 1.22 Illustrative fabrication of a highly stretchable, fiber-shaped supercapacitor with a coaxial structure.
Source: Reprinted from Chen, T., Peng, H., Durstock, M., Dai, L., 2014b. Sci. Rep. 4, 3612.

reactions or fluid pressure) into mechanical energy (displacement/force) (Janocha, 2004). Common examples of the actuators are solenoids, electrical motors, valves and fluid-based motors. Since the middle part of the 20th century, polymers and their derivatives as well as polymer hybrid materials have drawn increased attention worldwide as materials used to developed new types of electroactive actuators possessing high power-to-weight and power-to-volume ratios. Electroactive polymer (EAP) actuators are often called "artificial muscles" due to their similarities with mammalian muscle tissue such as compliance, noiseless and resilience actuation.

Depending on the structure and composition we can distinguish several types of artificial muscles such as: conducting polymer actuators, ionic polymer metal composite actuators, conducting polymer actuators, carbon nanotube–polymer composite actuators, liquid crystal elastomers and dielectric elastomer actuators (Mirfakhrai et al., 2007). All of these devices take advantage of polymeric materials and exhibit movement or dimensional changes while voltage is applied. In contrary to the conventional actuators, such as electric or combustion motors, EAP actuators usually produce motion by material bending or extension rather than typical rotational movement.

Among EAPs, a very promising class are ionic electroactive polymer actuators (iEAP) due to their ability to operate at low voltage (1–4 V). Their actuation mechanism is based on the movement of ions within the actuator strip cross-section while an electric field is applied resulting in the contraction of one side and expansion of the other (Jo et al., 2013). The iEAPs originally derived from a concept of ionic polymer metal composites (IPMC) reported in 1992 in which metal electrodes are fabricated onto ion exchange polymer membrane (e.g., Nafion) (Tiwari and Garcia, 2011; Shahinpoor, 1992; Segalman et al., 1992; Shahinpoor and Kim, 2001). Later on, the metal electrodes were replaced by carbon nanotube, graphene and polymer based electrodes as wells as their composites (Kong and Chen, 2014). In addition, instead of traditional Nafion/aqueous electrolytes many researchers started using ionic liquids and solid polymer electrolytes. In this way dry "sandwich" actuators usually laminated from three layers (electrode/electrolyte/electrode) by hot-pressing have been achieved. These modifications of original idea allowed for the actuation not only at low voltages but also for thousands of cycles in open air (Fukushima et al., 2005; Gendron et al., 2015b).

1.3.2.2 Polymer actuators based on carbon nanotubes

In 2003, by combining carbon nanotubes (CNT) and ionic liquids, a group of Japanese scientists revealed an interesting phenomenon—a gel that is created when single-walled carbon nanotubes (SWCNT) are mixed and grounded with imidazolium ion-based room-temperature ionic liquids (Fukushima et al., 2003). The authors named these materials "bucky gels of ionic liquids", which was later shortened to simply "bucky gel". These bucky gels are formed due to noncovalent $\pi-\pi$ stacking and cation$-\pi$ interactions between carbon nanotubes and the imidazolium ring (Fukushima et al., 2003; Wang et al., 2008). Once cast and dried, bucky gels can be used as actuators' electrodes. In 2005, for the first time, polymer-supported bucky gel electrodes were used to build low-voltage and operable in air, fully plastic trilayer actuators. More precisely, these devices were prepared through layer-by-layer casting and consisted of two outer bucky gel electrodes with inner solid polymer electrolyte layer between them (Fukushima et al., 2005). In the above-mentioned work, poly(vinylidenefluoride-co-hexafluoropropylene) (PVDF-HFP) allowed facile preparation and provided necessary mechanical robustness of the electrodes (Mukai et al., 2009). Fig. 1.23 shows a schematic view of the typical carbon nanotube based polymer actuator. The electrical model of carbon nanotube electroactive polymer actuators can be represented by a supercapacitor consisting of two double-layer capacitors and a resistor in series. Therefore, a wide selection of techniques devoted to characterization of the supercapacitors and described in a

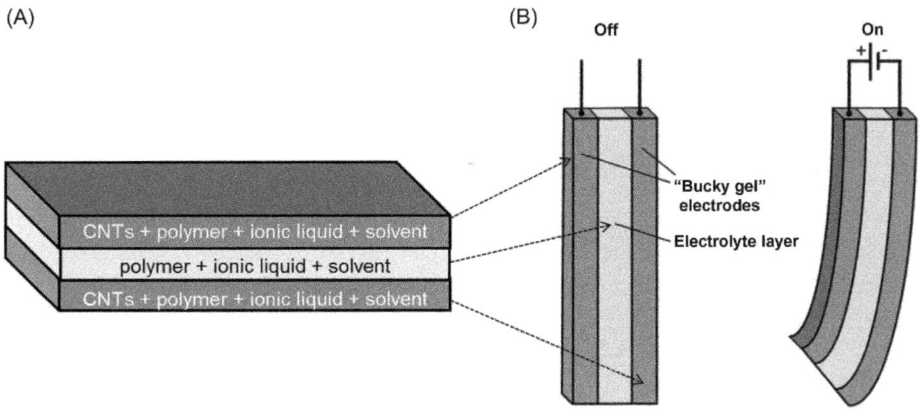

Figure 1.23 (A) An overview of the carbon nanotube-polymer based ionic actuator. The usual composition of the electrodes: IL + CNT + active or nonactive polymer; composition of the electrolyte layer: IL + polymer matrix + solvent, (B) a schematic representation of a "bucky gel actuator" during the OFF and ON states.

previous section can be used in the estimation of the capacitance, power, energy, and conductivity of CNT/polymer actuators. Typical figure of merits benchmarking the performance of bending actuators are displacement, strain and blocking force. The strain (i.e., strain difference) is defined as $\varepsilon = (2d\delta)/(L^2 + \delta^2)$, where L is the actuator free length, d is the thickness, and δ is the peak-to-peak displacement. The blocking force is the force measured at zero displacement (i.e., when actuator movement is blocked by a force sensor).

Since the first report of a polymer-supported bucky gel electrode, the fabrication procedure has been modified and typical preparation employs hot-pressing rather than original layer-by-layer casting in order to better control the thickness of each layer (Biso et al., 2012a). In general, addition of the polymer binder/supporting polymer to the original electrode preparation methods and employment of the solid polymer electrolytes enabled the development of lightweight, easy to store, pack and transport solid-state actuating devices. A relatively facile preparation of the actuators initiated an intensive process of modifications and improvements of both electrode and electrolyte layers.

Considering polymer derivatives used in the electrode preparation Terasawa et al. investigated different molecular weights of the PVDF and PVDF-HFP (Terasawa et al., 2011). It was demonstrated that the electrical conductivity of the bucky gel electrodes can be almost twofold higher (12–14 S cm^{-1} compared with 6–8 S cm^{-1}) when using optimal molecular weight and varying the type of supporting polymer for a specific ionic liquid. Interestingly, the gravimetric capacitance did not change significantly among PVDF and PVDF-HFP supported electrodes with different molecular weights. However, electrodes with the lowest molecular weight of PVDF-HFP (average $M_w = 212{,}000$) showed one order of magnitude lower Young's modulus than the one with the average M_w of 455,000, which can help in tuning the blocking force of the actuators. Overall, the authors concluded that HFP content should be low (or zero) as they obtained higher maximum strains for PVDF based actuators. The same group used poly(ethylene oxide) as a matrix polymer in the bucky gel electrodes in order to understand how the interactions between ionic liquids and oxygen atoms (of the polymer) can influence the properties of the electrodes and actuators (Terasawa et al., 2014). In this case, maximum strain value for the poly(ethylene oxide) electrode actuators was approximately 1.3 times larger than for the actuators containing PVdF(HFP) based electrodes.

In addition, electronically conducting polymers previously reported in the preparation of supercapacitor electrodes (Wang et al., 2012a; Khomenko et al., 2005) were investigated towards their use in the electrode composition for iEAPs. For instance, Sugino et al. (2011) investigated bucky gels enriched by 4–17 wt% of polyaniline (PANI). CNT/PANI based actuators showed more than two times higher maximum strains (Fig. 1.24) when compared to the additive free CNT actuators of the same kind, which was attributed mainly to the improved capacitance due to the conductive and capacitive nature of PANI.

Apart, from simply mixing conductive polymer with CNTs, ionic liquid, and supporting polymer, Biso et al. (2012b) carried out in situ polymerization of polypyrrole directly in a bucky gel slurry leading to the significant performance

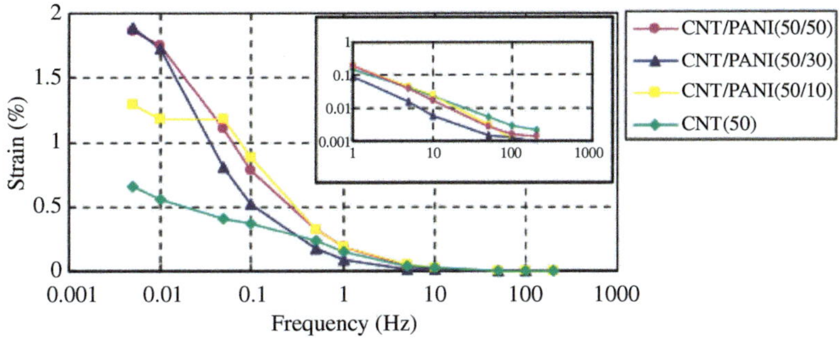

Figure 1.24 Strain–frequency dependence for iEAPs having PANI modified electrodes versus standard one based on PVDF-HFP, carbon nanotubes, and 1-ethyl-3-methylimidazoliumtetrafluoroborate.

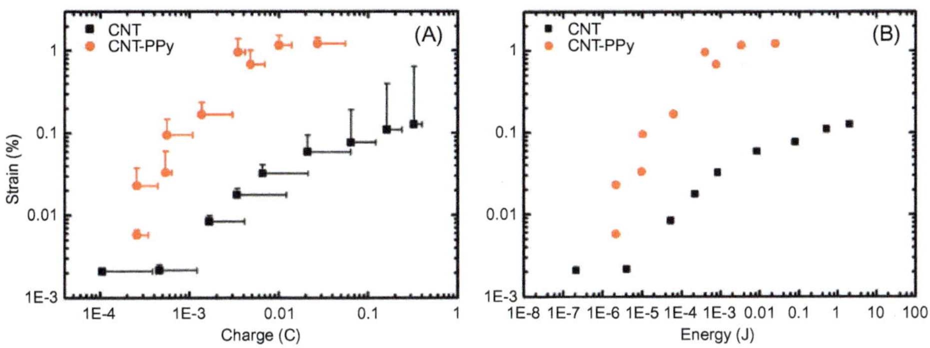

Figure 1.25 (A) Strain versus charge response of actuators composed of CNTs (squares) and CNT–polypyrrole (circles), (B) strain response versus energy stored considering the capacitance of the actuators.
Source: Reprinted from Biso M., Ansaldo, A., Picardo, E., Ricci, D., 2012b. Carbon, 4506–4511.

improvement in the actuation efficiency confirmed by higher strains obtained for a certain applied charge and stored energy (Fig. 1.25).

SEM was employed to examine changes in the electrode morphology of above-mentioned PPy and PANI enhanced polymeric composites (Fig. 1.26). Moreover, SEM cross-section of the actuator became a very useful technique allowing for a validation of the thickness and conformity of biocompatible protecting polymeric coating (Parylene C) (Bubak et al., 2015) as well as assessing the thickness of each layer and condition of solid polymer electrolyte after the lamination process (Sugino et al., 2011) (Fig. 1.27).

In addition, for polymeric composite electrodes enhanced by the additives in the form of metal particles (e.g., metal chalcogenides—WS_2, MoS_2) an energy dispersive X-ray spectroscopy (EDX) connected to the SEM could be a very convenient technique to estimate a homogeneity/dispersion of these additives in the actuator

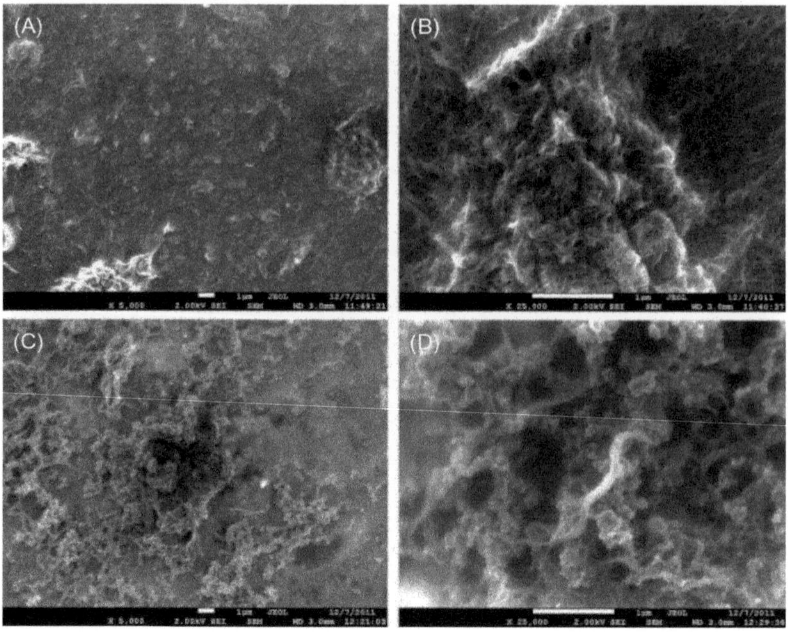

Figure 1.26 SEM images at 5000 × (A) and 25000 × (B) magnifications of the CNTs as received; SEM images at the same magnifications of the novel composite (C and D) containing CNT−PPy.
Source: Reprinted from Sugino, T., Kiyohara, K., Takeuchi, I., Mukai, K., Asaka, K., 2011. Carbon, 3560−3570.

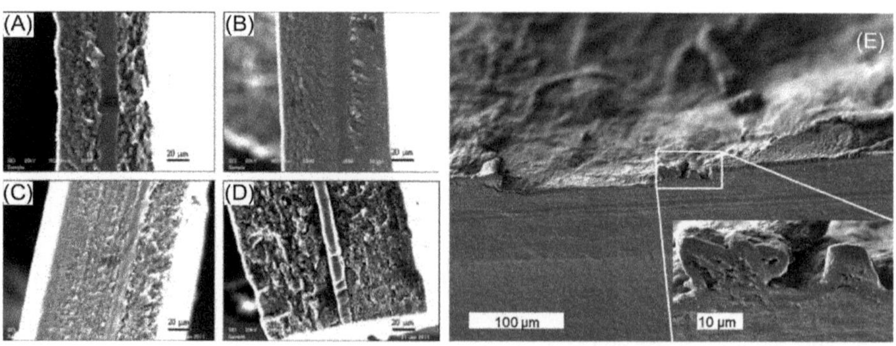

Figure 1.27 Cross-section SEM images of three-layered actuator elements: CNT(50) (A), CB(50) (B), CNT/CB(50/40) (C), and CNT/PANI(50/50) (D). (Magnification: 650 ×). Reprinted from (Sugino et al., 2011); and (E) details of the cross-section showing that the Parylene coating is conformal to the bucky gel electrode surface (inset).
Source: Reprinted from Bubak, G., Gendron, D., Ceseracciu, L., Ansaldo, A., Ricci, D., 2015. ACS Appl. Mater. Interfaces 7, 15542−15550.

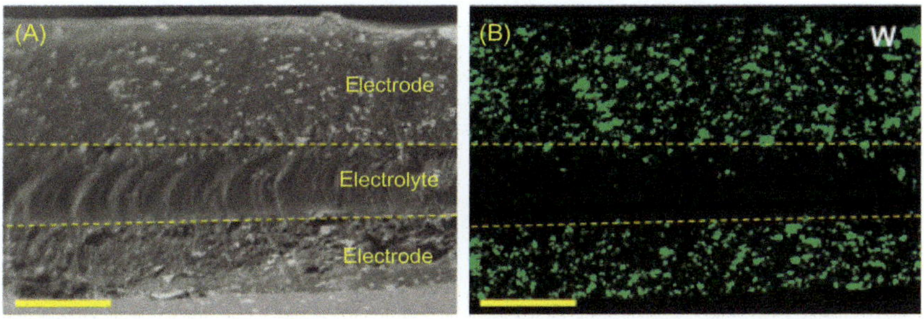

Figure 1.28 (A) Scanning electron micrograph (SEM) coupled with (B) energy dispersive X-ray spectroscopy (EDX) of W, in a typical cross-section of a carbon nanotube–metal chalcogenide actuator. In this example WS2(2-1). Scale bar = 50 μm.
Source: Reprinted from Gendron, D., Bubak, G., Ceseracciu, L., Ricciardella, F., Ansaldo, A., Ricci, D., 2016. Sensors Actuators B: Chem. 230, 673–683.

(Gendron et al., 2016) (Fig. 1.28). It could be also used to check whether those additives migrate within the composite after the actuation period or even after certain storage time.

Recently, bucky gel electrodes without a presence of ionic liquid were reported. By processing polymerized ionic liquid (PIL)—also called poly(ionic liquid)—with CNTs highly conductive and self-supported electrodes for actuating devices were obtained (Gendron et al., 2015a). Five different kinds of composites with three different ions were studied: bromide (Br^-), tetrafluoroborate (BF_4^-) and bis(trifluoromethane)sulfonamide ($TFSI^-$). Raman spectroscopy was employed to study carbon nanotubes integrity and their possible interaction with polymers showing that imidazolium based polymerized ionic liquids do not considerably alter carbon nanotubes (Fig. 1.29). On the contrary thermal stability of PIL-CNTs composites differs significantly as confirmed by thermogravimetric analysis (TGA), (Fig. 1.30). The degradation temperatures (T_d) for the polymerized ionic liquid containing the biggest anion among studied—$TFSI^-$ with the van der Waals volume of 147 Å—was the highest (313°C). For a bromide and tetrafluoroborate containing polymers lower T_d was recorded (165 and 196°C, respectively) showing that indeed the polymer anion in the PIL–CNT bucky gel composite has a significant effect on thermal properties. Nevertheless, all the composites had a T_d higher than typical temperature in which actuators are laminated ($\leq$130°C) (Gendron et al., 2015b; Mukai et al., 2009).

One of the most common mechanical parameters of the carbon nanotube–polymer composite electrodes is Young's modulus. Depends on the composition it can vary from tens of MPa for single-walled carbon nanotubes with PVDF(HFP) through hundreds of MPa for CNT/PANI and CNT/PVDF electrodes up to even few GPa for above-mentioned PIL/CNT composites without the presence of ionic liquid (Kong and Chen, 2014; Fukushima et al., 2005; Gendron et al., 2015a). Since according to the simple elastic bending model of electroactive polymer actuators (Ceseracciu et al., 2011) the generated strain α and blocking force depend on the Young's modulus of the electrodes it seems to be rational to seek for "stiffer" electrodes when higher actuation forces are sought.

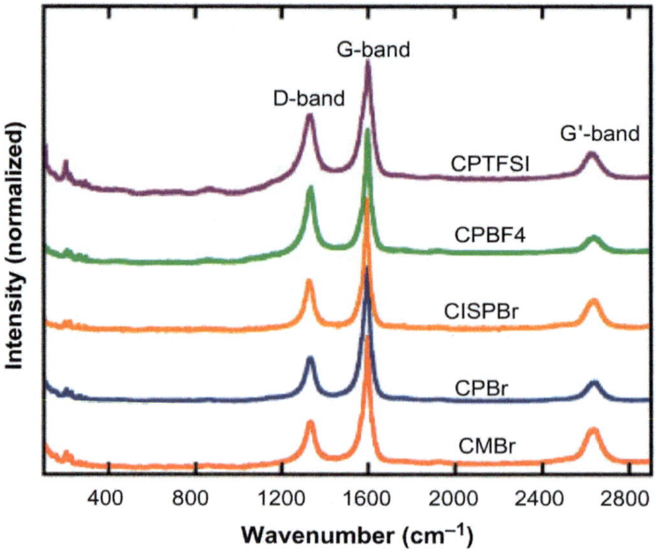

Figure 1.29 Raman spectra of composite CMBr, CISPBr, CPBr, CBF4, and CPTFSI. *Source*: Reprinted from Gendron, D., Ansaldo, A., Bubak, G., Ceseracciu, L., Vamvounis, G., Ricci, D., 2015a. Compos. Sci. Technol. 117, 364–370.

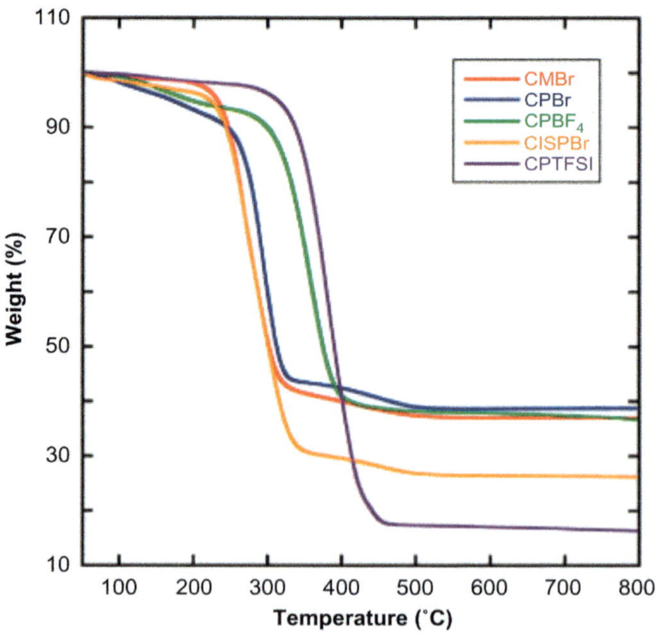

Figure 1.30 TGA thermograms of composites CMBr, CISPBr, CPBr, CPBF4, and CPTFSI. *Source*: Reprinted from Gendron, D., Ansaldo, A., Bubak, G., Ceseracciu, L., Vamvounis, G., Ricci, D., 2015a. Compos. Sci. Technol. 117, 364–370.

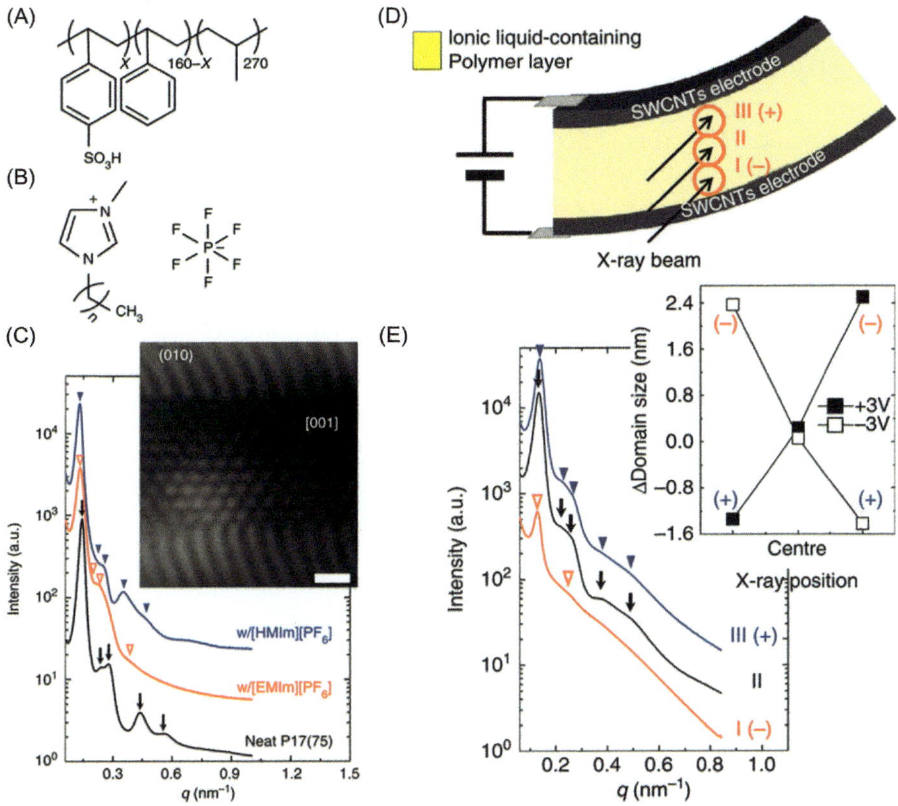

Figure 1.31 Molecular structures of (A) PSS-b-PMB block copolymer and (B) 1-alkyl-3-methyl imidazolium hexafluorophosphate. (C) SAXS profiles of P17(75) membranes with the addition of [HMIm][PF_6] and [EMIm][PF_6] at 25°C. The arrows for neat P17(75), the inverted open triangles of P17(75) integrated with 60 wt% [EMIm][PF_6] and the inverted filled triangles of P17(75) containing 60 wt% [HMIm][PF_6] represent Bragg peaks at q^*, $\sqrt{3}q^*$, $\sqrt{4}q^*$, $\sqrt{9}q^*$, and $\sqrt{16}q^*$ indicative of HEX morphology. The scattering profiles are vertically offset for clarity. A cross-sectional TEM image of neat P17(75) copolymer given in the inset figure of (C) shows the (010) plane and the [001] view of the HEX structure. Scale bar, 100 nm; the PSS phases were darkened by RuO_4 staining. (D) A schematic drawing of the measurement setup of in situ SAXS experiments. (E) Representative SAXS profiles obtained with the actuator comprising P17(75) and [HMIm][PF_6], acquired in regions I, II, and III under voltages of ± 3 V. The change in domain size of the HEX structure in each region upon switching the direction of the applied voltage is plotted in the inset figure.
Source: Kim O., Shin T.J. and Park M.J., Nat. Commun.4, 2013, 2208.

Considering the electrolyte layer of carbon based iEAPs, an elegant approach to boost actuator performance by using nanostructured layer of sulfonated polymers with covalently bonded nonionic polymethylbutylene (PSS-b-PMB block copolymer) (see Fig. 1.31) and ionic liquids instead of conventional PVDF based electrolytes or Nafion was reported by Kim et al. (2013). Thanks to shorter ion diffusion pathways along the nanoscale ionic channels, these actuators demonstrated strain of about 4% at ± 3 V making it one of the best results up to date. A well-defined hexagonal-cylindrical (HEX) morphology of this new kind of PSS-b-PMB based solid polymer electrolytes were investigated by small-angle X-ray scattering (SAXS) and cross-sectional TEM (Fig. 1.31C).

Interestingly, thanks to the cooperation of several research groups different kinds of ionic electroactive polymer actuators were also tested under laboratory imitated space conditions (Punning et al., 2014). In general, studies revealed that most of iEAP materials, mostly based on PVDF/PVDF-HFP/ionic liquid electrolytes are tolerant to vacuum lower than 1 mbar as well as freezing in 77 K and 4.22 K. Moreover, the performance of tested devices was not substantially affected by the exposure to UV radiation, X-rays, and ionizing γ-radiation.

References

Amb, C.M., Chen, S., Graham, K.R., Subbiah, J., Small, C.E., So, F., et al., 2011. J. Am. Chem. Soc. 133, 10062–10065.

Armin, A., Juska, G., Ullah, M., Velusamy, M., Burn, P.L., Meredith, P.A., et al., 2014. Adv. Energy Mater. 4, 1300954. Available from: http://dx.doi.org/10.1002/aenm.201300954.

Béguin, F., Presser, V., Balducci, A., Frackowiak, E., 2014. Adv. Mater. 26, 2219–2251.

Biso, M., Ansaldo, A., Ricci, D., 2012a. Phys. Status Solidi. 249, 2361–2364.

Biso, M., Ansaldo, A., Picardo, E., Ricci, D., 2012b. Carbon4506–4511.

Bredas, J.-L., Norton, J.E., Cornil, J., Coropceanu, V., 2009. Acc. Chem. Res. 42, 1691–1699.

Bubak, G., Gendron, D., Ceseracciu, L., Ansaldo, A., Ricci, D., 2015. ACS Appl. Mater. Interfaces. 7, 15542–15550.

Ceseracciu, L., Biso, M., Ansaldo, A., Futaba, D.N., Hata, K., Barone, A.C., et al., 2011. Sensors Actuators B Chem. 156, 949–953.

Chen, T., Yang, Z., Peng, H., 2013. ChemPhysChem. 14, 1777–1782.

Chen, Q., Li, X., Zang, X., Cao, Y., He, Y., Li, P., et al., 2014a. RSC Adv. 4, 36253.

Chen, T., Peng, H., Durstock, M., Dai, L., 2014b. Sci. Rep. 4, 3612.

Chen, Z., Cai, P., Chen, J., Liu, X., Zhang, L., Lan, L., et al., 2014c. Adv. Mater. 26, 2586–2591.

Cheng, Y.-J., Yang, S.-H., Hsu, C.-S., 2009. Chem. Rev. 109, 5868–5923.

Clarke, T.M., Peet, J., Nattestad, A., Drolet, N., Dennler, G., Lungenschmied, C., et al., 2012. Org. Electron. 13, 2639–2646.

Coffey, D.C., Reid, O.G., Rodovsky, D.B., Bartholomew, G.P., Ginger, D.S., 2007. Nano Lett. 7, 738–744.

DeLongchamp, D.M., Kline, R.J., Herzing, A., 2012. Energy Environ. Sci. 5, 5980–5993.

Dennler, G., Scharber, M.C., Brabec, C.J., 2009. Adv. Mater. 21, 1323–1338.

Fukushima, T., Kosaka, A., Ishimura, Y., Yamamoto, T., Takigawa, T., Ishii, N., et al., 2003. Science. 300, 2072–2074.

Fukushima, T., Asaka, K., Kosaka, A., Aida, T., 2005. Angew. Chem. Int. Ed. Engl. 44, 2410–2413.

Gao, H., Lian, K., 2014. RSC Adv. 4, 33091.

Gao, J., Dou, L., Chen, W., Chen, C.-C., Guo, X., You, J., et al., 2014. Adv. Energy Mater. 4, 1300739.

Gendron, D., Ansaldo, A., Bubak, G., Ceseracciu, L., Vamvounis, G., Ricci, D., 2015a. Compos. Sci. Technol. 117, 364–370.

Gendron, D., Bubak, G., Ceseracciu, L., Ansaldo, A., Ricci, D., 2015b. Sensors Actuators B Chem. 222, 1073–1082.

Gendron, D., Bubak, G., Ceseracciu, L., Ricciardella, F., Ansaldo, A., Ricci, D., 2016. Sensors Actuators B Chem. 230, 673–683.

Günes, S., Neugebauer, H., Sariciftci, N.S., 2007. Chem. Rev. 107, 1324–1338.

Hoppe, H., Sariciftci, N.S., 2004. J. Mater. Res. 19, 1924–1945.

Huang, Y., Zhang, M., Ye, L., Guo, X., Han, C.C., Li, Y., et al., 2012a. J. Mater. Chem. 22, 5700.

Huang, Y., Guo, X., Liu, F., Huo, L., Chen, Y., Russell, T.P., et al., 2012b. Adv. Mater. 24, 3383–3389.

Huang, Y., Liu, F., Guo, X., Zhang, W., Gu, Y., Zhang, J., et al., 2013. Adv. Energy Mater. 3, 930–937.

Huang, Y., Kramer, E.J., Heeger, A.J., Bazan, G.C., 2014. Chem. Rev. 114, 7006–7043.

Janocha, H., 2004. Actuators: Basics and Applications. Springer-Verlag Berlin Heidelberg, Berlin, Heidelberg.

Jo, C., Pugal, D., Oh, I.K., Kim, K.J., Asaka, K., 2013. Prog. Polym. Sci. 38, 1037–1066.

Kaempgen, M., Chan, C.K., Ma, J., Cui, Y., Gruner, G., 2009. Nano Lett. 9, 1872–1876.

Kang, Y.J., Yoo, Y., Kim, W., 2016. ACS Appl. Mater. Interfaces acsami.6b02690.

Khomenko, V., Frackowiak, E., Béguin, F., 2005. Electrochim. Acta. 50, 2499–2506.

Kim, O., Shin, T.J., Park, M.J., 2013. Nat. Commun. 4, 2208.

Kong, L., Chen, W., 2014. Adv. Mater. 26, 1025–1043.

Li, C., Liu, M., Pschirer, N.G., Baumgarten, M., Müllen, K., 2010. Chem. Rev. 110, 6817–6855.

Li, G., Shrotriya, V., Huang, J., Yao, Y., Moriarty, T., Emery, K., et al., 2005. Nat. Mater. 4, 864–868.

Li, H., Wang, J., Chu, Q., Wang, Z., Zhang, F., Wang, S., 2009. J. Power Sources. 190, 578–586.

Liang, Y., Yu, L., 2010. Acc. Chem. Res. 43, 1227–1236.

Lou, S.J., Szarko, J.M., Xu, T., Yu, L., Marks, T.J., Chen, L.X., 2011. J. Am. Chem. Soc. 133, 20661–20663.

Lu, L., Zheng, T., Wu, Q., Schneider, A.M., Zhao, D., Yu, L., 2015. Chem. Rev. 115, 12666–12731.

Meng, C., Liu, C., Chen, L., Hu, C., Fan, S., 2010. Nano Lett. 10, 4025–4031.

Mirfakhrai, T., Madden, J., Baughman, R., 2007. Mater. Today. 10, 30–38.

Mukai, K., Asaka, K., Sugino, T., Kiyohara, K., Takeuchi, I., Terasawa, N., et al., 2009. Adv. Mater. 21, 1582–1585.

Murphy, L., Hong, W., Aziz, H., Li, Y., 2013. Org. Electron. 14, 3484–3492.

Peet, J., Kim, J.Y., Coates, N.E., Ma, W.L., Moses, D., Heeger, A.J., et al., 2007. Nat. Mater. 6, 497–500.

Pingree, L.S.C., Reid, O.G., Ginger, D.S., 2009. Adv. Mater. 21, 19–28.

Punning, A., Kim, K.J., Palmre, V., Vidal, F., Plesse, C., Festin, N., et al., 2014. Sci. Rep. 4, 6913.

Qian, X., Gu, N., Cheng, Z., Yang, X., Wang, E., Dong, S., 2001. J. Solid State Electrochem. 6, 8–15.

Rivnay, J., Mannsfeld, S.C.B., Miller, C.E., Salleo, A., Toney, M.F., 2012. Chem. Rev. 112, 5488–5519.

Roberts, M.E., Wheeler, D.R., McKenzie, B.B., Bunker, B.C., 2009. J. Mater. Chem. 19, 6977.

Segalman, D.J., Witkowski, W.R., Adolf, D.B., Shahinpoor, M., 1992. Smart Mater. Struct. 1, 95–100.

Shahinpoor M., 1992. 1, 91–94.

Shahinpoor, M., Kim, K.J., 2001. Smart Mater. Struct. 10, 819–833.

Sugino, T., Kiyohara, K., Takeuchi, I., Mukai, K., Asaka, K., 2011. Carbon3560–3570.

Sylla, S., Sanchez, J.-Y., Armand, M., 1992. Electrochim. Acta. 37, 1699–1701.

Tang, C.W., 1986. Appl. Phys. Lett. 48, 183.

Terasawa, N., Ono, N., Hayakawa, Y., Mukai, K., Koga, T., Higashi, N., et al., 2011. Sensors Actuators B: Chem. 160, 161–167.

Terasawa, N., 2014. Sensors Actuators B. 202, 382–387. Available from: http://dx.doi.org/10.1016/j.snb.2014.05.096.

Thompson, B.C., Fréchet, J.M.J., 2008. Angew. Chem. Int. Ed. 47, 58–77.

Tiwari, R., Garcia, E., 2011. Smart Mater. Struct. 20, 083001.

van Bavel, S.S., Bärenklau, M., de With, G., Hoppe, H., Loos, J., 2010. Adv. Funct. Mater. 20, 1458–1463.

Wang, G., Zhang, L., Zhang, J., 2012a. Chem. Soc. Rev. 41, 797–828.

Wang, G., Lu, X., Ling, Y., Zhai, T., Wang, H., Tong, Y., et al., 2012b. ACS Nano. 6, 10296–10302.

Wang, J., Chu, H., Li, Y., 2008. ACS Nano. 2, 2540–2546.

Yang, C., Shen, J., Wang, C., Fei, H., Bao, H., Wang, G., 2014. J. Mater. Chem. A. 2, 1458–1464.

Yang, Z., Deng, J., Chen, X., Ren, J., Peng, H., 2013a. Angew. Chem. Int. Ed. 52, 13453–13457.

Yang, Z., Li, L., Luo, Y., He, R., Qiu, L., Lin, H., et al., 2013b. J. Mater. Chem. A. 1, 954–958.

You, J., Dou, L., Yoshimura, K., Kato, T., Ohya, K., Moriarty, T., et al., 2013. Nat. Commun. 4, 1446.

Yu, G., Gao, J., Hummelen, J.C., Wudl, F., Heeger, A.J., 1995. Science. 270, 1789–1791.

Yu, H., Wu, J., Fan, L., Xu, K., Zhong, X., Lin, Y., et al., 2011. Electrochim. Acta. 56, 6881–6886.

Yuan, C., Zhang, X., Wu, Q., Gao, B., 2006. Solid State Ionics. 177, 1237–1242.

Zhang, Q.T., Tour, J.M., 1998. J. Am. Chem. Soc. 120, 5355–5362.

Zhong, C., Deng, Y., Hu, W., Qiao, J., Zhang, L., Zhang, J., 2015. Chem. Soc. Rev. 44, 7484–7539.

Hybrid thermoplastic composites using nonwood plant fibers

2

Alireza Ashori
Department of Chemical Technologies, Iranian Research Organization for Science and Technology (IROST), Tehran, Iran

Chapter Outline

2.1 Introduction

Due to the concerns about environment and sustainability issues, this century has witnessed remarkable improvements in green materials in the field of polymer science through the development of biocomposites. In line with this objective, material engineers have conducted studies to replace current reinforced materials with plant fibers (Ashori, 2006; Nourbakhsh et al., 2008; Tabarsa et al., 2011; Begum and Islam, 2013; Jawaid et al., 2013; Ghanbari et al., 2014). It is important that these replacement materials exhibit similar capabilities compared with their counterparts while introducing other "green" characteristics. Plant fibers have been

Hybrid Polymer Composite Materials: Properties and Characterisation. DOI: http://dx.doi.org/10.1016/B978-0-08-100787-7.00002-0

utilized for material reinforcement for more than 3000 years ago (Taj et al., 2007). With the recent advancement in technology, these materials have been combined with polymers (Nourbakhsh et al., 2010; Azwa et al., 2013). Several types of non-wood plant fibers have been used for this purpose such as kenaf, roselle, jute, sugar palm, oil pump empty fruit bunch, sisal, pineapple leaf, rice husk, kapok, wood, barleys oat coir, and abaca (Nguong et al., 2013; Nadlene et al., 2016).

Fiber plastic composites (FPCs) are a composite material consisting of a polymer matrix embedded with high-strength plant fibers, like kenaf, oil palm, sisal, jute, and flax. Usually, polymers can be categorized into two categories, thermoplastics and thermosets (Nourbakhsh and Ashori, 2010). FPCs got considerable attention in numerous applications because of the good properties and superior advantages of plant fiber over synthetic fibers in term of its relatively low weight, low cost, less damage to processing equipment, good relative mechanical properties such as tensile modulus and flexural modulus, improved surface finish of molded parts composite, renewable resources, being abundant, flexibility during processing, biodegradability, and minimal health hazards (Ashori, 2008).

2.2 Natural fibers

Fibers are broadly classified as natural or man-made (synthetic) fibers (Fig. 2.1). Natural fibers have replaced synthetic fibers in many applications in the last few decades due to low cost, low density, and low tool wear. Wide use of natural fibers as reinforcements due to renewability and sustainability has brought several

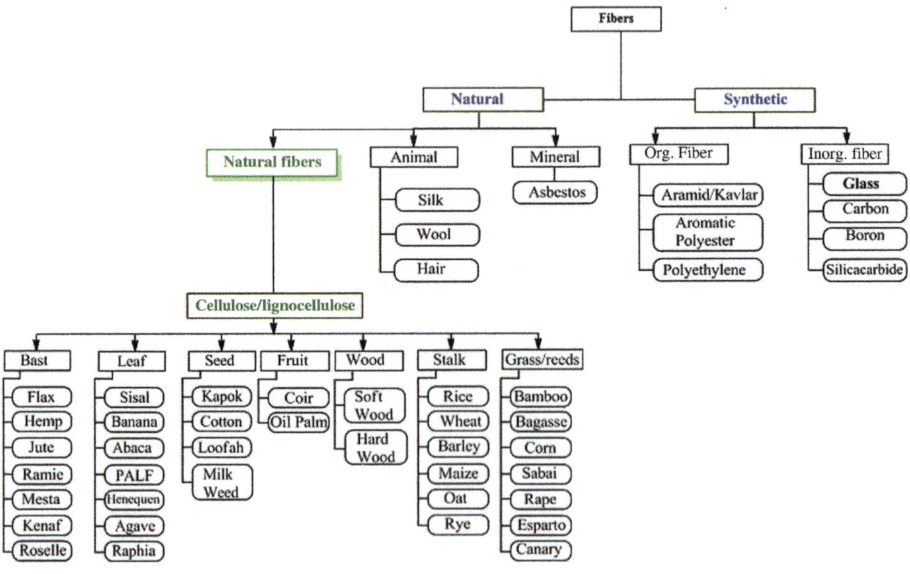

Figure 2.1 Classification of plant and synthetic fiber (Averous and Boquillon, 2004).

fibers into the composite field. Natural fiber is subdivided, based on their origins, viz. plants, animals, or minerals (Azwa et al., 2013; Gurunathan et al., 2015). All plant fibers are composed of cellulose while animal fibers consist of proteins (hair, silk, and wool). Plant fibers are categorized into two types based on their origins; wood plant fibers and nonwood plant fibers. Animal fibers and mineral fibers have not been widely used as reinforcement fibers. But several plant fibers have been used widely in biocomposites field for applications in the areas of automotive, marine, and construction. Regenerated cellulose fibers fall between plant and synthetic fibers which have been used as reinforcement fibers in recent times (Averous and Boquillon, 2004).

2.2.1 Wood plant fibers

Wood is a natural and complex polymeric composite, which essentially contains cellulose, hemicellulose, lignin, and extractives. A wide range of different substances is included under the extractives heading: flavonoids, lignans, stilbenes, tannins, inorganic salts, fats, waxes, alkaloids, proteins, simple and complex phenolics, simple sugars, pectins, mucilages, gums, terpenes, starch, glycosides, saponins, and essential oils. Extractives content in most temperate and tropical wood species are 4–10% and 20% of the dry weight, respectively (Ashori, 2006). Although extractives contribute merely a few percent to the entire wood composition, they have significant influence on its properties, such as mechanical strength, and the quality of wood can be affected by the amount and type of these extractives. Shebani et al. (2008) have noted that removing extractives improved the thermal stability of different wood species. Therefore using extracted wood for the production of wood–plastic composite would improve the thermal stability (Sheshmani et al., 2012).

2.2.2 Nonwood plant fibers

Forests, the major sources of wood fibers, are declining at the alarming rate of 13.0 million ha/year in developing countries. Due to the sharp ecological damage, the global demand for fibrous material and worldwide shortage of trees in many areas, there has been growing interest in the use of nonwood plant fibers as an alternative or supplementary fiber source for the 21st century (Ashori, 2006). Nonwood plant fibers are abundantly available in many countries and are the major source of fiber for papermaking in some developing countries, particularly China and India. Approximately 2.5 billion tons of nonwood raw materials are available each year worldwide, however, most of this raw material is currently untapped. Most nonwood plants are annual plants that develop full fiber potential in one growing season. Table 2.1 shows a wide variety of nonwood plant fibers that can be used (Ashori, 2008).

The fibers from the plants can be in the form of hairs (cotton, kapok), hard fibers (coir, sisal), and fiber sheaves (flax, hemp, jute). The plant fibers are classified depending on their utility such as primary and secondary. Plants to be used as fibers

Table 2.1 List of important plant fibers (Ashori, 2008)

Fiber source	Species	Origin
Abaca	*Musa textilis*	Leaf
Bagasse	–	Grass
Bamboo	(>1250 species)	Grass
Banana	*Musa indica*	Leaf
Broom root	*Muhlenbergia macroura*	Root
Cantala	*Agave cantala*	Leaf
Caroa	*Neoglaziovia variegate*	Leaf
China jute	*Abutilon theophrasti*	Stem
Coir	*Cocos nucifera*	Fruit
Cotton	*Gossypium* sp.	Seed
Curaua	*Ananas erectifolius*	Leaf
Date palm	*Phoenix dactylifera*	Leaf
Flax	*Linum usitatissimum*	Stem
Hemp	*Cannabis sativa*	Stem
Henequen	*Agave foourcrocydes*	Leaf
Isora	*Helicteres isora*	Stem
Istle	*Samuela carnerosana*	Leaf
Jute	*Corchorus capsularis*	Stem
Kapok	*Ceiba pentranda*	Fruit
Kenaf	*Hibiscus cannabinus*	Stem
Kudzu	*Pueraria thunbergiana*	Stem
Mauritius hemp	*Furcraea gigantea*	Leaf
Nettle	*Urtica dioica*	Stem
Oil palm	*Elaeis guineensis*	Fruit
Piassava	*Attalea funifera*	Leaf
Pineapple	*Ananus comosus*	Leaf
Phormium	*Phormium tenas*	Leaf
Roselle	*Hibiscus sabdariffa*	Stem
Ramie	*Boehmeria nivea*	Stem
Sansevieria	*Sansevieria*	Leaf
Sisal	*Agave sisilana*	Leaf
Sponge gourd	*Luffa cylindrica*	Fruit
Straw (cereal)	–	Stalk
Sun hemp	*Crotolaria juncea*	Stem
Cadillo/urena	*Urena lobata*	Stem

for primary utilities include hemp, jute, kenaf, etc., while the byproducts of plants such as coir, pineapple, etc., belong to the secondary group. There are six types of plant fibers namely bast fibers (flax, hemp, jute, kenaf, and ramie), leaf fibers (abaca, pineapple, and sisal), seed fibers (coir, cotton, and kapok), straw fibers (corn, rice, and wheat), grass fibers (bagasse and bamboo), and wood fibers (softwood and hardwood). Table 2.2 lists the physical and chemical properties of some nonwoods in comparison to those of wood. The dimensions of

Table 2.2 Comparison of physical and chemical properties of nonwood fibers with those of wood raw materials (Ashori, 2006)

Properties	Kenaf	Straw	Bagasse	Bamboo	Eucalyptus	Birch	Spruce
Physical							
Fiber length, mm	1.4	1.3	1.7	2.3	1.0	1.9	3.6
Fiber width, μm	29	12.9	20	14.4	18	25	35
Felting factor[a]	51	102	85	161	51	58	101
Chemical							
Holocellulose, %	77.2	78.1	77.8	76.6	74	81	71
Hemicellulose, %	28.1	24.1	27.9	19.5	18	40	27
Lignin, %	19.9	18.4	20.8	23.4	26	19	29

[a]The ratio of fiber length to fiber width.

nonwood fibers are between those of hardwoods and softwoods. The cellulose content of most of nonwoods listed in Table 2.2 is comparable to that of woods, while the lignin content is much lower than for woods (Ashori, 2006).

2.2.3 *Recycled fibers*

Shortage of fiber supply in the most countries made researchers both in industry and in academia to look out for new sources of fibrous materials. If a decision is taken to substitute a new material for an established one, care must be taken to ensure that all the characteristics of the new material are well understood. The main parameters that need to be examined for material alternative can be grouped as follows (Ashori and Nourbakhsh, 2009):

1. Technical performance advantage, as a result of introducing a stronger, stiffer, tougher, or lighter material.
2. Economic advantage over the total life cycle of the product. This can be achieved as a result of introducing cheaper material, more cost effective use of material, lower cost of processing, better recycle ability and lower cost of disposal, or lower running cost of the product.
3. Changing the character of the product by incorporating a material that is esthetically more attractive, with a different feel or that can provide more comfort to the user through sound or heat insulation for example.
4. Environmental and legislative considerations including less damage to the environment, better recycling or reuse, and compliance with environmental regulations.

The possibility of using recycled materials in the development of composites is very attractive, especially with respect to the large quantity of wood fiber/plastic waste generated daily. Recycled fibers such as wastepaper can meet all the requirements in order to replace inorganic fillers in thermoplastic composites (Ashori and Nourbakhsh, 2008). Advantages associated with biocomposite products include lighter weight and improved acoustic, impact, and heat reformability properties—all at a cost less than that of comparable products made from plastics alone. In addition, these composites may possibly be reclaimed and recycled for the production of second-generation composites. Interest is high for the use of recycled fibers in composites, thus providing cost and environmental benefits (Ashori, 2010).

2.2.4 Mechanical and physical properties of plant fibers

The mechanical and physical properties of plant fibers are very important for industrial applications and can contribute to the use of plant fibers in numerous applications. Lower density leads lower-weight structures in the automotive industry and aerospace applications. Mechanical properties such as tensile properties, flexural properties, and impact strength are strongly affected by fiber content (Nourbakhsh and Ashori, 2008). Compared with oil palm epoxy composites, the tensile properties of jute oil palm fiber hybrid composites are enhanced substantially with increased jute fiber content loading. The mechanical properties of jute fiber-reinforced composites are superior to those of sisal fiber-reinforced composites (Alkbir et al., 2016). Venkateshwaran et al. (2011) reported that hybridizations of sisal fibers with banana/ epoxy composites of up to 50 wt% enhance the mechanical properties and degrade the water absorption properties of these fibers. The overall tensile and flexural properties of plant fiber-reinforced polymer hybrid composites are highly dependent on the aspect ratio, moisture absorption tendency, morphology, and dimensional stability of the fibers used. Some of the physical properties of nonwood plant fibers are listed in Table 2.3. These fibers are low-cost fibers with low densities and specific properties that are comparable with those of synthetic fibers.

Table 2.3 Mechanical and physical properties of nonwood plant fibers (Alkbir et al., 2016)

Fibers	Tensile strength (MPa)	Young's modulus (GPa)	Elongation at break (%)	Density ($g\ cm^{-3}$)
Sisal	350–640	12.8–22	3.7	1.41–1.45
Oil palm	70.9–248	14–6.7	14–25	0.7–1.55
Bamboo	215–218	28–30	1.3	0.6–0.9
Banana	529–914	27–32	5–9	1.35
Coir	120–304	4–6	15–40	1.15–1.25
Cotton	287–800	5.5–12.6	3–10	1.51–1.6
Flax	345–1500	23.9–27.6	1.6–3.2	1.5
Hemp	690	60–70	1.6–4	1.4–1.5
Jute	393–780	13–30	1.9	1.3–1.45
Kenaf	284–1191	21–60	1.6–3.5	0.13–0.17
Pineapple	413–1627	60–82		1.07–2.4
Ramie	400–938	44–128	1.2–8	1–1.55

2.3 Composites

FPCs have attracted a significant interest in the last decades, due to the specific advantages they can grant in comparison with the classic mineral filler/plastic composites (Ashori, 2010). FPCs are widely used in many industries such as the aircraft, automobile, leisure, electronic, and medical industries. The composite industry always looks into alternative low-cost sources, which can decrease overall manufacturing costs and increase stiffness of the materials. Plant materials offer significant advantages, which justify their use. The potential advantages of plant fibers, apart from their environmental benefits, are the abundant availability of the raw materials from renewable resources, rather than fossil sources, and their low cost. Also, they have high specific strength due to their low density. Furthermore, it is possible to obtain a higher loading of plant fibers in FPCs than conventional inorganic fillers because of the softer nonabrasive nature of the plant fibers (Hamzeh et al., 2012).

FPCs include wood/fiber thermosetting composites and wood/fiber thermoplastic composites. The techniques of producing wood thermosetting composites have been well established since the 1900s. Phenol formaldehyde and urea formaldehyde are the most common thermosetting resins used as adhesives for wood composite products, such as plywood, particleboard, fiberboard, and oriented strand board. Fiber thermoplastic composites are manufactured by dispersing fibers or wood flour (WF) into molten plastics to form composite materials by processing techniques such as extrusion, thermoforming, and compression or injection molding (Nourbakhsh and Ashori, 2008). Although wood thermoplastic composites have been developed in the United States since 1980s, they have recently experienced dramatic growth. The abbreviation "FPC" most often represents fiber thermoplastic composite. FPCs take advantage of both wood and plastics. As mentioned earlier, the advantages of wood include low density, low-equipment abrasiveness, relatively low cost, and good biodegradability. However, plastics provide FPCs with good moisture and decay resistance. In addition, various surface optical effects can be obtained by adding different wood species and colored pigments. It is forecast that the demand for plant fiber/plastic composites will grow about 60% per year for building construction products and 50% per year for automotive applications (Ashori and Nourbakhsh, 2009).

In spite of the advantages, the use of wood in thermoplastics has been plagued by the thermal stability limitation of wood, and the difficulties in obtaining good filler dispersion and strong interfacial adhesion. This is because of the natural incompatibility between the hydrophilic, polar wood fibers and hydrophobic, nonpolar thermoplastics. Such phase incompatibility causes a weak interface between the wood filler and the matrix. Moreover, strong wood–wood interactions resulting from hydrogen bonding and physical entanglement impair the dispersion of the fillers in the viscous matrix. Thermoplastics commonly used for manufacturing FPCs include low- and high-density polyethylene (HDPE and LDPE), polypropylene (PP), polyvinyl polystyrene (PS), and chloride (PVC) (Table 2.4) (Sheshmani et al., 2012; Hamzeh et al., 2012; Nourbakhsh and Ashori, 2009; Ashori et al., 2013).

Table 2.4 Properties of commodity plastics

Property	LDPE	HDPE	PP	PS
Elongation (%)	90–800	20–1000	200–700	1.2–2.5
Tensile strength (68×10^6N m^{-2})	0.6–2.3	3.1–5.5	4.3–5.5	5–12
Tensile modulus (10^5 psi)	0.2–0.4	0.6–1.8	1.6–2.3	4–6
Impact strength (53.1 J m^{-1})	16	0.5–2.0	0.5–2.0	0.2–0.4
Burning rate	Very slow	Very slow	slow	slow
Effect of alkalis	resistance	resistance	resistance	Attacks
Effect of organic solvents	Resistance to below 80°C	Resistance to below 80°C	Resistance to below 80°C	Soluble
Clarity	Opaque	Opaque	Opaque	Transparent
Specific gravity	0.91–0.92	0.94–0.96	0.9–0.91	1.4–1.09

LDPE, low-density polyethylene; HDPE, high-density polyethylene; PP, polyethylene; PS, polystyrene.

2.4 Thermoplastic composites

2.4.1 Low-density polyethylene

LDPE can be polymerized using a high- or low-pressure method in the temperature range of 180–200°C. It is used for many applications that require low temperature flexibility, toughness, and durability. It has good corrosion resistance, low moisture permeability, and good resistance to stress cracking. LDPE has excellent resistance to dilute and concentrated acids, alcohols, bases, and esters, but is not recommended for use in contact with halogenated hydrocarbons. LDPE is widely used for manufacturing containers, dispensing bottles, wash bottles, tubing, and molded laboratory equipment. Its most common use is in packaging; food storage and laboratory containers are also made of it. It is used for making chemical-resistant tanks; six-pack soda can rings; and thermoformed products such as trays, end caps, and tops (Nourbakhsh and Ashori, 2009).

2.4.2 High-density polyethylene

HDPE has better tensile strength than LDPE due to stronger intermolecular forces. It is more opaque and can withstand somewhat higher temperatures (120°C for short periods, 110°C continuously). HDPE has superior corrosion resistance, good chemical resistance, and excellent stress cracking resistance. HDPE can be sterilized by boiling, and meets food and drug administration requirements for direct food contact applications. Rods, tubing, pipes, tapes, racks, trays, Tupperware containers, laundry detergent bottles, milk cartons, and fuel tanks are some general-purpose applications. It can be used for pipes for domestic water supply, chemical-resistant use, natural gas distribution, and geothermal heat transfer systems. Some of the industrial applications are for steel pipelines/wall coverings, laboratory equipment, lavatory partitions, and man-hole covers. There is a good scope of using HDPE with cheaply available WF in making composites. Wood–plastic composites are susceptible to weathering that affect appearance and mechanical properties of these composites. Addition of ultraviolet absorbers and pigments is effective in preventing the degradation due to weathering, as studied by Stark and Matuana (Stark and Matuana, 2004). Li and Yan (2007) prepared HDPE/ionomer blends by using WF as a filler up to 60% loading.

2.4.3 Polypropylene

The most useful polymers are those that have outstanding physical, chemical, mechanical, thermal, and electrical properties. Compared with LDPE or HDPE, PP has lower impact strength, but has superior working temperature and good tensile strength. Some of the general-purpose applications are for film, sheet, rolls, tape, rods, tubing, pipes, housewares, furniture, appliances, luggage, toys, battery cases, and other durable items for home, garden, or leisure uses. Other applications include hoods, orthopedic devices, structural covers, light tables, laboratory tables, and rinse and etch housing for electronics. PP is a tough and heat-resistant material, ideal for transfer of hot liquids or gases (Ashori and Nourbakhsh, 2008). One of its important applications is in automobiles for automotive interiors, monomaterial dashboard, bumpers, cladding, and exterior trim due to its low coefficient of thermal expansion, light weight, high chemical resistance, good weather ability, and proper impact/stiffness balance. Faruk et al. (2012) studied wood fiber-reinforced PP with different fiber contents (40%, 50%, and 60% by weight) and maleic anhydride grafted polypropylene (MAPP) as the compatibilizer. The Charpy impact strength increased with moisture content. PP reinforced with long wood fibers had better impact resistance than hardwood fiber-based composites (Ashori, 2013).

2.4.4 Polystyrene

General-purpose PS is brittle and transparent in nature. Most foods, drinks, and household fluids have no effect on PS, but it can be destroyed by citrus fruit rind oil, gasoline, turpentine, and lacquer thinner. Its continuous service temperature is well under 95°C and a major drawback is its brittleness (which can be improved by adding styrene butadiene rubber). Some of the important applications are for packaging, toys, housewares, bottles, lenses, electronic appliances, furniture, refrigeration, toys, utensils, display boxes, model aircraft kits, and ballpoint pen barrels. It is used as building/construction material and largely for packaging applications. Other applications are for cutlery, yogurt and cottage cheese containers, cups, clear salad bar containers, and video/audiocassette housings.

2.4.5 Polyvinyl chloride

PVC can be polymerized industrially by emulsion or suspension polymerization techniques. Its light weight, good mechanical strength, and toughness are key technical advantages for building and construction. Other properties of importance are resistance to weathering, chemical rotting, corrosion, shock, and abrasion. PVC is an excellent material in electrical applications for sheathing of cables. It performs better in terms of combustibility, flammability, flame propagation, and heat release compared with other plastics. PVC is the material of choice for scaffolding billboards, interior design articles, window frames, and fresh and waste water systems. Chlorinated PVC is ideally suitable for self-supporting constructions where good corrosion resistance at high temperatures is needed.

2.5 Hybrid composites

Reinforcement by two or more fibers into a single matrix leads to the development of hybrid composites with a great diversity of material properties. It appears that the behavior of hybrid composites is simply a weighted sum of the individual components so that there is a more favorable balance of properties in the resultant composite material (Ashori, 2010).

In recent years, hybrid composites have been developed by using more than one type, shape, or size of reinforcement. These composites have been developed to provide synergistic properties of the chosen fillers and matrix. They offer a range of properties that cannot be obtained with a single type of reinforcement. Hybridization may offset the disadvantages of one component by the addition of another fiber. A requisite for the occurrence of a hybrid effect is that the two fibers will vary by their mechanical properties and by the interfaces they form with the matrix (Ashori and Nourbakhsh, 2010). A combination of only two types of fibers would likely be most useful. For example, the hybridization of two types of short biofibers or lignocellulosic fibers having different lengths and diameters offers some advantages over each fiber alone being used in a single polymer matrix. Hybrid reinforcement with good fibers selection possibly will produce excellent properties and fulfill a current demand for polymer matrix composites. In recent years, there have been quite a number of studies on the hybridization of synthetic fiber and plant lignocellulose fiber. However, hybrid composites using plant fibers are less studied and the most published reports are limited to the hybrid composite consisting of one plant fiber and one synthetic fiber. For example, the use of plant/glass fibers and talc/carbon fiber has been reported (Ashori, 2010).

With regard to the environmental aspects it is very interesting that plant fibers could be used instead of synthetic fibers such as carbon, aramid, and glass fibers in composite materials. Synthetic fibers cause health problems to workers due to the skin irritations during processing and handling. Compared to the traditional synthetic fibers, plant fibers from renewable plant resources present lower density, less abrasiveness, lower cost and they are renewable and biodegradable. Much work has been done on virgin thermoplastic and lignocellulosic fiber composites, which have successfully proven their applicability to various fields of technical applications, especially for loadbearing application. However, work done on waste hybrid biocomposite systems are still limited (Ashori and Sheshmani, 2010).

2.6 Modification of plant fibers

Even though there is wide diversity in the chemical compositions of plant fibers, almost all plant organic fibers have a hydrophilic nature. This is due to the presence of hydroxyl (OH) and carboxylic acid (COOH) groups in heteropolysaccharides, such as hemicelluloses and pectins. Additionally, cellulose is rich in hydroxyl groups, but the high linearity and crystallinity of cellulose microfibrils reduce its hydrophilicity. However, free OH-groups on the surface of the microfibrils are still susceptible to absorb significant amounts of water (Nourbakhsh and Ashori, 2008; Väisänen et al., 2016). Lignin is the third major structural organic polymer of lignocellulosic materials. Unlike the carbohydrate-based fractions of the cell wall, particularly hemicelluloses, lignin has a hydrophobic structure. The majority of the matrix polymers used in FPCs are hydrophobic, which lead to the problem of poor surface wetting of plant fibers by the polymers (Adekunle, 2015). This can lead to impaired stress transfer and the formation of void spaces within the composite. Another concern related to the use of FPCs is the plant fibers' tendency for moisture absorption under fluctuating weather conditions and the consequent dimensional instability (Adhikary et al., 2008). The swelling of plant fibers may evoke a stress in the surrounding matrix, which may eventually result in composite damage, even its failure (Sombatsompop and Chaochanchaikul, 2004). In addition, the high moisture uptake increases the susceptibility to biodegradation due to microbial and fungal attack (Klyosov, 2007). There are several procedures intended to overcome the problems associated with FPCs. A typical goal of these approaches is to create a chemical bond between polymer matrix and reinforcement by the incorporation of an interacting substance. The treatments that aim to improve compatibility and adhesion between plant fibers and the polymer matrix can be roughly divided into chemical and physical treatments (Väisänen et al., 2016).

2.6.1 Chemical modification of plant fibers

To develop composites with good mechanical properties, chemical modification of fiber carried out to reduce the hydrophilic behavior of fibers and the absorption of moisture (Gassan and Bledzki, 1999; Rong et al., 2001; Cordeiro et al., 2013; Tran et al., 2014). The different surface treatments of advanced composites applications were reviewed by several researchers. The effects of different chemical treatments on cellulosic fibers that were employed as reinforcements for thermoplastics and thermoset were also examined. For the treatments, the different kinds of chemical treatment include silane, alkali, acrylation, benzoylation, maleated coupling agents, permanganate, acrylonitrile and acetylation grafting, stearic acid, peroxide, isocyanate, triazine, fatty acid derivate (oleoyl chloride), sodium chloride, and fungal. The main purpose of surface treatments of plant fibers to enhanced fiber/matrix interfacial bonding and stress transferability of the composites. Chemical modifications are applied to plant fibers in order to improve the matrix–fiber adhesion (Fig. 2.2). Some chemical modification could lead to reduced moisture absorption of plant fibers and their composites. Most of the chemical modifications of plant fiber involve silanization, alkalization (mercerization), acetylation, cyanoethylation, benzoylation, isocyanation, dewaxing, esterification, etherification, and graft copolymerization. Other modifications of plant fibers include crosslinking with formaldehyde, *p*-phenylenediamine and phthalic anhydride; nitration; dinitrophenylation and transesterification. Their chemical composition allows them to react with the fiber surface, which forms a bridge of chemical bonds between the fiber and matrix. Some of the most effective chemical modification strategies are briefly described as follows (Tran et al., 2014).

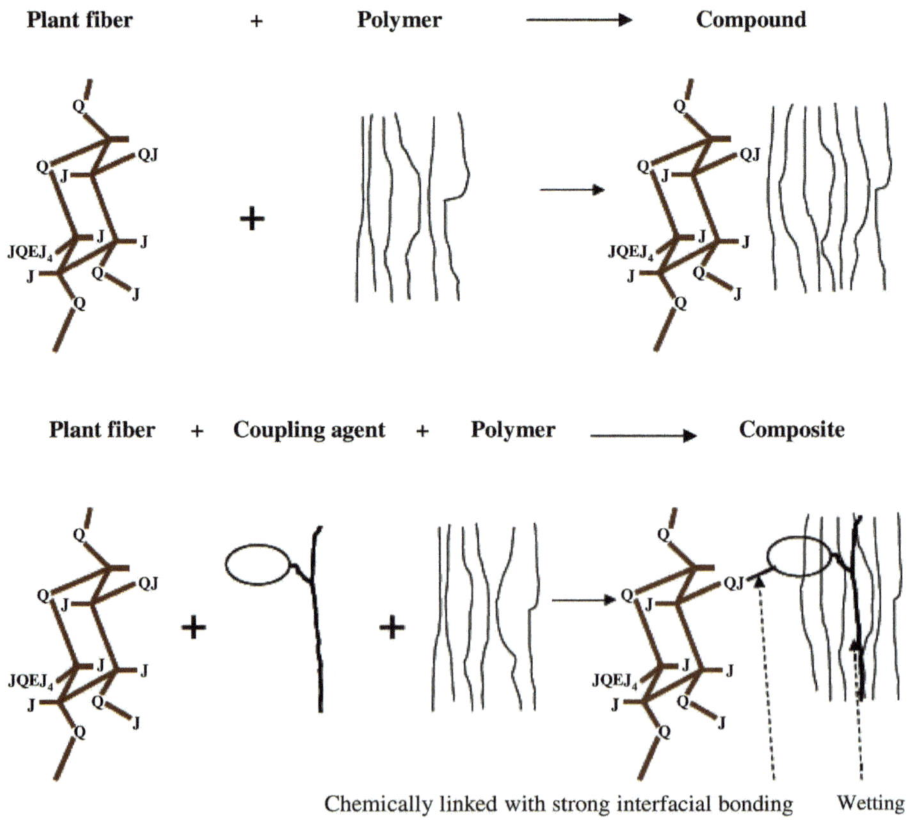

Figure 2.2 Mechanism of coupling agent between hydrophilic fibers and hydrophobic matrix polymer (Ashori, 2008).

Alkaline treatment or mercerization is one of the oldest, cost effective and most used chemical methods for plant fiber when they used to reinforce thermoplastics and thermosets. Its efficiency depends on the type and concentration of the alkaline solution, time of treatment, and the temperature used for modification. If the alkali concentration is higher than the optimum condition, the excess delignification of the fiber can take place, which results in weakening or damaging the fiber. This treatment removes lignin, hemicellulose, wax, and oils covering the surface of the fiber (Cordeiro et al., 2012). Addition of aqueous sodium hydroxide (NaOH) to plant fiber promotes the ionization of the hydroxyl group to the alkoxide plant fibers. The development of a rough surface and the enhanced aspect ratio result in a better mechanical interlocking that induces as improved the fiber/matrix interfacial adhesion in the resulting composites. Several studies conducted on alkali treatment of plant fibers and they reported that the mercerization led to an increase in the amount of amorphous cellulose while reduce the hydrogen bonding intensity. Ali et al. (2003) have studied the effect of fiber treatment on the tensile properties of Mater-Bi Y and Mater-Bi Z with 20% of short sisal fiber. They have showed that the alkaline treatment favored an increase fiber aspect ratio and improved mechanical properties of the Mater-Bi Z and Mater-Bi Y/sisal fiber biocomposites. The effects of alkali treatment of pineapple leaf fiber on the performance of pineapple leaf fiber/polylactic acid (PLA) biocomposites have been shown by Huda et al. (2008). It was found that the alkali-treated fiber-reinforced biocomposites offered superior mechanical properties compared to untreated fiber biocomposites. This study also suggested that the appropriate modification of plant fiber surface significantly contributes to improving the interfacial properties of the resulting biocomposites. Cao et al. (2006) investigated the effect of NaOH treatment of bagasse fiber on the mechanical properties of bagasse fiber-reinforced polyester biocomposites. Among, the various concentration of NaOH used, superior properties were obtained for the biocomposites made from 1% NaOH treated bagasse fiber.

Silane is a multifunctional molecule which is used as a coupling agent to modify fiber surfaces. It was found to be the most effective among many coupling agents for the plant fiber surface treatment. The uptake of silane is very much dependent on a number of factors including hydrolysis time, organofunctionalized of silane, temperature, and pH. It undergoes several stages of hydrolysis, condensation, and bond formation during the treatment process of the plant fiber. After hydrolysis, one end of silanol reacts with the cellulose hydroxyl groups (Si—O—cell-fiber) and the other end reacts (bond formation) with the matrix (Si-matrix) functional groups. After silane modification, hydrocarbon chains allow the fiber to absorb more water, which means that its chemical affinity to the polymer matrix is improved. Goriparthi et al. (2012) also studied jute fiber to reinforced PLA composites, but their focus was to improve the adhesion between the fiber-matrix by surface modification of jute fiber in the presence of alkali, permanganate, peroxide, and silane treatments. Combination of the special prepreg fabrication method along with surface treatment on their composite sample exhibits enhancement at least 45% on the tensile and flexural modulus. Other authors investigated the effect of silane treatments on nonwoven kenaf/PLA composites and the effect of a combined alkali and silane treatment on ramie fiber/PLA composites. They found a significant improvement of the overall mechanical properties, the alkali and combined alkali/silane treatments showed the best results. Acetylation treatment is known as esterification methods for plasticizing of plant fiber. Acetyl group ($-CH_3COO$) reacts with the hydrophilic hydroxyl groups of the fiber and takes out the existed moisture. As a result, hydrophilic nature of the fiber is reduced while improves the dimensional stability as well as dispersion of fiber into polymeric matrices. After acetylation, the moisture regains considerably reduced as the fiber became more hydrophobic due to the substitution of hydroxyl groups with acetyl groups. Plant fibers are acetylated with and without an acid catalyst to graft acetyl groups onto the cellulose structure. In general, acetic acid and acetic anhydride individually do not react sufficiently with the plant fiber (Gurunathan et al., 2015).

2.6.2 Physical methods

Physical methods include stretching, calendaring, thermo treatment, and the production of hybrid yarns for the modification of plant fibers. Physical treatments change structural and surface properties of the fiber and thereby influence the mechanical bonding of polymers. Physical treatments do not extensively change the chemical composition of the fibers. Therefore the interface is generally enhanced via an increased mechanical bonding between the fiber and the matrix.

Physical methods involve treatment by corona-discharge treatment, physico-chemical ones, such as steam explosion treatment, high energy ray radiation processing and autoclave treatment (Albanoa et al., 2002; Ragoubi et al., 2012). All these techniques aim to improve the fiber-matrix adhesion by reducing the difference between hydrophilic/hydrophobic characters of fiber and the matrix (Deepa et al., 2011). It is noteworthy that physical treatments appear as the most ecofriendly ones. Plasma treatment offers a unique approach to modifying the chemical and physical structures of both fiber and polymeric surfaces without altering the bulk structures and characteristics of resulting materials (Mahlberg et al., 1999). Plasma treatment is mainly applied for the cleaning, sterilization and surface etching of the films in food packaging application. In fact, surface hydrophilicity and adhesion ability of the films increase dramatically after plasma treatment because polar groups are formed on film surfaces. Therefore, further modification is possible through assembling hydrophilic substances with antimicrobial abilities onto the surface of the treated films. On the other hand, steam explosion results in improved properties of lignocellulosic materials, which include reduced stiffness, smoother surface, improved bending properties, and better distribution. Steam explosion process, a high-pressure steaming involves heating of lignocellulosic materials at high temperatures and pressures followed by mechanical disruption of the pretreated material by violent discharge (explosion) into a collecting tank. This process has been applied to many lignocellulosic materials to enhance dispersibility and adhesion with the polymer matrix (Foscher et al., 1998).

2.7 Conclusions

The environment issues and the shortage of wood fiber have spurred efforts to find alternative ways for green composites. Hence, new composites utilizing nonwood plant fibers have been developed. Several studies have shown that FPCs consisting of nonwood fibers have comparable properties with the traditional reinforced composite materials. There are many types of plant fibers with highly complex structures, the properties of FPCs can be tailored according to the different fiber types. Additionally, it is possible to modify the properties of FPCs with proper chemical and physical treatments.

FPCs offer multiple advantages over traditional thermoplastic composite materials with conventional reinforcements, such as E-glass, Kevlar or carbon fiber, e.g., reductions in processing temperature, cycle time and density. However, there are characteristic factors that affect the overall performance of FPCs, which may limit the applicability of FPCs. The inherent hydrophilicity of natural fibers deteriorates the bonding between the polymer matrix and the fiber reinforcement, which dictates the final properties of the composite. Moreover, the moisture and thermal instability of plant fibers are well-recognized drawbacks associated with FPCs. Nonetheless, one can safely predict that the development of FPCs modified or reinforced with organic waste and residues will be able to overcome these limitations. The replacement of petroleum-derived and ecologically unfriendly FPC constituents, such as polymers and additives, with more sustainable alternatives is the current trend of the modern polymer composite industry.

References

Adekunle, K.F., 2015. Surface treatments of natural fibers—a review: part 1. Open J. Org. Polym. Mater. 5, 41−46.

Adhikary, K.B., Pang, S., Staiger, M.P., 2008. Long-term moisture absorption and thickness swelling behavior of recycled thermoplastics reinforced with *Pinus radiata* sawdust. Chem. Eng. J. 142, 190−198.

Albanoa, C., Reyes, J., Ichazo, M., Gonzálezc, J., Britob, M., Moronta, D., 2002. Analysis of the mechanical, thermal and morphological behavior of polypropylene compounds with sisal fiber and wood flour irradiated with gamma rays. Polym. Degrad. Stab. 76, 191−203.

Ali, R., Iannace, S., Nicolais, L., 2003. Effect of processing conditions on mechanical and viscoelastic properties of biocomposites. J. Appl. Polym. Sci. 88, 1637−1642.

Alkbir, M.F.M., Sapuan, S.M., Nuraini, A.A., Ishak, M.R., 2016. Fiber properties and crashworthiness parameters of natural fiber-reinforced composite structure: A literature review. Compos. Struct. 148, 59−73.

Ashori, A., 2006. Non-wood fibers—a potential source of raw material in papermaking. Polym.-Plast. Technol. Eng. 45, 1133−1136.

Ashori, A., 2008. Wood-plastic composites as promising green-composites for automotive industries. Bioresour. Technol. 99, 4661−4667.

Ashori, A., 2010. Hybrid composites from waste materials. Polym. Environ. 18, 65−70.

Ashori, A., 2013. Effects of nanoparticles on the mechanical properties of rice straw/polypropylene composites. Compos. Mater. 47, 149–154.

Ashori, A., Nourbakhsh, A., 2008. A comparative study on mechanical properties and water absorption behavior of fiber-reinforced polypropylene composites prepared by OCC fiber and aspen fiber. Polym. Comp. 29, 574–578.

Ashori, A., Nourbakhsh, A., 2009. Polypropylene cellulose-based composites: the effect of bagasse reinforcement and polybutadiene isocyanate treatment on the mechanical properties. Appl. Polym. Sci. 111, 1684–1689.

Ashori, A., Nourbakhsh, A., 2010. Bio-based composites from waste agricultural residues. Waste Manage. 30, 680–684.

Ashori, A., Sheshmani, S., 2010. Hybrid composites made from recycled materials: Moisture absorption and thickness swelling behavior. Bioresour. Technol. 101, 4717–4729.

Ashori, A., Sheshmani, S., Farhani, F., 2013. Preparation and characterization of bagasse/high density polyethylene composite using multi-walled carbon nanotubes. Carbohydr. Polym. 92, 865–871.

Averous, L., Boquillon, N., 2004. Biocomposites based on plasticized starch: thermal and mechanical behaviors. Carbohydr. Polym. 56, 111–122.

Azwa, Z.N., Yousif, B.F., Manalo, A.C., Karunasena, W., 2013. A review on the degradability of polymeric composites based on natural fibers. Mater. Des. 47, 424–442.

Begum, K., Islam, M.A., 2013. Natural fiber as a substitute to synthetic fiber in polymer composites: a review. Res. J. Eng. Sci. 2, 46–53.

Cao, Y., Shibata, S., Fukumoto, I., 2006. Mechanical properties of biodegradable composites reinforced with bagasse fiber before and after alkali treatments. Compos. A. 37, 423–429.

Cordeiro, N., Ornelas, M., Ashori, A., Sheshmani, S., Norouzi, H., 2012. Investigation on the surface properties of chemically modified natural fibers using inverse gas chromatography. Carbohydr. Polym. 87, 2367–2375.

Cordeiro, N., Ashori, A., Hamzeh, Y., Faria, M., 2013. Effects of hot water pre-extraction on surface properties of bagasse fiber. Mater. Sci. Eng. C. 33, 613–617.

Deepa, B., Abraham, E., Cherian, B.M., Bismarck, A., Blaker, J.J., Pothan, L.A., et al., 2011. Structure, morphology and thermal characteristics of banana nano fibers obtained by steam explosion. Bioresour. Technol. 102, 988–1997.

Faruk, O., Bledzki, A.K., Fink, H.-P., Sain, M., 2012. Biocomposites reinforced with natural fibers: 2000–2010. Progr. Polym. Sci. 37, 1552–1596.

Foscher, B., Marzetti, A., Beltrame, P.L., Avella, M., 1998. Steam exploded biomass for the preparation of conventional and advanced biopolymer-based materials. Biomass Bioenergy. 14, 187–194.

Gassan, J., Bledzki, A.K., 1999. Possibilities for improving the mechanical properties of jute/epoxy composites by alkali treatment of fibers. Compos. Sci. Technol. 59, 1303–1309.

Ghanbari, A., Madhoushi, M., Ashori, A., 2014. Wood plastic composite panels; Influence of species, formulation variables and blending process on the density and withdrawal strength of fasteners. Polym. Environ. 22, 260–266.

Goriparthi, B.K., Suman, K.N.S., Mohan Rao, N., 2012. Effect of fiber surface treatments on mechanical and abrasive wear performance of polylactide/jute composites. Compos. A. 43, 1800–1808.

Gurunathan, T., Mohanty, S., Nayak, S.K., 2015. A review of the recent developments in biocomposites based on natural fibers and their application perspectives. Compos. A. 77, 1–25.

Hamzeh, Y., Ashori, A., Hojati Marvast, E., Rashedi, K., Mohammad Olfat, A., 2012. A comparative study on the effects of *Coriolus versicolor* on properties of HDPE/wood flour/paper sludge composites. Compos. B. 43, 2409–2414.

Huda, M.S., Drzal, L.T., Mohanty, A.K., Misra, M., 2008. Effect of chemical modifications of the pineapple leaf fiber surfaces on the interfacial and mechanical properties of laminated biocomposites. Compos. Interfaces. 15, 169–191.

Jawaid, M., Abdul Khalil, H.P.S., Hassan, A., Dungani, R., Hadiyane, A., 2013. Effect of jute fiber loading on tensile and dynamic mechanical properties of oil palm epoxy composites. Compos. B. 45, 619–624.

Klyosov, A., 2007. Water absorption by composite materials and related effects. In: Klyosov, A. (Ed.), Wood–Plastic Composites. John Wiley & Sons, Hoboken, NJ, US.

Li, T., Yan, N., 2007. Mechanical properties of wood flour/HDPE/ionomer composites. Compos. A. 38, 1–12.

Mahlberg, R., Niemi, H.M., Denes, F., Rowell, R., 1999. Application of AFM on the adhesion studies of oxygen–plasma-treated polypropylene and lignocellulosics. Langmuir. 15, 2985–2992.

Nadlene, R., Sapuan, S.M., Jawaid, M., Ishak, M.R., Yusriah, L., 2016. A review on roselle fiber and its composites. J. Nat. Fibers. 13, 10–41.

Nguong, C.W., Lee, S.N.B., Sujan, D., 2013. 73. A Review on Natural Fiber Reinforced Polymer Composites. World Academy of Science, Engineering and Technology, pp. 1123–1130.

Nourbakhsh, A., Ashori, A., 2008. Fundamental studies on wood–plastic composites: effects of fiber concentration and mixing temperature on the mechanical properties of poplar/PP composite. Polym. Compos. 29, 569–573.

Nourbakhsh, A., Ashori, A., 2009. Preparation and properties of wood plastic composites made of recycled HDPE. Compos. Mater. 43, 877–883.

Nourbakhsh, A., Ashori, A., 2010. Wood plastic composites from agro-waste materials: analysis of mechanical properties. Bioresour. Technol. 101, 2525–2528.

Nourbakhsh, A., Kokta, B.V., Ashori, A., Jahan-Latibari, A., 2008. Effect of a novel coupling agent, polybutadiene isocyanate, on mechanical properties of wood-fiber polypropylene composites. Reinf. Plast. Compos. 27, 1679–1687.

Nourbakhsh, A., Ashori, A., Ziaei Tabari, H., Rezaei, F., 2010. Mechanical and thermo-chemical properties of wood-flour polypropylene blends. Polym. Bull. 65, 691–700.

Ragoubi, M., George, B., Molina, S., Bienaimé, D., Merlin, A., Hiver, J.M., et al., 2012. Effect of corona discharge treatment on mechanical and thermal properties of composites based on miscanthus fibers and polylactic acid or polypropylene matrix. Compos. A. 43, 675–685.

Rong, M.Z., Zhang, M.Q., Liu, Y., Yang, G.C., Zeng, H.M., 2001. The effect of fiber treatment on the mechanical properties of unidirectional sisal-reinforced epoxy composites. Compos. Sci. Technol. 61, 1437–1447.

Shebani, A.N., van Reenen, A.J., Meincken, M., 2008. The effect of wood extractives on the thermal stability of different wood species. Thermochim. Acta. 471, 43–50.

Sheshmani, S., Ashori, A., Farhani, F., 2012. Effects of extractives on the performance properties of wood flour-polypropylene composites. Appl. Polym. Sci. 123, 1563–1567.

Sombatsompop, N., Chaochanchaikul, K., 2004. Effect of moisture content on mechanical properties, thermal and structural stability and extrudate texture of poly (vinyl chloride)/wood sawdust composites. Polym. Int. 53, 1210–1218.

Stark, N.M., Matuana, L.M., 2004. Surface chemistry changes of weathered HDPE/wood flour composites studied by XPS and FTIR spectroscopy. Polym. Degrad. Stab. 86, 1–9.

Tabarsa, T., Jahanshahi, S., Ashori, A., 2011. Mechanical and physical properties of wheat straw boards bonded with a tannin modified phenol-formaldehyde adhesive. Compos. B. 42, 176–180.

Taj S., M.A. Munawar, S. Khan, Natural fiber-reinforced polymer composites, In Proceedings Pakistan Academy Science 44 (2007) 129–144.

Tran, T.P.T., Bénézet, J.-C., Bergeret, A., 2014. Rice and Einkorn wheat husks reinforced poly(lactic acid) (PLA) biocomposites: effects of alkaline and silane surface treatments of husks. Ind. Crops Prod. 58, 111–124.

Väisänen, T., Haapala, A., Lappalainen, R., Tomppo, L., 2016. Utilization of agricultural and forest industry waste and residues in natural fiber-polymer composites: A review. Waste Manage. 54, 62–73.

Venkateshwaran, N., ElayaPerumal, A., Alavudeen, A., Thiruchitrambalam, M., 2011. Mechanical and water absorption behavior of banana/sisal reinforced hybrid composites. Mater. Des. 32, 4017–4021.

Epoxy resin based hybrid polymer composites

3

Naheed Saba and Mohammad Jawaid
Laboratory of Biocomposite Technology, Institute of Tropical Forestry and Forest Products (INTROP), Universiti Putra Malaysia, Serdang, Selangor, Malaysia

Chapter Outline

3.1 Introduction

3.1.1 Reinforcements

Reinforcement in polymer composites include materials either from renewable source such as plant fibers (bast, leaf, core fibers), fibers from recycled wood or waste paper, regenerated cellulose fibers (viscose/rayon), byproducts from food crops, bioagricultural wastes or synthetic/manmade fibers such as aramid, carbon, and glass fibers. In the composites, reinforced fibers provide stiffness and sufficient strength and govern the inherent properties of the final material. Fibers are basically classified in two major types: natural fibers and synthetic fibers, which are then further subclassified depending on their origin (Saba et al., 2014). Broad classification of fibers and their subclassifications are illustrated in Fig. 3.1. Currently, different types of natural fibers (bast, leaf, fruit, and core fibers) have been explored as a potential replacement of

Hybrid Polymer Composite Materials: Properties and Characterisation. DOI: http://dx.doi.org/10.1016/B978-0-08-100787-7.00003-2

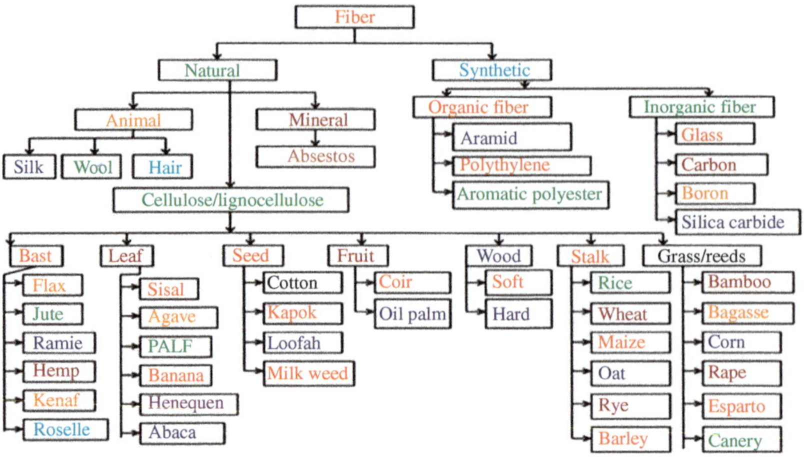

Figure 3.1 Fibers classifications and subclassifications.

synthetic fibers like glass and carbon fibers. These provide benefits to the environment with respect to the degradability and utilization of natural materials. Although natural fibers have relatively lower strength properties compared to the synthetic fibers, the specific modulus and elongation at break signifies the potentiality of these fibers to replace synthetic fibers in engineering polymer composites. Certain modifications by chemical treatments can overcome the limitations of natural fibers such as its hydrophilic tendency and poor compatibility with the matrix. The chemical treatment includes, silane, acetylation, alkaline, benzoylation peroxide, sodium chlorite, isocyanate, permanganate, triazine treatments, and maleated coupling agents are well established to change the fiber structure and surface morphology (Kabir et al., 2012). They also confer improvement in mechanical strength and dimensional stability of the resultant polymer composites.

3.1.1.1 Synthetic fibers

Synthetic fibers are made of polymers that do not occur naturally, and are produced entirely in the laboratory, most generally from petroleum byproducts. Synthetic fibers are made from different chemicals, having their own properties. Fiber produced from these polymers includes nylon, polyesters, acrylics, polyurethanes, etc. Synthetic fibers are more in length and are long lasting. The three most common synthetic fibers used in composites industries are Kevlar (aramid), carbon, and glass fibers, displayed in Fig. 3.2.

3.1.1.2 Kevlar fibers

The combination of para-phenylenediamine and terephthaloyl chloride, results the formation of aromatic polyamide (aramid) threads or Kevlar. Kevlar fibers are

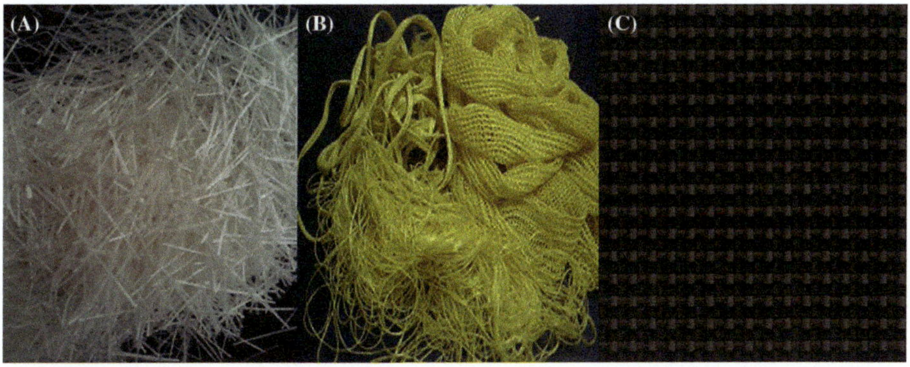

Figure 3.2 Displaying (A) glass fibers, (B) Kevlar fibers, and (C) carbon fibers.

highly expensive due to the costly manufacturing process and costly specific equipment. It exists in three main types as: Kevlar, Kevlar 29, Kevlar 49.

3.1.1.3 Carbon fibers

About 90% of the carbon fibers are prepared from polyacrylonitrile (PAN), while the remaining is from rayon or petroleum pitch. All of these materials are organic polymers, characterized by long strings of molecules bound together by carbon atoms. Currently, three types of precursor are commonly used in manufacturing process including, PAN precursors, rayon precursors, and pitch precursor. Among the three, PAN precursors are the major precursors for commercial carbon fibers, generating about 50% of original fiber mass, followed by pitch precursors yielding high carbon at relatively low cost.

3.1.1.4 Glass fibers

Glass fibers are the most versatile and cheap synthetic fibers compared to Kevlar and carbon, widely used in the polymer composites industries, having high percentage (50%) of silica content, along with different mineral oxides. Glass fibers are lightweight, less brittle, lesser stiff, extremely strong, and robust material. The comparative physical and mechanical properties of different glass fibers types are tabulated in Table 3.1. Glass reinforced polymer composites are used where the higher stiffness of carbon or aramid fibers are not required.

3.1.1.5 Comparison between synthetic fibers

The comparison between glass, carbon, Kevlar, and the most popular thermoset epoxy are tabulated in Table 3.2. Elastic modulus, strength and fatigue strength of Kevlar and carbon fibers are higher than glass fibers. Kevlar (aramid) and carbon fibers has a high strength-to-weight ratio compared to most commonly used E-glass fibers. Kevlar and carbon fibers are resistant to elevated temperatures, but Kevlar

Table 3.1 Different types of glass fibers, physical, and mechanical properties

Glass fibers type	Density ($g\ cm^{-3}$)	Tensile strength (MPa)	Modulus (GPa)	Elongation at break (%)
A-glass	2.44	3300	72	4.8
C-glass	2.56	3300	69	4.8
D-glass	2.11	2500	55	4.5
E-glass	2.54	3400	72	4.7
R-glass	2.52	4400	86	5.1
S-glass (also S-2 glass)	2.53	4600	89	5.2
ECR-glass	2.72	3400	80	4.3
AR-glass	2.7	1700	72	2.3

Table 3.2 Comparative between synthetic fibers and epoxy

Synthetic fibers	Young's modulus	Strength-to-weight	Fiber strength	Laminate strength
E-glass	30–40	564	3450	1500
Carbon fiber	125–181	1013	4127	1600
Kevlar	70.5–112.4	993	2757	1430
Epoxy	3	28	–	12–40

Source: http://www.christinedemerchant.com/carbon-kevlar-glass-comparison.html.

and glass do not conduct electricity (http://www.christinedemerchant.com/carbon-kevlar-glass-comparison.html).

3.1.1.6 Natural fibers

Natural fibers are the most promising reinforcements, substitute to synthetic fibers for fibers reinforced polymer composites, owing to nontoxic, nonabrasive, higher specific strength, lower density, minimal environmental impact, biodegradability besides desirable mechanical properties compared to synthetic fiber, such as glass, carbon, Kevlar fibers (Rajesh et al., 2016). The natural fibers with high content of lignin exhibit high char yield, high effective heat of combustion (EHC), high activation energy of combustion (E_a) and low CO/CO_2 ratio (Dorez et al., 2014; Saba et al., 2016). The physical and mechanical properties of some important natural fibers are listed in Table 3.3 (Ramamoorthy et al., 2015; Onuaguluchi and Banthia, 2016; Yan et al., 2016b).

Huge varieties of natural fibers have been used as reinforcing agent in thermoplastics and thermosets for the modification or improving the properties, from past

Table 3.3 Tabulated chemical compositions and mechanical properties of natural fibers

Fibers	Compositions (%) Cellulose/hemicellulose/lignin	Tensile strength (MPa)	Young Modulus (GPa)	Elongation at break (%)	Density (g cm^{-3})
Jute	33.4/22.7/28	393–773	26.5	1.5–1.8	1.3
Abaca	56–63/20–25/7–9	400	12	3–10	1.5
Alfa	45.4/38.5/38.5	35	22	5.8	0.89
Flax	71/18.6–20.6/2.2	345–1035	27.6	2.7–3.2	1.5
Bagasse	32–48/19–24/23–32	20–290	19.7–27.1	1.1	1.2
Hemp	68/15/10	690	70	1.6	1.48
Kenaf	72/20.3/9	280	53	1.6	1.2
Ramie	68.6–76.2/13–16/0.6–0.7	560	24.5	2.5	1.5
Bamboo	26–43/30/21–31	140–230	11–17	2.5–3.7	0.6–1.1
Banana	60–65/6–8/5–10	355	33.8	53	1.35
Sisal	73.11/13.33/11.0	511–635	9.4–22	2–2.5	1.5
Pineapple	70–82/18.0/5–12	400–627	1.44	14.5	0.8–1.6
Coir	36–43/0.15–0.25/41–45	175	4–6	30	1.2
Rice	28–36/23–28/12–16	–	–	–	–
OPEFB	65/–/29	248	3.2	25	0.7–1.55
Henequen	60/28/28	430–570	10.1–16.3	3.7–5.9	1.2
Cotton	89/4/0.75	287–800	5.5–12.6	3–10	1.5–1.6
Piassava	–/–/–	134–143	1.07–4.59	7.8–21.9	–

Source: (Ramamoorthy et al., 2015; Onuaguluchi and Banthia, 2016; Yan et al., 2016b).

Figure 3.3 Commonly used natural fibers as reinforcement in thermosets polymer.

decades. Some of the most widely used natural fibers in thermosets polymer composite industries are displayed in Fig. 3.3. Instead of the several advantages, natural fibers confer certain limitations due to physicochemical incompatibility between hydrophilic fibers and hydrophobic matrix, when used as reinforcement in polymeric matrices (Luna et al., 2016).

3.1.2 Thermoplastics and thermosets

Thermoplastics are linear polymers existing either as semicrystalline or amorphous glasses. On heating beyond the melting point of crystals or beyond the glass transition temperature, thermoplastics loose its structure due to the free segmental movement which consequences more flexible and deformable structure, resulting the possibilities of reshaping and reforming by simply heating and cooling. On the other hand, thermosets are cross-linked polymers and they remain in the solid state as long as the covalent chemical bonds are not destroyed. Thermoset polymers based on petroleum are highly flammable and combustible, such as epoxy, polyester, or vinyl ester resins. Thermoplastic polymers are ductile and tougher (Kabir et al., 2012), but they have lower stiffness and strength compared to thermoset polymers (Saba et al., 2015b). Thermoplastics have poor creep resistance and more susceptible to solvent than thermosets. Commonly used thermosets includes epoxy, polyester, vinyl ester, and phenol formaldehyde to manufactured fibers based composites.

3.1.2.1 Epoxy resin

Epoxy resins have better mechanical properties than polyesters and vinyl esters, and are the principal thermoset polymer used as matrix in aerospace composites. Epoxy resins describe a broad class of molecular structures having at least two oxirane groups as epoxide functional groups (shown in Fig. 3.4) in the polymer chain (Saba et al., 2015b), being first synthesized as early as 1891. The epoxide group is planar, with a three-membered ring composed of one oxygen and two carbon atoms, where the carbon atoms of the ring are electrophilic and highly reactive.

The most popular epoxy monomers are derived from the reaction of bis(4-hydroxy phenylene)-2,2 propane (bisphenol A) and 1-chloroprene,2-oxide (called epichlorohydrin), in the presence of base (sodium hydroxide). Some typical epoxy monomers are represented in Fig. 3.5. The reaction of epichlorohydrin with an aromatic amine, results other typical epoxy resins, such as tetraglycidylmethylenedianiline (TGMDA), that are extensively used in aerospace composites (Vidil et al., 2016). Polyglycidyl derivatives of phenolic prepolymers (phenolic resin) are also common epoxy resins with high glass transition temperature (T_g) and thermal stability. Additionally, cycloaliphatic resins also shows considerable interest as they have better weather

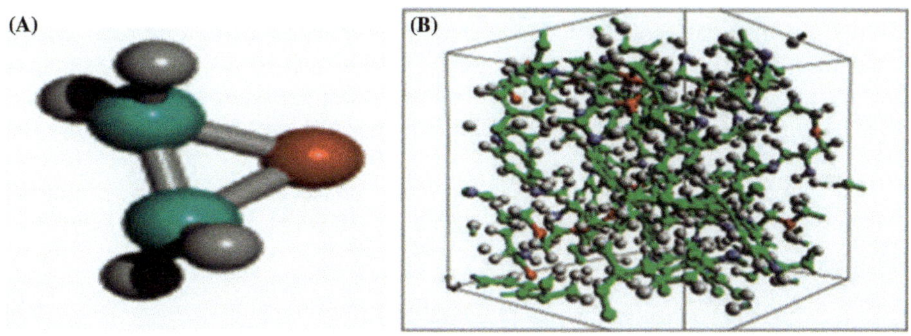

Figure 3.4 (A) Epoxy oxirane ring and (B) cross-linked structure of cured epoxy.

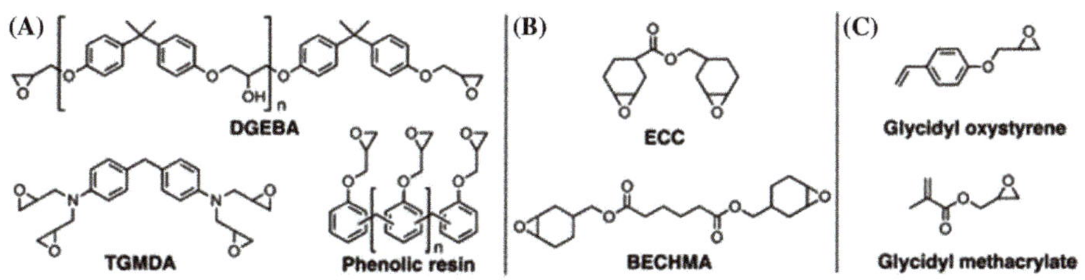

Figure 3.5 Typical epoxy monomers, (A), (B) and (C).

resistance and lesser tendency to get yellow compared to aromatic resins (Vidil et al., 2016). Epoxy monomers containing vinyl groups, like glycidyl (meth) acrylate, or glycidyl oxystyrene, can be used for the synthesis of functional oligomers.

Cured epoxies with tight three-dimensional molecular network structures have relatively high thermal stability and T_g. Cured epoxy exhibit inherent brittle fracture behavior and poor crack growth resistance that restrained its mechanical applications. The brittleness of epoxy resins can be reduced by the process of rubber toughening, involving the addition of a liquid rubber, such as butadiene–acrylonitrile, to the uncured resin. Toughened epoxy polymer composites are extensively applied for load-bearing applications (Gong et al., 2015). Epoxies possess higher curing time and are highly combustible and expensive with respect to phenolic. Epoxies on exposing to high temperatures (300–400°C), decomposes releasing smoke, heat, toxic volatiles, and soot from the organic matrix of the cured epoxy laminate. However by incorporating flame retardants (FRs) additives or by copolymerization with reactive FRs, its flame retardancy can be improved (Saba et al., 2015b). FR epoxies are widely used in printed wiring boards, electrical and electronics, construction materials, automobile, aerospace, marine, painting, coating, adhesive and in advanced engineered composites fabrication for high end applications on a large scale worldwide (Saba et al., 2015a).

3.2 Polymer composites

Composite materials composed of at least two constituents of different phase, in order to achieve combined properties that cannot be met by a single-phase material, and it can be classified in three subclasses, including particles, fibers, and structural based composites (Fig. 3.6) (http://textilelearner.blogspot.my/2012/09/glass-fiber-composites-properties-of.html).

The polymer composites have at least one phase as a polymer or matrix as binder and second phase includes reinforcing fibers or particles in a matrix to

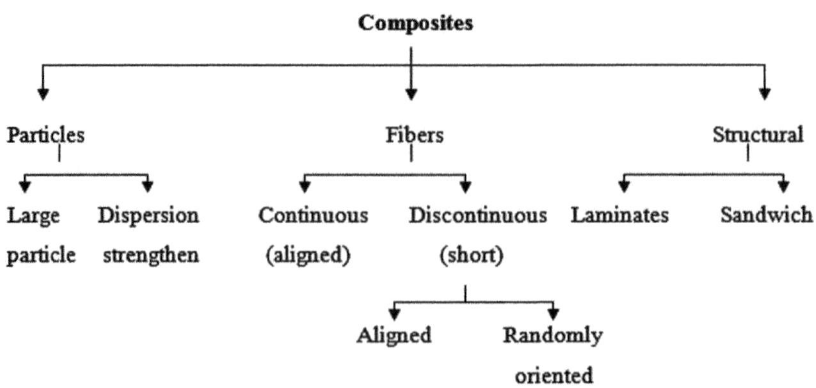

Figure 3.6 Classification of polymer composites.

improve dimensional and thermal stability, stiffness, toughness, and tensile strength. Polymer composites are able to meet diverse design requirements with significant weight savings as well as high strength-to-weight ratio.

3.3 Natural fibers polymer composites

Currently natural fibers are reinforced in a variety of thermoplastic and thermosets polymers for modification of their properties. Among thermosets such as polyester and phenolic, epoxy polymer is the most common polymer that is reinforced by natural fibers. Natural fibers reinforced polymer composites are of great interest due to the ease of fabrication, cost effectiveness, and remarkable structural rigidity. These composites show extensive applications, including constructional, aerospace, and automotive industries because they represent an ecological and inexpensive alternative to conventional petroleum derived materials (Väisänen et al., 2016).

3.4 Hybrid composites

Hybridization is the approach to meet competing and diverse design requirements in a more cost-effective way compared to traditional engineering materials. The term "hybrid composites" represents a composite having at least two different types of fibers reinforced in the single matrix, or blends of polymer reinforced with single fibers, in order to provide a synergistic effect such as enhanced mechanical properties (Kalantari et al., 2016). Hybrid composites offered balanced thermal stability, reduced weight/cost, balanced strength and stiffness, better fatigue resistance, impact resistance, fracture toughness besides reduced notch sensitivity (Ferrante et al., 2015). Several studies have been reported on the hybridization of epoxy reinforced with natural/synthetic fibers, fabricated by variety of techniques such as hand lay-up, compression molding, resin transfer molding (RTM), illustrate marked increase in mechanical, thermal, physical, and flame retardancy properties.

3.5 Epoxy based hybrid polymer composites

Huge varieties of epoxy hybrid composites have been fabricated from synthetic fibers, natural fibers, and a combination of both synthetic/natural fibers, and with natural fibers/natural fibers to widen the epoxy industrial applications.

3.5.1 Natural fibers/synthetic fibers based epoxy hybrid polymer composites

Some of the important and recent studies on the hybridization of epoxy with natural/synthetic fibers, along with characteristics improvement in properties are tabulated in Table 3.4. The advantages of hybridization are fully utilized to reduce the

Table 3.4 **Reported study on the natural/synthetic fibers based epoxy hybrid polymer composites**

Natural/synthetic fibers reinforced epoxy hybrid polymer composites			
Polymer	**Reinforcements**	**Hybridization effects**	**References**
Epoxy	Basalt/carbon fibers	The hybridization modifies the failure mechanisms depending on the stacking sequence	Ferrante et al. (2015)
Epoxy	Flax/carbon fibers	Hybridization improved the mechanical properties and reduced the vibration damping of the composites	Flynn et al. (2016)
Epoxy	Microsized red mud/glass fibers	Considerable improvements in mechanical properties, erosion wear performances and density are observed by the addition of red mud to the glass/epoxy composites	Biswas and Satapathy (2009)
Epoxy	Jute/E-glass fibers	Hybridization results improved mechanical properties.	Johnson et al. (2016)
Epoxy	Core–shell particles (CSP)/glass fibers	Hybridization of glass fibers with CSP particles results less structural defects with significant improvement in impact properties	Zhong and Joshi (2016)
Epoxy	Rice husk particulates/glass fibers	Hardness, tensile modulus, impact energy and erosion resistance of hybrid composites improved by the hybridization. Small declines in tensile and flexural properties are also observed	Rout and Satapathy (2012)
Epoxy	Basalt/carbon fibers	Mechanical analysis indicate that hybrid laminates with intercalated configuration have better impact energy absorption capability, while hybrid laminates with sandwich-like configuration displayed higher flexural behavior, fabricated by RTM	(Sarasini et al., 2014)
Epoxy	*Pennisetum purpureum*/E-glass	Higher tensile and flexural strengths were recorded for the hybrid composites with the 5% alkali-treated *Pennisetum purpureum* fibers, fabricated by vacuum infusion process	Ridzuan et al. (2016b)
Epoxy	Jute/glass fibers	Hybridization of jute fibers with glass fibers/epoxy composites results considerable increase in impact energy, density, tensile and flexural strength along with noticeable decrease in water absorption properties	Braga and Magalhaes (2015)

Epoxy	Coir pith/Nylon fabric	Durability, chemical and flame retardancy get improved on hybridization in moist conditions	Narendar et al. (2014)
Epoxy	NaOH treated woven kenaf/aramid fibers	Tensile and flexural properties of treated hybrid composites are better than nontreated hybrid composites. Hybrid composites with Kevlar as outer layers possess better mechanical properties among the rest	Yahaya et al. (2015)
Epoxy	Kenaf/glass fibers	Hybridization results enhancement in mechanical properties to use for car bumper beams as automotive structural components fabricated by modified SMC	Davoodi et al. (2010)
Epoxy	Banana/hemp/glass fibers	Banana/hemp/glass fibers reinforced hybrid epoxy composites exhibited superior properties and can use as potential material to replace synthetic fibers reinforced composites.	Bhoopathi et al. (2014)
Epoxy	Pennisetum purpureum/glass fibers	The incorporation of the glass fibers into the P. purpureum/epoxy composites enhanced the tensile, flexural strength, as well as their modulus, fabricated by vacuum infusion method	Ridzuan et al. (2016a)
Epoxy	Basalt woven fabric/glass fibers	Hybridization improves the impact energy absorption and damage tolerance tendency with respect to glass laminates fabricated by RTM	Sarasini et al. (2013)
Epoxy	Basalt/glass fibers	Increase in tensile and flexural properties are observed by the hybridization, compared to those of GFRP laminates	Fiore et al. (2011)
Epoxy	Sisal/jute/glass fibers	Hybridization of sisal fiber with glass improved the tensile properties. While Hybridization of jute fibers with glass fibers improves the flexural properties fabricated by hand lay-up	Ramesh et al. (2013)
Epoxy	Kenaf/aramid fibers	Hybridization of woven kenaf mat with aramid fibers yield composite material with high tensile strength and impact resistance properties	Yahaya et al. (2016)

Notes: Resin transfer molding (RTM), sheet molding compound (SMC).

use of synthetic fibers which are generally nonenvironmentally friendly (Yahaya et al., 2016).

3.5.2 Natural fibers/natural fibers based epoxy hybrid polymer composites

Incorporation of natural/natural fibers also results improved properties by the hybridization effects. Some of the reported findings are listed in Table 3.5.

3.5.3 Synthetic/synthetic fibers based epoxy hybrid polymer composites

Huge varieties of hybridized epoxy composites are fabricated by different techniques by the reinforcements of synthetic/synthetic fibers, resulting considerable improvement in thermal, mechanical, and flame retardancy properties of epoxy composites. Hybrid composites involving carbon and glass fibers allow to combine the advantages of both groups of fibers including, low price of glass fibers, low weight, high tensile strength and stiffness of carbon fibers, and to minimize their limitations including, high costs and low compressive strength of carbon fibers (Dai and Mishnaevsky, 2015). The hybridization involving synthetic/synthetic fibers reinforced in epoxy and its effects are tabulated in Table 3.6.

3.5.4 Epoxy based hybrid polymer nanocomposites

Currently, incorporation of nanosized fillers such as carbon nanotubes (CNTs), graphite, nanoclay has been extensively used along with natural fibers such as kenaf, jute or with synthetic fibers such as carbon and glass fibers. Some exclusive study on epoxy based hybrid nanocomposites with different nanofillers/synthetic or nanofillers/natural fibers are listed in Table 3.7. Hybrid epoxy nanocomposites possess marked enhanced properties compared with epoxy composites, at relatively lower concentrations.

3.6 Applications

3.6.1 Applications of epoxy based polymer composites

Epoxy based composites are extensively been used in making automobiles components including radiator supports, bumper beams, fenders, hoods, roof panels, deck lids, and a number other exterior and interior body components. Synthetic fibers reinforced epoxy polymer composites are being increasingly used for aircraft structures owing to their superior structural performance, such as long fatigue life, high stiffness, high strength, and low density (Chowdhury et al., 2016).

Table 3.5 **Reported study on the natural/natural fibers based epoxy hybrid polymer composites**

Natural fibers reinforced hybrid polymer composites			
Polymer	**Reinforcements**	**Properties improvement**	**References**
Epoxy	Jute/Sansevieria cylindrica fibers	Hybrid composites showed higher strength than untreated composites fabricated through hand lay-up technique	Kumar and Reddy (2014)
Epoxy	Plain woven/ Rib-knitted flax preforms	Thermal stability and tensile strength of flax hybrid preforms composites get improved fabricated by hand lay-up	Muralidhar (2013)
Epoxy	Jute/oil palm fibers	Hybridization increases the tensile and dynamic mechanical properties of the oil palm/epoxy composites manufactured by hand lay-up	Jawaid et al. (2013)
Epoxy	Flax/glass fibers	Flax/glass hybridization fabricated by compression molding, shows positive effect in a wet environment at low temperatures ($\sim 20°C$) for Young's modulus and tensile strength. While negative effect on the tensile strength and on the specific tensile strength are observed on hybridization	Saidane et al. (2016)
Epoxy	OPEFB fibers/ woven jute fibers	Hybridization increases the tensile and flexural properties, compared to independent composites	Jawaid et al. (2011)
Epoxy	Sisal/Jute fibers	The hybrid composites fabricated by hand lay-up technique followed by light compression molding technique, displayed higher storage and loss modulus values and lower value of damping parameter and water absorption properties	Gupta and Srivastava (2015)
Epoxy	Luffa fibers/ groundnut	The mechanical properties of hybridized luffa fibers/groundnut/ epoxy composites fabricated by hand lay-up get enhanced compared to luffa fibers/epoxy composites	Panneerdhass et al. (2014)
Epoxy	OPEFB/jute fibers	The flexural properties of hybrid composites are higher than that of pure OPEFB composites	Jawaid et al. (2010)
Epoxy	Jute/banana fibers	Hybridization of banana fibers in jute/ epoxy composites fabricated by Hand lay-up technique results in	Boopalan et al. (2013)

(*Continued*)

Table 3.5 (Continued)

Natural fibers reinforced hybrid polymer composites			
Polymer	**Reinforcements**	**Properties improvement**	**References**
		increasing the mechanical and thermal properties, along with a marked decrease in the moisture absorption properties	
Epoxy	Abaca/jute fibers	The abaca/jute/epoxy hybrid composites displayed better tensile and shear properties compared with single fiber composites. Hybridization also improves the ductility than the single type composites	Ramnath et al. (2013)
Epoxy	Flax/basalt fibers	Storage modulus of hybrid composites decreases after 15 days of aging. T_g increment is also higher for hybrid composites	Fiore et al. (2016)
Epoxy	Basalt/hemp fibers	The hybridization in a sandwich configuration markedly improved both post-impact residual properties and damage tolerance tendency, fabricated by combination of hand lay-up and compression molding techniques	Dhakal et al. (2015)

Note: Oil palm empty fruit bunch (OPEFB).

3.6.2 Applications of epoxy based hybrid polymer composites

Most promising applications of epoxy hybrid polymer composites are in making the components of coal dust carrying pipes, desert structures, industrial fans, low cost housing, false ceiling, partition boards and fishing boats/water-sports equipment (Chowdhury et al., 2016).

Currently, epoxy based hybrid polymer composites and epoxy based hybrid polymer nanocomposites are being extensively utilized in electrical and electronic components, automotive and military applications. Research study also illustrates the promising applications of epoxy hybrid composites in drug delivery, dentary fills, artificial limbs and in orthopedic trauma applications (Ramakrishna et al., 2001). Hybrid composites made of natural/natural fibers offer the opportunity for extensive applications in the fields of low cost construction and civil structures, domestic, and toilets accessories and in many other common applications where the prohibitive cost of reinforcements restricts the use of conventional lightweight reinforced plastics (Harish et al., 2009).

Table 3.6 Reported study on synthetic/synthetic fibers based epoxy hybrid polymer composites

Synthetic/synthetic fibers reinforced hybrid polymer composites			
Polymer	**Reinforcements**	**Hybridization effects**	**References**
Epoxy	Glass/kevlar fabrics	Kevlar/glass hybrid structure fabricated by hand lay-up, showed improvement in specific mechanical strength as well as bending and impact energy properties	Valença et al. (2015)
Epoxy	SiC/pitch-based carbon fibers	Thermal conductivity of SiC/pitch-CF/ epoxy composites increases 18.8 times to that of epoxy resin	Mun et al. (2015)
Epoxy	Unidirectional glass/carbon fibers	Unidirectional glass/carbon fibers/ epoxy hybrid composites possess maximum flexural strength and robustness under flexural loading	Kalantari et al. (2016)
Epoxy	Glass/carbon fibers	Hybridization increases the tensile strength and modulus of hybrid composites	Naresh et al. (2016)
Epoxy	E-glass/T700S carbon fibers	Hybridization potentially improves the flexural strength	Dong and Davies (2015)
Epoxy	Glass/ceramic whisker/solid lubricant filler	Incorporation of solid lubricant results in the improvement of both mechanical and tribological properties of composites, fabricated by the hand lay-up procedure followed by vacuum bagging technique	Sudheer et al. (2014)

Furthermore, the epoxy hybrid composites reinforced by natural/synthetic fibers are widely used in making the components of desert structures, low cost housing, fishing boats/water-sports equipment, false ceiling, partition boards, automotive (tails, wings, propellers), bicycle frames, boat hulls, fishing rods, storage tanks, baseball bats, ice skating boards, shelters, clothes, door panels, and in weapons construction (Sanjay et al., 2015). Carbon fibers based epoxy hybrid composites shows extensive applications in load-bearing structural materials in aerospace and automotive sectors along with sports and consumer goods, owing to an interesting combination of low weight, high strength, and excellent corrosion resistance (Yan et al., 2016a).

Currently, epoxy based hybrid polymer nanocomposites signify as most encouraging materials hence are receiving higher attention in the field of cosmetics, construction, food packaging, medical sciences, and other composite based industries, owing to distinctive features of incorporated nanofiller in enhancing the mechanical and barrier properties of epoxy polymer composites (Saba et al., 2014). Some of

Table 3.7 Exclusive research study on epoxy based hybrid polymer nanocomposites

Epoxy hybrid polymer nanocomposites			
Polymer	**Reinforcements**	**Hybridization effects**	**References**
Epoxy	Banana fiber/silica powder/	Hybridization of banana fibers with silica powder improves the bending strength	Singh et al. (2012)
Green Epoxy	Alkali-treated jute/nanojute fibers	Hybrid composites fabricated by hand lay-up method and compression molding technique display higher storage modulus and T_g values. Marked reduction in delta peak height also been realized	Jabbar et al. (2016)
Epoxy	Kevlar/nanoclay	The hybrid laminates manufactured by hand lay-up with epoxy resin/ 6 wt% of nanoclays display higher elastic recuperation and penetration threshold	Reis et al. (2013)
Epoxy	CNTs/GO	Hybrid composites fabricated by ultra-sonication followed hand lay-up technique with 0.5 phr MWCNTs and 0.1 phr GO, results increase in the friction coefficient and a reduction in the specific wear rate are observed	Reis et al. (2013)
Epoxy	Fe_2O_3/RGO nanoplatelets	TGA analysis revealed improvement in the thermal stability of Epoxy/RGO–Fe_2O_3 nanocomposites. Hybrid composites also possess improved dielectric and microwave properties, through in situ polymerization	Sharmila et al. (2016)
Epoxy	Kevlar fibers/CNT	The hybrid composites shows improved tendency to absorb mechanical shocks and effectively shield from electromagnetic interferences by the impact of metallic bullets fired at about 400 m s^{-1} and 1000 m s^{-1}	Micheli et al. (2016)
Epoxy	Silica coated with hybrid particles of MWCNTs	The rheology and electrical conductivity (conductivity $\sim 10^{-4}$ S m^{-1}) of epoxy resin	Wilkinson et al. (2016)

(*Continued*)

Table 3.7 (Continued)

Epoxy hybrid polymer nanocomposites			
Polymer	**Reinforcements**	**Hybridization effects**	**References**
		suspensions of particles found comparable to the neat resin	
Epoxy	CNT−Al_2O_3	The flexural strength, flexural modulus and dielectric constant of CNTs−Al_2O_3/epoxy hybrid composites shows significant improvement up to 30%, 35%, and 20%, respectively, compared to the epoxy composites	Zakaria et al. (2015)
Epoxy	MWCNTs/MnZn ferrite	3MWNCTs/1MnZn ferrite/epoxy hybrid nanocomposites fabricated by ultra-sonication followed by hand lay-up, displayed the highest effective electromagnetic-interference (EMI) and shielding effectiveness (SE). The EMISE of the hybrid composites are better than epoxy composites filled with single conductive filler and are comparable with that of commercial EMI absorber	Phan et al. (2016)
Epoxy	Nanosilica/ AgNWs	Epoxy/SNP/AgNWs hybrid nanocomposites displayed distinct improvements in thermal conductivity without degrading mechanical properties	Chen et al. (2016)
Epoxy	CNTs/NDs	Hybrid nanocomposites manufactured by ultra-sonication followed by hand lay-up having 0.2 wt% MWCNTs/0.2 wt% NDs showed 50% increase in hardness while tensile strength and modulus enhanced to 70% and 84%, respectively. Flexural strength and modulus also increases by 104% and 56%, respectively. Fracture strain also increased in both the tensile and flexural testing. The impact resistance or toughness of hybrid nanocomposites also get increased to 161%	Subhani et al. (2015)

(Continued)

Table 3.7 **(Continued)**

Epoxy hybrid polymer nanocomposites			
Polymer	**Reinforcements**	**Hybridization effects**	**References**
Epoxy	Al_2O_3/GNPs/ magnesium hydroxide	Hybridization by the incorporation of layered GNPs efficiently increases the thermal conductivity of epoxy/Al_2O_3 composites with considerable flame retardancy	Guan et al. (2016)
Epoxy	CTBN-rubber/ GNPs	Hybridization enhanced the fracture toughness and thermal conductivity of the epoxy composites with the addition of 5 μm (GNP-5) to the CTBN/ epoxy composites	Wang et al. (2016)
Epoxy	Graphene-functionalized with POSS	Hybridization results reduce in the dielectric constant of epoxy composite materials with ultra-low filler content	Yu et al. (2014)
Epoxy	GNPs/carbon fiber	Hybridization results an increment in the overall mechanical properties by the addition of GNPs to the carbon fiber/epoxy composites	Hadden et al. (2015)
Epoxy	CNTs/Al_2O_3 particles	Addition of CNT−Al_2O_3 hybrid compound to the epoxy composites exhibit an enhancement of 117% and 148% in compressive strength and compressive modulus respectively. Thermal stability also improved for the CNT−Al_2O_3 hybrid compound/ epoxy composites	Zakaria et al. (2016)
Epoxy	Reactive liquid rubber/Silica nanoparticles	Improvement in mechanical properties by the addition of surface-modified silica nanoparticles of 20 nm size to epoxy was perceived	Sprenger et al. (2014)
Epoxy	Nano-SiO_2/short carbon fiber	The hybrid composites with 4 wt% nano-SiO_2/6 wt% carbon fiber/ epoxy hybrid nanocomposites delivered the highest improvement of the tribological performance	Guo et al. (2009)

(Continued)

Table 3.7 **(Continued)**

Epoxy hybrid polymer nanocomposites			
Polymer	**Reinforcements**	**Hybridization effects**	**References**
Epoxy	Nanocopper particles/ MWCNTs	Heat transfer performance as a thermal interface material (TIM) of hybrid nanocomposites gets improved by the incorporation of MWCNTs and nanocopper particles into epoxy	Zhang et al. (2014)
Epoxy	CNTs/carbon fibers	Hybridization of CNTs and carbon fibers to the epoxy shows positive effect on erosive wear response	Papadopoulos et al. (2016)
Epoxy	Bentonites/silica modified with POSS	Hybridization results an improvement in mechanical properties with particular increase in tensile strength by 44%, and Charpy impact strength by 93% for hybrid nanocomposites	Oleksy et al. (2014)
Epoxy	Kenaf/silica nanoparticles	Inclusion of hydrophobic silica nanoparticles had a detrimental effect on the mechanical properties of hybrid nanocomposites including flexural modulus, flexural strength, compressive strength and compressive modulus, fabricated by vacuum infusion	Bajuri et al. (2016)
Rubbery epoxy	Carbon nanofiber/ BN	The developed hybrid nanocomposites are thermally conducting and electrically insulating TIMs	Raza et al. (2015)

Notes: Polyhedral oligomeric silsesquioxane (POSS), Multiwall carbon nanotubes (MWCNTs), Iron oxide (Fe_2O_3), Reduced graphene oxide nanoplatelets (RGO), Manganese zinc ferrite (MnZn ferrite), Carboxyl terminated butadiene acrylonitrile (CTBN), Carbon nanotube (CNTs), Silicon carbide (SiC), Oil palm empty fruit bunches (OPEFB), Thermal interface materials (TIMs), Alumina (Al_2O_3), Graphene nanoplatelet (GNPs), Nanodiamonds (NDs), Graphene oxide (GO), Boron nitride (BN).

the most important and exclusive applications of epoxy hybrid polymer composites and epoxy hybrid nanocomposites are listed in Table 3.8.

3.7 Conclusion

Epoxy delivers diverse applications from adhesives to coatings; however its pervasive applications in advanced engineering are restricted due to its delamination, low

Table 3.8 Applications of epoxy based hybrid polymer composites and hybrid nanocomposites in different sectors

Epoxy hybrid polymer composites	Fields of applications	References
Epoxy/CNT/coated silica particles	As electrically conductive composites	Wilkinson et al. (2016)
Epoxy/carbon/glass fibers	As advanced pseudoductile unidirectional thin-ply	Czél et al. (2016)
Epoxy/Kevlar fibers/CNT	In aerospace structures for low energy range of potential mechanical shocks	Micheli et al. (2016)
Epoxy/PDMS-OH	As protective agents for stone surface	Xu et al. (2015)
Epoxy/basalt/glass mat fibers	In real ship component and marine applications	Fiore et al. (2011)
Epoxy/flax fiber/carbon fibers	In controlling vibration damping applications	Flynn et al. (2016)
Epoxy/Al_2O_3, GNPs/magnesium hydroxide	As thermal conductive hybrid epoxy nanocomposites with satisfactory flame retardancy	Guan et al. (2016)
Rubbery epoxy/carbon nanofibers/BN	As interfacial thermal transport at thick bond lines	Raza et al. (2015)
Epoxy/graphene/POSS	Applications in low-dielectric epoxy composites	Yu et al. (2014)
Epoxy/carbon/glass fibers	As hybrid laminate reinforcement and in the repairing of aeronautic structures	Guermazi et al. (2014)
Epoxy/kenaf/aramid fibers	Military vehicle's spall-liner applications	Yahaya et al. (2016)
Epoxy/nanosilica/AgNWs	As electronic packaging hybrid materials with thermally conductive and electrically insulating properties	Chen et al. (2016)
Epoxy/carbon/basalt, glass fibers	As components in wind turbine blades and wind energy generation	Chikhradze et al. (2015)
Epoxy/carbon/flax fibers	In long bone fracture fixation for replacing clinically used metal plates	Bagheri et al. (2013)
Epoxy/graphene/graphite oxide	As electrically conductive composites in electrical and electronic industries	Pokharel and Truong (2014)
Epoxy/CNTs/carbon/glass fibers	Wind energy applications	Dai and Mishnaevsky (2015)
Epoxy/carbon fiber/flax fibers	As bone fracture plate in orthopedic trauma	Bagheri et al. (2015)
Epoxy/unidirectional glass/carbon fibers	Robust designing	Kalantari et al. (2016)

impact resistance, low fracture toughness behavior, inherent brittleness, and inferior thermal stability. Modification of epoxy by reinforcing natural fibers and synthetic fibers or combination of both (natural/synthetic fibers) results superior physical, mechanical, thermal, wear, flame retardancy, and electrical properties. Currently reinforcing of nanofiller at relatively lower concentration in epoxy are receiving more responses in the fabrication of high performance engineering materials.

The present review provides the valuable research and analysis that has been carried out in the area of the epoxy based hybrid polymer composites reinforced with natural fibers, synthetic fibers and nanofillers, for further investigations. From the literature it seemed that a wide variety of research studies have been conducted focusing on improving the physical, mechanical (tensile, flexural, impact strength), thermal, and electrical properties of cured epoxy through hybridization. The developed epoxy hybrid composites displayed extensive and wide applications in areas such as aircraft, automotive components, sporting goods, building industry, and biomedical science.

The future research study will be the hybridization of epoxy by the introduction of nanofillers exclusively derived from agriculture wastes such as coconut pith, groundnut husk, wheat straw, flax straw, sunflower leaves, used tea leaves, and newspapers which are still underutilized along with natural fibers to modify and enhance the properties of cured epoxies, would be the keen interests areas for researchers to yield epoxy bionanocomposites.

Acknowledgments

All authors acknowledge Universiti Putra Malaysia (UPM) for providing access throughout to complete this review article.

References

Bagheri, Z.S., El Sawi, I., Schemitsch, E.H., Zdero, R., Bougherara, H., 2013. Biomechanical properties of an advanced new carbon/flax/epoxy composite material for bone plate applications. J. Mech. Behav. Biomed. Mater. 20, 398–406.

Bagheri, Z.S., Giles, E., El Sawi, I., Amleh, A., Schemitsch, E.H., Zdero, R., et al., 2015. Osteogenesis and cytotoxicity of a new carbon fiber/flax/epoxy composite material for bone fracture plate applications. Mater. Sci. Eng. C. 46, 435–442.

Bajuri, F., Mazlan, N., Ishak, M.R., Imatomi, J., 2016. Flexural and compressive properties of hybrid kenaf/silica nanoparticles in epoxy composite. Proc. Chem. 19, 955–960.

Bhoopathi, R., Ramesh, M., Deepa, C., 2014. Fabrication and property evaluation of banana-hemp-glass fiber reinforced composites. Proc. Eng. 97, 2032–2041.

Biswas, S., Satapathy, A., 2009. Tribo-performance analysis of red mud filled glass–epoxy composites using Taguchi experimental design. Mater. Des. 30, 2841–2853.

Boopalan, M., Niranjanaa, M., Umapathy, M., 2013. Study on the mechanical properties and thermal properties of jute and banana fiber reinforced epoxy hybrid composites. Compos. B: Eng. 51, 54–57.

Braga, R., Magalhaes, P., 2015. Analysis of the mechanical and thermal properties of jute and glass fiber as reinforcement epoxy hybrid composites. Mater. Sci. Eng. C. 56, 269–273.

Chen, C., Wang, H., Xue, Y., Xue, Z., Liu, H., Xie, X., et al., 2016. Structure, rheological, thermal conductive and electrical insulating properties of high-performance hybrid epoxy/nanosilica/AgNWs nanocomposites. Compos. Sci. Technol. 128, 207–214.

Chikhradze, N.M., Marquis, F.D., Abashidze, G.S., 2015. Hybrid fiber and nanopowder reinforced composites for wind turbine blades. J. Mater. Res. Technol. 4, 60–67.

Chowdhury, N.M., Wang, J., Chiu, W.K., Chang, P., 2016. Static and fatigue testing bolted, bonded and hybrid step lap joints of thick carbon fibre/epoxy laminates used on aircraft structures. Compos. Struct. 142, 96–106.

Czél, G., Jalalvand, M., Wisnom, M.R., 2016. Design and characterisation of advanced pseudo-ductile unidirectional thin-ply carbon/epoxy–glass/epoxy hybrid composites. Compos. Struct. 143, 362–370.

Dai, G., Mishnaevsky Jr., L., 2015. Carbon nanotube reinforced hybrid composites: computational modeling of environmental fatigue and usability for wind blades. Compos. B: Eng. 78, 349–360.

Davoodi, M., Sapuan, S.M., Ahmad, D., Ali, A., Khalina, A., Jonoobi, M., 2010. Mechanical properties of hybrid kenaf/glass reinforced epoxy composite for passenger car bumper beam. Mater. Des. 31, 4927–4932.

Dhakal, H.N., Sarasini, F., Santulli, C., Tirillò, J., Zhang, Z., Arumugam, V., 2015. Effect of basalt fibre hybridisation on post-impact mechanical behaviour of hemp fibre reinforced composites. Compos. A: Appl. Sci. Manuf. 75, 54–67.

Dong, C., Davies, I.J., 2015. Flexural strength of bidirectional hybrid epoxy composites reinforced by E glass and T700S carbon fibres. Compos. B: Eng. 72, 65–71.

Dorez, G., Ferry, L., Sonnier, R., Taguet, A., Lopez-Cuesta, J.M., 2014. Effect of cellulose, hemicellulose and lignin contents on pyrolysis and combustion of natural fibers. J. Anal. Appl. Pyrolysis. 107, 323–331.

Ferrante, L., Tirillo, J., Sarasini, F., Touchard, F., Ecault, R., Urriza, M.V., et al., 2015. Behaviour of woven hybrid basalt-carbon/epoxy composites subjected to laser shock wave testing: preliminary results. Compos. B: Eng. 78, 162–173.

Fiore, V., Di Bella, G., Valenza, A., 2011. Glass–basalt/epoxy hybrid composites for marine applications. Mater. Des. 32, 2091–2099.

Fiore, V., Calabrese, L., Di Bella, G., Scalici, T., Galtieri, G., Valenza, A., et al., 2016. Effects of aging in salt spray conditions on flax and flax/basalt reinforced composites: Wettability and dynamic mechanical properties. Compos. B: Eng. 93, 35–42.

Flynn, J., Amiri, A., Ulven, C., 2016. Hybridized carbon and flax fiber composites for tailored performance. Mater. Des. 102, 21–29.

Gong, L.X., Zhao, L., Tang, L.C., Liu, H.Y., Mai, Y.W., 2015. Balanced electrical, thermal and mechanical properties of epoxy composites filled with chemically reduced graphene oxide and rubber nanoparticles. Compos. Sci. Technol. 121, 104–114.

Guan, F.L., Gui, C.X., Zhang, H.B., Jiang, Z.G., Jiang, Y., Yu, Z.Z., 2016. Enhanced thermal conductivity and satisfactory flame retardancy of epoxy/alumina composites by combination with graphene nanoplatelets and magnesium hydroxide. Compos. B: Eng. 98, 134–140.

Guermazi, N., Haddar, N., Elleuch, K., Ayedi, H., 2014. Investigations on the fabrication and the characterization of glass/epoxy, carbon/epoxy and hybrid composites used in the reinforcement and the repair of aeronautic structures. Mater. Des. 56, 714–724.

Guo, Q.B., Rong, M.Z., Jia, G.L., Lau, K.T., Zhang, M.Q., 2009. Sliding wear performance of nano-SiO 2/short carbon fiber/epoxy hybrid composites. Wear. 266, 658–665.

Gupta, M., Srivastava, R., 2015. Effect of sisal fibre loading on dynamic mechanical analysis and water absorption behaviour of jute fibre epoxy composite. Mater. Today: Proc. 2, 2909–2917.

Hadden, C.M., Klimek-McDonald, D.R., Pineda, E.J., King, J.A., Reichanadter, A.M., Miskioglu, I., et al., 2015. Mechanical properties of graphene nanoplatelet/carbon fiber/epoxy hybrid composites: multiscale modeling and experiments. Carbon. 95, 100–112.

Harish, S., Michael, D.P., Bensely, A., Lal, D.M., Rajadurai, A., 2009. Mechanical property evaluation of natural fiber coir composite. Mater. Charact. 60, 44–49.

Jabbar, A., Militký, J., Kale, B.M., Rwawiire, S., Nawab, Y., Baheti, V., 2016. Modeling and analysis of the creep behavior of jute/green epoxy composites incorporated with chemically treated pulverized nano/micro jute fibers. Ind. Crops Prod. 84, 230–240.

Jawaid, M., Khalil, H.A., Bakar, A.A., 2010. Mechanical performance of oil palm empty fruit bunches/jute fibres reinforced epoxy hybrid composites. Mater. Sci. Eng.: A. 527, 7944–7949.

Jawaid, M., Khalil, H.A., Bakar, A.A., 2011. Woven hybrid composites: tensile and flexural properties of oil palm-woven jute fibres based epoxy composites. Mater. Sci. Eng.: A. 528, 5190–5195.

Jawaid, M., Khalil, H.A., Hassan, A., Dungani, R., Hadiyane, A., 2013. Effect of jute fibre loading on tensile and dynamic mechanical properties of oil palm epoxy composites. Compos. B: Eng. 45, 619–624.

Johnson, S., Kang, L., Akil, H.M., 2016. Mechanical behavior of jute hybrid bio-composites. Compos. B: Eng. 91, 83–93.

Kabir, M., Wang, H., Lau, K., Cardona, F., 2012. Chemical treatments on plant-based natural fibre reinforced polymer composites: an overview. Compos. B: Eng. 43, 2883–2892.

Kalantari, M., Dong, C., Davies, I.J., 2016. Multi-objective analysis for optimal and robust design of unidirectional glass/carbon fibre reinforced hybrid epoxy composites under flexural loading. Compos. B: Eng. 84, 130–139.

Kumar, M.A., Reddy, G.R., 2014. Mechanical & thermal properties of epoxy based hybrid composites reinforced with jute/sansevieria cylindrica fibres. Int. Lett. Chem. Phys. Astron. 19, 191.

Luna, P., Lizarazo-Marriaga, J., Mariño, A., 2016. Guadua angustifolia bamboo fibers as reinforcement of polymeric matrices: an exploratory study. Constr. Build. Mater. 116, 93–97.

Micheli, D., Vricella, A., Pastore, R., Delfini, A., Giusti, A., Albano, M., et al., 2016. Ballistic and electromagnetic shielding behaviour of multifunctional Kevlar fiber reinforced epoxy composites modified by carbon nanotubes. Carbon. 104, 141–156.

Mun, S.Y., Lim, H.M., Lee, D.J., 2015. Thermal conductivity of a silicon carbide/pitch-based carbon fiber-epoxy composite. Thermochim. Acta. 619, 16–19.

Muralidhar, B., 2013. Study of flax hybrid preforms reinforced epoxy composites. Mater. Des. 52, 835–840.

Narendar, R., Dasan, K.P., Nair, M., 2014. Development of coir pith/nylon fabric/epoxy hybrid composites: mechanical and ageing studies. Mater. Des. 54, 644–651.

Naresh, K., Shankar, K., Rao, B., Velmurugan, R., 2016. Effect of high strain rate on glass/carbon/hybrid fiber reinforced epoxy laminated composites. Compos. B: Eng.

Oleksy, M., Szwarc-Rzepka, K., Heneczkowski, M., Oliwa, R., Jesionowski, T., 2014. Epoxy resin composite based on functional hybrid fillers. Materials. 7, 6064–6091.

Onuaguluchi, O., Banthia, N., 2016. Plant-based natural fibre reinforced cement composites: a review. Cement Concrete Compos. 68, 96–108.

Panneerdhass, R., Gnanavelbabu, A., Rajkumar, K., 2014. Mechanical properties of luffa fiber and ground nut reinforced epoxy polymer hybrid composites. Proc. Eng. 97, 2042–2051.

Papadopoulos, A., Gkikas, G., Paipetis, A., Barkoula, N.M., 2016. Effect of CNTs addition on the erosive wear response of epoxy resin and carbon fibre composites. Compos. A: Appl. Sci. Manuf. 84, 299–307.

Phan, C., Mariatti, M., Koh, Y., 2016. Electromagnetic interference shielding performance of epoxy composites filled with multiwalled carbon nanotubes/manganese zinc ferrite hybrid fillers. J. Magn. Magn. Mater. 401, 472–478.

Pokharel, P., Truong, Q.T., 2014. Multi-step microwave reduction of graphite oxide and its use in the formation of electrically conductive graphene/epoxy composites. Compos. B: Eng. 64, 187–193.

Rajesh, M., Pitchaimani, J., Rajini, N., 2016. Free vibration characteristics of banana/sisal natural fibers reinforced hybrid polymer composite beam. Proc. Eng. 144, 1055–1059.

Ramakrishna, S., Mayer, J., Wintermantel, E., Leong, K.W., 2001. Biomedical applications of polymer-composite materials: a review. Compos. Sci. Technol. 61, 1189–1224.

Ramamoorthy, S.K., Skrifvars, M., Persson, A., 2015. A review of natural fibers used in biocomposites: plant, animal and regenerated cellulose fibers. Polym. Rev. 55, 107–162.

Ramesh, M., Palanikumar, K., Reddy, K.H., 2013. Comparative evaluation on properties of hybrid glass fiber-sisal/jute reinforced epoxy composites. Proc. Eng. 51, 745–750.

Ramnath, B.V., Kokan, S.J., Raja, R.N., Sathyanarayanan, R., Elanchezhian, C., Prasad, A. R., et al., 2013. Evaluation of mechanical properties of abaca–jute–glass fibre reinforced epoxy composite. Mater. Des. 51, 357–366.

Raza, M., Westwood, A., Stirling, C., Ahmad, R., 2015. Effect of boron nitride addition on properties of vapour grown carbon nanofiber/rubbery epoxy composites for thermal interface applications. Compos. Sci. Technol. 120, 9–16.

Reis, P., Ferreira, J., Zhang, Z., Benameur, T., Richardson, M., 2013. Impact response of Kevlar composites with nanoclay enhanced epoxy matrix. Compos. B: Eng. 46, 7–14.

Ridzuan, M., Majid, M.A., Afendi, M., Azduwin, K., Amin, N., Zahri, J., et al., 2016a. Moisture absorption and mechanical degradation of hybrid Pennisetum purpureum/glass–epoxy composites. Compos. Struct. 141, 110–116.

Ridzuan, M., Majid, M.A., Afendi, M., Mazlee, M., Gibson, A., 2016b. Thermal behaviour and dynamic mechanical analysis of Pennisetum purpureum/glass-reinforced epoxy hybrid composites. Compos. Struct. 152, 850–859.

Rout, A.K., Satapathy, A., 2012. Study on mechanical and tribo-performance of rice-husk filled glass–epoxy hybrid composites. Mater. Des. 41, 131–141.

Saba, N., Paridah, M.T., Jawaid, M., 2014. A review on potentiality of nano filler/natural fiber filled polymer hybrid composites. Polymers. 6, 2247–2273.

Saba, N., Jawaid, M., Paridah, M.T., Alothman, O.Y., 2015a. A review on flammability of epoxy polymer, cellulosic and non-cellulosic fiber reinforced epoxy composites. Polym. Adv. Technol.

Saba, N., Jawaid, M., Alothman, O.Y., Paridah, M.T., Hassan, A., 2015b. Recent advances in epoxy resin, natural fiber-reinforced epoxy composites and their applications. J. Reinf. Plastics Compos.0731684415618459.

Saba, N., Jawaid, M., Asim, M., 2016. Recent advances in nanoclay/natural fibers hybrid composites. Nanoclay Reinf. Polym. Compos., Springer. pp. 1−28.

Saidane, E.H., Scida, D., Assarar, M., Sabhi, H., Ayad, R., 2016. Hybridisation effect on diffusion kinetic and tensile mechanical behaviour of epoxy based flax−glass composites. Compos. A: Appl. Sci. Manuf. 87, 153−160.

Sanjay, M., Arpitha, G., Yogesha, B., 2015. Study on mechanical properties of natural-glass fibre reinforced polymer hybrid composites: a review. Mater. Today: Proc. 2, 2959−2967.

Sarasini, F., Tirillò, J., Valente, M., Valente, T., Cioffi, S., Iannace, S., et al., 2013. Effect of basalt fiber hybridization on the impact behavior under low impact velocity of glass/basalt woven fabric/epoxy resin composites. Compos. A: Appl. Sci. Manuf. 47, 109−123.

Sarasini, F., Tirillò, J., Ferrante, L., Valente, M., Valente, T., Lampani, L., et al., 2014. Drop-weight impact behaviour of woven hybrid basalt−carbon/epoxy composites. Compos. B: Eng. 59, 204−220.

Sharmila, T.B., Antony, J.V., Jayakrishnan, M., Beegum, P.S., Thachil, E.T., 2016. Mechanical, thermal and dielectric properties of hybrid composites of epoxy and reduced graphene oxide/iron oxide. Mater. Des. 90, 66−75.

Singh, V., Gope, P., Chauhan, S., Bisht, D., 2012. Mechanical behavior of banana fiber based hybrid bio composites. J. Mater. Environ. Sci. 3, 185−194.

Sprenger, S., Kothmann, M.H., Altstaedt, V., 2014. Carbon fiber-reinforced composites using an epoxy resin matrix modified with reactive liquid rubber and silica nanoparticles. Compos. Sci. Technol. 105, 86−95.

Subhani, T., Latif, M., Ahmad, I., Rakha, S.A., Ali, N., Khurram, A.A., 2015. Mechanical performance of epoxy matrix hybrid nanocomposites containing carbon nanotubes and nanodiamonds. Mater. Des. 87, 436−444.

Sudheer, M., Hemanth, K., Raju, K., Bhat, T., 2014. Enhanced mechanical and wear performance of epoxy/glass composites with PTW/graphite hybrid fillers. Proc. Mater. Sci. 6, 975−987.

Väisänen, T., Haapala, A., Lappalainen, R., Tomppo, L., 2016. Utilization of agricultural and forest industry waste and residues in natural fiber-polymer composites: a review. Waste Manage.

Valença, S.L., Griza, S., de Oliveira, V.G., Sussuchi, E.M., de Cunha, F.G.C., 2015. Evaluation of the mechanical behavior of epoxy composite reinforced with Kevlar plain fabric and glass/Kevlar hybrid fabric. Compos. B: Eng. 70, 1−8.

Vidil, T., Tournilhac, F., Musso, S., Robisson, A., Leibler, L., 2016. Control of reactions and network structures of epoxy thermosets. Prog. Polym. Sci..

Wang, F., Drzal, L.T., Qin, Y., Huang, Z., 2016. Enhancement of fracture toughness, mechanical and thermal properties of rubber/epoxy composites by incorporation of graphene nanoplatelets. Compos. A: Appl. Sci. Manuf. 87, 10−22.

Wilkinson, A.N., Kinloch, I.A., Othman, R.N., 2016. Low viscosity processing using hybrid CNT-coated silica particles to form electrically conductive epoxy resin composites. Polymer. (Guildf).

Xu, F., Wang, C., Li, D., Wang, M., Xu, F., Deng, X., 2015. Preparation of modified epoxy−SiO_2 hybrid materials and their application in the stone protection. Prog. Org. Coat. 81, 58−65.

Yahaya, R., Sapuan, S.M., Jawaid, M., Leman, Z., Zainudin, E., 2015. Effect of layering sequence and chemical treatment on the mechanical properties of woven kenaf−aramid hybrid laminated composites. Mater. Des. 67, 173−179.

Yahaya, R., Sapuan, S.M., Jawaid, M., Leman, Z., Zainudin, E., 2016. Effect of fibre orientations on the mechanical properties of kenaf–aramid hybrid composites for spall-liner application. Defence Technol. 12, 52–58.
Yan, H., Lu, C.X., Jing, D.Q., Chang, C.B., Liu, N.X., Hou, X.I., 2016a. Recycling of carbon fibers in epoxy resin composites using supercritical 1-propanol. New Carbon Mater. 31, 46–54.
Yan, L., Kasal, B., Huang, L., 2016b. A review of recent research on the use of cellulosic fibres, their fibre fabric reinforced cementitious, geo-polymer and polymer composites in civil engineering. Compos. B: Eng. 92, 94–132.
Yu, W., Fu, J., Dong, X., Chen, L., Shi, L., 2014. A graphene hybrid material functionalized with POSS: synthesis and applications in low-dielectric epoxy composites. Compos. Sci. Technol. 92, 112–119.
Zakaria, M.R., Akil, H.M., Kudus, M.H.A., Kadarman, A.H., 2015. Improving flexural and dielectric properties of MWCNT/epoxy nanocomposites by introducing advanced hybrid filler system. Compos. Struct. 132, 50–64.
Zakaria, M.R., Akil, H.M., Kudus, M.H.A., Othman, M.B.H., 2016. Compressive properties and thermal stability of hybrid carbon nanotube-alumina filled epoxy nanocomposites. Compos. B: Eng. 91, 235–242.
Zhang, P., Li, Q., Xuan, Y., 2014. Thermal contact resistance of epoxy composites incorporated with nano-copper particles and the multi-walled carbon nanotubes. Compos. A: Appl. Sci. Manuf. 57, 1–7.
Zhong, Y., Joshi, S.C., 2016. Environmental durability of glass fiber epoxy composites filled with core–shell polymer particles. Mater. Des. 92, 866–879.

Mechanical properties of hybrid polymer composite

4

Hai Nguyen[1], *Wael Zatar*[1] *and Hiroshi Mutsuyoshi*[2]
[1]College of Information Technology & Engineering, Marshall University, Huntington, WV, United States, [2]Department of Civil & Environmental Engineering, Saitama University, Saitama, Japan

Chapter Outline

4.1 Introduction

Fiber-reinforced composites have received great attention from researchers and scientists worldwide due to their attractive material characteristics. They have been widely used for various applications such as aerospace, automobile, civil infrastructure, and marine. Fibers (in the forms of roving, yarn, woven, etc.) reinforced with polymer matrix are known to result in enhanced mechanical properties of the composites. There are two basic types of fibers including natural and man-made fibers (a.k.a. synthetic or artificial fibers). Natural fibers such as flax, jute, and sisal have low densities compared to synthetic counterparts. They have acceptable specific strength/stiffness and relatively high elongations at breaking. The advantages of natural-fiber composites (a.k.a. green or fully-biocomposites) over synthetic fiber-based composites include low cost, light weight, abundantly available from renewable resources, environmentally friendly, biodegradability, recyclability, and renewability. On the other hand, synthetic fibers such as aramid, basalt, carbon, glass, and nylon are more durable and stronger than most natural fibers. Synthetic-fiber composites have thus been used in high-performance applications such as automobile and aircraft industries (Ramamoorthy et al., 2015). The major disadvantages of synthetic-fiber composites (synthetic fibers

Hybrid Polymer Composite Materials: Properties and Characterisation. DOI: http://dx.doi.org/10.1016/B978-0-08-100787-7.00004-4

are mostly derived from petroleum) are their low melting temperature, high cost, and nondegradable, while those of natural-fiber composites (natural fibers are made from plants, animals, and mineral sources) are their poor fiber-matrix adhesion and relatively high moisture sorption. The combined use of natural and synthetic fibers in polymer hybrid composites may take advantages of both fibers while minimizing their shortcomings. This chapter reviews the most recent studies on natural fiber-based hybrid composites with the main emphasis on their mechanical properties.

4.2 Polymer matrix composites (PMCs)

4.2.1 Reinforcing fibers

4.2.1.1 Natural fibers

Natural fibers have numerous advantages over traditional reinforcing synthetic fibers (e.g., glass and carbon) such as low density, low cost, high toughness, acceptable specific strength, renewability, biodegradability, ease of separation, lower energy requirements for processing, and worldwide availability (Saw et al., 2012; Lee and Wang, 2006). All fibers which come from natural sources (plants, animals, etc.) and do not require fiber formation or reformation are defined as natural fibers (Needles, 2001; Jacob et al., 2004). There are three basic types of natural fibers according to their origin. They are classified as the following:

- Plant fibers (referred to as cellulosic or lignocellulosic fibers): Plant fibers are categorized into six types including:
  - Bast or stem fibers (e.g., flax, hemp, isora, jute, kenaf, kudzu, mesta, nettle, okra, paper mulberry, roselle hemp, ramie, rattan, urena, wisteria)
  - Leaf fibers (e.g., abaca, agave, banana, cantala, caroa, curaua, date palm, fique, henequen, istle, Mauritius hemp, piassava, pineapple, phormium, raphia, sansevieria, sisal)
  - Seed/fruit fibers (e.g., coconut, coir, cotton, kapok, milkweed hairs, loofah, oil palm, sponge gourd)
  - Wood fibers (softwood and hardwood)
  - Stalk fibers (derived from stalks of barley, maize, oat, rice, wheat, and other crops)
  - Cane, grass, and reed fibers (e.g., albardine, bamboo, bagasse, canary, corn, esparto, rape, papyrus, sabai).
- Animal fibers: Animal fibers generally compose of proteins such as collagen, keratin, and fibroin. They are classified as animal wool or hairs (e.g., alpaca, angora wool, bison, camel, cashmere, mohair, goat hair, horse hair, lamb's wool, qiviut, yak wool, etc.), silk fibers (e.g., mulberry silk cocoons, tussah silkmoths, spider silk), and keratin fiber (e.g., bird and chicken feathers).
- Mineral fibers: Mineral fibers include the asbestos group (chrysotile, amosite, crocidolite, tremolite, anthophyllite, and actinolite), fibrous brucite, and wollastonite.

4.2.1.2 Man-made fibers

Man-made fibers are fibers in which either the basic chemical units have been formed by chemical synthesis followed by fiber formation or the polymers from natural sources have been dissolved and regenerated after passage through a spinneret to form fibers. Those fibers made by chemical synthesis are often called synthetic fibers, while fibers regenerated from natural polymer sources are called regenerated fibers or natural polymer fibers (Needles, 2001). The regenerated fibers include viscose/cuprammonium rayon (the fiber is mainly cellulose), cellulose ester, protein, and miscellaneous natural polymer fibers. Synthetic fibers can be classified according to their chemical structure as follows: polyamides, polyesters, polyvinyl derivatives, polyolefins, polyurethanes, and miscellaneous synthetic fibers (Gordon Cook, 1984).

4.2.1.3 Nanofillers

Nanofillers are defined as nano-objects with one, two, or three external dimensions in the size range from approximately 1–100 nm (i.e., nanoscale). According to the International Organization for Standardization (ISO) technical specification ISO/TS 80004-2:2015, nanofillers can be classified as three different types: (1) nanoplate (a nano-object with one external dimension at the nanoscale); (2) nanofiber (a nano-object with two external dimensions at the nanoscale) [e.g., hollow nanofiber–nanotube; rigid nanofiber–nanorod; and electrically conducting nanofiber–nanowire]; and (3) nanoparticle (a nano-object with three external dimensions in the nanoscale). Nanofillers play important role in modifying and improving physical, mechanical, optical, electrical, and thermal properties of polymer-based composites (Saba et al., 2014). The most commonly used nanofillers are nanoclays (morphology of layered silicate), nano-oxides, carbon nanotubes (CNT), polyhedral oligomeric sislesquioxanes (POSS), expanded graphite, carbon black, and fullerenes (Tables 4.1 and 4.2).

4.2.2 Polymer matrices

The important functions of polymer matrices are to bond fibers together and to transfer loads to the fibers. The polymer matrices can also provide a good surface finish quality of the composites and protect reinforcing fibers against chemical attack. They are classified as either thermosetting or thermoplastic resins.

4.2.2.1 Thermosetting resins

Thermosetting resins undergo chemical reactions (curing process) that crosslink the polymer chains and thus connect the entire matrix together in a three-dimensional network. Once cured, they cannot be remelted or reformed. Thermosetting resins tend to have high dimensional stability, high-temperature resistance, and good resistance to solvents because of their three-dimensional cross-linked structure (U.S. Congress, Office of Technology Assessment, June 1988). The most frequently used thermosetting resins are polyesters, vinylesters, epoxies, phenolics, polyamides (PA), and bismaleimides (BMI).

Table 4.1 Chemical composition of some important natural fibers (Mohanty et al., 2000; Jawaid and Abdul Khalil, 2011; Faruk et al., 2012)

Fiber	Cellulose (wt%)	Hemicellulose (wt%)	Lignin (wt%)	Pectin (wt%)
Bast				
Flax	71	18.6–20.6	2.2	2.3
Hemp	70.2–74.4	17.9–22.4	3.7–5.7	0.9
Jute	61–71.5	13.6–20.4	12–13	0.2
Kenaf	31–39	21.5	15–19	–
Ramie	68.6–76.2	13.1–16.7	0.6–0.7	1.9
Leaf				
Abaca	56–63	20–25	7–9	–
Banana	60–65	19	5–10	–
Curaua	73.6	9.9	7.5	–
Henequen	77.6	4–8	13.1	–
PALF	70–82	–	5–12	–
Sisal	67–78	10.0–14.2	8.0–11.0	10.0
Seed				
Cotton	82.7	5.7	–	–
Fruit				
Coir	36–43	0.15–0.25	41–45	3–4
Oil palm	65	–	29	–
Wood				
Hardwood	31–64	25–40	14–34	–
Softwood	30–60	20–30	21–37	–
Stalk				
Wheat straw	38–45	15–31	12–20	–
Rice husk	35–45	19–25	20	–
Rice straw	41–57	33	8–19	–
Cane/grass				
Bagasse	55.2	16.8	25.3	–
Bamboo	26–43	30	21–31	–

Table 4.2 Mechanical properties of natural and man-made fibers (Ramamoorthy et al., 2015; Mohanty et al., 2000; Jawaid and Abdul Khalil, 2011; Hyer, 2009)

Fiber	Density (g cm^{-3})	Diameter (μm)	Elongation (%)	Tensile strength (MPa)	Young's modulus (GPa)
Bast					
Flax	1.4–1.5	5–38	1.2–3.2	345–1500	27.6–80
Hemp	1.48	10–51	1.6	550–900	70
Jute	1.3–1.46	5–25	1.5–1.8	393–800	10–30
Kenaf	1.2	12–36	2.7–6.9	295	–
Ramie	1.5	18–80	2.0–3.8	220–938	44–128
Leaf					
Abaca	1.5	–	3.0–10	400	12
Banana	1.35	13.16	5.3	355	33.8
Curaua	1.4	–	3.7–4.3	500–1150	11.8
Henequen	1.4	–	3–4.7	430–580	–
PALF	1.5	20–80	1–3	170–1627	82
Sisal	1.33–1.5	7–47	2.0–3.0	400–700	9–38
Seed					
Cotton	1.5–1.6	12–35	3.0–10.0	287–597	5.5–12.6
Fruit					
Coir	1.2	–	15.0–30.0	175–220	4–6
Oil palm EFB	0.7–1.55	19.1–25.0	2.5	248	3.2
Wood					
Softwood kraft pulp	1.5	33	4.4	1000	40
Cane/glass					
Bagasse	1.2	10–34	1.1	20–290	19.7–27.1
Bamboo	0.6–1.1	–	–	140–230	11–17
Man-made					
PAN-based Carbon (IM)	1.78–1.82	8–9	1.0	2410–2930	228–276
PAN-based Carbon (HM)	1.67–1.9	7–10	0.5	2070–2900	331–400
PAN-based Carbon (UHM)	1.86	7–10	0.3–0.4	1720	517

(Continued)

Table 4.2 (Continued)

Fiber	Density ($g\ cm^{-3}$)	Diameter (μm)	Elongation (%)	Tensile strength (MPa)	Young's modulus (GPa)
Rayon	1.53–1.66	6.5	1.5–2.5	620–2200	41–393
E-glass	2.54	8–14	1.8–3.2	3450	72.4
S-glass	2.49	10	5.7	4590	85.5
Aramid (Kevlar-29)	1.44	12	3–4	2760	62
Aramid (Kevlar-49)	1.48	12	2.2–2.8	2800–3792	131

Note: EFB, empty-fruit bunches; PALF, pineapple leaf fiber; PAN, polyacrylonitrile; IM, intermediate modulus; HM, high modulus; UHM, ultra-high modulus.

4.2.2.2 Thermoplastic resins

Unlike thermosetting resins, thermoplastic molecules do not crosslink and they can be melted by heating and solidified by cooling, which render them capable of repeated reshaping and reforming. They are, in general, ductile and tougher than thermosetting resins and are widely used for nonstructural applications without reinforcements and fillers (Mallick, 2007). Thermoplastic resins offer attractive mechanical properties such as excellent tensile strength and stiffness, good compression and fatigue strength, high dimensional stability, and excellent durability and damage tolerance. In addition, they have good wear-resistant and flame-retardant characteristics, which are suitable for various applications especially aerospace (McKague, 2001). Typical thermoplastic resins include polypropylene (PP), polyvinylidene fluoride (PVDF), polymethyl methacrylate (PMMA, also known as acrylic), polyphenylene sulfide (PPS), polyether etherketone (PEEK), polyetherimide (PEI), and polyetherketone ketone (PEKK). Comparisons on qualitative characteristics of thermoplastic and thermosetting resins are shown in Table 4.3.

4.2.3 Manufacturing processes for PMCs

Manufacturing processes of PMCs can be grouped into three categories: short-fiber suspension methods; squeeze flow methods; and porous media methods (Astrom, 2001). Short-fiber suspension methods involve the transport of fibers (usually short discontinuous fibers) and resin (either thermosetting or thermoplastic) as a suspension into a mold or through a die to form the composite. Injection molding, compression molding, and extrusion processes are included in this category. Squeeze flow methods include fibers (usually continuous or long discontinuous fibers) partially or fully preimpregnated with thermoplastic resin. Pultrusion, thermoforming (thermoplastic sheet forming), and tape winding processes fall under

Table 4.3 Qualitative comparisons of thermoplastic and thermosetting resins (McKague, 2001)

Characteristic	Thermoplastics	Thermosets
Tensile properties	Excellent	Excellent
Stiffness properties	Excellent	Excellent
Compression properties	Good	Excellent
Compression strength after impact	Good to excellent	Fair to excellent
Bolted joint properties	Fair	Good
Fatigue resistance	Good	Excellent
Damage tolerance	Excellent	Fair to excellent
Durability	Excellent	Good to excellent
Maintainability	Fair to poor	Good
Service temperature	Good	Good
Dielectric properties	Good to excellent	Fair to good
Environmental weakness	None, or hydraulic fluid	Moisture
NBS smoke test performance	Good to excellent	Fair to good
Processing temperature, °C (°F)	343–427 (650–800)	121–315 (250–600)
Processing pressure, MPa (psi)	1.38–2.07 (200–300)	0.59–0.69 (85–100)
Lay-up characteristics	Dry, boardy, difficult	Tack, drape, easy
Debulking, fusing, or heat tacking	Every ply if part is not flat	Typically every 3 or more plies
In-process joining options	Co-fusion	Co-cure, co-bond
Post-process joining options	Fastening, bonding, fusion	Fastening, bonding
Manufacturing scrap rates	Low	Low
Ease of prepregging	Fair to poor	Good to excellent
Volatile-free prepreg	Excellent	Excellent
Prepreg shelf life and out time	Excellent	Good
Health/safety	Excellent	Excellent

Note: NBS, National Bureau of Standards.

this category. Porous media methods compose continuous fibers impregnated with thermosetting resin (due to its low viscosity) to form the composite in an open or a closed mold. Liquid composite molding, thermoset pultrusion, filament winding, and autoclave processes are belonging to this category. Fig. 4.1 shows an overview of manufacturing processes for PMCs.

4.3 Hybrid composites and their mechanical properties

4.3.1 Introduction

Hybrid composites are defined as composite materials consisting of two or more different reinforcing fibers impregnated in the same matrix. The purpose of

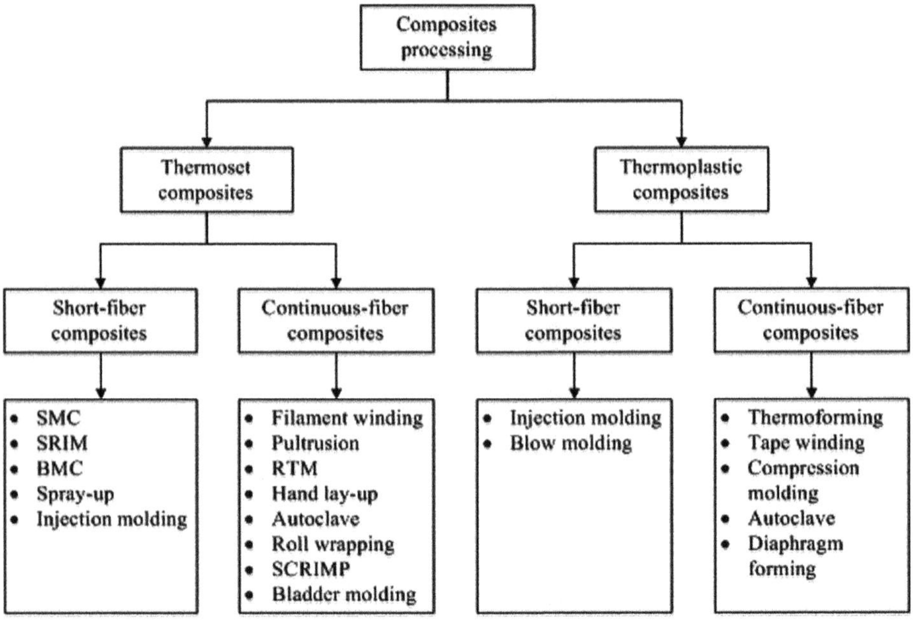

Figure 4.1 Manufacturing Processes for PMCs (Mazumdar, 2001).

hybridization is to achieve tailor-made properties of polymer composites and to take advantages of constituent materials in the composites. There are several types of hybrid composites depending on the way the constituent materials are mixed (Chamis and Lark, 1977; Fukuda, 1984; Pegoretti et al., 2004; Wang et al., 2008). According to Chamis and Lark (1977), there are four general categories of hybrid composites including: (1) interply hybrids; (2) intraply hybrids; (3) interply/intraply hybrids; and (4) superhybrids. The interply hybrids consist of plies from two or more unidirectional composites stacked in a specified sequence. Intraply hybrids include two or more different fibers mixed in the same ply. Interply/intraply hybrids compose of plies of intraply and interply hybrids stacked in a specified sequence. Superhybrids compose of metal foils or metal composite plies stacked in a specified sequence. The term "hybrid" or "synergistic" effect is usually used to imply that the initial failure strain of a hybrid composite (corresponding to the failure of low--elongation fibers in a hybrid) is greater than failure strain of a low-elongation, non-hybrid composite (Fukuda, 1984). Brittle inorganic fibers and ductile organic fibers are often combined to make hybrid composites such as aramid/glass, palm/glass, mineral fiber/glass, etc. (Wang et al., 2008). Hybrid biocomposites are defined as the combination of two or more different natural fibers (biofibers) in a matrix or a combination of biofibers and synthetic fibers in a matrix. Glass fibers are generally used to improve the mechanical properties of natural-fiber composites. The following sections discuss mechanical properties of various hybrid natural/synthetic fiber composites.

4.3.2 Hybrid natural fiber-reinforced composites

4.3.2.1 Hybrid bagasse/jute fiber-reinforced composites

Jute is one of the most well-known plant (vegetable) fibers, largely found in Asian countries like Bangladesh, China, India, Nepal, and Thailand (they produce about 95% of the global production of jute fibers) (Alves et al., 2010). It is a lignocellulosic bast fiber having inherent advantages such as renewable nature, biodegradability (associated with environmentally friendly), high strength and initial modulus over other fibers (Saw et al., 2012). Sugar-cane bagasse (generally called "bagasse") is one of the largest cellulosic agro-industrial byproducts. It is a lingo-cellulosic residue (byproduct) of the sugar industry and is major used by the sugar factories as fuel for the boilers. Bagasse offers many advantages over other crop residues (e.g., rice and wheat straw) and agricultural residues because of its low ash contents (Pandey et al., 2000). It is typically found in tropical countries such as Brazil, India, China, and Thailand (Table 4.4).

Saw and Datta (2009) studied mechanical properties of hybrid polymer composites reinforced with short bagasse fiber (BF) and short jute fiber (JF) bundles. Epoxidized phenolic novolac (EPN) was used as the resin matrix. Different fiber ratios and fiber surface treatments were investigated. JF bundles were treated by alkali solution (a.k.a. sodium hydroxide—NaOH) while BF bundles were either untreated or modified by chlorine dioxide (ClO_2) and furfuryl alcohol ($C_5H_6O_2$). The purpose of fiber surface modification was to create quinones in the lignin portions of BF bundles. The quinones reacted with the furfuryl alcohol to improve adhesion ability of the modified BF bundles. The results showed that the hybridization of the modified BF and alkali-treated JF bundles in the EPN resin matrix resulted in higher tensile, flexural, and impact properties in comparison to those of the unmodified BF bundles. The optimal mechanical properties were obtained when the BF/JF ratio was 50:50 (Table 4.5).

4.3.2.2 Hybrid bamboo fiber-reinforced composites

Bamboo is known as one of the most attractive biofibers because it has several advantages such as small environmental load, renewability, rapid growth, and relatively high strength compared to other natural fibers (e.g., jute and cotton) (Takagi and Ichihara, 2004). Asian countries such as China and India produce over 80% of the worldwide availability of bamboo fiber (Han et al., 2008).

Okubo et al. (2009) developed novel hybrid biocomposites consisting of a biodegradable poly-lactic acid (PLA) matrix reinforced with bamboo fiber bundles and microfibrillated cellulose (MFC). MFC is a cellulosic material with expanded high-volume cellulose and usually consists of aggregates of cellulose microfibrils. Its diameter is in the range of 20–60 nm and it has a length of several micrometers (Lavoine et al., 2012). Various terms are used to describe MFC in the literature including microfibril, microfibril aggregates, microfibrillar cellulose, nanofibril, nanofiber, nanofibrillar cellulose, and fibril aggregates (Siró and Plackett, 2010). Okubo et al. (2009) investigated the influence of MFC dispersion on the properties

Table 4.4 **Hybrid biocomposites and their manufacturing processes**

Hybrid biocomposites	Resin	Chemical treatments	Manufacturing processes	References	Year
All natural fibers					
Bagasse/jute	EPN	Chlorine dioxide and furfuryl alcohol (bagasse); Alkali (jute)	Hand lay-up and compression molding	Saw and Datta (2009)	2009
Bamboo/MFC	PLA	Alkali (bamboo)	Injection molding	Okubo et al. (2009)	2009
Banana/kenaf	UP	Alkali or sodium lauryl sulfate (SLS)	Compression molding	Thiruchitrambalam et al. (2009)	2009
Banana/sisal	Epoxy	Untreated	Hand lay-up	Venkateshwaran et al. (2011)	2011
Coconut/cork	HDPE	Untreated	Twin-screw extrusion and compression molding	Fernandes et al. (2013)	2013
Coir/silk	UP	Alkali treatment	Hand lay-up	Khanam et al. (2009)	2009
Corn husk/kenaf	PLA	Untreated	Injection molding	Kwon et al. (2014)	2014
Cotton/jute	Phenolic novolac	Untreated	Compression molding	De Medeiros et al. (2005)	2005
Cotton/kapok	UP	Alkali treatment	Hydraulic compression molding	Mwaikambo and Bisanda (1999)	1999
Cotton/ramie	UP	Untreated	Compression molding	Paiva Júnior et al. (2004)	2004
Jute/oil palm EFB	Epoxy	Untreated	Compression molding	Jawaid et al. (2011)	2011
Kenaf/PALF	HDPE	Untreated	Compression molding	Aji et al. (2011)	2011
Roselle/sisal	UP	Untreated	Hydraulic compression molding	Athijayamani et al. (2009)	2009
Silk/sisal	UP	Alkali treatment	Hand lay-up	Khanam et al. (2007)	2007
Synthetic/natural fibers					
Aramid/coir	Epoxy	Untreated	Hand lay-up	Rashid et al. (2011)	2011
Aramid/kenaf	Epoxy	Untreated	Hand lay-up	Yahaya et al. (2016)	2016
Aramid/sisal	Phenolic	Untreated	Hot press	Zhong et al. (2011)	2011

Basalt/flax-hemp	Epoxy	Untreated	Vacuum infusion	Petrucci et al. (2013)	2013
Basalt/flax-glass	Epoxy	Untreated	Vacuum infusion	Petrucci et al. (2013)	2013
Basalt/glass-hemp	Epoxy	Untreated	Vacuum infusion	Petrucci et al. (2013)	2013
Carbon/basalt-flax	Epoxy	Untreated	Hand lay-up and vacuum bagging	Nisini et al. (2016)	2016
Carbon/flax	Epoxy	Others	Vacuum bagging	Fiore et al. (2012)	2012
Carbon/sisal	UP	Alkali treatment	Hand lay-up	Khanam et al. (2010)	2010
Glass/abaca	Orthophthalic	Untreated	Hand lay-up	Venkatasubramanian and Raghuraman (2015)	2015
Glass/abaca-banana	Orthophthalic	Untreated	Hand lay-up	Venkatasubramanian and Raghuraman (2015)	2015
Glass/banana	Orthophthalic	Untreated	Hand lay-up	Venkatasubramanian and Raghuraman (2015)	2015
Glass/bamboo	PP	MAPP	Injection molding	Thwe and Liao (2003)	2003
Glass/coir	UP	PVA	Hand lay-up	Jayabal et al. (2011)	2011
Glass/curaua	UP	AAP	Hydraulic press	Almeida Júnior et al. (2012)	2012
Glass/jute	Polyester	Untreated	Hand lay-up	Ahmed Sabeel and Vijayarangan (2008)	2008
Glass/kapok	Polyester	Alkali treatment	Hand lay-up	Venkata Reddy et al. (2008)	2008
Glass/kenaf	Epoxy	Untreated	Modified SMC	Davoodi et al. (2010)	2010
Glass/PALF	Polyester	Untreated	Hydraulic press	Mishra et al. (2003)	2003
Glass/palmyra	Rooflite	Untreated	Hydraulic compression molding	Velmurugan and Manikandan (2007)	2007
Glass/silk	Epoxy	Untreated	Hand lay-up	Priya and Rai (2006)	2006
Glass/sisal	Polyester	Alkali, cyanoethylation, and acetylation treatments	Hydraulic press	Mishra et al. (2003)	2003

Note: EFB, empty-fruit bunches; PLA, poly-lactic acid; UP, unsaturated polyester; HDPE, high-density polyethylene; PALF, pineapple leaf fiber; MFC, microfibrillated cellulose; EPN, epoxidized phenolic novolac; PP, Polypropylene; MAPP, maleic anhydride polypropylene; PVA, polyvinyl acetate; AAP, acetyl acetone peroxide; SMC, sheet molding compound.

Table 4.5 **Mechanical properties of hybrid biocomposites**

Hybrid biocomposites	Fiber ratio (by weight or volume)	Flexural modulus (GPa)	Flexural strength (MPa)	Tensile modulus (GPa)	Tensile strength (MPa)	Impact strengt (kJ m^{-2})	References
Natural fibers							
Bagasse/jute	Bagasse fiber bundles (untreated) and jute fiber bundles (treated)						Saw and Datta (2009)
	0:100	0.645	31.15	0.302	11.45	6.90	
	20:80	0.789	36.46	0.356	16.02	7.46	
	35:65	1.101	45.32	0.420	19.45	9.53	
	50:50	1.480	55.63	0.492	23.07	10.66	
	65:35	1.311	51.19	0.399	21.15	8.33	
	100:0	0.502	26.78	0.227	9.87	6.67	
	Bagasse fiber bundles (treated) and jute fiber bundles (treated)						
	20:80	1.178	42.72	0.526	18.72	10.00	
	35:65	1.484	54.57	0.635	22.57	13.33	
	50:50	1.748	65.22	0.753	26.77	15.93	
	65:35	1.518	60.12	0.704	23.54	10.93	
	100:0	0.632	30.78	0.286	11.20	8.66	
Bamboo/MFC	MFC/PLA composites (milled to 5 μm)						Okubo et al. (2009)
	1 wt% of MFC	–	–	4.61 ± 0.27	45.9 ± 4.1	–	
	2 wt% of MFC	–	–	3.95 ± 0.14	51.7 ± 2.3	–	
Banana/kenaf	50:50, nonwoven hybrid						Thiruchitrambalam et al. (2009)
	10% NaOH treatment	–	57.2	–	44	13	
	10% SLS treatment		60.8		50	16	
	50:50, woven hybrid						
	10% NaOH treatment	–	62.0	–	50	18	
	10% SLS treatment		68.0		54	21	

Banana/sisal	100:0	8.920	57.33	0.642	16.12	13.25	Venkateshwaran et al. (2011)
	75:25	9.025	58.51	0.662	17.39	15.57	
	50:50	9.130	59.69	0.682	18.66	17.90	
	25:75	9.235	60.87	0.703	19.93	20.22	
	0:100	9.340	62.04	0.723	21.20	22.54	
Coconut/cork	10:44:44:2 (wt% of coconut/cork/HDPE/coupling agent)	–	–	0.599 ± 0.02	20.4 ± 0.3	–	Fernandes et al. (2013)
Coir/silk	Alkali treatment						Khanam et al. (2009)
	10 mm fiber	–	39.53	–	15.01	–	
	20 mm fiber	–	45.07	–	17.24	–	
	30 mm fiber	–	42.02	–	16.14	–	
Corn husk/kenaf	0:30 (PLA 70 wt%)	–	–	2.117	–	–	Kwon et al. (2014)
	15:15 (PLA 70 wt%)	–	–	1.547	–	–	
	30:0 (PLA 70 wt%)	–	–	1.221	–	–	
Cotton/jute	23.7:76.3 (jute fabric type III)						De Medeiros et al. (2005)
	Test angle, 0°	9.9 ± 0.8	136.7 ± 4.0	7.1 ± 0.3	59.4 ± 1.7	9.3 ± 0.9	
	Test angle, 45°	8.4 ± 0.7	84.6 ± 4.7	4.6 ± 0.1	21.1 ± 1.4	7.5 ± 1.0	
	Test angle, 90°	7.2 ± 0.7	58.3 ± 5.4	4.1 ± 0.1	14.6 ± 0.5	5.5 ± 1.0	

(Continued)

Table 4.5 (Continued)

Hybrid biocomposites	Fiber ratio (by weight or volume)	Flexural modulus (GPa)	Flexural strength (MPa)	Tensile modulus (GPa)	Tensile strength (MPa)	Impact strengt (kJ m^{-2})	References
Cotton/kapok	3:2						Mwaikambo and Bisanda (1999)
	Untreated ($V_f = 60\%$)	–	–	0.884	55.70	110.53	
	Alkali treatment ($V_f = 43\%$)	–	–	1.635	52.87	119.25	
	Non-accelerated weather condition ($V_f = 46.6$ %)	0.709	52.40	–	–	–	
	Accelerated weather condition (V_f=46.6 %)	0.703	39.55	–	–	–	
Cotton/ramie (ramie fibers placed longitudinally to the mould length)	10.8:41.1 (0° composite)	–	–	–	90.9 ± 12.7	–	Paiva Júnior et al. (2004)
	11.9:45.5 (0° composite)	–	–	–	117.3 ± 13.3	–	
	11.9:45.1 (0° composite)	–	–	–	118.0 ± 6.5	–	
Jute/OPEFB	1:4						Jawaid et al. (2011)
	OPEFB/Jute/OPEFB	–	–	2.39	25.53	–	
	Jute/OPEFB/Jute	–	–	2.59	27.41	–	
	Pure OPEFB	–	–	2.23	22.61		
	Pure jute	–	–	3.89	45.55		

Kenaf/PALF	1:1 (At 0.25 mm fiber length and 60% fiber loading)	4.114	34.01	0.874	32.24	6.167	Aji et al. (2011)
Roselle/sisal	1:1						Athijayamani et al. (2009)
	Dry condition, fiber length = 15 cm	–	76.5	–	58.7	1.30	
	Wet condition, fiber length = 15 cm	–	62.9	–	44.9	1.28	
Silk/sisal	1:1, fiber length = 20 mm						Khanam et al. (2007)
	Untreated	–	46.18	–	18.95	–	
	Alkali treatment	–	54.74	–	23.61	–	
Natural/synthetic fibers							
Aramid/coir	Coir (warp) + Kevlar (weft)	–	16.70	–	–	66.82	Azrin Hani Abdul et al. (2011)
	Kevlar (warp) + Coir (weft)	–	25.16	–	–	61.12	
Aramid/kenaf	Fiber volume fraction ratio						Yahaya et al. (2016)
	21.2:10.46 (woven)	–	94.21	3.337	145.8	51.41	
	16.78:16.51 (unidirectional)	–	100.30	2.368	115.36	41.24	
	21.39:9.57 (mat)	–	35.82	1.888	101.56	24.64	
Aramid/sisal	20:80 (Degree of surface microfibrillation of sisal fiber = 32 SR)	–	–	–	26.9	–	Zhong et al. (2011)
Basalt/flax-hemp	7.85:5.57:9.11 (V_f ratio)	7.45 ± 0.67	128.46 ± 29.14	7.69 ± 0.63	115.97 ± 3.77	–	Petrucci et al. (2013)
Basalt/flax-glass	7.16:11.72:2.30 (V_f ratio)	8.02 ± 0.68	137.95 ± 19.85	6.64 ± 0.49	153.16 ± 17.41	–	Petrucci et al. (2013)
Basalt/glass-hemp	11.38:2.59:8.56 (V_f ratio)	5.90 ± 0.42	126.22 ± 13.63	8.11 ± 0.60	128.84 ± 8.70	–	Petrucci et al. (2013)

(Continued)

Table 4.5 (Continued)

Hybrid biocomposites	Fiber ratio (by weight or volume)	Flexural modulus (GPa)	Flexural strength (MPa)	Tensile modulus (GPa)	Tensile strength (MPa)	Impact strengt (kJ m^{-2})	References
Carbon/ basalt-flax	12:14:27 (wt%)						Nisini et al. (2016)
	Laminate N1	16.42 ± 0.55	256.08 ± 9.79	16.20 ± 0.52	189.23 ± 3.75	–	
	Laminate N2	17.08 ± 1.00	286.67 ± 15.26	16.89 ± 0.31	185.24 ± 5.66	–	
Carbon/flax	51.1 ± 3.3 (total fiber content)	23.84 ± 0.74	160.42 ± 10.46	6.48 ± 0.32	288.03 ± 30.23	–	Fiore et al. (2012)
Carbon/sisal	18% NaOH treatment						Khanam et al. (2010)
	0:100	5.32	138.78	1.96	78.22	–	
	25:75	6.52	140.89	1.99	84.74	–	
	50:50	8.69	158.31	2.17	93.97	–	
	75:25	11.33	169.14	2.78	107.51	–	
	100:0	13.47	176.53	2.98	122.11	–	
Glass/abaca	60% fiber + 40% resin	0.621	68.23	0.750	93.29	1.458	Venkatasubramanian and Raghuraman (2015)
Glass/abaca-banana	60% fiber + 40% resin	0.222	82.85	0.567	97.28	1.090	Venkatasubramanian and Raghuraman (2015)
Glass/banana	60% fiber + 40% resin	0.235	139.66	0.750	96.00	1.315	Venkatasubramanian and Raghuraman (2015)

Glass/bamboo	1:7	–	–	4.8	24.4	–	Thwe and Liao (2003)
Glass/coir	Glass/glass/coir Glass/coir/glass Coir/glass/glass	2.358 2.881 2.361	77 65 71	1.349 1.453 1.373	51 47 52	144 101 140	Jayabal et al. (2011)
Glass/curaua	$V_f = 40\%$						Almeida Júnior et al. (2012)
	0:100 70:30 100:0	– – –	– – –	– – –	– – –	32.6 ± 1.7 149 ± 17 153.9 ± 19.7	
Glass/jute	60:40	12.38	159.85	12.46	124.44	–	Ahmed Sabeel and Vijayarangan (2008)
Glass/kapok	Untreated hybrid composites						Venkata Reddy et al. (2008)
	0:100 25:75 50:50 75:25 100:0	– – – – –	– – – – –	0.975 1.133 1.182 1.229 2.469	67.34 78.05 82.11 102.55 112.87	– – – – –	
	Alkali-treated hybrid composites						
	0:100 25:75 50:50 75:25 100:0	– – – – –	– – – – –	1.426 1.605 1.645 2.363 2.469	79.1 94.1 98.6 107.6 112.8	– – – – –	

(Continued)

Table 4.5 (Continued)

Hybrid biocomposites	Fiber ratio (by weight or volume)	Flexural modulus (GPa)	Flexural strength (MPa)	Tensile modulus (GPa)	Tensile strength (MPa)	Impact strengt (kJ m^{-2})	References
Glass/kenaf	–	12.0	223.6	7.5	151.8		Davoodi et al. (2010)
Glass/PALF	8.6:16.4	–	99	–	71	–	Mishra et al. (2003)
Glass/palmyra	Randomly mixed glass/palmyra hybrid composites, $V_f = 55\%$ (by weight), Palmyra fiber length = 50 mm	3.54	59.19	1.515	42.65	60.5	Velmurugan and Manikandan (2007)
Glass/silk	0:100	1.503	60.81	0.844	58.35	–	Priya and Rai (2006)
	10:90	1.847	94.31	0.891	60.99	–	
	20:80	3.015	97.31	0.922	64.87	–	
	30:70	4.221	106.5	0.944	70.12	–	
	40:60	5.251	108.2	0.992	77.81	–	
	50:50	5.440	114.5	1.008	84.04	–	
Glass/sisal	5.7:24.3 (wt%)	–	138	–	98	–	Mishra et al. (2003)

Note: OPEFB, oil palm empty-fruit bunches; PALF, pineapple leaf fiber; MFC, microfibrillated cellulose; PLA, poly-lactic acid; SLS, sodium lauryl sulfate; SR, Schopper–Riegler; V_f, fiber volume fraction.

of bamboo fiber-reinforced composites. MFC was dispersed in a PLA polymer matrix using a calendering process (usually employed to smooth or compress a material) with a three-roll mill. The purpose of using PLA, a bio-based and biodegradable polymer matrix, is to enhance interfacial bonding with the MFC. Bamboo fiber bundles have diameters of about 200 μm whereas MFC has much smaller diameters of a few microns. The MFC/PLA mixture was processed in the three-roll mill at progressively decreasing gap settings of 70, 50, 35, 25, 15, 10, and 5 μm. Significant increase in fracture energy (nearly 200%) was achieved when 1 wt% of MFC was added to the PLA matrix and the MFC/PLA composite was milled at the minimum gap setting of 5 μm. The hybrid bamboo/MFC/PLA composite, including the bamboo fiber and the PLA matrix reinforced with 1 wt% of MFC, was found to effectively prevent sudden crack path through the reinforcing bamboo fiber and result in substantial fracture strength improvements.

4.3.2.3 Banana/kenaf and banana/sisal hybrid composites

Banana fiber (extracted from the bark of banana trees) is a potential reinforcing material for various polymer composites. It has superior mechanical properties such as good tensile strength and modulus, resulting from its high cellulose content and low microfibrillar angle (Liu et al., 2009). According to FAOStat (FAOStat, September 12, 2016), the five largest countries of banana production in 2013–2014 are India, China (mainland), Philippines, Brazil, and Ecuador.

Kenaf fiber is extracted from bast fiber of kenaf plants. It is a promising reinforcement element for polymer composites because of its excellent mechanical properties, renewability, and ecofriendly. On the other hand, sisal is known as a durable fiber and one of the toughest reinforcing materials. Its composites have high impact strength and moderate tensile and flexural properties compared to other natural fiber-reinforced composites. It has been used for various applications such as marine and agriculture to make ropes, twines, cords, bagging and rugs, etc. (Jacob et al., 2004). The main disadvantage associated with natural fibers, including sisal and kenaf fibers, is their poor interfacial bonding with a polymer matrix (Akil et al., 2011).

Thiruchitrambalam et al. (2009) investigated woven and nonwoven hybrid banana/kenaf fiber reinforced with unsaturated polyester matrix. The fiber contents were kept constant at 40% with 50:50 ratio of banana and kenaf fiber-reinforced composites. The fibers were 30-minute treated with either 10% of NaOH solution or 10% of sodium lauryl sulfate (SLS). The specimens with SLS treatment showed better improvement in the mechanical properties than the ones with alkali treatment. The SLS treatment resulted in enhanced tensile, flexural, and impact strength of both woven and nonwoven hybrid banana/kenaf composites (Table 4.5).

Mechanical properties of hybrid banana/sisal fiber reinforced with epoxy matrix were evaluated by Venkateshwaran et al. (2011). The hybridization of banana and sisal fibers in the epoxy composite resulted in 16% increase in tensile strength, 4% increase in flexural strength, and 35% increase in impact strength. The fiber ratio of 50:50 by weight was found to enhance the mechanical properties of the banana/sisal hybrid composite while decreasing its moisture uptake.

4.3.2.4 Hybrid coconut/cork fiber-reinforced composites

Coconut fiber (a.k.a. "coir") is a natural fiber extracted from coconut trees, which mainly grows in tropical regions in Asia countries such as India, Vietnam, and Thailand. Cork fiber is harvested from the bark of a specific species of cork oak trees (*Quercus suber*). The cork oak tree naturally regrows its new cork bark, making it a renewable resource.

Fernandes et al. (2013) prepared hybrid composites from high-density polyethylene (HDPE) reinforced with cork powder and randomly distributed short coconut fibers. Coupling agent (CA) based on maleic anhydride was used to improve the compatibility and interfacial bonding between the fiber and matrix. The coconut/HDPE/cork hybrid composites resulted in 27% increase in elastic modulus and 47% increase in the tensile strength as compared with the cork/HDPE composite. In addition, the use of CA enhanced the elongation at break and tensile properties of the hybrid composites. The addition of 10 wt% of short coconut fibers and 2 wt% of CA was recommended for the better mechanical performance of the cork-based composites.

4.3.2.5 Hybrid coir/silk fiber-reinforced composites

Silk is a light, soft, thin, and continuous protein fiber, which is produced by various insects. Silk fiber is synthesized by the silkworm and spun in the form of a silk cocoon. The silkworm produces massive amount of silk proteins (fibroin and sericin, which are major components of silk cocoons) during the final stage of larval development (Mondal, 2007). Silk fiber is known as the strongest natural material with high specific strength and stiffness. It has excellent drape and wonderful luster but possesses a poor resistance to sunlight exposure.

Khanam et al. (2009) investigated the hybrid composites of coir/silk fiber reinforced with unsaturated polyester matrix. Different fiber lengths (10 mm, 20 mm, and 30 mm) were studied. Coir fibers were treated with NaOH solution. The purpose of the NaOH treatment was to remove hemicellulose and lignin from the coir fiber, which may result in a better fiber-matrix bonding. The 20 mm fiber length composites were found to have higher flexural and tensile strength compared to the 10 mm and 20 mm fiber length counterparts. The NaOH-treated coir/silk hybrid composites were proved to have significant improvement in compressive, flexural, and tensile strength properties, resulting from the enhanced interfacial bonding between the coir fiber interface and the polyester matrix.

4.3.2.6 Hybrid corn husk/kenaf fiber-reinforced composites

Agricultural wastes (e.g., rice husk, rice straw, and corn husk) produce large amount of raw natural fibers, which can be used as reinforcing materials in polymer composites. Corn husks are thin, leafy sheaths that cover the corn cobs and contains cellulose-rich fibers (Mahalaxmi et al., 2010). Kenaf is an important source of fiber for paper industry and other sectors.

Kwon et al. (2014) prepared hybrid biocomposites composed of kenaf fiber and corn husk flour reinforced with poly-lactic acid (PLA) matrix. The ratio of fiber/matrix by weight was fixed at 30:70 while various ratios of kenaf fiber and corn husk flour were evaluated. The influence of the aspect ratios of kenaf fibers (measured before and after passing through extrusion process) to the mechanical properties was investigated. The results indicated that the aspect ratio determined after extrusion did not influence the predicted values obtained by the Halpin–Tsai equation. It should be noted that the Halpin–Tsai model for the prediction of elastic behavior of composite materials is based on the geometry/orientation and elastic properties of the fibers and matrix. It assumes no interaction between the fiber and matrix in the composite. The difference of Young's modulus of fibers was found to affect the stress transfer from matrix to fiber. It was reported that a scale ratio between reinforcements of different aspect ratios may be a controlling factor in optimizing the mechanical properties of a hybrid biocomposite.

4.3.2.7 Hybrid cotton fiber-reinforced composites

Cotton fibers are unbranched, unicellular (single-cell) seed hairs (or seed trichomes) and being among the longest plant cells ever characterized (they can elongate up to approximately 3 cm). Unlike many plant secondary cell walls, the cotton fiber wall contains no lignin (Kim and Triplett, 2001). Cotton fibers are considered the world's most important fibers and widely used in textile industry. They have rich cellulose content and possess many advantages such as good strength, excellent drape, and high absorbency. According to FAOStat (FAOStat, September 12, 2016), the top 4 countries with largest production of cottonseed in 2013–2014 include China, India, United States, and Pakistan.

De Medeiros et al. (2005) investigated mechanical properties of hybrid cotton/jute woven fabrics reinforced with novolac type phenolic matrix. The results indicated that the mechanical properties of the hybrid cotton/jute fabric composites were strongly dependent on fiber orientation, fiber content, fiber-matrix adhesion, and fabric characteristics. The anisotropy of the composites depended upon the characteristics of fiber roving/fabric and increased with the increase of the test angle. The mechanical properties were found to be inversely proportional to the test angle as the specimens tested at zero degree with respect to the jute roving direction showed best overall performance. The composites tested at 45 and 90 degrees with respect to the jute fiber direction exhibited a controlled brittle failure while those tested at zero degree to the longitudinal direction displayed a catastrophic failure without control. Jute fiber was found to be a strong reinforcing material and the combination of jute and cotton in the fabric composites can avoid catastrophic failure mode.

Mwaikambo and Bisanda (1999) prepared hybrid cotton-kapok fiber fabric incorporated with unsaturated polyester matrix with varying fiber volume fraction (V_f). The fabric composites were either untreated or treated with 5% NaOH to improve fiber-matrix bonding. Mechanical properties of the cotton-kapok composites subjected or not subjected to accelerated weathering condition were evaluated. It was found that the composites with untreated fibers exhibited higher V_f values than those with alkali treatments. The untreated fibers improved the tensile strength of the composites while the alkali-treated fibers enhanced the composites' tensile modulus. The increase of V_f resulted in decreasing the impact strength for both treated and untreated composites. The composites subjected to accelerated weather conditions showed reductions in flexural strength and modulus.

Ramie fibers are obtained from the bast/stem of the ramie plant (*Boehmeria nivea*) of the nettle family, Utricaceae (Lodha and Netravali, 2002). Ninety-nine percent of ramie plants are cultivated in Asian countries such as China, Laos, Philippines, and Republic of Korea and one percent is grown in Americas such as Brazil. Hybrid composites of ramie/cotton plain weave fabric reinforced with unsaturated polyester resin were investigated by Paiva Júnior et al. (2004). The results indicated that ramie fibers have a great potential as reinforcing fibers in polymer composites. The contribution of the cotton fibers was negligible because of their poor alignments in the composites and the weak fiber-matrix interfacial bonding.

4.3.2.8 Hybrid jute/oil palm EFB fiber-reinforced composites

Oil palm (*Elaeis guineensis*) is one of the most economical perennial oil crops for its valuable oil-producing fruits in tropical regions such as West/Southwest Africa and Southeast Asia. In the oil extraction process, the fruits or nuts are first stripped from fruit bunches, leaving behind the empty-fruit bunches as waste (Law et al., 2007). The oil palm industries generate abundant amount of biomass which can be a waste disposal challenge if not properly used. Oil palm fibers are derived from two sources of oil palm tree including oil palm empty-fruit bunches (OPEFB) and mesocarp. OPEFB fibers are the most commonly used for composite materials because they contain highest composition of hemicellulose compared to coir, pineapple, banana, and even soft and hardwood fibers (Hassan et al., 2010).

Jawaid et al. (2011) prepared three-ply hybrid composites of jute/OPEFB fibers reinforced with epoxy resin. The ratio by weight of the jute/OPEFB composites was fixed at 1:4. The chemical resistance, void content, and tensile properties of the hybrid composites were investigated. The results indicated that the jute/OPEFB/jute and OPEFB/jute/OPEFB composites were strongly resistant to the following chemicals: benzene (C_6H_6), toluene (C_7H_8), carbon tetrachloride (CCl_4), water (H_2O), hydrochloric acid (HCl), 40% nitric acid (HNO_3), 5% acetic acid (CH_3COOH), 10% sodium hydroxide (NaOH), 20% sodium carbonate (Na_2CO_3), and 10% ammonium hydroxide (NH_4OH). The jute/OPEFB/jute composites showed less void content compared to the pure OPEFB and OPEFB/jute/OPEFB composites. This was attributed to the fact that the jute fiber mats were tightly packed and more compatible towards the epoxy resin. The high-strength jute fibers at the outer ply were able to withstand the tensile stress while the OPEFB fiber core absorbed the stresses and evenly distributed them in the composites. As a result, the hybrid jute/OPEFB had higher tensile strength and modulus compared to the pure OPEFB composite (Table 4.5). The hybrid composites also exhibited better adhesion to the matrix than the pure OPEFB composite.

4.3.2.9 Hybrid kenaf/PALF fiber-reinforced composites

Pineapple (*Ananas comosus*) is a tropical plant, a member of the bromeliad family (Bromeliaceae) native to South America, and the third most important tropical fruit crop after banana and mango (in terms of total global production of fruit weight) (Davis et al., 2015). Pineapple leaf fibers (PALFs) are extracted from pineapple leaves, which are a waste product of pineapple cultivation. PALFs show excellent mechanical properties due to its high cellulose content (70–82%) and high degree of crystallinity (44–60%) (Reddy and Yang, 2005). Kenaf fiber has superior mechanical properties such as excellent flexural and tensile strength and the combination of kenaf and PALF in a polymer composite may yield robust materials for various applications.

Aji et al. (2011) investigated the effect of fiber size and fiber loading on the mechanical properties of hybridized kenaf/PALF fibers reinforced with high-density polyethylene (HDPE). All tested specimens were prepared at kenaf/PALF fiber ratio of 1:1. Four types of fiber lengths (0.25, 0.5, 0.75, and 2 mm) were evaluated at varying percentages of fiber loadings (ranged from 10–70%). The 0.25 mm fiber size showed the best tensile and flexural properties while the 0.75 and 2 mm fiber sizes exhibited enhanced impact strength. The increase of fiber length resulted in reduction in some mechanical properties, which was attributed to fiber entanglement rather than fiber attrition. Tensile and impact strengths were found to be inversely proportional while flexural strength generally satisfied the rule of mixture. Hybridization effect (resulting from the synergistic strengthening of kenaf and PALF fibers) was clearly observed. Scanning electron microscopy (SEM) was used to evaluate the composites' surface and the results showed good adhesion between the matrix and fibers.

4.3.2.10 Hybrid sisal fiber-reinforced composites

Sisal (*Agave sisalana*) is a member of the Agavaceae family, which are hard-fiber plants originally from Central America and Mexico but widely cultivated and naturalized in many tropical countries in Americas, Africa, and Asia. Sisal fibers are extracted from the leaves of sisal plants. According to FAOStat (FAOStat, September 12, 2016), world production of sisal fibers in 2011 is about 411,102 tons. Top three producers of sisal fibers include Brazil, Mexico, and Tanzania. Sisal fibers are tough and strong and being widely used in composite materials as well as in paper/plastic industries.

Roselle (*Hibiscus sabdariffa*) is a species of Hibiscus native to West Africa. The Roselle plant is found in abundance in nature and primarily used for its bast fibers and its fruit. Roselle fibers have been widely used in composite materials and textile industry as their mechanical properties are comparable to other natural fibers such as kenaf and jute. Athijayamani et al. (2009) investigated the effect of moisture absorption (under wet conditions) on mechanical properties of short sisal and roselle fibers (with sisal/roselle fiber ratio of 1:1) reinforced with unsaturated polyester resin. Different fiber lengths and contents were considered. The results revealed that the tensile and flexural strength of the hybrid sisal/roselle composites increased with the increase of the fiber length and the fiber content at the dry condition. On the other hand, at the wet condition, significant strength reductions were observed for both tensile and flexural properties. The impact strength was found to be inversely proportional to the fiber content and fiber length at both wet and dry conditions.

Khanam et al. (2007) prepared polyester based hybrid composites of sisal and silk fibers. Sisal/silk fiber ratio was at 1:1 and different fiber lengths were evaluated. It was found that the composites with 20 mm fiber length had higher tensile, compressive, and flexural strength than those with 10 mm and 30 mm fiber lengths. Significant improvements in mechanical properties (tensile, compressive, and flexural strength) were observed for the hybrid composites with alkali-treated fibers.

4.3.3 Hybrid natural/synthetic fiber-reinforced composites

The natural and synthetic fibers can be combined in the same matrix to produce hybrid composites that offer a range of properties that cannot be obtained with a single kind of reinforcement (Khanam et al. 2009). The following sections discuss mechanical properties of hybrid composites of some common synthetic fibers (aramid, basalt, carbon, and glass) and natural fibers.

4.3.3.1 Hybrid aramid fiber-based composites

Rashid et al. (2011) investigated mechanical properties of hybrid coir/Kevlar reinforced epoxy composites. Kevlar is the registered trademark of the E.I. du Pont de Nemours and Company (a.k.a. DuPont) for their para-aramid fibers. Kevlar has a unique combination of high strength, high modulus, toughness and thermal stability. Coconut or coir fibers have been increasingly used as a reinforcing material due to their low cost and good mechanical properties. It was found that the coir/woven Kevlar composites exhibited highest impact strength while their flexural strength was lowest. The results showed that the hybrid composites of woven coir yarn (warp) and Kevlar yarn (weft) had the flexural and impact strength of 16.7 MPa and 66.82 kJ m^{-2}, respectively (Table 4.5). The results suggested that coir fibers are promising reinforcements for high-impact resistant application such as body armors.

Yahaya et al. (2016) presented an evaluation on the effect of kenaf fiber orientation on the mechanical properties of hybrid aramid/kenaf reinforced epoxy composites for military application. The effect of kenaf structure including woven, nonwoven unidirectional (UD), and mat fabrics was investigated. Aramid fabric (Kevlar 129) was the plain weaved structure. It was found that the nonwoven mat kenaf/Kevlar hybrid composite had relatively low density because of its high void contents. The tensile and Charpy impact strength properties of the woven kenaf/

Kevlar composite were higher compared with those of other hybrid composites. On the other hand, the flexural strength of the hybrid composites with the UD kenaf was slightly higher than that of the woven kenaf hybrid composite. The scanning electron micrograph revealed that the mat kenaf hybrid composites exhibited higher void content than the woven and UD kenaf composites.

Zhong et al. (2011) investigated the effect of surface microfibrillation of sisal fiber on the mechanical properties of hybrid aramid/sisal fiber-reinforced phenolic composites. The results showed that surface microfibrillation of sisal fibers significantly influenced the mechanical properties of the hybrid aramid/sisal composites. Microfibrils and aggregates formed on the sisal-fiber surface resulted in a larger contact area between sisal fibers and the phenolic matrix, thus producing stronger mechanical interlocking strength. In addition, the microfibrils and aggregates inhibited the formation of spontaneous cracks in the composites. As a result, the compression, tensile and fiber/matrix interfacial bonding strengths and wear resistance of the hybrid composites were significantly enhanced.

4.3.3.2 Hybrid basalt fiber-based composites

Petrucci et al. (2013) evaluated mechanical properties of hybrid basalt fiber-based composite laminates manufactured by vacuum infusion process. Basalt fibers are made from basalt, a type of igneous rock formed by volcanic lava. The basalt fibers were combined with either flax, hemp, or glass fibers in the composites. The test results suggested that the hybrid basalt/flax-glass exhibited best general performance among all investigated composites. The hybrid composites with hemp fibers showed relatively low layer-interface quality. SEM observations of the tested hybrid composite laminates exhibited the diffuse presence of fiber pull-out in hemp and flax fibers and all laminates showed a brittle failure.

4.3.3.3 Hybrid carbon fiber-based composites

Nisini et al. (2016) investigated mechanical and impact properties of ternary hybrid composite laminates with carbon, basalt, and flax fibers. All laminates were fabricated by hand lay-up technique and then consolidated by vacuum bagging process. Basalt and flax fiber-layers were sandwiched between carbon-fiber layers on the outer faces. It was found that the intercalation of basalt with flax fiber layers resulted in enhanced flexural and interlaminar strength. Two laminates with different stacking sequences of basalt and flax fiber layers exhibited insignificant improvement in impact performance.

Fiore et al. (2012) studied mechanical behavior of hybrid carbon/flax/epoxy composite for structural applications. Two different bidirectional flax fabrics were used to produce flax fabric reinforced plastic (FFRP) laminates using vacuum bagging process. The test results showed that the addition of one external carbon-fiber layer in the FFRP composites remarkably increased their mechanical properties. The hybrid carbon/flax composites were recommended for several applications such as nautical and automobile.

Khanam et al. (2010) prepared hybrid composites of carbon/sisal fibers reinforced with unsaturated polyester matrix. Tensile, flexural, and chemical resistance properties were evaluated. The tensile and flexural strength of the hybrid carbon/sisal composites increased with the increase of the carbon fiber loading. Significant improvement in tensile and flexural properties were observed for the hybrid composites with alkali treatment of sisal fibers. The chemical resistance test results indicated that all hybrid composites were strongly resistant to all chemicals except carbon tetrachloride (CCl_4).

4.3.3.4 Hybrid glass fiber-based composites

Venkatasubramanian and Raghuraman (2015) evaluated the mechanical behavior of hybrid composites consisting of abaca/banana and glass fibers reinforced with orthophthalic resins. The hybrid banana–abaca/glass composites showed higher tensile strength than the abaca/glass and banana/glass composites. Flexural strength of the banana/glass composites was found to be highest, attributable to the good adhesion properties of the banana fiber. The abaca/glass composites exhibited highest impact strength, resulting from the high strength and stiffness of the abaca fiber.

Thwe and Liao (2003) investigated durability of bamboo fiber-reinforced polypropylene (BFRP) composites and hybrid bamboo/glass fiber-reinforced polypropylene (BGRP) composites. The results indicated that both tensile strength and tensile modulus of BFRP and BGRP decreased after exposing to water (25°C and 75°C) for prolonged period. The level of reductions in strength and stiffness depended upon the exposed time and water temperature. BGRP specimens exhibited a better resistance to the exposed environment in terms of retention of tensile strength and stiffness. The tensile strength and stiffness were enhanced by the incorporation of maleic anhydride polypropylene (MAPP) as a coupling agent in the polypropylene matrix, resulting in an improved interfacial bonding. The hybridization of high-durable glass fiber and bamboo fiber was found to be an effective way to enhance the durability of natural-fiber composites subject to environmental aging. The hybrid glass/bamboo composites showed better fatigue behavior than all bamboo fiber-reinforced composites.

Jayabal et al. (2011) developed hybrid composites incorporating woven coir/glass fabric fiber preimpregnated with the resin matrix consisting of unsaturated orthophthalic polyester, cobalt octoate accelerator, and methyl ethyl ketone peroxide (MEKP) catalyst in the ratio of 1:0.015:0.015. Polyvinyl acetate release agent was applied to the laminates' surface before placing in the mold. Different laminates' stacking sequences were considered to evaluate mechanical properties of the hybrid coir/glass composites. It was found that the glass/glass/coir and coir/glass/glass composites showed highest tensile, flexural, and impact strength. The hybrid composites with two plies of glass fibers (glass/glass/coir and coir/glass/glass) exhibited higher breaking resistance than the coir/glass/coir composites with a single glass ply. The coir fibers failed faster than the glass fibers and the incorporation of the glass woven fabric in the coir-fiber composites enhanced their mechanical properties. The glass fibers were found to have better interfacial bond with the polyester matrix than the coir fibers.

Curaua (*Ananas erectifolius*) plants are native to Brazilian Amazon region and belong to Bromeliaceae family. Curaua fibers exhibit excellent properties such as good breaking elongation, high specific strength, and low density (Almeida et al., 2013). Almeida Júnior et al. (2012) investigated thermal, mechanical, and dynamic mechanical properties of hybrid curaua/glass composites. The results showed that the density of the hybrid curaua/glass composites increased with the increase of the glass fiber content and overall fiber volume fraction. The incorporation of glass fibers in the curaua composites resulted in significant improvement in impact strength and hardness. This was attributed to the intrinsic characteristics of the glass fiber such as stronger interfacial bond to the resin matrix and higher energy dissipation compared to the curaua fiber. Dynamic mechanical properties exhibited an increase in storage modulus whereas the glass transition temperature showed no significant change with the intermingled glass fibers. It was found that the hybrid composites with 30% of curaua fibers showed similar properties compared to the pure glass fiber-reinforced composites.

The mechanical properties of other glass fiber-based hybrid composites including glass/jute (Ahmed Sabeel and Vijayarangan, 2008), glass/kapok (Venkata Reddy et al., 2008), glass/kenaf (Davoodi et al., 2010), glass/PALF (Mishra et al., 2003), glass/palmyra (Velmurugan and Manikandan, 2007), glass/silk (Priya and Rai, 2006), and glass/sisal (Mishra et al., 2003) are listed in Table 4.5.

4.4 Conclusions

Mechanical characterizations of various hybrid composites were reviewed in this chapter. Hybrid composites of all natural fibers generally exhibit satisfactory strength and can be potentially used for various applications. Hybridizations of natural and synthetic fibers in polymer composite effectively enhance mechanical properties (e.g., flexural, tensile, and impact strength) of all natural-fiber composites. The natural/synthetic fiber hybrid composites are thus promising for high-performance structural applications. Hybrid composites with chemical treatments or modifications of fibers generally show better mechanical properties compared with untreated composites, resulting from the improved fiber-matrix bonding in treated composites.

References

Ahmed Sabeel, K., Vijayarangan, S., 2008. Tensile, flexural and interlaminar shear properties of woven jute and jute–glass fabric reinforced polyester composites. J. Mater. Process. Technol. 207 (1), 330–335.

Aji, I.S., Zainudin, E.S., Khalina, A., Sapuan, S.M., Khairul, M.D., 2011. Studying the effect of fiber size and fiber loading on the mechanical properties of hybridized kenaf/PALF–reinforced HDPE composite. J. Reinf. Plastics Compos. 30 (6), 546–553.

Akil, H.M.L., Omar, M.F., Mazuki, A.A.M., Safiee, S.Z.A.M., Mohd Ishak, Z.A., Abu Bakar, A., 2011. Kenaf fiber reinforced composites: a review. Mater. Des. 32 (8), 4107–4121.

Almeida, J.H.S., Amico, S.C., Botelho, E.C., Rico Amado, F.D., 2013. Hybridization effect on the mechanical properties of curaua/glass fiber composites. Compos. B: Eng. 55, 492–497.

Almeida Júnior, J.H.S., Ornaghi Júnior, H.L., Amico, S.C., Rico Amado, F.D., 2012. Study of hybrid intralaminate curaua/glass composites. Mater. Des. 42, 111–117.

Alves, C., Silva, A.J., Reis, L.G., Freitas, M., Rodrigues, L.B., Alves, D.E., 2010. Ecodesign of automotive components making use of natural jute fiber composites. J. Clean Product. 18 (4), 313–327.

Astrom, B.T., 2001. Introduction to Manufacturing of Polymer-Matrix Composites. ASM International, Materials Park, OH, pp. 421–422.

Athijayamani, A., Thiruchitrambalam, M., Natarajan, U., Pazhanivel, B., 2009. Effect of moisture absorption on the mechanical properties of randomly oriented natural fibers/ polyester hybrid composite. Mater. Sci. Eng.: A. 517 (1), 344–353.

Chamis, C.C., R.F. Lark. Hybrid composites, state-of-the-art review: analysis, design, application and fabrication, 1977.

Gordon Cook, J., 1984. Handbook of Textile Fibres: Man-Made Fibres. Elsevier.

Davis, S.C., Ming, R., LeBauer, D.S., Long, S.P., 2015. Toward systems-level analysis of agricultural production from crassulacean acid metabolism (CAM): scaling from cell to commercial production. New Phytol. 208 (1), 66–72.

Davoodi, M.M., Sapuan, S.M., Ahmad, D., Ali, A., Khalina, A., Jonoobi, M., 2010. Mechanical properties of hybrid kenaf/glass reinforced epoxy composite for passenger car bumper beam. Mater. Des. 31 (10), 4927–4932.

De Medeiros, E.S., Agnelli, J.A.M., Joseph, K., de Carvalho, L.H., Mattoso, L.H.C., 2005. Mechanical properties of phenolic composites reinforced with jute/cotton hybrid fabrics. Polym. Compos. 26 (1), 1–11.

FAOStat. Food and Agriculture Organization of the United Nations, <http://faostat3.fao.org/ browse/Q/QC/E> [accessed 12.09.16].

Faruk, O., Bledzki, A.K., Fink, H.-P., Sain, M., 2012. Biocomposites reinforced with natural fibers: 2000–2010. Prog. Polym. Sci. 37 (11), 1552–1596.

Fernandes, E.M., Correlo, V.M., Mano, J.F., Reis, R.L., 2013. Novel cork–polymer composites reinforced with short natural coconut fibres: effect of fibre loading and coupling agent addition. Compos. Sci. Technol. 78, 56–62.

Fiore, V., Valenza, A., Di Bella, G., 2012. Mechanical behaviour of carbon/flax hybrid composites for structural applications. J. Compos. Mater.0021998311429884.

Fukuda, H., 1984. An advanced theory of the strength of hybrid composites. J. Mater. Sci. 19 (3), 974–982.

Han, G., Lei, Y., Wu, Q., Kojima, Y., Suzuki, S., 2008. Bamboo–fiber filled high density polyethylene composites: effect of coupling treatment and nanoclay. J. Polym. Environ. 16 (2), 123–130.

Hassan, A., Salema, A.A., Ani, F.N., Abu Bakar, A., 2010. A review on oil palm empty fruit bunch fiber-reinforced polymer composite materials. Polym. Compos. 31 (12), 2079–2101.

Hyer, M.W., 2009. Stress Analysis of Fiber-Reinforced Composite Materials. DEStech Publications, Inc.

Jacob, M., Thomas, S., Varughese, K.T., 2004. Mechanical properties of sisal/oil palm hybrid fiber reinforced natural rubber composites. Compos. Sci. Technol. 64 (7), 955–965.

Jawaid, M., Abdul Khalil, H.P.S., 2011. Cellulosic/synthetic fibre reinforced polymer hybrid composites: a review. Carbohydr. Polym. 86 (1), 1–18.

Jawaid, M., Abdul Khalil, H.P.S., Abu Bakar, A., Noorunnisa Khanam, P., 2011. Chemical resistance, void content and tensile properties of oil palm/jute fibre reinforced polymer hybrid composites. Mater. Des. 32 (2), 1014–1019.

Jayabal, S., Natarajan, U., Sathiyamurthy, S., 2011. Effect of glass hybridization and staking sequence on mechanical behaviour of interply coir–glass hybrid laminate. Bull. Mater. Sci. 34 (2), 293–298.

Paiva Júnior, C.Z., De Carvalho, L.H., Fonseca, V.M., Monteiro, S.N., d'Almeida, J.R.M., 2004. Analysis of the tensile strength of polyester/hybrid ramie–cotton fabric composites. Polym. Test. 23 (2), 131–135.

Khanam, P.N., Mohan Reddy, M., Raghu, K., John, K., Venkata Naidu, S., 2007. Tensile, flexural and compressive properties of sisal/silk hybrid composites. J. Reinf. Plastics Compos. 26 (10), 1065–1070.

Khanam, N., Ramachandra Reddy, G., Raghu, K., Venkata Naidu, S., 2009. Tensile, flexural and compressive properties of coir/silk fiber reinfroced hybrid composites. J. Reinf. Plastics Compos.

Khanam, P.N., Abdul Khalil, H.P.S., Jawaid, M., Ramachandra Reddy, G., Surya Narayana, C., Venkata Naidu, S., 2010. Sisal/carbon fibre reinforced hybrid composites: tensile, flexural and chemical resistance properties. J. Polym. Environ. 18, 727–733.

Kim, H.J., Triplett, B.A., 2001. Cotton fiber growth in planta and in vitro. Models for plant cell elongation and cell wall biogenesis. Plant Physiol. 127 (4), 1361–1366.

Kwon, H.-J., Sunthornvarabhas, J., Park, J.-W., Lee, J.-H., Kim, H.-J., Piyachomkwan, K., et al., 2014. Tensile properties of kenaf fiber and corn husk flour reinforced poly (lactic acid) hybrid bio-composites: role of aspect ratio of natural fibers. Compos. B: Eng. 56, 232–237.

Lavoine, N., Desloges, I., Dufresne, A., Bras, J., 2012. Microfibrillated cellulose – its barrier properties and applications in cellulosic materials: a review. Carbohydr. Polym. 90 (2), 735–764.

Law, K.-N., Daud, W.R.W., Ghazali, A., 2007. Morphological and chemical nature of fiber strands of oil palm empty-fruit-bunch (OPEFB). BioResources. 2 (3), 351–362.

Lee, S.-H., Wang, S., 2006. Biodegradable polymers/bamboo fiber biocomposite with bio-based coupling agent. Compos. A: Appl. Sci. Manuf. 37 (1), 80–91.

Liu, H., Wu, Q., Zhang, Q., 2009. Preparation and properties of banana fiber-reinforced composites based on high density polyethylene (HDPE)/Nylon-6 blends. Bioresour. Technol. 100 (23), 6088–6097.

Lodha, P., Netravali, A.N., 2002. Characterization of interfacial and mechanical properties of "green" composites with soy protein isolate and ramie fiber. J. Mater. Sci. 37 (17), 3657–3665.

Mahalaxmi, Y., Sathish, T., Subba Rao, Ch, Prakasham, R.S., 2010. Corn husk as a novel substrate for the production of rifamycin B by isolated *Amycolatopsis* sp. RSP 3 under SSF. Process Biochem. 45 (1), 47–53.

Mallick, P.K., 2007. Fiber-Reinforced Composites: Materials, Manufacturing, and Design. CRC Press.

Mazumdar, S., 2001. Composites Manufacturing: Materials, Product, and Process Engineering. CRC Press.

McKague, L., 2001. Thermoplastic Resins. ASM International, Materials Park, OH, pp. 132–140.

Mishra, S., Mohanty, A.K., Drzal, L.T., Misra, M., Parija, S., Nayak, S.K., et al., 2003. Studies on mechanical performance of biofibre/glass reinforced polyester hybrid composites. Compos. Sci. Technol. 63 (10), 1377–1385.

Mohanty, A.K., Misra, M., Hinrichsen, G., 2000. Biofibres, biodegradable polymers and biocomposites: an overview. Macromol. Mater. Eng. 276 (1), 1–24.

Mondal, M., 2007. The silk proteins, sericin and fibroin in silkworm, *Bombyx mori* Linn. – a review. Caspian J. Environ. Sci. 5 (2), 63–76.

Mwaikambo, L.Y., Bisanda, E.T.N., 1999. The performance of cotton–kapok fabric–polyester composites. Polym. Test. 18 (3), 181–198.

Needles, H.L., 2001. Textile Fibers, Dyes, Finishes, and Processes. Standard Publishers Distributors.

Nisini, E., Santulli, C., Liverani, A., 2016. Mechanical and impact characterization of hybrid composite laminates with carbon, basalt and flax fibres. Compos. B: Eng.

Okubo, K., Fujii, T., Thostenson, E.T., 2009. Multi-scale hybrid biocomposite: processing and mechanical characterization of bamboo fiber reinforced PLA with microfibrillated cellulose. Compos. A: Appl. Sci. Manuf. 40 (4), 469–475.

Pandey, A., Soccol, C.R., Nigam, P., Soccol, V.T., 2000. Biotechnological potential of agro-industrial residues. I: sugarcane bagasse. Bioresour. Technol. 74 (1), 69–80.

Pegoretti, A., Fabbri, E., Migliaresi, C., Pilati, F., 2004. Intraply and interply hybrid composites based on E-glass and poly (vinyl alcohol) woven fabrics: tensile and impact properties. Polym. Int. 53 (9), 1290–1297.

Petrucci, R., Santulli, C., Puglia, D., Sarasini, F., Torre, L., Kenny, J.M., 2013. Mechanical characterisation of hybrid composite laminates based on basalt fibres in combination with flax, hemp and glass fibres manufactured by vacuum infusion. Mater. Des. 49, 728–735.

Priya, S.P., Rai, S.K., 2006. Mechanical performance of biofiber/glass-reinforced epoxy hybrid composites. J. Ind. Text. 35 (3), 217–226.

Ramamoorthy, S.K., Skrifvars, M., Persson, A., 2015. A review of natural fibers used in biocomposites: plant, animal and regenerated cellulose fibers. Polym. Rev. 55 (1), 107–162.

Rashid, A.H.A., Roslan, A., Jaafar, M., Roslan, M.N., Ariffin, S., 2011. Mechanical properties evaluation of woven coir and kevlar reinforced epoxy composites. Advanced Materials Research, vol. 277. Trans Tech Publications.

Venkata Reddy, G., Venkata Naidu, S., Shobha Rani, T., 2008. Kapok/glass polyester hybrid composites: tensile and hardness properties. J. Reinf. Plastics Compos.

Reddy, N., Yang, Y., 2005. Biofibers from agricultural byproducts for industrial applications. Trends Biotechnol. 23 (1), 22–27.

Saba, N., Md Tahir, P., Jawaid, M., 2014. A review on potentiality of nano filler/natural fiber filled polymer hybrid composites. Polymers. 6 (8), 2247–2273.

Saw, S.K., Datta, C., 2009. Thermo mechanical properties of jute/bagasse hybrid fibre reinforced epoxy thermoset composites. BioResources. 4 (4), 1455–1475.

Saw, S.K., Sarkhel, G., Choudhury, A., 2012. Effect of layering pattern on the physical, mechanical, and thermal properties of jute/bagasse hybrid fiber-reinforced epoxy novolac composites. Polym. Compos. 33 (10), 1824–1831.

Siró, I., Plackett, D., 2010. Microfibrillated cellulose and new nanocomposite materials: a review. Cellulose. 17 (3), 459–494.

Takagi, H., Ichihara, Y., 2004. Effect of fiber length on mechanical properties of" green" composites using a starch-based resin and short bamboo fibers. JSME Int. J. Ser. A. 47 (4), 551–555.

Thiruchitrambalam, M., Alavudeen, A., Athijayamani, A., Venkateshwaran, N., Elaya Perumal, A., 2009. Improving mechanical properties of banana/kenaf polyester hybrid composites using sodium laulryl sulfate treatment. Mater. Phys. Mechan. 8 (2), 165–173.

Thwe, M.M., Liao, K., 2003. Durability of bamboo-glass fiber reinforced polymer matrix hybrid composites. Compos. Sci. Technol. 63 (3), 375–387.

U.S. Congress, Office of Technology Assessment, 1988. Advanced Materials by Design, OTAE-351. U.S. Government Printing Office, Washington, DC.

Velmurugan, R., Manikandan, V., 2007. Mechanical properties of palmyra/glass fiber hybrid composites. Compos. A: Appl. Sci. Manuf. 38 (10), 2216–2226.

Venkatasubramanian, H., Raghuraman, S., 2015. Mechanical behaviour of abaca-glass-banana fibre reinforced hybrid composites. J. Eng. Sci. Technol. 10 (8), 958–971.

Venkateshwaran, N., ElayaPerumal, A., Alavudeen, A., Thiruchitrambalam, M., 2011. Mechanical and water absorption behaviour of banana/sisal reinforced hybrid composites. Mater. Des. 32 (7), 4017–4021.

Wang, X., Hu, B., Feng, Y., Liang, F., Mo, J., Xiong, J., et al., 2008. Low velocity impact properties of 3D woven basalt/aramid hybrid composites. Compos. Sci. Technol. 68 (2), 444–450.

Yahaya, R., Sapuan, S.M., Jawaid, M., Leman, Z., Zainudin, E.S., 2016. Effect of fibre orientations on the mechanical properties of kenaf–aramid hybrid composites for spall-liner application. Def. Technol. 12 (1), 52–58.

Zhong, L.X., Fu, S.Y., Zhou, X.S., Zhan, H.Y., 2011. Effect of surface microfibrillation of sisal fibre on the mechanical properties of sisal/aramid fibre hybrid composites. Compos. A: Appl. Sci. Manuf. 42 (3), 244–252.

Physical properties of hybrid polymer/clay composites

5

Ayesha Kausar
Nanoscience and Technology Department, National Center For Physics, Quaid-i-Azam University, Islamabad, Pakistan

Chapter Outline

5.1 Introduction

In recent research and development efforts, nanotechnology is the most competent field. One of the major areas of recent research in nanotechnology is polymer matrix-based nanocomposite. Composite material is usually made up of two or more components. Generally, composite material is a solid multiphase material with unique physical, structural, and chemical properties (Amina et al., 2013). In composites, the main constituent is *continuous phase*, i.e., matrix. The other phase is *filler/reinforcement* material, i.e., dispersed phase. The composite material can be classified into polymer, metal, and ceramic composites. The host phase (matrix) is usually present in large quantity, whereas the filler material is implanted in the host matrix. Polymer nanocomposites have several advantageous properties relative to metals and ceramics such as light weight, low cost, surface area/volume ratio, and several improved physical properties (Shah et al., 2015). In other words, polymeric nanocomposites are multiphase materials in which polymers are hardened by the nanoscale filler material. The polymer/clay nanocomposite is a distinctive intance of nanotechnology. In this regard, the widely

Hybrid Polymer Composite Materials: Properties and Characterisation. DOI: http://dx.doi.org/10.1016/B978-0-08-100787-7.00005-6

used nanofiller materials are clay fillers. The most important nanoclay filler material is montmorillonite (MMT), i.e., based on smectite class of aluminum silicate clays. MMT clay has high-surface area and aspect ratio, so can be used in various polymer/layered silicate nanocomposite systems. Clays are naturally occuring materials which are available in large quantities, and are environment friendly (Nguyen and Baird, 2006). The intent of clay addition to polymers is to form polymer/clay nanocomposites with demanded characteristic relevance because of improved material properties. Even at very low filler, the polymer/clay nanocomposite may offer considerable improvement in physical and engineering characteristics (Fornes and Paul, 2003). The most important clay fillers are smectite-type clays, including mica, hectorite, and MMT. To enhance the final properties of polymer nanocompsoite, individual clay aggregates are needed to break down into fine particles. Various methods have been utilized to prepare polymer/clay nanocomposite including in situ polymerization, solution intercalation, and melt processing. In situ polymerization involve dispersing clay layers in polymer matrix through introducing polymer precursor between clay layers (Yano et al., 1993). Polymer/clay nanocomposites have gained attention because of wide range of advanced properties such as strength, toughness, modulus, and barrier properties relative to conventional composite materials. Polymer nanocomposites also have advantages of easy processability and low density. Owing to nanoscale dispersion of filler within polymer matrix, fine interfacial interaction may occur between the polymer and inorganic filler. The interfacial interaction may consequence in enhanced characteristics properties than the matrix phase. In polymer nanocomposites, the properties may be varied to large extent due to decrease in the dimensions of filler particles to nanoscale (Kumar et al., 2009). Consequently, advances in materials science and engineering have resulted in the production of light, strong, thin, and cheap materials.

5.2 An overture to clay as reinforcement

On the basis of origin and structural chemistry, clay minerals are divided into various categories. Clay minerals are mainly divided into four major groups relying on the changes in sheet-like layered structure. The major group of clay minerals encompasses kaolinite, smectite, illite, and chlorite group. The kaolinite group further consists of kaolinite, nacrite, and dickite divisions. The main formula is $Al_2Si_2O_5(OH)_4$, relating variation is the layered structure. The kaolinite, nacrite, and dickite are structurally polymorphs, i.e., same molecular formulae but different structure. In all of these classes, aluminum oxide/hydroxide layers ($Al_2(OH)_4$) are bonded to tetrahedrally and octahedrally arranged silicate sheets (Si_2O_5). The kaolinite group is mainly used in plastics, paper, paint, rubber, and ceramics. Another large group is smectite clay group. It comprises of five important members such as MMT, talc, pyrophyllite, saponite, and nontronite. The most widely used one is MMT owing to wide range of applications. The general formula for smectite clay group is (Ca,Na,H) $(Al,Mg,Fe,Zn)_2$ $(Si,Al)_4O_{10}$ $(OH)_2$ XH_2O. The main difference between kaolinite and smectite clay group can be seen in their chemical struture. In smectite group, the layered structure is composed of two tetrahedral silicate layers sandwiching an octahedral aluminum oxide/hydroxide layer ($Al_2(OH)_4$). This class of clay mineral is widely used in paint, rubber, plasticizer, as well as in electrical and heat resistant materials. Illite group also forms an important category of clay minerals. The general formula of this group is (K,H) $Al_2(Si,Al)_4O_{10}$ $(OH)_2$ XH_2O. Its layered structure is just like smectite group in which the silicate layers may sandwich the aluminum oxide/hydroxide layer in structure. The chlorite group is another large group, however it is not delibrated as clay in terms of polymer fillers. The members of this group are cookeite, amesite, chamosite, and daphnite. Another classification of clay minerals is porcelain, plastic, and fire clays. The porcelain clay is pure kaolin which burns to white color product. The plastic clay contains more impurities and burns to yellow-red color. It is used for making ordinary clay pots. Fire clay contains larger amount of iron and silica. The commonly employed nanofillers for polymer nanocomposite are bentonite, kaolin, talc, and mica. These nanofillers render particular properties such as thermal stability, electrical insulation, hardness, opacity and brightness to the polymers (Zhang et al., 2015). MMT is usually obtained by volcanic ash. MMT corresponds to phyllosilicate group and is a soft form of clay. It is a member of smectite family. The platelets of MMT consist of octahedral sheets covered by tetrahedral sheets on both sides. It forms a sandwich like structure. Its basic structure is $MgOAl_2O_2 \cdot SiO_2 \cdot nH_2O$ (Kumar et al., 2014). Bentonite is natural clay mineral, which is impure clay comprising MMT. Bentonite has large interlayer spacing, good swelling capacity, water absorption, and cation exchange capability (Macter and Broussean, 1990). For polymer/clay nanocomposite, the clay minerals may be classified as 2:1 type, 1:1 type, and layered silicic acids. The 2:1 type clays with crystal structure consist of nanometer platelet (aluminum octahedron) squeezed between silicon tetrahedron sheets (Fig. 5.1). The 1:1 type clay consists of layers of

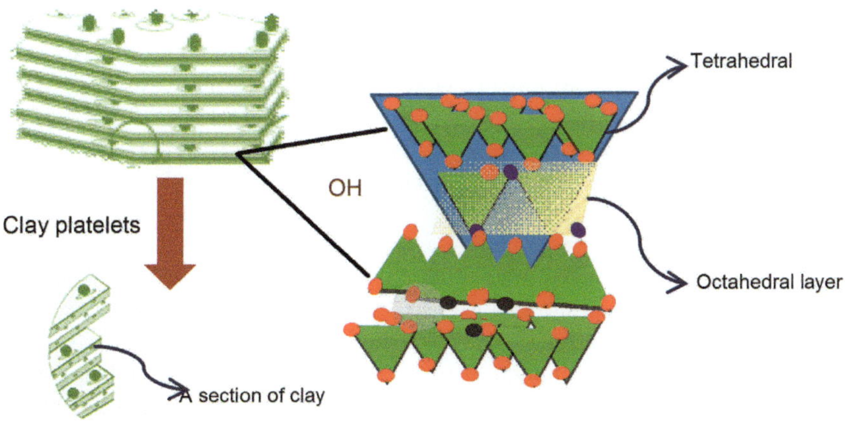

Figure 5.1 Nanaoclay structure.

Table 5.1 Smectite-type clays

Smectite	Chemical formula
Montmorillonite	$Mx(Al_4\text{-}xMgx)Si_8O_{20}(OH)_4$
Hectorite	$Mx(Mg_6\text{-}xLix)Si\ _8O_{20}(OH)_4$
Saponite	$MxMg_6(Si_8\text{-}xAlx)O_{20}(OH)_4$

silicon tetrahedron and aluminum octahedron sheet. The hydroxyl groups in octahedral sheets and oxygen group in tetrahedral sheets are capable of forming hydrogen bonding between the layers. The basic structure is made up of layered silicate networks and interlayer hydrated alkali metal cations (Zeng et al., 2005). The smectite-type clays and their formula are given in Table 5.1.

5.3 Surface modification of nanoclay

Various methods have been developed to modify the layered silicate structure of nanoclays (Pavlidou and Papaspyrides, 2008). In this regard, cation exchange method has been employed as simple and appropriate route to modify the nanoclays (Gul et al., 2015). Initially, quaternary alkylammonium salts have been used for the modification of MMT clay. Later, different types of alkylammonium cations have been developed and employed for organic modification of nanoclays. Onium ions have also been employed for nanoclay modification. The ammonium ion (NH_4^+) and hydronium ion (H_3O^+) belongs to class of positive ions called *onium ions*. Epoxy/clay nanocomposite reinforced with onium ion-modified clay has been reported (Pavlidou and Papaspyrides, 2008). Reinforcement properties of nanocomposite have been explored via interfacial effects of polymer and clay (Shi et al., 1996). Polypropylene (PP) and stearyl amine modified MMT nanocomposite has been reported (Hasegawa et al., 1998). The organomodification of nanoclay was used to enhance the compatibility of polymer with hydrophillc nanoclay. Melt compounding was used to disperse nanoclay in PP matrix. Mechanical properties of the resulting nanocomposite were found to enhance with nanoclay inclusion. Smectite nanoclays were modified using quaternary ammonium cations, i.e., diethyl methyl ammonium chloride $[(C_2H_5)_2(CH_3)N_1(O{-}i\text{Pr})_{25}]Cl_2$ and methyl trioctyle-ammonium chloride $[CH_3(C_8II_{17})_3N_1]Cl_2$. The in situ polymerization technique was used to form poly (methyl methacrylate) (PMMA)/organomodified clay and polystyrcne (PS)/modified clay nanocomposite (Okamoto and Morita, 2000). Physical characteristics, mechanical properties, and storage modulus of the resulting nanocomposite were investigated. Melt extrusion technique was used to form polylactide (PLA)/MMT nanocomposite. Na-MMT was modified by octadecylammonium cation and used for reinforcement. The PLA/MMT nanocomposite revealed improved properties compared with the neat matrix material (Ray et al., 2002). Na-MMT modified with stearyl ammonium ions has been employed for the fabrication of nylon 6/clay nanocomposite via melt compounding. The nanocomposites showed enhanced strength, modulus, and heat distortion temperature than neat nylon 6 (Hasegawa et al., 2003). MMT modified with bis(2-hydroxyethyl) quaternary ammonium salt was also blended with epoxy matrix. Flame retardant properties of styrene butadiene rubber have been improved by incorporation of Na-MMT modified with quaternary ammonium ion. Quaternary ammonium cation modified cloisite clay was reinforced in high impact PS to attain improved gas barrier, heat resistance, and mechanical properties (Hwang et al., 2008). The bentonite clay has been modified using intercalating agents such as *N*-acetyl-*N*,*N*,*N*-trimethyl ammonium bromide, tetrabutyl ammonium chloride, and hexadecyl trimethyl ammonium chloride (Sonawane and Chaudhari, 2008). Styrene-(ethylene-co-butylene)-styrene triblock copolymer was used as matrix to disperse the modified filler. The mechanical, morphological, and thermal properties of nanocomposite were investigated. Melt mixing route was used to form polyolefinic system and modified clay nanocomposite. The mechanical and thermal properties of the resultant nanocomposite were studied. In this case, the modifers used were dialquildimethylammonium chloride (DADMA), diestearildimethylammonium chloride (DEDMA), ditallowalkyldimethylammonium chloride (DTADMA), alquildimethylbenzylammonium chloride (ADMBA), hexadecyltrimethylammonium chloride (HDTMA), and fettalkyldimethylhydroxiethyl-ammonium chloride (FADMHEA) (Delbem and Valera, 2010). In another attempt, PS/modified MMT nanocomposite was prepared. MMT was modified using 1-methyl-3(4-vinylbenzyl)imidazoliumchloride (VBIMCl), 1-dodecyl-3-(4-vinylbenzyl) imidazolium chloride (VBIMCl), and 1-hexyl-3 (4-vinylbenzyl) imidazoliumchloride (VBIMCl). The in situ polymerization technique was employed to form PS and organomodified nanoclay nanocomposite (Pucci and Liuzzo, 2012). Thermal stability and morphology of the nanocomposite were studied (Corres and Zubitur, 2013). High-density polyethylene/nanokaolinite clay nanocomposite was prepared through melt intercalation method. The nanocomposites exhibited better thermomechanical properties relative to neat polymer (Anjana and Krishnan, 2014). Thermoplastic polyurethane/clay nanocomposites have also been reported. Commercially modified cloisite nanoclay was used in the system. The physical properties including morphology, thermal stability, and chemical resistance were explored (Ornaghi and Pistor, 2015).

5.4 Matrices for clay filler

Polyamide (PA) is an important thermoplastic with amide linkage (−NH−C=O) in the polymer backbone. Polyamides are classified into various categories depending on the arrangement and chemical nature of monomers. Aromatic, cycloaliphatic, and aliphatic polyamides are important types of polyamides. Polyamide 6,6 (PA 66) and polyamide 6 (PA 6) are imperative types of polyamide. As these are aliphatic polyamides, they are also known as *nylons* (Steingruber, 2000; Marchildon, 2011). Nylon 6 and nylon 66 comprises about 50% of industrial polyamide consumption

(Diamond et al., 2014). Greater number of carbon atom in aliphatic chains of nylon 6 may decrease the polarity of polyamide. Such polyamides have properties comparable to that of polyolefins. In this regard, PA 1818 has been reported in literature (Bennett et al., 2009). PA 624 and PA 634 have also been reported with long chain polyamide sequence (Ehrenstein et al., 2000). Hoshino et al. have prepared polyamide using cyclic diamine monomer (Inoue and Hoshino, 1980). The *para*-bis(4-aminocyclohexyl)methane (PACM) and *para*-bis(4-amino-3-methylcyclohexyl)methane (MACM) have been frequently used as cyclic diamines for cycloaliphatic polyamide (Kircher, 2014). Aromatic polyamide or aramid is the polyamide which is made up of monomer precursors containing aromatic rings in their structure. The aromatic–aliphatic polyamide possesses both aromatic and aliphatic characteristics due to the presence of aliphatic chain along with aromatic rings. Another form of aromatic–aliphatic polyamide consists of aromatic rings separated by methylene group. This group is also referred as *polyarylamide* (Saotome and Komoto, 1966). Epoxy matrix has also been widely employed for the reinforcement of organically modified smectitic clay in nanocomposite (Park and Jana, 2003). Mechanical properties of epoxy resins have been enhanced with nanoclay loading. In case of pure epoxy, there are several processing draw backs including poor resistant to crack propagation affecting toughness and impact strength (Xu et al., 1994). Both organic as well as inorganic nanofiller have been utilized to improve toughness, impact strength, and other mechanical properties of the epox resins. Nanoclays are also reinforced in epoxy blend matrices of polyamide, poly(phenylene-ether), poly(ether-sulfone), poly(ether-imide), and polycarbonates (Ok and Choe, 2013). In such systems, epoxy resin is blended with thermoplastic phase before curing process. However, these sysytems may possess phase separated morphology. Epoxy and thermoplastic blends have high tensile strength and resistance towards crack propagation. These materials have high potential in aerospace field (Thoppul et al., 2009).

5.5 High performance nanoclay reinforced polymeric hybrid

A nanocomposite material is combination of two or more phases which may have different composition and struture. Nanocomposites have a range of diverse advantages relative to conventional composite owing to small size and high surface-to-volume ratio (Motawie et al., 2014). By definition, in nanocomposite structure at least one of the dimensions of the dispersed phase must be in nanometer range. Generally, inorganic fillers have been incorporated in polymer matrices to enhance the mechanical and physical properties. In clay nanofiller, platelets are nanometer in thickness and width, therefore may form polymer–surface interaction in resulting nanocomposite (Fig. 5.2) (Giannelis et al., 1999).

As clays are *layered silicates*, they can be employed as fillers by intercalating in organic molecules. Generally, properties of nanocomposite are determined by the

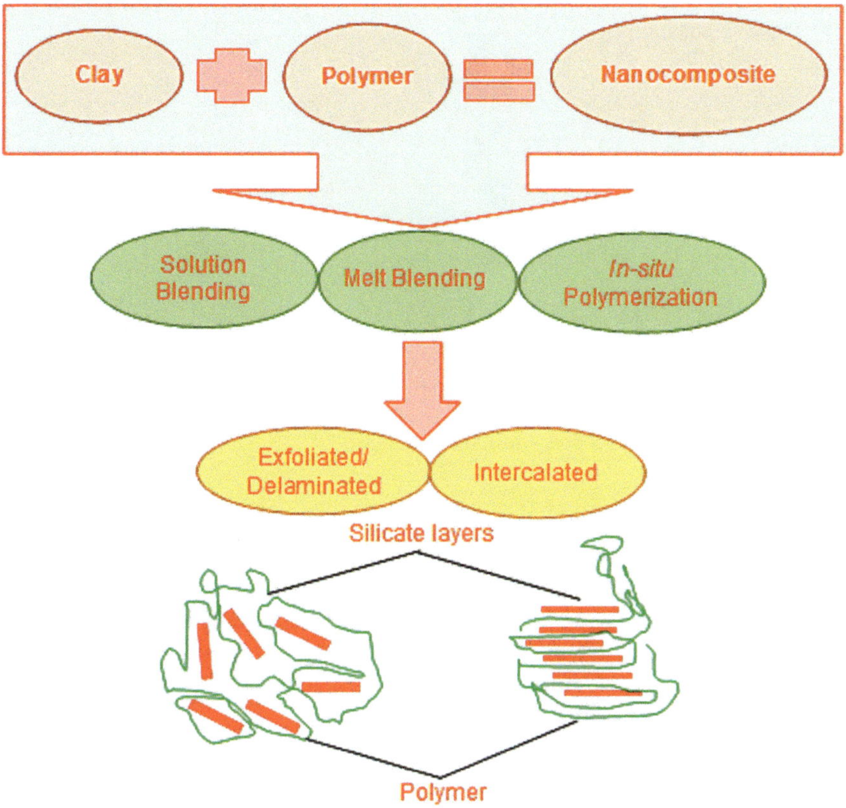

Figure 5.2 Various methods for the preparation of polymer/clay nanocomposite.

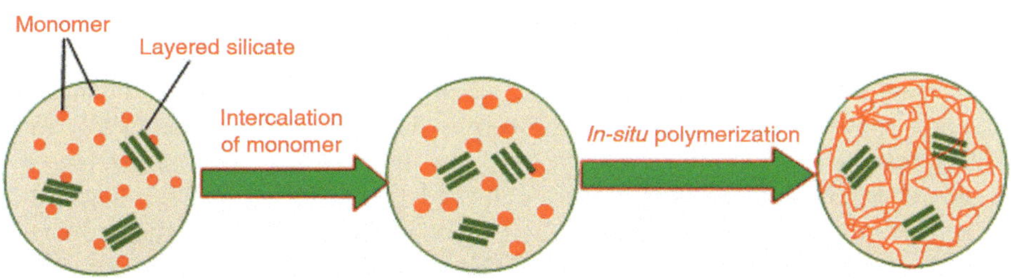

Figure 5.3 Preparation of nanocomposite by in situ polymerization.

extent of nanoclay dispersion in polymer matrix. Interaction between polymer matrix and clay has defined two categories of polymer/clay nanocomposite such as (i) intercalated and (ii) exfoliated (Masenelli-Varlot et al., 2002). However, there are different methods for the synthesis of nanocomposite. In situ polymerization involves polymerization of monomer inside the layered silicate platelets and layered silicate may swell up (Fig. 5.3).

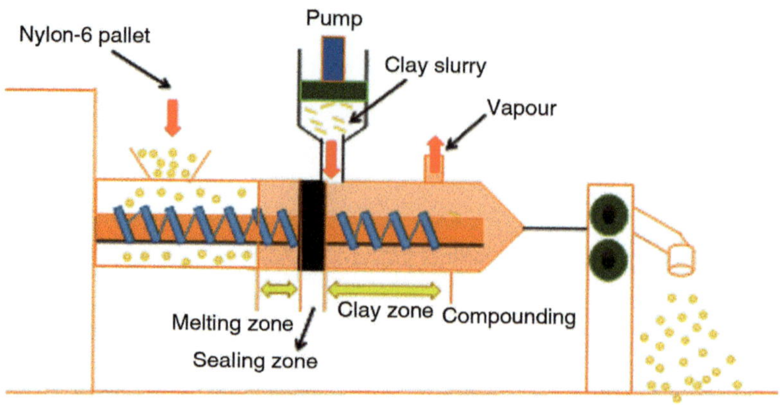

Figure 5.4 Compounding process using the clay slurry.

Solution intercalation of polymer involves the intercalation of polymer molecules from solution causing swelling of silicate layers. Melt intercalation method involves softening of polymer above its melting temperature and clay intercalation (Fig. 5.4). This method is simple, economical, and environmentally friendly. The polymer chain may diffuse into clay galleries to form intercalated structure on heating. The multilayered structure formed by melt technique may have improved mechanical, thermal, and physical properties.

5.5.1 Mechanical strength

Improvement of mechanical properties is an important objective to attain technical applications of polymer/clay nanocomposite. Wan et al. (2003) have studied the mechanical properties of polyvinyl chloride (PVC)/MMT nanocomposite with 0–5 wt% nanofiller loading. Toughness and impact strength of these nanocomposites were enhanced with 0.5–3 wt% nanofiller. The PVC/MMT intercalated structure showed better mechanical properties than the neat polymer. PVC/MMT nanocomposite was synthesized by in situ intercalative polymerization and improved mechanical properties were achieved (Gong et al., 2004b). The MMT layers were exfoliated in polymer to form exfoliated nanostructure. The addition of organic modified montmorillonite (OMMT) in PVC matrix improved the tensile strength and Young's modulus as compare to neat PVC. The OMMT content was varied from 1 to 5 wt%. The nanocomposite with 5 wt% nanoclay showed the tensile strength of 83.5 MPa and Young's modulus of 1.80 GPa, i.e., 31% and 54% increase correspondingly. The elongation at break of nanocomposite showed slightly decreasing trend (Table 5.2). Series of PVC/clay nanocomposite were synthesized by melt blending of polymer with organically modified clay with and without (2-ethylhexyl)phthalate (DOP) (Wang et al., 2001). In the absence of DOP, the clay acted as a plasticizer for PVC. The tensile strength was found to be the

Table 5.2 **Effect of OMMT content on the tensile properties of the nanocomposites (Gong et al., 2004b)**

Sample	Tensile strength (MPa)	Elongation at break (%)	Young's modulus (GPa)	Work of fracture (kJ m^{-2})
PVC	63.7	174	1.17	2249
PVC-OMMT1	75.0	182	1.64	2835
PVC-OMMT3	79.2	181	1.71	3173
PVC-OMMT5	83.5	165	1.80	3041

function of clay content in PVC/clay nanocomposite. In the absence of DOP, there was 20–25% increase in tensile strength. The thermal and mechanical properties of these nanocomposites were varied depending upon the clay and DOP content. Nylon 6 and clay hybrid (NCH) depicted higher strength, modulus, and heat distortion temperature relative to nylon 6 (Kojima et al., 1993).

5.5.2 Thermal stability

Poly(methyl methacrylate) (PMMA)/clay nanocomposites have been fabricated through in situ polymerization and emulsion polymerization (Meneghetti and Qutubuddin, 2006). The resulting nanocomposites were somewhat exfoliated and intercalated. Thermogravimetric analysis (TGA) has shown two main reaction steps during degradation of PMMA-based nanocomposites. The first step was around 160–240°C due to the decomposition of weak head-to-head linkages between matrix and filler. The second step between 300°C and 400°C represented irregular scission of the polymer chains. PMMA/clay nanocomposites showed improved thermal stability relative to neat polymer. PVC/MMT nanocomposites have been reported by in situ intercalative polymerization. Thermal properties were studied by TGA (Fig. 5.5) and differential scanning calorimetry (DSC) (Fig. 5.6) (Gong et al., 2004a). The nanocomposites showed higher glass transition temperature relative to pure PVC. The presence of organophilic MMT improved the decomposition temperature and reduced the maximum decomposition rate. The dense char layered formed was further investigated using Fourier transform infrared spectroscopy (FTIR).

Solution-compounding method was used to form polyvinyl alcohol/silica (PVA/ SiO_2) nanocomposite.

Thermal degradation mechanism of nanocomposite was studied (Peng and Kong, 2007). The thermal degradation of nanocomposite was increased with SiO_2 nanoparticle addition, relative to neat PVA. The PVA/SiO_2 nanocomposite degraded in two-steps in the range of 300–450°C and 450–550°C. The degradation products were recognized by pyrolysis-gas chromatography/mass spectrometric analysis (Py-GC/Ms) and FTIR/TGA. The results suggested that the first

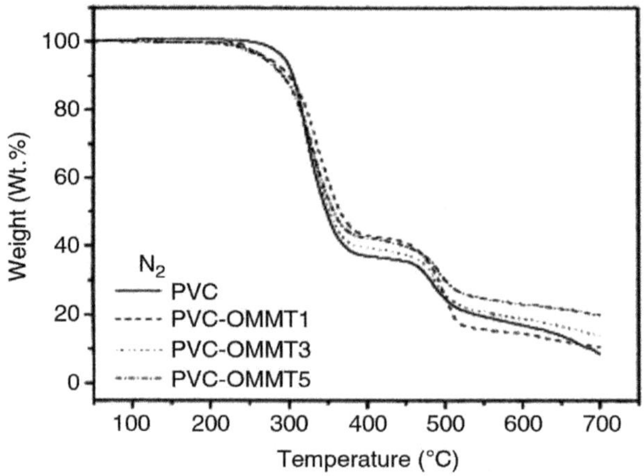

Figure 5.5 Thermograms of pure PVC and its nanocomposite (Gong et al., 2004a).

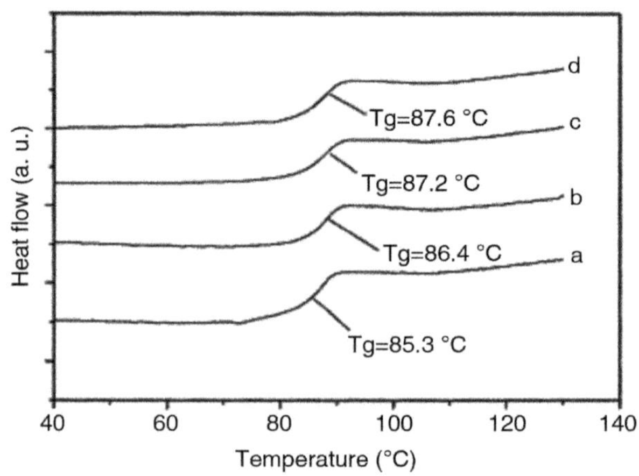

Figure 5.6 DSC thermograms of (a) PVC, (b) PVCeOMMT1, (c) PVCeOMMT3 and (d) PVCeOMMT5 (Gong et al., 2004a).

degradation step of nanocomposite was due to H_2O removal, acetate group exclusion, and little chain-scission reaction. The second degradation step was due to cyclization and major chain-scission reactions (Tyan et al., 1999). Thermally stable polyimide/clay nanocomposites were prepared from reactive organoclay and poly(amic acid). The polyimide and poly(amic acid) were formed from pyromellitic dianhydride-4,4′-oxydianiline (PMDA-ODA). The reactive organoclay was formed by using *p*-phenylenediamine as a swelling agent for MMT.

The decomposition temperature of PMDA-ODA nanocomposite increased with the organoclay content.

5.5.3 Morphology

The morphology study of polymer intercalated clay nanocomposite is important to understand the texture, nature, fracture mechanism, and toughening properties of polymer/clay nanocomposite (Zerda and Lesser, 2001). Aliphatic diamine has been used as curing agent in epoxy/intercalated clay nanocomposite. The fracture-surface topology was studied by using scanning electron and tapping-mode atomic force microscopy. In 5 wt% clay nanocomposite, the fracture energy of nanocomposite was increased by 100%. Epoxy has been modified with polyhedral oligomeric silsesquioxane (POSS) (Liu et al., 2005). The 30 wt% phenyltrisilanol POSS, [$Ph_7Si_7O_9(OH)_3$, POSS-triol], loaded Bisphenol A diglycidyl ether (DGEBA) epoxy sample was prepared. The organic–inorganic hybrids were prepared by in situ polymerization of epoxy monomer in POSS-triol. The morphology of organic–inorganic hybrids was studied by scanning electron microscopy (SEM). The composites showed phase separated and heterogeneous morphology with higher glass transition temperature (T_g) than the polymer. The spherical particles of POSS-triol (<0.5 μm) were found evenly dispersed in epoxy matrix. However, the size of POSS particles was found to decrease with increasing POSS-triol content. The optical microscopic study of PP/clay composite showed that the material was crystallized at higher temperature than pure PP (Kodgire et al., 2001). The morphology was altered due to the presence of silicate layers. Plasticized poly(L-lactide) (PLA) and Na-MMT nanocomposite was obtained by melt blending (Paul et al., 2003). X-ray diffraction study has shown the formation of MMT intercalated nanostructure.

5.5.4 Flame retardancy

Various polymer/clay nanocomposites have been studied using cone calorimetry and other flammability techniques. The cone calorimeter is one of most effectual technique for studying flammability characteristics of materials. Heat release rate (HRR) and peak HRR (PHRR) have been considered as significant parameters to elucidate fire resistance. Zhang et al. have reported polyethylene (PE)/clay nanocomposite by melt blending method (Zhang and Wilkie, 2003). The flammability parameters HRR, PHRR, specific extinction area (SEA), and mass loss rate (MLR) were measured using cone calorimetry (Fig. 5.7). According to results, PHRR of PE nanocomposite revealed showed 30–40% reduction relative to pure PE. However, the HRR of nanocomposite was same as for virgin PE.

The nonflammability properties of nylon-6/clay and nylon-12/clay nanocomposite have shown significant improvement with nanofiller loading (Gilman, 1999). The cone calorimetry data revealed that the HRR and PHRR were significantly abridged for the clay intercalated nanocomposite. The nylon-6 nanocomposite showed 63% lowering in HRR relative to neat nylon 6. The lowering in flammability was accredited to the formation of carbonaceous char in condensed phase (Gilman, 1999). Cone calorimetry has been employed to determine HRR and other flammability properties of PP and PS nanocomposite. Incident heat flux of $35\ kW\ m^{-2}$ was

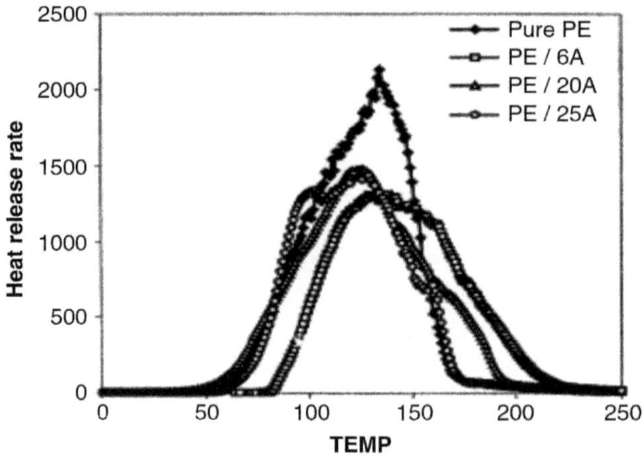

Figure 5.7 Comparison of the heat release rate (HRR) plots for pure PE and PE/clay nanocomposites at 35 kW m^{-2} heat flux (Zhang and Wilkie, 2003).

used (Gilman et al., 2000). The 0.1 wt% clay nanocomposite decreased the rate of heat release of PS (Zhu and Wilkie, 2000). The PHRR showed 50% decreased for all the nanocomposites. The overall nonflammability of nanocomposite was improved than the neat PS. PS/clay nanocomposites have also been fabricated by bulk polymerization method (Zhu et al., 2001). The clay loading has improved the degradation temperature by 50°C, while PHRR was reduced by 27−58%. The char layer formed acted as barrier to the mass and energy transport, thus increasing the nonflammability of polymer/clay nanocomposite.

5.5.5 *Crystallinity*

Nylon 6 and nylon 6/MMT nanocomposites have been studied for crystallization behavior (Maiti and Okamoto, 2003). Relative to neat nylon 6, the crystallization rate of nylon 6/clay nanocomposite was found to be higher. Owing to epitaxial crystallization, nylon 6 usually crystallizes in γ-form. Transmission electron microscopy (TEM) images have shown sandwiched structure of the crystallized nylon 6 sample. The clay particles may also improve the crystallization rate of nylon 6. Two type of crystal structure may appear in pure polyamide at low temperature i.e. monoclinic (α) and pseudo hexagonal (γ) crystal structure. However, the γ-form may disappear with increasing crystallization temperature (T_c). In the case of nylon 6 nanocomposite, only γ-form was produced. Melt compounding was used to prepare PP/sepiolite (Sep) nanocomposite in miniextruder apparatus (Bilotti et al., 2008). The PP was semicrystalline in nature; however its crystal structure was modified with the presence of nanoclay. Organic modifiers have been reported to affect crystallinity and nanomechanical properties of polymer/clay nanocomposite (Sikdar et al., 2008). The crystallinity of polymer/clay nanocomposite was

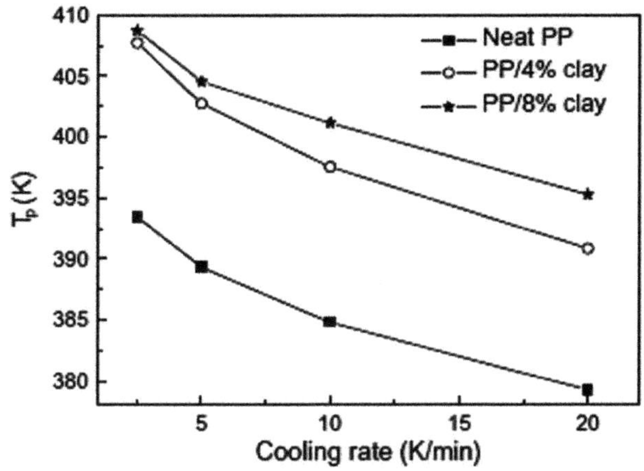

Figure 5.8 Crystallization peak temperature T_p versus cooling rate for PP−clay nanocomposites (Yuan et al., 2006).

dependent on (i) nature of organic modifier and (ii) interaction between polymer and nanoclay. The nonisothermal crystallization behavior of neat PP and melt intercalated PP/clay nanocomposite has also been studied (Yuan et al., 2006). According to kinetic studies, nanoclay addition in PP consequenced in decrease in half-time for crystallization cooling rates. The nucleating effect of clay nanoparticles was seemed to be responsible for decreasing crystallization cooling rates. The activation energy for nonisothermal crystallization of PP and PP/clay nanocomposite was determined by Kissinger method. The nucleating effect of clay on PP crystallization was observed by lower value of activation energy for PP/clay nanocomposite (Fig. 5.8).

5.6 Application of polymer/clay hybrid

It is intricate to convolute all the applications of polymer/clay hybrids in this brief section; however they are highlighted duly keeping in view their attractive characteristics. Industrial applications of nanoclay-based nanocomposite range from aerospace/automotive (interior and exterior panels, flame retardant panels, gas tanks, and bumpers) to construction/building, textile, and packaging (Gacitua et al., 2005). Major use of polymer/clay hybrids is in automotive, aeronautical, and packaging industry i.e. up to 80%. Industrial sector is paying interest and investment towards the development of nanoclay materials and polymer/nanoclay nanocomposite (Zeng et al., 2005). Flame retardancy and weight issues of space industry can be solved by using organic polymers such as thermoset with dispersed inorganic nanomaterial. The nanoclay-based materials can sustain the harsh space environment and are light in weight (Njuguna and Pielichowski, 2003). Polymer/clay materials have been studied for flame retardancy and thermal stability to attain austere aerospace requirements. Antennas in aerospace have also been made using nanocomposite materials with improve physical and structural properties such as coefficient of thermal expansion (Gorinevsky and Hyde, 2002). Polymer/clay hybrids have also replaced the conventional plastics in food packaging owing to fine barrier properties towards gases (oxygen, carbon dioxide) and water vapors. The conventional plastics used lack environmental concerns. In addition to simple food packaging, these nanocomposites may act as active and intelligent packaging material due to ability to control active molecules release rate. Nanoscience developments have also focused the construction industry. Recently, polymer/clay nanocomposite has been employed in structurally significant areas of construction industry such as thermal and sound insulation, adhesives, coatings, plastics, concrete and asphalt strength (Majeed et al., 2013). However, research is still in dearth regarding the use of nanocomposite in construction. Modified structural adhesives have been developed from nanomaterials (Feldman, 2014). The building and civil engineering applications have successfully employed nanoparticle incorporated adhesives. Moreover, nanoclay modified epoxy adhesive have been used for strengthening concrete-based carbon fiber reinforced polymer. Polymer/nanoclay nanocomposites have also gained significance in the field of medicine and biomedical commodities (Ma, 2004). Polymer/clay nanocomposites have been explored for tissue engineering and recovery of lost organs. Another essential application regarding medical facet is the use in drug delivery system. The nanocomposite has been for the treatment of certain diseases due to hydrophobicity and short life of drugs (Urbańska et al., 2014; Aguzzi et al., 2007).

5.7 Conclusion

Polymer/clay nanocomposite has fascinated noteworthy interest in technological research and development worldwide. Polymer/clay nanocomposite can be simply prepared by dispersing nanoclay in polymer matrix. Clay minerals are effective nanofiller due to distinctive structure and properties. The polymer/clay nanocomposites have been synthesized using different approaches such as in situ polymerization, solution exfoliation, and melt intercalation. The monomer can be in situ polymerized in clay galleries by initiator, heat, or other techniques. Solution exfoliation involves a soluble polymer and exfoliation of layered clay in solution. The melt intercalation involves direct mixing of polymer and layered silicate in the molten state. The structure and properties of polymer/clay nanocomposite have been characterized by variety of techniques such as TEM, SEM, FTIR, XRD, TGA, and DSC. The polymer/clay nanocomposite possesses improved physical properties relative to conventional composite. Moreover, the properties are improved at very low nanoclay loading level. The industrial applications of these high performance nanocomposites may be found in automobile, packaging, coating, construction, and paint and pigment industries.

References

Aguzzi, C., Cerezo, P., Viseras, C., Caramella, C., 2007. Use of clays as drug delivery systems: possibilities and limitations. Appl. Clay Sci. 36, 22–36.

Amina, A., Kandila, H., Ismaila, M.N., 2013. The effect of presence of glycopolymer grafts in poly *N*-isopropylacrylamide/clay nanocomposite hydrogels. Polym.-Plast. Technol. Eng. 52, 1034–1042.

Anjana, R., Krishnan, A.K., 2014. Design of experiments for thermo-mechanical behaviour of polypropylene/high-density polyethylene/ nanokaolinite clay composites. Polym. Bull. 71, 315–335.

Bennett, C., Kaya, E., Sykes, A.M., Jarrett, W.L., Mathias, L.J., 2009. Synthesis and characterization of nylon 18 18 and nylon 18 adamantane. J. Polym. Sci. Part A. Polym. Chem. 47, 4409–4419.

Bilotti, E., Fischer, H.R., Peijs, T., 2008. Polymer nanocomposites based on needle-like sepiolite clays: effect of functionalized polymers on the dispersion of nanofiller, crystallinity, and mechanical properties. J. Appl. Polym. Sci. 107, 1116–1123.

Corres, M.A., Zubitur, M., 2013. Thermal decomposition of phenoxy/clay nanocomposites: effect of organoclay microstructure. Polym. Degrad. Stab. 98, 818–828.

Delbem, M.F., Valera, T.S., 2010. Modification of a Brazilian smectite clay with different quaternary ammonium salts. Quim. Nova. 33, 309–315.

Diamond, G., Murphy, V., Boussie, T.R., 2014. Applications of high throughput experimentation to the production of commodity chemicals from renewable feedstocks. Modern Appl. High Throughput R&D Heterogen. Catal.288–309.

Ehrenstein, M., Dellsperger, S., Kocher, C., Stutzmann, N., Weder, C., Smith, P., 2000. New polyamides with long alkane segments: nylon 6.24 and 6.34. Polymer. 41, 3531–3539.

Feldman, D., 2014. Polymer nanocomposites in building, construction. J. Macromol. Sci. 51, 203–209.

Fornes, T.D., Paul, D.R., 2003. Modeling properties of nylon 6/clay nanocomposites using composite theories. Polymer. 44, 4993–5013.

Gacitua, W.E., Ballerini, A.A., Zhang, 2005. Polymer nanocomposites: synthetic and natural fillers a review. Maderas Cienciay Tecnol. 7, 159–178.

Giannelis, E.P., Krishnamoorti, R., Manias, E., 1999. Polymer-silicate nanocomposites: model systems for confined polymers and polymer brushes. Adv. Polym. Sci. 138, 107–147.

Gilman, J.W., 1999. Flammability and thermal stability studies of polymer layered-silicate (clay) nanocomposites. Appl. Clay Sci. 15, 31–49.

Gilman, J.W., Jackson, C.L., Morgan, A.B., Harris, R., 2000. Flammability properties of polymer – layered-silicate nanocomposites: polypropylene and polystyrene nanocomposites. Chem. Mater. 12, 1866–1873.

Gong, F., Feng, M., Zhao, C., Zhang, S., Yang, M., 2004a. Thermal properties of poly(vinyl chloride)/montmorillonite nanocomposites. Polym. Degrad. Stab. 84, 289–294.

Gong, F., Feng, M., Zhao, C., Zhang, S., Yang, M., 2004b. Particle configuration and mechanical properties of poly(vinyl chloride)/montmorillonite nanocomposites *via in-situ* suspension polymerization. Polym. Test. 23, 847–853.

Gorinevsky, D., Hyde, T.T., 2002. Adaptive membrane for large lightweight space telescopes. Astronom. Telescop. Instrum.1–9.

Gul, S., Kausar, A., Muhammad, B., 2015. Research progress on properties and applications of polymer/clay nanocomposite. Polym.-Plast. Technol. Eng.. Available from: http://dx.doi.org/10.1080/03602559.2015.1098699.

Hasegawa, N., Kawasumi, M., Kato, M., Usuki, A., Okada, A., 1998. Preparation and mechanical properties of polypropylene-clay hybrids using a maleic anhydride-modified polypropylene oligomer. J. Appl. Polym. Sci. 67, 87–92.

Hasegawa, N., Okamoto, H., Kato, M., Usuki, A., Sato, N., 2003. Nylon 6/Na–montmorillonite nanocomposites prepared by compounding Nylon 6 with Na–montmorillonite slurry. Polymer. 44, 2933–2937.

Hwang, S.J., Joo, Y.L., Lee, S.J., 2008. Properties of high-impact polystyrene/organoclay nanocomposites synthesized via in situ polymerization. J. Appl. Polym. Sci. 110, 1441–1450.

Inoue, K., Hoshino, S., 1980. Crystal structure of polyamides containing 1,3-cyclohexane rings. J. Polym. Sci. Polym. Phys. Ed. 18, 697–706.

Kircher, M., 2014. The emerging bioeconomy: industrial drivers, global impact, and international strategies. Ind. Biotechnol. 10, 11–18.

Kodgire, P., Kalgaonkar, R., Hambir, S., Bulakh, N., Jog, J.P., 2001. PP/clay nanocomposites: effect of clay treatment on morphology and dynamic mechanical properties. J. Appl. Polym. Sci. 81, 1786–1792.

Kojima, Y., Usuki, A., Kawasumi, M., Okada, A., Fukushima, Y., 1993. Mechanical properties of nylon-6 hybrids. J. Mater. Res. 8, 1185–1189.

Kumar, A.P., Depan, D., Tomer, N.S., Singh, R.P., 2009. Nanoscale particles for polymer degradation and stabilization: trends and future perspectives. Prog. Polym. Sci. 34, 479–515.

Kumar, M.S.S., Raju, N.M.S., Sampath, P.S., Jayakumari, L.S., 2014. Effects of nanomaterials on polymer composites. Rev. Adv. Mater. Sci. 38, 40–54.

Liu, H., Zheng, S., Nie, K., 2005. Morphology and thermomechanical properties of organic – inorganic hybrid composites involving epoxy resin and an incompletely condensed polyhedral oligomeric silsesquioxane. Macromolecules. 38, 5088–5097.

Ma, P.X., 2004. Scaffolds for tissue fabrication. Materials. 7, 30–40.

Macter, W.J., Broussean, R., 1990. The A.C. response of hardened cement paste. Cem. Concr. Res. 20, 891–900.

Maiti, P., Okamoto, M., 2003. Crystallization controlled by silicate surfaces in nylon 6-clay nanocomposites. Macromol. Mater. Eng. 288, 440–445.

Majeed, K., Jawaid, M., Hassan, A., 2013. Potential materials for food packaging from nanoclay/natural fibres filled hybrid composites. Mater. Des. 46, 391–410.

Marchildon, K., 2011. Polyamides – still strong after seventy years. Macromol. React. Eng. 5, 22–54.

Masenelli-Varlot, K., Reynaud, E., Vigier, G., Varlet, J., 2002. Mechanical properties of clay-reinforced polyamide. J. Polym. Sci. B: Polym. Phys. 40, 272–283.

Meneghetti, P., Qutubuddin, S., 2006. Synthesis, thermal properties and applications of polymer-clay nanocomposites. Thermochim. Acta. 442, 74–77.

Motawie, A.M., Madany, M.M., El-Dakrory, A.Z., Osman, H.M., Ismail, E.A., Badr, M.M., et al., 2014. Physico-chemical characteristics of nano-organo bentonite prepared using different organo-modifiers. Egypt. J. Petrol. 23, 331–338.

Nguyen, Q.T., Baird, D.G., 2006. Preparation of polymer–clay nanocomposites and their properties. Adv. Polym. Technol. 25, 270–285.

Njuguna, J., Pielichowski, K., 2003. Polmer nanocoposites for aerospace. Adv. Eng. Mater. 5, 769–778.

Ok, S., Choe, Y., 2013. Impact optimized performance of epoxy/polyamide/csr(core shell rubber)/anhydride blends at low temperature. Mol. Cryst. Liq. Cryst. 579, 55–61.

Okamoto, M., Morita, S., 2000. Synthesis and structure of smectic clay/poly(methyl methacrylate) and clay/polystyrene nanocomposites via in situ intercalative polymerization. Polymer. 41, 3887–3890.

Ornaghi, F.G., Pistor, V., 2015. Thermoplastic polyurethane synthesis with modified montmorillonite prepared by torque rheometry: investigation of morphological, thermal, chemical, and physical properties. Appl. Polym. Sci.1–10.

Park, J.H., Jana, S.C., 2003. Mechanism of exfoliation of nanoclay particles in epoxy–clay nanocomposites. Macromolecules. 36, 2758–2768.

Paul, M.-A., Alexandrea, M., Degee, P., Henristc, C., Rulmontc, A., Dubois, P., 2003. New nanocomposite materials based on plasticized poly(L-lactide) and organo-modified montmorillonites: thermal and morphological study. Polymer. 44, 443–450.

Pavlidou, S., Papaspyrides, C.D., 2008. A review on polymer-layered silicate nanocomposites. Prog. Polym. Sci. 33, 1119–1198.

Peng, Z., Kong, L.X., 2007. A thermal degradation mechanism of polyvinyl alcohol/silica nanocomposites. Polym. Degrad. Stab. 92, 1061–1071.

Pucci, A., Liuzzo, V., 2012. Polymerizable ionic liquids for the preparation of polystyrene/clay composites. Polym. Int. 61, 426–433.

Ray, S., Maiti, P., Okamoto, M., 2002. New polylactide/layered silicate nanocomposites. 1. Preparation, characterization, and properties. Macromolecules. 35, 3104–3110.

Saotome, K., Komoto, H., 1966. Polyamides having long methylene chain units. J. Polym. Sci A-1. 4, 1463–1473.

Shah, R., Kausar, A., Muhammad, B., Shah, S., 2015. Progression from graphene and graphene oxide to high performance polymer-based nanocomposite: a review. Polym.-Plast. Technol. Eng. 54, 173–183.

Shi, H., Lan, T., Pinnavaia, T., 1996. Interfacial effects on the reinforcement properties of polymer–organoclay nanocomposites. Chem. Mater. 8, 1584–1587.

Sikdar, D., Katti, D.R., Katti, K.S., 2008. The role of interfacial interactions on the crystallinity and nanomechanical properties of clay–polymer nanocomposites: a molecular dynamics study. J. Appl. Polym. Sci. 107, 3137–3148.

Sonawane, S., Chaudhari, P., 2008. Combined effect of ultrasound and nanoclay on adsorption of phenol. Ultrason. Sonochem. 15, 1033–1037.

Steingruber, E., 2000. Indigo and indigo, colorants. Steingruber, E., 2004. Indigo and indigo colorants. Ullmann's Encyclopedia of Industrial Chemistry. Available from: http://dx.doi.org/10.1002/14356007.114-149.

Thoppul, S.D., Finegan, J., Gibson, R.F., 2009. Mechanics of mechanically fastened joints in polymer–matrix composite structures-a review. Compos. Sci. Technol. 69, 301–329.

Tyan, H.-L., Liu, Y.-C., Wei, K.-H., 1999. Thermally and mechanically enhanced clay/polyimide nanocomposite via reactive organoclay. Chem. Mater. 11, 1942–1947.

Urbańska, J., Karewicz, A., Nowakowska, M., 2014. Polymeric delivery systems for dexamethasone. Life. Sci. 96, 1–6.

Wan, C., Qiao, X., Zhang, Y., Zhang, Y., 2003. Effect of different clay treatment on morphology and mechanical properties of PVC–clay nanocomposites. Polym. Test. 22, 453–461.

Wang, D., Paiuow, D., Yao, Q., Wilkie, C.A., 2001. PVC–clay nanocomposites: preparation, thermal and mechanical properties. J. Vinyl Additiv. Technol. 7, 204–213.

Xu, L., Fu, J.H., Schlup, J.R., 1994. In situ near-infrared spectroscopic investigation of epoxy resin-aromatic amine cure mechanisms. J. Am. Chem. Soc. 116, 2821–2826.

Yano, K., Usuki, A., Okada, A., Kurauchi, T., Kamigaito, O., 1993. Synthesis and properties of polyimide–clay hybrid. J. Polym. Sci: A, Polym. Chem. 31, 2493–2498.

Yuan, Q., Awate, S., Misra, R.D.K., 2006. Nonisothermal crystallization behavior of polypropylene–clay nanocomposites. Eur. Polym. J. 42, 1994–2003.

Zeng, Q.H., Yu, A.B., Lu, G.Q., Paul, D.R., 2005. Clay-based polymer nanocomposites: research and commercial development. J. Nanosci. Nanotechnol. 5, 1574–1592.

Zerda, A.S., Lesser, A.J., 2001. Intercalated clay nanocomposites: morphology, mechanics, and fracture behavior. J. Polym. Sci. B: Polym. Phys. 39, 1137–1146.

Zhang, J., Wilkie, C.A., 2003. Preparation and flammability properties of polyethylene–clay nanocomposites. Polym. Degrad. Stab. 80, 163–169.

Zhang, Y., Stringer, J., Grainger, R., Smith, P.J., Hodzic, A., 2015. Fabrication of patterned thermoplastic microphases between composite plies by inkjet printing. J. Compos. Mater. 49, 1907–1913.

Zhu, J., Wilkie, C.A., 2000. Thermal and fire studies on polystyrene–clay nanocomposites. Polym. Int. 49, 1158–1163.

Zhu, J., Morgan, A.B., Lamelas, F.J., Wilkie, C.A., 2001. Fire properties of polystyrene–clay nanocomposites. Chem. Mater. 13, 3774–3780.

Carbon nanotube hybrid polymer composites: recent advances in mechanical characterization

6

Wagner Mauricio Pachekoski[1], *Sandro Campos Amico*[2], *Sergio Henrique Pezzin*[3] *and Jose Roberto Moraes d'Almeida*[4]

[1]Federal University of Santa Catarina (UFSC), Joinville, Brazil, [2]PPGEM, Federal University of Rio Grande do Sul (UFRGS), Porto Alegre, Brazil, [3]Center of Technological Sciences, Santa Catarina State University (UDESC), Joinville, Brazil, [4]Materials Engineering Department, Pontifical Catholic University of Rio de Janeiro (PUC), Rio de Janeiro, Brazil

Chapter Outline

6.1 Introduction

Carbon nanotube (CNT) hybrid polymer composite materials are hierarchical hybrid systems generally comprised of three constituents: stiff fibers (micrometer scale) organized in tows or woven, a thermoset polymer resin (the matrix), and CNT (nanometer scale) (Thostenson et al., 2002; Herceg et al., 2016). The polymer resin binds the micron-sized fibers and the CNT together. In such composites, CNT can be grown in situ onto the surface of the fibers (Thostenson et al., 2002; De Riccardis et al., 2006; Bekyarova et al., 2007a; Song et al., 2013; Garcia et al., 2008a,b) or dispersed within the array of fibers in the cloth material by different mixing methods (Gojny et al., 2005; Fan et al., 2008).

The CNT reinforce the polymer matrix between the fibers, providing enhanced strength and toughness as well as an electrically conductive pathway. CNT also carry the promise of enhancing intrinsically poor out-of-plane mechanical performance of regular fiber-reinforced polymer composites. The resulting structure can be described

Hybrid Polymer Composite Materials: Properties and Characterisation. DOI: http://dx.doi.org/10.1016/B978-0-08-100787-7.00006-8

as a hybrid composite laminate, or a "fuzzy" fiber-reinforced plastic (FFRP) (Garcia et al., 2008a; Wicks et al., 2010; Wicks et al., 2014).

In recent years, there is a growing interest in the development of such hybrid thermoset matrix composites. The high-strength fibers, such as carbon, glass and aramid, are responsible for the gross mechanical properties of the composites whereas the nanofillers add toughness and a variety of other properties, such as reduced permeability to fluids, yielding multifunctional materials. When CNT are under consideration toughness is the most impacted mechanical characteristic. However, whenever nanoparticles are employed, questions regarding their homogeneous distribution over the entire composite will always be of prime importance. Agglomerates must be avoided and any discussion about the mechanical properties and characterization of composites with nanofillers must consider the way the nanoparticles were incorporated into the polymeric matrix and/or onto the fiber preform.

As in conventional nanocomposites, the mechanical properties of hybrid nanocomposites are not determined only by the individual properties of the matrix and the dispersed constituents, but also by the interaction between the different phases. From this interaction, the most relevant parameter regarding mechanical strength regards the ability of the interface between the different phases to transfer stress. In order to evaluate the mechanical performance of a nanocomposite material, different approaches can be used: (i) tests in which the components are analyzed separately; (ii) tests in the nanocomposite itself (Van Hemelrijck and Anastassopoulos, 1996). The interface between the fiber and the matrix can be evaluated through several methods, including fiber pull-out, single-fiber fragmentation, and indentation/nanoindentation procedures (Bannister et al., 1995).

On this context, this chapter describes some general mechanical characterization techniques used for composites, followed by a more focused discussion of CNT hybrid composites regarding the effect of manufacturing on properties and also toughness-related issues.

6.2 General mechanical characterization of composites

The testing techniques generally applied to CNT hybrid polymer composite materials are mostly the same used for regular composites. They can be classified according to the application of the load in static, dynamic and constant load. In static tests the load is applied sufficiently slowly, leading to a succession of equilibrium states, as in tensile, compression and bending tests. In dynamic tests, the load is quickly or cyclically applied during the test, as in impact and fatigue tests. Those tests are standardized by regulatory bodies such as the International Organization for Standardization (ISO) and the American Society for Testing and Materials (ASTM), and some of the most used ones are briefly cited below.

a. *Tensile testing*: The tensile test is the most commonly used for materials characterization and enables the determination of properties such as longitudinal and transverse tensile strengths, longitudinal and transverse elasticity moduli, and Poisson's ratios in the main material directions. Due to the composite anisotropy characteristics, the tensile tests are commonly performed in the longitudinal and transversal fiber directions.

Important information about their behavior under the action of a load can be retrieved, e.g., separation or delamination can provide evidence on how the crack starts and develops. The nature of the failure can also be observed, i.e., it can occur in a brittle and unexpected fashion, or be preceded by visual and sound clues. All this information can be vital for their proper study when selecting new applications (Hodgki, 2000). It is worth mentioning that the properties of a composite are mainly associated with the fiber type and orientation, with an important reduction in strength and stiffness as the main fiber direction deviates from the direction in which the load is applied. The tensile test for fiber-reinforced polymeric composites should follow the guidance of regulatory institutions such as ASTM D3039 or ISO 527 (Part 4 or 5). ASTM standards are generally preferred for tensile tests on composites, as they enable a better control over test conditions that can cause variations in the results.

b. *Compression testing*: The elasticity modulus and compressive strength of a composite material are critical parameters for many structural uses. The factors that determine the compressive strength of a composite material are complex, and compressive strength can be much smaller than tensile strength. A compression test needs to apply a compressive load to the sample, while avoiding its bending or buckling. For that, compression devices must have good axial alignment and high lateral stiffness.

There are three force application methods for compression tests: (i) load is applied at both ends of the sample; (ii) shear load is applied on the sides of the samples; (iii) a combination of these loads is applied. Tests in which load is applied to the sample's ends are the simplest and most common ones. However, they are unsuitable for high-strength composites, because these materials tend to have relatively low transverse strength and fail due to brooming (crushing of the ends). The main standards for this type of test are ASTM D695 and ISO 14126.

Compressive strength tests in which a shear load is applied require special care during sample placement in order to provide high friction interface between the grips and the sample surface, while preventing the grips from damaging the tested material. This seems to be the method that provides the most reliable compressive strength results for a composite material. The main standards that fit into this category are ASTM D3410 and ISO 14126. Initially developed by Boeing, combined load tests try to reduce the shear stress generated during the test to a minimum, significantly reducing the possibility of brooming. The main standard for this type of compression test is ASTM D6641, which has become very popular in the last 10 years due to the efficiency of the load and its performance, and it will probably replace the other tests in a few years (Schultheisz and Waas, 1996; Waas and Schultheisz, 1996; Carlsson et al., 2014).

c. *Shear testing*: The shear behavior of composites is a particularly important area of continuous study and many shear testing methods have been introduced over the years. For a complete understanding of the shear behavior of composites, it is mandatory to evaluate in-plane and interlaminar properties (Almeida Jr. et al., 2015). There are several in-plane shear testing methods available, such as tensile test with fibers oriented at $\pm 45°$ (ASTM D3518), Iosipescu (ASTM D5379) and two-rail shear (ASTM D4255). The Iosipescu shear test is perhaps the most used, mainly due to its versatility and accuracy in obtaining the shear properties. Nevertheless, Adams et al. (2007) developed the V-notched rail shear test (ASTM D7078) for unidirectional, multidirectional, isotropic and for thin laminates. The specimen length was the same, but the width was considerably increased. In addition, the two end regions on either side of V-notched specimen were shortened for face loading instead of edge loading. This fixture was designed to produce a uniform state of shear stress across the specimen gage section, allowing accuracy in determining the shear properties and providing acceptable failures in the gage section. In addition, the reduced cross-sectional area increases shear stresses in the gage section (Adams and Busse, 2004).

Regarding out-of-plane shear methods, short beam test (ASTM D2344) is largely used for several types of composites. The advantages include fixture simplicity, easy specimen geometry, fast testing, low cost and good testing efficiency in acceptable failures (Adams and Busse, 2004). Nevertheless, this test presents drawbacks related to the lack of a uniform shear stress state and to the low span-to-thickness (s:t) ratio, which may cause failure by crushing (for s:t < 4:1) or due to vertical (for s:t > 6:1) cracks (typical of bending failures). The double-notched shear (DNS) test (ASTM D3846) was developed targeting unquestionable delamination failure to evaluate interlaminar shear strength (ILSS) of a composite. The failure must occur in the gage section (between notches) since the straight sides contain half-thickness and are flat-bottomed on opposing surfaces (Kedward, 1972).

d. *Flexural testing*: The flexural test is performed by placing a test sample over two separated supporting pins and applying a load on the central point. ASTM D7264 and ASTM D790 are usually employed to define the procedure for the three-point flexural test of polymeric matrix composite materials, whereas ASTM D6272 is used for the four-point flexural test, in which the load is applied off-center, on two points equally distant from it. According to the standard, the flexural stress cannot be determined for samples that do not fail visibly within 5% strain. This standard also establishes that samples that do not fail using the three-point method can be tested using the four-point method. To further ensure the perfect positioning of the sample on the supports, the total sample length must be 20% greater than the space between support pins. The standard also suggests special relations between the length and thickness of the test sample for different types of composites aiming at reducing the resulting shear stress, which mainly occurs in the three-point test. A 16:1 ratio is recommended when the ratio between tensile strength (with fibers oriented parallel to the support pins) and in-plane shear strength is less than or equal to 8. For composites that present low shear strength and high tensile strength parallel to the support pins, 32:1 or 40:1 relationships are suggested and when determination of the tangent flexural modulus is required, a 60:1 ratio is suggested to reduce the effects caused by shear. Flexural test on composites also require special attention due to several possible failure modes (Kretsis, 1987; Hodgki, 2000).

e. *Impact testing*: Composite materials with polymeric matrices have, in general, high stiffness and high mechanical strength, but low toughness compared to metallic materials. Unlike metals, that possess high ductility, most composites can only absorb energy through elastic strain or damage mechanisms, not through plastic strain. It is therefore important to know the characteristics and limitations of such materials, particularly relative to impact speed, load application time and perceived damage. Impact resistance tests normally used to simulate impact events belong to the low-speed category ($1-10\ \mathrm{m\ s^{-1}}$), or large mass, which is characterized by a wide damaged area and a global structure response. Most publications for polymeric matrix composites involve low energy impact tests (in the range of $1-5$ J), with low test sample damage. The main effect of this impact is the failure of the polymeric matrix evidenced by cracking, and the loss of adhesion between fiber and matrix.

The most used impact characterization tests are the Charpy/Izod (ASTM D256 and D4812) and the drop-weight test (ASTM D6264 and D7136). The Charpy test submits a sample to a three-point flexural impact load induced by a swinging pendulum. In general, the test sample contains a 2 mm-deep notch at 45 degree angle in order to induce failure. The Charpy test of composite materials, however, has limitations because the failure mode is commonly of a more complex nature, being difficult to relate to the measured energy, thus such test is becoming progressive less used in the characterization of that property. In the drop-weight impact test, a mass is raised to a known height and then released, impacting the test sample. This is the basis of ASTM D7136, the most common toughness characterization method for polymeric matrix composite materials. To measure the compressive residual strength of tested samples, they can undergo a post-impact compression test according to ASTM D7137 (Belingardi and Vadori, 2002; Cantwell and Morton, 1991; Richardson and Wisheart, 1996; Shyr and Pan, 2003; Winkel and Adams, 1985).

f. *Interlaminar characteristics*: Failure associated with the interface between the layers of a polymeric composite, called delamination, progresses mostly within the polymeric matrix, usually presenting low strength compared to the composite strength. ASTM D2344 was the first experimental procedure used to evaluate interlaminar strength of composites. However, due to the complexity of the failure modes found in different composites, interpretation of the results was found to be considerably complex. New tests were then developed, for a more precise characterization of failure by delamination.

Tests have been proposed for the three modes of propagation of interlaminar fracture. In Mode I, or opening delamination tests, the necessary tensile load for a cantilever beam to promote sample delamination from an initial crack is measured. ASTM D5528 and ISO 1524 present the related test conditions and this is the most commonly used delamination characterization test. In Mode II, delamination takes place by the sliding of two fracture surfaces, i.e., in-plane shear. The Mode II Interlaminar Fracture Toughness test described in ASTM D7905 is very recent, and used for qualitatively comparing composite materials, being useful for design and analysis. Alternatively, the Mixed Mode I-Mode II Interlaminar Fracture Toughness of Unidirectional Fiber Reinforced Polymer Matrix Composites (ASTM D6671) is a more established standard. On the other hand, Mode III tests, for out-plane shear or twisting, have been the toughest challenge in composite delamination characterization, since it is very difficult to ensure a pure torsional mode. The main methods are the split cantilever beam and the edge torsion crack test but standardized tests for them are yet to be reported (Brunner et al., 2008; Ding, 1999; Sela and Ishai, 1989).

g. *Fatigue testing*: The common standards that determine the experimental procedures for fatigue tests for composite materials with polymeric matrix are ASTM D3479 and ISO 13003. The test is about subjecting a sample to an axial load at a specific frequency and determining the number of cycles that it stands before failing. A fundamental problem in the use of fiber-reinforced polymeric matrix composites regards the difficulty to determine their mechanical strength under the action of cyclic stresses. Composite materials have complex fatigue failure mechanisms due to their anisotropic characteristics. Also, there are four possible failure mechanisms resulting from a material fatigue process: crack propagation through the polymeric matrix, delamination, fiber breakage, and interfacial detachment. The different failure modes, combined with the inherent anisotropy, complex stress fields and the nonlinear behavior of composites severely limit the understanding of the nature of fatigue in this type of material (Degrieck and Van Paepegem, 2001; Konur and Matthews, 1989).

h. *Fiber pull-out and single-fiber fragmentation testing*: The single-fiber pull-out test is an experimental method to evaluate failure through load transfer between fiber and matrix. In this test, one end of a single fiber is encapsulated in the polymeric matrix.

The encapsulated end is held stationary while the free end is pulled. In the experiment, the force required to detach the fiber from the polymer is a function of the encapsulated fiber length, from which the interfacial shear strength can be calculated. This test may be useful for quantitative studies, such as to compare surface treatments and fiber interaction with different matrices. However, this method has several limitations when a more thorough interfacial characterization is desired, especially regarding sample preparation and handling, temperature changes and sample loading conditions.

An attempt to acquire more fundamental results of interfacial properties is to characterize the interfacial failure from the perspective of fracture mechanics. Using such interpretation, the fiber pull-out test can be seen as a test that initiates and propagates a crack, i.e., interfacial debonding along the fiber/matrix interface. Like so, it is possible to calculate the necessary energy for crack growth as a function of crack length (Bannister et al., 1995; DiFrancia et al., 1996a,b; Hampe et al., 1995; Nairn et al., 2001).

Another common method for the characterization of the interfacial stress between fiber and matrix is the single-fiber fragmentation test. In this test, a single fiber is encapsulated in a polymeric matrix test sample and a tensile load is applied to it. When stress is applied to the matrix, it is transferred to the fiber by means of a shear stress. With the increase in load, the fiber's maximum strain will eventually exceed causing its fracture. Continuous increase in stress, leads to fiber fractures into ever smaller lengths until the shear stress that is transferred through the interface becomes insufficient to cause more fractures.

For transparent polymers, the measurement of the minimum fiber length, which can be used to determine critical fiber length, can be performed with an optical microscope. For opaque samples, this measurement must be performed ex-situ, and may not exactly replicate the sample fragmentation situation when encapsulated in the matrix, presenting a technical limitation. The average interfacial shear strength can be estimated based on the model proposed by Kelly and Tyson. However, due to the random nature of fiber fractures and fiber size measurements that tend to be somewhat inaccurate, the use of statistical distribution models is necessary to acquire more reliable values. The most commonly used model to predict the average fiber tensile strength is the Weibull distribution for probability of failure (Tripathi and Jones, 1998; Lee et al., 1998; Feih et al., 2004).

i. *Indentation testing*: Instrumented indentation is also very common for the mechanical characterization for composites. The method consists in pressing a diamond tip into the material, controlling and recording values of applied load and penetration depth using sensors. The maximum load is kept constant for a few seconds before it is removed. The time, in seconds, is controlled in three stages: during loading, while at maximum load, and during unloading. Currently, technological advancements enable indentations of tenths of a nanometer and loads near nano-Newtons, making this technique an important tool for nanocomposite materials. Several nanomechanical properties can be obtained from the tested area, such as elasticity modulus, elastic and plastic strain, hardness and wear resistance. If an atomic-force microscope is used in conjunction with the nanoindenter, in situ information about material strain and crack propagation during nanoindentation can be obtained.

Methods such as continuous stiffness measurement (CSM) enable the study of materials in which mechanical properties vary according to the depth of penetration, such as multilayer nanocomposites. The test is based on the principle of overlapping a small oscillation amplitude of the displacement signal (or load) of the indenter and measuring the load response (or displacement) in the tested material. One of the advantages of this technique is the possibility of measuring stiffness at any point of the curve and not only at the unloading point, as performed in conventional nanoindentation measurements. The CSM method enables measurements at nanoscale of the displacement and stress relaxation in creep tests and high frequency fatigue tests (Cakmak et al., 2012; Díez-Pascual et al., 2015; Gibson, 2014; Nardi et al., 2015).

j. *Dynamic mechanical analysis*: The dynamic mechanical analysis (DMA), also known as dynamic mechanical thermal analysis (DMTA), evaluates the material response to an oscillatory strain of certain frequency, as a function of a certain temperature. An oscillating stress is applied to the sample to create a series of stress–strain curves at a particular frequency during a certain time. The displacement (strain) and the resulting amplitude of this applied force are measured so that the elasticity and viscosity moduli of the material can be calculated. DMA results are composed of three major parameters: (i) storage modulus (G'), that defines the elastic behavior of the sample; (ii) loss modulus (G''), that defines its viscous behavior; and (iii) tan δ, which evaluates energy dissipation in the material.

One of the main goals of this technique is to correlate the macroscopic properties, such as thermomechanical properties, with molecular relaxations, which may be caused by conformational changes and/or microscopic strains caused by molecular movement (due to temperature change, frequency or time). From this test, various characteristics of a material such as the ability to lose energy in the form of heat (damping), strain recovery ability (elasticity), and transitions of molecular mobility, such as glass transition temperature (T_g), can be obtained. DMA data is used not only to determine the mechanical behavior of materials (polymers, in particular), but also their structure, morphology and viscoelastic behavior. For composites, there are some standard recommendations controlling DMA analysis, regarding the observation method and the calculation to determine the parameters of a material such as glass transition temperature (T_g) and the curing point. ASTM D7028 has specifications for the determination of T_g by DMA using the flexure method. In this method, fiber orientation must follow the sample length and T_g is determined by the storage modulus curve, E', as a function of temperature (Guo et al., 2013; Lorandi et al., 2016; Menard, 1999).

6.3 Influence of manufacturing on properties of CNT/fiber hybrid composites

A number of works have reported that the addition of CNT brings only marginal improvements in mechanical properties of fiber/thermoset composites (Tehrani et al., 2013a; Sanchez et al., 2013; Godara et al., 2009; Iwahori et al., 2005), while others report significant enhancements, especially in interlaminar fracture toughness and matrix-dominated properties (Garcia et al., 2008a; Garcia et al., 2008b; Wicks et al., 2014). This seems to be strongly related to the manufacturing process chosen and the dispersion state of the nanotubes in the interlaminar region.

Chandrasekaran et al. (2010) addressed the role of processing on the final properties of epoxy/glass fiber/multiwalled carbon nanotube (MWCNT) hybrid composites. Sonication was used to prepare MWCNT/epoxy suspensions. When epoxy/glass fiber hybrid composites modified with 0.5% MWCNT were produced by injection double vacuum-assisted resin transfer molding (IDVARTM) process, no improvement in ILSS was observed. The authors suggest that any improvements in the ILSS for hybrid composites may be the result of sonication itself, which changes the viscosity and surface tension of the epoxy system, instead of the presence of nanotubes. The decrease in viscosity and surface tension of the epoxy resin system due to sonication would result in a better resin infiltration into the preform and an enhanced fiber–matrix interface, improving ILSS. On the other hand, when the samples were prepared by flow flooding chamber (FFC), a variation of the standard vacuum-assisted resin transfer molding (VARTM) technique, the average ILSS increased 21%.

The influence of fabrication on the performance of a hybrid CNT-fiber composite was also reported by Fan et al. (2008). MWCNT–epoxy suspensions were injected into glass fiber mats preforms following two different routes. Short beam and compression shear tests were performed, and the results showed an increase of up to 33% when the hybrid composite is compared to the neat epoxy/glass fiber composite. Increase of interlaminar shear properties was attributed to the distribution of MWNT through and in the thickness direction of the composites. The special spatial distribution of aligned MWNT across the thickness of the composites was obtained using a variation of VARTM. The experimental configuration used was reported to achieve a uniform flow of the MWNT–epoxy suspension across the thickness of the composite, differently from when ordinary VARTM technique is used where a gradient of nanofillers may occur due to restriction of the flow caused by preform compaction. Although both short beam and compression shear tests showed increase in ILSS of the hybrid composites, a complex failure mode occurred when short beam tests were performed. The use of compression shear test was reported as more reliable, minimizing the occurrence of different failure modes.

Sanchez et al. (2013) manufactured epoxy/carbon fibers/CNT by vacuum-assisted resin infusion molding (VARIM), dispersing pristine and amine-functionalized CNT (0.1–0.3 wt%) with calendaring. A small improvement in flexural strength and ILSS, around 10%, was observed for composites filled with amine-CNT. These results were attributed to a less effective dispersion of CNT, leading to the presence of microagglomerates that reduced the potential strengthening effect. The somewhat higher values obtained for amino-CNT in comparison to pristine nanotubes were related to the formation of a stronger interface with the epoxy matrix. Therefore, even for randomly oriented nanotubes, the method of dispersion of CNT is important.

A 'wet' method to deposit carboxylated and hydroxylated MWCNT onto carbon fiber surface, without the removal of the commercial sizing, has been proposed by Li et al. (2013). The functionalized MWCNT adhered to the carbon fiber (uniformly distributed with random orientation) and lead to an increase in surface roughness compared to the original fiber. Single-fiber fragmentation tests using epoxy resin showed greater interfacial shear strength after the deposition of CNT, especially for carboxylated CNT. Tensile strength and ILSS of the CNT-deposited laminates were also improved. The authors suggest that the increased interfacial bonding is related to resin toughening near the interphase, interfacial friction and to chemical bonding (between the epoxy in the sizings and carboxylic or hydroxyl groups of the CNT).

Khan and Kim (2011) reviewed many works and also concluded that the nanofillers incorporation method plays a definite role on the final mechanical performance of the hybrid composite. Very different outcomes were reported even when the same fiber, matrix and nanofiller were used. For example, hybrid composites of epoxy matrix reinforced by alumina fibers and MWCNT obtained 0–69% increase in ILSS (Garcia et al., 2008a, Abot et al., 2008), the difference being attributed to the method of incorporation of the nanofiller into the composite. A similar trend was observed for interlaminar fracture toughness.

In order to eliminate the need to disperse CNT in thermoset resins, various methods of controlled growth in substrates have been reported. One of these methods is based on the vapor-deposition chemical growth of carbon nanofilaments on the surface of carbon fibers. However, the required temperature for the growth of nanotubes (between 650°C and 1100°C) is high enough to destroy the carbon fibers. An alternative to such process is the growth of these nanotubes on substrates of silica and alumina. For instance, the method called graphitic structures by design (GSD) allows the growth of graphitic complex structures at low temperatures and at specific spots of the substrate, preventing agglomeration (Tehrani et al., 2013b). The technique of grafting CNT onto the surface of carbon fibers has been found not only to modify mechanical properties but also electrochemical properties (Islam et al., 2016).

Even though most of the studies involving incorporation of CNT into hybrid fiber composites deal with epoxy matrix composites, some are devoted to other thermoset polymers. Zhu et al. (2007) incorporated CNT at glass fiber-reinforced vinyl ester matrix composites and obtained an increase of up to 45% in interfacial shear strength. An increase in interfacial shear strength of up to 90% was reported for CNT grown directly onto the surface of carbon fibers by chemical vapor deposition (CVD) technique in hybrid composites with polyester matrix (Agnihotri et al., 2011) for optimum CNT length and density. Hybrid phenolic matrix-carbon fiber composites obtained by growing CNT onto the surface of carbon fibers by CVD where manufactured using several different carbon fiber substrates (Mathur et al., 2008). Tensile and flexural properties were improved when the CNT fraction was higher than ca. 5 wt%, depending on the substrate, and greater increase was obtained for a bidirectional carbon fiber cloth. Scanning electron microscopy (SEM) analysis of the fracture surface indicated that the spatial arrangement of nanotubes grown at warp and weft crossover points helped promoting a better interlock between fibers and the matrix, reducing shear delamination.

The prepreg technique was also studied as an alternative way to reduce problems with resin flow through closely packed fibers and to avoid reagglomeration of CNT, which are common in the processing of three phase hybrid composites via resin transfer molding (RTM) (Godara et al., 2009). This technique reduces CNT mobility during curing since a high viscosity resin is employed. A positive effect of CNT (0.5 wt%) on some mechanical properties of unidirectional (UD) carbon fiber/epoxy composites was found, such as considerable improvement in crack initiation and propagation energy. This effect was attributed to crack bridging by the CNT, and a further improvement was achieved by modifying the three-component system with a compatibilizer, which increased the crack initiation energy (G_{IC}) by 75%.

The use of blended thermoplastic/thermosetting matrices is another alternative, as in the work of Song (2007), who used an epoxy/PEO blend. The CNT content was 3 wt%, and the composite was manufactured by compression molding to obtain thin films of CNT dispersed into PEO followed by vacuum-assisted resin transfer molding (VARTM) to blend PEO and CNT into an epoxy matrix and to impregnate the aramid fiber preform. The results showed no statistical variation in both flexural modulus and strength between the hybrid CNT/aramid and aramid fiber composites. However, impact properties increased by 8%, which was attributed to crack deviation at the homogeneously dispersed CNT.

6.4 Interlaminar, toughness and damping characteristics of CNT/fiber hybrid composites

The structural failure of CNT hybrid polymer composites normally occurs by microcracking and/or delamination (Grant and Rousseau, 2000; Herceg et al., 2016). Thus, the most useful methods for their mechanical characterization are ILSS, commonly obtained using short beam shear test (ASTM D2344) (Garcia et al., 2008a; Herceg et al., 2016), and interlaminar fracture toughness in Mode I (ASTM D5528) (Almuhammadi et al., 2014; Falzon et al., 2013).

Khan and Kim (2011) reported that despite the large bounds found in the literature regarding the magnitude of the effect of incorporation of nanofillers into fiber-reinforced composites, the results clearly show that the use of CNT increases through-the-thickness properties of these composites. This positive effect is attributed to the interaction of the dispersed nanofillers with the crack front, including crack deflection (Yokozeki et al., 2007a), crack bridging (Karapappas et al., 2009) and pull-out of CNT (Karapappas et al., 2009; Wicks et al., 2010), which increase fracture toughness. Kepple et al. (2008) studied the in situ functionalization of carbon fiber woven (CF) with CNT to test the hypothesis that growing CNT on CF would enhance the properties of polymer-matrix composites, specifically epoxy–carbon composites that are used in aerospace applications, and achieved a significant improvement in fracture toughness (ca. 50%) with no loss in structural stiffness. This increase in the fracture toughness indicates that the addition of CNT is beneficial for the damage tolerance of a composite structure.

The increase in fracture toughness and interlaminar properties of hybrid composites is partially attributed to modifications of the resin matrix by the nanofillers. In fact, when nanofillers are homogeneously distributed in the resin matrix, damage tolerance to failure mechanisms, such as matrix microcracking, is enhanced (Gojny et al., 2004, Lubineau and Rahaman, 2012). It was also demonstrated that the presence of a layer of resin rich in nanofillers at the fiber/matrix interface can increase damage tolerance, reducing fiber breakage at low stress levels (Siddiqui et al., 2009). If matrix microcracking and fiber breakage increase in number and/or coalesce transverse cracks are formed, the interlaminar properties of the composite are reduced. Therefore, inhibiting these mechanisms contribute to improve in-plane properties of hybrid composites. In this context, Yokozeki et al. (2007b) studied the effect of CNT incorporation on fracture toughness of a hybrid carbon fiber–epoxy composite. The onset strain at which transverse cracks begin to form increased by 21% and fracture toughness increased by 37%.

Several works demonstrated that interlaminar properties can be improved if the proper strategy is used to incorporate the nanofillers. Therefore, strong increase in critical energy release rate for both Modes I and II (G_{Ic} and G_{IIc}) were obtained when nanotubes were grafted on the surface of the fibers. Improvements as high as 348% for Mode I, and 54% for Mode II of crack growth were obtained using double-cantilever beam and end notch flexure tests (Veedu et al., 2006). The results obtained by Zhou et al. (2016) demonstrate that the higher the fraction of CNT incorporated onto the surface of the fibers, the greater the delamination resistance. Similar results were obtained by Storck et al. (2011) for epoxy matrix composites reinforced by glass or carbon fibers with CNT grown by CVD onto their surface. High density of short CNT, with lengths smaller than twice the diameter of the fibers, improved the interlaminar strength by ca. 30%.

The variation in fracture toughness of vinyl ester matrix hybrid composites was also reported by Seyhan et al. (2008). Mode I interlaminar fracture toughness (G_{Ic}) determined by DCB test and Mode II (G_{IIc}) determined by the end notched flexure test (ENF) showed different trends. While for Mode I the incorporation of CNT into the resin matrix did not affect the result, for Mode II an increase of about 8% was found in relation to the value of the composite manufactured without CNT. The ILSS of the composite modified with CNT was 11% higher than the value of the unmodified composite. The authors indicate that it is necessary to obtain a uniform distribution of CNT and to achieve a stronger interfacial bonding to develop hybrid composites with better performance.

It is important to bear in mind that due to the anisotropic properties of graphite, the physical properties of CNT are also anisotropic. Geometrical considerations suggest that CNT are less rigid in radial directions than in directions along the tube axis. The alignment of the CNT also greatly influences the mechanical properties of nanocomposites. CNT are uniaxially strong, just like the carbon fibers, and the alignment in a uniaxial direction should lead to improved properties. Some work using randomly oriented carbon nanofibers and CNT placed between woven plies by spraying (Thakre et al., 2006) or unaligned CNT mixed between tows (Kim et al., 2007) showed almost no mechanical reinforcement, probably due to agglomeration and poor dispersion. However, a successful approach has been achieved with nanoscale modification of the interface between composite plies, by placing random CNT at low volume fraction on fibers (including at the ply interface) (Zhu et al., 2007; Bekyarova et al., 2007b). Functionalized single-wall carbon nanotubes (SWCNT) sprayed on the surface of glass cloth showed an increase of up to 45% in short beam (interlaminar) shear strength (Zhu et al., 2007).

When aligned and organized around the fibers, CNT have the potential to offer both interlaminar and intralaminar modification of the composite performance.

Garcia et al. (2008a) described a hybrid composite architecture consisting of an epoxy matrix, alumina fiber cloth woven in a 0°/90° satin-weave, and CNT. Aligned CNT were grown directly on the surface of the alumina fibers, enabling multifunctional performance, as demonstrated by 69% increase in ILSS and 10^6 (in-plane) and 10^8 (through-thickness) increase in electrical conductivity. The authors suggest a capillarity-driven mechanism to explain the wetting of the aligned CNT in the interior of the laminate by the thermoset epoxy resin. The reinforcement of the polymer matrix by aligned CNT has been also confirmed using novel nanocompression techniques (Garcia et al., 2007). And when aligned CNT are arranged in the interlaminar region of a hybrid composite, an increment of up to 200% in the Mode II fracture toughness was reported (Garcia et al., 2008b; Wicks et al., 2014).

A recent fractographic study in woven (alumina cloth) composite laminates reinforced with aligned CNT revealed toughening mechanisms for epoxy laminates at different length scales, from the pull-out of CNT induced tow breakage (Wicks et al., 2014). The results show that the toughening behavior and the magnitude of improvement are dependent on the type of epoxy (more brittle infusion aerospace resin or tougher hand lay-up marine resin). Only modest improvement in steady-state toughness was achieved for more brittle matrices, while tougher marine epoxy allowed large improvement with CNT by significant microfiber breakage and CNT pull-out along with epoxy fracture. Crack propagation path and subsequent interlaminar toughness is also affected by CNT length. Growth of CNT on the surface of glass fibers was also reported (Rahmanian et al., 2013), and enhancement of CNT-coated fiber adhesion with the polymeric matrix was verified by SEM.

Although fiber-reinforced polymer composites (FRPC) can offer good mechanical performance, they have limitations. One of these limiting factors is the low vibration damping. CNT are able to dissipate energy very efficiently when dispersed in polymeric matrices, expanding the number of applications when added to the FRPC. However, due to the great difficulty of dispersing CNT in a polymeric matrix, this phenomenon has a limited effect. (Tehrani et al., 2013b). The effect of adding MWCNT to the epoxy matrix of a carbon fiber-reinforced composite (CFRP) on its impact properties (damping performance, impact resistance and impact damage progression) has been addressed by Tehrani et al. (2013a). CFRP specimens were mechanically tested under tensile, quasi-static punch test, vibration, and intermediate velocity impact (IVI) environments. Tensile failure strain increased with the addition of MWCNT, while tensile modulus and strength remained almost unaffected. The samples containing MWCNT also retained higher damping in dynamic-mechanical analysis, indicating their greater capability to attenuate impact shocks. The energy dissipation capacity of the composites was evaluated by out-of-plane impact tests (at $\sim 100\ m\ s^{-1}$), which revealed that the presence of MWCNT in the CFRP increased the absorbed impact energy by 21%. Punch tests, used to evaluate damage mechanics of the ballistic impact of composite panels, and X-ray radiography measurements showed that the addition of MWCNT to the matrix of a CFRP improves its inter- and intra-laminar mechanical performance yielding a better impact resistance.

From the discussion above, it is clear that the key aspects regarding the effects caused by the incorporation of CNT on the mechanical properties of hybrid composites are related to their interaction with the polymeric matrix. Several techniques can be used to investigate CNT–polymer interaction, such as contact angle measurements using several approaches, including drop-on-fiber analysis (Tran et al., 2008) and atomic-force microscopy (AFM) (Barber et al., 2005). However, a large variation in values is common even when the same polymer/CNT is being investigated. Therefore, wetting experiments can be regarded as a qualitative way to verify if there is a surface tension match between CNT and the polymeric matrix, which can suggest a strong interaction at the interface. Raman spectroscopy is a powerful tool to investigate CNT/polymer interaction. This technique can be used not only to verify the efficacy of surface treatments by identifying specific chemical groups at the CNT/polymer interface (Martinez-Rubi et al., 2007), but it can also give information about nanofiller dispersion (Bassil et al., 2005) and about the stress transfer from the matrix to CNT/fiber reinforcement (Chang et al., 2005, Brownlow et al., 2010). Another technique that can be used to access modifications at the interface is Fourier transform infrared spectroscopy (FTIR) (Brownlow et al., 2010).

6.5 Concluding remarks

Mechanical tests for the characterization of hybrid composites obtained with CNT and common reinforcement fibers, e.g., carbon fibers, into a single polymeric matrix were briefly presented and discussed, with emphasis on more recent and relevant methods. Hybrid CNT/fiber-reinforced thermoset composites may show good performance when the nanoparticles are introduced into the matrix or grown onto the surface of the reinforcing fibers. Both architectures may improve matrix-dominated properties, such as interlaminar shear and interlaminar toughness. Crack deflection at the many nanofiller/polymer matrix interfaces is an important fracture mechanism. However, the effectivity of this mechanism is strongly dependent on the homogeneous distribution of the nanofillers, since agglomerates will significantly reduce the CNT surface area available to interact with the crack front. The size of the CNT and their alignment can also affect the final mechanical properties. Therefore, a key point to effectively enhance, the mechanical properties of hybrid composites is to guarantee adequate dispersion of the nanofillers, indicating that a proper manufacturing route must, at first, be dominated to guarantee reproducibility and confidence in the results.

Acknowledgments

The authors would like to thank CNPq agency, Brazil. Also, Mariana Lima for her help with the references.

References

Abot, J.L., Song, Y., Schulz, M.J., Shanov, V.N., 2008. Novel carbon nanotube array-reinforced laminated composite materials with higher interlaminar elastic properties. Compos. Sci. Technol. 68, 2755.

Adams, D.F., Busse, J.M. Suggested modifications of the short beam shear test method. In: 49th International SAMPE Conference, Long Beach/California, 2004, EUA; 2004. p. 1–14.

Adams, D.O., Moriarty, J.M., Gallegos, A.M., Adams, D.F., 2007. The V-notched rail shear test. J. Compos. Mater. 41, 281.

Agnihotri, P., Basu, S., Kar, K.K., 2011. Effect of carbon nanotube length and density on the properties of carbon nanotube-coated carbon fiber/polyester composites. Carbon. 49, 3098.

Almeida Jr., J.H.S., Angrizani, C.C., Botelho, E.C., Amico, S.C., 2015. Effect of fiber orientation on the shear behavior of glass fiber/epoxy composites. Mater. Des. 65, 789.

Almuhammadi, K., Alfano, M., Yang, Y., Lubineau, G., 2014. Analysis of interlaminar fracture toughness and damage mechanisms in composite laminates reinforced with sprayed multi-walled carbon nanotubes. Mater. Des. 53, 921.

Bannister, D.J., Andrews, M.C., Cervenka, A.J., Young, R.J., 1995. Analysis of the single-fiber pull-out test by means of Raman-spectroscopy. 2. Micromechanics of deformation for an aramid/epoxy system. Compos. Sci. Technol. 53, 411.

Barber, A.H., Cohen, S.R., Wagner, H.D., 2005. External and internal wetting of carbon nanotubes with organic liquids. Phys. Rev. B. 71.

Bassil, A., Puech, P., Landa, G., Bacsa, W., Barrau, S., Demont, P., et al., 2005. Spectroscopic detection of carbon nanotube interaction with amphiphilic molecules in epoxy resin composites. J. Appl. Phys. 97.

Bekyarova, E., Thostenson, E.T., Yu, A., Itkis, M.E., Fakhrutdinov, D., Chou, T.W., et al., 2007a. Functionalized single-walled carbon nanotubes for carbon fiber–epoxy composites. J. Phys. Chem. C. 111, 17865.

Bekyarova, E., Thostenson, E.T., Yu, A., Kim, H., Gao, J., Tang, J., et al., 2007b. Multiscale carbon nanotube–carbon fiber reinforcement for advanced epoxy composites. Langmuir. 23, 3970.

Belingardi, G., Vadori, R., 2002. Low velocity impact tests of laminate glass–fiber–epoxy matrix composite material plates. Int. J. Impact Eng. 27, 213.

Brownlow, S.R., Moravsky, A.P., Kalugin, N.G., Majumdar, B.S., 2010. Probing deformation of double-walled carbon nanotube (DWNT)/epoxy composites using FTIR and Raman techniques. Comp. Sci. Technol. 70, 1460.

Brunner, A.J., Blackman, B.R.K., Davies, P., 2008. A status report on delamination resistance testing of polymer matrix composites. Eng. Fract. Mech. 75, 2779.

Cakmak, U.D., Schoberl, T., Major, Z., 2012. Nanoindentation of polymers. Meccanica. 47, 707.

Cantwell, W.J., Morton, J., 1991. The impact resistance of composite materials – a review. Composites. 22, 347.

Carlsson, L.A., Adams, D.F., Pipes, R.B., 2014. Experimental Characterization of Advanced Composite Materials. CRC Press.

Chandrasekaran, V.C.S., Advani, S.G., Santare, M.H., 2010. Role of processing on interlaminar shear strength enhancement of epoxy/glass fiber/multi-walled carbon nanotube hybrid composites. Carbon. 48, 3692.

Chang, T.E., Jensen, L.R., Kisliuk, A., Pipes, R.B., Pyrz, R., Sokolov, A.P., 2005. Microscopic mechanism of reinforcement in single-wall carbon nanotube/polypropylene nanocomposite. Polymer. 46, 439.

De Riccardis, M.F., Carbone, D., Makris, T.D., Giorgi, R., Lisi, N., Salernitano, E., 2006. Anchorage of carbon nanotubes grown on carbon fibres. Carbon. 44, 671.

Degrieck, J., Van Paepegem, W., 2001. Fatigue damage modeling of fibre-reinforced composite materials: Review. Appl. Mech. Rev. 54, 279.

Díez-Pascual, A.M., Gómez-Fatou, M.A., Ania, F., Flores, A., 2015. Nanoindentation in polymer nanocomposites. Prog. Mater. Sci. 67, 1.

DiFrancia, C., Ward, T.C., Claus, R.O., 1996a. The single-fibre pull-out test. 1. Review and interpretation. Compos. A: Appl. Sci. Manuf. 27, 597.

DiFrancia, C., Ward, T.C., Claus, R.O., 1996b. The single-fibre pull-out test. 2. Quantitative evaluation of an uncatalysed TGDDM/DDS epoxy cure study. Compos. A: Appl. Sci. Manuf. 27, 613.

Ding, W., 1999. Delamination Analysis of Composite Laminates, 161.

Falzon, B.G., Hawkins, S.C., Huynh, C.P., Radjef, R., Brown, C., 2013. An investigation of Mode I and Mode II fracture toughness enhancement using aligned carbon nanotubes forests at the crack interface. Compos. Struct. 106, 65.

Fan, Z.H., Santare, M.H., Advani, S.G., 2008. Interlaminar shear strength of glass fiber reinforced epoxy composites enhanced with multi-walled carbon nanotubes. Compos. A: Appl. Sci. Manuf. 39, 540.

Feih, S., Wonsyld, K., Minzari, D., Westermann, P., Lilholt, H., 2004. Testing procedure for the single fiber fragmentation test. Forskningscenter Risø1483.

Garcia, E.J., Hart, A.J., Wardle, B.L., Slocum, A.H., 2007. Fabrication and nanocompression testing of aligned carbon−nanotube−polymer nanocomposites. Adv. Mater. 19, 2151.

Garcia, E.J., Wardle, B.L., Hart, A.J., 2008a. Joining prepreg composite interfaces with aligned carbon nanotubes. Compos. A: Appl. Sci. Manuf. 39, 1065.

Garcia, E.J., Wardle, B.L., Hart, A.J., Yamamoto, N., 2008b. Fabrication and multifunctional properties of a hybrid laminate with aligned carbon nanotubes grown in situ. Compos. Sci. Technol. 68, 2034.

Gibson, R.F., 2014. A review of recent research on nanoindentation of polymer composites and their constituents. Compos. Sci. Technol. 105, 51.

Godara, A., Mezzo, L., Luizi, F., Warrier, A., Lomov, S.V., van Vuure, A.W., et al., 2009. Influence of carbon nanotube reinforcement on the processing and the mechanical behaviour of carbon fiber/epoxy composites. Carbon. 47, 2914.

Gojny, F., Wichmann, M., Fiedler, B., Bauhofer, U., Schulte, K., 2005. Influence of nano--modification on the mechanical and electrical properties of conventional fibre-reinforced composites. Compos. A: Appl. Sci. Manuf. 36, 1525.

Gojny, F.H., Wichmann, M.H.G., Köpke, U., Fiedler, B., Schulte, K., 2004. Carbon nanotube-reinforced epoxy-composites: enhanced stiffness and fracture toughness at low nanotube content. Comp. Sci. Technol. 64, 2363.

Grant, P., Rousseau, C.Q., 2000. Composite Structures: Theory and Practice, ASTM STP 1383, West Conshohocken, PA.

Guo, C., Zhou, L., Lv, J., 2013. Effects of expandable graphite and modified ammonium polyphosphate on the flame-retardant and mechanical properties of wood flour-polypropylene composites. Polym. Polym. Compos. 21, 449.

Hampe, A., Kalinka, G., Meretz, S., Schulz, E., 1995. An advanced equipment for single-fiber pull-out test designed to monitor the fracture process. Composites. 26, 40.

Herceg, T.M., Abidin, M.S.Z., Greenhalgh, E.S., Shaffer, M.S.P., Bismarck, A., 2016. Thermosetting hierarchical composites with high carbon nanotube loadings: en route to high performance. Compos. Sci. Technol. 127, 134.

Hodgki, J.M., 2000. Mechanical Testing of Advanced Fibre Composites. CRC Press.

Islam, M.S., Deng, Y., Tong, L.Y., Faisal, S.N., Roy, A.K., Minett, A.I., et al., 2016. Grafting carbon nanotubes directly onto carbon fibers for superior mechanical stability: towards next generation aerospace composites and energy storage applications. Carbon. 96, 701.

Iwahori, Y., Ishiwata, S., Sumizawa, T., Ishikawa, T., 2005. Mechanical properties improvements in two-phase and three-phase composites using carbon nano-fiber dispersed resin. Compos. A: Appl. Sci. Manuf. 36, 1430.

Karapappas, P., Vavouliotis, A., Tsotra, P., Kostopoulos, V., Paipetis, A., 2009. Enhanced fracture properties of carbon reinforced composites by the addition of multi-wall carbon nanotubes. J. Compos. Mater. 43, 977.

Kedward, K.T., 1972. On the short beam test method. Fibre Sci. Technol. 5, 85.

Kepple, K.L., Sanborn, G.P., Lacasse, P.A., Gruenberg, K.M., Ready, W.J., 2008. Improved fracture toughness of carbon fiber composite functionalized with multi walled carbon nanotubes. Carbon. 46, 2026.

Khan, S.U., Kim, J.K., 2011. Impact and delamination failure of multiscale carbon nanotube-fiber reinforced polymer composites: a review. Int. J. Aeronaut. Space Sci. 12, 115.

Kim, Y.A., Kamio, S., Tajiri, T., Hayashi, T., Song, S.M., Endo, M., et al., 2007. Enhanced thermal conductivity of carbon fiber/phenolic resin composites by the introduction of carbon nanotubes. Appl. Phys. Lett. 90.

Konur, O., Matthews, F.L., 1989. Effect of the properties of the constituents on the fatigue performance of composites: a review. Composites. 20, 317.

Kretsis, G., 1987. A review of the tensile, compressive, flexural and shear properties of hybrid fiber-reinforced plastics. Composites. 18, 13.

Lee, S., Nguyen, T., Chin, J., Chuang, T.J., 1998. Analysis of the single-fiber fragmentation test. J. Mater. Sci. 33, 5221.

Li, M., Gu, Y.Z., Liu, Y.N., Li, Y.X., Zhang, Z.G., 2013. Interfacial improvement of carbon fiber/epoxy composites using a simple process for depositing commercially functionalized carbon nanotubes on the fibers. Carbon. 52, 109.

Lorandi, N.P., Cioffi, M.O.H., Ornaghi Jr, H., 2016. Dynamic mechanical analysis (DMA) of polymeric composite materials. Sci. Cum. Ind. 4, 48.

Lubineau, G., Rahaman, A., 2012. A review of strategies for improving the degradation properties of laminated continuous-fiber/epoxy composites with carbon-based nanoreinforcements. Carbon. 50, 2377.

Martinez-Rubi, Y., Guan, J., Lin, S., Scriver, C., Sturgeon, R.E., Simard, B., 2007. Rapid and controllable covalent functionalization of single-walled carbon nanotubes at room temperature. Chem. Commun. 48, 5146.

Mathur, R.B., Chatterjee, S., Singh, B.P., 2008. Growth of carbon nanotubes on carbon fibre substrates to produce hybrid/phenolic composites with improved mechanical properties. Compos. Sci. Technol. 68, 1608.

Menard, K.P., 1999. Dynamic Mechanical Analysis: A Practical Introduction. CRC Press.

Nairn, J.A., Liu, C., Mendels, D., Zhandarov, S., 2001. Fracture mechanics analysis of the single-fiber pull-out test and the microbond test including the effects of friction and thermal stresses. Proc. 16th Ann. Tech. Conf. of the Amer. Soc. Composites, 9–12.

Nardi, T., Hammerquist, C., Nairn, J.A., Karimi, A., Månson, J.A.E., Leterrier, Y., 2015. Nanoindentation of functionally graded polymer nanocomposites: assessment of the strengthening parameters through experiments and modeling. Front. Mater. 2 (August), 1–10.

Rahmanian, S., Thean, K.S., Suraya, A.R., Shazed, M.A., Salleh, M.A.M., Yusoff, H.M., 2013. Carbon and glass hierarchical fibers: influence of carbon nanotubes on tensile, flexural and impact properties of short fiber reinforced composites. Mater. Des. 43, 10.

Richardson, M.O.W., Wisheart, M.J., 1996. Review of low-velocity impact properties of composite materials. Compos. A: Appl. Sci. Manuf. 27, 1123.

Sanchez, M., Campo, M., Jimenez-Suarez, A., Urena, A., 2013. Effect of the carbon nanotube functionalization on flexural properties of multiscale carbon fiber/epoxy composites manufactured by VARIM. Compos. B.: Eng. 45, 1613.

Schultheisz, C.R., Waas, A.M., 1996. Compressive failure of composites, part 1: testing and micromechanical theories. Prog. Aerosp. Sci. 32, 1.

Sela, I.V., Ishai, O., 1989. Interlaminar fracture toughness and toughening of laminated composite materials: a review. Composites. 20, 423.

Seyhan, A.T., Tanoglu, M., Schulte, K., 2008. Mode I and mode II fracture toughness of E-glass non-crimp fabric/carbon nanotube (CNT) modified polymer based composites. Eng. Fract. Mech. 75, 5151.

Shyr, T.-W., Pan, Y.-H., 2003. Impact resistance and damage characteristics of composite laminates. Compos. Struct. 62, 193.

Siddiqui, N.A., Sham, M.L., Tang, B.Z., Munir, A., Kim, J.K., 2009. Tensile strength of glass fibres with carbon nanotube–epoxy nanocomposite coating. Compos. A: Appl. Sci. Manuf. 40, 1606.

Song, K.A., Zhang, Y.Y., Meng, J.S., Green, E.C., Tajaddod, N., Li, H., et al., 2013. Structural polymer-based carbon nanotube composite fibers: understanding the processing–structure–performance relationship. Materials. 6, 2543.

Song, Y.S., 2007. Multiscale fiber-reinforced composites prepared by vacuum-assisted resin transfer molding. Polym. Comp. 28, 458.

Storck, S., Malecki, H., Shah, T., Zupan, M., 2011. Improvements in interlaminar strength: a carbon nanotube approach. Compos. B: Eng. 42, 1508.

Tehrani, M., Safdari, M., Boroujeni, A.Y., Razavi, Z., Case, S.W., Dahmen, K., et al., 2013a. Hybrid carbon fiber/carbon nanotube composites for structural damping applications. Nanotechnology. 24.

Tehrani, M., Boroujeni, A.Y., Hartman, T.B., Haugh, T.P., Case, S.W., Al-Haik, M.S., 2013b. Mechanical characterization and impact damage assessment of a woven carbon fiber reinforced carbon nanotube-epoxy composite. Compos Sci. Technol. 75, 42.

Thakre P.R., Lagoudas, D.C., Zhu, J., Barrera, E.V., Gates, T., 2006. Processing and characterization of epoxy-SWCNT-woven fabric composites. 47th AIAA structures, dynamics, and materials conference proceedings, Newport, RI, May 1–4; 2006 [AIAA-2006-1857-212].

Thostenson, E.T., Li, W.Z., Wang, D.Z., Ren, Z.F., Chou, T.W., 2002. Carbon nanotube/carbon fiber hybrid multiscale composites. J. Appl. Phys. 91, 6034.

Tran, M.Q., Cabral, J.T., Shaffer, M.S.P., Bismarck, A., 2008. Direct measurement of the wetting behavior of individual carbon nanotubes by polymer melts: the key to carbon nanotube–polymer composites. Nano Lett. 8, 2744.

Tripathi, D., Jones, F.R., 1998. Single fibre fragmentation test for assessing adhesion in fibre reinforced composites. J. Mater. Sci. 33, 1.

Van Hemelrijck, D., Anastassopoulos, A., 1996. Non Destructive Testing. A.A. Balkema Press.

Veedu, V.P., Cao, A., Li, X.S., Ma, K.G., Soldano, C., Kar, S., et al., 2006. Multifunctional composites using reinforced laminae with carbon-nanotube forests. Nat. Mater. 5, 457.

Waas, A.M., Schultheisz, C.R., 1996. Compressive failure of composites. 2. Experimental studies. Prog. Aerosp. Sci. 32, 43.

Wicks, S.S., de Villoria, R.G., Wardle, B.L., 2010. Interlaminar and intralaminar reinforcement of composite laminates with aligned carbon nanotubes. Comp. Sci. Technol. 70, 20.

Wicks, S.S., Wang, W.N., Williams, M.R., Wardle, B.L., 2014. Multi-scale interlaminar fracture mechanisms in woven composite laminates reinforced with aligned carbon nanotubes. Compos. Sci. Technol. 100, 128.

Winkel, J.D., Adams, D.F., 1985. Instrumented drop weight impact testing of cross-ply and fabric composites. Composites. 16, 268.

Yokozeki, T., Iwahori, Y., Ishiwata, S., 2007a. Matrix cracking behaviors in carbon fiber/epoxy laminates filled with cup-stacked carbon nanotubes (CSCNT). Compos. A: Appl. Sci. Manuf. 38, 917.

Yokozeki, T., Iwahori, Y., Ishiwata, S., Enomoto, K., 2007b. Mechanical properties of CFRP laminates manufactured from unidirectional prepregs using CSCNT-dispersed epoxy. Compos. A: Appl. Sci. Manuf. 38, 2121.

Zhou, H.W., Mishnaevsky, L., Yi, H.Y., Liu, Y.Q., Hu, X., Warrier, A., et al., 2016. Carbon fiber/carbon nanotube reinforced hierarchical composites: Effect of CNT distribution on shearing strength. Compos. B: Eng. 88, 201.

Zhu, J., Imam, A., Crane, R., Lozano, K., Khabashesku, V.N., Barrera, E.V., 2007. Processing a glass fiber reinforced vinyl ester composite with nanotube enhancement of interlaminar shear strength. Compos. Sci. Technol. 67, 1509.

Low-velocity impact behaviour of hybrid composites

7

Fabrizio Sarasini
Department of Chemical Engineering Materials Environment, University of Rome "La Sapienza", Rome, Italy

Chapter Outline

7.1 Introduction

In recent years composite materials experienced an increased usage in different industries ranging from aerospace, wind energy, transportation, naval, and defense structures. According to the market report by MarketsandMarkets (2016), the global market size of composites is projected to grow from USD 69.50 billion in 2015 to USD 105.26 billion by 2021 at a CAGR of 7.04% between 2016 and 2021. This massive use has triggered the need for improved understanding of dynamic deformation and fracture of composites as such components are subjected in service to both quasi-static and dynamic loading conditions. This is particularly of interest to the aerospace structures for which dynamic events are quite common (bird strikes, hail, runway debris) but it is destined to be a major issue also in the automotive sector due to the increasing applications of composites in critical components.

In this framework, the aim of the present chapter is to provide the reader with an overview of the response of composite laminates to low-velocity impacts and with

Hybrid Polymer Composite Materials: Properties and Characterisation. DOI: http://dx.doi.org/10.1016/B978-0-08-100787-7.00007-X

a sound description of how the impact damage and related damage tolerance can be tailored through a judicious hybridization.

7.1.1 Why hybrid materials?

Hybrid materials can be effectively defined as combinations of two or more materials, or of materials and space, assembled in such a way as to have attributes not offered by any one material alone (Ashby, 2005). Examples of the former include composite materials while the latter class is well represented by cellular structures. Ashby has grouped the materials properties in the well-known "Material Property Charts" with the aim of helping engineers and designers in the selection process of materials (Ashby and Bréchet, 2003). In these charts there are parts that are crowded with materials but other parts are empty. Some of these holes cannot be populated for fundamental reasons whilst others can be potentially filled. Hybrid materials have the ability to broaden the occupied areas of material-property space, thus offering materials with properties tailored to meet specific requirements and needs (Ashby, 2011).

7.1.2 Classification of hybrid composites

Composites are themselves hybrid materials but the term "hybrid composites" refers to composites containing more than one type of fiber. In this context the term "hybrid" is used to describe the incorporation of two different types of fibers into one single matrix, the level of mixing being either on a small scale (fibers, tows) or on a large scale (layers). There are several types of hybrid composites that are based on how the constituent materials are mixed. According to Kretsis (1987), these types of hybrid composites include: (i) sandwich structures, where one material is sandwiched between two layers of another, (ii) interply (layer-by-layer) hybrids, in which alternate layers of two (or more) fibers are stacked in a regular manner; (iii) intraply (yarn-by-yarn) hybrids, where tows of two (or more) fiber types are mixed in a regular or random manner, (iv) intimately (fiber-by-fiber) mixed hybrids, in which the constituent fibers are mixed as randomly as possible so that no concentrations of either type are present in the material. Historical research on hybrid composites started several decades ago and was mainly focused on carbon and glass fibers and led to the awareness that the use of both fibers could allow properties better than those expected from the application of the rule of mixtures. In particular, the incorporation of glass fibers in carbon fiber reinforced laminates was reported to improve impact properties and to increase the strain to failure of carbon fibers in tension (Summerscales and Short, 1978). This latter effect is well known in the open literature as "hybrid effect" and it has been the subject of controversial discussions since its very first occurrence as reported in 1972 by Hayashi (1972) who described a 40% increase of the failure strain of the carbon fiber layers in a carbon/glass hybrid composite compared to the reference carbon fiber composite. The most basic definition of the hybrid effect is the apparent failure strain enhancement of the low elongation fiber in a hybrid composite compared to the failure strain of a low elongation fiber-reinforced nonhybrid composite. A more accepted and general definition is as follows: a positive deviation of a certain property from the rule of mixtures. This effect has been debated for a long time in literature but now its existence is well established even though not thoroughly understood.

Hybrid composites have been the subject of several reviews during 1970s and 1980s (Kretsis, 1987; Short and Summerscales, 1980, 1979) but since then there has been a resurgent interest in such materials which culminated in a very recent review by Swolfs et al. (2014).

7.2 Overview of the impact behavior of fiber reinforced composites

One of the major threats to the extensive use of composite materials is represented by the effect of foreign objects impacts as such impacts can be expected to occur during the life of a composite structure. These concerns resulted in the publication of thousands of articles and books (Abrate, 2005, 1994, 1991; Cantwell and Morton, 1991; Reid and Zhou, 2000; Richardson and Wisheart, 1996; Silberschmidt, 2016). The importance of impacts can be ascribed to three factors: (i) damage is more likely to be induced in laminated composites than in similar metallic structures, (ii) damage can grow under load, and the strength and stiffness of the structure can be reduced in a significant way, and (iii) impact damage often can go undetected by traditional visual inspection.

7.2.1 Why are composites prone to impact damage?

Impact damage does not represent a major concern in metallic structures because their ability to dissipate the incident kinetic energy is completely different to that of composites. First of all impact damage in metals can be easily detected as it starts at the impacted surface whilst in composites it often begins on the nonimpacted surface or in the form of internal delaminations that can go easily unnoticed but can severely degrade the structural integrity of the component. This form of damage is referred to as Barely Visible Impact Damage. Secondly, for low to intermediate incident energies metals can absorb energy through elastic and mainly plastic deformation, therefore with minimal consequences on the load carrying ability of the component. On the contrary, most composite materials are characterized by an inherent brittleness with very limited plastic deformation that results in energy absorption through elastic deformation and irreversible damage mechanisms with serious consequences on the strength and stiffness of the composite structure. Most impacts on composite plates are in the transverse direction but due to the lack of through-thickness reinforcement, transverse damage resistance is particularly poor. Interlaminar stresses—shear and tension—are often the ones that cause first failure because of the correspondingly low interlaminar strengths. For this reason composite laminates are particularly prone to impact damage.

Two main strategies have been proposed over the years in order to increase the damage resistance of composite laminates: (i) toughness and (ii) through-thickness mechanical properties enhancement. As regards the first strategy, one of the most researched approach includes toughening of the polymer matrix by tailoring the polymer chemistry or by rubbers, thermoplastics or nanoscale reinforcements. The toughness can be also significantly increased if brittle fibers are replaced by ductile fibers. In this regard, metal fibers, with high stiffness and large failure strain, appear to be the right choice but their use is hindered by their high densities. Polymer fibers, on the other hand, exhibit low densities and can be ductile, but are limited by their low stiffness and poor temperature resistance and stability (Swolfs et al., 2014). The most common through-thickness reinforcement techniques are 3D weaving, stitching, braiding, embroidery, tufting and z-pinning (Isa et al., 2011; Mouritz, 2007). These techniques were found to be effective at increasing the delamination resistance and impact damage tolerance of the resulting composites but at the expense of more complicated manufacturing processes and, above all, of detrimental effects on the in-plane mechanical properties. The drawbacks of these strategies have triggered a resurgent interest in "hybridization" which is the subject of the present chapter.

7.2.2 Impact test techniques for composite materials

Impact can be defined as the relatively sudden application of an impulsive force to a limited volume of material or part of a structure. The problem in this definition lies in that there is no general consensus on the exact meaning of "relatively sudden" and therefore a host of interpretations can be found in literature (Reid and Zhou, 2000). Impulsive problems are usually categorized, depending on the speed of the impactor, into high- or low-velocity impacts. In a low-velocity impact, the stress waves reach the boundary of the structure and reflect several times during the impact; for this reason the response of the structure is global and is controlled by its geometry and boundary conditions. The upper limit of velocity is in the range from 1 to 10 $\mathrm{m\,s^{-1}}$, depending upon target stiffness, material properties, and the impactor's mass and stiffness (Cantwell and Morton, 1991; Sjoblom et al., 1988). The impact energy in low-velocity impact is absorbed by a composite specimen primarily in the form of strain energy, in addition to that dissipated through various failure modes such as matrix cracking, fiber breakage, and delamination (Abrate et al., 2013). In high-velocity impacts, the response of the structure is controlled by the stress-wave propagation through the thickness of the laminate, resulting in quite localized damage with the structure that has not enough time to respond. In this case the boundary-condition effects are of scant importance, since the impact event finishes before the initiated stress waves reach the boundary. No clear limit between high- and low-velocity impacts has been established, although several definitions are found to apply. One widely used defines 100 $\mathrm{m\,s^{-1}}$ as the minimum velocity for which an impact is considered a high-velocity impact (Abrate et al., 2013). Robinson and Davies (1992) suggested a simple equation, $\varepsilon_c = v_0/c$, to estimate the transition impact velocity above which stress waves are dominant. In this equation, v_0 is the impact velocity, c is the stress-wave propagation velocity, and ε_c is the failure strain of the composite. This equation for epoxy-matrix composites defines a transition velocity of about 10–20 m s^{-1}. In a composite structure the characteristics of the high-velocity impact process are controlled by the mechanical properties of the projectile and structure, the impact conditions (i.e., angle of the trajectory of the projectile), laminate configuration, geometry of the projectile, and plate thickness. Cantwell and Morton (1991) found that high-velocity impacts are more detrimental to carbon fiber reinforced polymer laminates than low-velocity drop-weight impacts. In high-velocity impacts, where the impact duration is in the order of microseconds compared to milliseconds of low-velocity impacts, it is important to estimate the projectile velocity that results or not in the complete perforation of the target. This velocity, called ballistic-limit velocity, is generally expressed in terms of probability as v_{50}, meaning that there is a 50% of probability of perforation and is primarily dependent on laminate thickness and secondarily on projectile shape.

While high-velocity impact tests are usually performed using high pressure gas guns, several techniques have been proposed and used to simulate the low-velocity impact response of composite materials which include the Charpy and Izod pendulums, the falling weight fixtures such as the Gardner and drop dart tests which are specifically designed to perform out-of-plane testing at velocities up to 10 m s^{-1}. Charpy and Izod pendulums were historically developed for testing metals and obtaining information about brittle/ductile transition temperature but were soon after successfully adopted by the plastics industry. The reason for this choice lies in the fact that the Charpy and Izod pendulums are both simple to use and can be instrumented so that it is possible to get information about the process of energy absorption and dissipation in composites. These techniques suffer from several drawbacks including the specimen geometry (a short and thick beam) that is not representative of actual component dimensions and the destructive nature of the test that causes failure modes that are not necessarily observed under low-velocity impact loadings on composite structures. These issues allow these techniques to be best suited for qualitatively ranking the impact response of composites.

A technique that has gained wide acceptance to simulate the impact response of composite laminates is the drop-weight impact test where a weight is allowed to fall from a predetermined height to strike the test specimen supported in the horizontal plane. Usually the impactor is instrumented, thus enabling the force/time characteristics to be determined whilst the incident velocity of the impactor can be obtained through optical sensors located just above the target. The impactor's displacement and energy dissipation during the impact event are determined from the equations of motion. The force that is read by the force tup is the total force on the composite panel (the mass times the acceleration of the force head). The acceleration of the force head is the acceleration of gravity, g, minus the acceleration that the composite panel exerts on the impactor itself. The force that the composite experiences is thus given as

$$F = mg - ma \tag{7.1}$$

where the force on the composite is the force read from the force tup and mg is the force due to gravity of the force head. In Eq. 7.1, the only unknown is the acceleration of the tup, a. Rearranging the equation, the acceleration can be solved for as:

$$a(t) = g - \frac{F(t)}{m} \tag{7.2}$$

By double integration with the suitable boundary conditions it is possible to derive expressions for the velocity and displacement as follows (Eq. 7.3):

$$\begin{aligned} &v(t) = v_0, at \quad t = 0 \\ &x(t) = 0, at \quad t = 0 \\ &v(t) = v_0 + \int_0^t \left(g - \frac{F(t')}{m} \right) dt' \\ &x(t) = 0 + \int_0^t v(t')dt' \end{aligned} \tag{7.3}$$

Usually three different cases can occur during a drop-weight impact test, namely (i) free fall, stop and rebound; (ii) free fall and stop (the corresponding energy threshold is called penetration energy), and (iii) free fall and perforation (the corresponding energy threshold is called perforation energy). In the first case the energy absorbed by the specimen is not high enough and a rebound occurs. Fig. 7.1 shows typical curves for force versus displacement and energy versus time.

The force versus displacement curve (Fig. 7.1A) shows a closed loop and the area inside the loop refers to the energy absorbed during the impact whilst the area under the descending path of the curve represents the elastic energy stored during the contact of the dart with the specimen. This energy is the one responsible for the rebounding of the falling dart. In Fig. 7.1B it can be noticed that energy

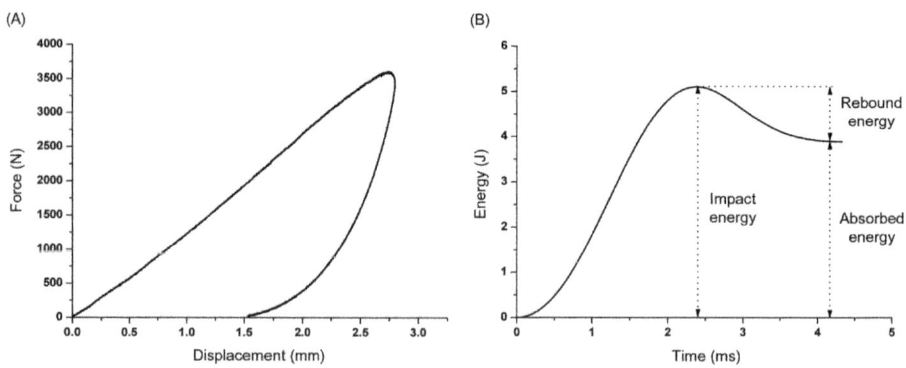

Figure 7.1 (A) Typical force versus displacement and (B) energy versus time response for a nonpenetrating impact event.

progressively grows until the maximum displacement is reached (the maximum energy level is equal to the initial kinetic energy of the impactor) then it decreases until the dart detaches from the plate. From this point on the specimen does not dissipate energy any further, only releasing the residual part of the elastic energy stored during the impact of the dart.

In the second case, the impactor stops without rebounding. When this happens, the specimen has dissipated the whole amount of energy available and penetration has been achieved. In this case the force versus displacement curve does not show a closed loop but it is open and the area enclosed represents the energy absorbed during the impact. In the third case (iii) the dart displacement grows continuously and the velocity does not change sign. The force versus displacement graph (Fig. 7.2) does not show a closed loop any more whilst the energy appears to grow further which is due to the friction of the edges of the perforation hole against the lateral surface of the dart (Belingardi and Vadori, 2002).

One of the main problems associated with this kind of impact test is the lack of established testing parameters for low-velocity impact of composite materials despite the existence of some specific standards. In fact, much of the work published in the literature has been conducted, especially in the past, on purpose-built machines using convenient specimen geometries with parameters rarely in accordance with those suggested by the relevant standards. As a result, direct comparisons between different material systems and researches are often very difficult and immediate conclusions are usually hard to draw.

For the sake of completeness, the determination of the damage resistance of polymer-matrix composite laminated plates subjected to a drop-weight impact event is described in the ASTM D7136 "Standard Test Method for Measuring the Damage Resistance of a Fiber-Reinforced Polymer Matrix Composite to a Drop-Weight Impact Event". This test method can be used to screen materials for damage resistance but it is mainly used to induce damage into a specimen for subsequent

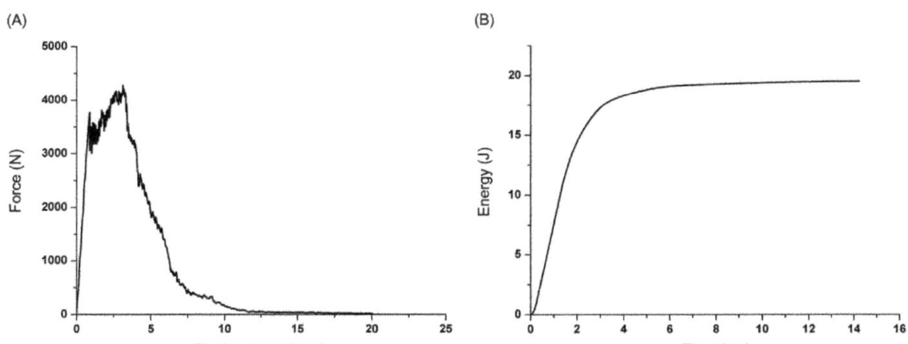

Figure 7.2 (A) Typical force versus displacement and (B) energy versus time response for a perforating impact event.

damage tolerance testing which is usually carried out in compression in accordance with ASTM D7137 "Test Method for Compressive Residual Strength Properties of Damaged Polymer Matrix Composite Plates". The overall test sequence is commonly referred to as the Compression After Impact (CAI) method. In ASTM D7136, due to the subsequent compression test, the specimens of 100 mm × 150 mm are subjected to an out-of-plane, concentrated impact (perpendicular to the plane of the laminated plate) using a drop-weight with a hemispherical striker tip (Fig. 7.3A). The impact support fixture (Fig. 7.3B) has a cut-out in the plate of 75 ± 1 mm by 125 ± 1 mm. The impactor shall have a mass of 5.5 ± 0.25 kg and shall have a smooth hemispherical striker tip with a diameter of 16 ± 0.1 mm. Results are strongly affected by the support fixture cut-out dimensions, material, fixture bending rigidity. The location of the clamps, clamp geometry, clamping force, shape and size of the impactor can all affect the deformation of the specimen during impact in addition to parameters related to the material itself such as thickness, stacking sequence and fiber volume fraction. Another widely used test configuration is the one adapted from the standards related to the puncture tests of rigid plastics (ASTM D5628 "Standard Test Method for Impact Resistance of Flat, Rigid Plastic Specimens by Means of a Falling Dart (Tup or Falling Mass)" and ISO6603 "Determination of puncture impact behavior of rigid plastics. Part 2—Instrumented puncture testing"). These standards use as support fixture the one represented in Fig. 7.3C which, compared to the one in Fig. 7.3b, allows to test specimens of different geometries and dimensions. The relevant standards suggest different dimensions of the support ring and different shapes and dimensions of the impactor. Typical parameters are as follows: impactor with hemispherical tip of diameter 20.0 ± 0.2 mm and inside diameter of the support ring equal to 40.0 ± 2 mm.

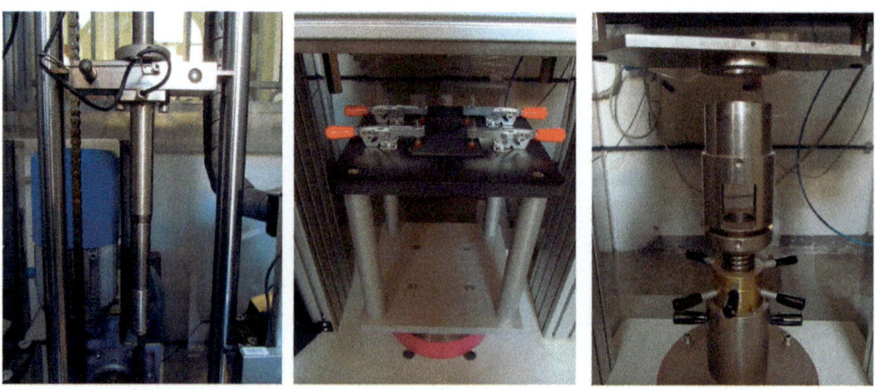
(A) (B) (C)

Figure 7.3 (A) Typical impactor with hemispherical tip; (B) Impact support fixture according to ASTM D7136; (C) typical support ring for instrumented puncture tests.

7.3 Low-velocity impact response of hybrid composites

This paragraph deals with the impact resistance of laminates, which can be broadly defined as the study of damage caused by foreign object impact in a laminate, with a special focus on the role played by hybridization. The understanding of impact damage development and the related failure modes has been achieved through a host of experimental studies along with the effect of several factors that can affect the damage extension. When dealing with low-velocity impact, one should distinguish between nonperforating and perforating impacts. For the former case, experimental investigations point out that damage usually occurs via matrix cracking, delaminations, and fiber failure (in tension fiber breakage and in compression fiber buckling) due to the highly heterogeneous and anisotropic nature of composite laminates (Abrate, 2005; Richardson and Wisheart, 1996).

7.3.1 Modes of failure in low-velocity impact

Among the different failure modes occurring during an impact event delaminations, cracks running in the resin-rich area between plies of different fiber orientation and not between laminae belonging to the same ply group, are of utmost concern as they significantly reduce the strength of the laminate because the laminate is actually subdivided into thinner sublaminates with lower buckling load. This supports the choice of the compression test after impact to assess the residual properties of impacted laminates. In composites the delaminated area has been reported to exhibit a characteristic oblong or "peanut" shape with its major axis coincident with the fiber direction of the layer below the interface. The delaminations are always associated to matrix cracks that represent the first damage appearing after an impact event. Usually two types of cracks are reported: tensile and shear cracks. Shear cracks are created by the high transverse shear stress (linked to the contact force and contact area) through the material, and are inclined at approximately 45 degrees. Once these cracks reach the lower interface, they create an opening force at the interface (mainly mode I) and the delamination grows due to interlaminar out-of-plane shear stress. These delaminations at ply interfaces in turn induce matrix cracking in adjacent plies and therefore the damage progresses from the top (impacted side) to the bottom of the laminate thus leading to what is well known as a "pine-tree pattern". This damage scenario is well summarized in Fig. 7.4 where the traditional pine-tree pattern is readily visible along with the presence of shear cracks and associated delaminations. It is also possible to observe the residual indentation on the impacted side of the laminate.

This scenario is typical of thick laminates where the localized contact stresses are high enough to induce the formation of shear cracks under the point of impact. In thin and flexible laminates, bending stresses cause tensile failure of the bottom ply in the transverse direction, because membrane effects are predominant. This preliminary failure induces debonding of the first interface followed by a sequence of matrix cracks and delaminations that this time propagate upwards, thus leading

Figure 7.4 Micro-CT of a carbon fiber/epoxy composite ($[(0/90)_3/0]_s$) impacted at 10 J.

to a reversed pine-tree pattern. Usually unidirectional composites are not suited for applications where a high impact resistance is desired because they tend to fail by splitting at very low energies and in these case woven fabrics are the material of choice, nonetheless is the strain energy absorbing capacity of the fibers to define the impact resistance of composite laminates. Therefore it can be envisaged the fundamental role played by fiber hybridization in tailoring the impact resistance of the ensuing composites.

7.3.2 Impact resistance and damage tolerance of hybrid composite materials

Table 7.1 reports a literature survey on the impact response of hybrid composites that is meant not to be exhaustive. Two different considerations can be drawn from the table: first of all, the testing parameters differ from one study to another and secondly considerable data have been generated for hybrid composites based on glass fibers because they offer a good compromise in terms of cost, availability and ease of processing. Over the last years other fibers have gained acceptance such as natural fibers, both of mineral (basalt), and vegetal origin (jute, flax, hemp, sisal, etc.).

In hybrid composites the positioning and dispersion of fibers represent two important issues not only for the quasi-static behavior but also for the low-velocity impact response because they can significantly affect the energy absorption mode (Peijs et al., 1990). Usually the hybridization procedure involves the combination of a high strain at failure fiber (H) with a low strain at failure fiber (L), therefore it is not surprising the considerable data on glass (H) and carbon (L) hybrid laminates. A first distinction is to be made between asymmetric and symmetric configurations. Sayer et al. (2010) investigated the impact behavior of asymmetric carbon/glass hybrid composites in order to assess the impact response of both surfaces with glass fibers or with carbon fibers. The impact energies were chosen from approximately 25–75 J, up to the complete perforation of the samples. From the data, it was concluded that the energy absorption capability of hybrid composite impacted on carbon layers was smaller than the hybrid composite impacted on glass layers and that the perforation threshold of the former hybrid composite was approximately 30% higher than that of the latter hybrid composite. The same conclusions were found by Park and Jang (2001) who performed impact tests on aramid/glass hybrid composites with a four-layer asymmetric configuration. In this case the impact velocity was fixed at 4.0 m s^{-1} and the incident impact energy was only 160 J. They found that impact energy and delaminated area of hybrid composites depended on the position of the aramid layer. In particular, when aramid layer (H) was at the back

Table 7.1 Literature survey on low-velocity impact behavior of hybrid composites

Types of fibers	Reinforcement architecture	Shape and dimension of impactor	Type of specimen support	Reference
E-glass/carbon	Unidirectional	Hemispherical ($\varphi = 12.5$ mm)	n/a	Sayer et al. (2010)
S2-glass/carbon	Plain weave fabric	Hemispherical ($\varphi = 16$ mm)	Circular clamping fixture ($\varphi = 76.2$ mm)	Sevkat et al. (2009)
S2-glass/carbon	Plain weave fabric	Charpy-straight line Hemispherical ($\varphi = 25.4$ mm) Hemispherical ($\varphi = 12.7$ mm) 10 mm flat-ended cylindrical	Circular clamping fixture ($\varphi = 76.2$ mm)	Sevkat et al. (2013)
S2-glass/carbon	Satin woven fabric	Hemispherical ($\varphi = 12.7$ mm)	Unsupported area of 127×127 mm^2	Enfedaque et al. (2010)
E-glass/carbon	Plain weave fabric (stitched with Kevlar 49)	Hemispherical ($\varphi = 12.7$ mm)	Circular clamping fixture (φ = n/a)	Onal and Adanur (2002)
E-glass/carbon	Plain and twill weave fabrics	n/a	Unsupported area of 127×127 mm^2	Naik et al. (2001)
S2-glass/carbon	Twill weave fabric	Hemispherical ($\varphi = 12.7$ mm)	Circular clamping fixture ($\varphi = 75$ mm)	Hosur et al. (2004)
E-glass/carbon	Pultruded laminates	n/a	n/a	Kowsika and Mantena (1999)
S2-glass/carbon	Plain and twill weave fabrics	Hemispherical ($\varphi = 12.7$ mm)	Circular clamping fixture ($\varphi = 75$ mm)	Hosur et al. (2005)
E-glass/carbon	Unidirectional	Hemispherical ($\varphi = 12.5$ mm)	n/a	Sayer et al. (2012)
Kevlar 29/spectra	Plain weave fabric	Hemispherical ($\varphi = 17.6$ mm)	Circular clamping fixture ($\varphi = 75$ mm)	Park and Jang (2000)
Kevlar 29/S2-glass	Plain weave fabric	Hemispherical ($\varphi = 17.6$ mm)	Circular clamping fixture ($\varphi = 75$ mm)	Park and Jang (2001)

(*Continued*)

Table 7.1 (Continued)

Types of fibers	Reinforcement architecture	Shape and dimension of impactor	Type of specimen support	Reference
Kevlar 49/carbon	Plain, Satin and Twill weave fabrics	Hemispherical ($\varphi = 12.7$ mm)	Circular clamping fixture ($\varphi = 76.2$ mm)	Gustin et al. (2005)
Nylon, PE, carbon, aramid, PET and E-glass	Plain weave fabric	Hemispherical ($\varphi = 12.7$ mm)	Circular clamping fixture ($\varphi = 75$ mm)	Jang et al. (1989)
E-glass/PE, nylon, PET	Noncrimp fabrics	Hemispherical ($\varphi = 20$ mm)	Circular clamping fixture ($\varphi = 40$ mm)	Thanomsilp and Hogg (2003)
Basalt/nylon 6	Plain weave fabric	Hemispherical ($\varphi = 12.7$ mm)	Unsupported area of 125×75 mm^2	Dehkordi et al. (2010)
Basalt/nylon 6	Plain weave fabric	Hemispherical ($\varphi = 12.7$ mm)	Unsupported area of 125×75 mm^2	Tehrani Dehkordi et al. (2013)
Basalt/E-glass	Plain weave fabric	Hemispherical ($\varphi = 12.7$ mm)	Circular clamping fixture ($\varphi = 40$ mm)	Sarasini et al. (2013b)
Basalt/carbon	Plain weave fabric	Hemispherical ($\varphi = 12.7$ mm)	Circular clamping fixture ($\varphi = 40$ mm)	Sarasini et al. (2014)
Basalt/Twaron	Plain weave fabric	Hemispherical ($\varphi = 12.7$ mm)	Circular clamping fixture ($\varphi = 40$ mm)	Sarasini et al. (2013a)
E-glass/jute	Plain weave fabric	Hemispherical ($\varphi = 12.7$ mm)	Unsupported area of 125×125 mm^2	Ahmed et al. (2007)
E-glass/jute	Plain weave fabric	Hemispherical ($\varphi = 12.7$ mm)	Circular clamping fixture ($\varphi = 40$ mm)	De Rosa et al. (2009)
Flax/carbon	Unidirectional	Hemispherical ($\varphi = 16$ mm)	Unsupported area of 125×75 mm^2	Sarasini et al. (2016)

surface, the composites exhibited higher total impact energy due to the increase of propagation energy, thus implying that the degree of deformation at back surface is a major factor in controlling the absorption mechanism of impact energy. This improvement almost disappeared when aramid fibers were surface treated to increase the adhesion to the vinylester matrix through oxygen plasma treatment and silane coupling agent (γ-methacryloxypropyltrimethoxysilane). This was ascribed to the restriction in deformation of aramid fiber. Jang et al. (1989) investigated the role of various fiber properties, such as strength and ductility, on penetration resistance of hybrid composites. An impact velocity of 4.5 m s^{-1} and an incident energy of 110 J were used throughout this study. Different fibers with a plain weave architecture were used, namely carbon, glass, nylon, aramid, polyester, and polyethylene. For two-layer asymmetric configurations, the results of Jang et al. (1989) seem to contradict the conclusions reported by Sayer et al. (2010) and Park and Jang (2001). It was found that the impact resistance is better when the hybrid laminates are impacted on the surface containing the most ductile fibers, as in the case for carbon/aramid, carbon/polyethylene, and carbon/nylon composites. As regards five layers hybrids, they concluded that the asymmetric hybrids have better impact properties than their alternating sequence counterparts.

As regards symmetric configurations, Enfedaque et al. (2010) and Sevkat et al. (2009) confirmed that symmetric carbon/glass hybrids exhibited better penetration resistance with glass fibers facing the impactor. Enfedaque et al. tested incident impact energies in the range 30–245 J with impact velocity in the range 3.2–4.0 m s^{-1}. Hybridization with S2-glass fibers increased the maximum load under impact (by approximately 10%) and the energy dissipated (by approximately 25% for 21 vol.% of glass fibers). These improvements were mainly ascribed by the authors to the higher ductility of the S2 glass fiber plies located near the top and bottom laminate surfaces. These plies were in fact able to experience higher deformations before fracture and hindered the propagation of damage to the inner plies from the broken plies on the bottom and top surfaces. In addition, S2 glass fibers helped to sustain higher deformations before final failure by the percolation of a through-thickness crack, thus significantly improving the energy dissipated under impact. Sevkat et al. (2009) performed impact tests with energy ranging from 47 to 122 J and velocity from 3.9 to 6.3 m s^{-1}. Hybrid composites were found to delaminate more than nonhybrid ones and delamination in hybrid composites where glass fibers were used as the outer skin was more pronounced compared to that in the other hybrid composite where carbon fibers were used as the face sheet. Among the four lay-up sequences tested, only glass fiber composites showed highest impact resistance, while only carbon fiber composites had the least resistance to impact. Both hybrid laminates performed better than only carbon composites but not as well as only glass composites and, between the two types of hybrids, the ones with glass-skin/carbon-core type performed slightly better compared to the carbon-skin/glass-core type. This behavior was also confirmed by the same authors (Sevkat et al., 2013) when assessing the effect of various impactor geometries on the same hybrid sandwich-like structures. In terms of load sustainment, structures with skins made of glass fibers sustained more load during all tests with blunt impactors such as Charpy and 25.4 mm spherical, and also with impacts by 12.7 mm spherical and 10 mm cylindrical impactors. Also Onal and Adanur (2002) reported that as glass volume fraction increased in the hybrid structures and as glass plies came close to composite surface, impact energies at maximum load increased. The positive effect played by the positioning of high strain fibers in the outer skins of an impacted laminate was also verified in case of repeated impacts (Sevkat et al., 2010). The authors demonstrated that damage accumulation rate could be slowed down through hybridization and that hybrid specimens with glass/epoxy skins survived the double number of successive impacts compared to hybrid specimens with graphite/epoxy skins. These findings appear to be supported also by numerical investigations. Lee et al. (1997) investigated the response of symmetric hybrid composite plates subjected to low-velocity impact using shear deformation theory and concluded that the constituent having good impact resistance should be laid on the impacted surface in hybrid plates. Sarasini et al. (2016) showed that the introduction of flax fiber laminates to be stacked together with carbon fiber ones, apart from environmental benefit at end-of-life, resulted in lower sensitivity of hybrids to nonpenetrating impact events compared to carbon fiber laminates and that better mechanical and impact absorption performances could be reached by specimens with outer flax skins. These improvements were ascribed by the authors to the higher compliance of flax skins that were able to hinder the propagation of cracks generated in the carbon core. Naik et al. (2001) evaluated the impact response of glass/carbon fibers arranged symmetrically in three stacking sequences, namely two clustered configurations (sandwich-like) with glass or carbon outer skins, and one dispersed configuration (sequence of alternating glass/carbon plies). The impact tests were performed at only one impact energy of around 20 J. They found that hybrids exhibited less notch sensitivity compared to reference configurations but it is interesting to note that in this study clustered configurations with outer carbon plies outperformed the other configurations in terms of lower notch sensitivity and higher damage tolerance, which was evaluated in compression. As composite materials are particularly prone to impact damage, a significant loss in both residual strength and structural integrity is to be expected. This aspect is well described in terms of damage tolerance that refers to a system's ability to perform postimpact. Residual properties are usually evaluated in tension, bending and compression, being the last one the most used and also standardized (i.e., ASTM D7137). In an attempt to explore the effect of fiber dispersion on the impact response of composite laminates, Sarasini et al. tested several fiber combinations including basalt/aramid, basalt/carbon and basalt/glass (Sarasini et al., 2014, 2013a,b). Four different configurations were investigated, namely two nonhybridized configurations and two hybrids with a sandwich-like and a well dispersed (intercalated) arrangements. In the sandwich-like lay-up the principle of using the most impact resistant ply as outer skin was adopted. The authors found that for nonpenetrating impacts (5, 12.5, and 25 J), well dispersed configurations offered higher energy absorption and postimpact flexural strength compared to the reference configurations due to multiple small delaminations between dissimilar layers instead of extensive fiber breakages or main delamination in the compression side for basalt/glass hybrids, or main delamination in the tension side at the interface between basalt and aramid layers or transverse crack propagation in carbon "core" of sandwich laminate and secondly main delaminations at the basalt-skin/carbon-core interface. This damage partition mechanism was reported also by Peijs et al. (1990) who investigated hybrids with different degree of hybridization but with a constant total fiber volume fraction and a constant carbon/polyethylene fiber ratio. Despite the different impact testing technique, namely Charpy impact test, they demonstrated an improvement in penetration impact resistance at higher degrees of dispersion of polyethylene and carbon fibers. Park and Jang (2000) compared aramid/polyethylene fiber hybrid composites with interply and intraply arrangements and reported interply hybrid composites to exhibit higher impact energy absorption than intraply hybrid composites due to extensive delamination. On the other hand, the intraply hybrid composites showed lower delaminated area that should lead to a higher damage tolerance.

As a general conclusion, an increased dispersion appears to improve nonpenetrating impact resistance and particularly the residual properties in hybrid composites whilst positioning the fibers with the highest ductility on the outside allows the hybrid composite to absorb more energy.

7.4 Identification of further research areas

Due to the susceptibility of composite laminates to low-velocity impacts, a number of strategies have been proposed and implemented over the years to improve impact resistance and related damage tolerance. A well exploited approach involves the hybridization of two or more fibers in the same matrix which has proven to be promising. However, standardized information and data on impact over a range of strain rates, operating conditions and constituent materials are largely inadequate. This aspect is further complicated by the appearance of new materials with different constituent content and properties as well as architecture. Each of these variations exhibits different impact properties that are difficult to understand and predict. It is therefore not unusual to find in literature quite conflicting or nonconclusive views that are caused most of the time by the lack of standardized procedures and differences in the damage mechanisms, which are generated by dissimilar material properties and fiber/matrix interfacial adhesion. The understanding of this relationship is the challenge to be addressed in the next future if optimization of hybrid composites to impact loading is to be achieved.

References

Abrate, S., 1991. Impact on laminated composite materials. Appl. Mech. Rev. 44, 155–190. Available from: http://dx.doi.org/10.1115/1.3119500.

Abrate, S., 1994. Impact on laminated composites: recent advances. Appl. Mech. Rev. 47, 517–544. Available from: http://dx.doi.org/10.1115/1.3111065.

Abrate, S., 2005. Impact on composite structures. Cambridge University Press.

Abrate, S., Castanié, B., Rajapakse, Y.D.S. (Eds.), 2013. Dynamic failure of composite and sandwich structures, solid mechanics and its applications. Springer, Netherlands, Dordrecht. Available from: http://dx.doi.org/10.1007/978-94-007-5329-7.

Ahmed, K.S., Vijayarangan, S., Kumar, A., 2007. Low velocity impact damage characterization of woven jute glass fabric reinforced isothalic polyester hybrid composites. J. Reinf. Plast. Compos. 26, 959–976. Available from: http://dx.doi.org/10.1177/0731684407079414.

Ashby, M., 2011. Hybrid materials to expand the boundaries of material-property space. J. Am. Ceram. Soc. 94, s3–s14. Available from: http://dx.doi.org/10.1111/j.1551-2916.2011.04559.x.

Ashby, M.F., 2005. Hybrids to fill holes in material property space. Philos. Mag. 85, 3235–3257.

Ashby, M.F., Bréchet, Y.J.M., 2003. Designing hybrid materials. Acta Mater. 51, 5801–5821, doi:10.1016/S1359-6454(03)00441-5.

Belingardi, G., Vadori, R., 2002. Low velocity impact tests of laminate glass-fiber-epoxy matrix composite material plates. Int. J. Impact Eng. 27, 213–229, doi:10.1016/S0734-743X(01)00040-9.

Cantwell, W.J., Morton, J., 1991. The impact resistance of composite materials—a review. Composites. 22, 347–362, doi:10.1016/0010-4361(91)90549-V.

De Rosa, I.M., Santulli, C., Sarasini, F., Valente, M., 2009. Post-impact damage characterization of hybrid configurations of jute/glass polyester laminates using acoustic emission and IR thermography. Compos. Sci. Technol. 69, 1142–1150. Available from: http://dx.doi.org/10.1016/j.compscitech.2009.02.011.

Dehkordi, M.T., Nosraty, H., Shokrieh, M.M., Minak, G., Ghelli, D., 2010. Low velocity impact properties of intra-ply hybrid composites based on basalt and nylon woven fabrics. Mater. Des. 31, 3835–3844. Available from: http://dx.doi.org/10.1016/j.matdes.2010.03.033.

Enfedaque, A., Molina-Aldareguia, J.M., Galvez, F., Gonzalez, C., LLorca, J., 2010. Effect of glass fiber hybridization on the behavior under impact of woven carbon fiber/epoxy laminates. J. Compos. Mater. 44, 3051–3068. Available from: http://dx.doi.org/10.1177/0021998310369602.

Gustin, J., Joneson, A., Mahinfalah, M., Stone, J., 2005. Low velocity impact of combination Kevlar/carbon fiber sandwich composites. Compos. Struct. 69, 396–406. Available from: http://dx.doi.org/10.1016/j.compstruct.2004.07.020.

Hayashi, T., 1972. On the improvement of mechanical properties of composites by hybrid composition, In: 8th International Reinforced Plastics Conference. pp. 149–152.

Hosur, M.V., Abdullah, M., Jeelani, S., 2004. Manufacturing and low-velocity impact characterization of hollow integrated core sandwich composites with hybrid face sheets. Compos. Struct. 65, 103–115. Available from: http://dx.doi.org/10.1016/j.compstruct.2003.10.013.

Hosur, M.V., Adbullah, M., Jeelani, S., 2005. Studies on the low-velocity impact response of woven hybrid composites. Compos. Struct. 67, 253–262. Available from: http://dx.doi.org/10.1016/j.compstruct.2004.07.024.

Isa, M.D., Feih, S., Mouritz, A.P., 2011. Compression fatigue properties of z-pinned quasi-isotropic carbon/epoxy laminate with barely visible impact damage. Compos. Struct. 93, 2269–2276. Available from: http://dx.doi.org/10.1016/j.compstruct.2011.03.015.

Jang, B.Z., Chen, L.C., Wang, C.Z., Lin, H.T., Zee, R.H., 1989. Impact resistance and energy absorption mechanisms in hybrid composites. Compos. Sci. Technol. 34, 305–335, doi:10.1016/0266-3538(89)90002-X.

Kowsika, M.V., Mantena, P.R., 1999. Static and low-velocity impact response characteristics of pultruded hybrid glass-graphite/epoxy composite beams. J. Thermoplast. Compos. Mater. 12, 121–132. Available from: http://dx.doi.org/10.1177/089270579901200203.

Kretsis, G., 1987. A review of the tensile, compressive, flexural and shear properties of hybrid fibre-reinforced plastics. Composites. 18, 13–23, doi:10.1016/0010-4361(87)90003-6.

Lee, Y.-S., Kang, K.-H., Park, O., 1997. Response of hybrid laminated composite plates under low-velocity impact. Comput. Struct. 65, 965–974, doi:10.1016/0045-7949(95)00298-7.

MarketsandMarkets, 2016. Composites Market by Type (Carbon Fiber Composites, Glass Fiber Composites and Others), Resin Type (Thermoplastic Composites and Thermosetting Composites), Manufacturing Process, Application and by Region – Global Trends and Forecasts to 2021.

Mouritz, A.P., 2007. Review of z-pinned composite laminates. Compos. A Appl. Sci. Manuf. 38, 2383–2397. Available from: http://dx.doi.org/10.1016/j.compositesa.2007.08.016.

Naik, N., Ramasimha, R., Arya, H., Prabhu, S., ShamaRao, N., 2001. Impact response and damage tolerance characteristics of glass–carbon/epoxy hybrid composite plates. Compos. B Eng. 32, 565–574, doi:10.1016/S1359-8368(01)00036-1.

Onal, L., Adanur, S., 2002. Effect of stacking sequence on the mechanical properties of glass-carbon hybrid composites before and after impact. J. Ind. Text. 31, 255–271. Available from: http://dx.doi.org/10.1106/152808302028713.

Park, R., Jang, J., 2000. Effect of laminate geometry on impact performance of aramid fiber/polyethylene fiber hybrid composites. J. Appl. Polym. Sci. 75, 952–959, doi:10.1002/(SICI)1097-4628(20000214)75:7<952::AID-APP12>3.0.CO;2-V.

Park, R., Jang, J., 2001. Impact behavior of aramid fiber/glass fiber hybrid composite: evaluation of four-layer hybrid composites. J. Mater. Sci. 36, 2359–2367. Available from: http://dx.doi.org/10.1023/A:1017545528304.

Peijs, A.A.J.M., Venderbosch, R.W., Lemstra, P.J., 1990. Hybrid composites based on polyethylene and carbon fibres, part 3: impact resistant structural composites through damage management. Composites. 21, 522–530, doi:10.1016/0010-4361(90)90425-V.

Reid, S.R., Zhou, G. (Eds.), 2000. Impact Behaviour of Fibre-Reinforced Composite Materials and Structures. Woodhead Publishing Limited.

Richardson, M.O.W., Wisheart, M.J., 1996. Review of low-velocity impact properties of composite materials. Compos. A Appl. Sci. Manuf. 27, 1123–1131, doi:10.1016/1359-835X(96)00074-7.

Robinson, P., Davies, G.A.O., 1992. Impactor mass and specimen geometry effects in low velocity impact of laminated composites. Int. J. Impact Eng. 12, 189–207, doi:10.1016/0734-743X(92)90408-L.

Sarasini, F., Tirillò, J., Valente, M., Ferrante, L., Cioffi, S., Iannace, S., et al., 2013a. Hybrid composites based on aramid and basalt woven fabrics: Impact damage modes and residual flexural properties. Mater. Des. 49, 290–302. Available from: http://dx.doi.org/10.1016/j.matdes.2013.01.010.

Sarasini, F., Tirillò, J., Valente, M., Valente, T., Cioffi, S., Iannace, S., et al., 2013b. Effect of basalt fiber hybridization on the impact behavior under low impact velocity of glass/basalt woven fabric/epoxy resin composites. Compos. A Appl. Sci. Manuf. 47, 109–123. Available from: http://dx.doi.org/10.1016/j.compositesa.2012.11.021.

Sarasini, F., Tirillò, J., Ferrante, L., Valente, M., Valente, T., Lampani, L., et al., 2014. Drop-weight impact behaviour of woven hybrid basalt-carbon/epoxy composites. Compos. B Eng. 59, 204–220. Available from: http://dx.doi.org/10.1016/j.compositesb.2013.12.006.

Sarasini, F., Tirillò, J., D'Altilia, S., Valente, T., Santulli, C., Touchard, F., et al., 2016. Damage tolerance of carbon/flax hybrid composites subjected to low velocity impact. Compos. B Eng. 91, 144–153. Available from: http://dx.doi.org/10.1016/j.compositesb.2016.01.050.

Sayer, M., Bektaş, N.B., Sayman, O., 2010. An experimental investigation on the impact behavior of hybrid composite plates. Compos. Struct. 92, 1256–1262. Available from: http://dx.doi.org/10.1016/j.compstruct.2009.10.036.

Sayer, M., Bektaş, N.B., Demir, E., Çallioğlu, H., 2012. The effect of temperatures on hybrid composite laminates under impact loading. Compos. B Eng. 43, 2152–2160. Available from: http://dx.doi.org/10.1016/j.compositesb.2012.02.037.

Sevkat, E., Liaw, B., Delale, F., Raju, B.B., 2009. Drop-weight impact of plain-woven hybrid glass–graphite/toughened epoxy composites. Compos. A Appl. Sci. Manuf. 40, 1090–1110. Available from: http://dx.doi.org/10.1016/j.compositesa.2009.04.028.

Sevkat, E., Liaw, B., Delale, F., Raju, B.B., 2010. Effect of repeated impacts on the response of plain-woven hybrid composites. Compos. B Eng. 41, 403–413. Available from: http://dx.doi.org/10.1016/j.compositesb.2010.01.001.

Sevkat, E., Liaw, B., Delale, F., 2013. Drop-weight impact response of hybrid composites impacted by impactor of various geometries. Mater. Des. 52, 67–77. Available from: http://dx.doi.org/10.1016/j.matdes.2013.05.016.

Short, D., Summerscales, J., 1979. Hybrids – a review. Part 1. Techniques, design and construction. Composites. 10, 215–222, doi:10.1016/0010-4361(79)90022-3.

Short, D., Summerscales, J., 1980. Hybrids – a review. Part 2. Physical properties. Composites. 11, 33–38, doi:10.1016/0010-4361(80)90019-1.

Silberschmidt, V.V., 2016. Dynamic Deformation, Damage and Fracture in Composite Materials and Structures. Woodhead Publishing Limited.

Sjoblom, P.O., Hartness, J.T., Cordell, T.M., 1988. On low-velocity impact testing of composite materials. J. Compos. Mater. 22, 30–52. Available from: http://dx.doi.org/10.1177/002199838802200103.

Summerscales, J., Short, D., 1978. Carbon fibre and glass fibre hybrid reinforced plastics. Composites. 9, 157–166, doi:10.1016/0010-4361(78)90341-5.

Swolfs, Y., Gorbatikh, L., Verpoest, I., 2014. Fibre hybridisation in polymer composites: a review. Compos. A Appl. Sci. Manuf. 67, 181–200. Available from: http://dx.doi.org/10.1016/j.compositesa.2014.08.027.

Tehrani Dehkordi, M., Nosraty, H., Shokrieh, M.M., Minak, G., Ghelli, D., 2013. The influence of hybridization on impact damage behavior and residual compression strength of intraply basalt/nylon hybrid composites. Mater. Des. 43, 283–290. Available from: http://dx.doi.org/10.1016/j.matdes.2012.07.005.

Thanomsilp, C., Hogg, P., 2003. Penetration impact resistance of hybrid composites based on commingled yarn fabrics. Compos. Sci. Technol. 63, 467–482, doi:10.1016/S0266-3538(02)00233-6.

Hybrid carbon nanotube/fiber thermoplastic composites: mechanical, thermal, and electrical characterization

8

Ana M. Díez-Pascual
Analytical Chemistry, Physical Chemistry and Chemical Engineering Department, Faculty of Biology, Environmental Sciences and Chemistry, Alcalá University, Madrid, Spain

Chapter Outline

8.1 Introduction

Fiber-reinforced polymer composites (FRPs) possess outstanding mechanical properties, including high modulus and strength, which combined with their corrosion resistance, easy of handling and their light weight make them useful for a wide range of applications in aeronautic (Irving and Soutis, 2015), sporting goods (Mallick, 2008), automobile (Marsh, 2003), marine (Mouritz et al., 2001), civil infrastructures (Hollaway, 2003) and the energy sector, e.g., wind turbine blades (Brondsted et al., 2005) and underground oil-drilling (Moritis and Hub, 2003). Despite these composites exhibit optimal in-plane and fiber dominant properties, their out-of-plane properties are restricted by the poor matrix properties and the weak fiber–matrix interfacial adhesion (Khan and Kim, 2011). In order to improve the through-the-thickness properties and reduce the risk of crack initiation and

Hybrid Polymer Composite Materials: Properties and Characterisation. DOI: http://dx.doi.org/10.1016/B978-0-08-100787-7.00008-1

propagation along the laminar structure, which may result in a catastrophic failure of the whole structure, several approaches have been developed, such as toughening the matrix resin with elastomers (Kroll et al., 2013) and interleaving with tough polymers, short fibers, or metal layers (Hojo et al., 2006; Sadighi et al., 2012). Besides, delamination can be reduced by incorporating different through-thickness reinforcing strategies such as 3D textiles (Tong et al., 2002), z-pinning (Mouritz, 2007), stitching (Mouritz et al., 1997), embroidery (Tong et al., 2002) braiding (Ayranci and Carey, 2008; Fujihara et al., 2004), etc. Nonetheless, such strategies do not enhance the fiber–matrix interfacial adhesion, and improve the interlaminar fracture toughness and impact damage tolerance at the cost of reducing the in-plane mechanical properties by damaging the fibers and inducing fiber waviness (Khan and Kim, 2011). Conversely, the fiber–matrix interface can be strengthened via different techniques like fiber surface treatments (Mallick, 2008; Jones, 2010; Shubhra et al., 2011), sizings (Jones, 2010), and coupling agents (Shubhra et al., 2011; Ji-Nian and Ai-Qin, 2013), although these procedures are expensive since have to be particularly optimized prior to the design step and may be difficult to scale up. Therefore, there is growing interest in the development of hierarchical (or hybrid or nanostructured or multiscale) composites, in which a nanoscale carbon nanotube (CNT) reinforcement is used together with conventional microscale reinforcing fibers.

CNTs, one dimensional carbon-based nanomaterials (Iijima, 1991), are one of the most efficient nanofillers to reinforce polymer matrices. They possess extraordinary high Young's modulus and tensile strength, combined with very large aspect ratio, high flexibility and low density (Popov, 2004). Compared with other nanoscale reinforcements like SiC nanowhiskers or inorganic nanoparticles, which have been grown on fibers for improving the fiber/matrix interface (Wu et al., 2004), CNTs possess smaller diameters, lower density, better alignment and mechanical properties, higher surface area and a lower coefficient of thermal expansion (CTE). CNTs can be divided into two groups: single-walled carbon nanotubes (SWCNTs), consisting of a single graphite sheet wrapped into a cylindrical tube with a diameter in the range of 0.7–3 nm, and multiwalled carbon nanotubes (MWCNTs), composed of more than two coaxial cylinders, each rolled out of single sheets, with diameters between 2 and 40 nm. Five methods of synthesis of CNTs have been reported (Dervishi et al., 2009): arc-discharge, laser ablation, chemical vapour deposition (CVD), the HIPCO process and surface mediated growth of vertically aligned tubes. CVD materials usually contain residual catalyst particles, while the main contaminants in the arc-discharge and laser ablation process are carbonaceous impurities. CNTs synthesized by the CVD technique typically have a large number of defects while those produced by the laser process possess higher quality (few defects, high crystallinity, very high aspect ratio). The purity, quality, aspect ratio and nature of impurities have a profound effect on the final properties of CNT-reinforced composites (Díez-Pascual et al., 2009).

The addition of CNTs to FRPs reduces the limitations associated with the matrix-dominated properties. CNTs provide both intra- and interlaminar reinforcement, thus improving delamination resistance and through-thickness properties. A number of methods have been reported to incorporate CNTs into FRPs: infusion/impregnation of a CNT–resin mixture into the primary fiber assembly via resin transfer molding (RTM) or vacuum-assisted resin transfer molding (VARTM) (Gojny et al., 2005; Qiu et al., 2007), direct growth of CNTs onto fibers by chemical or thermal vapour deposition (CVD and TVD) (Rahmanian et al., 2013) or via graphitic structures by design (GSD) (Tehrani et al., 2013), electrophoretic deposition (EPD) onto the surface of fabric layers (Qian et al., 2010a; Bekyarova et al., 2007), direct placement of CNTs between layers of the preform (Abot et al., 2008), reactions between functionalized CNTs and fibers (Qian et al., 2010a), coating of primary fibers with CNT-modified sizing agents (Barber et al., 2004; Mäder et al., 2008; Rausch and Mäder, 2010a; Rausch and Mäder, 2010b) and electrostatic assembly of oxidized CNTs onto functionalized fibers (Zhang et al., 2012). Using the aforementioned approaches, the delamination resistance and out-of-plane properties were found to improve through interactions between the propagating cracks and the CNTs (Gojny et al., 2005). Besides, the addition of CNTs generally increases the glass transition temperature (T_g) and decrease the CTE (Warrier et al., 2010).

The abovementioned methods have been applied mainly to develop carbon fiber (CF)-reinforced composites with CNTs, even if some studies have also been published on glass fiber (GF)-reinforced plastics (Zhang et al., 2012; Barber et al., 2004; Mäder et al., 2008; Rausch and Mäder, 2010a; Rausch and Mäder, 2010b; Wichmann et al., 2006; Díez-Pascual et al., 2011; Ashrafi et al., 2012; Shen et al., 2009; Díez-Pascual et al., 2012, 2013; Zhu et al., 2007). Moreover, only a few studies (Bekyarova et al., 2007; Barber et al., 2004; Díez-Pascual et al., 2011, 2012, 2013; Ashrafi et al., 2012; Zhu et al., 2007; Thakre et al., 2011) have employed SWCNTs in hierarchical composites, given that they are more difficult to disperse and more expensive than MWCNTs. Most of the articles dealing with hierarchical composites comprise a thermoset resin, while works with a thermoplastic matrix are still limited (Rahmanian et al., 2013; Zhang et al., 2012; Barber et al., 2004; Mäder et al., 2008; Rausch and Mäder, 2010a; Rausch and Mäder, 2010b; Díez-Pascual et al., 2011, 2012, 2013; Ashrafi et al., 2012; Shen et al., 2009; Qian et al., 2010b, 2010c; Suraya et al., 2009; Hapuarachchi and Peijs, 2010; Li and Bai, 2011; Zhang, 2011; Meszaros et al., 2011; Ruiya et al., 2011) because processing is a foremost challenge due to their high viscosity. In this chapter, recent literature on hierarchical thermoplastic-based composites incorporating both CNTs and conventional fibers will be summarized, and their morphology, thermal, mechanical and electrical properties are presented. Examples have been chosen to show the importance of mixing different scale constituents for developing new materials with enhanced properties due to synergistic effects. Finally, potential applications and future perspectives of this new generation of multifunctional high-performance composites are discussed.

8.2 Manufacture of multiscale composites based on a CNT-reinforced thermoplastic matrix

The simplest and more widely used method to manufacture hierarchical composites is by mechanically or ultrasonically shear-mixing CNTs into the matrix and then infusing or impregnating the primary fiber stack using conventional techniques such as RTM or VARTM (Gnidakouong et al., 2013). This approach is scalable at an industrial level but it is limited to short CNTs at low volume fractions (≤2 vol.%). Higher volume fractions may cause a strong increase in the viscosity of the matrix and void formation, which may result in incomplete fiber wetting (Gojny et al., 2005), or CNT agglomeration/depletion in different areas of the fabric. The development of hierarchical composites by impregnating microscale primary fibers with CNT-modified thermoplastics is restricted to polymers with low viscosity. Moreover, matrix flow during impregnation tends to align the CNTs parallel to the fiber direction, which hardly improves the through-thickness properties. In the case of high-viscosity polymers, hot-press processing is generally used to impregnate the fibers with the matrix/CNT mixture (Díez-Pascual et al., 2011; Shen et al., 2009; Hapuarachchi and Peijs, 2010). This procedure has been applied to fabricate GF-reinforced poly(ether ether ketone) (PEEK)/SWCNT laminates (Díez-Pascual et al., 2011; Ashrafi et al., 2012) incorporating polyetherethersulfone (PEES) as a compatibilizing agent. Thin PEEK/SWCNT films (~500 μm thick) were prepared in a hot-press (Fig. 8.1A) and subsequently 4 of these films were alternatively placed within 5 GF fabric plies (Fig. 8.1B). Consolidation of the composite was carried out in a hot-press under high pressure (Fig. 8.1C). Analogous procedure was employed by Shen et al. (2009) for the preparation of polyamide-6 (PA-6)/MWCNT/GF composites.

In order to prepare composites with higher CNT volume fractions, CNTs can be grown or grafted directly onto the surfaces of the primary fibers. The grafting

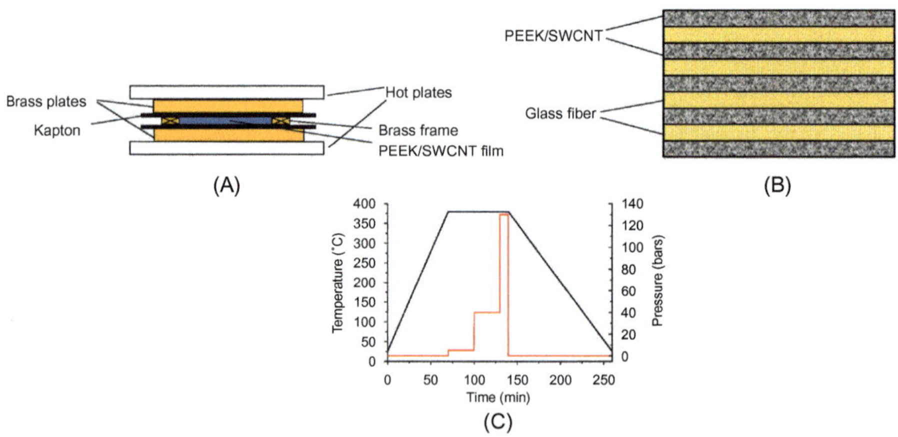

Figure 8.1 (A) Schematic representation of the hot-press used for the preparation of the PEEK/SWCNT films. (B) Scheme of the lay-up stacking sequence of the PEEK/SWCNT/GF hierarchical composites. (C) Consolidation cycle for composite manufacturing.
Source: From Díez-Pascual, A.M., Ashrafi, B., Naffakh, M., González-Domínguez, J.M., Johnston, A., Simard, B., et al., 2011. Influence of carbon nanotubes on the thermal, electrical and mechanical properties of poly(ether ether ketone)/glass fiber laminates. Carbon 49, 2817–2833, with permission from Elsevier.

process can be carried out via CVD (Qian et al., 2010b, 2010c; Agnihotri et al., 2011) in the presence of a metal catalyst (i.e., Ni or Fe) that can be deposited by different methods like solution impregnation, electron beam, sputtering, etc. The CNTs are grown in a quartz tube reactor at low-temperature CVD (550–900°C) to yield MWCNTs or high-temperature (900–1200°C) reaction that favors SWCNT growth, using C_2H_2 as the hydrocarbon source and employing a H_2/Ar mixture as the carrier gas. Using this procedure, Qian et al. prepared poly(methyl methacrylate) (PMMA) based composites incorporating CNTs grown onto CFs (Qian et al., 2010b) and silica fibers (Qian et al., 2010c), and explored the influence of the CNT grafting on the wetting between CFs and the polymer matrix. Agnihotri et al. (2011) manufactured CNT-coated CF/polyester composites, investigating the effect of the run time of the CVD reactor on the length and quantity of grown CNTs. Similarly, Rahmanian et al. (2013) grew MWCNTs onto short CFs and GFs by feeding a vaporized solution of benzene and ferrocene into a CVD reactor, and the CNT-grafted fibers were blended with polypropylene (PP) in a mixer, and Naito grew CNTs on CFs using ferrocene as catalyst and deposited a polyimide (PI) nanolayer coating on the surface of the CFs by high-temperature vapor deposition polymerization (VDPH) (Naito, 2014). This CVD approach enables to introduce high loadings of aligned CNTs although is difficult to scale up and the use of metal catalysts is not desirable. More importantly, temperatures higher than 750°C can induce CF degradation. Recently, injection CVD (ICVD) technique has been employed to grow the CNTs on the fibers via a pyrolysis of solutions containing a catalyst precursor and a hydrocarbon source (Mathur et al., 2008). The ICVD technique leads to better orientation and growth of longer CNTs compared to the conventional CVD approach.

In order to reduce the negative effect of the elevated temperatures on the CFs during the CVD process, a variant of the former method named Graphitic Structure by Design (GSD), has been developed for growing CNTs over CFs at 550°C under atmospheric pressure (Tehrani et al., 2013). The process is rapid, inexpensive and scalable, in which CNTs are grown from fuel mixtures using nickel particles as catalyst instead of toxic hydrocarbons or catalysts commonly employed in the CVD approach. GSD also allows growing the CNTs in determined locations in which the catalyst is pre-deposited, whereas with the CVD technique, CNTs grow everywhere.

Another approach is the dispersion of the CNTs in the sizing of the fibers, which protects the fiber surface and improves stress transfer at the interface. Following this method, Rausch and Mäder fabricated MWCNT-coated GF yarns embedded in a polypropylene (PP) matrix (Mäder et al., 2008; Rausch and Mäder, 2010a; Rausch and Mäder, 2010b). The coated GFs were embedded in PP by compression moulding or by simultaneous melt-spinning and co-mingling. With the same technique, Barber et al. (2004) prepared isotactic polypropylene (iPP)/SWCNT/GF composites. GFs were submerged into a sizing-SWCNT solution, and subsequently placed between two iPP sheets and hot-pressed. The major drawbacks of this procedure are the lack of control of the CNT orientation and the weak attachment of the CNTs to the fibers due to the absence of chemical bonding.

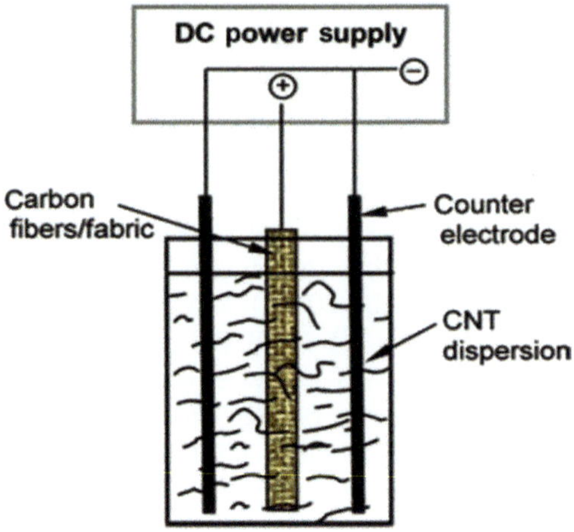

Figure 8.2 Schematic representation of the electrophoretic deposition (EPD) process.

A different method consists in the selective deposition of CNTs onto the surface of fabric layers by EPD (Bekyarova et al., 2007). In this approach CNTs are dispersed in an aqueous media between two negative electrodes to charge the CNTs (Fig. 8.2). The carbon fabric is then immersed in the CNT doped media and held between two steel plates connected to a positive charge. Due to the electric potential, the CNTs are deposited onto the carbon fabric and the CNT-CF preforms are then infused with the matrix (i.e., via VARTM) or staked with the polymer layers and then consolidated by hot-compression. This is a practical, scalable and cost-effective method albeit presents similar shortcomings to the fiber sizing approach. Otherwise, negatively charged CNTs synthesized through oxidation treatment are deposited on positively charged functionalized GFs by electrostatic assembly (Zhang et al., 2012), and the thickness of the adsorbed CNT layer can be tailored by adjusting the process conditions. Another approach consists in spraying CNTs directly onto the fibers followed by VARTM or hot-press processing (Zhu et al., 2007). The drawback of this technique is the lack of control over the CNT orientation.

8.3 Characterization of the multiscale composites

8.3.1 Surface morphology

The morphological characterization of this type of composites is usually performed by imaging techniques such as scanning electron microscopy (SEM), which provides information about the dispersion and orientation of the CNTs, and can be indicative of the wetting of the fibers by the matrix. The majority of the studies did not report the CNT orientation, though a random arrangement is the most frequent. SEM has been used to assess the surface morphology of noncompatibilized and compatibilized PEEK/SWCNT (1.0 wt%)/GF laminates (Fig. 8.3A–D)

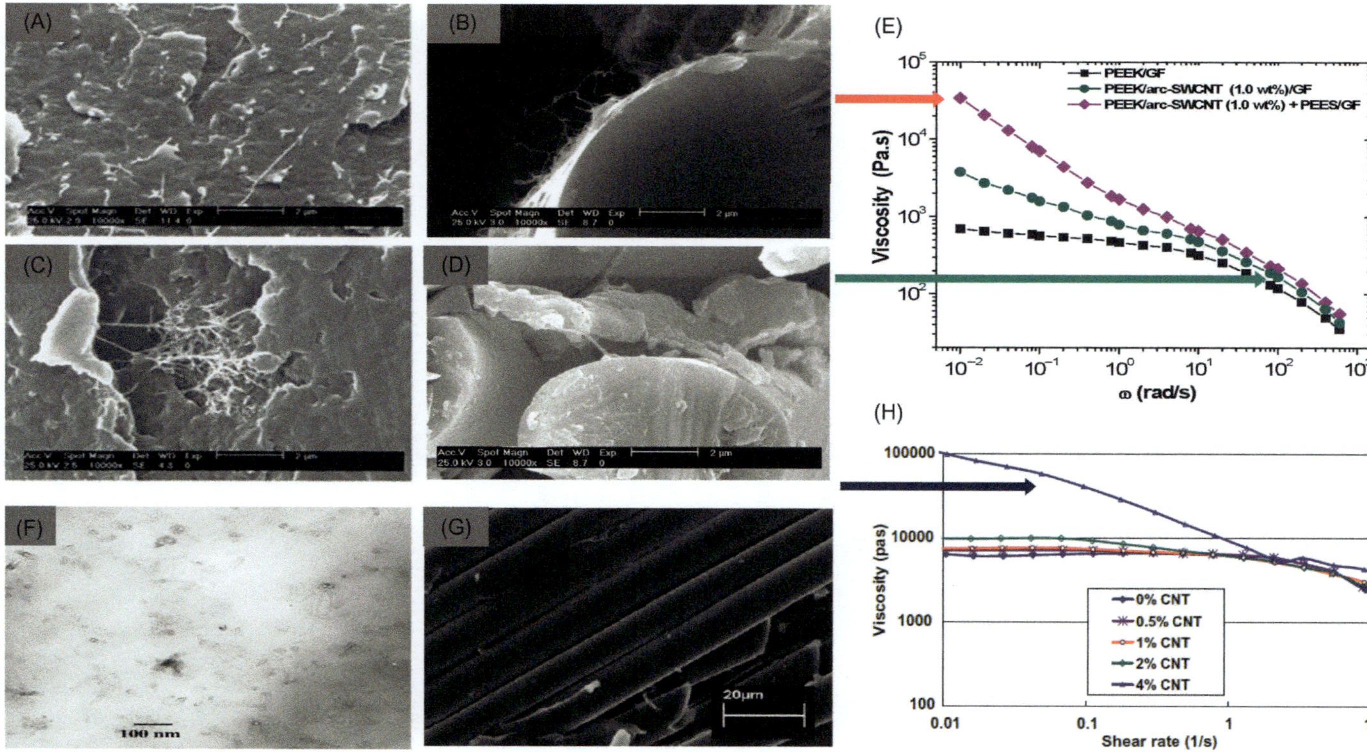

Figure 8.3 SEM micrographs of the cross-section of: (A) and (B) PEEK/laser-SWCNT (1.0 wt%) + PEES (compatibilizer)/GF; (C) and (D): PEEK/laser-SWCNT (1.0 wt%)/GF. (A) and (C) correspond to a matrix-rich region, whilst (B) and (D) to a region within the fiber tows. *Source*: Adapted from Ashrafi, B., Díez-Pascual, A.M., Johnson, L., Genest, M., Hind, S., Martinez-Rubi, Y., et al., 2012, with permission from Elsevier. (E) Melt-viscosity of compatibilized and noncompatibilized PEEK based multiscale laminates. *Source*: Adapted from Díez-Pascual, A.M., González-Domínguez, J.M., Martinez, M.T., Gomez-Fatou, M.A., 2013, with permission from Elsevier. (F) TEM photo of PA-6/MWCNT (4.0 wt%) composite. (G) SEM image of PA-6/MWCNT (4.0 wt%)/GF laminate. (H): Melt-viscosity of PA-6/MWCNT nanocomposites. *Source*: Adapted from Shen, Z., Bateman, S., Wu, D.Y., McMahon, P., Dell'Olio, M., Gotama, J., 2009, with permission from Elsevier.

(Ashrafi et al., 2012). A homogeneous SWCNT dispersion in PEEK and a good fiber impregnation was found for composites with the compatibilizing agent (Figs. 8.3A and B), whereas SWCNT aggregates that did not penetrate the fiber tows were observed in the noncompatibilized composites (Figs. 8.3C and D). In another work (Díez-Pascual et al., 2013), the authors correlated the morphology with the melt-viscosity (η) of the composites (Fig. 8.3E). The addition of SWCNTs to PEEK/GF caused an increase in η, the rise being stronger for the compatibilized laminates. However, these composites showed better fiber wetting, likely because the nanotubes are acid functionalized, hence can interact both the GF (through H-bonding) and with the matrix (via $\pi-\pi$ stacking), and these interactions overweigh the rise in viscosity.

The morphology of PA-6/MWCNT (0.5−4.0 wt%)/GF laminates was also investigated by SEM and transmission electron microscopy (TEM) (Shen et al., 2009). A good MWCNT dispersion was observed for all the composites (Fig. 8.3F). Samples with low MWCNT loadings (≤2.0 wt%) exhibited complete fiber impregnation, whilst that with 4.0 wt% displayed poor wetting (Fig. 8.3G), ascribed to the high viscosity of this composite (Fig. 8.3H) that hindered the flow of the matrix during processing. In this study unmodified MWCNTs were used as nanoreinforcement, which do not interact with the GF nor with the matrix. Therefore, it can be concluded that the interfacial interactions between the laminate components play an important role on the fiber wetting, hence on the final composite properties.

The topography of PP/MWCNT (0.5 wt%)/GF composites has also been examined by SEM (Fig. 8.4) (Rausch and Mäder, 2010a). The GF surface is inhomogenously covered by the CNTs dispersed in a PP-film former, showing an "island-in-the-sea" topography with cluster-like structures (Fig. 8.4A and B). To attain good electrical properties, a more homogenous coating layer is required, which can be attained by annealing. Upon annealing the GF surfaces at 200°C for 15 min, these are appropriately wetted by the molten PP (Fig. 8.4C). The high magnification image (Fig. 8.4D) reveals that the MWCNTs are homogeneously distriuted within the matrix, resulting in the formation of a network which could be used as a strain sensor to measure the GF−matrix interfacial adhesion.

8.3.2 Thermal properties

8.3.2.1 Crystallization and melting behavior

The crystallization process greatly influences on the fabrication of thermoplastic-based composites and plays a key role in the final properties of the material. However, very scarce information regarding the crystallization of hierarchical laminates have been published. Fig. 8.5 presents DSC thermograms of PEEK/SWCNT/GF laminates (Díez-Pascual et al., 2011). Composites with 1.0 wt% SWCNTs showed lower crystallization temperature (T_c) and degree of crystallinity (X_c) than the reference PEEK/GF, and this drop was more pronounced in composites with the compatibilizer PEES (an amorphous polymer). Conversely, both parameters are similar to those of the reference in the composite with 0.5 wt%

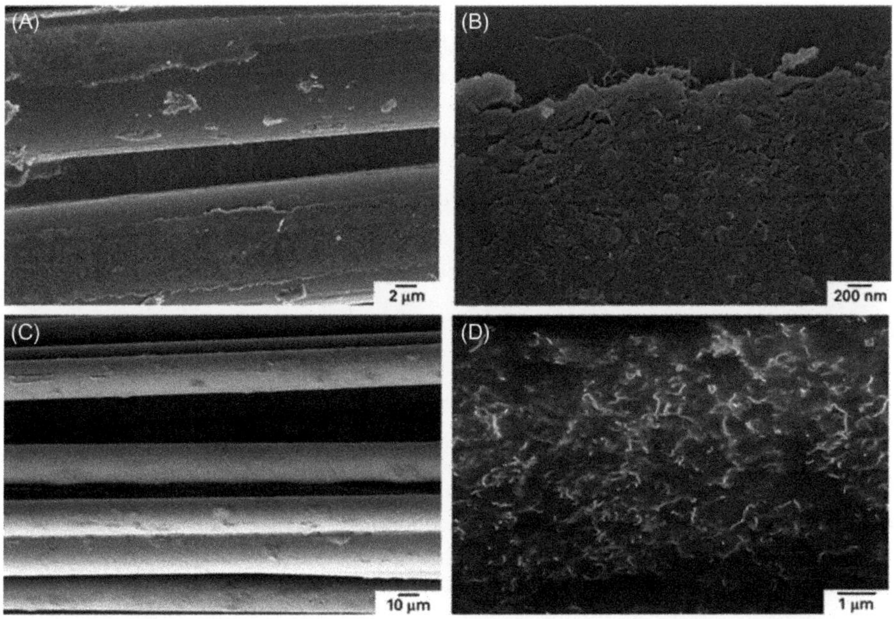

Figure 8.4 SEM images of GF coated with MWCNTs (0.5 wt%) dispersed within PP-film former: (A) as-coated low magnification micrograph; (B) as-coated high magnification image; (C) after annealing at 200°C for 15 min; (D) high magnification charge contrast micrograph after annealing.
Source: Adapted from: Rausch, J., Mäder, E., 2010a. Health monitoring in continuous glass fiber reinforced thermoplastics: Manufacturing and application of interphase sensors based on carbon nanotubes. Compos. Sci. Technol. 70, 1589–1596, with permission from Elsevier.

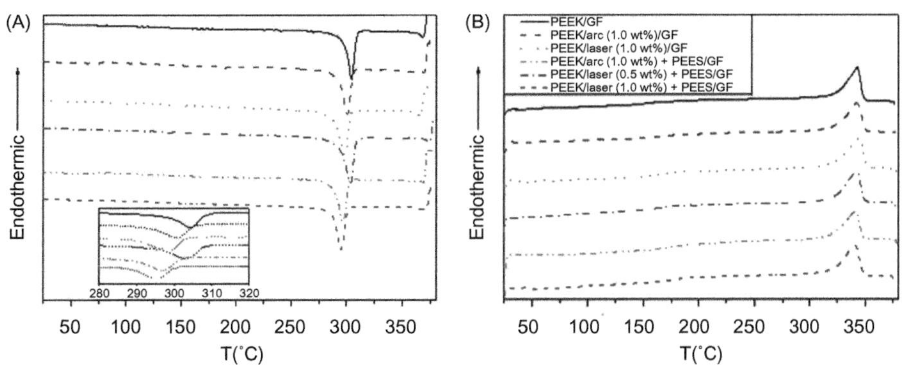

Figure 8.5 DSC thermograms of PEEK/SWCNT/GF laminates: (A) cooling scans; (B) heating scans.
Source: From Díez-Pascual, A.M., Ashrafi, B., Naffakh, M., González-Domínguez, J.M., Johnston, A., Simard, B., et al., 2011. Influence of carbon nanotubes on the thermal, electrical and mechanical properties of poly(ether ether ketone)/glass fiber laminates. Carbon 49, 2817–2833, with permission from Elsevier.

wrapped SWCNTs. Two factors are known to influence the crystallization of polymer composites (Díez-Pascual et al., 2009): the nucleating effect of the fillers, that causes an increase in T_c and X_c, and their barrier effect, since they can obstruct the diffusion and migration of the polymer segments, thereby hampering crystallization. In the case of composites with 0.5 wt% SWCNTs, the nucleating effect should compensate for the restrictions in chain mobility, so T_c and X_c remain almost unchanged. Nonetheless, at higher concentrations, the formation of a SWCNT network confined the crystal growth, and these restrictions in mobility overweigh the nucleating effect, provoking a drop in both parameters. Moreover, it was found that the confinement effect was stronger in the composites with PEES, given that the nanotubes wrapped in the compatibilizer can interact strongly with the matrix. Regarding the effect of the CNT type, composites with arc-discharge SWCNTs exhibited somewhat higher crystallinity than those with laser-grown SWCNTs, ascribed to the higher degree of disentanglement and debundling of the arc-SWCNTs (Díez-Pascual et al., 2011). One the other hand, very small changes were observed in the melting temperature (T_m) upon addition of the SWCNTs to PEEK/GF.

In contrast to the aforementioned results, an increase in the crystallization rate of the matrix was found for PP/MWCNT (0.04–0.2 wt%)/GF (Mäder et al., 2008) hybrids prepared by dispersing the MWCNTs in the fiber sizing, since the onset of crystallization was shifted to shorter times. An increase in the degree of crystallinity was also observed, indicating a change in the composite morphology. Nevertheless, the MWCNT concentration had very small influence on the crystallization kinetics, pointing that a saturation of the nucleation density is achieved at very low MWCNT contents. The MWCNTs acted as heterogeneous nucleation sites on the fiber surface, hindering the lateral growth of crystallites. This is an effective way to provoke the growth of a transcrystalline layer around the GFs, which has been related to better elastic/mechanical composite properties.

8.3.2.2 *Thermal conductivity*

Taking into account that CNTs exhibit a high thermal conductivity (λ) of 200–3000 W mK^{-1} at room temperature (Yang et al., 2004), their incorporation into polymers should lead to enhanced thermal conductivity leading to better heat transport, making them suitable for applications such as printed circuit boards, connectors, thermal interface materials, heat sinks, etc. Fig. 8.6 compares room temperature λ values for CNT-reinforced PEEK (Díez-Pascual et al., 2011) and PA-6 (Shen et al., 2009), multiscale composites manufactured by extrusion and hot-compression. It is worthy to mention that the increments in λ in the laminates are smaller than those of the corresponding polymer/CNT composites. λ of PEEK/GF was found to be ~0.22 W m^{-1} K^{-1}, similar to that of the pure resin. The addition of 1.0 wt% nonwrapped arc and laser-grown SWCNTs increased this value by about 20 and 55%, respectively, whereas for the composites with the same SWCNT content wrapped in PEES the increments were ~33 and 93%. This confirms that the

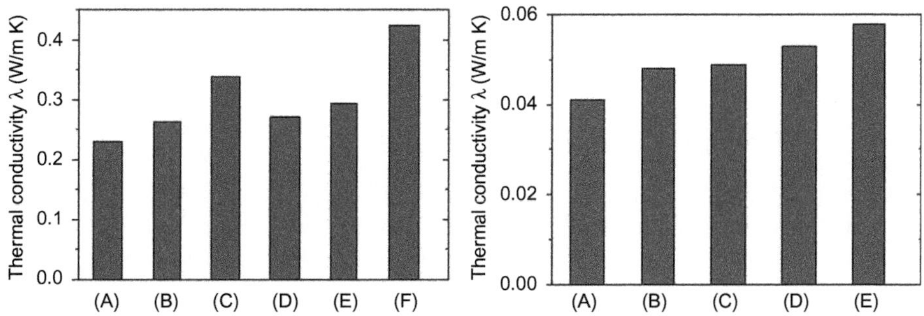

Figure 8.6 *Left*: Room temperature out-of-plane thermal conductivity of PEEK laminates: (A) PEEK/GF; (B) PEEK/arc (1.0 wt%)/GF; (C) PEEK/laser (1.0 wt%)/GF; (D) PEEK/laser (0.5 wt%) + PEES/GF; (E) PEEK/arc (1.0 wt%) + PEES/GF; (F) PEEK/laser (1.0 wt%) + PEES/GF. *Source*: From Ref. Díez–Pascual, A.M., Ashrafi, B., Naffakh, M., González-Domínguez, J.M.,Johnston, A., Simard, B., et al., 2011, with permission from Elsevier. *Right*: thermal conductivity of PA-6 laminates with different MWCNT content (wt%): (A): 0; (B) 0.5; (C) 1.0; (D): 2.0; (E): 4.0. *Source*: From Ref. Shen, Z., Bateman, S., Wu, D.Y., McMahon, P., Dell'Olio, M., Gotama, J., 2009, with permission from Elsevier.

thermal conductivity depends on the type and quality (defect content, metallic impurities) of the nanotubes and their state of dispersion within the matrix. With regard to the SWCNT type, laminates with laser-grown SWCNTs exhibited higher conductivity due to their higher quality (few defects, high crystallinity). Regarding the compatibilizer, it wraps the nanotubes, hampering their direct contact, which should decrease λ. However, the better SWCNT dispersion in the presence of the compatibilizer overweighs this effect, leading to higher conductivity. Further, λ rises with increasing SWCNT concentration, causing ~50% enhancement when the concentration was doubled.

The λ increments found for PA-6/MWCNT/GF laminates were considerably lower than those of PA-6/MWCNT binary nanocomposites (Shen et al., 2009). This behavior can be explained considering that the laminates present some voids that restrict the heat transfer (λ_{air} ~0.024 W m^{-1} K^{-1}). Further, the laminates comprise 10 plies of GF alternated with 11 plies of PA-6/MWCNT films, hence the weight fraction of MWCNTs in the laminates is approximately half of that in the nanocomposites. In addition, the orientation of the CNTs could have changed during the fabrication of the laminates, and this factor influences the thermal conductivity. However, for both the binary composites and the laminates, λ rose gradually with increasing MWCNT content. Thus, the PA-6/GF laminate had a conductivity of 0.041 W m^{-1} K^{-1}, which increased up to 0.058 W m^{-1} K^{-1} (~42% increment) at 4.0 wt% MWCNT. Thermal conductivity enhancements have also been described for multiscale thermoset-based composites incorporating CNTs. For instance, improvements of 67% and 150% have been reported on addition of 1.0 and 3.0 wt % MWCNTs, respectively, to polyester/vinyl ester resin/GF composites (Wang and Qiu, 2010). Enhancements of ~60% have been found for epoxy/MWCNT/GF composites with 1.2 wt% MWCNT (Assael et al., 2009) and phenolic resin/CF composites with 7.0 wt% MWCNTs (Kim et al., 2007).

8.3.2.3 Thermal stability and flammability

The thermal stability of hierarchical CNT-reinforced composites can be evaluated by thermogravimetric analysis (TGA). Measurements can be performed under both inert and oxidative conditions. Fig. 8.7 shows TGA curves in air atmosphere for PEEK/SWCNT/GF laminates (Díez-Pascual et al., 2011). All the composites exhibited two degradation steps, similar to PEEK/GF, the first attributed to the random scission of the polymer chains and the second to the degradation of the polymer aromatic rings. The initial degradation temperature (T_i) of PEEK/GF is ~516°C and its maximum rates of degradation (T_{max1} and T_{max2}) correspond to 547 and 602°C, respectively. An important grow in the degradation temperatures was found upon incorporation of SWCNTs, especially those with the compatibilizer. For instance, compared to PEEK/GF, T_i, T_{max1}, and T_{max2} increased by ~34, 30, and 40°C, respectively in the compatibilized laminate with 1.0 wt% laser SWCNTs, while in a similar laminate without compatibilizer the increments were smaller, 9, 12, and 22°C, respectively. The thermal stability improvement was ascribed to the barrier effect of the CNTs, which delay the diffusion of degradation products from interior of the composite to the exterior. The compatibilizer enhances the CNT dispersion and their adhesion with the matrix, making the barrier effect stronger, which

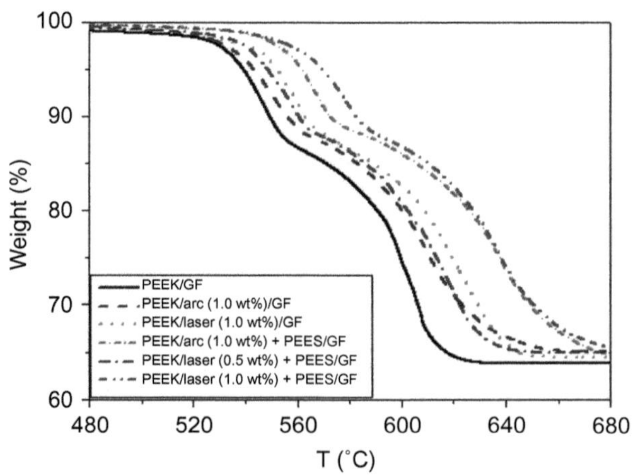

Figure 8.7 TGA curves under oxidative atmosphere for PEEK/SWCNT/GF laminates. *Source*: From Díez-Pascual, A.M., Ashrafi, B., Naffakh, M., González-Domínguez, J.M., Johnston, A., Simard, B., et al., 2011. Influence of carbon nanotubes on the thermal, electrical and mechanical properties of poly(ether ether ketone)/glass fiber laminates. Carbon 49, 2817–2833, with permission from Elsevier.

leads to higher degradation temperatures. Regarding the nanotube type, for the same CNT loading, the increments found in T_i, T_{max1}, and T_{max2} for laminates with laser SWCNTs were on average 2, 6, and 10°C higher, respectively, than those with arc-SWCNTs, since laser-grown SWCNTs exhibit less defect content, hence higher thermal conductivity that promotes heat dissipation. The increments in thermal stability found in these laminates are larger than those of binary PEEK/SWCNT (1.0 wt%) nanocomposites (Diez-Pascual et al., 2010a), in which T_i and T_{max} increased by 5 and 8°C, respectively. The greater enhancements in the laminates can be attributed to a synergistic stabilization effect due to the presence of both fillers, given that the GFs also holdup the degradation process of the polymer.

Thermal stability improvements have also been reported for PA-6/MWCNT/GF laminates (Shen et al., 2009). The thermal degradation of the polymer occurs via chain scission and radical formation, and the MWCNTs are believed to act as radical scavengers, delaying the degradation. The increments in the degradation temperatures of these laminates are noticeably smaller than those reported for PEEK/SWCNT/GF laminates, which is in agreement with the lower λ improvements of the former laminates, as mentioned earlier. Besides, the MWCNTs do not interact with PA-6, whilst the SWCNTs are able to interact with PEEK via $\pi-\pi$ stacking, leading to a stronger interfacial adhesion, and consequently better barrier effect.

PA-6/MWCNT/GF composites also exhibited better fire resistance than those without CNTs, as determined from the heat release rate (HRR) curves obtained from cone test measurements (Shen et al., 2009). The ignition time and peak HRR values increased progressively with the MWCNT loading. The ignition time of the laminate with 4 wt% MWCNT was 31% longer than the reference PA-6/GF and the peak HRR occurring time also showed an increase of 118%, although the values of peak HRR increased only slightly. This is the first report of cone test on laminates in the literature. The delayed ignition time and the time to achieve the peak HRR found for these laminates have been ascribed to the better thermal conductivity of the samples since the CNTs form a heat conductive network that transports and dissipates the heat rapidly through the bulk of the composite, hence it delays the ignition point at the surface.

Natural fiber-reinforced composites such as polylactic acid (PLA) filled with hemp, flax, or kenaf (John and Thomas, 2008) have recently become interesting for certain applications such as transport industry due to their reduced weight, lower fuel consumption, and consequently emissions compared to traditional fiber-reinforced plastics. Nonetheless, to apply these biomaterials at an industrial level it is important to enhance their flame retardancy. In this regard, MWCNTs and sepiolite nanoclays were incorporated into PLA/hemp (Hapuarachchi and Peijs, 2010), and an initial rise in HRR was found for the hemp/ternary nanocomposites compared to PLA/hemp. However, at longer times, the HRR of the ternary composites decreased. It was proposed that the CNTs formed a network with the clay layers, which reduced the amount of volatile gases evading from the degrading polymer and made the oxygen entrance difficult, consequently diminishing the HRR.

8.3.2.4 Heat distortion temperature and thermal expansion coefficient

The heat distortion temperature (HDT), temperature at which a polymeric material deforms under a specific load, is a parameter widely used for the design of materials. However, very scarce literature dealing with the HDT of fiber-reinforced laminates has been reported (Wu et al., 2001). The HDT is expected to rise upon incorporation of nanofillers, although the HDT of PA-6/GF (49 wt%) hardly changed on addition of 0.5–4.0 wt% MWCNTs (Shen et al., 2009). Conversely, the addition of 3.0 wt% nanoclay to PA-6/GF (10–30 wt%) laminates caused increases between 11 and 18% in the HDT of the matrix suggesting a synergistic effect of both fillers (Wu et al., 2001). The GF to nanofiller ratio was believed to be the reason for the different behavior between these two types of laminates. In the laminates reinforced with MWCNTs this ratio was high, hence the GFs predominated in the HDT properties, and no cooperative effects between both fillers were observed. On the other hand, the HDT of CF-reinforced polyimide (PI) composites increased by 9 and 22% upon incorporation of 1.0 wt% raw or surface treated CNTs, respectively (Li and Bai, 2011). The addition of polymer treated CNTs on the CF surface improved the fiber–matrix interface adhesion, leading to a larger HDT improvement.

The thermal expansion coefficient (CTE) must be taken into account when important changes in dimensions due to temperature variations are expected. The CTE of PEI/CNTs, PEI/CF, and PEI/CF/CNT composites has been comparatively investigated (Rios et al., 2013). The addition of 1.75 wt% and 3.5 wt% MWCNTs to PEI hardly reduced the CTE and did not cause important anisotropy. In contrast, the addition of 7.5 wt% CFs strongly decreased the CTE in the direction perpendicular to the polymer flow, leading to a highly anisotropic sample. When 3.5 wt% MWCNTs were added to PEI/CF, the CTE in the direction of the polymer flow diminished and the material become less anisotropic. Similar conclusions were drawn for polycarbonate (PC)/CF/MWCNT composites. In summary, it was found that the addition of CNTs to thermoplastic/CF composites reduces the anisotropy of thermal expansion.

8.3.3 Mechanical properties

8.3.3.1 Dynamic mechanical properties

Very scarce data dealing with the mechanical properties of multiscale composites versus temperature and/or frequency have been published. Dynamic mechanical analysis (DMA) is a technique that gives information about the stiffness of material as a function of temperature, as well as about the relaxation processes of the matrix. The change in the storage modulus (E') of PEEK/SWCNT/GF laminates as a function of temperature was measured by DMA (Díez-Pascual et al., 2011) (Fig. 8.8A), and a very strong drop was found between 150 and 200°C, range that corresponds to the glass transition (T_g) of the composites. The addition of 1.0 wt% raw arc- and

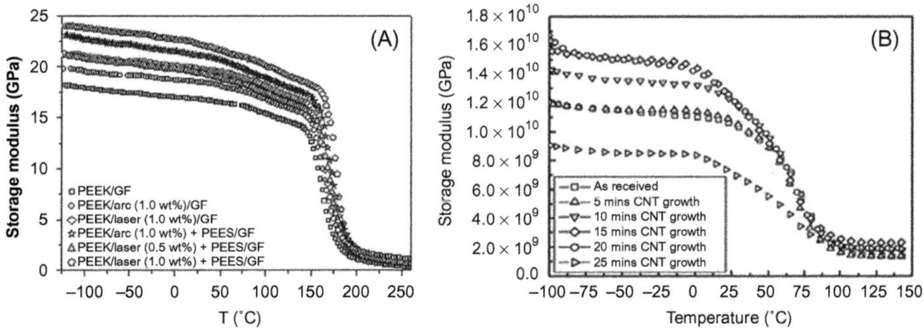

Figure 8.8 Storage modulus versus temperature for PEEK/SWCNT/GF laminates (A), *Source*: Adapted from Díez–Pascual, A.M., Ashrafi, B., Naffakh, M., González-Domínguez, J.M., Johnston, A., Simard, B., et al., 2011, and polyester/CF/CNT composites with different growth times (B), *Source*: Taken from Agnihotri, P., Basu, S., Kar, K.K., 2011, with permission from Elsevier.

laser-SWCNTs to PEEK/GF caused about 5 and 8% increase in E' at 25°C, respectively, while the same amount of SWCNTs wrapped in the compatibilizer increased E' by ~18 and 21%. The same increasing tendency was found in all the temperature range, though the percentage of increase in E' was larger at higher temperatures, suggesting that the reinforcement effect was stronger above the T_g. The higher modulus of the samples with the compatibilizer was ascribed to a more homogenous distribution of the CNTs and a better polymer–nanotube stress transfer. On the other hand, enhanced modulus was reported for laminates incorporating laser-grown SWCNTs in comparison to those with arc-purified SWCNTs, which was explained considering that the acid treatment could have shortened the tubes and induced sidewall defects, thus reducing their reinforcing efficiency. As expected, E' increased with the SWCNT content albeit following a nonlinear trend where composites incorporating 1.0 wt% loading had ~10% higher modulus than those with 0.5 wt%.

An improvement of ~30% in the room temperature E' has been found for a PA-6/MWCNT (0.18 wt%)/GF laminate prepared by electrostatic assembly (Zhang et al., 2012). This extraordinary enhancement obtained with such a small amount of MWCNTs likely arises from the fiber–matrix interconnecting effect. Accordingly, the oxidized MWCNTs adsorbed on the GFs formed a porous and interconnected structure that could be simply interpenetrated by the polymer, resulting in strong fiber–matrix interfacial adhesion. The percentage of increase in E' was stronger at temperatures <60°C, while become insignificant above 120°C, in contrast to the trend reported for PEEK/GF laminates.

Fig. 8.8B shows the variation of the storage modulus with temperature for CNT-coated CF/polyester composites with CNT growth times of 5, 10, 15, 20, and 25 min (Agnihotri et al., 2011). For growth times between 5 and 20 min, E' raised gradually with increasing time. Nonetheless, for a growth time of 25 min, E'

dropped strongly, albeit both the density and length of the CNTs augmented. The increase in E' up to a growth time of 20 min was ascribed to a larger contact area and mechanical interlocking between the fiber and matrix due to the presence of CNTs on the fiber surface. Conversely, when the growth time rose beyond 20 min, the CNTs surface density increased to such degree that the interlocking of the CNTs with the polyester was not feasible, thus weakening the polymer-CNT interaction, and consequently the composite properties. Further, the CNTs in this case were very long, and on pulling out, they were prone to break from the point of tethering on the fiber surface, leading to failure.

DMA experiments also provide information about the transition temperatures of the matrix from the plot of the ratio of the loss to storage modulus (tan δ) versus temperature. Most T_g data reported on CNT-reinforced composites are for CNT/polymer systems rather than multiscale composites. Moreover, contradictory data on changes in T_g have been found for CNT/polymer systems (Allaoui and Bounia, 2009). For instance, the T_g of polyester/CF/CNT composites was not influenced by the growth time of CNTs (Agnihotri et al., 2011), while in the case of PEEK based laminates, an increase of $\sim 20°C$ in relation to that of PEEK/GF was found when 1.0 wt% laser-grown SWCNTs dispersed in PEES were incorporated (Díez-Pascual et al., 2011). The addition of SWCNTs reduces the free volume and hampers the mobility of the PEEK chains, which leads to higher T_g values. The noteworthy T_g improvement found in the compatibilized multiscale composites was ascribed to the high T_g of PEES together with the improved matrix-nanofiller adhesion and a joint effect between the micro- and nano-scale fillers on restricting the expansion of the polymer chains, consistent with the behavior reported for other hierarchical laminates (Wichmann et al., 2006; Green et al., 2009). On the contrary, the T_g of PA-6/GF increased only slightly upon incorporation of 0.18 wt% MWCNTs (Zhang et al., 2012), despite SEM observations indicated strong fiber–matrix adhesion in the presence of the CNTs. Higher loading of MWCNTs would be necessary to successfully restrict the mobility of the PA-6 chains.

8.3.3.2 Static mechanical properties

Table 8.1 collects the percentage of variation reported in the mechanical properties of CNT-reinforced hierarchical composites compared to the corresponding binary matrix/fiber samples. The data hint that, in general, the increments in the fiber-dominated properties (i.e., Young's modulus and tensile strength) are smaller than those attained in the matrix-dominated properties (i.e., flexural and interlaminar shear strength).

Tensile and flexural

On the whole, the incorporation of CNTs has a beneficial effect on the tensile and flexural properties of fiber-reinforced thermoplastic composites, although the improvements are strongly influenced by the type, purity, aspect ratio, concentration, orientation, and degree of dispersion of the nanotubes as well as on the nature of the matrix. In the case of high-performance and engineering thermoplastics such

Table 8.1 Improvement in the mechanical properties (in %) of CNT-reinforced thermoplastic hierarchical composites

Matrix	Fiber (wt%)	CNT (wt%)	E (%)	σ_y (%)	G (%)	E_f (%)	σ_{fM} (%)	IFSS (%)	ILSS (%)	Reference
PA-6	GF (49)	MWCNTs (0.5)				16.5	35.0			Shen et al. (2009)
PA-6	GF (49)	MWCNTs (1.0)				21.0	32.0			Shen et al. (2009)
PA-6	GF (49)	MWCNTs (2.0)				15.0	15.0			Shen et al. (2009)
PA-6	GF (49)	MWCNTs (4.0)				2.2	14.2			Shen et al. (2009)
PA-6	BF[a] (30)	MWCNTs (0.5)	12.2	9.0		35.4	41.0			Meszaros et al. (2011)
PA-6	BF (30)	MWCNTs (1.0)	12.2	8.5		37.4	41.0			Meszaros et al. (2011)
PA-6	BF (30)	MWCNTs (1.5)	8.1	6.5		29.1	37.0			Meszaros et al. (2011)
PA-6	BF (30)	MWCNTs (2.0)	6.2	5.0		32.5	37.0			Meszaros et al. (2011)
PA-6	GF (5)	MWCNTs (0.12)		12.8						Zhang et al. (2012)
PA-6	GF (5)	MWCNTs (0.18)		14.7						Zhang et al. (2012)
PEEK	GF (64)	Arc-SWCNTs (1.0)	2.6	−2.8	−6.2	5.2	7.7		3.8	Diez-Pascual et al. (2011)
PEEK	GF (64)	Laser SWCNTs (1.0)	6.4	−2.0	−3.8	12.8	9.0		−21.6	Diez-Pascual et al. (2011)
PEEK	GF (64)	PEES-wrapped laser SWCNTs (0.5)	9.6	8.4	12.5	14.5	13.5		11.9	Diez-Pascual et al. (2011)
PEEK	GF (64)	PEES-wrapped laser SWCNTs (1.0)	16.0	7.6	10.0	32.6	17.9		64.4	Diez-Pascual et al. (2011)
PEEK	GF (64)	PEES-wrapped arc-SWCNTs (1.0)	14.1	9.2	7.5	28.5	17.0		−25.5	Diez-Pascual et al. (2011)
PP	GF[b] (68)	SWCNTs (0.04)						36.6		Mäder et al. (2008)
PP	GF (5)	MWCNTs	40.0	39.2	24.0	36.0	43.0			Rahmanian et al. (2013)
PP	CF (5)	MWCNTs	57.0	37.3	34.0	51.0	35.0			Rahmanian et al. (2013)
PP	CF (15)	MWCNTs (1.0)			−32.0	60.0				Kim et al. (2011)
PP	CF (15)	MWCNTs (2.0)			−35.0	99.6				Kim et al. (2011)

(Continued)

Table 8.1 (Continued)

Matrix	Fiber (wt%)	CNT (wt%)	E (%)	σ_y (%)	G (%)	E_f (%)	σ_{fM} (%)	IFSS (%)	ILSS (%)	Reference
PP	CF (12)	CVD CNTs	133.0	57.0						Suraya et al. (2009)
PI	CF (15)	PE/PAM[c]-modified CVD MWCNTs (1.0)							65.0	Li and Bai (2011)
PI	CF (30)	MWCNTs (15)	33.5	125.0	75.0	36.0	29.6			Zhang (2011)
PI	CF (25)[d]	CNT (1.0)[d]	1.8	1.7						Ruiya et al. (2011)
PI	CF (25)[d]	CNT (3.0)[d]	3.0	8.7						Ruiya et al. (2011)
PI	CF (25)[d]	CNT (5.0)[d]	6.0	17.4						Ruiya et al. (2011)
PI	CF[a] (61)[e]	CVD CNT	0.3	0				15.6		Naito (2014)
PMMA	CF[e]	MWCNTs						26.0		Qian et al. (2010b)
PMMA	Silica[e]	Short MWCNTs						150.0		Qian et al. (2010c)
		Long MWCNTs						80.0		

E: Young's modulus; σ_y: tensile strength at yield; G: impact strength; E_f: flexural modulus; σ_{fM}: flexural strength; IFSS: interfacial shear strength; ILSS: interlaminar shear strength.
[a]Basalt fiber.
[b]Glass fiber yarns.
[c]Polyethylene/polyamine.
[d]Content in vol.%.
[e]Composites with CNT-grafted fibers.

as PEEK and PA-6, respectively, important improvements in flexural properties (up to 41% (Meszaros et al., 2011)) have been reported, ascribed to a stiffer CNT-reinforced matrix, enhanced fiber–matrix interfacial adhesion and/or strong synergetic effect between nano and microscale fillers. Nonetheless, only moderate increments in the in-plane Young's modulus (E) and tensile strength (σ_y) (up to 16 and 9%, respectively (Díez-Pascual et al., 2011)) have been found. For instance, the incorporation of 1.0 wt% arc- and laser-SWCNTs wrapped in PEES to PEEK/GF raised the flexural modulus (E_f) by about 29 and 33%, respectively, whilst the increments in E were considerably lower, around 14 and 16%. However, E and E_f of composites without compatibilizing agent were on average 10 and 20% lower than the corresponding compatibilized samples, and their σ_y was even lower than that of PEEK/GF, attributed to a weaker nanotube–matrix interfacial bonding and the existence of nanotube agglomerates. The percentage of increment in both moduli found in these hierarchical composites were higher than those reported for binary PEEK/SWCNT nanocomposites (Díez-Pascual et al., 2010b), owed to a cooperative effect of both fillers on enhancing the stiffness of the polymer. A comparable synergistic effect was reported for high-performance PI/MWCNT/CF composites (Zhang, 2011), where the mechanical properties of the hybrids were always higher than those of PI/MWCNT and PI/CF composites. CNTs could behave as bridges between the polar fibers and the nonpolar matrix, resulting in a better fiber interlock. Contrary to the behavior observed for other high-performance composites, an unprecedented rise in σ_y of ~125% was reported upon addition of 15 wt% MWCNTs to PI/CF, considerably larger than the 30% augment found for the flexural strength (σ_{fM}). This suggests that the main effect of the nanofillers is a strengthening of the fiber–matrix interface instead of reinforcing the thermoplastic matrix. Conversely, hardly improvement in E or σ_y was found for CNT-grafted CF/PI composites (Naito, 2014). On another point, both tensile and flexural properties were found to be considerably dependent on the CNT type. For instance, the increments obtained upon incorporation of high-quality laser-grown SWCNTs to PEEK/GF were notably higher than those attained with the same amounts of arc-purified SWCNTs (Díez-Pascual et al., 2011). Regarding the CNT loading, the improvements in E_f and E of PEEK/GF attained at 1.0 wt% SWCNT loading were approximately double those found with 0.5 wt%. Likewise, a quasilinear raise in the tensile properties of PI/CF was described on increasing CNT content (Ruiya et al., 2011), whereas comparable E_f and E values were reported for basalt fiber (BF)-reinforced PA-6/MWCNT composites with 0.5 and 1.0 wt% CNT content (Meszaros et al., 2011). Similar E_f and σ_{fM} increments at 1.0 and 0.5 wt% MWCNT contents were also found for PA-6/GF/MWCNT laminates (Shen et al., 2009), which decreased considerably at higher loadings, though the absolute values were still higher than those of samples without MWCNTs. Conversely, for binary PA-6/MWCNT composites (Shen et al., 2009) both parameters augmented regularly with the MWCNT loading, reaching 20 and 16% improvement, respectively, at 4 wt% MWCNT. Several factors such as the CNT dispersion, the viscosity of the matrix and the corresponding wetting with the GF were believed to be the reason for the smaller improvements observed in the laminates. This concern can be worked out by fabricating the laminates via electrostatic assembly, as described for PA-6 hybrid composites (Zhang et al., 2012). As a result, the addition of MWCNT contents ≤ 0.2 wt% increased σ_y of PA-6/GF by $\sim 15\%$.

Focusing on commodity plastics such as PP, strong improvements in both tensile and flexural properties of hierarchical composites have been attained (Rahmanian et al., 2013; Suraya et al., 2009; Kim et al., 2011). In particular, E and E_f of samples incorporating 5.0 wt% MWCNT-coated short CFs increased by up to 57 and 51%, respectively, and those of composites with the same amount of MWCNT-coated short GFs increased by ~ 40 and 36% compared to neat short fiber composites (Rahmanian et al., 2013). The improvement was even more important at higher fiber contents, up to 133% improvement in E upon addition of 12.0 wt% coated CFs (Suraya et al., 2009). This behavior indicates a more efficient fiber–matrix stress transfer due to the CNT coating and the existence of synergistic effects of both fillers on reinforcing the matrix. More interestingly, about 100% E_f increase was found on addition of 2.0 wt% MWCNTs to PP/CF (Kim et al., 2011), attributed to the reinforcing effect of the well dispersed nanofillers and a strong PP-CF interface.

The elongation at break of polymer/fiber composites is also influenced by the addition of CNTs, since these restrict the extent of plastic deformation of the matrix, as reported for binary nanocomposites. This is the case of PEEK/GF laminates (Díez-Pascual et al., 2011), in which composites with 1.0 wt% raw SWCNTs displayed $\sim 40\%$ reduction in strain at break, owed to the nanotube agglomerates that delay the ductile flow of the polymer chains. However, the addition of a compatibilizer decreased the stress concentrations at the polymer–nanotube interface, thus improving the composite ductility, and the compatibilized laminates showed comparable elongation at break to that of the reference PEEK/GF. On the other hand, the addition of MWCNTs does not significantly modify the elongation at break of hierarchical composites, as found for BF/PA-6/MWCNT hybrids (Meszaros et al., 2011), given that MWCNTs are easier to disperse than SWCNTs. Accordingly, the elongation at break can be indicative of the quality of the CNT dispersion in thermoplastic composites.

Very scarce studies modeling the mechanical properties of CNT-based hierarchical composites have been published (Kim et al., 2009; Tsai et al., 2011). Their tensile properties can be calculated using a combination of the Halpin–Tsai equations (Halpin and Kardos, 1976) with a micromechanical approach (Kim et al., 2009) that is generally applied for traditional fiber-reinforced composites. Another choice is to apply a two-phase model (Diez-Pascual et al., 2011). Firstly, the theoretical values for polymer/CNT mixtures can be estimated according to the well-known rule of mixtures. Given that CNTs are not completely extended when dispersed in a polymer matrix, a decreased shape factor (1/5) must be taken into account. The CNT-reinforced polymer is then considered as a new matrix phase and the fibers as the filler phase. The rule of mixtures can be then applied to determine the laminate modulus: $E_c = \zeta_E V_f E_f + (1 - V_f)E_m$, where ζ_E is the fiber efficiency factor (~ 0.5 for continuous bidirectional fibers), E_f and E_m are the fiber and matrix modulus, respectively, and V_f is the fiber volume fraction.

By means of this approach, the modulus of PEEK/SWCNT (1.0 wt%)/GF laminates was calculated to be 13–27% higher than the measured data (Diez-Pascual et al., 2011). The discrepancies between experimental and calculated values were attributed to several factors, mostly the very high viscosity of the PEEK/SWCNT mixture that prevents perfect fiber wetting, and the fact that the SWCNTs are assembled into small bundles and shear slippage of individual nanotubes within the bundle may occur, thus limiting stress transfer from the matrix to the nanofillers. Note that the rule of mixtures is an upper boundary expected in the case of perfect CNT dispersion and adhesion to the matrix. Novel analytical models should be developed that can provide a more real approximation of the laminates' modulus.

Impact strength

A number of factors determine the impact strength of reinforced composites (Thomas, 1973), including a diminishing effect of the fillers owed to a drop in the elongation at break, hence a reduction in ductility. Moreover, stress concentrations can be formed around the filler ends, an area of poor filler–matrix adhesion. On the other hand, the fillers have a beneficial effect since reduce the crack propagation rate by forcing cracks to get around them. Furthermore, the boost of fracture resistance is the result of energy-dissipating mechanisms based on filler bridging, debonding, slip, and pull-out. The net effect depends on the competition of these factors. Focusing on laminates including CNTs, the toughness is sensitive to their size, state of dispersion and interfacial adhesion with the matrix. Regarding SWCNT-reinforced PEEK/GF composites (Diez-Pascual et al., 2011), a slight decrease in the impact resistance was found on addition of 1.0 wt% raw arc- and laser-SWCNTs, while the same amount of SWCNTs dispersed in the compatibilizer raised the impact strength by ~9%. The laminate with 0.5 wt% wrapped laser-grown SWCNTs showed the highest impact strength, ~13% higher than that of PEEK/GF. The drop in toughness of the former samples was ascribed to the presence of small nanotube agglomerates that act as stress concentration sites favoring the formation of dimples and nucleating cracks, consequently leading to a premature failure. The addition of a compatibilizer improved the toughness owed to a better nanofiller dispersion that reduces the number of stress concentration nuclei, and results in an enhanced CNT–matrix interfacial adhesion, which provides an effective barrier for pinning and bifurcation of the advancing cracks. Stronger improvements in impact strength have been reported for PP composites containing CNT-coated fibers. Thus, the addition of MWCNTs to CF and GF-reinforced PP increased the impact resistance by 34 and 24%, respectively (Rahmanian et al., 2013), attributed to an enhanced fiber–matrix interfacial region and the combined reinforcement effect of CNTs and fibers. Conversely, a decrease in the impact strength of PP/CF composites was found on addition of MWCNTs (Kim et al., 2011), since the CF length was significantly decreased owed to the increased torque caused by the MWCNTs during the extrusion processing. On the other hand, an extraordinary enhancement of ~75% has been reported for PI/CF composites upon addition of 15 wt% MWCNTs (Zhang, 2011), attributed to the development of a 3D network upon addition of the CNTs that avoids the beginning of shear failure in the composites. Failure typically involves the pull-out, rupture or bridging of the nanofillers, which is believed to consume additional energy and thus contributes to the superior toughness.

Interlaminar and interfacial shear strength

The interlaminar shear strength (ILSS) provides information regarding the fiber/matrix interfacial adhesion and can be evaluated using short beam shear (SBS) or compression shear tests (CST). Several factors are known to play a role in the laminates' ILSS, namely the level of porosity, interlaminar layer thickness, and crystallinity. Depending on the CNT type, concentration, and surface chemistry, the ILSS of hierarchical composites containing nanotubes can be increased in the interlaminar region in the range of 2–45% (Zhu et al., 2007). For instance, the use of functionalized CNTs resulted in larger improvements over pristine CNTs (Qiu et al., 2007) due to stronger interfacial bonding between the nanotubes and the matrix. The ILSS of PEEK/SWCNT/GF composites has been assessed using SBS (Ashrafi et al., 2012), and strong increments have been attained, by ~64 and 12% increase for SWCNT loadings wrapped in a compatibilizing agent of 1.0 and 0.5 wt%, respectively, attributed to the low void content of the laminates combined with homogenous SWCNT distribution inside the matrix. A mixture of different mechanisms such as compression failure, inelastic deformation, fiber failure, and shear failure was believed to be responsible for the fracture of these nanomodified laminates. Analogously, a 65% raise in the ILSS of PI/CF composites has been found on incorporation of polyethylene-polyamine surface treated CNTs (Li and Bai, 2011), attributed to a better matrix strength, an increase in the CF surface roughness and a stronger fiber–matrix interfacial bonding.

A large number of studies have investigated the interfacial properties of CNT-grafted CF multiscale composites by measuring the interfacial shear strength (IFSS) via single fiber fragmentation tests (Bekyarova et al., 2007; Qian et al., 2010b; Qian et al., 2010c; Agnihotri et al., 2011). IFSS increments between 11 and 470% have been described depending on the CNT type, concentration, length, and alignment. Qian et al. found a 26% augmentation in the IFSS of PMMA/CF composites upon growing MWCNTs onto the fibers using the CVD method (Qian et al., 2010b), ascribed to the superior wettability of the CNT-grafted fibers, and the mechanical interlocking between MWCNTs and the matrix. They also measured the IFSS of PMMA/MWCNT/silica fiber composites (Qian et al., 2010c) and found that the extent of enhancement was dependent on the length and morphology of the grafted CNTs: Shorter MWCNTs symmetrically located around the fibers caused ~150% in the IFSS, whilst longer asymmetrical situated MWCNTs only led to an 80% enhancement.

A noticeable increase in the IFSS of polyester/CNT-coated CF composites was found on increasing CNT growing time onto the fiber (Fig. 8.9), ascribed to the raise in the CNT length and density, similarly to the behavior found for the storage modulus (Agnihotri et al., 2011). Thus, at an optimum growth time of 20 min, the IFSS increased by about 88%, albeit drop strongly at higher times. For growth

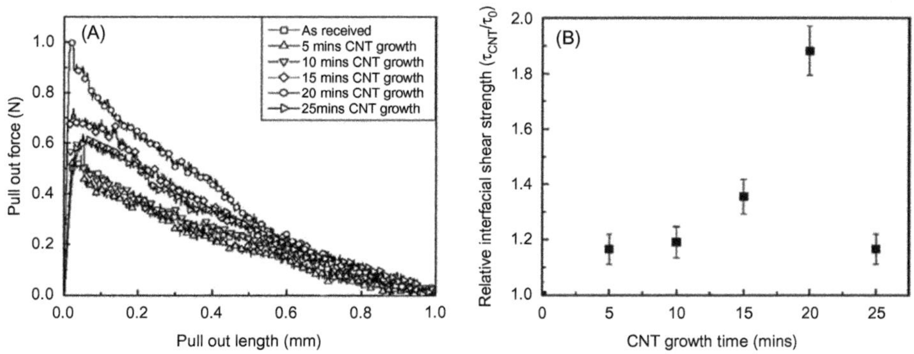

Figure 8.9 Interfacial properties of polyester/CF/CNT composites: (A) variation of pull-out force with the pull-out length for different CNT growing times. (B) Relative IFSS versus CNT growth time. τ_0 and τ_{CNT} are the interfacial shear strengths of the as-received and CNT-coated carbon fiber, respectively.
Source: Adapted from Agnihotri, P., Basu, S., Kar, K.K., 2011. Effect of carbon nanotube length and density on the properties of carbon nanotube-coated carbon fiber/polyester composites. Carbon 49, 3098–3106, with permission from Elsevier.

times ≤20 min, the CNT surface density is low and they are fairly short, which enables the polymer molecules to penetrate into the spaces between the CNTs and form a strong interface. In this case, failure in the pull-out tests takes place by fracture in the matrix. However, for longer times, the CNTs are denser and longer, hence polymeric chains cannot penetrate within the nanofillers and failure occurs at the tethering points in which the CNTs experiment large forces due to their length. Conversely, only a moderate rise (~16%) in the IFSS of PI/CF composites was found on grafting CNTs onto the fibers (Naito, 2014) using growing times of 15 min. The discrepancies could be related to the different matrix stiffness. Thus, when a fiber breaks, the stress is transferred back to a neighboring fiber by the surrounding matrix, hence is governed by the stiffness of the surrounding material. If the surrounding matrix is very stiff, the stress is transferred over a small distance, consequently the stress concentration nearby he fiber would be very high, which significantly increases the possibility that one fiber fracture can produce an unstable sequence of neighboring fiber fractures, eventually resulting in complete failure. These considerations may explain the small IFSS improvement for CNT-grafted CF-reinforced high-performance polymers like PI.

Another work (Diez-Pascual et al., 2012) assessed the interphase properties of SWCNT-reinforced PEEK/GF laminates using nanoindentation. The article shows how the polymer–fiber interface can be simply detected using the continuous stiffness measurement (CSM) technique, and proves that the extent of the interphase rises when the matrix is reinforced with SWCNTs, particularly those wrapped in a compatibilizing agent. On the whole, the multiscale reinforcement is a suitable approach to develop polymer–fiber laminates with improved interphases.

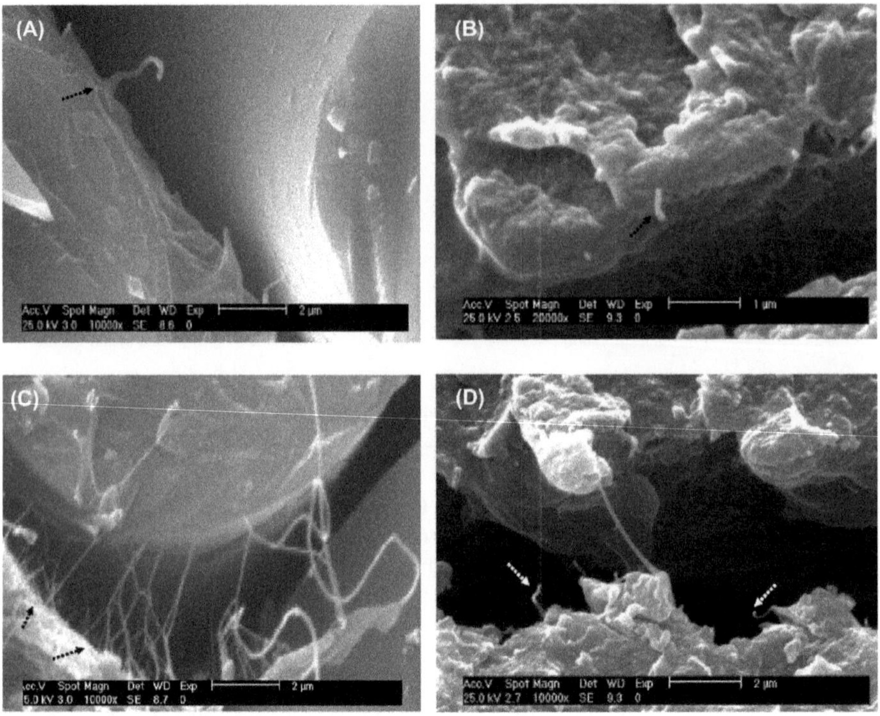

Figure 8.10 SEM images from fractured surfaces of flexural specimens: (A) and (B) PEEK/laser (1.0 wt%)/GF; (C) and (D) PEEK/laser (1.0 wt%) + PEES/GF. (A) and (C) were taken from a region nearby the fiber tows, whereas (B) and (C) correspond to matrix-rich areas. The dashed arrows point out SWCNTs pulled-out of the matrix.
Source: Adapted from Díez-Pascual, A.M., Ashrafi, B., Naffakh, M., González-Domínguez, J.M., Johnston, A., Simard, B., et al., 2011. Influence of carbon nanotubes on the thermal, electrical and mechanical properties of poly(ether ether ketone)/glass fiber laminates. Carbon 49, 2817–2833, with permission from Elsevier.

Fractographic analysis

Fractographic study is a crucial tool for investigating the failure mode of hierarchical composites. Examination of the fractured surfaces of PEEK/SWCNT/GF laminates after flexural tests using SEM (Fig. 8.10) revealed that fiber fracture dominated the failure mechanism, together with resin yielding at the fracture zone (Diez-Pascual et al., 2011). Furthermore, CNT pull-out instead of CNT breaking was found to be the result of flexural deformation, which improved the delamination resistance compared to the reference binary laminate. In the composites with raw CNTs (Fig. 8.10A and B) only a few SWCNTs were found to stick out of the matrix, whilst in the composites with a compatibilizing agent (Fig. 8.10C and D) a number of SWCNTs were pulled-out of the matrix and others were observed at the matrix–fiber interface extending into the surrounding matrix, which is believed to stiffen the matrix.

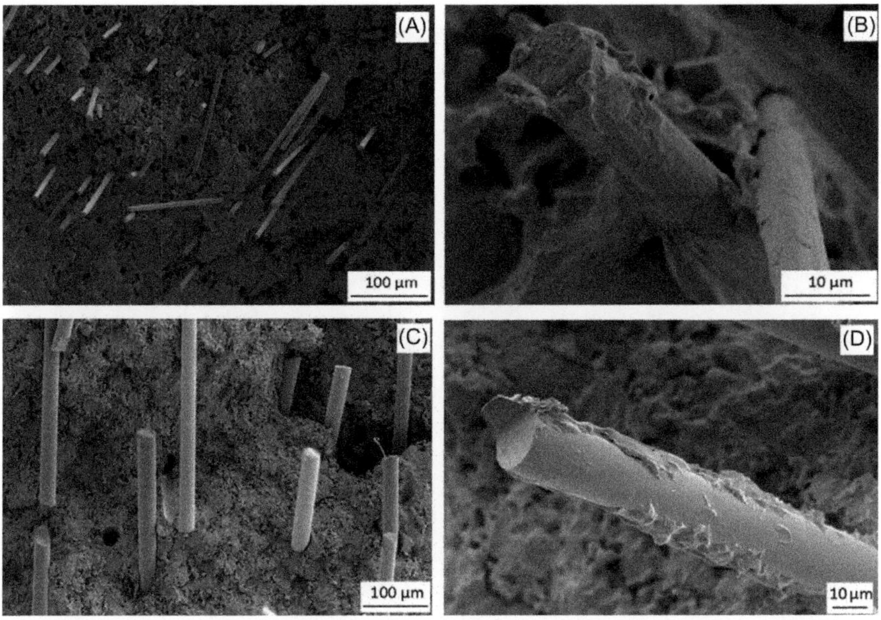

Figure 8.11 SEM micrographs of the fracture surfaces of PMMA/silica fiber composites (A), (B) without and (C), (D) with CNTs grafted onto the fibers. The micrographs in (B) and (D) are a magnification of (A) and (C), respectively.
Source: From Qian, H., Bismarck, A., Greenhalgh, E.S., Shaffer, M.S.P., 2010c. Carbon nanotube grafted silica fibres: Characterising the interface at the single fibre level. Compos. Sci. Technol. 70, 393–399, with permission from Elsevier.

The cross-sections of PMMA/MWCNT/silica composites fractured in flexion were also characterized by SEM (Fig. 8.11) (Qian et al., 2010c). The fracture plane was found at the fiber/matrix interface for both composites with and without CNTs. Nevertheless, composites with raw silica fibers (Fig. 8.11A and B) exhibited debonding at the interface, indicating a weak fiber–matrix interfacial adhesion, while in the composites with CNT-grafted fibers (Fig. 8.11C and D) there was no debonding, suggesting the existence of a strong interface. The failure took place at or close to the root of the CNTs, hinting that the CNT-fiber shear strength was higher than the matrix–fiber one. Besides, the CNTs were fully embedded in the polymer matrix demonstrating very good wettability by the PMMA. Analogously, the fracture morphology of PMMA/MWCNT/CF composites was assessed by SEM (Qian et al., 2010b), revealing that the failure mode was not influenced by the CNT grafting. The CF diameter hardly changed after the fracture, corroborating the absence of a cohesive mechanism.

The tensile fractured surfaces of PI/CF/CNT composites were also analyzed (Fig. 8.12) (Naito, 2014). All the composites failed due to extensive longitudinal splitting along the whole length of the specimen, leading to a brush-like fracture

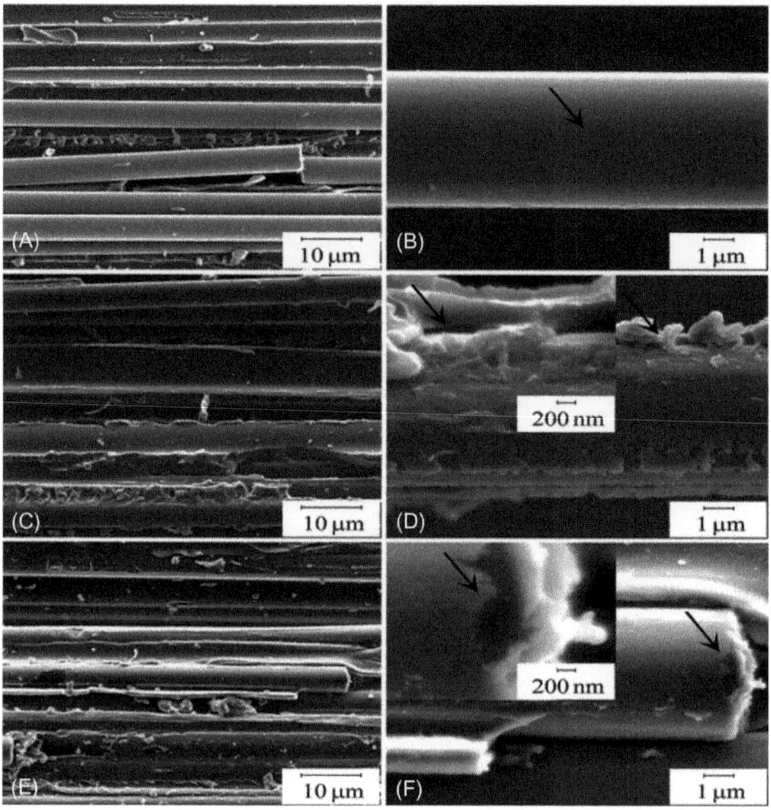

Figure 8.12 SEM micrographs at different magnifications of the tensile fracture surfaces of as-received (A and B), CNT-grafted (C and D) and PI-coated CF composites (E and F). *Source*: Reprinted from Naito, K., 2014. Tensile properties of polyimide composites incorporating carbon nanotubes-grafted and polyimide-coated carbon fibers. J. Mater. Eng. Perf. 23, 3245–3256, with permission from Springer.

surface, which is indicative of a fiber-dominated behavior. However, there were some differences between the morphology of binary and ternary composites. Composites with CNT-grafted fibers showed rough fractured surfaces with the original surface features and some adhesion of the resin (arrow in Fig. 8.12D), while the surface of the binary composites was relatively smooth with PI-coated fiber surface features (arrow in Fig. 8.12F). The as-received fibers also exhibited a smooth surface (Fig. 8.12B). Similar observations were reported for the fractured surfaces of PP/CNT-coated fiber composites after impact tests (Rahmanian et al., 2013). Composites with neat CFs and GFs showed a smooth fiber surface, without indication of interfacial fiber–matrix interactions, whereas composites with CNTs showed a rough surface and some PP residue, suggesting an effective CNT grafting onto the fibers and a CNT-PP micromechanical coupling. Analogously, the fiber surface of PA-6/MWCNT/GF composites (Zhang et al., 2012) was irregular and coarse after tensile tests, and most of the fibers remained embedded in the matrix, indicating enhanced fiber–matrix adhesion, whilst for binary composites the surface was smoother and many fibers protuded out of the matrix. In the hierarchical lamiantes, the intense CNT-fiber and CNT–matrix interactions increased the fiber–matrix debonding energy, the energy for crack growth and for fiber pull-out, leading to outstanding improvements in impact strength.

8.3.4 Electrical conductivity

Multiscale composites with CNTs display superior electrical conductivity (Gojny et al., 2005; Qiu et al., 2007) since the CNTs form a network above a concentration named as the percolation threshold. Their conductive properties are dependent on the concentration, aspect ratio, and distribution of the CNTs, as well as on filler–matrix and filler–filler interfacial interactions. The CNT type also influences the electrical behavior, and composites with raw CNTs have better conductivity than those with functionalized nanotubes (Ashrafi et al., 2012), given that the oxidation and sonication processed employed for the functionalization can damage and shorten the CNTs, diminishing their conductivity. Another factor that influences conductivity is the polymer wrapping (Diez-Pascual et al., 2011), which has a twofold effect: it precludes the formation of a conductive network but it improves the nanotube dispersion, hence the electron charge transfer. Overall, composites with raw or polymer-wrapped CNTs displayed almost the same conductivity. Besides, the electrical conductivity is an anisotropic property, and in the case of GF-reinforced composites such as those based on a PEEK matrix (Diez-Pascual et al., 2011), the in-plane conductivity was about an order of magnitude higher than the out-of-plane one. This behavior has not been found in composites with CFs, since the high conductivity of the fibers camouflages the anisotropic effect of the CNTs.

Fig. 8.13 shows the electrical conductivity of CNT-coated GF yarns embedded in a PP matrix (Rausch and Mäder, 2010a; Rausch and Mäder, 2010b). The behavior found resembles that reported for binary polymer/CNT nanocomposites, in which the volume resistivity drops with increasing CNT concentration, showing an abrupt change in the region of the percolation threshold. The inset of Fig. 8.13 presents a log–log plot of the conductivity as a function of the CNT concentration, which was used to determine the percolation threshold: 0.09 wt% MWCNTs, value lower than that found for melt-extruded PP/CNT nanocomposites, which corroborates the potential of electrically conductive coatings. On the other hand, the conductivity was found to rise as the GF yarn length decreased (Rausch and Mäder, 2010a), and for the same length it increased with the CNT weight fraction. For coating contents below 10 wt% the GF were not homogeneously covered, leading to lower conductivity values.

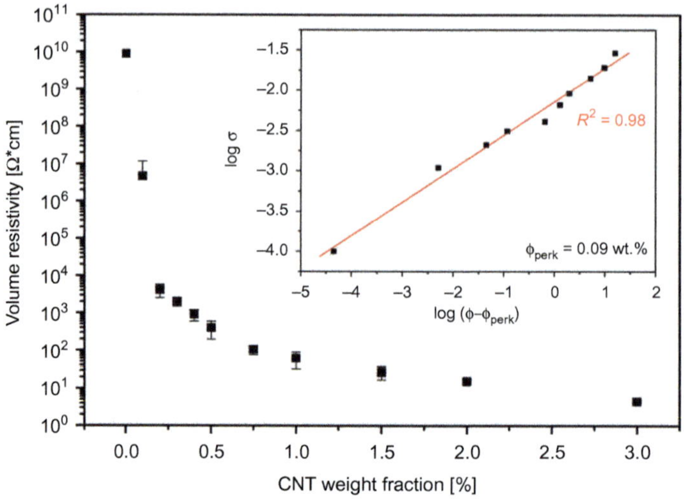

Figure 8.13 Volume resistivity of MWCNT-modified-GF yarns embedded into a PP matrix versus CNT weight fraction. The inset presents the best linear fit of the experimental data to determine the percolation threshold, $\varphi_c = 0.09$ wt%.
Source: Adapted from Rausch, J, Mäder, E., 2010a. Health monitoring in continuous glass fibre reinforced thermoplastics: Manufacturing and application of interphase sensors based on carbon nanotubes. Compos. Sci. Technol. 70, 1589–1596, with permission from Elsevier.

8.4 Concluding remarks and future trends

This chapter has reviewed the latest studies on multiscale composites developed by adding CNTs to conventional fiber-reinforced thermoplastics. These hierarchical materials can be fabricated either by dispersing the CNTs into the polymer matrix or by growing or depositing them onto the fabric layers. A wide number of factors condition the properties of this type of composites, namely the CNT type, quality, concentration, degree of dispersion and interfacial adhesion with the matrix. The combination of micro- and nano-scale fillers has been found to lead to synergistic effects that improve the mechanical, thermal and electrical properties of the matrix. For instance, noticeable improvements in thermal conductivity, thermal stability and flammability have been found on incorporation of SWCNTs to conventional fiber-reinforced composites, as well as simultaneous improvements in stiffness, strength, and toughness. On the whole, outstanding improvements in the matrix-dominated properties like flexural and interlaminar shear strength were reported, while only small increases in the fiber-dependent properties such as Young's modulus and tensile strength were found. Composites including CNT-grafted fibers display noticeably enhanced interfacial shear strength due to the enhanced wettability of the fibers, offering a controlled method for introducing high loadings of oriented CNTs into the matrix surrounding the fibers.

Unfortunately, the study of multiscale CNT-modified fiber-reinforced thermoplastic composites is still at its infancy, and commercial applications are not available yet. Further development and optimization is necessary prior to their use at a large-scale. The key confront that delays the application of these novel composites is related with the impregnation of the reinforcing fibers by very viscous CNT containing polymer melts. Besides, scalable and cost-effective fabrication processes that enable an efficient production such as displaced foam dispersion (DFD) or powder impregnation approaches should be employed. It is expected that in the near future CNTs would be used at an industrial level as hybrid fillers in conventional composites, mainly in the aeronautics, automobile, and energy sectors. In particular, they can considerably decrease the weight in traditional structural composites, being well suited for aircraft parts. They could also be employed in wind turbine blades, bridges, solar panels, and are perfect candidates for energy-dependant applications, such as access panels for aircrafts, laptops, and mobiles. More importantly, CNTs embedded in a sizing or polymer matrix can be used as health sensors to monitor damage. The formation of CNT networks on the fibers enables a local strain mapping within the composite interphase, which can be used for health monitoring of fiber-reinforced composites in a nondestructive way. This is a novel route towards the development of easy sensors and integrative switches at the micrometer-scale activated by strain or temperature.

Acknowledgments

AD acknowledges the MINECO for a Ramón y Cajal senior postdoctoral research fellowship cofinanced by the EU.

References

Abot, J.L., Song, Y., Schulz, M.J., Shanov, V.N., 2008. Novel carbon nanotube array-reinforced laminated composite materials with higher interlaminar elastic properties. Compos. Sci. Technol. 68, 2755–2760.

Agnihotri, P., Basu, S., Kar, K.K., 2011. Effect of carbon nanotube length and density on the properties of carbon nanotube-coated carbon fiber/polyester composites. Carbon. 49, 3098–3106.

Allaoui, A., Bounia, N.E., 2009. How carbon nanotubes affect the cure kinetics and glass transition temperature of their epoxy composites? – A review. eXPRESS Polym. Lett. 3, 588–594.

Ashrafi, B., Díez-Pascual, A.M., Johnson, L., Genest, M., Hind, S., Martinez-Rubi, Y., et al., 2012. Processing and properties of PEEK/glass fiber laminates: effect of addition of single-walled carbon nanotubes. Compos. A Appl. Sci. Manuf. 43, 1267–1279.

Assael, M.J., Antoniadis, K.D., Metaxa, I.N., 2009. Measurement of the enhancement of the thermal conductivity of an epoxy resin reinforced with glass fiber and carbon multi-walled nanotubes. J. Chem. Eng. Data. 54, 2365–2370.

Ayranci, C., Carey, J., 2008. 2D braided composites: a review for stiffness critical applications. Compos. Struct. 85, 43–58.

Barber, A.H., Zhao, Q., Wagner, H.D., Baillie, C.A., 2004. Characterization of E-glass-propylene interfaces using carbon nanotubes as strain sensors. Compos. Sci. Technol. 64, 1915–1919.

Bekyarova, E., Thostenson, E.T., Yu, A., Kim, H., Gao, J., Tang, J., et al., 2007. Multiscale carbon nanotube – carbon fiber reinforcement for advanced epoxy composites. Langmuir. 23, 3970–3974.

Brondsted, P., Lilholt, H., Lystrup, A., 2005. Composite materials for wind power turbine blades. Annu. Rev. Mater. Res. 35, 505–538.

Dervishi, E., Li, Z., Xu, Y., Saini, V., Biris, A.R., Lupu, D., et al., 2009. Carbon nanotubes: synthesis, properties, and applications. Part. Sci. Technol. 27, 107–125.

Díez-Pascual, A.M., Naffakh, M., Gomez, M.A., Marco, C., Ellis, G., Martinez, M.T., et al., 2009. Development and characterization of PEEK/carbon nanotube composites. Carbon. 47, 3079–3090.

Díez-Pascual, A.M., Naffakh, M., González-Domínguez, J.M., Anson, A., Martinez-Rubi, Y., Martinez, M.T., et al., 2010a. High performance PEEK/carbon nanotube composites compatibilized with polysulfones-I. Structure and thermal properties. Carbon. 48, 3485–3499.

Díez-Pascual, A.M., Naffakh, M., González-Domínguez, J.M., Anson, A., Martinez-Rubi, Y., Martinez, M.T., et al., 2010b. High performance PEEK/carbon nanotube composites compatibilized with polysulfones-II. Mechanical and electrical properties. Carbon. 48, 3500–3511.

Díez-Pascual, A.M., Ashrafi, B., Naffakh, M., González-Domínguez, J.M., Johnston, A., Simard, B., et al., 2011. Influence of carbon nanotubes on the thermal, electrical and mechanical properties of poly(ether ether ketone)/glass fiber laminates. Carbon. 49, 2817–2833.

Díez-Pascual, A.M., Gomez-Fatou, M.A., Ania, F., Flores, A., 2012. Nanoindentation assessment of the interphase in carbon nanotube-based hierarchical composites. J. Phys. Chem. C. 116, 24193–241200.

Díez-Pascual, A.M., González-Domínguez, J.M., Martinez, M.T., Gomez-Fatou, M.A., 2013. Poly(ether ether ketone)-based hierarchical composites for tribological applications. Chem. Eng. J. 218, 285–294.

Fujihara, K., Teo, K., Gopal, R., Loh, P.L., Ganesh, V.K., Ramakrishna, S., et al., 2004. Fibrous composite materials in dentistry and orthopaedics: review and applications. Compos. Sci. Technol. 64, 775–788.

Gnidakouong, J.R.N., Kim, M., Park, H.W., Park, Y.B., 2013. Electromagnetic interference shielding of composites consisting of a polyester matrix and carbon nanotube-coated fiber reinforcement. Compos. A Appl. Sci. Manuf. 50, 73–80.

Gojny, F.H., Wichmann, M.H.G., Fiedler, B., Bauhofer, W., Schulte, K., 2005. Influence of nano-modification on the mechanical and electrical properties of conventional fiber-reinforced composites. Compos. A Appl. Sci. Manuf. 36, 1525–1535.

Green, K.J., Dean, D.R., Vaidya, U.K., Nyairo, E., 2009. Multiscale fiber reinforced composites based on a carbon nanofiber/epoxy nanophased polymer matrix: synthesis, mechanical, and thermomechanical behaviour. Compos. A Appl. Sci. Manuf. 40, 1470–1475.

Halpin, J.C., Kardos, J.L., 1976. The Halpin–Tsai equations: a review. Polym. Eng. Sci. 16, 344–352.

Hapuarachchi, T.D., Peijs, T., 2010. Multiwalled carbon nanotubes and sepiolite nanoclays as flame retardants for polylactide and its natural fibre reinforced composites. Compos. A Appl. Sci. Manuf. 41, 954–963.

Hojo, M., Ando, T., Tanaka, M., Adachi, T., Ochiai, S., Endo, Y., 2006. Modes I and II interlaminar fracture toughness and fatigue delamination of CF/epoxy laminates with self-same epoxy interleaf. Int. J. Fatigue. 28, 1154–1165.

Hollaway, L.C., 2003. The evolution of and the way forward for advanced polymer composites in the civil infrastructure. Construct. Built. Mater. 17, 365–378.

Iijima, S., 1991. Helical microtubules of graphitic carbon. Nature. 354, 56–58.

Irving, P.E., Soutis, C., 2015. Polymer Composites in the Aerospace Industry. Woodhead Publising & Elsevier, Cambridge.

Ji-Nian, Y., Ai-Qin, X., 2013. Phase structures and mechanical properties of SGF/PP composites with and without compatilizers. Adv. Mater. Res. 668, 33–36.

John, M.J., Thomas, S., 2008. Biofibres and biocomposites – a review. Carbohydr. Polym. 71, 343–364.

Jones, F.R., 2010. A review of interphase formation and design in fibre-reinforced composites. J. Adh. Sci. Technol. 24, 171–202.

Khan, S.U., Kim, J.K., 2011. Impact and delamination failure of multiscale carbon nanotube-fiber reinforced polymer composites: a review. Int. J. Aeronaut. Space Sci. 12, 115–133.

Kim, M., Park, Y.-B., Okoli, O., Zhang, C., 2009. Processing, characterization, and modeling of carbon nanotubes-reinforced multiscale composites. Compos. Sci. Technol. 69, 335–342.

Kim, S. B., Nam, B. U., Park, J. S., 2011. Properties of polypropylene/carbon fiber /multi-walled carbon nanotube nanocomposites. In: 18th International Conference on Composite Materials, Korea, P3-19-IK1906.

Kim, Y.A., Kamio, S., Tajiri, T., Hayashi, T., Song, S.M., Endo, M., et al., 2007. Enhanced thermal conductivity of carbon fiber/phenolic resin composites by the introduction of carbon nanotubes. Appl. Phys. Lett. 90, 093125–093127.

Kroll, M., Langer, B., Grellmann, W., 2013. Toughness optimization of elastomer-modified glass–fiber reinforced PA6 materials. J. Appl. Polym. Sci. 127, 57–66.

Li, J., Bai, T., 2011. The addition of polyethylene/polyamine surface treated carbon nanotube on the interfacial adhesion of carbon fiber reinforced polyimide composites. Polym. Plast. Technol. Eng. 50, 1393–1397.

Mäder, E., Rausch, J., Schmidt, N., 2008. Commingled yarns – processing aspects and tailored surfaces of polypropylene/glass composites. Compos. A Appl. Sci. Manuf. 39, 612–623.

Mallick, P.K., 2008. Fiber Reinforced Composites: Materials, Manufacturing and Design. 3rd ed. CRC Press, New York.

Marsh, G., 2003. Next step for automotive materials. Mater. Today 6, 36–43.

Mathur, R.B., Chatterjee, S., Singh, B.P., 2008. Growth of carbon nanotubes on carbon fibre substrates to produce hybrid/phenolic composites with improved mechanical properties. Compos. Sci. Technol. 68, 1608–1615.

Meszaros, L., Gali, I.M., Czigany, T., Czvikovszky, T., 2011. Effect of nanotube content on mechanical properties of basalt fibre reinforced polyamide 6. Plast. Rub. Compos. 40, 289–293.

Moritis, G., Hub, 2003. New technologies provide access to more deepwater reserves. Oil Gas J. 101, 54–61.

Mouritz, A.P., 2007. Review of z-pinned composite laminates. Compos. A Appl. Sci. Manuf. 38, 2383–2397.

Mouritz, A.P., Leong, K.H., Hersberg, I., 1997. A review of the effect of stitching on the in-plane mechanical properties of fibre-reinforced polymer composites. Compos. A Appl. Sci. Manuf. 28A, 979–991.

Mouritz, A.P., Gellert, E., Burchill, P., Challis, K., 2001. Review of advanced composite structures for naval ships and submarines. Compos. Struct. 53, 21–42.

Naito, K., 2014. Tensile properties of polyimide composites incorporating carbon nanotubes-grafted and polyimide-coated carbon fibers. J. Mater. Eng. Perf. 23, 3245–3256.

Popov, V.N., 2004. Carbon nanotubes: properties and application. Mater. Sci. Eng. R. 43, 61–102.

Qian, H., Greenhalgh, E.S., Shaffer, M.S.P., Bismarck, A., 2010a. Carbon nanotube-based hierarchical composites: a review. J. Mater. Chem. 20, 4751–4762.

Qian, H., Bismarck, A., Greenhalgh, E.S., Shaffer, M.S.P., 2010b. Carbon nanotube grafted carbon fibres: A study of wetting and fibre fragmentation. Compos. A Appl. Sci. Manuf. 41, 1107–1114.

Qian, H., Bismarck, A., Greenhalgh, E.S., Shaffer, M.S.P., 2010c. Carbon nanotube grafted silica fibres: Characterising the interface at the single fibre level. Compos. Sci. Technol. 70, 393–399.

Qiu, J.J., Zhang, C., Wang, B., Liang, R., 2007. Carbon nanotube integrated multifunctional multiscale composites. Nanotechnology 18, 275708–275711.

Rahmanian, S., Thean, K.S., Suraya, A.R., Shazed, M.A., Salleh, M.A.M., Yusoff, H.M., 2013. Carbon and glass hierarchical fibers: Influence of carbon nanotubes on tensile, flexural and impact properties of short fiber reinforced composites. Mater. Des. 43, 10–16.

Rausch, J., Mäder, E., 2010a. Health monitoring in continuous glass fibre reinforced thermoplastics: Manufacturing and application of interphase sensors based on carbon nanotubes. Compos. Sci. Technol. 70, 1589–1596.

Rausch, J., Mäder, E., 2010b. Health monitoring in continuous glass fibre reinforced thermoplastics: Tailored sensitivity and cyclic loading of CNT-based interphase sensors. Compos. Sci. Technol. 70, 2023–2030.

Rios, P.F., Kenig, S., Cohen, R., Shechter, A., 2013. Injection-molded hybrid polymer composites with improved thermal isotropy. Soc. Plast. Eng. doi:10.2417/spepro.005040 (3p).

Ruiya, R., Zhenhua, L., Yunxuan, L., Chilan, C., 2011. Mechanical and tribological properties of polyimide composite reinforced with carbon fibers and carbon nanotube. Adv. Mater. Res. 311–313, 189–192.

Sadighi, M., Alderliesten, R.C., Benedictus, R., 2012. Impact resistance of fiber-metal laminates: a review. Int. J. Impact. Eng. 49, 77–90.

Shen, Z., Bateman, S., Wu, D.Y., McMahon, P., Dell'Olio, M., Gotama, J., 2009. The effects of carbon nanotubes on mechanical and thermal properties of woven glass fibre reinforced polyamide-6 nanocomposites. Compos. Sci. Technol. 69, 239–244.

Shubhra, Q.T.H., Alam, A.K.M.M., Quaiyyum, M.A., 2011. Mechanical properties of polypropylene composites: a review. J. Thermoplast. Compos. Mater. 26, 362–391.

Suraya, A.R., Yunus, R., Nor Azowa, I., 2009. Growth of carbon nanotubes on carbon fibres and the tensile properties of resulting carbon fibre reinforced polypropylene composites. J. Eng. Sci. Technol. 4, 400–408.

Tehrani, M., Safdari, M., Boroujeni, A.Y., Razavi, Z., Case, S.W., Dahmen, K., Garmestani, H., Al-Haik, M.S., 2013. Hybrid carbon fiber/carbon nanotube composites for structural damping applications. Nanotechnology 24, 155704–155715.

Thakre, P.R., Lagoudas, D.C., Riddick, J.C., Gates, T.S., Frankland, S.J.V., Ratcliffe, J.G., et al., 2011. Investigation of the effect of single wall carbon nanotubes on interlaminar fracture toughness of woven carbon fiber-epoxy composites. J. Compos. Mater. 45, 1091–1107.

Thomas, W.F., 1973. Factors affecting the impact strength of glass-fibre-reinforced polyester composites. Composites 4, 105–110.

Tong, L., Mouritz, A.P., Bannister, M.K., 2002. 3D Fibre reinforced polymer composites. Elsevier, Boston, pp. 1–12.

Tsai, C.-H., Zhang, C., Jack, D.A., Liang, R., Wang, B., 2011. The effect of inclusion waviness and waviness distribution on elastic properties of fiber-reinforced composites. Compos. B Eng. 42, 62–70.

Wang, S., Qiu, J., 2010. Enhancing thermal conductivity of glass fiber/polymer composites through carbon nanotubes incorporation. Compos. B Eng. 47, 533–536.

Warrier, A., Godara, A., Rochez, O., Mezzo, L., Luizi, F., Gorbatikh, L., et al., 2010. The effect of adding carbon nanotubes to glass/epoxy composites in the fibre sizing and/or the matrix. Compos. A Appl. Sci. Manuf. 41, 532–538.

Wichmann, M.H.G., Sumfleth, J., Gojny, F.H., Quaresimin, M., Fiedler, B., Schulte, K., 2006. Glass-fibre-reinforced composites with enhanced mechanical and electrical properties – benefits and limitations of a nanoparticle modified matrix. Eng. Fract. Mech. 73, 2346–2359.

Wu, S.-H., Wang, F.-Y., Ma, C.-C.M., Chang, W.-C., Kuo, C.-T., Kuan, H.-C., et al., 2001. Mechanical, thermal and morphological properties of glass fiber and carbon fiber reinforced polyamide-6 and polyamide-6/clay nanocomposites. Mater. Lett. 49, 327–333.

Wu, Y.J., Qin, W., Yang, Z.X., Wu, J.S., Zhang, Y.F., 2004. Preparation of high-quality ß-SiC nanowhiskers by using carbon fibres as carbon source. J. Mater. Sci. 39, 5563–5565.

Yang, D.J., Wang, S.G., Zhang, Q., Sellin, P.J., Chen, G., 2004. Thermal and electrical transport in multi-walled carbon nanotubes. Phys. Lett. A. 329, 207–213.

Zhang, J.G., 2011. The effect of carbon fibers and carbon nanotubes on the mechanical properties of polyimide composites. Mech. Compos. Mater. 47, 447–450.

Zhang, L., Su, D., Jin, L., Li, C., 2012. Polyamide 6 composites reinforced with glass fibers modified with electrostatically assembled multiwall carbon nanotubes. J. Mater. Sci. 47, 5446–5454.

Zhu, J., Imam, A., Crane, R., Lozano, K., Khabashesku, V.N., Barrera, E.V., 2007. Processing a glass fiber reinforced vinyl ester composite with nanotube enhancement of interlaminar shear strength. Compos. Sci. Technol. 67, 1509–1517.

Hybrid bast fiber reinforced thermoset composites

9

M.R. Nurul Fazita[1], *H.P.S. Abdul Khalil*[1,2], *Tham Mun Wai*[1], *E. Rosamah*[3] *and N.A. Sri Aprilia*[4]

[1]Schools of Industrial Technology, Universiti Sains Malaysia, Penang, Malaysia, [2]Cluster for Polymer Composites, Science and Engineering Research Center, University Sains Malaysia, Penang, Malaysia, [3]Faculty of Forestry, Mulawarman University, East Kalimantan, Indonesia, [4]Department of Chemical Engineering, Syiah Kuala University, Banda Aceh, Indonesia

Chapter Outline

Hybrid Polymer Composite Materials: Properties and Characterisation. DOI: http://dx.doi.org/10.1016/B978-0-08-100787-7.00009-3

9.1 Introduction

Plant fibers extracted from the stem are called bast fibers (Table 9.1). Bast fibers are produced in the layer between the xylem and the epidermis where they are surrounded by phloem. These fibers occur in bundles/aggregates in many dicotyledonous plants, running parallel to the stem between nodes (Rials and Wolcott, 1996) up to over 2 m long depending on the length of the plant (Meshram and Palit, 2013) to provide structural rigidity to the stems. The bundles consist of 10–25 elementary fibers (Fig. 9.1) (2–5 mm length and 10–50 μm diameter) where both the bundles and the elementary fibers are glued together by lignin and pectin to form a three dimensional network (Munder et al., 2005).

During bast fibers extraction process, the lignin and pectin are removed through a separation process such as retting, degumming or decortication to yield bast fibers (Zimniewska et al., 2011) with required strength, fineness, length, and purity (Batog et al., 2006). The bast fibers obtained are used as raw materials not only in the production of textiles but also for composites used for various applications such as for the car interior (Holbery and Houston, 2006; Fimmm and Rosemaund, 2009; Shinoj et al., 2011) and exterior (Mussig et al., 2006) in the automative industry, building and construction (Drzal et al., 2002), aerospace (Subash and Pillai, 2015), sports, and more (Chand and Fahim, 2008).

9.2 Natural bast fibers

9.2.1 Flax

Flax (*Linum usitatissimum*) under genus *Linum* in the Linaceae family is an annual plant that grows to 1.2 m tall with a stem diameter of 3 mm or more (Zimniewska et al., 2011) in temperate regions (Elzebroek and Wind, 2008) and is endemic to the area from the eastern Mediterranean to India (Sen and Reddy, 2011). It is now grown in a lot of countries including Canada, USA, China, India, and throughout Europe for its fibers and seeds (Joshi, 2015; Hall, 2016). The seeds are converted to linseed oil for health supplements, paints, and other industrial products (Jhala and Hall, 2010).

The fibers are used to produce textiles (Pallesen, 1996), pulp and paper (Aracri et al., 2010) and as reinforcements for polymeric composites (Bos et al., 2002). Its fibers are lustrous (Sen and Reddy, 2011) with a good length at 0.6–1.4 m and a diameter of 40–80 μm (Cutter, 2011) while its elementary fibers have lengths between 0.2 and 0.5 m and diameters ranging from 10 to 25 μm (Bos et al., 2002). It is also soft and flexible (Sen and Reddy, 2011; Nair and Joseph, 2014) but this contradicted the statements made by Cutter (2011) and Yan et al. (2014) who found it to be the stiffest and strongest plant fibers due to the substantial degree of crystallinity in its structure.

Table 9.1 Tabular presentation of bast fibers

Fiber	Botanical name	Plant photo	Longitudinal view of fibers	Image of fiber cross section
Flax	*Linum usitatissimum*			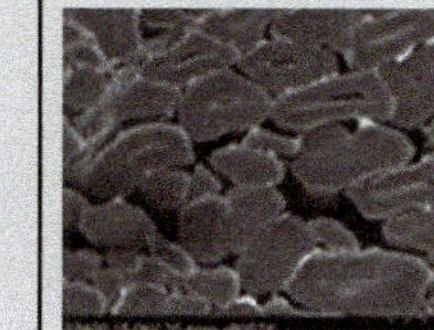
Hemp	*Cannabis sativa*			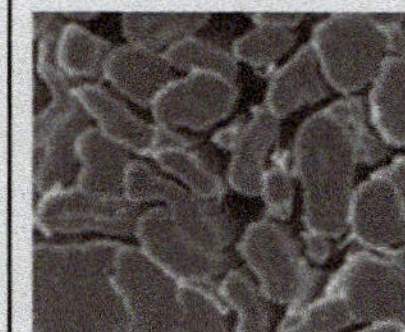
Kenaf	*Hibiscus cannabinus*			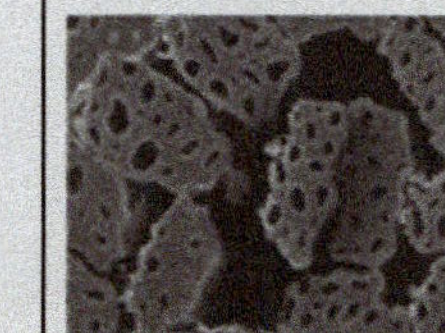
Jute	*Corchorus capsularis*			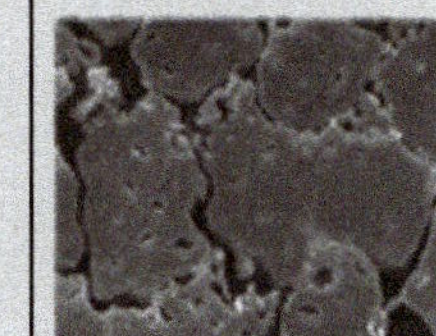

Source: Zimniewska, M., Wladyka-Przybylak, M., Mankowski, J. 2011. Cellulosic bast fibers, their structure and properties suitable for composite applications. In: Kalia, S., Kaith, B.S. and Kaur, I. (Eds.), Cellulose Fibers: Bio- and Nano-polymer Composites: Green Chemistry and Technology, Springer, Berlin, Germany, pp. 97–120.

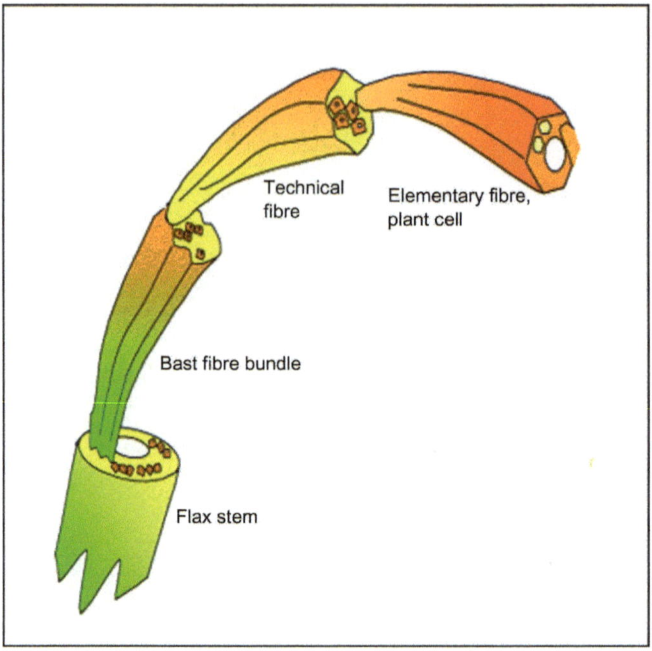

Figure 9.1 Schematic representation of a flax fiber from stem to elementary fiber.
Source: Bos, H.L., Mussig, J., van den Oever, M.J.A., 2006. Mechanical properties of short-flax-fibre reinforced compounds. Compos. A: Appl. Sci. Manuf. 37(10), 1591–1604.

Kers et al. (2010) also discovered that the flax bast fibers are strong where they are two times stronger than cotton and five times as strong as wool. Moreover, flax fibers have the added advantage in its ability to absorb up to 12% of its own weight when in contact with water which increases its strength by 20% (Murthy, 2015) unlike other bast fibers. Other than that, it also dries up quickly and is antistatic (Tahir et al., 2011) but exhibits relatively low longitudinal extension to failure when subjected to tensile loads (Cutter, 2011).

9.2.2 Jute

Jute [*Corchorus capsularis* (white jute) and *Corchorus olitorius* (dark jute)] (Summerscales et al., 2010) classified in the genus *Corchorus*, family Malvaceae (Ashby, 2012) are annual plants capable of growing up to 2–4.5 m in height (Lewington, 2003) with a stem diameter of 2–3 cm (Cutter, 2011). It originates from the Mediterranean (Cook, 1984) and thrives in hot humid environments like the tropical lowland areas which has a humidity of 60–90% (Yumnam et al., 2015) without needing fertilizers or pesticides. Today, it is widely grown in India, Bangladesh, Thailand, China, Brazil, and Indonesia (Summerscales et al., 2010; Cutter, 2011).

Jute is grown exclusively for its long, soft and shiny off-whitish to brownish in color fibers (1–4 m length and 17–20 μm diameter) unlike flax and hemp (Cutter, 2011). Its fibers are strong and resilient to microorganism's attack (Murthy, 2015) but are fairly brittle (Ganguly et al., 1999), sensitive to chemical and photochemical attack (Ghosh, 2003), exhibit low elongation at break due to its high lignin content (Cutter, 2011) and has a lower tensile strength than other bast fibers like hemp and flax (Biagiotti et al., 2004).

Despite its shortcomings, their advantages far outweigh its shortcomings (John and Thomas, 2008) for it to be used in sackings, carpets, wrapping fabrics, and construction fabric manufacturing industries dating back to ancient times (Savastano Jr et al., 2009). Besides that, the fibers are used by itself or combined with other types of fibers to make twine and rope (Sreenath et al., 1996). Presently, it is also used as raw materials to produce pulp and paper (Sahin, 2003), textiles (Liu et al., 2010), polymer composites in the automotive (Punyamurthy et al., 2014) as well as in the building and construction industry (Roul, 2009).

9.2.3 Hemp

Hemp (*Cannabis sativa* L.) is an annual plant belonging to the Moraceae family which can only grow in temperate regions (Faruk et al., 2012) and are indigenous to central Asia including China where it was grown over 12,000 years ago and later reached central Europe (Shahzad, 2012). At present, it can be found in the EU, central Asia, Philippines, China, Chile, France, Korea, and Spain with China cultivating and producing almost half of the world's industrial hemp supply (Sanjay et al., 2016).

Hemp can grow up to 5 m in height (Kymalainen and Sjoberg, 2008) and a stem diameter of 4–20 mm in approximately 140–145 days (Batra, 2006; Zimniewska et al., 2011) with a bast fiber content in the range of 28–46% (Cutter, 2011). However, its growth rate can be sped up to reach a height of 4 m in just 84 days when grown in a suitably warm condition (Summerscales et al., 2010). Additionally, hemp has the added advantage of being very resilient compared to other fiber crops, requiring little to no herbicides, fungicides, pesticides, and fertilizers (Summerscales et al., 2010; Cutter, 2011).

The hemp bast fibers obtained from its stem have a length of 1.0–2.5 m (Summerscales et al., 2010) while its elementary fibers length averages at 12–25 mm (Cutter, 2011). Both fibers are fine, strong, lustrous, and light in color where their color and cleanliness greatly differs depending on the fiber extraction method used (Tahir et al., 2011). Their fibers are also highly resistant to moisture degradation as it rots very slowly in water although they are hygroscopic in nature similar to natural fibers of other sources (van Rijswijk et al., 2001).

Apart from that, hemp fibers also possess low elongation to break due to its low cellular microfibril angle (Bowen et al., 1994; Cutter, 2011) but excellent mechanical properties like specific strength and stiffness comparable to glass fibers (Shahzad, 2013). The above mentioned latter characteristics make hemp fibers an enticing material as reinforcements to replace synthetic fibers in the production of thermoset polymeric composites (Wotzel et al., 1999; Mwaikambo and Ansell, 2006).

9.2.4 Kenaf

Kenaf (*Hibiscus cannabinus* L.) of Malvaceae family is an annual plant originating from central Africa and is also commonly found as a wild plant in tropical and subtropical Asia (Cheng et al., 2004). It grows 1.5–3.5 m tall and 1–5 cm stem diameter (Nishino, 2013; Zimniewska et al., 2011) but other sources claimed that it can reach heights of 2.4–6.0 m in 4–5 months (Summerscales et al., 2010; Cutter, 2011). The stems have two distinct types of fibers, long bast fibers and woody like short core fibers in a 30:70 (w/w) ratio (Sanadi et al., 1996; Saba et al., 2015a) where these fibers differ in their appearance and anatomical structure (Voulgaridis et al., 2000).

However, both type of fibers are characteristically similar to wood fibers unlike jute, hemp, and flax fibers (Tahir et al., 2011). According to Tahir et al. (2011) and Tahir et al. (2014), kenaf dry fibers yield was reported to be 5–6% of fresh stems equivalent to 18–22% of dry plant which is higher than jute, hemp, and flax. Moreover, kenaf fibers yield was even found to be greater by 3–5 fold than southern pine with a production of 5–10 tons of dry fiber/acre (Sen and Reddy, 2011). This makes kenaf a cost effective and an attractive raw material for its natural fibers.

Kenaf bast fiber bundles consist of short elementary fibers with an average of 2.5 mm length (Saba et al., 2015a) while Cutter (2011) stated that the elementary fibers length is in the range from 1.5 to 6 mm. These fibers are thus too short for textile production (Calamari et al., 1997). Besides being too short, its fibers are brittle, coarse due to its striated surface, nonuniform as a result of its irregular shape making them difficult to process using existing textile equipment (Cutter, 2011; Tahir et al., 2011). Nevertheless, it possesses good mechanical properties similar to those of jute but with a relatively lower specific gravity as it has a lower cellulose content (Cutter, 2011; Sen and Reddy, 2011). Its favorable mechanical properties enables it to be used for pulp and paper (Ashori, 2006) as well as reinforcements for polymeric composites (Yousif et al., 2012).

9.2.5 Cell wall architecture of bast fibers

Bast fibers are defined as tightly joint fiber cells in bundles present in plant stems such as hemp, jute, flax, and kenaf (Haugan and Holst, 2013). Each of these bast fiber cells have an empty space called lumen surrounded by a cell wall (Bos et al., 2006) comprising of cellulose, hemicellulose, and lignin (Fig. 9.2) (Naik and Fronk, 2013). Its cell wall gradually thickens up till the point the lumen will appear as if it has almost disappeared as the bast fiber cells reaches maturity during plant development (Raven et al., 2005). The cell wall has two main layers, the primary (outermost layer) and secondary (innermost layer) wall with the secondary wall

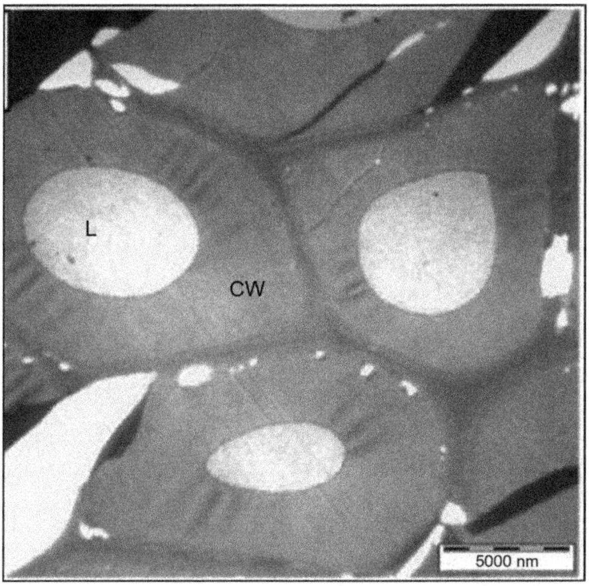

Figure 9.2 Transverse section of kenaf bast fibers where L and CW indicates lumen and cell wall, respectively.
Source: Abdul Khalil, H.P.S., Ireana Yusra, A.F., Bhat, A.H., Jawaid, M., 2010. Cell wall ultrastructure, anatomy, lignin distribution and chemical composition of Malaysian cultivated kenaf fiber. Ind. Crops Prod. 31(1), 113–121.

thicker than the primary wall and is only formed deposited inside the primary wall after the primary wall is completely developed. The secondary wall is made up of another three distinct layers known as S1, S2, and S3 with S2 being the thickest layer, thus playing a significant role in the bast fiber cells' mechanical properties (Meshram and Palit, 2013). On the contrary, Romhany et al. (2003) and Blake et al. (2008) discovered that secondary cell walls consist of multiple layers after examining the micrographs of bast fiber cross-sections and a speculative model of the bast fiber cell wall structure were drawn and shown in Fig. 9.3.

Every cell wall layers are reinforced with cellulosic macrofibrils (made up of cellulosic microfibrils) and microfibrils embedded in a hemicellulose, lignin and pectin matrix (Naik and Fronk, 2013). The cellulosic microfibrils are arranged differently in the primary and secondary wall (Hughes, 2004). In the primary wall, cellulosic microfibrils are generally randomly arranged in the longitudinal direction while the cellulosic microfibrils are arranged in a helical manner, winding around the fibers longitudinal axis in the secondary wall (Beck, 2005). In each of the three distinct secondary wall layers, the cellulosic microfibrils winds and twists in different directions. However, the twist orientation of the thickest layer, S2 is used to designate the overall cellulosic microfibrillar

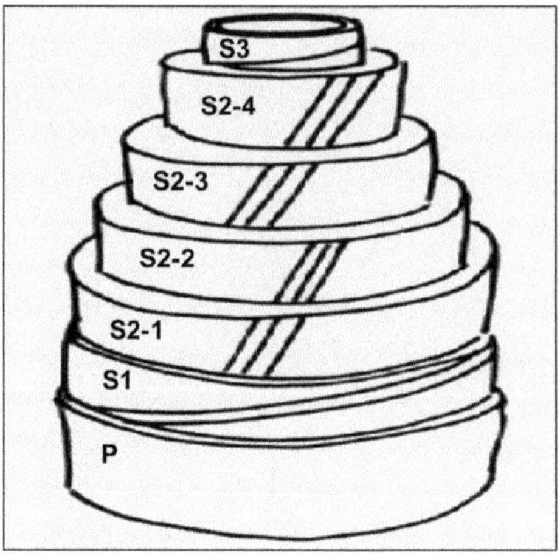

Figure 9.3 Schematic of possible bast fiber cell wall structure.
Source: Eder, M., Burgert, I., 2010. Natural fibres – function in nature. In: Mussig, J. (Ed.), Industrial Applications of Natural Fibres: Structure, Properties and Technical Applications, Wiley, Hoboken, New Jersey, pp. 23–40.

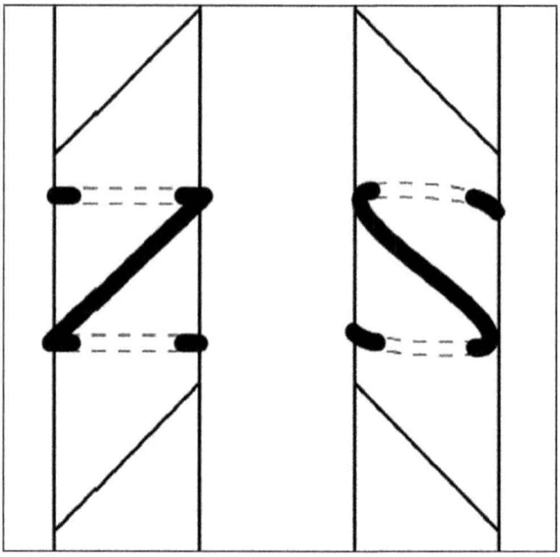

Figure 9.4 Cellulosic microfibrillar orientation—Z-twist and S-twist.
Source: Skoglund, G., Nockert, M., Holst, B., 2013. Viking and early middle ages northern Scandinavian textiles proven to be made with hemp. Sci. Rep. 3, <http://dx.doi.org/10.1038/srep02686> (accessed 15.04.16).

orientation of a bast fiber cell wall as either Z (right-handed) or S (left-handed) twist (Fig. 9.4). The S2 twist angle is also otherwise known as the fibrillar angle (φ). For example, the S1 is Z-twist and S2 is S-twist in flax and ramie but the S3 is Z-twist for flax while ramie has cellulosic microfibrillar orientation almost parallel to the bast fiber axis in S3. Accordingly, the overall cellulosic microfibrillar orientation of flax and ramie is S-twist with fibrillary angles (φ) of 6.5 and 7.0 degrees, respectively based on their S2 (Haugan and Holst, 2013).

9.3 Characterization of the bast fibers

9.3.1 Chemical composition

Bast fiber cells are chemically simple materials consisting of three main chemical constituents, cellulose, hemicellulose, and lignin with smaller amounts of pectin (Summerscales et al., 2010). All of these chemical constituents are distributed throughout the cell wall including the primary and secondary cell wall layers (Faruk et al., 2012). In the bast fiber cell wall layers, in which approximately half of it consists of cellulose bundles held together by roughly a quarter of hemicellulose and lignin matrix (Biagiotti et al., 2004). The remaining quarter of chemical constituents including pectin, protein, mineral substances, resin, tannin, dye, wax, and fat also helps in cementing the cellulose bundles (Bogoeva-Gaceva et al., 2007). Table 9.2 shows some of the main chemical constituents for a few bast fibers.

9.3.2 Physical properties

Natural fibers (fibers bundles or unit fibers) physical properties include their fineness, density, length, width, lumen diameter, cell wall thickness, microfibrillar angle, crystallinity, moisture content, and absorption (Table 9.3) (Franck, 2005; Goda and Cao, 2007; Celino et al., 2013). These physical properties of natural fibers including bast fibers vary considerably even for a specific given fiber like hemp, jute, kenaf, or flax fibers (Celino et al., 2013). For example, the hemp fibers' physical property, density was reported to be 1.48 g cm^{-3} by Faruk et al. (2012) but this data obtained differ a bit with the findings of Biagiotti et al. (2004) who found that it is supposed to be in the range of 1.40–1.50 g cm^{-3}. Another example will be the flax fibers physical property, moisture content which was revealed to be 8–12% (Thiruchitrambalam

Table 9.2 Chemical constituents of bast fibers

Bast fibers	Chemical composition (wt%)			
	Cellulose	Hemicelloluse	Lignin	Pectin
Jute	51.0–72.0	12.0–20.4	5.0–13.0	0.2
Hemp	70.0–78.0	17.9–22.0	3.7–5.0	0.9
Flax	60.0–81.0	14.0–18.6	2.0–3.0	1.8–2.3
Kenaf	36.0	21.0	18.0	2.0
Ramie	68.6–76.0	13.1–15.0	0.6–1.0	1.9–2.0

Source: Biagiotti, J., Puglia, D. and Kenny, J.M. 2004. A review on natural fibre-based composites—part 1: structure, processing and properties of vegetable fibres. J. Nat. Fibers 1(2), 37–68.

Table 9.3 Physical properties of bast fibers

Bast fibers	Density (g cm^{-3})	Specific gravity (g cm^{-3})	Diameter (μm)	Microfibrillar angle (degree)	Moisture content (%)
Jute	1.30–1.50	1.30	10–200	8.0	10.0–13.7
Hemp	1.40–1.50	1.30	16–2000	2.0–6.2	6.2–12.0
Flax	1.40–1.50	1.20	20–620	5.0–10.0	8.0–12.0
Kenaf	1.19–1.40	1.04	17.7–100	—	12.0
Ramie	1.50–1.55	1.16	10–34	7.5	7.5–17.0

Sources: Bledzki, A.K., Gassan, J. 1999. Composites reinforced with cellulose based fibres. Prog. Polym. Sci. 24(2), 221–274; Mohanty, A.K., Misra, M., Hinrichsen, G., 2000. Biofibres, biodegradable polymers and biocomposites: an overview. Macromol. Mater. Eng. 276–277(1), 1–24; Biagiotti, J., Puglia, D., Kenny, J.M., 2004. A review on natural fibre-based composites – part 1: structure, processing and properties of vegetable fibres. J. Nat. Fibers 1(2), 37–68; Bogoeva-Gaceva, G., Avella, M., Malinconico, M., Buzarovska, A., Grozdanov, A., Gentile, G., et al., 2007. Natural fiber eco-composites. Polym. Compos. 28(1), 98–107; Goda, K., Cao, Y., 2007. Review paper: research and development of fully green composites reinforced with natural fibers. J. Solid Mech. Mater. Eng. 1(9), 1073–1084; Li, X., Tabil, L.G., Panigrahi, S., 2007. Chemical treatments of natural fiber for use in natural fiber-reinforced composites: a review. J. Polym. Environ. 15(1), 25–33; Cutter, A.G., 2011. Development and Characterization of Renewable Resource-Based Structural Composite Materials. ProQuest, Ann Arbor, Michigan; Faruk, O., Bledzki, A.K., Fink, H.P., Sain, M., 2012. Biocomposites reinforced with natural fibers: 2000–2010. Prog. Polym. Sci. 37(11), 1552–1596; Thiruchitrambalam, M., Alavudeen, A., Venkateshwaran, N., 2012. Review on kenaf fiber composites. Rev. Adv. Mater. Sci. 32(2), 106–111.

et al., 2012) but Mohanty et al. (2000) obtained a value of 10% for it. These differences occur due to the fibers chemical composition, plant size and maturity, environmental conditions during the plants growth (Bourmaud et al., 2013), different testing environmental conditions, and testing methods used (Placet et al., 2012) as well as different fiber extraction methods employed (Mohanty et al., 2000).

9.3.3 Mechanical properties

Bast fibers (fibers bundles or unit fibers) like other natural fibers exhibit good specific mechanical properties especially when it comes to their strength and stiffness which is comparable to synthetic fibers like glass fibers but with the added advantage of a lower density (Li et al., 2007; Cutter, 2011). Table 9.4 lists some of the bast fibers main mechanical properties including tensile strength, Young's modulus, elongation at break, fracture stress, specific modulus, and specific strength (Bogoeva-Gaceva et al., 2007; Goda & Cao, 2007; Faruk et al., 2012). These mechanical properties are of great importance as they are usually evaluated when considering bast fibers suitability for a certain application (Bledzki and Gassan, 1999). However, one major drawback of bast fibers are their varying mechanical properties even among similar plant type fibers like hemp bast fibers mechanical properties may differ from one fiber to another (Celino et al., 2013). One such example is the considerable range of hemp bast fibers tensile strength at 580–1110 MPa or the flax bast fibers tensile strength from 343 to 1035 MPa (Biagiotti et al., 2004). These differences occur due to the very same factors that

Table 9.4 **Mechanical properties of bast fibers**

Bast fibers	Tensile strength (MPa)	Specific Strength (GPa $cm^3 g^{-1}$)	Young's Modulus (GPa)	Specific Modulus (GPa $cm^3 g^{-1}$)	Fracture Stress (MPa)	Elongation at Break (%)
Jute	187–800	303.10	3–55	2.0–42.3	393–800	1.16–3.10
Hemp	550–1110	609.20	3–90	2.0–60.0	270–900	1.30–4.70
Flax	343–2000	1.60	27–85	19.0–71.0	345–1500	1.20–3.20
Kenaf	295–930	430.80	22–53	18.0–38.0	–	1.50–6.90
Ramie	220–938	482.80	44–128	21.1	–	1.20–3.80

Sources: Bledzki, A.K., Gassan, J. 1999. Composites reinforced with cellulose based fibres. Prog. Polym. Sci. 24(2), 221–274; Biagiotti, J., Puglia, D., Kenny, J.M., 2004. A review on natural fibre-based composites – part 1: structure, processing and properties of vegetable fibres. J. Nat. Fibers 1(2), 37–68; Bogoeva-Gaceva, G., Avella, M., Malinconico, M., Buzarovska, A., Grozdanov, A., Gentile, G., et al., 2007. Natural fiber eco-composites. Polym. Compos. 28(1), 98–107; Goda, K., Cao, Y., 2007. Review paper: research and development of fully green composites reinforced with natural fibers. J. Solid Mech. Mater. Eng. 1(9), 1073–1084; Li, X., Tabil, L.G., Panigrahi, S., 2007. Chemical treatments of natural fiber for use in natural fiber-reinforced composites: a review. J. Polym. Environ. 15(1), 25–33; Mohanty, A.K., Misra, M., Hinrichsen, G., 2000. Biofibres, biodegradable polymers and biocomposites: an overview. Macromol. Mater. Eng. 276–277(1), Cutter, A.G., 2011. Development and Characterization of Renewable Resource-Based Structural Composite Materials. ProQuest, Ann Arbor, Michigan; Faruk, O., Bledzki, A.K., Fink, H.P., Sain, M., 2012. Biocomposites reinforced with natural fibers: 2000–2010. Prog. Polym. Sci. 37(11), 1552–1596.

affect the bast fibers physical properties which also affects its mechanical properties (Celino et al., 2013). Besides that, the bast fibers physical properties also influence its mechanical properties (Bogoeva-Gaceva et al., 2007). For example, the fibers mechanical properties, strength, and stiffness are mainly influenced by its chemical composition, cellulose content and also its physical property, microfibrillar angle where a higher cellulose content and a lower microfibrillar angle increases its mechanical properties (Mohanty et al., 2000).

9.4 Hybrid bast fibers reinforced thermoset composites

The term hybrid originates from Greek-Latin and is often used in the polymer composites field to refer to two or more reinforcing and filling materials being incorporated into a single matrix leading to the formation of hybrid composites (Jawaid and Abdul Khalil, 2011). The reinforcing materials can either consist of a mixture of bast fibers and other cellulosic fibers or bast fibers and synthetic fibers while the matrix used is a thermoset matrix (Sathishkumar et al., 2014). The incorporation of hybrid fibers was shown to enhance the physical and mechanical properties of the hybrid composites produced where the advantages of one type of fiber offsets what are lacking in the other fiber type and vice versa (Shahzad, 2011). Furthermore, the hybridization of fibers as reinforcements for thermosets aids in achieving a balance between performance and cost of composites, hence propelling the utilization of hybrid composites prospects in a variety of applications including higher load bearing structural applications (Al-Harbi, 2001). For example, the hybridization of synthetic fibers and bast fibers into thermosets not only offsets the bast fibers inferior mechanical properties (Sanjay and Yogesha, 2016) but also reduces the composites moisture absorption due to the lower content of bast fibers (Salman et al., 2015). Similarly, the addition of bast fibers with synthetic fibers into thermosets lessen the composites production cost as well as the composites weight while maintaining the composites desired mechanical properties (Salman et al., 2015) and its biodegradability although at a lower level (Shahzad, 2011).

9.4.1 Potential and challenges in development of hybrid composites

9.4.1.1 Fiber–polymer matrix interface

The fiber–polymer matrix interface may be a distinct phase or in some cases a planar region consisting of only a few atoms in thickness as a result of a reaction between the fiber and the polymer matrix where its properties differ from both the fiber and the polymer matrix. Despite its thinness, the bonding occurring in the fiber–polymer matrix interface significantly influences the composites mechanical properties as it assists in transferring stress from the polymer matrix to the fiber (Jawaid and Abdul Khalil, 2011). Thus, there must be a strong bonding between the fiber and the polymer matrix to produce composites with desirable mechanical properties, strength, and stiffness. However, an interface bonding that is too strong results in the composites being too brittle (low resistance to fracture) whereas a weak interface bonding produces composites exhibiting low stiffness and strength (Park and Seo, 2011). Hence, there is a need to achieve optimum interfacial bonding. In order to achieve optimum interfacial bonding, there must be an intimate contact between the fiber and the polymer matrix. This is affected by wettability defined as the extent to which a polymer matrix spreads over the fiber. A good wettability means that the polymer matrix liquid spreads and covers every bumps and dips of the fiber rough surface displacing air in the process. This promotes intimate contact between both the fiber and the polymer matrix. A poor wettability produces composites with interfacial defects acting as stress concentrators in it (Pickering et al., 2016).

There are a few types of interfacial bonding mechanism such as mechanical interlocking, chemical bonding, and interdiffusion bonding where one or more types of bonding mechanism may occur at the same time in the composites interface (Pickering et al., 2016). Mechanical interlocking is more effective with rougher fiber surface, increasing the composites shear strength considerably but has very little strengthening effect on the transverse tensile strength. Unlike mechanical bonding, chemical bonding strength does not depend on the fiber surface roughness but on the reactivity of the chemical groups on the fiber surface with the polymer matrix chemical groups to form chemical bonds per unit area and on the type of chemical bonds formed. Interdiffusion bonding occurs from the two components (fiber and polymer matrix) molecules interdiffusion and intertwining in which its bonding strength are affected by the distance and degree of molecules intertwined and also the number of molecules per unit area at the interface (Park and Seo, 2011). Unfortunately, weak interfacial bonding occurs between the fiber and the polymer matrix due to the incompatibility of fiber (hydrophilic) and polymer matrix (hydrophobic) (John and Anandjiwala, 2008). To improve this, extensive research had been carried out and reported that mechanically or chemically treating the fibers helps to enhance the weak interfacial bonding. Mechanical treatments include corona, plasma, ultraviolet (UV), heat treatments electron radiation and fiber beating while some of the chemical treatments available are alkali, acetyl, silane, acrylonitrile, and maleated anhydride grafted coupling agent (Adekunle, 2015).

9.4.1.2 Moisture content of bast fibers

Bast fibers constitute primarily of cellulose in its cell walls similar to other types of natural fibers and usually have moisture contents (5–15%) (Jawaid and Abdul Khalil, 2011) in its voids as well as in its noncrystalline/amorphous regions (Bledzki and Gassan, 1999). The bast fibers chemical constituent, cellulose contains many hydroxyl groups forming hydrogen bonds between its macromolecules within the bast fiber cell walls. These hydrogen bonds will break as soon as moisture (water molecules) from the surrounding environment comes into contact with cellulose, thereby freeing the cellulose macromolecules hydroxyl groups. The cellulose macromolecules free hydroxyl groups will next form new hydrogen bonds with water molecules instead (Kabir et al., 2012). Thus, resulting in bast fibers being hydrophilic in nature (Saheb and Jog, 1999). This causes polymer composites reinforced with bast fibers to exhibit a high moisture absorption when exposed to water even with the presence of hydrophobic polymer matrix. The moisture absorbed into the composites induces the bast fibers within it to swell leading to diminishing bonding strength at its interface and ultimately results in the composites microcracking, dimensional instability, and poorer mechanical properties (Biagiotti et al., 2004).

One such example was reported by Raghavendra et al. (2015) who found that jute fiber reinforced epoxy composites exhibited depreciating mechanical properties, tensile and flexural strength after being immersed in water. In contrast, Phani and Bose (1987) had discovered the added benefits of jute fibers incorporation into glass fiber reinforced thermoset composites where jute fibers assisted in lowering its composites degradation rate after being exposed to moisture for more than 70 hours. This was either probably due to the swollen jute fiber layers accommodation of the resin swelling strain or the swollen jute fibers protection to a central glass fiber layer. Despite the findings of Phani and Bose (1987), it is evidently essential to remove moisture from the bast fibers before its used as a reinforcement in polymers. A few other researchers had also suggested subjecting bast fibers to chemical treatments which removes its hydrophilic hydroxyl groups to reduce its moisture absorption (Wang et al., 2007). Besides that, bast fiber chemical treatments also enhance the bast fibers reinforced polymer composites interfacial adhesion as a good interfacial adhesion lowers the rate and amount of water absorbed by the composites. Other alternatives include coating the bast fibers reinforced polymer composites with surface barriers but this might be too costly (Biagiotti et al., 2004).

9.4.1.3 Dispersion of bast fibers in the matrix

Bast fibers for the most part are made up of cellulose which imparts polarity and hydrophilicity to the fibers due to the high density of hydroxyl groups on its surface at the cell walls (Saheb and Jog, 1999). These hydroxyl groups on its surface has a tendency to form hydrogen bonding with hydroxyl groups occurring on the surface of other bast fibers adjacent to it while having less interaction with the polymer matrix (Wang, 2008). Thus resulting in agglomeration/entanglement of the bast fibers or otherwise known as weak dispersion of bast fibers when used as reinforcements in nonpolar and hydrophobic polymer matrix unlike polar polymer matrix. Bhatnagar (2004) discovered that the dispersion of natural fibers in thermoplastic polymer matrix, polyvinyl alcohol (PVA) was not an issue due to the polarity and hydrophilicity of the polymer matrix used. However, the majority of polymers are nonpolar and hydrophobic in nature. The findings of Bhatnagar (2004) was supported by Barkoula and Peijs (2011) who explained that the occurrence of a strong interaction between natural fibers and polymer matrix is a result of their similar polarity.

This in turn promotes natural fibers wetting by molten polymer matrix, thus leads to a good dispersion of natural fibers in the polymer matrix. Various methods and treatments can be employed to change the natural fibers polarity to make it less polar, hence less hydrophilic as this will increase its compatibility with the hydrophobic polymer matrix. Some of the treatments include physical and chemical treatments (Rahman et al., 2015). Other than that, Nando and Gupta (1996) had suggested the wetting of natural fibers to prevent hydrogen bonding between fibers before mixing them with the polymer matrix but this will induce the formation of pores in the composites formed. Apart from that, increased mixing time assists in fibers dispersion in polymer matrix rapidly but only up to a certain extent as it was found that fibers dispersion gradually slows down with mixing time. Shorter natural fibers were also recommended for use as longer fibers are more inclined to agglomerate (Pickering et al., 2016).

9.4.1.4 Thermal stability

Bast fibers thermal stability is defined as the bast fibers resistance to decomposition/degradation at higher temperatures up to a certain extent. The fibers thermal stability is influenced by their chemical constituents, cellulose, hemicellulose, lignin and pectin. Each of these chemical constituents are sensitive to a different range of temperatures, hence the different stages of fibers decomposition resulting in the fibers weight loss (Saheb and Jog, 1999). Hemp fibers for example, starts to decompose and suffer from weight loss at 50°C due to the evaporation of its moisture content. Once the temperature reaches above 160°C, hemp fibers binding material, lignin begins to soften causing physical and chemical changes within the fibers. At about 270°C, the hemp fibers weight loss is attributed to the decomposition of hemicellulose or pectin while the weight loss at 360°C are associated with cellulose decomposition (Shahzad, 2011). Troedec et al. (2008) reported that the hemp fibers hemicellulose and pectin degradation corresponds to 320–370°C while its cellulose degradation occurs at 390–420°C. Generally, natural fibers starts degrading at 200°C (Jawaid and Abdul Khalil, 2011).

These researchers revealed that the fibers goes through changes in its physical and chemical properties after exposure to high temperatures. Studies showed that the changes in the fibers physical and chemical properties also affects its mechanical properties negatively. This was supported by Sridhar et al. (1982) who observed a 60% decrease in jute fibers tensile strength after heating under vacuum at 300°C for 2 hours. Mohanty et al. (2000) also had the same findings in their review paper where ramie fibers were reported to suffer from nearly 10% reduction in its tensile strength after being exposed to a high temperature at about 200°C for a duration of only 10 minutes. The reduction in the natural fibers mechanical properties will in turn affect the natural fiber reinforced polymer composites strength. Therefore, there is a need to consider the temperature and duration in the processing of the polymer composites to avoid thermal degradation of the natural fibers within it (Summerscales et al., 2010). Besides that, there are also ways to improve the natural fibers thermal stability by grafting monomers onto the fibers evidently shown by Saheb and Jog (1999).

9.4.1.5 Biodegradability

Bast fibers are easily biodegraded by microorganisms, bacteria, and fungi as these organisms are capable of recognizing the lignocellulosic materials such as cellulose, hemicellulose, and lignin present in its cell walls (Saheb and Jog, 1999). Next, the bacteria and fungi produces the necessary specific enzymes to break down the lignocellulosic materials in the fibers cell wall into smaller units to be assimilated into the microorganisms for its nutritional needs and growth (Mohanty et al., 2000). Out of the two microorganisms, bacteria were found to degrade lignocellulosic materials at a slower rate than fungi (Kuhad et al., 1997). As a result of the natural fibers biodegradation, the fibers losses its mechanical strength and this will in turn drastically lower the natural fibers reinforced polymer composites mechanical strength and its service life (Walentowska and Kozlowski, 2012). Additionally, the incorporation of biodegradable natural fibers into either biodegradable or nonbiodegradable polymers enhances the polymers biodegradability. Thus, limiting these polymer composites from outdoor applications (Jawaid and Abdul Khalil, 2011). However, certain measures can be taken to reduce the fibers biodegradability by altering its cell wall chemistry through chemical treatments (Joseph et al., 1999).

9.5 Hybrid bast fiber reinforced thermoset composites processing

There are a few methods frequently used by researchers to produce hybrid bast fiber reinforced thermoset composites such as hand lay-up, compression moulding and pultrusion where some methods are favoured over the other (Table 9.5). Out of all

Table 9.5 Reported works on hybrid bast fibers reinforced thermoset composites processing

Fibers	Matrix	Fiber fraction (wt%/vol%)	Processing	Tensile strength	Tensile modulus	Flexural Strength	Flexural modulus	Impact strength	References
Flax/basalt/glass	Epoxy	21.18 vol% (flax: basalt:glass, 11.72:7.16:2.30)	Vacuum infusion	153.16 ± 17.41 MPa	8.11 ± 0.60 GPa	137.95 ± 19.85 MPa	8.02 ± 0.68 GPa	–	Petrucci et al. (2013)
Hemp/basalt/ glass		22.53 vol% (hemp: basalt:glass, 8.56:11.38:2.59)		128.84 ± 8.70 MPa	6.64 ± 0.49 GPa	126.22 ± 13.63 MPa	5.90 ± 0.42 GPa		
Flax/hemp/ basalt		21.18 vol% (flax: hemp:basalt, 9.11:7.85:5.57)		115.97 ± 3.77 MPa	7.69 ± 0.63 GPa	128.46 ± 29.14 MPa	7.45 ± 0.67 GPa		
Jute/Glass *Note*: i. Jute (J) ii. Glass fibers (GF) iii. Glass chopped strand mat (GCSM) iv. Surface veil (SV)	Unsaturated polyester	*Dry composites*	Pultrusion	*Dry composites*				–	Akil et al. (2014)
		70 vol% (J:GF, 1:1)		261.22 ± 8.06 MPa	25.70 ± 0.82 GPa	366.38 ± 21.05 MPa	23.40 ± 0.46 GPa		
		70 vol% (J:GF:GCSM, 1:1:1)		266.22 ± 17.85 MPa	27.50 ± 0.69 GPa	343.32 ± 23.19 MPa	24.60 ± 0.31 GPa		
		[70 vol % (J:GF, 1:1)] + SV		122.71 ± 7.03 MPa	15.90 ± 1.02 GPa	136.80 ± 9.89 MPa	15.40 ± 0.78 GPa		
		Soaked composites (4076 h)		*Soaked composites (4076 h)*					
		70 vol % (J:GF, 1:1)		152.40 ± 9.86 MPa	18.30 ± 1.06 GPa	297.54 ± 15.61 MPa	19.50 ± 0.86 GPa		
		70 vol % (J:GF:GCSM, 1:1:1)		166.49 ± 10.95 MPa	20.10 ± 0.69 GPa	276.35 ± 7.04 MPa	21.50 ± 0.58 GPa		
		[70 vol % (J:GF, 1:1)] + SV		64.48 ± 8.31 MPa	11.70 ± 0.88 GPa	119.61 ± 9.60 MPa	11.90 ± 0.61 GPa		
Kenaf/Glass	Unsaturated polyester	–	Hand lay-up & Compression molding	–	*Dry composites* Day 1 till Week 4 8.36 GPa *Distilled water soaked* Day 1–6.53 GPa Week 1–6.66 GPa Week 2–6.46 GPa Week 3–6.83 GPa	–	–	–	Ghani et al. (2012)

					Week 4–5.64 GPa *Rain water soaked* Day 1–8.27 GPa Week 1–6.65 GPa Week 2–6.92 GPa Week 3–6.74 GPa Week 4–6.06 GPa *Sea water soaked* Day 1–7.19 GPa Week 1–7.15 GPa Week 2–6.37 GPa Week 3–6.64 GPa Week 4–5.80 GPa				
Jute (J)/banana (B)	Epoxy	Weight ratio J:B, 100:0	Hand lay-up & Compression molding	16.62 MPa	664 MPa	57.22 MPa	8956 MPa	13.44 kJ/m^2	Boopalan et al. (2013)
		Weight ratio J:B, 75:25		17.89 MPa	682 MPa	58.60 MPa	9065 MPa	15.81 kJ/m^2	
		Weight ratio J:B, 50:50		18.96 MPa	724 MPa	59.84 MPa	9170 MPa	18.23 kJ/m^2	
		Weight ratio J:B, 25:75		18.25 MPa	720 MPa	59.30 MPa	9056 MPa	17.89 kJ/m^2	
		Weight ratio J:B, 0:100 *(Fiber fraction not stated)*		17.92 MPa	718 MPa	58.06 MPa	9048 MPa	16.92 kJ/m^2	
Flax/carbon	Epoxy	2 cross ply flax fiber + *unidirectional ply carbon fiber*	Compression molding	284.80 ± 0.02 MPa	11.90 ± 0.01 GPa	145.00 ± 0.18 MPa	9.71 ± 0.34 GPa	–	Dhakal et al. (2013)
		2 unidirectional ply flax fiber + *unidirectional ply carbon fiber* *(Fiber fraction not stated)*		–	–	318.83 ± 0.22 MPa	28.83 ± 0.29 GPa		
Jute/Palmyra palm leaf stalk (PPLS)	Unsaturated polyester	30 wt % (jute:PPLS, 0:100)	Compression molding	56.90 ± 1.60 MPa	2.28 ± 0.34 GPa	105.49 ± 2.41 MPa	15.32 ± 1.53 GPa	36.38 ± 8.14 kJ/m^2	Shanmugam & Thiruchi trambalam (2013)
		30 wt % (jute:PPLS, 25:75)		60.30 ± 1.59 MPa	2.84 ± 0.34 GPa	116.83 ± 15.20 MPa	16.83 ± 0.64 GPa	34.87 ± 6.12 kJ/m^2	
		30 wt % (jute:PPLS, 50:50)		64.30 ± 1.95 MPa	2.45 ± 0.40 GPa	145.66 ± 9.35 MPa	17.95 ± 0.53 GPa	27.01 ± 4.13 kJ/m^2	
		30 wt % (jute:PPLS, 75:25)		83.30 ± 5.13 MPa	3.78 ± 0.60 GPa	164.00 ± 12.14 MPa	18.23 ± 0.95 GPa	26.02 ± 3.34 kJ/m^2	
		30 wt % (jute:PPLS, 100:0)		77.10 ± 3027 MPa	5.07 ± 0.20 GPa	176.00 ± 3.50 MPa	19.26 ± 1.29 GPa	24.71 ± 3.09 kJ/m^2	

(Continued)

Table 9.5 (Continued)

Fibers	Matrix	Fiber fraction (wt%/vol%)	Processing	Tensile strength	Tensile modulus	Flexural Strength	Flexural modulus	Impact strength	References
Flax (F)/ Glass (G)	Phenolic	67 vol % (F:G, 50:50) stacked GF 67 vol % (F:G, 50:50) stacked GGFF 67 vol % (F:G, 50:50) stacked GGGGFFFF	Compression molding	450.1 ± 16.5 MPa 412.5 ± 12.7 MPa 392.5 ± 20.0 MPa	40.1 ± 1.7 GPa 40.8 ± 1.4 GPa 39.7 ± 0.6 GPa	–	–	–	Zhang et al. (2013)
Hemp/Glass Hemp/Banana/ Glass	Epoxy	–	Hand lay-up	37.5 MPa 28.0 MPa	–	0.29 kN 0.51 kN	–	5.33 J 8.66 J	Bhoopathi et al. (2014)
Flax (F)/ Banana (B)/ Glass (G) Flax (F)/ Glass (G)	Epoxy	[40 vol % (F layer, B layer, F layer)] + G surface veil [40 vol % (3 F layers)] + G surface veil	Hand lay-up	30 MPa 32 MPa	–	13.54 MPa 11.59 MPa	–	16 J 11 J	Srinivasan et al. (2014)
Jute/Vetiver Jute/Vetiver/ Glass	Vinyl ester	*Treated vetiver* 34 wt % (jute:vetiver, 17:17) 34 wt % (jute:vetiver, 24:10) 34 wt % (jute:vetiver, 10:24) 34 wt % (jute:vetiver: glass, 13:13:8) 34 wt % (jute:vetiver: glass, 10:10:14)	Hand lay-up	71.73 MPa 63.30 MPa 64.53 MPa 74.14 MPa 70.96 MPa	–	133.11 MPa 114.79 MPa 121.31 MPa 131.90 MPa 137.60 MPa	–	11.00 J 10.33 J 11.67 J 15.33 J 18.33 J	Vinayagamoorthy & Rajeswari (2014)

the fabrication methods, hand lay-up is often used as it is the simplest method (Gupta and Srivastava, 2016) but it still requires the necessary workmanship skills to ensure the polymer composites uniformity (in terms of thickness), fiber to polymer matrix ratio and void content throughout the polymer composites produced (Kabir et al., 2012). This method basically requires the manual placement and arrangement of fibers in woven fabric form or in chopped form in the mould. Next, a mixture of molten thermosetting polymer and hardener is poured onto the fibers in the mould to wet the fibers. A brush is later used to evenly spread the molten polymer while a roller is used in the removal of air as well as excess molten polymer before the mould is closed and left to cure or pressed at room temperature (Sreenivasan et al., 2013; Ghani et al., 2012).

Other than hand lay-up, compression moulding are also often used to form hybrid bast fiber reinforced thermoset composites by placing the fibers, fiber mats (Abdellaoui et al., 2015) or loosely chopped fibers either randomly oriented (Gopinath et al., 2014) or aligned (Coroller et al., 2013) together with molten polymer matrix into the mould cavity. The mould is later closed and placed in a compression molder for compression moulding at the required temperature, pressure, and time before the composites formed are removed from the mold. In order to produce good quality polymer composites, the temperature and time needs to be carefully controlled (Pickering et al., 2016) so as to prevent the bast fibers degradation from occurring when exposed to temperatures above 200°C for a long duration of time. Hence reducing its fibers strength significantly. This was supported by Herrmann et al. (1998) who observed a 10% reduction in natural fibers strength after being heated at 200°C for only 10 minutes. With careful control of temperature and time, researchers had found that this method has the added advantage of producing polymer composites with a low number of air voids in it and the possible incorporation of various fibers length from short to long (Sreenivasan et al., 2013).

The other less frequently used method are pultrusion in the production of hybrid bast fiber reinforced thermoset composites (Sathishkumar et al., 2014; Saba et al., 2015b). Pultrusion is a continuous process involving the pulling of continuous fibers (cords or strands) through a polymer resin bath to impregnate the fibers with polymer resin. This is followed by a separate preforming system where it is shaped and rid of excessive polymer resin before it is guided through a heated die to allow the polymer resin to cure (Anandjiwala and Blouw, 2007). The final product obtained a constant cross-sectional shaped composite which exhibits good mechanical properties and also dimensional stability (Sreenivasan et al., 2013). This was shown by Akil et al. (2011) who proved that the pultrusion fabricated composites achieved a good flexural strength at 250 MPa with 70% fiber composition which can only be attained through this method. However, there are also some drawbacks to this method where its composites good mechanical properties and dimensional stability are only limited to the cross-sections. Besides that, the composites fabricated shape is also limited to cylindrical form (Sreenivasan et al., 2013).

9.6 Physical and mechanical properties of hybrid bast fibers reinforced thermoset composites

9.6.1 Epoxy based hybrid composites

The physical and mechanical properties of epoxy based bast fiber hybrid composites had been studied by several researchers as reported by Saba et al. (2015b) and Gupta and Srivastava (2016) in their review papers. Some of the researchers include Ramesh and Nijanthan (2016) who fabricated kenaf-glass fiber reinforced epoxy composites with two different fiber orientations at 0 and 90 degrees. These composites were tested for their mechanical properties, tensile strength, impact strength, and flexural strength. They concluded that the kenaf-glass fiber reinforced epoxy composites mechanical properties with 0 degree fiber orientation are better than the 90 degrees. Finite element analysis (FEA) carried out also validated the experimental results obtained as the experimental results are very close to the FEA model results. Similar study was also conducted by Gujjala et al. (2014) on the mechanical properties of woven jute–glass fibers reinforced epoxy composites where the woven jute and glass mat were stacked in a different sequence. Karahan and Karahan (2015) also investigated the tensile and impact properties as well as the water absorption of jute–carbon woven fabric epoxy composites. They found that the hybridized jute–carbon woven fabric epoxy composites showed significant improvement in its tensile and impact strength compared with the jute woven fabric epoxy composites. The hybrid epoxy composites also exhibited reduced moisture absorption when soaked in water for 2 and 24 hours, respectively.

9.6.2 Polyester based hybrid composites

Ramesh et al. (2013) investigated the mechanical property of sisal–jute–glass fiber reinforced polyester composites in comparison with jute–glass fiber reinforced polyester composites. In terms of tensile strength, they reported that jute–glass fiber reinforced composites exhibits higher tensile strength whereas sisal–jute–glass fiber reinforced composites showed superior flexural and impact strength. Unlike Ramesh et al. (2013), Scutaru and Baba (2014) only focused on hybridizing two type of fibers, carbon, and hemp fabric as polyester composite fillers to determine the composites impact strength with different impact speed and falling heights. They concluded that the hybridized composites produced had a good stiffness. This is in agreement with the findings made by Ahmed et al. (2007) who observed an increased in their fabricated hybrid composites stiffness. In their study, Ahmed et al. (2007) produced woven jute–glass fiber reinforced polyester composites where its stiffness was enhanced with increasing glass fiber content in the hybrid composites. The hybrid composites stiffness enhancement of Scutaru and Baba (2014) and Ahmed et al. (2007) is the result of the glass and carbon fibers high stiffness. Ahmed et al. (2007) also discovered that the addition of glass fibers into woven jute reinforced composites increased the composites tensile strength due to the greater extensibility of glass fibers.

9.6.3 Phenolic based hybrid composites

Ozturk (2010a) is among one of the few researches to report on the influence of synthetic fibers hybridization into phenolic based bast fiber composites. In his research, he hybridized rockwool fibers into jute fiber reinforced phenol formaldehyde composites. He found that the addition of rockwool fibers to jute reinforced phenol formaldehyde composites increases the hybrid composites flexural strength but also decreases its tensile and impact strength. Another researcher by the same last name, Ozturk (2005) investigated the hybridization of basalt fibers into hemp phenol formaldehyde composites. His findings are almost similar in which the hybridization of basalt fibers decreases not only the tensile and impact strength but also the flexural strength. These findings are in agreement with his recent work where he also observed a decreasing trend for the hybrid composites mechanical properties with the addition of fibrefrax fiber into kenaf reinforced phenol formaldehyde composites (Ozturk, 2010b). All of these results are attributed to the weak adhesion between the synthetic fibers (rockwool fiber, fibrefrax fiber, and basalt fiber) with the phenol formaldehyde polymer matrix. Medeiros et al. (2005) on the other hand studied the addition of cellulosic fibers, cotton to jute reinforced phenolic composites by weaving both cotton with different jute roving textures together into a fabric as a reinforcement for phenolic polymer matrix. Next, the hybrid composites mechanical properties were tested in relation with the different test angles and jute roving textures. They concluded that the hybrid composites mechanical properties are strongly dependent on the test angles and jute roving textures. The best overall mechanical properties were the ones tested along the jute roving direction and these properties are reduced with increasing test angles.

9.6.4 Unsaturated polyester based hybrid composites

The physical and mechanical properties of unsaturated polyester resin based bast fiber hybrid composites had been studied by a few researchers already. One of them is Zamri et al. (2011) who had conducted an experiment to determine the effect of water absorption on the jute–glass fiber reinforced unsaturated polyester composites. The hybrid composites were immersed in three different types of water: distilled water, sea water, and an acidic solution for up till 3 weeks to determine the composites water absorption. From the results obtained, they concluded that the water absorption pattern follows the non-Fickian behavior. The composites also exhibited the highest values for diffusion coefficient and maximum moisture content when immersed in distilled water followed by acidic solution with sea water having the lowest value. Besides that, the incorporation of synthetic glass fibers into jute reinforced unsaturated polyester composites were observed to lower the composites moisture absorption and at the same time enhance its mechanical properties, flexural, and compression strength. Similarly, Salleh et al. (2012) had also conducted a research on the water absorption effect on long kenaf-woven glass fiber unsaturated polyester composites where their findings are in agreement with Zamri et al. (2011). Others like Kafi et al. (2006) and Lai et al. (2008) had conducted their research focusing on enhancing the mechanical properties of bast fiber hybrid reinforced unsaturated polyester composites either by chemical or physical fiber treatments. Unlike the others, Hashemi et al. (2015) instead focused on the relationship between the kenaf fiber volume fraction and void volume fraction of kenaf/glass fiber unsaturated polyester composites.

9.6.5 Vinyl ester based hybrid composites

Vinayagamoorthy and Rajeswari (2014) investigated the hybrid woven jute–vetiver–glass fiber reinforced vinyl ester composites mechanical performances in terms of tensile, compressive, flexural and impact strength. In this study, only the vetiver fibers were chemically treated with alkaline solution before undergoing heat treatment to enhance the hybrid composites properties with differing hybrid fiber weight ratios. They found that the vetiver fiber treatments, an increase in glass fiber weight ratio up till 15% and hybridization substantially enhanced the hybrid composites mechanical properties. They also discovered that it is possible to replace the fiber glass in the hybrid woven jute–vetiver–glass fiber reinforced composites with other natural fibers without affecting the hybrid composites mechanical properties negatively except for its impact strength. Li et al. (2015) who studied the hybridization of jute–ramie fiber reinforced polyester composites also came to a similar conclusion that hybridization enhances the mechanical properties only when the ramie fiber fractions are increased. They also discovered that by increasing the jute fiber fractions, this will in turn improved the hybrid composites permeability. Kannan et al. (2015) like Vinayagamoorthy and Rajeswari (2014) also studied the effect of alkaline treatment on natural fibers, jute and banana fibers where both jute and banana fibers were later used together as fillers for vinyl ester polymer matrix. These alkaline treated and untreated hybrid jute–banana fiber reinforced composites were tested for their mechanical properties. In contrast to the findings of Vinayagamoorthy and Rajeswari (2014), Kannan et al. (2015) inferred that the untreated hybrid jute–banana fiber reinforced composites exhibits stronger mechanical properties compared with the treated ones.

9.7 Applications of hybrid bast fibers reinforced thermoset composites

Atiqah et al. (2014) attempted the development of hybrid kenaf-glass reinforced unsaturated polyester composites for structural applications in their research. First, they fabricated the hybrid kenaf-glass reinforced composites where the different fractions of kenaf fibers used were either alkaline treated or untreated before subjecting the hybrid composites for mechanical testing. They discovered that the hybrid composites with 15% volume of treated kenaf fibers possess the highest flexural, tensile and impact strength due to good interfacial bonding between the kenaf fibers and the polymer matrix. Based on their findings, they came to a conclusion that the 15% volume treated kenaf-glass hybrid reinforced composites is suitable for structural applications. This finding is in agreement with Burgueno et al. (2005a), Ray and Rout (2005), Dittenber and GangaRao (2012) and Alam et al. (2015). These researchers concluded that the hybridization of natural fibers with glass and/or carbon fibers significantly reduces the thermoset composites water absorption properties and also enhances the composites mechanical and thermal properties. These improvements allow the hybridized natural fibers–glass fibers or natural fibers–carbon fibers reinforced thermoset composites to compete with conventional structural materials in their application. Similarly, Burgueno et al. (2005b) also investigated the applicability of hybrid biofibre based composites for structural applications, specifically on structural cellular plates. They hybridized chopped natural fibers (hemp and flax) with synthetic (glass and carbon) or natural (jute) based fabrics as fillers for unsaturated polyester polymer matrix. These hybridized composites were tested and also analyzed with micromechanics and sandwich analysis. Results indicated that they are a viable alternative to conventional structural materials for current and future applications.

Besides the utilization of hybrid bast fibers reinforced thermoset composites on structural and building materials, it can also be applied for car structural components. This was done by Davoodi et al. (2010) where they fabricated hybrid kenaf-glass fiber reinforced epoxy composites and evaluated its mechanical properties. The hybrid composites mechanical properties were compared with glass mat thermoplastic (GMT) which is the material used to produce passenger car bumper beam. The hybrid composites had comparable mechanical properties with GMT except for the hybrid composites lower impact strength but based on its overall properties, it has the potential to be used as a material for the production of car bumper beams. Devireddy and Biswas (2016) focused their research on the applicability of hybrid banana-jute fiber reinforced epoxy composites in relation with its physical and thermal properties for not only car components but also building materials. They also came to the same conclusion that hybrid bast fibers reinforced thermoset composites are suitable for building materials and car components. Other researchers with similar findings on hybrid bast fiber thermoset composites applicability as automotive components includes Yahaya et al. (2016), Mansor et al. (2013) and Suresh et al. (2015). Other than building and car materials, hybrid bast fiber reinforced thermoset composites were also found to possess the potential for use as curved pipes where three hybrid fibers, kenaf-glass fiber, flax-glass fiber and hemp-glass fiber were evaluated (Cicala et al., 2009). Another utilization is its development as a combat armor made from hybrid ramie-kevlar 29 polyester composites as it meets the ballistic threats equivalent to the third level of protective ballistic limits in the National Institute of Justice (NIJ) standards (Radif et al., 2011).

9.8 Conclusion

Bast fibers like jute, flax, hemp and kenaf had been thoroughly investigated by numerous researchers up till today where most of them had focused on the fibers chemical constituents, physical and mechanical properties as well as its cell wall architecture. These properties of bast fibers makes it a suitable reinforcement for thermoset polymers with the added benefits of being environmentally friendly and also possessing high specific strength comparable to synthetic fibers. Despite the advantages, bast fibers also lack behind in certain properties like moisture absorption. In order to offset what bast fibers are lacking, hybridization of bast fibers with other natural fibers or synthetic fibers were done and the results were promising as hybridization enhances the reinforced thermoset composites. To further improve the hybrid bast fiber thermoset composites, researchers had also conducted several studies on its fiber-matrix interface, bast fiber thermal stability, moisture content, biodegradability, and dispersion in the matrix. This had led to the development of hybrid bast fiber thermoset composites exhibiting properties comparable to synthetic thermoset composites with the exception of one or two lower properties for the hybrid composites. Due to these comparable properties, the hybrid composites has the potential for application in building and structural materials, automotive components, piping and body armor. However, there is still a need for future research to broaden its application to other utilizations by further improving the bast fibers moisture absorption, thermal stability and durability, allowing them to completely replace synthetic fibers one day.

References

Abdul Khalil, H.P.S., Ireana Yusra, A.F., Bhat, A.H., Jawaid, M., 2010. Cell wall ultrastructure, anatomy, lignin distribution and chemical composition of Malaysian cultivated kenaf fiber. Ind. Crop. Prod. 31 (1), 113–121.

Abdellaoui, H., Bensalah, H., Echaabi, J., Bouhfid, R., Qaiss, A., 2015. Fabrication, characterization and modelling of laminated composites based on woven jute fibres reinforced epoxy resin. Mater. Des. 68, 104–113.

Adekunle, K.F., 2015. Surface treatments of natural fibers – a review. Open J. Polym. Chem. 5 (3), 41–46.

Ahmed, K.S., Vijayarangan, S., Kumar, A., 2007. Low velocity impact damage characterization of woven jute glass fabric reinforced isothalic polyester hybrid composites. J. Reinf. Plast. Compos. 26 (10), 959–976.

Akil, H.M., Zamri, H., Shahidan, H.A., Arifin, M.I.Z., Bakar, A.A., 2011. Pultruded natural fibre reinforced composites: preparation, properties and application. The 19th Annual International Conference on Composites/Nano Engineering – ICCE-19, Shanghai 24th – 30th July 2011. University of New Orleans, New Orleans, USA.

Akil, H.M., Santulli, C., Sarasini, F., Tirillo, J., Valente, T., 2014. Environmental effects on the mechanical behaviour of pultruded jute/glass fibre-reinforced polyester hybrid composites. Compos. Sci. Technol. 94, 62–70.

Alam, M.A., Alriyami, K., Jumaat, M.Z., Noori, K., 2015. Development of hybrid jute fibre composite plates for potential application in shear retrofitting of RC structure. Ind. J. Sci. Technol. 8 (30), <http://dx.doi.org/10.17485/ijst/2015/v8i1/72172> (accessed 15.04.16).

Al-Harbi, K.M.A.S., 2001. Application of the AHP in project management. Int. J. Proj. Manage. 19 (1), 19–27.

Anandjiwala, R.D., Blouw, S., 2007. Composites from bast fibres – prospects and potential in the changing market environment. J. Nat. FiberFiber. 4 (2), 91–109.

Aracri, E., Fillat, A., Colom, J.F., Gutierrez, A., del Rio, J.C., Martinez, A.T., et al., 2010. Enzymatic grafting of simple phenols on flax and sisal pulp fibres using laccases. Bioresour. Technol. 101 (21), 8211–8216.

Ashby, M.F., 2012. Materials and the Environment: Eco-informed Material Choice. 2nd ed. Butterworth-Heinemann, Oxford, UK.

Ashori, A., 2006. Pulp and paper from kenaf bast fibers. Fiber Polym. 7 (1), 26–29.

Atiqah, A., Maleque, M.A., Jawaid, M., Iqbal, M., 2014. Development of kenaf-glass reinforced unsaturated polyester hybrid composite for structural applications. Compos. B: Eng. 56, 68–73.

Barkoula, N.M., Peijs, T., 2011. Interface engineering through matrix modification in natural fibres composites. In: Zafeiropoulos, N.E. (Ed.), Interface Engineering of Natural Fibre Composites for Maximum Performance. Woodhead Publishing, Sawston, Cambridge, pp. 43–81.

Batog, J., Konczewicz, W., Kozlowski, R., Muzyczek, M., Sedelnik, N., Tanska, B., 2006. Survey and recent report on enzymatic processing of bast fibres. J. Nat. FiberFiber. 3 (2–3), 113–129.

Batra, S.K., 2006. Other long vegetable fibres: abaca, banana, sisal, henequen, flax, ramie, hemp, sunn and coir. In: Lewin, M. (Ed.), Handbook of Fiber Chemistry, 3rd ed. CRC Press, Boca Raton, Florida, pp. 453–520.

Beck, C.B., 2005. An Introduction to Plant Structure and Development. Cambridge University Press, Cambridge, UK.

Bhatnagar, A., 2004. Isolation of Cellulose Nanofibers from Renewable Feed Stocks and Root Crops. University of Toronto, Toronto, Canada.

Bhoopathi, R., Ramesh, M., Deepa, C., 2014. Fabrication and property evaluation of banana--hemp-glass fiber reinforced composites. Proc. Eng. 97, 2032–2041.

Biagiotti, J., Puglia, D., Kenny, J.M., 2004. A review on natural fibre-based composites–part 1: structure, processing and properties of vegetable fibres. J. Nat. Fiber. 1 (2), 37–68.

Blake, A.W., Marcus, S.E., Copeland, J.E., Blackburn, R.S., Knox, J.P., 2008. In situ analysis of cell wall polymers associated with phloem cells in stems of hemp, *Cannabis sativa* L. Planta. 228, 1–13.

Bledzki, A.K., Gassan, J., 1999. Composites reinforced with cellulose based fibres. Prog. Polym. Sci. 24 (2), 221–274.

Bogoeva-Gaceva, G., Avella, M., Malinconico, M., Buzarovska, A., Grozdanov, A., Gentile, G., et al., 2007. Natural fiber eco-composites. Polym. Compos. 28 (1), 98–107.

Boopalan, M., Niranjanaa, M., Umapathy, M.J., 2013. Study on the mechanical properties and thermal properties of jute and banana fiber reinforced epoxy hybrid composites. Compos. B: Eng. 51, 54–57.

Bos, H.L., van den Oever, M.J.A., Peters, O.C.J.J., 2002. Tensile and compressive properties of flax fibres for natural fibre reinforced composites. J. Mater. Sci. 37 (8), 1683–1692.

Bos, H.L., Mussig, J., van den Oever, M.J.A., 2006. Mechanical properties of short-flax-fibre reinforced compounds. Compos. A: Appl. Sci. Manuf. 37 (10), 1591–1604.

Bourmaud, A., Morvan, C., Bouali, A., Placet, V., Perre, P., Baley, C., 2013. Relationships between micro-fibrillar angle, mechanical properties and biochemical composition of flax fibers. Ind. Crop. Prod. 44, 343–351.

Bowen, D.H., Signorelli, R.A., de Luca, P.L., Williams, J.C., Kelly, A., Greer, A.L., et al., 1994. M – Manufacturing methods for composites: an overview. In: Kelly, A. (Ed.), *Concise Encyclopedia of Composite Materials, Revised Edition*. Pergamon Press, Oxford, UK, pp. 175–199.

Burgueno, R., Quagliata, M.J., Mehta, G.M., Mohanty, A.K., Misra, M., Drzal, L.T., 2005a. Sustainable cellular biocomposites from natural fibers and unsaturated polyester resin for housing panel applications. J. Polym. Environ. 13 (2), 139–149.

Burgueno, R., Quagliata, M.J., Mohanty, A.K., Mehta, G., Drzal, L.T., Misra, M., 2005b. Hybrid biofiber-based composites for structural cellular plates. Compos. A: Appl. Sci. Manuf. 36 (5), 581–593.

Calamari, T.A., Tao, W., Goynes, W.R., 1997. A preliminary study of kenaf fiber bundles and their composite cells. TAPPI J. 80 (8), 149–154.

Celino, A., Freour, S., Jacquemin, F., Casari, P., 2013. The hygroscopic behaviour of plant fibers: a review. Front. Chem. 1, <http://dx.doi.org/10.3389/fchem.2013.00043> (accessed 15.04.16).

Chand, N., Fahim, M., 2008. Tribology of Natural Fiber Polymer Composites. Woodhead Publishing, Sawston, Cambridge.

Cheng, Z., Lu, B.R., Sameshima, K., Fu, D.X., Chen, J.K., 2004. Identification and genetic relationships of kenaf (*Hibiscus cannabinus* L.) germplasm revealed by AFLP analysis. Genet. Resour. Crop. Evol. 51 (4), 393–401.

Cicala, G., Cristaldi, G., Recca, G., Ziegmann, G., El-Sabbagh, A., Dickert, M., 2009. Properties and performances of various hybrid glass/natural fibres composites for curved pipes. Mater. Des. 30 (7), 2538–2542.

Cook, J.C., 1984. Handbook of Textile Fibres: Volume 1 – Natural Fibres. Woodhead Publishing, Sawston, Cambridge.

Coroller, G., Lefeuvre, A., Duigou, A.L., Bourmaud, A., Ausias, G., Gaudry, T., et al., 2013. Effect of flax fibres individualisation on tensile failure of flax/epoxy unidirectional composite. Compos. A: Appl. Sci. Manuf. 51, 62–70.

Cutter, A.G., 2011. Development and Characterization of Renewable Resource-Based Structural Composite Materials. ProQuest, Ann Arbor, Michigan.

Davoodi, M.M., Sapuan, S.M., Ahmad, D., Ali, A., Khalina, A., Jonoobi, M., 2010. Technical report: mechanical properties of hybrid kenaf/glass reinforced epoxy composite for passenger car bumper beam. Mater. Des. 31 (10), 4927–4932.

Devireddy, S.B., Biswas, S., 2016. Physical and thermal properties of unidirectional banana-jute hybrid fiber-reinforced epoxy composites. J. Reinf. Plast. Compos.<http://dx.doi.org/10.1177/0731684416642877> (accessed 15.04.16).

Dhakal, H.N., Zhang, Z.Y., Guthrie, R., MacMullen, J., Bennett, N., 2013. Development of flax/carbon fibre hybrid composites for enhanced properties. Carbohydr. Polym. 96 (1), 1–8.

Dittenber, D.B., GangaRao, H.V.S., 2012. Review: critical review of recent publications on use of natural composites in infrastructure. Compos. A: Appl. Sci. Manuf. 43 (8), 1419–1429.

Drzal, L.T., Mohanty, A.K., Mehta, G., Misra, M., 2002. Low-cost, bio-based composite materials for housing applications. In: Anson, M., Ko, J.M., Lam, E.S.S. (Eds.), *Advances in Building Technology, 1st Volume*. Elsevier Science, Philadelphia, Pennsylvania, pp. 225–232.

Eder, M., Burgert, I., 2010. Natural fibres – function in nature. In: Mussig, J. (Ed.), Industrial Applications of Natural Fibres: Structure, Properties and Technical Applications. Wiley, Hoboken, New Jersey, pp. 23–40.

Elzebroek, A.T.G., Wind, K., 2008. Guide to Cultivated Plants. CABI, Wallingford, Oxfordshire, UK.

Faruk, O., Bledzki, A.K., Fink, H.P., Sain, M., 2012. Biocomposites reinforced with natural fibers: 2000–2010. Prog. Polym. Sci. 37 (11), 1552–1596.

Fimmm, B.C.M., Rosemaund, A. 2009. Industrial fibres: recent and current developments. In: Discover natural fibers 2009: Proceedings of the symposium on natural fibres, Rome 20th October 2008, Common Fund for Commodities 2009, Amsterdam, Netherlands.

Franck, R.R., 2005. Overview. In: Franck, R.R. (Ed.), Bast and Other Plant Fibres. Woodhead Publishing, Sawston, Cambridge, pp. 1–23.

Ganguly, P.K., Sengupta, S., Samajpati, S., 1999. Mechanical behaviour of jute and polypropylene blended needle-punched fabrics. Ind. J. Fibre Text. Res. 24 (1), 34–40.

Ghani, M.A.A., Salleh, Z., Hyie, K.M., Berhan, M.N., Taib, Y.M.D., Bakri, M.A.I., 2012. Mechanical properties of kenaf/fiberglass polyester hybrid composite. Proc. Eng. 41, 1654–1659.

Ghosh, P., 2003. Fibre Science and Technology. Tata McGraw-Hill, New Delhi, India.

Goda, K., Cao, Y., 2007. Review paper: research and development of fully green composites reinforced with natural fibers. J. Solid Mech. Mater. Eng. 1 (9), 1073–1084.

Gopinath, A., Kumar, M.S., Elayaperumal, A., 2014. Experimental investigations on mechanical properties of jute fiber reinforced composites with polyester and epoxy resin. Proc. Eng. 97, 2052–2063.

Gujjala, R., Ojha, S., Acharya, S.K., Pal, S.K., 2014. Mechanical properties of woven jute-glass hybrid-reinforced epoxy composite. J. Compos. Mater. 48 (28), 3445–3455.

Gupta, M.K., Srivastava, R.K., 2016. Mechanical properties of hybrid fibers-reinforced polymer composite: a review. Polym.-Plast. Technol. Eng. 55 (6), 626–642.

Hall, L.M., 2016. Flax (*Linum usitatissimum* L.). In: McKeon, T., Hayes, D., Hildebrand, D., Weselake, R. (Eds.), Industrial Oil Crops, *1st ed.* Academic Press and AOCS Press, UK and USA, pp. 157–194.

Hashemi, F., Tahir, P.M., Madsen, B., Jawaid, M., Majid, D.L., Brancheriau, L., et al., 2015. Volumetric composition and shear strength evaluation of pultruded hybrid kenaf/glass fiber composites. J. Compos. Mater.<http://dx.doi.org/10.1177/0021998315602948> (accessed 15.04.16).

Haugan, E., Holst, B., 2013. Determining the fibrillary orientation of bast fibres with polarized light microscopy: the modified Herzog test (red plate test) explained. J. Microsc. 252 (2), 159–168.

Herrmann, A.S., Nickel, J., Riedel, U., 1998. Construction materials based upon biologically renewable resources – from components to finished parts. Polym. Degrad. Stab. 59 (1–3), 251–261.

Holbery, J., Houston, D., 2006. Overview low-cost composites in vehicle manufacture: Natural-fiber-reinforced polymer composites in automotive applications. J. Miner. Met. Mater. Soc. (TMS). 58 (11), 80–86.

Hughes, M., 2004. Quantifying the range of properties in natural raw material origin polymers and fibres. In: Tucker, N., Johnson, M.G. (Eds.), Low Environmental Impact Polymers. Smithers Rapra Technology, Shrewsbury, UK, pp. 67–104.

Jawaid, M., Abdul Khalil, H.P.S., 2011. Cellulosic/synthetic fibre reinforced polymer hybrid composites: a review. Carbohydr. Polym. 86 (1), 1–18.

Jhala, A.J., Hall, L.M., 2010. Flax (*Linum usitatissimum* L.): Current uses and future applications. Austr. J. Basic Appl. Sci. 4 (9), 4304–4312.

John, M.J., Thomas, S., 2008. Review: biofibres and biocomposites. Carbohydr. Polym. 71 (3), 343–364.

John, M.J., Anandjiwala, R.D., 2008. Recent developments in chemical modification and characterization of natural fiber-reinforced composites. Polym. Compos. 29 (2), 187–207.

Joseph, P.V., Joseph, K., Thomas, S., 1999. Effect of processing variables on the mechanical properties of sisal-fiber-reinforced polypropylene composites. Compos. Sci. Technol. 59 (11), 1625–1640.

Joshi, M., 2015. Textbook of Field Crops. PHI, New Delhi, India.

Kabir, M.M., Wang, H., Lau, K.T., Cardona, F., 2012. Chemical treatments on plant-based natural fibre reinforced polymer composites: An overview. Compos. B: Eng. 43 (7), 2883–2892.

Kafi, A.A., Abedin, M.Z., Beg, M.D.H., Pickering, K.L., Khan, M.A., 2006. Study on the mechanical properties of jute/glass fiber-reinforced unsaturated polyester hybrid composites: Effect of surface modification by ultraviolet radiation. J. Reinf. Plast. Compos. 25 (6), 575–588.

Kannan, K., Harikannan, S., Arun, V., Anusundar, R., Ashokraj, S., 2015. Analysis of mechanical properties of jute and banana fibre reinforced vinyl ester composites. Int. J. Innov. Res. Sci. Eng. Technol. 4 (6), 1630–1638.

Karahan, M., Karahan, N., 2015. Investigation of the tensile properties of natural and natural/synthetic hybrid fiber woven fabric composites. J. Reinf. Plast. Compos. 34 (10), 795–806.

Kers, J., Peetsalu, P., Saarna, M., Viikna, A., Krumme, A., Menind, A., 2010. Preliminary investigation into tensile characteristics of long flax fibre reinforced composite material. Agron. Res. 8 (1), 107–114.

Kuhad, R.C., Singh, A., Eriksson, K.E.L., 1997. Microorganisms and enzymes involved in the degradation of plant fiber cell walls. In: Eriksson, K.E.L., Babel, W., Blanch, H.W., Cooney, C.L., Enfors, S.O., Eriksson, K.E.L., Fiechter, A., Klibanov, A.M., Mattiasson, B., Primrose, S.B., Rehm, H.J., Rogers, P.L., Sahm, H., Schugerl, K., Tsao, G.T., Venkat, K., Villadsen, J., von Stocker, U., Wandrey, C. (Eds.), *Advances in Biochemical Engineering Biotechnology: Biotechnology in the Pulp and Paper Industry, 57th Volume*. Springer, Berlin, Germany, pp. 45–125.

Kymalainen, H.R., Sjoberg, A.M., 2008. Flax and hemp fibres as raw materials for thermal insulations. Build. Environ. 43 (7), 1261–1269.

Lai, W.L., Mariatti, M., Mohamad Jani, S., 2008. The properties of woven kenaf and betel palm (*Areca catechu*) reinforced unsaturated polyester composites. Polym.-Plast. Technol. Eng. 47 (12), 1193–1199.

Lewington, A., 2003. Plants for People. Transworld Publishers, London, UK.

Li, X., Tabil, L.G., Panigrahi, S., 2007. Chemical treatments of natural fiber for use in natural fiber-reinforced composites: a review. J. Polym. Environ. 15 (1), 25–33.

Li, Y., Xie, L., Ma, H., 2015. Permeability and mechanical properties of plant fiber reinforced hybrid composites. Mater. Des. 86, 313–320.

Liu, L., Wang, Q., Xia, Z., Yu, J., Cheng, L., 2010. Mechanical modification of degummed jute fibre for high value textile end uses. Ind. Crop. Prod. 31 (1), 43–47.

Mansor, M.R., Sapuan, S.M., Zainudin, E.S., Nuraini, A.A., Hambali, A., 2013. Hybrid natural and glass fibres reinforced polymer composites material selection using Analytical Hierarchy Process for automotive brake lever design. Mater. Des. 51, 484–492.

Medeiros, E.S.D., Agnelli, J.A.M., Joseph, K., Carvalho, L.H.D., Mattoso, L.H.C., 2005. Mechanical properties of phenolic composites reinforced with jute/cotton hybrid fabrics. Polym. Compos. 26 (1), 1–11.

Meshram, J.H., Palit, P., 2013. Review article: Biology of industrial bast fibers with reference to quality. J. Nat. Fiber. 10 (2), 176–196.

Mohanty, A.K., Misra, M., Hinrichsen, G., 2000. Biofibres, biodegradable polymers and biocomposites: an overview. Macromol. Mater. Eng. 276–277 (1), 1–24.
Munder, F., Furll, C., Hempel, H., 2005. Processing of bast fiber plants for industrial applications. In: Mohanty, A.K., Misra, M., Drzal, L.T. (Eds.), Natural Fibers, Biopolymers and Biocomposites. CRC Press, Boca Raton, Florida, pp. 109–140.
Murthy, H.V.S., 2015. Introduction to Textile Fibres. Woodhead Publishing, India, New Delhi, India.
Mussig, J., Schmehl, M., von Buttlar, H.B., Schonfeld, U., Arndt, K., 2006. Exterior components based on renewable resources produced with SMC technology – Considering a bus component as example. Ind. Crop. Prod. 24 (2), 132–145.
Mwaikambo, L.Y., Ansell, M.P., 2006. Mechanical properties of alkali treated plant fibres and their potential as reinforcement materials. I. hemp fibres. J. Mater. Sci. 41 (8), 2483–2496.
Naik, D.L., Fronk, T.H., 2013. Effective properties of cell wall layers in bast fiber. Comput. Mater. Sci. 79, 309–315.
Nair, A.B., Joseph, R., 2014. Eco-friendly bio-composites using natural rubber (NR) matrices and natural fiber reinforcements. In: Kohjiya, S., Ikeda, Y. (Eds.), Chemistry, Manufacture and Applications of Natural Rubber. Woodhead Publishing, Sawston, Cambridge, pp. 249–283.
Nando, G.B., Gupta, B.R., 1996. Short fibre-thermoplastic elastomer composites. In: De, S. K., White, J.R. (Eds.), Short Fibre-Polymer Composites. CRC Press, Boca Raton, Florida, pp. 84–115.
Nishino, T., 2013. Preparation, microstructure and properties of biofibers. In: Thomas, S., Joseph, K., Malhotra, S.K., Goda, K., Sreekala, M.S. (Eds.), *Polymer Composites: Biocomposites, 3rd Volume*. Wiley-VCH, Weinheim, Germany, pp. 109–132.
Ozturk, S., 2005. The effect of fibre content on the mechanical properties of hemp and basalt fibre reinforced phenol formaldehyde composites. J. Mater. Sci. 40 (17), 4585–4592.
Ozturk, B., 2010a. Hybrid effect in the mechanical properties of jute/rockwool hybrid fibres reinforced phenol formaldehyde composites. Fiber Polym. 11 (3), 464–473.
Ozturk, S., 2010b. Effect of fiber loading on the mechanical properties of kenaf and fiberfrax fiber-reinforced phenol-formaldehyde composites. J. Compos. Mater. 44 (19), 2265–2288.
Pallesen, B.E., 1996. The quality of combine-harvested fibre flax for industrial purposes depends on the degree of retting. Ind. Crop. Prod. 5, 65–78.
Park, S.J., Seo, M.K., 2011. Interface Science and Composites. Academic Press, Cambridge, UK.
Petrucci, R., Santulli, C., Puglia, D., Sarasini, F., Torre, L., Kenny, J.M., 2013. Technical report: Mechanical characterisation of hybrid composite laminates based on basalt fibres in combination with flax, hemp and glass fibres manufactured by vacuum infusion. Mater. Des. 49, 728–735.
Phani, K.K., Bose, N.R., 1987. Hydrothermal aging of jute glass-fiber hybrid composites – An acoustoultrasonic study. J. Mater. Sci. 22 (6), 1929–1933.
Pickering, K.L., Efendy, M.G.A., Le, T.M., 2016. A review of recent developments in natural fibre composites and their mechanical performance. Compos. A: Appl. Sci. Manuf. 83, 98–112.
Placet, V., Trivaudey, F., Cisse, O., Gucheret-Retel, V., Boubakar, M.L., 2012. Diameter dependence of the apparent tensile modulus of hemp fibres: A morphological, structural or ultrastructural effect? Compos. A: Appl. Sci. Manuf. 43 (2), 275–287.
Punyamurthy, R., Sampathkumar, D., Ranganagowda, R.P., Bennehalli, B., Badyankal, P., Venkateshappa, S.V., 2014. Surface modification of abaca fiber by benzene diazonium

chloride treatment and its influence on tensile properties of abaca fiber reinforced polypropylene composites. Cienc. Tecnol. Mater. 26 (2), 142–149.

Radif, Z.S., Ali, A., Abdan, K., 2011. Development of a green combat armour from ramie-kevlar-polyester composite. Pertanika J. Sci. Technol. 19 (2), 339–348.

Raghavendra, G., Kumar, K.A., Kumar, M.H., Raghukumar, B., Ojha, S., 2015. Moisture absorption behaviour and its effect on the mechanical properties of jute-reinforced epoxy composite. Polym. Compos.<http://dx.doi.org/10.1002/pc.23610> (accessed 15.04.16).

Rahman, M.A., Parvin, F., Hasan, M., Hoque, M.R., 2015. Introduction to manufacturing of natural fibre-reinforced polymer composites. In: Salit, M.S., Jawaid, M., Yusoff, N.B., Hoque, M.E. (Eds.), Manufacturing of Natural Fibre Reinforced Polymer Composites. Springer, Berlin, Germany, pp. 17–44.

Ramesh, M., Nijanthan, S., 2016. Mechanical property analysis of kenaf-glass fibre reinforced polymer composites using finite element analysis. Bull. Mater. Sci. 39 (1), 147–157.

Ramesh, M., Palanikumar, K., Reddy, K.H., 2013. Mechanical property evaluation of sisal-jute-glass fiber reinforced polyester composites. Compos. B: Eng. 48, 1–9.

Raven, P., Evert, R., Eichhorn, S., 2005. Biology of plants, 7th ed., W.H. Freeman and Co, New York.

Ray, D., Rout, J., 2005. Thermoset biocomposites. In: Mohanty, A.K., Misra, M., Drzal, L.T. (Eds.), Natural Fibers, Biopolymers and Biocomposites. CRC Press, Boca Raton, Florida, pp. 291–345.

Rials, T.G., Wolcott, M.P., 1996. Physical and mechanical properties of agro-based fibers. In: Rowell, R.M., Young, R.A., Rowell, J.K. (Eds.), Paper and Composites from Agro-Based Resources. CRC Press, Boca Raton, Florida, pp. 63–82.

Romhany, G., Karger-Kocsis, K., Czigany, T., 2003. Tensile fracture and failure behaviour of technical flax fiber. J. Appl. Polym. Sci. 90 (13), 3638–3645.

Roul, C., 2009. The International Jute Commodity System. Northern Book Centre, New Delhi, India.

Saba, N., Paridah, M.T., Jawaid, M., Abdan, K., Ibrahim, N.A., 2015a. Potential utilization of kenaf biomass in different applications. In: Hakeem, K.R., Jawaid, M., Alothman, O. (Eds.), Agricultural Biomass Based Potential Materials. Springer, Berlin, Germany, pp. 1–34.

Saba, N., Jawaid, M., Alothman, O.Y., Paridah, M.T., Hassan, A., 2015b. Recent advances in epoxy resin, natural fiber-reinforced epoxy composites and their applications. J. Reinf. Plast. Compos.<http://dx.doi.org/10.1177/0731684415618459> (accessed 15.04.16).

Saheb, S.N., Jog, J.P., 1999. Natural fiber polymer composites: a review. Adv. Polym. Technol. 18 (4), 351–363.

Sahin, H.T., 2003. Base-catalyzed organosolv pulping of jute. J. Chem. Technol. Biotechnol. 78 (12), 1267–1273.

Salleh, Z., Taib, Y.M., Hyie, K.M., Mihat, M., Berhan, M.N., Ghani, M.A.A., 2012. Fracture toughness investigation on long kenaf/woven glass hybrid composite due to water absorption effect. Proc. Eng. 41, 1667–1673.

Salman, S.D., Leman, Z., Sultan, M.T.H., Ishak, M.R., Cardona, F., 2015. Kenaf/synthetic and kevlar®/cellulosic fiber-reinforced hybrid composites: A review. Bioresour. 10 (4), 8580–8603.

Sanadi, A.R., Caulfield, D.F., Jacobson, R.E., 1996. Agro-fibre/thermoplastic composites. In: Rowell, R.M., Young, R.A., Rowell, J.K. (Eds.), Paper and Composites from Agro-Based Resources. CRC Press, Bora Raton, Florida, pp. 377–402.

Sanjay, M.R., Yogesha, B., 2016. Studies on mechanical properties of jute/E-glass fiber reinforced epoxy hybrid composites. J. Miner. Mater. Charact. Eng. 4 (1), 15–25.

Sanjay, M.R., Arpitha, G.R., Naik, L.L., Gopalakrisha, K., Yogesha, B., 2016. Applications of natural fibers and its composites: An overview. Nat. Resour. 7 (3), 108–114.

Sathishkumar, T.P., Naveen, J., Satheeshkumar, S., 2014. Hybrid fiber reinforced polymer composites – a review. J. Reinf. Plast. Compos. 33 (5), 454–471.

Savastano Jr, H., Santos, S.F., Agopyan, V., 2009. Sustainability of vegetable fibres in construction. In: Khatib, J. (Ed.), Sustainability of Construction Materials, *1st ed.* Woodhead Publishing, Sawston, Cambridge, pp. 55–81.

Scutaru, M.L., Baba, M., 2014. Investigation of the mechanical properties of hybrid carbon-hemp laminated composites used as thermal insulation for different industrial applications. Adv. Mech. Eng. 6, <http://dx.doi.org/10.1155/2014/829426> (accessed 15.04.16).

Sen, T., Reddy, H.N.J., 2011. Various industrial applications of hemp, kenaf, flax and ramie natural fibres. Int. J. Innov. Manage. Technol. 2 (3), 192–198.

Shahzad, A., 2011. Hemp fiber and its composites – a review. J. Compos. Mater. 46 (8), 973–986.

Shahzad, A., 2012. Hemp fiber and its composites – a review. J. Compos. Mater. 46 (8), 973–986.

Shahzad, A., 2013. A study in physical and mechanical properties of hemp fibres. Adv. Mater. Sci. Eng.accessed 15 April 2016, <http://dx.doi.org/10.1155/2013/325085>.

Shanmugam, D., Thiruchitrambalam, M., 2013. Static and dynamic mechanical properties of alkali treated unidirectional continuous Palmyra palm leaf stalk fiber/jute fiber reinforced hybrid polyester composites. Mater. Des. 50, 533–542.

Shinoj, S., Visvanathan, R., Panigrahi, S., Kochubabu, M., 2011. Oil palm fiber (OPF) and its composite: A review. Ind. Crop. Prod. 33 (1), 7–22.

Skoglund, G., Nockert, M., Holst, B., 2013. Viking and early middle ages northern Scandinavian textiles proven to be made with hemp. Sci. Rep. 3, <http://dx.doi.org/10.1038/srep02686> (accessed 15.04.16).

Sreenath, H.K., Shah, A.B., Yang, V.W., Gharia, M.M., Jeffries, T.W., 1996. Enzymatic polishing of jute/cotton blended fabrics. J. Ferment. Bioeng. 81 (1), 18–20.

Sreenivasan, S., Sulaiman, S., Baharudin, B.T.H.T., Ariffin, M.K.A., Abdan, K., 2013. Recent developments of kenaf fibre reinforced thermoset composites: Review. Mater. Res. Innov. 17, s2–s11.

Sridhar, M.K., Basavarajappa, G., Kasturi, S.G., Balasubramaniam, N., 1982. Evaluation of jute as a reinforcement in composites. Indian J. Text. Res. 7, 87–92.

Srinivasan, V.S., Boopathy, S.R., Sangeetha, D., Ramnath, B.V., 2014. Technical report: Evaluation of mechanical and thermal properties of banana-flax based natural fibre composite. Mater. Des. 60, 620–627.

Subash, T., Pillai, S.N., 2015. Bast fibers reinforced green composites for aircraft indoor structures applications: Review. National Conference on Recent Trends and Developments in Sustainable Green Technologies. Anna University, Chennai, India.

Summerscales, J., Dissanayake, N.P.J., Virk, A.S., Hall, W., 2010. Review: A review of bast fibres and their composites. Part 1 – Fibre as reinforcements. Compos. A: Appl. Sci. Manuf. 41 (10), 1329–1335.

Suresh, B.K., Somashekar, H.K., Padmanabha, S., 2015. An evaluation of jute-epoxy-hybrid composite materials for automotive frontal crash safety applications. Int. J. Res. Eng. Technol. 4 (7), 315–321.

Tahir, P.M., Ahmed, A.B., Azry, S.O.A.S., Ahmed, Z., 2011. Retting process of some bast plant fibres and its effect on fibre quality: a review. Bioresour. 6 (4), 5260–5281.

Tahir, P.M., Zaini, L.H., Jonoobi, M., Abdul Khalil, H.P.S., 2014. Preparation of nanocellulose from kenaf (*Hibiscus cannabinus* L.) via chemical and chemo-mechanical processes. In: Pandey, J.K., Takagi, H., Nakagaito, A.N., Kim, H.J. (Eds.), Handbook of Polymer Nanocomposites. Processing, Performance and Application – Volume C: Polymer Nanocomposites of Cellulose Nanoparticles. Springer, Berlin, Germany, pp. 119–144.

Thiruchitrambalam, M., Alavudeen, A., Venkateshwaran, N., 2012. Review on kenaf fiber composites. Rev. Adv. Mater. Sci. 32 (2), 106–111.

Troedec, M.L., Sedan, D., Peyratout, C., Bonnet, J.P., Smith, A., Guinebretiere, R., et al., 2008. Influence of various chemical treatments on the composition and structure of hemp fibres. Compos. A: Appl. Sci. Manuf. 39 (3), 514–522.

van Rijswijk, K., Brouwer, W.D., Beukers, A., 2001. Application of Natural Fibre Composites in the Development of Rural Societies. Delft University of Technology, Delft, Netherlands.

Vinayagamoorthy, R., Rajeswari, N., 2014. Mechanical performance studies on *Vetiveria zizanioides*/jute/glass fiber-reinforced hybrid polymeric composites. J. Reinf. Plast. Compos. 33 (1), 81–92.

Voulgaridis, E., Passialis, C., Grigoriou, A., 2000. Anatomical characteristics and properties of kenaf stems (*Hibiscus cannabinus*). IAWA J. 21 (4), 435–442.

Walentowska, J., Kozlowski, R.M., 2012. Prevention of fungal growth in natural fibres. In: Kozlowski, R.M. (Ed.), *Handbook of Natural Fibres: Types, Properties and Factors Affecting Breeding and Cultivation, 1st Volume*. Woodhead Publishing, Sawston, Cambridge, pp. 532–549.

Wang, B., 2008. Dispersion of Cellulose Nanofibers in Biopolymer Based Nanocomposites. University of Toronto, Toronto, Canada.

Wang, B., Panigrahi, S., Tabil, L., Crerar, W., 2007. Pre-treatment of flax fibres for use in rotationally molded biocomposites. J. Reinf. Plast. Compos. 26 (5), 447–463.

Wotzel, K., Wirth, R., Flake, R., 1999. Life cycle studies on hemp fiber reinforced components and ABS for automotive parts. Die Angew. Makromolekulare Chem. 272 (1), 121–127.

Yahaya, R., Sapuan, S.M., Jawaid, M., Leman, Z., Zainudin, E.S., 2016. Effect of fibre orientations on the mechanical properties of kenaf-aramid hybrid composites for spall-liner application. Def. Technol. 12 (1), 52–58.

Yan, L., Chouw, N., Jayamaran, K., 2014. Flax fibre and its composites – A review. Compos. B: Eng. 56, 296–317.

Yousif, B.F., Shalwan, A., Chin, C.W., Ming, K.C., 2012. Flexural properties of treated and untreated kenaf/epoxy composites. Mater. Des. 40, 378–385.

Yumnam, S., Sawarkar, A., Mukherjee, S., Sarkar, K.K., 2015. Stability performance of tossa jute (*Corchorus olitorius* L.) under waterlogged condition. Bioscan. 10 (3), 977–980.

Zamri, M.H., Akil, H.M., Bakar, A.A., Ishak, Z.A.M., Leong, W.C., 2011. Effect of water absorption on pultruded jute/glass fibre-reinforced unsaturated polyester hybrid composites. J. Compos. Mater.accessed 15 April 2016, <http://dx.doi.org/10.1177/0021998311410488>.

Zhang, Y., Li, Y., Ma, H., Yu, T., 2013. Tensile and interfacial properties of unidirectional flax/glass fiber reinforced hybrid composites. Compos. Sci. Technol. 88, 172–177.

Zimniewska, M., Wladyka-Przybylak, M., Mankowski, J., 2011. Cellulosic bast fibers, their structure and properties suitable for composite applications. In: Kalia, S., Kaith, B.S., Kaur, I. (Eds.), Cellulose Fibers: Bio- and Nano-Polymer Composites: Green Chemistry and Technology. Springer, Berlin, Germany, pp. 97–120.

Influence of interphase and inclusion waviness on stiffness properties of a nanoenhanced matrix

10

Mohammad Rouhi[1] *and Masoud Rais-Rohani*[2]
[1]Concordia University, Montreal, QC, Canada, [2]University of Maine, Orono, ME, United States

Chapter Outline

10.1 Introduction

Nanoreinforcements, such as carbon nanotubes (CNT) and carbon nanofibers (CNF), can be used as fillers to enhance the overall stiffness properties of the base polymer matrix. When conventional reinforcing fibers are combined with a nanoenhanced matrix, the nanoreinforcements can improve the interfacial shear strength (Thostenson et al., 2002; Garg et al., 2008) and other mechanical properties of the hybrid composite material (Chisholm et al., 2005; Gojny et al., 2005; Zhou et al., 2008).

Several models have been developed to predict the properties of nanocomposite materials. Molecular dynamics (MD) simulation (Odegard et al., 2002; Odegard et al., 2003) and finite element analysis (FEA) have been used successfully in modeling small (10^{-9} m) and large (10^{-1} m) length-scale problems, respectively. In the latter case, homogenized properties are often used to capture the effect of nanoenhancements on macroscopic responses. Hybrid methods that use both FEA and micromechanical models are found to have good accuracy and computational

Hybrid Polymer Composite Materials: Properties and Characterisation. DOI: http://dx.doi.org/10.1016/B978-0-08-100787-7.00010-X

performance at different length scales (Chisholm and Brinson, 2003; Bradshaw et al., 2003; Rouhi and Rais-Rohani, 2013; Rouhi et al., 2013).

As a result of manufacturing-process imperfections, variability is introduced in the CNT/CNF geometric properties such as length, cross-sectional area, and waviness. Previous experimental and analytical investigations have explored the effects of CNT waviness (Fisher, 2002) and CNF waviness and aspect ratio (Yu et al., 2011) on stiffness properties of enhanced matrix. Generally, the stiffness properties tend to diminish in a nonlinear manner as waviness increases. However, the effect of waviness on stiffness has not been quantitatively compared with those of volume fraction and interphase.

Microscopic images of fiber-reinforced polymer materials show the presence of a three-dimensional region surrounding the fiber inside a matrix (Gao and Mader, 2002). The so-called interphase region has unique properties that are different from those of the fiber and the matrix. There is considerable variability in the shape and size of the interphase. Also, the mechanical properties of the interphase are found to inhomogeneous (i.e., vary from one location to another).

Because of the extremely small length scales involved, it is very difficult to perform an experimental study of the interphase region in nanoreinforced materials. Recent MD simulations have revealed the existence of interphase between CNF and the surrounding polymer matrix (Nouranian et al., 2011). Yu et al. (2011) performed a parametric study to investigate the effects of the interphase properties and thickness as well as CNF waviness and morphology (hollow vs solid cross section) on elastic properties of nanoreinforced vinyl ester matrix. They concluded that the development of a nanofiber–matrix interphase can have a profound effect on the overall elastic properties of the material. In addition, small degrees of nanofiber waviness can result in a significant decrease in the effective composite properties. Liu et al. (2016) proposed an extended micromechanics method to include the interacting interphases in polymer nanocomposites. The interphase properties, however, were assumed to be constant in all distances from the surface of the nanoparticles. Ansari et al. (2016) used a simplified unit cell approach to incorporate the existence of waviness and homogeneous interphase and studied their effects on the overall elastic properties of a CNT/polymer nanocomposite material. They also considered the interphase as a homogeneous region between the nanoinclusion and its surrounding matrix.

Although the addition of nanoreinforcements has been shown to always improve the stiffness properties of a nanoenhanced matrix, the same cannot be said about its strength properties. For example, Zhou et al. (2008) found improvements of about 18% in the Young's modulus and 9% in strength properties of CNF/epoxy nanocomposite with 3% CNF weight fraction, whereas at 2% CNF weight fraction, strength improvement was 14.5%. The reduction in strength improvement with higher CNF weight fraction was attributed to the formation of weak spots in the epoxy resin due to agglomeration of CNF.

Simulation codes such as GENOA (Garg et al., 2008; Rouhi et al., Apr 2010) or MAC/GMC (Arnold et al., 1999; Micromechanics Analysis Code) have been used to estimate the stiffness and strength properties of the enhanced matrix based on a building block approach that relies on the results of physical experiments at various length scales. However, the three-dimensional interphase region is not captured in these and other similar hybrid micromechanical/FE modeling codes.

In this chapter, through numerical modeling and simulation, the influence of interphase and CNF waviness on stiffness properties of a thermoset polymer matrix material is investigated. The effects of geometric attributes (e.g., aspect ratio and waviness) and volume fraction on the stiffness properties of enhanced matrix are examined using a micromechanical approach. Also, the multi-inclusion approach is used with different representations of the interphase as a functionally graded material to investigate the effect of interphase and its profile on the stiffness properties of the enhanced matrix.

10.2 Stiffness properties of nanoenhanced matrix

Eshelby's results (Eshelby, 1957; Mura, 1987) show that if an elastic homogeneous ellipsoidal inclusion, surrounded by an infinite linear elastic matrix, is subjected to a uniform unconstrained strain $\boldsymbol{e}^{\mathrm{T}}$ (also known as "eigenstrain"), a uniform strain $\boldsymbol{e}^{\mathrm{C}}$ is induced in the constrained inclusion, which is proportional to the induced unconstrained strain according to the relationship

$$\boldsymbol{e}^{C} = \mathbb{S}\,\boldsymbol{e}^{T} \tag{10.1}$$

where $\mathbb{S}$ is the interior point Eshelby fourth-order tensor that depends solely on the geometry of the inclusion and the Poisson's ratio of the surrounding isotropic linear elastic matrix. The concept of equivalent homogeneous inclusions was introduced to handle the mean field description of a composite material with inhomogeneities. Eshelby's solution resulted in the calculation of the effective properties of a two-phase composite. Due to limitation of Eshelby's method to only low volume fraction of inclusions, Hill (1965) and Budiansky (1965) developed a self-consistent method for nondilute concentrations that, unlike the Eshelby's prediction, approaches the limiting value of the inclusion properties as the inclusion volume fraction approaches one. Mori and Tanaka (1973) accounted for the interaction of the inclusions by defining a second transformation matrix (strain concentration tensor) to match the limiting values of the effective properties at both low and high volume fractions of the inclusion. In addition, all of the five elastic constants match with the elasticity solution for fibrous composites. A detailed discussion on different types of self-consistent methods can be found in (Gramoll et al., 1991; Qu and Cherkaoui, 2006). Based on the Eshelby's solution, the Mori–Tanaka (M-T) scheme (Mori and Tanaka, 1973) estimates the effective stiffness tensor as

$$\mathbb{L}^{MT} = \mathbb{L}_0 + c_1\{(\mathbb{L}_1 - \mathbb{L}_0)\mathbb{T}\}[c_0\mathbb{I} + c_1\{\mathbb{T}\}]^{-1} \tag{10.2}$$

where indices 0 and 1 represent the matrix and the inhomogeneity, respectively, $\mathbb{L}$ is the fourth-order stiffness tensor, c is the volume fraction of the inhomogeneity, $\mathbb{I}$ is the fourth-order identity tensor, and $\mathbb{T}$ is a fourth-order tensor that relates the uniform strain in an inhomogeneity embedded in an effective continuum to the average matrix strain. The curly brackets { } denote physical properties averaged

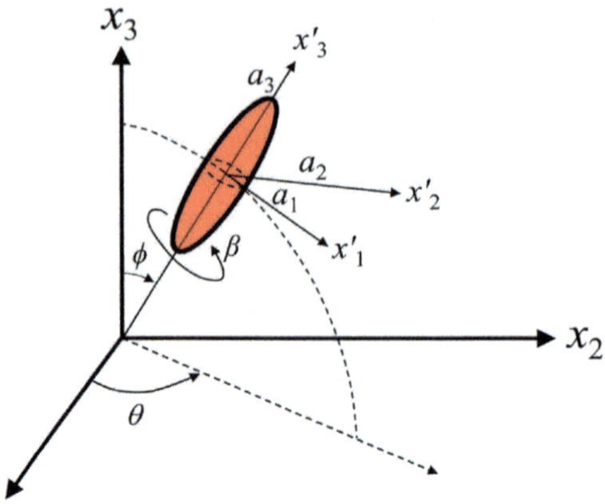

Figure 10.1 A spatially oriented ellipsoidal inhomogeneity.

over all possible inhomogeneity orientations. For an ellipsoidal inhomogeneity, $\mathbb{T}$ is calculated as

$$\mathbb{T} = [\mathbb{I} + \mathbb{S}\mathbb{L}_0^{-1}(\mathbb{L}_1 - \mathbb{L}_0)]^{-1} \tag{10.3}$$

where $\mathbb{S}$ only depends on the geometry of the inhomogeneity and Poisson's ratio of the matrix. With c, $\mathbb{L}$, and $\mathbb{S}$ known, one may implement the micromechanical model based on Eqs. (10.2) and (10.3).

When the inhomogeneities are in the form of nonaligned (randomly oriented) discontinuous fibers, the stiffness tensor is averaged over all possible fiber orientations. Fig. 10.1 shows a rotated ellipsoidal inhomogeneity and the corresponding orientation angles used in averaging the stiffness tensor over all possible orientations using (Mura, 1987; Huang, 2001; Schjodt-Thomsen and Pyrz, 2001)

$$\{\mathbb{L}_{ijkl}\} = \int_0^{2\pi}\int_0^{2\pi}\int_0^{\pi} g(\phi)a_{ip}a_{jq}a_{kr}a_{ls}\mathbb{L}_{ijkl}\ \sin(\phi)d\phi d\theta d\beta \tag{10.4}$$

where $g(\phi)$ is the orientation distribution function, and a_{ij} are angular functions found as

$$a_{ij} = \begin{cases} a_{11} = \cos(\phi)\cos(\theta)\cos(\beta) - \sin(\theta)\sin(\beta) \\ a_{12} = -\cos(\phi)\cos(\theta)\sin(\beta) - \sin(\theta)\cos(\beta) \\ a_{13} = \sin(\phi)\cos(\theta) \\ a_{21} = \cos(\phi)\sin(\theta)\cos(\beta) + \cos(\theta)\sin(\beta) \\ a_{22} = -\cos(\phi)\sin(\theta)\sin(\beta) + \cos(\theta)\cos(\beta) \\ a_{23} = \sin(\phi)\sin(\theta) \\ a_{31} = -\sin(\phi)\cos(\beta) \\ a_{32} = \sin(\phi)\sin(\beta) \\ a_{33} = \cos(\phi) \end{cases} \tag{10.5}$$

Maekawa et al. (1989) used an incomplete Beta function to represent $g(\phi)$ in order to account for partially aligned fibers. When fibers are randomly oriented, $g(\phi)$ is constant with its value found to be (Mura, 1987):

$$g(\phi) = \frac{1}{8\pi^2} \tag{10.6}$$

Tandon and Weng (1984) developed a simplified approach based on the M-T homogenization scheme to calculate the effective properties of an enhanced matrix with unidirectional discontinuous inhomogeneities. They treated the enhanced matrix as a transversely isotropic material with aligned inhomogeneities of known aspect ratio and elastic properties in a two-dimensional space. They showed that for a small volume fraction of inhomogeneities (e.g., CNF), the effective or homogenized elastic constants in the principal material directions can be calculated analytically as:

$$\begin{aligned}
\frac{E_{11}}{E_0} &= \frac{1}{1+f(A_1+2\nu_0 A_2)/C_A} \\
\frac{E_{22}}{E_0} &= \frac{1}{1+f[(-2\nu_0 A_3+(1-\nu_0)A_4+(1+\nu_0)A_5 C_A]/C_A} \\
\frac{G_{12}}{G_0} &= 1+\frac{f}{\dfrac{G_0}{G_1-G_0}+2(1-f)S_{1212}} \\
\frac{G_{23}}{G_0} &= 1+\frac{f}{\dfrac{G_0}{G_1-G_0}+2(1-f)S_{2323}}
\end{aligned} \tag{10.7}$$

where E_{ij} and G_{ij} represent the elastic and shear moduli, respectively, of the nanoenhanced matrix material with $G_{23} = G_{13}$, f is the inhomogeneity volume fraction, E_0, G_0, and ν_O are the elastic modulus, shear modulus, and Poisson's ratio of the neat matrix, respectively, whereas A_1, A_2, A_3, A_4, A_5, and C_A are constants calculated from the geometric properties of the inhomogeneity and elastic properties of both the inhomogeneity and matrix materials as shown in (Tandon and Weng, 1984). The S_{ijkl}'s are the components of the Eshelby tensor calculated using the aspect ratio of the inhomogeneity and the Poisson's ratio of the neat matrix (Eshelby, 1957; Mura, 1987).

Garg et al. (2008) developed a technique based on Eq. (10.7) to approximate the in-plane properties of a matrix with randomly distributed nanoreinforcements. They treated the matrix as an equivalent quasi-isotropic $[0/45/-45/90]_s$ laminate made of multiple layers of aligned nanoenhanced matrix material, where the ply angles represent the orientation angles of the aligned nanoreinforcements in the individual layers. By assuming the out-of-plane elastic properties are equal to the in-plane values, they approximated the effect of three-dimensional random orientation of CNF in the matrix.

The above methods can predict the effective properties of a dilute composite with relatively high accuracy. For a composite with a finite concentration of inhomogeneities, a repetitive process is exploited for the effective properties in terms of the volume fractions. In the so-called differential schemes, the effective properties of a composite made of a matrix ($\mathbb{L}_0$) and a small amount of inhomogeneities (Δc) is calculated using any of the above methods ($\mathbb{L}_1$). In the next step, $\mathbb{L}_0$ is replaced by $\mathbb{L}_1$ and another small amount of inhomogeneities (Δc) is added to the matrix and the new effective properties are calculated ($\mathbb{L}_2$). This process can repetitively continue until the volume fraction of the inhomogeneities reaches to the desired finite value (c). This is the basic idea of all types of the differential schemes.

It is worth noting that all of the above methods assume uniform distribution for the inhomogeneities. In addition, Eshelby tensor $\mathbb{S}$ is shape dependent, but not size dependent. In other words, the effective properties predicted by these methods will not depend on the size of the inhomogeneities and length scales are not considered. However, the shapes and orientations of the inhomogeneities are taken into account and the interactions among inhomogeneities are considered differently by different methods. In general, the Eshelby method works only for very dilute concentration, whereas the other methods are applicable to somewhat higher concentrations.

The practical amount of CNF volume fraction in the epoxy matrix is in the range that M-T approach can predict the stiffness properties accurately. Table 10.1 shows the calculated effective Young's modulus based on the general M-T approach as described by Eqs. (10.1)–(10.5) and the approximate method by Garg et al. (2008) when CNF ($E_1 = 450$ GPa, and $\nu_1 = 0.3$) nanoreinforcements are mixed with vinyl ester resin ($E_0 = 3.5$ GPa, $\nu_0 = 0.35$). The effects of CNF aspect ratio and volume fraction are captured. Only at very low volume fractions, the approximate method produces a value for the elastic modulus that is in reasonable agreement with that from the M-T approach. The closest values are at volume fraction of 0.01 and aspect ratio of one. The difference between the two approaches increases as CNF

Table 10.1 Predicted elastic modulus of CNF reinforced vinyl ester for different CNF aspect ratios and volume fractions using the general M-T and approximate (Garg et al., 2008) approaches

Elastic modulus (GPa)	CNF aspect ratio	CNF volume fraction				
		0.01	0.05	0.1	0.2	0.4
$E_{\text{M-T}}$	1	3.57	3.88	4.30	5.29	8.23
E_{App}		3.51	3.53	3.55	3.59	3.66
$E_{\text{M-T}}$	10	3.67	4.36	5.32	7.55	14.09
E_{App}		3.88	4.97	6.47	9.91	19.69
$E_{\text{M-T}}$	100	4.23	7.25	11.30	20.60	46.21
E_{App}		4.90	10.36	17.27	31.42	61.31
$E_{\text{M-T}}$	1000	4.29	7.56	11.95	21.99	49.46
E_{App}		5.02	10.94	18.37	33.40	64.37

volume fraction or aspect ratio is increased. For aspect ratio of one, the approximate values are less than those of M-T, with the difference increasing to as much as 55% as the CNF volume fraction reaches 0.4. For all the other aspect ratios, the approximate values overestimate those of M-T by as much as 54%.

10.3 Interphase model

Since in the Mori–Tanaka approach the fiber and matrix are assumed to have an interface with perfect bonding, a different technique is needed for modeling the effect of interphase. Nemat-Nasser and Hori (1993) developed the double-inclusion and multi-inclusion models in an attempt to capture the distinct three-dimensional interphase region whose properties may be considerably different from those of the inclusion and matrix materials. Fig. 10.2 shows a multi-inclusion model consisting of an ellipsoid with separate phases (a nested series of smaller ellipsoids Ω_α ($\alpha = 1, 2, \ldots, N$) such that $\Omega_1 \subset \Omega_2 \subset \ldots \subset \Omega_N \equiv V$) in an infinite domain of the matrix material. The overall elasticity of the multiphase composite region is treated as the volumetric average of properties of the individual constituent materials, and for the aligned inclusion case, it is found as:

$$\bar{\mathbb{L}}^{MI} = \mathbb{L}\left\{\mathbb{I} + \sum_{\alpha=1}^{n} f_\alpha(\mathbb{S}^\Omega - \mathbb{I})(\mathbb{A}^\alpha - \mathbb{S}^\Omega)^{-1}\right\}\left\{\mathbb{I} + \sum_{\alpha=1}^{n} f_\alpha \mathbb{S}^\Omega(\mathbb{A}^\alpha - \mathbb{S}^\Omega)^{-1}\right\}^{-1} \tag{10.8a}$$

$$\mathbb{A}^\alpha \equiv (\mathbb{L} - \mathbb{L}^\alpha)^{-1}\mathbb{L} \tag{10.8b}$$

where $\mathbb{L}$ and $\mathbb{I}$ are the fourth-order stiffness and identify tensors, respectively. f_α is the volume fraction of the α interphase region and $\mathbb{S}$ is the Eshelby tensor that, in

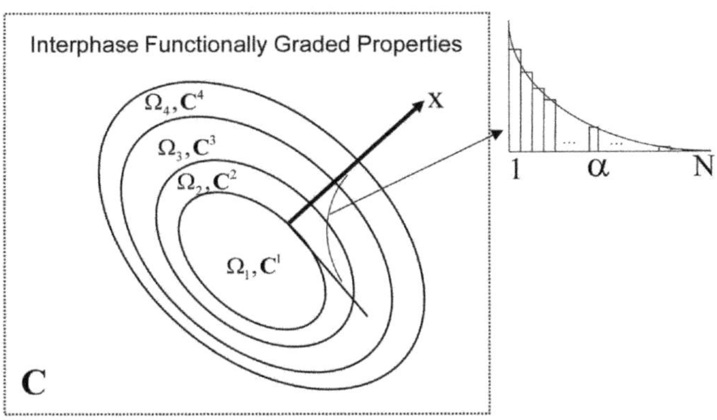

Figure 10.2 A multi-inclusion model with piecewise constant (functionally graded) interphase representation.

this case, also depends on the geometry of the multi-inclusion model. For the case of randomly distributed inclusions, the stiffness properties have to be averaged similar to that in Eqs. (10.4) and (10.5).

The interphase region may be inhomogeneous with properties that vary as a function of distance from the inclusion surface (Gao and Mader, 2002; Nouranian et al., 2011). Hence, the multi-inclusion model can be used to approximate the varying properties of the interphase. As shown in Fig. 10.2, an approximate model based on piecewise constant properties is used to model the interphase region, where the properties are assumed to vary similar to a functionally graded material (Reddy, 2000; Ruhi et al., 2005) and approximated as

$$P = P_{in} + (P_{out} - P_{in})\left(\frac{x}{L}\right)^n \tag{10.9}$$

where P represents the property of interest in the interphase at distance x away from the surface of the inclusion, with P_{in} and P_{out} representing the interphase property at $x = 0$ and $x = L$, respectively. The parameter n is a constant power determining the nonlinearity of the graded property. By changing the value of n, the rate of property variation from P_{in} to P_{out} can be adjusted as shown in Fig. 10.2.

Since in the multi-inclusion model each domain is assumed to have constant properties, Eq. (10.9) is modified to

$$P = P_{in} + (P_{out} - P_{in})\left(\frac{\alpha - 1}{N}\right)^n \tag{10.10}$$

where α represents the interphase subdivision in the range of 1 to N. Eq. (10.10) represents an approximation of Eq. (10.9) with the assumption of piecewise constant properties in each interphase subdivision.

Table 10.2 shows the effective elastic modulus of a CNF reinforced vinyl ester matrix with different elastic properties of a homogeneous interphase assuming a

Table 10.2 Effective elastic modulus of CNF reinforced vinyl ester with homogeneous interphase for $E_f = 450$ GPa, $E_0 = 3.5$ GPa, AR = 100, $V_f = 0.01$, $N = 1$, and $n = 0$

Interphase modulus, E_I (GPa)	ITR[a]	Effective modulus, E_c (GPa)
0 (No interphase)	0.0	4.20
2.0	0.1	4.21
2.0	0.5	4.19
2.0	1.0	4.15
100	0.1	4.26
100	0.5	4.47
100	1.0	4.83

[a]ITR: interphase thickness ratio = interphase thickness/fiber radius.

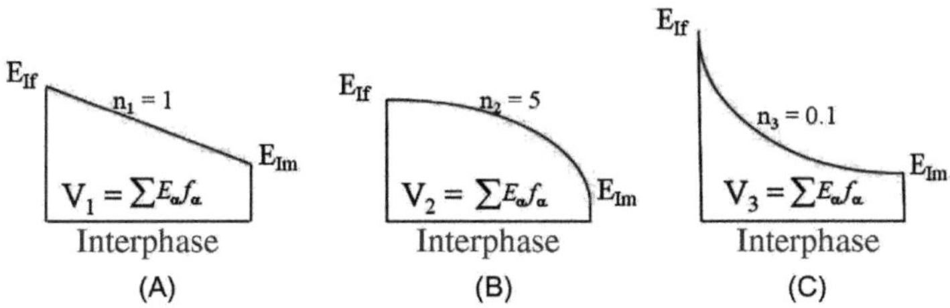

Figure 10.3 Three different graded profiles for the interphase region with $V_1 = V_2 = V_3$.

double-inclusion representation ($N = 1$) where the interphase properties are held constant along its thickness ($n = 0$). As to be expected, when the homogeneous interphase has greater elastic modulus than the matrix, the overall stiffness of the composite material increases by increasing the thickness of the interphase; otherwise, the overall stiffness decreases. For an order of magnitude increase in ITR, the effective modulus decreases by ~1.4% for $E_I = 2$ GPa and increases by ~13% for $E_I = 100$ GPa. As expected, when the interphase has higher modulus than the matrix, the effective elastic property of the resulting composite material is improved.

In the case of inhomogeneous interphase, the effect of graded profile can be found from Eqs. (10.8)–(10.10), where the overall stiffness of the composite material is a weighted average of the stiffness properties of the individual layers of the interphase region based on their corresponding volume fractions. Three different profiles are considered as shown in Fig. 10.3 with the interphase elastic modulus varying from E_{If} in the fiber vicinity to E_{Im} at the matrix boundary. The profiles are varied based on the selected value for parameter n in Eq. (10.10) such that the weighted volume fraction of the interphase region ($\sum E_\alpha f_\alpha$) remains constant for all the profiles.

For a nanoreinforced material system, the effect of n is captured in Table 10.3 for $N = 20$. To make the weighted volume fraction of the interphase equal in different profiles, the E_{If} and E_{Im} values are adjusted as shown in Table 10.3. As the results show, as long as the weighted interphase volume fraction is kept constant, the profile shape of the interphase region does not affect the overall elastic properties of the composite material. With no interphase, the overall elastic modulus of the composite material using the general M-T approach is $E_c = 6.24$ GPa.

Tables 10.4–10.6 show the elastic modulus of the CNF reinforced vinyl ester with different interphase parameters using the multi-inclusion approach and a linear interphase profile with $n = 1$. The results in Table 10.4 show how the increase in ITR can affect the interphase volume fraction (IVF), weighted interphase volume fraction, and effective modulus while holding the CNF volume fraction fixed. In this case, the E_{If} and E_{Im} values are equal to elastic moduli of the adjacent materials. Essentially, the interphase is gradually getting wider and wider such that it finally becomes twice as wide as the CNF radius. The plot of E_c versus the weighted interphase volume fraction is shown in Fig. 10.4A indicating a mild nonlinear relationship.

Table 10.3 Effective elastic modulus of CNF reinforced vinyl ester with inhomogeneous interphase for $E_f = 450$ GPa, $E_0 = 3.5$ GPa, AR = 100, $V_f = 0.03824$, $N = 20$, and $\sum E_\alpha f_\alpha = 23.19$ GPa

E_{If}, E_{Im} (GPa)	E_c (GPa)		
	$n = 1$	$n = 5$	$n = 0.1$
450, 3.5	10.47		
350, 84	10.54		
250, 164	10.57		
150, 244	10.56		
3.5, 362	10.50		
254, 3.5		10.54	
190, 250		10.57	
138, 450		10.53	
450, 182			10.57
150, 207			10.57
3.5, 219			10.57

Table 10.4 Effective elastic modulus of CNF reinforced vinyl ester with different ITR values for $E_f = E_{If} = 450$ GPa, $E_0 = E_{Im} = 3.5$ GPa, AR = 100, $V_f = 0.0382$, $N = 20$, and $n = 1$

ITR	IVF	$\sum E_\alpha f_\alpha$(GPa)	E_c (GPa)
0	0	0.0	6.24
0.2	0.0169	3.72	6.89
0.4	0.0368	7.89	7.64
0.6	0.0597	12.52	8.48
0.8	0.0857	17.62	9.42
1.0	0.1148	23.19	10.46
1.2	0.1469	29.22	11.61
1.4	0.1821	35.73	12.87
1.6	0.2203	42.70	14.26
1.8	0.2616	50.13	15.78
2.0	0.306	58.03	17.45

The results in Table 10.5 show how changes in E_{If} and E_{Im} values can affect the effective elastic modulus. The bounds are taken between the selected upper and lower bounds irrespective of the moduli of the adjacent materials. The purpose here is to simply examine the effect of linear variation in the interphase properties from one extreme to another.

Table 10.5 Effective elastic modulus of CNF reinforced vinyl ester with different E_{If} and E_{Im} values for $E_f = 450$ GPa, $E_0 = 3.5$ GPa, AR = 100, $V_f = 0.0382$, $N = 20$, ITR = 1, and $n = 1$ (IVF = 0.1148 for all cases)

E_{If} (GPa)	E_{Im} (GPa)	$\sum E_\alpha f_\alpha$(GPa)	E_c (GPa)
450	3.50	23.19	10.46
405.35	48.15	23.76	10.59
360.70	92.80	24.33	10.70
316.05	137.45	24.90	10.81
271.40	182.10	25.46	10.90
226.75	226.75	26.03	10.99
182.10	271.40	26.60	11.08
137.45	316.05	27.17	11.15
92.80	360.70	27.74	11.22
48.15	405.35	28.30	11.28
3.50	450.00	28.87	11.33

Table 10.6 Effective elastic modulus of CNF reinforced vinyl ester with different E_{If} and E_{Im} values for $E_f = 450$ GPa, $E_0 = 3.5$ GPa, AR = 100, $V_f = 0.0382$, $N = 20$, ITR = 2, and $n = 1$ (IVF = 0.306 for all cases)

E_{If} (GPa)	E_{Im} (GPa)	$\sum E_\alpha f_\alpha$(GPa)	E_c (GPa)
450	3.50	58.03	17.45
405.35	48.15	60.30	17.97
360.70	92.80	62.57	18.40
316.05	137.45	64.84	18.81
271.40	182.10	67.12	19.19
226.75	226.75	69.39	19.55
182.10	271.40	71.66	19.90
137.45	316.05	73.93	20.22
92.80	360.70	76.20	20.53
48.15	405.35	78.47	20.81
3.50	450.00	80.74	21.05

The results in Table 10.6 are found in the same way as those in Table 10.5 except that the ITR is increased from 1 to 2, which also causes an increase in the IVF value. The graph of E_c versus the weighted interphase volume fraction (IVF) is shown in Fig. 10.4.

The results in Tables 10.4–10.6 and Fig. 10.4 indicate that the weighted volume fraction of the interphase through changes in ITR has the greatest influence on the effective elastic modulus of the CNF enhanced matrix material.

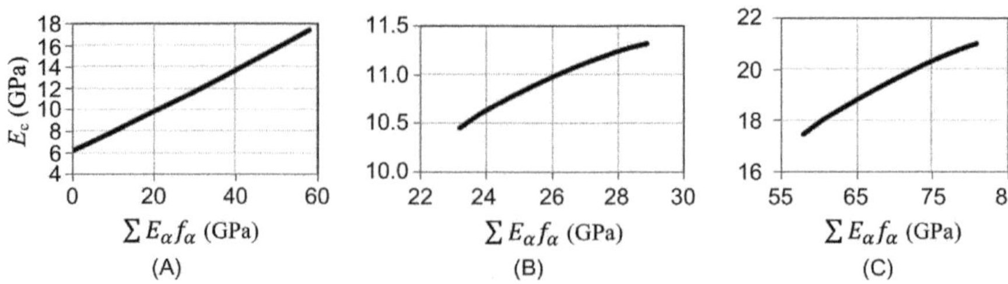

Figure 10.4 Variation of the effective elastic modulus as a function of weighted interphase volume fraction for the case defined in (A) Table 10.4, (B) Table 10.5, and (C) Table 10.6.

10.4 Waviness model

Scanning electron microscope and transmission electron microscope images show that CNTs can be highly curved when dispersed in a polymer (Shaffer and Windle, 1999; Vigolo et al., 2000; Qian et al., 2000). Given the relatively high aspect ratio and low bending stiffness of CNT and CNF, these findings are to be expected.

Different analytical and hybrid approaches have been used in modeling fiber waviness. For example, in an analytical approach (e.g., Fisher, 2002; Shady and Gowayed, 2010), a wavy fiber is isolated and its effective reinforcing modulus (E_{ERM}) is calculated based on Castigliano's second theorem. Given the assumed sinusoidal geometry of a wavy fiber as described in Fig. 10.5, one may calculate the internal axial force (T), shear force (V) and bending moment (M) in the fiber as a function of coordinate z using (Fisher, 2002)

$$
\begin{aligned}
T(z) &= F\cos(\theta) = \frac{F}{\sqrt{1+\dfrac{4\pi^2a^2}{\lambda^2}\sin^2\left(\dfrac{2\pi z}{\lambda}\right)}} \\
V(z) &= F\sin(\theta) = \frac{-\dfrac{2a\pi}{\lambda}F\sin\left(\dfrac{2\pi z}{\lambda}\right)}{\sqrt{1+\dfrac{4\pi^2a^2}{\lambda^2}\sin^2\left(\dfrac{2\pi z}{\lambda}\right)}} \qquad (10.11)\\
M(z) &= F[a-y(z)] = Fa\left[1-\cos\left(\dfrac{2\pi z}{\lambda}\right)\right]
\end{aligned}
$$

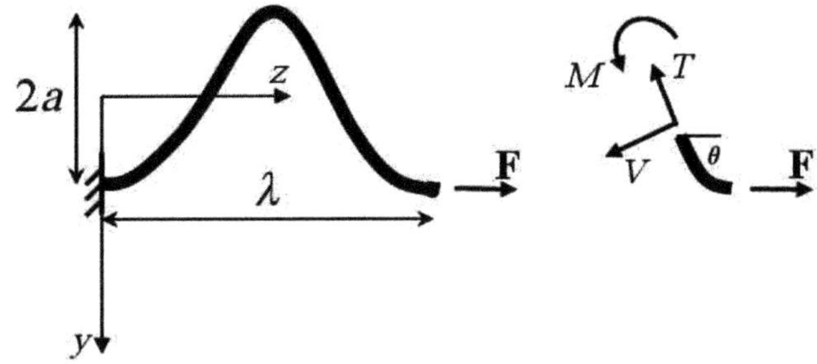

Figure 10.5 Schematic of a clamped-free sinusoidal wavy fiber.

The total displacement is found by considering the contributions of the internal loads over a wavelength of the sinusoidal fiber given by

$$\delta_F = \delta_T + \delta_V + \delta_M = \int \frac{T}{EA}\frac{\partial T}{\partial F}ds + \int \frac{kV}{GA}\frac{\partial V}{\partial F}ds + \int \frac{M}{EI}\frac{\partial M}{\partial F}ds \tag{10.12}$$

where δ_T, δ_V, and δ_M represent the deflections due to T, V, and M, respectively, with ds as an increment of arc length along the fiber in Fig. 10.5. In the shear contribution term, k is the correction factor for the shear strain energy (equal to 1.33 for a solid circular cross-section) and G is the shear modulus. The terms A and I are the cross-sectional area and the moment of inertia for a circular cross-section, respectively. Equating the displacement found from Eq. (10.12) with the displacement of an equivalent straight fiber of the same length will result in the effective reinforcing modulus of a free-standing wavy fiber found as

$$E_{ERM} = \frac{\sigma}{\varepsilon} = \frac{F/A}{\delta_F/L} = \frac{FL}{A\delta_F} \tag{10.13}$$

The effective reinforcing modulus is a representative value that accounts for the reduction in reinforcement provided by the wavy fiber as compared to that by a straight fiber (of modulus E_f) (Chisholm and Brinson, 2003; Bradshaw et al., 2003). Hence, E_f represents the Young's modulus of the fiber material, whereas E_{ERM} ($E_{ERM} \leq E_f$) represents the effective modulus considering the fiber waviness.

Using the model described by Eqs. (10.11)–(10.13), Fig. 10.6A shows how the planar sinusoidal waviness of a fiber drastically decreases E_{ERM} of the fiber. This effective property can then be used in any micromechanical model as an equivalent value for the true CNF/CNT modulus. Fisher (2002) showed that the analytical model is only useful when the modulus of the straight nanotube E_{NT} is much larger than the modulus of the matrix.

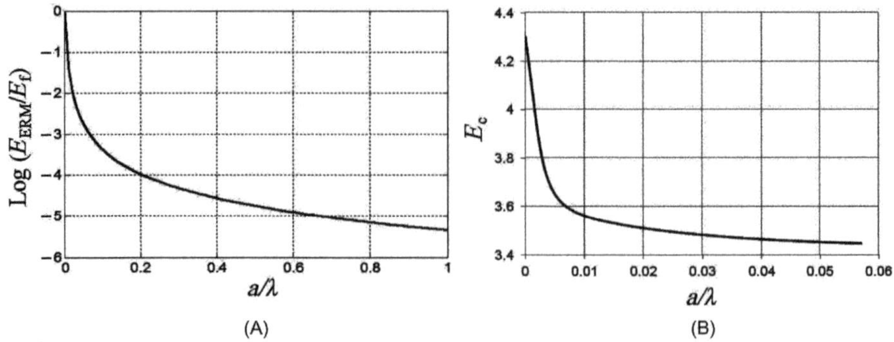

Figure 10.6 Variation of effective modulus of (A) a free-standing wavy fiber and (B) wavy nanofiber reinforced matrix using analytical method for $\lambda/d = 100$.

The rapid decrease in effective modulus of nanotube/nanofiber as shown in Fig. 10.6A results in a sharp decrease in modulus of the nanoenhanced matrix as shown in Fig. 10.6B. It is worth noting that this analytical method will underpredict the effective modulus of the nanoenhanced matrix as shown in Fig. 10.6B, and that the prediction is even lower than the modulus of pure matrix ($E_m = 3.5$ GPa) after waviness extends beyond a small value ($a/\lambda > 0.025$) which seems to be unrealistic.

When the criterion of large relative modulus is not satisfied (i.e., $E_{NT}/E_{matrix} < 1000$), a new model/analysis is needed to account for the lateral constraint (bonding between the inclusion and the matrix) imposed on the CNT by the surrounding matrix.

10.5 Summary and conclusion

The stiffness properties of a thermoset polymer matrix material with CNF reinforcement were calculated using two micromechanics-based approaches, one based on the general three-dimensional Mori–Tanaka (M-T) scheme and another based on a quasi-isotropic laminate representation of the nanoenhanced matrix. The effects of CNF aspect ratio and volume fraction on the overall stiffness properties of the CNF reinforced polymer matrix were studied. It was shown that the results of the approximate approach are consistent with those of the M-T scheme for very low nanoreinforcement volume fractions, but the accuracy of approximation drops with increased CNF volume fraction and aspect ratio.

A piecewise constant approximation along with multi-inclusion approach was used to model the interphase region as a functionally graded material. Different profile geometries were considered with the results showing that the profile has no impact on the predicted effective Young's modulus. Instead, the volume fraction of the interphase weighted by its elastic modulus (interphase effective volume fraction) was shown to be effective on the overall elastic properties of the composite.

Fiber waviness was modeled using analytical techniques. The results showed that any level of waviness causes a considerable reduction in the overall stiffness properties of the nanoenhanced matrix due to diminishing the effective (equivalent straight fiber) stiffness of the fibers in the matrix.

Acknowledgment

This material is based on the work partially supported by the US Department of Energy under Award Numbers DE-FC26-06NT42755 and DE-EE0002323.

References

Ansari, R., Hassanzadeh-Aghdam, M., Mahmoodi, M., 2016. Three-dimensional micromechanical analysis of the CNT waviness influence on the mechanical properties of polymer nanocomposites. Acta Mech. 1−21. Available from: http://dx.doi.org/10.1007/s00707-016-1666-6.

Arnold, S., Bednarcyk, B., Wilt, T., Trowbridge, D., 1999. Micromechanical Analysis Code with Generalized Method of Cells (MAC/GMC), User Guide: Version 3.0, NASA/TM-1999-209070.

Bradshaw, R.D., Fisher, F.T., Brinson, L.C., 2003. Fiber waviness in nanotube-reinforced polymer composites—II: Modelling via numerical approximation of the dilute strain concentration tensor. Compos. Sci. Technol. 63 (11), 1705−1722.

Budiansky, B., 1965. On the elastic moduli of some heterogeneous materials. J. Mech. Phys. Solids 13, 223−227.

Chisholm, N., Brinson, L.C., 2003. Fiber waviness in nanotube-reinforced polymer composites—I: modulus predictions using effective nanotube properties. Compos. Sci. Technol. 63 (11), 1689−1703.

Chisholm, N., Mahfuz, H., Rangari, V.K., Ashfaq, A., Jeelani, S., 2005. Fabrication and mechanical characterization of carbon/SiC−epoxy nanocomposites. Compos. Struct. 67 (1), 115−124.

Eshelby, J.D., 1957. The determination of the elastic field of an ellipsoidal inclusion, and related problems. Proc. R. Soc. Lond. A. 241, 376−396.

Fisher, F., 2002. Nanomechanics and Viscoelastic Behavior of Carbon Nanotube Reinforced Polymers [PhD Dissertation]. Northwestern University.

Gao, S.L., Mader, E., 2002. Characterization of interphase nanoscale property variations in glass fiber reinforced polypropylene and epoxy composites. Compos. A. 33, 559−576.

Garg M., Abdi F., McHugh S., 2008. Analyzing interlaminar shear strength of multi-scale composite via combined finite element and progressive failure analysis approach, SAMPE Fall Technical Conference on Multifunctional Materials, September 2008, Memphis, TN, USA.

Gojny, F.H., Wichmann, M.H.G., Fiedler, B., Bauhofer, W., Schulte, K., 2005. Influence of nano-modification on the mechanical and electrical properties of conventional fibre-reinforced composites. Compos. A. 36 (11), 1525−1535.

Gramoll, K.C., Walker, K.P., Freed, A.D., 1991. An overview of Self-Consistent Methods for Fiber Reinforced Composites. NASA TM 103713.

Hill, R., 1965. A self-consistent mechanics of composite materials. J. Mech. Phys. Solids 13, 213–225.

Huang, J.H., 2001. Some closed-form solutions for effective moduli of composites containing randomly oriented short fibers. Mater. Sci. Eng. A. 315, 11–20.

Liu, Z., Moore, J.A., Liu, W.K., 2016. An extended micromechanics method for probing interphase properties in polymer nanocomposites. J. Mech. Phys. Solids. 95, 663–680.

Maekawa, Z.-I., Hamada, H., Yokoyama, A., 1989. Lamination Theory of Composite Material with Complex Fiber Orientation Distribution, Proc. Int. Conf. on Conceptual Structures (ICCS), pp. 701–714.

Micromechanics Analysis Code, Generalized Method of Cells. Point of Contact: Dr. Steven M. Arnold, NASA Glenn Research Center, <http://www.grc.nasa.gov/WWW/LPB/mac/index.html>.

Mori, T., Tanaka, K., 1973. Average stress in matrix and average elastic energy of materials with misfitting inclusions. Acta Metal. 21, 571–574.

Mura, T., 1987. Micromechanics of defects in solids. Martinus Nijhoff Publishers.

Nemat-Nasser, S., Hori, M., 1993. Micromechanics: Overall Properties of Heterogeneous Solids. Elsevier, Amsterdam.

Nouranian, S., Jang, C., Lacy, T., Gwaltney, S.R., Toghiani, H., Pittman, C., 2011. Molecular dynamics simulations of vinyl ester resin monomer interactions with a pristine vapor-grown carbon nanofiber and their implications for composite interphase formation. Carbon 49 (10), 3219–3232.

Odegard, G.M., Gates, T.S., Nicholson, L.M., Wise, K.E., 2002. Equivalent-continuum modeling of nano-structured materials. Compos. Sci. Technol. 62 (14), 1869–1880.

Odegard, G.M., Harik, V.M., Wise, K.E., Gates, T.S., 2003. Constitutive modeling of nanotube-reinforced polymer composites. Compos. Sci. Technol. 63, 1671–1687.

Qian, D., Dickey, E.C., Andrews, R., Rantell, T., 2000. Load transfer and deformation mechanisms in carbon nanotube-polystyrene composites. Appl. Phys. Lett. 76 (20), 2868–2870.

Qu, J., Cherkaoui, M., 2006. Fundamentals of Micromechanics of Solids. John Wiley & Sons Inc., Hoboken, NJ.

Reddy, J.N., 2000. Analysis of functionally graded plates. Int. J. Numer. Meth. Eng. 47, 663–684.

Rouhi, M., Rais-Rohani, M., 2013. Modeling and probabilistic design optimization of a nanofiber-enhanced composite cylinder for buckling. Compos. Struct. 95, 346–353.

Rouhi, M., Angoshtari, A., Naghdabadi, R., 2005. Thermoelastic Analysis of Thick-walled Finite-length Cylinders of Functionally Graded Materials. J. Therm. Stress 28, 391–408.

Rouhi, M., Rais-Rohani, M., Lacy, T., Garg, M., Abdi, F., 2010. Mechanical Characterization of a Nanofiber Enhanced Polymer with Application to Composite Crush Tubes, 51st AIAA/ASME/ASCE/AHS/ASC Structures, Structural Dynamics, and Materials Conference, April 2010, Orlando, FL, USA.

Rouhi, M., Rais-Rohani, M., Najafi, A., 2013. Probabilistic analysis and optimization of energy absorbing components made of nanofiber enhanced composite materials. Compos. Struct. 100, 144–153.

Schjodt-Thomsen, J., Pyrz, R., 2001. The Mori–Tanaka stiffness tensor: diagonal symmetry, complex fiber orientations and non-dilute volume fractions. Compos. Sci. Technol. 33, 531–544.

Shady, E., Gowayed, Y., 2010. Effect of nanotube geometry on the elastic properties of nano-composites. Compos. Sci. Technol. 70, 1476–1481.

Shaffer, M.S.P., Windle, A.H., 1999. Fabrication and characterization of carbon nanotube/ poly (vinyl alcohol) composites. Adv. Mater. 11 (11), 937–941.

Tandon, G.P., Weng, G.J., 1984. The effect of aspect ratio of inclusions on the elastic properties of unidirectionally alligned composites. Polym. Compos. 5, 327–333.

Thostenson, E.T., Li, W.Z., Wang, D.Z., Ren, Z.F., Chou, T.W., 2002. Carbon nanotube/ carbon fiber hybrid multiscale composites. J. Appl. Phys. 91 (9), 6034–6037.

Vigolo, B., Penicaud, A.P., Couloun, C., Sauder, C., Pailler, R., Journet, C., et al., 2000. Macroscopic fibers and ribbons of oriented carbon nanotubes. Science 290, 1331–1334.

Yu, J., Lacy, T., Toghiani, H., Pittman, C., Hwang, Y., 2011. Classical micromechanics modeling of nanocomposites with carbon nanofibers and interphase. J. Compos. Mater. 45 (23), 2401–2413.

Zhou, Y., Pervin, F., Jeelani, S., Mallick, P.K., 2008. Improvement in mechanical properties of carbon fiber–epoxy composite using carbon nanofibers. J. Mater. Process. Technol. 198 (1–3), 445–453.

Properties and characterization of fiber metal laminates

11

Jarosław Bieniaś, Patryk Jakubczak and Barbara Surowska
Department of Materials Engineering, Faculty of Mechanical Engineering, Lublin University of Technology, Lublin, Poland

Chapter Outline

11.1 Introduction

Dynamic technological progress necessitates the need of intensive development in the field of modern, innovative materials and technologies. Hybrid materials are one of the most prospective material groups in terms of research and application. The aircraft industry appears to be the area with the highest demand for new materials and technologies. These materials are seen as modern materials applicable for current and future new-generation aircraft structures. Such view can be linked to high expectations regarding exploitation, safety and reliability of performance, high quality of composite structure, and efficiency of structure condition monitoring.

The development of polymer materials prompted a significant progress considering composite materials which, due to a number of beneficial properties, are now a noticeably popular and differentiated construction material group.

The interest gathered around composites may be due to their excellent mechanical and strength parameters as well as low specific weight, the combination of which is characteristic virtually for composites only. In research and implementation works,

Hybrid Polymer Composite Materials: Properties and Characterisation. DOI: http://dx.doi.org/10.1016/B978-0-08-100787-7.00011-1

the greatest importance among composite materials is usually attached to glass continuous fiber-reinforced laminates with a polymer matrix. The main advantage of laminate is the possibility to modify their properties to a certain extent, depending on their purpose. Element design allows not only to choose an existing material but also design its properties by selecting particular components, the percentage of components (based on mass or volume), the configuration (orientation) of reinforcement in the composite structure, with the option to differentiate the number of particular layers. Composite material provides not only optimal usage of given properties of all components, but it also enables differentiation of properties, depending on the processing techniques used. In the case of composite materials, it is thus possible to modify their properties at the design and production stage.

So far, from a wide range of materials, mainly light metals, such as aluminum, magnesium and titanium alloys and, especially, continuous fiber-reinforced polymer matrix composites, are applied in the aircraft industry.

In the recent years, the usage of composite materials has been growing from year to year. Composite materials are widely used for primary and secondary aircraft structures, at times also for critical elements. The basic criteria regarding composite application are their strength/density rates. They are the potential target material as, e.g., skin elements, fuselage, spars, blades, landing gear elements, stabilizers, hatches, and many more. Initially, the use of composites in aircraft structures reached several percent (military aircraft and structures). Now, the percentage of composite materials oscillates around 20–30%. The leading products employing composite materials, apart from military aircraft and space industry, are Airbus A380 (with the composite percentage of a dozen or so) and Boeing 787—with the percentage of composites of over 50%. The availability of input materials, their considerable modifications, lower production costs and well-known composite production process contribute to the fact that these materials constitute a group with a strong development potential.

Although metals and composites have a number of beneficial properties, they also have some disadvantages limiting their potential applications. Composites, compared to metal alloys, display high mechanical properties: static strength, high stiffness, low density, chemical and corrosive resistance. The disadvantages of composites include low formability, cold cracking and velocity impact resistance, moisture absorption, and relatively low operating temperature of composite materials.

Therefore, it is a challenge to produce a material retaining the features of both, metal component and fiber-reinforced polymer composite.

Fiber metal laminates (FML) are hybrid materials complying to the above mentioned requirements (Vlot and Gunnink, 2001; Vogelesang and Vlot, 2000; Wu and Yang, 2005; Botelho et al., 2006; Asundi and Choi, 1997). FMLs are materials consisting of alternating layers of metal sheets and fiber-reinforced polymer matrix composites (Fig. 11.1).

FMLs are characterized by new or improved properties, compared to traditional composites. The combination of two metal components and a fiber-reinforced polymer matrix composite as a hybrid composite leads to obtaining a material of high strength, great stiffness and strength to density ratio, flame and corrosive resistance, high fatigue and dynamic load resistance, good damping, and insulation properties and high durability.

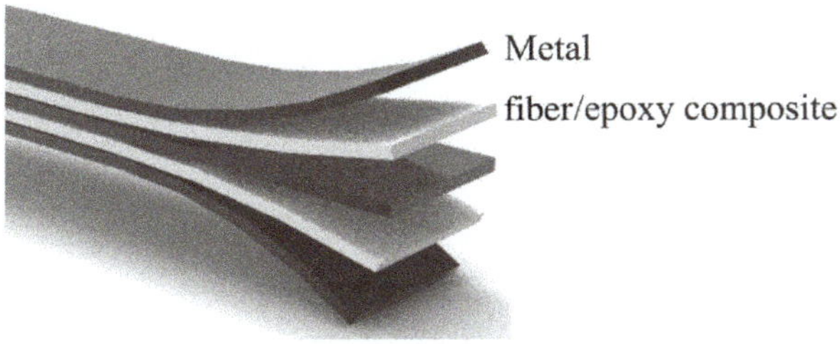

Figure 11.1 Schematic drawing of a 3/2 FML (3 metal layers and 2 fiber-reinforced polymer matrix interlayers).

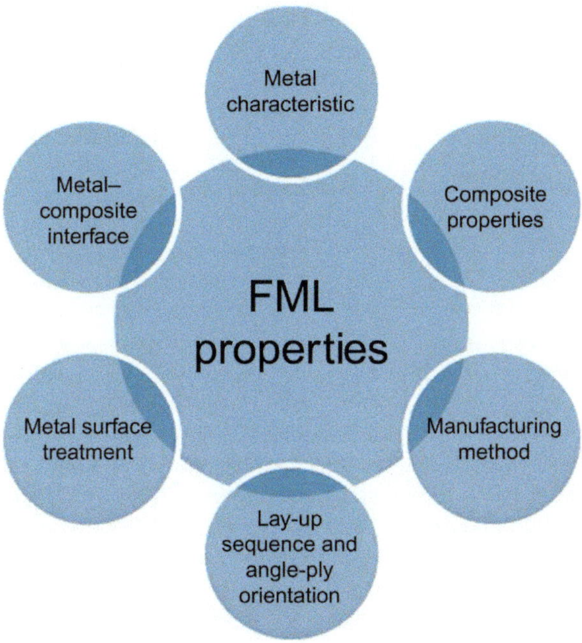

Figure 11.2 Characteristic factors influencing FML properties.

11.2 Mechanical behavior

FMLs consist of alternating thin layers of metal sheets and fiber-reinforced polymer matrix composites. Strength properties are thus determined chiefly by the characteristics of the properties of individual FML components (Wu and Yang, 2005). The combination of metal layers and a fiber-reinforced polymer matrix composite exerts a synergistic impact on numerous properties. The key characteristics influencing FML properties are shown schematically in Fig. 11.2.

Mechanical properties of FMLs are different and improved to individual metal alloy and composite material properties. Different combinations of metal layers (e.g.,

Table 11.1 Selected FML constructions on the example of GLARE laminates (Vlot and Gunnink, 2001; Wu and Yang, 2005)

Grade	Sub	Metal sheet thickness [mm] and alloy	Fiber layer orientation[1]	Characteristics	Density [g cm^{-3}][2]
Glare 1		0.3–0.4 7475-T6	0/0	Fatigue, strength, yield stress	2.52
Glare 2	2A	0.2–0.5 2024-T3	0/0	Fatigue, strength	2.52
	2B	0.2–0.5 2024-T3	90/90	Fatigue, strength	
Glare 3		0.2–0.5 2024-T3	0/90	Fatigue, impact	2.52
Glare 4	4A	0.2–0.5 2024-T3	0/90/0	Fatigue, strength in 0 degree direction	2.45
	4B	0.2–0.5 2024-T3	90/0/90	Fatigue, strength in 90 degrees direction	
Glare 5		0.2–0.5 2024-T3	0/90/90/0	Impact	2.38
Glare 6	6A	0.2–0.5 2024-T3	+45/-45	Shear, off-axis properties	2.52
	6B	0.2–0.5 2024-T3	−45/ + 45	Shear, off-axis properties	

aluminum, magnesium, titanium alloys) and glass, aramid or carbon fiber-reinforced polymer matrix composites, the possibility to orient individual layers, as well as to create laminate packs with certain thickness and a specific number of layers (both metal and polymer composite), but also various surface treatment methods allow the appropriate and desired modifying of particular FML strength properties. In order to obtain the desired composite properties, it is important to ensure the proper matrix-reinforcement interface and metal–composite interface surface.

The only FML materials used so far on an industrial scale are GLARE laminates (Glass LAminates REinforced). At least six GLARE laminate types are used (Table 11.1). The types vary in the number and orientation of prepreg layers. All the types are based on high strength glass continuous S2 fiber prepregs in an epoxy Cytec FM 94 matrix and aluminum 2024-T3 or 7475-T6 sheet prepregs (Vlot and Gunnink, 2001; Vogelesang and Vlot, 2000; Wu and Yang, 2005).

11.2.1 Static properties

11.2.1.1 Tensile strength

Tensile FML properties are a resultant of and depend on the properties of particular components: fiber-reinforced polymer matrix composite, metal material, and on the composite layer orientation in the laminate (Vlot and Gunnink, 2001; Vogelesang and Vlot, 2000; Wu and Yang, 2005).

Mechanical properties of laminate components have been analyzed and are available in the literature on nonferrous metals and polymer matrix fiber-reinforced composites. Selected properties of particular FML components can be seen in chapter "Structure and chemistry of Fiber Metal Laminates", Volume 1.

Generally, FMLs have high strength properties. They are mostly based on the strength properties and type of the composite material, especially the properties of fiber reinforcement, fiber orientation, and the type of metal. Also, fiber orientation of individual layers determines isotropic or anisotropic properties of a material (Botelho et al., 2006; Sinke, 2006).

Unidirectional FMLs display strong orientation properties. Fiber-reinforced polymer composite is the dominant factor in obtaining the proper strength and modulus of elasticity in the longitudinal direction (the direction of fibers), while metal layers influence the tensile strength of laminate as regards width. As a result, tensile strength of unidirectional FMLs is higher in comparison to the metal material used. However, as regards width, the FML properties are slightly lower than those of the metal material. Nonetheless, cross-ply FMLs allow to achieve similar strength values regarding both, length, and width.

The stress–deformation relationships in FMLs are a typical combination of high stiffness and strength of a composite and good plasticity of metal layers (Vlot and Gunnink, 2001; Wu and Yang, 2005; Botelho et al., 2006).

Selected strength properties of several examples of FMLs can be seen in Table 11.2.

Table 11.2 Selected tensile strength properties of FMLs (Majerski et al., 2014; Bieniaś et al., 2013a)

Laminates[a]	Fiber orientation	Tensile ultimate strength (MPa)	Tensile elastic modulus (GPa)	Tensile ultimate strain (%)
Al/glass	Unidirectional	919	62.9	4.2
	0/90	603	58.7	3.8
	±45	366	52.3	6.6
Al/carbon	Unidirectional	1071	91.4	1.7
	0/90	619	107	1.1
	±45	327	52.9	3.6
Ti/glass	Unidirectional	920	95	6.2
	0/90	601	83	7.1
Ti/carbon	Unidirectional	1115	123	3.9
Ti/carbon	0/90	665	105	3,7
Stop 2024	–	472	62.5	15
Ti grade 2	–	502	108	19,6
Glass Fibre Reinforced Polymer	Unidirectional	1534	46	4.6
Carbon Fibre Reinforced Polymer	Unidirectional	2344	136	1.6

[a]Laminates: 2/1 lay-up; metal thickness 0.5 mm; composite thickness for each layer 0.25 mm.

Experience from literature data indicates that in experimental and theoretical studies on FML composite properties, matrix-reinforcement interface and metal–composite interface surface are of vital importance. They play the key role in stress transmission between laminate components. It has been noted that the quality of interface between particular components influences to a great extent static and fatigue properties of laminate (Botelho et al., 2006; Ostapiuk et al., 2012)

An indicator determining the metal volume fraction (MVF) (Vlot and Gunnink, 2001; Vries, 2001) in laminate can be incorporated into the assessment of FML properties and relationships between components. It is defined as the ratio of the sum of individual metal layers thickness values to the total thickness of laminate

FML strength properties can be predicted with the classical lamination theory, using the rule of mixtures and MVF (Vlot and Gunnink, 2001; Bieniaś et al., 2013a; Wu and Wu, 1994). The estimation of a given static property can be calculated on the basis of the following equation:

$$Laminate\ properties = MVF * X + (1 - MVF) * Y$$

where:

X—value of a given metal property (e.g., Young's modulus)
Y—value of a given composite property (e.g., Young's modulus)

The above mentioned equation is applicable for determining the mechanical properties of laminates subjected to static strength tests for stress in the specimen plane (tensile, compressive, shear strength, Young's modulus calculation) and density. Generally, a high level of agreement between experimentally and theoretically calculated strength properties can be attained with the use of the above mentioned equation (Vlot and Gunnink, 2001; Zhu and Chai, 2012).

Property prediction is particularly important in the design and production process of new laminates consisting of different metal materials and fiber-reinforced polymer matrix composites.

The character and mechanisms of damage for FMLs are extremely complex and multiform. Main damage mechanisms include polymer composite matrix cracks, matrix–fiber interface degradation, fiber cracks, delamination between layers with different fiber orientation and delamination at the metal–composite interface surface.

11.2.1.2 Compressive strength

The comparison of compressive and tensile strength of GLARE laminates shows that the Young's modulus in tension and compression tests is similar. Any plastic limit determined during compression will be lower than during tension (Wu and Yang, 2005). Table 11.3 shows selected compressive properties of GLARE laminates.

As regards monolithic composite structures and FMLs, tensile strength of both types of materials is determined chiefly by the fiber orientation. Fiber buckling greatly affects

Table 11.3 Selected compressive properties of GLARE laminates (Wu and Yang, 2005)

Properties	Test direction	Glare 1	Glare 2	Glare 3	Glare 4	2024-T3
0.2% Compressive yield strength (MPa)	Longitudinal	424	414	309	365	304
	Transverse	403	236	306	285	345
Compressive elastic modulus (GPa)	Longitudinal	67	67	60	60	74
	Transverse	51	52	60	54	74

strength during unidirectional composite structure compression. Aluminum layer buckling, on the other hand, usually occurs earlier and results in delaminations at the metal–composite interface. The buckling level during compression greatly influences the load capacity of a construction. FMLs do not show tendency of aramid fibers to suffer microbuckling under compressive stresses (Hull and Clyne, 1996; Remmers and de Borst, 2001).

11.2.1.3 In-plane shear behavior

The available literature implies that the shear modulus and shear yield strength of various GLARE laminates (Glare 4, 5/4, 3/2, 2/1) are only about 50% of that of the monolithic 2024-T3 aluminum alloy (Wu and Yang, 2005; Kawai et al., 1998; Woerden et al., 2003; Botelho et al., 2007b; Botelho et al., 2005; Botelho et al., 2008).

11.2.1.4 Buckling

Currently advanced construction materials, including hybrid FML, appear mostly as thin-wall composite structures. Such thin-walled materials may undergo various types of buckling. The assessment of thin-walled structures is chiefly determined by their durability, not strength only. Current research on FML proves that it can effectively replace traditional aluminum alloys in light constructions in the aircraft industry, as regards buckling and strength properties (Mania et al., 2015; Banat et al., 2016)

The literature implies that buckling load for FMLs (GLARE type) is about three times higher than for columns made of traditional, glass fiber-reinforced epoxy matrix laminates. Moreover, breaking load for FMLs is twice as high as for traditional laminates (Kubiak and Mania, 2016).

11.2.2 Fatigue properties

One of the most characteristic features of fiber–metal laminates is their fatigue strength. FMLs exhibit a very high level of fatigue strength. Literature data indicate a high tolerance to the development of cracks in GLARE type structures compared to monolithic aluminum alloy (Vlot and Gunnink, 2001; Vogelesang and Vlot, 2000; Rhymer and Johnson, 2002).

The process of cyclic load-related change in FMLs, as in the case of monolithic metals, may be conventionally divided into three stages (Schijve, 2009). The first one is the period of crack initiation, which is a substantial part of the life cycle of monolithic materials. Crack initiation is of lesser importance in FMLs. Under cyclic loads, cracks appear in the metal layer. Initially, their development depends on the properties of the monolithic material. Initiation of fatigue cracks is contingent on the level of cyclic stress in the metal layer and does not depend on the presence of notches (Schijve, 2009; Homan, 2006; Spronk et al., 2015).

The moment when a fatigue crack may be observed macroscopically is considered as the beginning of the second stage, i.e., the crack propagation phase. Throughout that period, FMLs behave quite differently from monolithic materials. In metals, a propagating crack does not encounter any limits, hence the rapid length growth. Its propagation period is relatively short compared to the initiation period. In fiber–metal laminates, however, crack growth is hampered by the presence of the composite material, which transfers the tension from the cracked metal layer. For this reason, FMLs show a much longer period of crack propagation in comparison to metal (Huang et al., 2015; Khan et al., 2011).

In FMLs, crack growth is inhibited by the fiber-bridging effect (Fig. 11.3) (Alderliesten and Homan, 2006; Abdullah et al., 2009; Marissen, 1987). It involves transferring the stress from the damaged metal to intact composite fibers. Due to

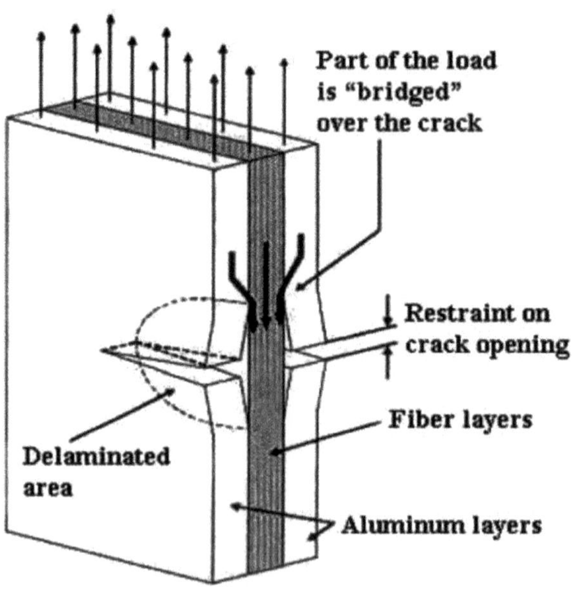

Figure 11.3 Bridging of fatigue cracks in FMLs (Wu and Yang, 2005; Alderliesten and Homan, 2006).

the propagating crack in the metal layer of FML, it does not transfer extending loads at the sites of discontinuity. However, adhesive connection between metal and composite enables the stress transfer from the damaged metal layer to the intact composite layer.

Stress transfer from the damaged metal layer onto composite fibers occurs in the adhesive connection, subjected to cyclic shear loads (Alderliesten, 2007b). Cyclic shear stresses in the adhesive layer lead to its degradation, encompassing the growth of delamination between metal and composite. As the length of the fatigue crack increases, so does the surface area and the length of delaminations. This increases the number of composite fibers which transfer the bridging stress from the aluminum layer to the composite layer (Alderliesten, 2007b; Khan et al., 2010; Alderliesten, 2007a; Alderliesten, 2005; Alderliesten et al., 2006).

In comparison to the monolithic metal materials, GLARE laminates type exhibit an almost constant, slow growth of cracks. In real load conditions, the growth cracks in GLARE laminates occur at a 10–100 times slower rate compared to monolithic aluminum components (Fig. 11.4). GLARE laminates therefore are often employed in aircraft applications, in critical fatigue loads (Vlot and Gunnink, 2001; Homan, 2006).

The final stage is the destruction phase. Depending on the load size, transverse cracks through the metal layers of FML lead to two types of behaviors observed: immediate and complete destruction of the entire remaining fiber composite structure or further work of cyclically loaded composite layer.

Phenomenological, analytical, and numerical prediction models for fatigue crack growth in fiber–metal laminates are found in the available literature (Alderliesten, 2007b; Shim et al., 2003; Johnson and Hammond, 2008). Most of the proposed

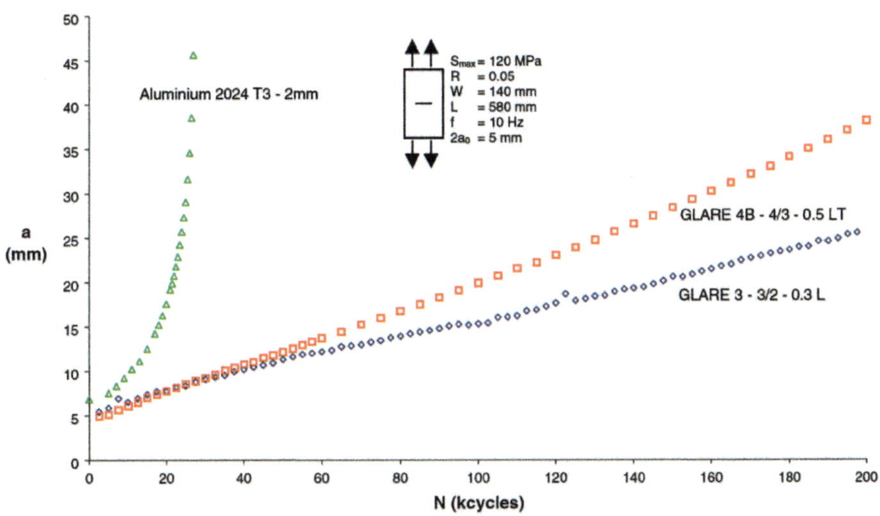

Figure 11.4 Degree of fatigue crack growth in selected GLARE laminates and aluminum alloy (Vlot and Gunnink, 2001; Alderliesten and Homan, 2006).

models of fatigue crack propagation in fiber–metal laminates concern laminates comprising only two materials. They have been tested on GLARE laminates in particular. There is a prediction model for the development of cracks in laminates of a complex structure, e.g., Al/CFRP/Al/GFRP/Al (Wilson et al., 2013). This model, however, has not been verified in an actual test, but only by the means of computing techniques based on the finite element method. Another model presented concerned predicting the angle of propagating crack in FMLs comprising composite layers of different orientation (Zaal, 1993).

However, the outcome of the research on laminates of a complex structure point to disparities of crack propagation angles predicted in theory and by the means of experimental studies (Gupta et al., 2013).

The analytical prediction model for crack propagation in FMLs (Alderliesten, 2007a; Alderliesten, 2005) suggests that the velocity of fatigue crack propagation depends on the stress intensity in the metal layer, whereas the stress intensity on the top of the crack is reduced by the stress intensity resulting from the bridging effect. This leads to a significant improvement in fatigue resistance of FMLs compared to monolithic materials.

Despite the existing theoretical models describing the cracking of FMLs, the mechanisms of the bridging effect have not been fully explained yet, while experimental studies repeatedly show results which differ from the forecasts. Literature data indicate that the damage development due to cyclic loads in FMLs leads to reduced stiffness and lower structure capacity overall (Alderliesten et al., 2003; Burianek and Spearing, 2002).

Strength reduction increases with the number of load cycles. The objective of the ongoing theoretical and experimental research on fatigue strength is mainly to develop effective methods for predicting the rate and directions of damage development in FMLs.

11.2.3 Impact

One of the key issues in the conventional composite materials, in particular the fiber-reinforced polymer–matrix composites is their high susceptibility to impact destruction (Richardson and Wisheart, 1996; Sjoblom et al., 1988; Abrate, 1991; Shyr and Pan, 2003; Abrate, 1998; Guan and Yang, 2002).

Impact damage may be caused by low- and high-velocity sources, including situations such as ground handling of aircraft, maintenance work, take-off and landing (solid objects being thrown up by the wheels of the aircraft and lifted by the wind, bird strikes, hail) and collisions with (ballistic) items moving at high velocity, e.g., missiles or bullets (Vogelesang and Vlot, 2000; Cantwell and Morton, 1991; Richardson and Wisheart, 1996; Sjoblom et al., 1988; Abrate, 1998; Sohn et al., 2000; Ardakani et al., July 2009).

Compared to standard composite materials, FMLs show higher resistance to impact. This is largely due to the protective role of metal layers. These materials are capable of transferring greater loads compared to metal alloys such as aluminum. Due to its plasticity range, aluminum alloy requires higher energy of impacts to undergo perforation, while the extent of the damage is significantly lower than in conventional composites. Excellent resistance to impact is also linked to the high value of the deformation related to the durability of glass fibers, as well as to the relatively high damage deformation (Sadighi et al., 2012a; Vlot, 1993; Wu et al., 2007; Liu and Malvern, 1987; Cortes and Cantwell, 2007; Lawcock et al., 1997; Bieniaś et al., 2015b; Beaumont et al., 1974; Seyed Yaghoubi et al., 2011; Caprino et al., 2007; Vlot, 1996; Chai and Manikandan, 2014; Fan et al., 2011).

It is often observed in the composite structures subjected to low-velocity or low-energy impact that the internal structure of the laminate is damaged. The developing site of damage is compact and sometimes difficult to identify. Subjected to impact, FMLs are not susceptible to the formation of large damage sites, in contrast to the standard laminates. The inner damage site covers a small area near the point of impact. It is usually smaller than the size of the plastic deformation observed in the fiber–metal laminate. However, internal damage may significantly reduce the durability and the stiffness of composite structures (Chai and Manikandan, 2014; Bienias et al., 2016a; Jakubczak et al., 2014a; Sadighi et al., 2012a; Liaw et al., 2001; Ardakani et al., 2008; Zhu and Chai, 2012; Caprino et al., 2004; Guo-Cai et al., 2015; Bienias et al., 2016b; Greenhalgh, 2009; Sadighi et al., 2012b; Seyed Yaghoubi et al., 2012; Laliberté et al., 2014; Bagnoli et al., 2009; Morinière et al., 2012; Nakatani et al., 2011; Bienias et al., 2015a).

On the other hand, high-velocity and high-energy impact produce visible changes in the laminate, which often leads to total destruction together with perforation of the element (Sadighi et al., 2012a).

The nature of the damage caused by impact in this type of hybrid structures is very complex. Dominant damage forms include transverse cracks in the composite material (both fibers and matrix), delaminations between the particular composite layers, delaminations at the boundary of metal and composite phases, plastic deformation of metal layers, cracks in metal layers, and laminate perforation (Sadighi et al., 2012a).

Postimpact fatigue strength and residual strength, e.g. of GLARE laminates, exceed the strength of standard composites and aluminum alloys. Compressive strength studies show a 2% drop in dent durability and a 10% drop in the durability of laminates with transverse cracks. Studies do not show buckling delaminations critical for laminates (Jakubczak et al., 2014b; Sanchez-Saez et al., 2008).

The relatively good resistance to impact of FMLs also regards high-velocity loads, whose effects on the material are considered in terms of terminal ballistics (Lemanski et al., 2007; Lemanski et al., 2006; Langdona et al., 2007; Varatharajoo et al., 2012; Vo et al., 2012).

Literature data indicate that the thickness of the metal layer affects the deformation and eventual destruction of the laminate. Similarly as in the case of low-velocity impact, this may comprise delaminations at the metal–composite boundary, as well as cracking of the matrix, fibers, and metal. The surface damage is visibly greater for high-velocity impact. It has also been demonstrated that delaminations are the main mechanism of laminate damage due to high-velocity impact (Lemanski et al., 2006).

As the studies have shown, resistance to such impact is contingent on laminate thickness, as this affects its stiffness. Stiffer laminates exhibit lower levels of deformation and higher levels of delamination (Varatharajoo et al., 2012).

Fiber—metal laminates with a high resistance to impact may therefore be considered as a potential alternative to monolithic light metal alloys employed thus far.

11.3 Durability

11.3.1 Environmental effect

FMLs have been known to contain composite interlayers with polymer matrix, prone to moisture absorption. In the case of FMLs, however, moisture absorption is very limited due to the protective metal layers (Vlot and Gunnink, 2001; Botelho et al., 2006; Botelho et al., 2005; Borgonje and Ypma, 2003; Botelho et al., 2007c). Moisture absorption of FMLs and standard composites is presented in Fig. 11.5.

The problem may arise in unprotected edges (side surfaces), e.g., after manual processing. The presence of moisture in the composite layers may contribute to delaminations at the metal/composite boundary, more often in distilled water or salt solution than in humid air and more significant at high temperatures. This may, in consequence, lower the level of fatigue strength (Botelho et al., 2006).

It is clear that the connection between the fiber and epoxy matrix is of great importance in the stress transfer in the composite. Unfortunately, fiber/matrix

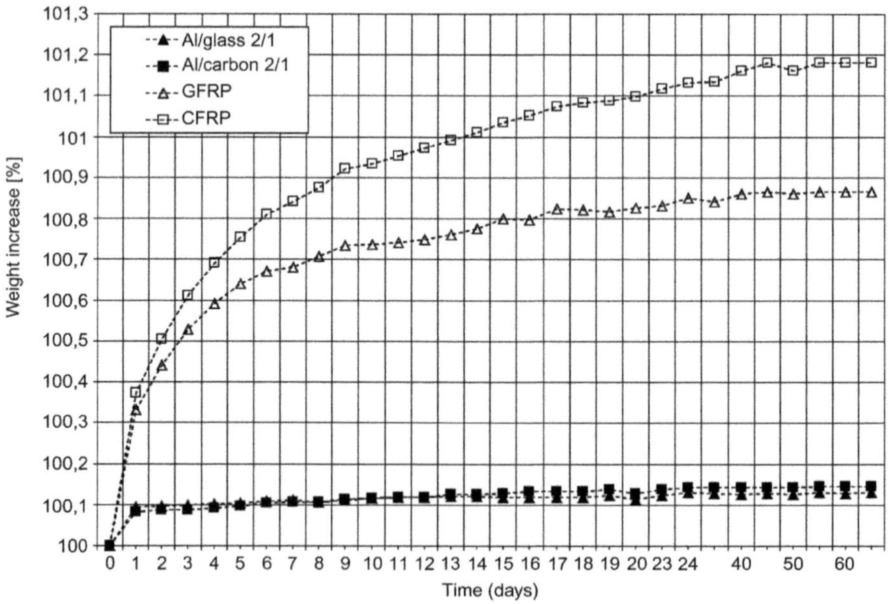

Figure 11.5 Moisture absorption of FMLs compared to standard composites.

interface is susceptible to chemical effects of moisture. Moisture absorption in composites occurs through surface, largely increasing the matrix plasticity. This leads to a reduction in effective stress distribution. Due to the barrier in the form of outer metal layers in FML composites, moisture absorption develops at a slower rate compared to polymer composites, even in relatively harsh conditions (Vlot and Gunnink, 2001; Botelho et al., 2006).

According to the literature, an increase in temperature leads to a decrease in mechanical properties, which may be connected with a reduction of matrix stiffness of the composite as the temperature increases (Borgonje and Ypma, 2003). It may therefore inhibit the effective stress distribution. Furthermore, an increase in ambient temperature to approximately 70°C reduces the strength of glass fibers by about 5%. In lower temperatures, in turn, stiffness increases, potentially leading to increased construction stiffness and degradation, e.g., as a result of delaminations and cracks.

11.3.2 Corrosion

One of the most important features of composite materials, including FMLs, is their corrosion resistance. In general, the corrosion resistance of FML composites is very high. Fiber metal laminates appear to have a comparable corrosion resistance to that found in conventional monolithic materials (e.g., aluminum). There are several distinctive features of fiber metal laminates with respect to their susceptibility to corrosion (Wu and Yang, 2005; Borgonje and Ypma, 2003; Bieniaś et al., 2013b).

Similarly to other properties of laminates, their corrosion resistance is dependent on the individual components of a fiber metal laminate. Both the metal used and the composite material can have a significant influence on the susceptibility to corrosion of this type of materials.

Standard anticorrosion treatment of fiber metal laminates consists in an appropriate surface treatment of the metal layers: anodisation of the metal surface followed by the application of primers containing corrosion inhibitors. Additionally, to increase corrosion resistance, it is beneficial to use composites reinforced with glass fibers, which are electrical insulators (Vlot and Gunnink, 2001).

Due to the use of a typical aluminum alloy in fiber metal laminates, the corrosion resistance of the outer metal layer is the same or even slightly higher than that of a monolithic sheet. If corrosion occurs, it is limited to the outer layers since the prepreg layers (a fiber-reinforced polymer composite) constitute a barrier against corrosion environment (Vlot and Gunnink, 2001; Wu and Yang, 2005; Alexopoulos et al., 2012).

The use of thin metal layers is beneficial with respect to the corrosion resistance of fiber metal laminates. The use of aluminum in fiber metal laminates (0.3–0.5 mm) results in higher corrosion resistance, compared to sheets of conventional thickness (2–4 mm). This is a result of the advantageous microstructure of the alloy, which is due to small thickness and, hence, fast heat removal during quenching after heat treatment (Alexopoulos et al., 2012).

Due to the composite layers that serve as a barrier against the effects of corrosion and corrosion being limited to the outer layers in FMLs, corrosion processes have a limited impact on mechanical properties, especially when combined with an increased number of layers.

Corrosion processes in fiber metal laminates can occur as a result of improper bonding or due to other factors causing delaminations. The penetration of the environment into the delamination can cause corrosion and degradation of the surface of the metal–composite interface.

Corrosion tests of panels made from a GLARE fiber metal laminate have revealed that areas with mechanical damages (impacts) exhibit increased susceptibility to corrosion. It has been concluded that, with respect to corrosion resistance, the use of anodizing surface treatment is beneficial, compared to laminates not subjected to surface treatment (Vlot and Gunnink, 2001).

Literature data also indicate high fatigue strength and residual strength after corrosion processes. Accelerated corrosion tests (175 hours) have shown that samples from a conventional monolithic aluminum alloy degraded after 48–62 fatigue cycles, whereas samples from fiber metal laminates were stopped after 100 cycles without any damage. Moreover, the residual strength of the tested fiber metal laminates was at approx. 340 MPa, compared to 390 MPa in the case of samples not subjected to the corrosion process (Vlot and Gunnink, 2001).

The differences between these conventional monolithic alloys and fiber metal laminates are found also in the tendency with regard to the intensity of particular types of corrosion. Stress corrosion is more common with FMLs due to the use of rolled sheets, while the intensity of pitting corrosion is at the same time lower than in the case of conventional aluminum alloys (Borgonje and Ypma, 2003).

However, FML hybrid composites can exhibit higher susceptibility to corrosion compared to monolithic materials. Their higher susceptibility to corrosion can be caused by the use of a more precious reinforcing phase compared to the metal layers, the structure of the metal/composite interface area and defects in the material, among others. The use of electrically-conductive fibers can significantly change the corrosion behavior of these materials and lead to an increased sensitivity to corrosion. This is one of the reasons why carbon fibers are not used in FMLs with aluminum layers (Vlot and Gunnink, 2001; Wu and Yang, 2005; Bieniaś et al., 2013b; Alexopoulos et al., 2012; Bieniaś, 2012; Wang et al., 2007; Hihara and Latanision, 1992). The results of corrosion resistance tests (mass decrement) of selected fiber metal laminates are presented in Fig. 11.6.

In fiber metal laminates with a carbon fiber-reinforced polymer composite as the interlayer, the formation of galvanic pairs between the composite and the metal layers is considered one of the main factors of the decrease of corrosion resistance. The essential issue is the presence of high electrical conductivity phase-carbon fibers at the metal/composite interface. Therefore, contact between carbon fiber composites and metals with similar properties in an electrolyte such as rain or seawater will be extremely undesirable. In normal service conditions, such corrosion is often unnoticeable. It is manifested only when damage is serious enough to make further use impossible (Vlot and Gunnink, 2001; Alexopoulos et al., 2012; Bieniaś, 2012; Wang et al., 2007; Gebhard et al., 2010; Peng and Nie, 2012). The surface of selected fiber metal laminates after corrosion resistance tests is shown in Fig. 11.7.

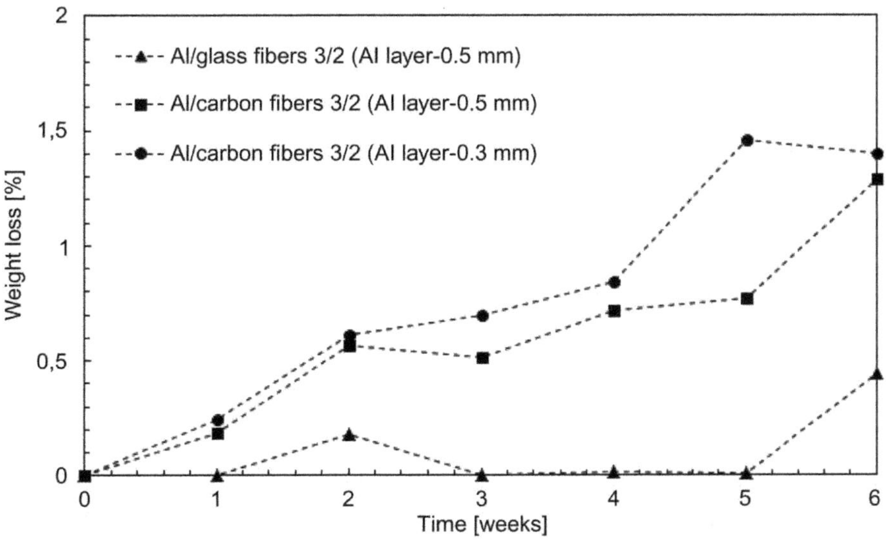

Figure 11.6 Dependence of mass decrement in time (corrosion immersion test) for laminates with aluminum layers and a composite with glass and carbon fibers.

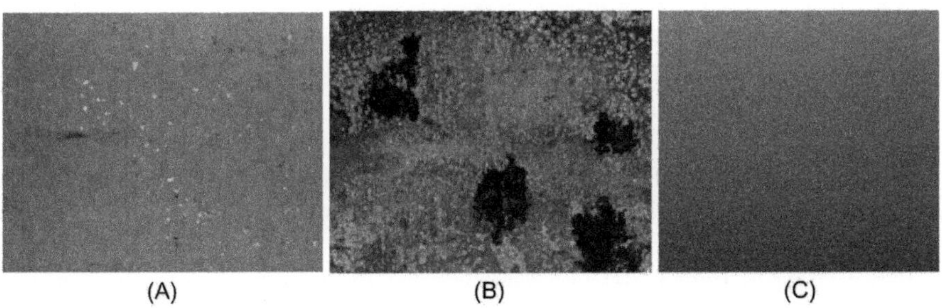

Figure 11.7 The surface of selected fiber metal laminates after corrosion resistance tests: (A) Al-glass, (B) Al-carbon, and (C) Ti-carbon.

Protection of the aluminum-composite interface (isolating carbon fibers from the aluminum) is an important issue for corrosion behavior. A thermoplastic coating like PolyEtherImide (PEI), isolating the carbon prepreg with glass prepreg on both sides and hybrid sol-gel coating can improve the corrosion resistance of FMLs (Vlot and Gunnink, 2001; Wang et al., 2007).

Any use of insulating interlayers, such as PEI, and layers with glass fibers can change the mechanical properties of fiber metal laminates (Vlot and Gunnink, 2001).

The authors' own research has indicated that corrosion processes are not found in laminates with titanium as the metal layer (even with carbon fibers) (Bieniaś et al., 2013b).

11.4 The application of FMLs

Fiber metal laminates have been developed for the aircraft industry at the Delfth University of Technology (Netherlands). The first group of FMLs were ARALL laminates (ARamid ALuminum Laminate) with aramid fibers. These laminates were used for commercial purposes for the cargo hatch in the C17 military aircraft. The next researched group of laminates was GLARE—GLAss REinforced Aluminum (Vlot and Gunnink, 2001; Vogelesang and Vlot, 2000; Botelho et al., 2006; Asundi and Choi, 1997; Vlot, 2001; Gunnink et al., 2002).

The unique properties of hybrid fiber metal laminates make these materials suitable for use in vital aircraft structures. A flagship application of GLARE laminates is Airbus A380 (Vlot and Gunnink, 2001; Wu and Yang, 2005; Botelho et al., 2006; Vlot, 2001; Gunnink et al., 2002; Pleitner, 2006; Sinke, 2003).

The first structural application of GLARE fiber metal laminates in commercial airlines is the structure of the upper skin on the fuselage of Airbus A380. Approximately 380 m^2 of GLARE laminates were used in every Airbus A380. Provided that the density of GLARE laminates is about 10% lower than that of aluminum, the mass is reduced by about 794 kg. GLARE laminates are characterized by high properties with respect to fatigue, damage, and flame resistance (Vlot and Gunnink, 2001; Wu and Yang, 2005; Gunnink et al., 2002; Sinke, 2003; http://www.fokker.com/innovations, July 2016c). An example of the application of GLARE composites is given in Fig. 11.8.

GLARE fiber metal laminates are also used for the leading edges of the vertical and horizontal tail units in Airbus A380. The use of FMLs for leading edges ensures one of their essential functionalities, i.e., protection against impacts of foreign bodies.

Other possible applications in the future include: floors and lining of aircraft cargo holds, explosion-proof luggage containers (Vlot and Gunnink, 2001; http://www.fmlc.nl/your-market/, July 2016b).

The creators of fiber metal laminates and the manufacturers of aircraft structures believe that, apart from exercising due caution during the production of parts and structures from fiber metal laminates, there are no restrictions as to the possible applications of this and similar FML materials.

NASA has developed a one-step technique implementing RTM or VARTM infiltration and infusion to manufacture fiber metal laminates which does not require the use of an autoclave or preimpregnated layers (http://technology.nasa.gov/patent/LAR-TOPS-6, July 2016a). Its advantages also include the possibility to manufacture large structures with complex shapes, high-quality structures, no necessity to use complex equipment and, ultimately, reduced manufacturing costs.

The fiber metal laminates manufactured using the process developed by NASA have properties comparable to the traditional fiber metal laminates manufactured using the autoclave technique. NASA reports that these laminates are used for aircraft and space constructions (fuselages, floor panels, cargo containers), pressure vessels and storage tanks, automotive structures, ballistic protection (explosives, bomb containment), among others.

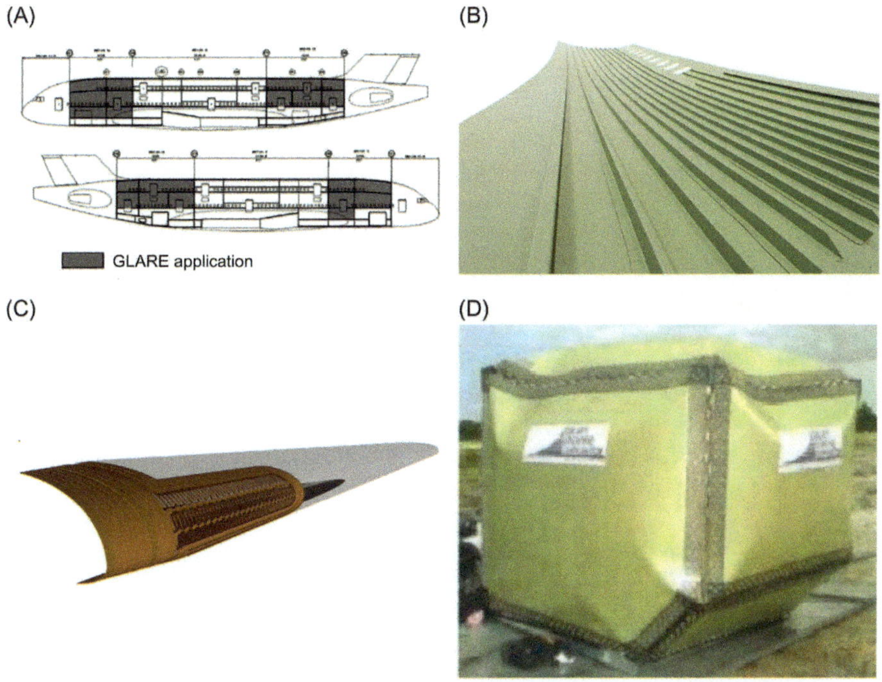

Figure 11.8 An example of the application of GLARE composites in the construction of Airbus A380; (A,B) fuselage (Woerden et al., 2003; http://www.fokker.com/innovations, July 2016c), (C) heated leading edges (http://www.fmlc.nl/your-market/, July 2016b), (D) container (Vlot and Gunnink, 2001).

The automotive industry may become a potential area for the application of FMLs which can contribute to mass reduction, compared with the currently used materials (http://www.fmlc.nl/your-market/, July 2016b). Another potential application of fiber metal laminates may be roofs and floors in trains, trucks and cars, as well as fire barriers, the production of which may capitalize on the low density and flame resistance of fiber metal laminates (http://www.fmlc.nl/your-market/, July 2016b).

11.5 Future trends in FMLs

The research studies currently conducted on FML composites, apart from comprehensive research on the properties of GLARE laminates, focus on new types of fiber metal laminates and manufacturing techniques (Vlot and Gunnink, 2001; Cortes and Cantwell, 2007; Antipov et al., 2012; Li et al., 2015; Poodts et al., 2015; Khalili et al., 2005; Reyes and Gupta, 2009; Li et al., 2016b; Li et al., 2016a; Pärnänen et al., 2008; Múgica et al., 2014).

It appears that the use of carbon fibers in fiber metal laminates may prove to be beneficial. These materials are characterized by superior high-cycle fatigue strength, low density, high strength and resistance to dynamic unloading. However, susceptibility to corrosion is an issue in this type of materials. Therefore, it is important to develop appropriate methods of the preparation of individual components or anticorrosion protection for a fiber metal laminate as a whole (Vlot and Gunnink, 2001; Botelho et al., 2007a; Sinmazçelik et al., 2011).

Glass or especially carbon fiber-reinforced titanium/polymer composite laminates – HTCL (*Hybrid Titanium Composite Laminates*)—appear to be another current trend (Rhymer and Johnson, 2002; Nakatani et al., 2011; Molitor and Young, 2002; Cortes and Cantwell, 2006b).

Despite higher density, typical of titanium, and technological difficulties, obtaining better operational properties of laminates with titanium layers is expected. The combination of these two materials as a hybrid composite, in comparison with FML-GLARE or polymer composites, makes it possible to obtain materials with high strength, excellent ratios of stiffness and strength to density, corrosion resistance, high fatigue strength, resistance to damage and impact, as well as high durability, both in low and elevated temperatures (even up to 300°C). The use of titanium in HTCL composites also contributes to the protection of a polymer composite against the effects of the external environment, such as oxidation or ingress of moisture. Moreover, the density of HTCL laminates (with carbon fibers) can be lower than that of traditional GLARE laminates, reducing the mass of the final element (Kolesnikov et al., 2008; Yamaguchi et al., 2009; Jakubczak et al., 2016).

A highly innovative solution may be the use of titanium in fiber metal laminates as metal layers and carbon fiber-reinforced composites with a PEEK thermoplastic matrix (Vlot and Gunnink, 2001).

One of the possible directions in the development of fiber metal laminates is the possibility of using magnesium alloys. However, the static and fatigue properties of magnesium-based fiber metal laminates are significantly lower than those of the currently used fiber metal laminates. Nevertheless, magnesium-based FMLs can be used for typical aircraft structures determined by compression buckling (Pärnänen et al., 2008; Alderliesten et al., 2008; Cortés and Cantwell, 2006a; Bernhardt et al., 2007; Asaee et al., 2015).

A new direction in the research on the production of laminates is also the use of prepregs with a thermoplastic matrix. Such laminates have a number of advantages, including short production time, ease of molding, high chemical resistance, very good maintainability and interlaminar fracture toughness properties. Research on a glass fiber-reinforced polypropylene FML has shown that this system offers an excellent resistance to low- and high-velocity impact loading conditions (Reyes and Gupta, 2009; Lee et al., 2014; Cortes and Cantwell, 2006b; Abdullah and Cantwell, 2006; Reyes and Kang, 2007; Carrillo and Cantwell, 2009; Iriondo et al., 2015; Kalyanasundaram et al., 2013).

The use of other methods of molding fiber metal laminates (apart from the autoclave technique) is also taken into consideration, e.g., the RTM and VARTM processes developed by NASA which make it possible to mold various combinations of individual components and reduce manufacturing costs. Automation of the production of fiber metal laminates is a promising concept as well (http://technology.nasa.gov/patent/LAR-TOPS-6, July 2016a).

Acknowledgements

Authors own research were financially supported of Structural Funds in the Operational Programme—Innovative Economy (IE OP) financed from the European Regional Development Fund—Project No POIG.0101.02-00-015/08 is gratefully acknowledged (until 2015) and in 2016 with help of the research project UMO-2014/15/B/ST8/03447 funded by the National Science Centre in Poland.

References

Abdullah, M.R., Cantwell, W.J., 2006. The impact resistance of polypropylene-based fibre–metal laminates. Compos. Sci. Technol. 66, 1682–1693.

Abdullah, S., Fahruddin, A., Syarif, J., Omar, M.Z., Zulkifli, R., 2009. Fatigue crack growth modelling of fibre metal laminate (FML) composite. Eur. J. Sci. Res. 35 (1), 43–53.

Abrate, S., 1991. Impact on laminated composite materials. Appl. Mech. Rev. 44 (4), 155–190.

Abrate, S., 1998. Impact on composite structures. Chaper 4, Low-Velocity Impact Damage. Cambridge University Press, pp. 135–160.

Alderliesten, R., Rans, C., Benedictus, R., 2008. The applicability of magnesium based Fibre Metal Laminates in aerospace structures. Compos. Sci. Technol. 28, 2983–2993.

Alderliesten, R.C., 2005. Fatigue Crack Propagation and Delamination Growth in Glare. Ph. D. Thesis. Delft University of Technology, Delft.

Alderliesten, R.C., 2007a. Analytical prediction model for fatigue crack propagation and delamination growth in glare. Int. J. Fatigue. 29, 628–646.

Alderliesten, R.C., 2007b. On the available relevant approaches for fatigue crack propagation prediction in Glare. Int. J. Fatigue. 29, 289–304.

Alderliesten, R.C., Homan, J.J., 2006. Fatigue and damage tolerance issues of Glare in aircraft structures. Int. J. Fatigue. 28, 1116–1123.

Alderliesten, R.C., Hagenbeek, M., Homan, J.J., Hooijmeijer, P.A., De Vries, T.J., Vermeeren, C.A.J.R., 2003. Fatigue and damage tolerance of Glare. Appl. Compos. Mater. 10, 223–242.

Alderliesten, R.C., Schijve, J., van der Zwaag, S., 2006. Application of the energy release rate approach for delamination growth in Glare. Eng. Fract. Mech. 73, 697–709.

Alexopoulos, N.D., Dalakouras, C.J., Skarvelis, P., Kourkoulis, S.K., 2012. Accelerated corrosion exposure in ultra thin sheets of 2024 aircraft aluminium alloy for GLARE applications. Corrosion Science. 55, 289–300.

Antipov, V.V., Senatorova, O.G., Beumler, T., Ijpma, M., 2012. Investigation of a new fibre metal laminate (FML) family on the base of Al-Li-Alloy with lower density. Materwiss. Werksttech. 43 (4), 350–355.

Ardakani, M.A., Khatibi, A.A., Ghazavi, S.A., 2008. A study on the manufacturing of glass-fiber-reinforced aluminum laminates and the effect of interfacial adhesive bonding on the impact behavior, In: Proceedings of the XIth International Congress and Exposition, June 2–5, Orlando, Florida USA.

Ardakani M.A., Khatibi, A.A., Parsaiyan, H., 2009. An experimental study on the impact resistance of glass-fiber-reinforced aluminum (Glare) laminates, In: 17th International Conference on Composite Materials, At Edinburgh, UK.

Asaee, Z., Shadlou, S., Taheri, F., 2015. Low-velocity impact response of fiberglass/magnesium FMLs with a new 3D fiberglass fabric. Compos. Struct. 122, 155–165.

Asundi, A., Choi, A.Y.N., 1997. Fiber metal laminates: an advanced material for future aircraft. J. Mater. Process. Technol. 63, 384–394.

Bagnoli, F., Bernabei, M., Figueroa-Gordon, D., Irving, P.E., 2009. The response of aluminium/GLARE hybrid materials to impact and to in-plane fatigue. Mater. Sci. Eng. A523, 118–124.

Banat, D., Kolakowski, Z., Mania, R.J., 2016. Investigations of profile buckling and post-buckling behaviour under axial compression. Thin-Wall. Struct. 107, 335–344.

Beaumont, P.W.R., Riewald, P.G., Zweben, C., 1974. Methods for Improving the Impact Resistance of Composite Materials, in Foreign Object Impact Damage to Composites, ASTM STP 568, American Society for Testing and Materials, pp. 134–158.

Bernhardt, S., Ramulu, M., Kobayashi, A.S., 2007. Low-velocity impact response characterization of a hybrid titanium composite laminate. J. Eng. Mater. Technol. 129, 220–226.

Bieniaś, J., 2012. Corrosion behaviour of carbon Fibre Metal Laminates—aspects of metal/carbon fibre interface, In: International Conference on Composite Interfaces, Kyoto, Japan, pp. 1–2.

Bieniaś, J., Majerski, K., Surowska, B., Jakubczak, P., 2013a. The mechanical properties and failure analysis of selected fiber metal laminates. Compos. Theory Pract. 3, 220–224.

Bieniaś J., P. Jakubczak, K. Majerski, B. Surowska, C. Antolak, 2013b. Corrosion studies of selected fibre metal laminates with carbon and glass fibres, In: 19th International Conference on Composite Materials, Montreal, Canada, pp. 7810–7811.

Bienias, J., Surowska, B., Jakubczak, P., 2015a. Influence of repeated impact on damage growth in fibre reinforced polymer composites. Eksploat. Niezawodn. – Maint. Reliab. 17 (2), 194–198.

Bieniaś, J., Jakubczak, P., Surowska, B., Dragan, K., 2015b. Low-energy impact behaviour and damage characterization of carbon fibre reinforced polymer and aluminium hybrid laminates. Arch. Civil Mech. Eng. 4 (15), 925–932.

Bienias, J., Surowska, B., Jakubczak, P., 2016a. The comparison of low velocity impact resistance of aluminum/carbon and glass fibres metal laminates. Polym. Compos. 4 (37), 1056–1063.

Bienias, J., Jakubczak, P., Dadej, K., 2016b. Low-velocity impact resistance of aluminium glass laminates – experimental and numerical investigation. Compos. Struct. 152, 339–348.

Borgonje, B., Ypma, M.S., 2003. Long term behaviour of Glare. Appl. Compos. Mater. 10, 243–255.

Botelho, E.C., Pardini, L.C., Rezende, M.C., 2005. Hygrothermal effects on damping behavior of metal/glass fiber/epoxy hybrid composites. Mater. Sci. Eng. A399, 190–198.

Botelho, E.C., Silva, R.A., Pardini, L.C., Rezende, M.C., 2006. A review on the development and properties of continuous fiber/epoxy/aluminum hybrid composites for aircraft structures. Mater. Res. 9 (3), 247–256.

Botelho, E.C., Pardini, L.C., Rezende, M.C., 2007a. Evaluation of hygrothermal effects on the shear properties of Carall composites. Mater. Sci. Eng. A. 452-453, 292–301.

Botelho, E.C., Almeida, R.S., Pardini, L.C., Rezende, M.C., 2007b. Elastic properties of hygrothermally conditioned glare laminate. Int. J. Eng. Sci. 45, 163–172.

Botelho, E.C., Almeida, R.S., Pardini, L.C., Rezende, M.C., 2007c. Influence of hygrothermal conditioning on the elastic properties of Carall laminates. Appl. Compos. Mater. 14 (3), 209–222.

Botelho, E.C., Rezende, M.C., Pardini, L.C., 2008. Hygrotermal effects evaluation using the Iosipescu shear test for glare laminates. J. Brazil. Soc. Mech. Sci. Eng. 30 (3), 213–220.

Burianek, D.A., Spearing, S.M., 2002. Fatigue damage in titanium-graphite hybrid laminates. Compos. Sci. Technol. 62, 607–617.

Cantwell, W.J., Morton, J., 1991. The impact resistance of composite materials – a review. Compos. 22 (5), 347–362.

Caprino, G., Spatarob, G., Del Luongo, S., 2004. Low-velocity impact behaviour of fibreglass–aluminium laminates. Compos. A. 35, 605–616.

Caprino, G., Lopresto, V., Iaccarino, P., 2007. A simple mechanistic model to predict the macroscopic response of fibre glass–aluminum laminates under low-velocity impact. Compos. A. 38, 290–300.

Carrillo, J.G., Cantwell, W.J., 2009. Mechanical properties of a novel fiber–metal laminate based on a polypropylene composite. Mech. Mater. 41 (7), 828–838.

Chai, G.B., Manikandan, P., 2014. Low velocity impact response of fibre-metal laminates – a review. Compos. Struct. 107, 363–381.

Cortés, P., Cantwell, W.J., 2006a. The fracture properties of a fibre–metal laminate based on magnesium alloy. Compos. B. 37 (2–3), 163–170.

Cortes, P., Cantwell, W.J., 2006b. The prediction of tensile failure in titanium-based thermoplastic fibre–metal laminates. Compos. Sci. Technol. 66 (13), 2306–2316.

Cortes, P., Cantwell, W.J., 2007. The impact properties of high-temperature fiber–metal laminates. J. Compos. Mater. 41, 613–632.

Fan, J., Guan, Z.W., Cantwell, W.J., 2011. Numerical modelling of perforation failure in fibre metal laminates subjected to low velocity impact loading. Compos. Struct. 93, 2430–2436.

Gebhard, A., Bayerl, T., Schlarb, A.K., Friedrich, K., 2010. Increased wear of aqueous lubricated short carbon fiber reinforced polyetheretherketone (PEEK/SCF) composites due to galvanic fiber corrosion. Wear. 268, 871–876.

Greenhalgh, E.S., 2009. Failure Analysis and Fractography of Polymer Composites. Woodhead Harsh Environment Publishing Limited, Cambridge, UK.

Guan, Z., Yang, Ch, 2002. Low-velocity impact and damage process of composite laminates. J. Compos. Mater. 36, 851–871.

Gunnink, J.W., Vlot, A., de Vries, T.J., van der Hoeven, W., 2002. Glare technology development 1997–2000. Appl. Compos. Mater. 9, 201–219.

Guo-Cai, Y., Lin-Zhi, W., Li, M., Jian, X., 2015. Low velocity impact of carbon fiber aluminum laminates. Compos. Struct. 119, 757–766.

Gupta, M., Alderliesten, R.C., Benedictus, R., 2013. Crack paths in fibre metal laminates: role of fibre bridging. Eng. Fract. Mech. 108, 183–194.

Hihara, L.H., Latanision, R.M., 1992. Galvanic corrosion of aluminium matrix composites. Corrosion. 48 (7), 546–552.

Homan, J.J., 2006. Fatigue initiation in fibre metal laminates. Int. J. Fatigue. 28, 366–374.

http://technology.nasa.gov/patent/LAR-TOPS-6, July 2016a.

http://www.fmlc.nl/your-market/, July 2016b.

http://www.fokker.com/innovations, July 2016c.

Huang, Y., Liu, J., Huang, X., Zhang, J., Yue, G., 2015. Delamination and fatigue crack growth behavior in Fiber Metal Laminates (Glare) under single overloads. Int. J. Fatigue. 78, 53–60.

Hull, D., Clyne, T.W., 1996. An Introduction to Composite Materials. 2nd ed. Cambridge University Press.

Iriondo, J., Aretxabaleta, L., Aizpuru, A., 2015. Characterisation of the elastic and damping properties of traditional FML and FML based on a self-reinforced polypropylene. Compos. Struct. 131, 47–54.

Jakubczak, P., Bienias, J., Majerski, K., Ostapiuk, M., Surowska, B., 2014a. The impact behavior of aluminum hybrid laminates. Aircraft Eng. Aerosp. Technol. 86 (4), 287–294.

Jakubczak, P., Bieniaś, J., Dadej, K., Zawiejski, W., 2014b. The issue of residual strength tests of thin fibre metal laminates. Compos. Theory Pract. 14 (3), 134–138.

Jakubczak, P., Surowska, B., Bieniaś, J., 2016. Evaluation of force-time changes during impact of hybrid laminates made of titanium and fibrous composite. Arch. Metall.Mater. 2A (61), 689–694.

Johnson, W.S., Hammond, M.W., 2008. Crack growth behavior of internal titanium plies of a fiber metal laminate. Compos. A. 39, 1705–1715.

Kalyanasundaram, S., DharMalingam, S., Venkatesan, S., Sexton, A., 2013. Effect of process parameters during forming of self-reinforced – PP based Fiber Metal Laminate. Compos. Struct. 97, 332–337.

Kawai, M., Morishita, M., Tomura, S., Takumida, K., 1998. Inelastic behavior and strength of fiber-metal hybrid composite: GLARE. Int. J. Mech. Sci. 40 (2–3), 183–198.

Khalili, S.M.R., Mittal, R.K., Gharibi Kalibar, S., 2005. A study of the mechanical properties of steel/aluminium/GRP laminates. Mater. Sci. Eng. A. 412 (1–2), 137–140.

Khan, S.U., Alderliesten, R.C., Rans, C.D., Benedictus, R., 2010. Application of a modified Wheeler model to predict fatigue crack growth in Fibre Metal Laminates under variable amplitude loading. Eng. Fract. Mech. 77, 1400–1416.

Khan, S.U., Alderliesten, R.C., Benedictus, R., 2011. Delamination in Fiber Metal Laminates (GLARE) during fatigue crack growth under variable amplitude loading. Int. J. Fatigue. 33, 1292–1303.

Kolesnikov, B., Herbeck, L., Fink, A., 2008. CFRP/titanium hybrid material for improving composite bolted joints. Compos. Struct. 83, 368–380.

Kubiak, T., Mania, R.J., 2016. Hybrid versus FR laminate channel section columns – buckling and postbuckling behaviour. Thin-Wall. Struct. 107, 335–344.

Laliberté J.F., C. Poon, P.V. Straznicky, 2014. Low Velocity Impact Damage in Glare© Fibre Metal Laminates ICCM, <http://www.iccm-central.org/Proceedings/ICCM12proceedings/site/papers/pap421.pdf>.

Langdona, G.S., Lemanski, S.L., Nurick, G.N., Simmons, M.C., Cantwell, W.J., Schleyer, G. K., 2007. Behaviour of fibre–metal laminates subjected to localised blast loading: Part I —Experimental observations. Int. J. Impact. Eng. 34 (7), 1202–1222.

Lawcock, G.D., Ye, L., Mai, Y.W., Sun, C.T., 1997. Effects of fibre/matrix adhesion on carbon-fibre-reinforced metal laminates-II. Impact behavior. Compos. Sci. Technol. 57, 1621–1628.

Lee, B.E., Park, E.T., Kim, J., Kang, B.S., Song, W.J., 2014. Analytical evaluation on uniaxial tensile deformation behavior of fiber metal laminate based on SRPP and its experimental confirmation. Compos. B. 67, 154–159.

Lemanski, S.L., Nurick, G.N., Langdon, G.S., Simmons, M.S., Cantwell, W.J., Schleyer, G. K., 2006. Understanding the behaviour of fibre metal laminates subjected to localised blast loading. Compos. Struct. 76, 82–87.

Lemanski, S.L., Nurick, G.N., Langdon, G.S., Simmons, M.C., Cantwell, W.J., Schleyer, G. K., 2007. Behaviour of fibre metal laminates subjected to localized blast loading—part II: quantitative analysis. Int. J. Impact. Eng. 34, 1223–1245.

Li, H., Hu, Y., Xu, Y., Wang, W., Zheng, X., Liu, H., et al., 2015. Reinforcement effects of aluminum–lithium alloy on the mechanical properties of novel fiber metal laminate. Compos. B. 82, 72–77.

Li, H., Hu, Y., Liu, C., Zheng, X., Liu, H., Tao, J., 2016a. The effect of thermal fatigue on the mechanical properties of the novel fiber metal laminates based on aluminum–lithium alloy. Compos. A. 84, 36–42.

Li, H., Hu, Y., Fu, X., Zheng, X., Liu, H., Tao, J., 2016b. Effect of adhesive quantity on failure behavior and mechanical properties of fiber metal laminates based on the aluminum–lithium alloy. Compos. Struct. 152, 687–692.

Liaw B.M., Y.X. Liu, E.A. Villars, 2001. Impact Damage Mechanisms in Fiber Metal Laminates, Proceedings of the SEM Annual Cenference on Experimental and Applied Mechanics, Portland, Oregon, June 4–6.

Liu, D., Malvern, L.E., 1987. Matrix cracking in impacted glass/epoxy plates. J. Compos. Materials. 21, 594–609.

Majerski, K., Surowska, B., Bieniaś, J., Jakubczak, P., Ostapiuk, M., 2014. Experimental investigation on the influence of fibers orientation on the failure of carbon hybrid laminates. Aircraft Eng. Aerosp. Technol. 86 (4), 307–311.

Mania, R.J., Kolakowski, Z., Bienias, J., Jakubczak, P., Majerski, K., 2015. Comparative study of FML profiles buckling and postbuckling behavior under axial loading. Compos. Struct. 134, 216–225.

Marissen, R., 1987. Fatigue mechanisms in ARALL, a fatigue resistant hybrid aluminum aramid composite material. Fatigue. 87, 1271–1279.

Molitor, P., Young, T., 2002. Adhesives bonding of a titanium alloy to a glass fibre reinforced composite material. Int. J. Adhesion Adhes. 22, 101–107.

Morinière, F.D., Alderliesten, R.C., Yarmohammad Tooski, M., Benedictus, R., 2012. Damage evolution in GLARE fibre–metal laminate under repeated low velocity impact tests. Cent. Eur. J. Eng. 2 (4), 603–611.

Múgica, J.I., Aretxabaleta, L., Ulacia, I., Aurrekoetxea, J., 2014. Impact characterization of thermoformable fibre metal laminates of 2024-T3 aluminium and AZ31B-H24 magnesium based on self-reinforced polypropylene. Compos. A. 61, 67–75.

Nakatani, H., Kosaka, T., Osaka, K., Sawada, Y., 2011. Damage characterization of titanium/ GFRP hybrid laminates subjected to low-velocity impact. Compos. A. 42, 772–781.

Ostapiuk, M., Surowska, B., Bieniaś, J., 2012. Interface analysis of fiber metal laminates. Compos. Interfaces. 4 (21), 309–318.

Pärnänen, T., Alderliesten, R., Rans, C., Brander, T., Saarela, O., 2008. Applicability of AZ31B-H24 magnesium in fibre metal laminates – an experimental impact research. Compos. Struct. 83, 247–258.

Peng Z., Nie, X., Galvanic corrosion property of contacts between carbon fiber cloth materials and typical metal alloys in an aggressive environment, Surface Coat. Technol., http://dx.doi.org/10.1016/j.surfcoat.2012.08.098.

Pleitner, J., Airbus Costumer Benefit from Fiber Metal Laminates – ESAS – Ref. L53PR0605135, 2006.

Poodts, E., Ghelli, D., Brugo, T., Panciroli, R., Minak, G., 2015. Experimental characterization of a fiber metal laminate for underwater applications. Compos. Struct. 129, 36–46.

Remmers, J.J.C., de Borst, R., 2001. Delamination buckling of fibre–metal laminates. Compos. Sci. Technol. 61, 2207–2213.

Reyes, G., Kang, H., 2007. Mechanical behavior of lightweight thermoplastic fiber–metal laminates. J. Mater. Process. Technol. 186 (1–3), 284–290.

Reyes, G., Gupta, S., 2009. Manufacturing and mechanical properties of thermoplastic hybrid laminates based on DP500 steel. Compos. A. 40 (2), 176−183.

Rhymer, D.W., Johnson, W.S., 2002. Fatigue damage mechanisms in advanced hybrid titanium composite laminates. Int. J. Fatigue. 24, 995−1001.

Richardson, M.O.W., Wisheart, M.J., 1996. Review of low-velocity impact properties of composite materials. Compos. A. 27, 1123−1131.

Sadighi, M., Alderliesten, R.C., Benedictus, R., 2012a. Impact resistance of fiber−metal laminates: a review. Int. J. Impact. Eng. 49, 77−90.

Sadighi, M., Parnanen, T., Alderliesten, R.C., Sayeaftabi, M., Benedictus, R., 2012b. Experimental and numerical investigation of metal type and thickness effects on the impact resistance of fiber−metal laminates. Appl. Compos. Mater. 19 (3), 545−559.

Sanchez-Saez, S., Barbero, E., Navarro, C., 2008. Compressive residual strength at low temperatures of composite laminates subjected to low-velocity impacts. Compos. Struct. 85, 226−232.

Schijve, J., 2009. Fatigue of Structures and Materials. Springer Science Business Media, B. V.

Seyed Yaghoubi, A., Liu, Y.X., Liaw, B.M., 2011. Drop-weight impact studies of GLARE 5 fiber−metal laminates. Exp. Appl. Mech. 6, 267−279.

Seyed Yaghoubi, A., Liu, Y., Liaw, B., 2012. Stacking sequence and geometrical effects on low-velocity impact behaviors of Glare5 (3/2) fiber-metal laminates. Journal of Thermoplastic Composite Materials. 25, 223−247.

Shim, D.J., Alderliesten, R.C., Spearing, S.M., Burianek, D.A., 2003. Fatique crack growth prediction in GLARE hybrid laminates. Composites Science and Technology. 63, 1759−1767.

Shyr, T.W., Pan, Y.H., 2003. Impact resistance and damage characteristics of composite laminates. Composite Structures. 62, 193−203.

Sinke, J., 2003. Manufacturing of GLARE, Parts and Structures. Applied Composite Materials. 10, 293−305.

Sinke, J., 2006. Development of Fibre Metal Laminates: concurrent multi-scale modeling and testing. Journal of Materials Science. 41, 6777−6788.

Sinmazçelik, T., Avcu, E., Bora, M.O., Çoban, O., 2011. A review: fibre metal laminates, background, bonding types and applied test methods. Mater. Des. 32, 3671−3685.

Sjoblom, P.O., Hartness, J.T., Cordell, T.M., 1988. On low-velocity impact testing of composite materials. J. Compos. Mater. 22, 30−52.

Sohn, M.S., Hu, X.Z., Kimb, J.K., Walker, L., 2000. Impact damage characterization of carbon fbre/epoxy composites with multi-layer reinforcement. Compos. B. 31, 681−691.

Spronk, S.W.F., Sen, I., Alderliesten, R.C., 2015. Predicting fatigue crack initiation in fibre metal laminates based on metal fatigue test data. Int. J. Fatigue. 70, 428−439.

Varatharajoo, R., Abdullah, E.J., Majid, D.L., Romli, F.I., Mohd Rafie, A.S., Ahmad, K.A., 2012. The behaviour of fibre-metal laminates under high velocity impact loading with different stacking sequences of Al alloy. Appl. Mechan. Mater. 225, 213−218.

Vlot, A., 1993. Impact properties of fibre metal laminates. Compos. Eng. 3 (10), 911−927.

Vlot, A., 1996. Impact loading on fiber−metal laminates. Int. J. Impact. Eng. 18, 291−307.

Vlot, A., 2001. GLARE − history of the development of a new aircraft material. Kluwert Academic Publishers, Dordrecht.

Vlot, A., Gunnink, J.W., 2001. Fiber Metal Laminates: An Introduction. Kluwer Academic Publishers, Dordrecht.

Vo, T.P., Guan, Z.W., Cantwell, W.J., Schleyer, G.K., 2012. Low-impulse blast behaviour of fibre-metal laminates. Compos. Struct. 94 (3), 954−965.

Vogelesang, L.B., Vlot, A., 2000. Development of fibre metal laminates for advanced aerospace structures. J. Mater. Process. Technol. 103, 1–5.
Vries, T.J., 2001. Blunt and Sharp Notch Behaviour of GLARE Laminates. PhD thesis. Delft University of Technology, The Netherlands.
Wang W.X., Takao, Y., Matsubara, T., 2007. Galvanic corrosion-resistant carbon fiber metal laminates, In: 16th International Conference on Composite Materials, Kyoto, Japan.
Wilson, G.S., Alderliesten, R.C., Benedictus, R., 2013. A generalized solution to the crack bridging problem of fiber metal laminates. Eng. Fract. Mech. 105, 65–85.
Woerden, H.J.M., Sinke, J., Hooijmeijer, P.A., 2003. Maintenance of Glare structures and Glare as riveted or bonded repair material. Appl. Compos. Mater. 10, 307–329.
Wu, G., Yang, J.M., 2005. The mechanical behavior of GLARE laminates for aircraft structures. JOM. 75, 72–79.
Wu, G., Yang, J.M., Hahn, H.T., 2007. The impact properties and damage tolerance and of bi-directionally reinforced fibre metal laminates. J. Mater. Sci. 42, 948–957.
Wu, H., Wu, L.L., 1994. Use of rule of mixtures and metal volume fraction for mechanical property predictions of fibre-reinforced aluminium laminates. J. Mater. Sci. 29, 4583–4591.
Yamaguchi, T., Okabe, T., Yashiro, S., 2009. Fatigue simulation for titanium/CFRP hybrid laminates using cohesive elements. Compos. Sci. Technol. 69, 1968–1973.
Zaal K.J.J.M., 1993. An experimental study of the angled crack in GLARE 3. T.U. Delft.
Zhu, S., Chai, G.B., 2012. Low-velocity impact response of fibre–metal laminates – experimental and finite element analysis. Compos. Sci. Technol. 72, 1793–1802.

Impact resistance and damage of fiber metal laminates

12

Patryk Jakubczak, Jarosław Bieniaś and Barbara Surowska
Department of Materials Engineering, Faculty of Mechanical Engineering, Lublin University of Technology, Lublin, Poland

Chapter Outline

12.1 Impact resistance and damage of fiber metal laminates

Fiber metal laminates (FMLs) display a number of beneficial properties stemming directly from their hybrid structure (Vlot and Gunnink, 2001; Wu and Yang, 2005; Vogelesang and Vlot, 2000). One such FML property is high resistance to impact, regardless of the structure and type of laminate components, compared with traditional fiber–polymer composite structures. There are many studies available where authors examine the response of FMLs to high- (Vlot and Krull, 1997; Zhu and Chai, 2012; Caprino et al., 2004; Fan et al., 2011; Bienias et al., 2016a) and low-velocity impact (Langdona et al., 2007; Varatharajoo et al., 2012). The need to broaden the knowledge on this subject results from current and potential FML uses, mainly in aircraft structures. As regards the use of FMLs as vast aircraft fabric covering pieces, impact resistance is one of the vital elements of the safety of their exploitation. There have been instances in the history of aircraft where the consequences of an impact on the covering led directly to a plane crash (among others: Space Shuttle Columbia, STS-107, 01.02.2003; Concorde Air France lot 4590,

Hybrid Polymer Composite Materials: Properties and Characterisation. DOI: http://dx.doi.org/10.1016/B978-0-08-100787-7.00012-3

25.07.2000). Impact, regardless of impact energy, causes permanent changes (damage) in composite structures and in FMLs, which, above all, deteriorate durability parameters of these materials (Lee et al., 2009; Sanchez-Saez et al., 2008; Alderliesten and Homan, 2006). Compared to fiber–polymer composite materials, FMLs appear superior in terms of impact resistance because of adhesive connection of elastic–plastic metal and stiff fibrous composite provides excellent energy absorption and relatively low damage especially after low-velocity impact (Lawcock et al., 1997; Bieniaś et al., 2015b; Vlot, 1993). Depending on the impact velocity, which to a large extent influences the amount of energy transferred from the impactor to the material, FML damage may take different forms (from a minor surface deformation, through matrix cracks, extensive delaminations, to a total perforation (Bienias et al., 2016a; Jakubczak et al., 2014b; Sadighi et al., 2012a; Chai and Manikandan, 2014). A recent surge in studies focused on defining relationships between impact energy and degradation mechanisms of laminates with different layer configurations regarding low- and high-velocity impact has been observed. Unfortunately, the experiments are carried out on different materials and interpreted by different criteria (Sadighi et al., 2012a; Song et al., 2010; Bagnoli et al., 2009; Ardakani et al., 2008; Nakatani et al., 2011; Cortés and Cantwell, 2006). A summary published by Sadighi et al. (2012a) shows that out of 40 works, about a half concerned low-velocity impact resistance of GLARE (GLass Aluminium REinforcment) laminates of different configurations. The least examined, though potentially the most crucial, group consists of CARAL (CArbon Reinforcment ALuminum) laminates.

Regardless of the aircraft fabric covering type (composite or FML type), low-velocity impact resistance is one of the important issues in composite structures, particularly in the aerospace sector. Impact damage occurs during preflight and taxiing operations, in case of runway debris, hail or bird strikes, maintenance damage (e.g., dropped tools), collision between service cars or cargo and the structure, ice from propellers striking the fuselage, engine debris and tire shrapnel from tread separation, and tire rupture (Vogelesang and Vlot, 2000; Sohn et al., 2000). Vogelesang and Vlot (2000) quoted that 13% repairs of primary structure in Boeing 747 aircraft are caused by the impact damage.

Determining the degradation mechanisms of different FML types resulting from impact is of key importance for design, application, and exploitation of these materials.

12.1.1 Low-velocity impact: definitions and test procedures

12.1.1.1 Definitions

In general, impact is a phenomenon which exerts high shear stress on structures, especially coating, which then causes degradation in the case of laminar–fibrous composites and FMLs (Fan et al., 2010).

Impact can be divided into low-velocity impact ($\leq 10\ \mathrm{m\ s^{-1}}$) (Cantwell and Morton, 1991; Richardson and Wisheart, 1996), high-velocity impact ($>10\ \mathrm{m\ s^{-1}}$)

and hyper-velocity impact (>100 m s^{-1}) (Richardson and Wisheart, 1996). There are no specified velocity thresholds which would define or differentiate types of impact as far as composite material impact is concerned (Richardson and Wisheart, 1996). The choice of criteria is individual and should be based on appropriate literature. Many authors present and defend the criteria they have chosen. Sjoblom et al. (1988) and Shivakumar et al. (1985) have described the relationship between impact velocity and material properties, weight, and indenter size. They argued that low-velocity impact ranges from quasi-statistical to several meters per second. The velocity thresholds are specified on the basis of contact time between the impactor and material. They claimed that at lower velocity, most of the impact energy is absorbed for elastic deformation, which determines the extent of final damage. Abrate (1991) assumed that low-velocity impact does not exceed 100 m/s. A different approach was proposed by Liu and Malvern (1987), who linked the impact classification with dominant composite damage mechanisms. They believed that fiber perforation and cracking follow high-velocity impact, while delamination and matrix cracking follow low-velocity impact. In general, the most popular division of impact based on velocity marks out low-velocity impact with velocity of impactor below 10 m s^{-1}.

12.1.1.2 Procedures

The assessment of FML impact damage vulnerability is a major academic and practical issue. The traditional Charpy and Izod impact tests used in research on metal and plastic with layer structure have proved to be unhelpful for FML impact resistance assessment (Bełzowski et al., 2002). Distinctive properties of FMLs necessitate a search for new methods of assessing their suitability to work in conditions at risk of, for example, low-velocity impact. The aspect of material is fundamental, because the effects of the laminate compounds on its response to impact and its degradation mechanism are vital for designing and exploitation of aircraft structures responsible, for example, for carrying the coating weight. Moreover, initial studies conducted on epoxy-glass and epoxy-carbon and metal laminates proved that the FML impact resistance does not correspond to the sum of impact resistances of individual components (Sohn et al., 2000; Zhou and Davies, 1995).

At present, there are no definitive procedures of measuring the FML low-velocity impact resistance. In such investigations, the authors of numerous works (Lawcock et al., 1997; Bieniaś et al., 2015b; Guo-Cai et al., 2015; Bienias et al., 2016b; Seyed Yaghoubi et al., 2011; Léonard et al., 2014) implemented methodologies of impact tests designed for traditional fiber–polymer composite materials. These include normalized procedures (e.g., ASTM D 7136, ISO 18352, EN 6038 standards) and procedures developed by aircraft manufacturers (e.g., Boeing BSS 7260, AITM 1-0010).

In the available literature (Bieniaś et al., 2015b, 2016b; Léonard et al., 2014), one method of examining low-velocity impact resistance of FMLs is the methodology described in the ASTM D7136 norm, where the suggested specimen size and support conditions allow postimpact CAI (compressive after impact) durability estimations. A flat, rectangular composite plate is subjected to an out-of-plane, concentrated impact using a drop-weight device with a hemispherical impactor. The potential energy of the drop-weight, as defined by the mass and drop height of the impactor, is specified prior to test (ASTM D 7136/D 7136M, 2007). The impact occurs in the central area of a 150 mm × 100 mm plate. The norm defines requirements regarding the quality of specimen's surface, parameters measured by the device and methods used for obtaining them, and requirements regarding support size and indenters. The damage resistance properties generated by this test method are highly dependent upon several factors, which include specimen geometry, layup, impactor geometry, impactor mass, impact force, impact energy, and boundary conditions. Thus, results are generally not scalable to other configurations, and are particular to the combination of geometric and physical conditions tested (ASTM D 7136/D 7136M, 2007). Even though the procedure is specified rather strictly, it allows deviations (different indenter sizes, wide scope of impact energy) which should be defined and included in data analysis. Finally, the impact resistance test procedure features material damage at the point of impact test methods, including macroscopic and nondestructive evaluation. The norm also defined the most commonly observed damage modes from out-of-plane drop-weight impact.

12.1.2 Measurements of FML impact resistance

There are no specific and measurable parameters of FML impact resistance assessment. The available literature provides only repeatable methods of measuring the response of FMLs and fiber–polymer composite structures to impact.

12.1.2.1 Force–time curves

Force is one of the measurable impact-derivative parameters in particular stages of impact (Caprino et al., 2004; Song et al., 2010; Bagnoli et al., 2009; Sohn et al., 2000; Manikandan and Chai, 2014). However, one should note that the use of the value of force exerted on material by the indenter when comparing the resistance of composite materials requires retaining the same test method with similar boundary conditions for all tested material types (indenter diameter, impact energy, number of impact repetitions). More details on the impact conditions influence on the parameters of FML impact resistance assessment have been presented in the literature (Sadighi et al., 2012a; Rajkumar et al., 2012a; Liu and Liaw, 2004; Dadej et al., 2015; Bienias et al., 2015a). The course of forces at the time of impact allows observation and quantitative comparison of changes in laminate connected, e.g., with the change of stiffness during the impact caused by damage initiation and propagation. Fig. 12.1 depicts an exemplary graph of change of force in time (f–t) of aluminum FML/carbon-epoxy composite after low-velocity impact with 20 J energy.

The basic measurable values describing the impact response of FMLs are the force value of the response of a laminate to contact with an object and the contact duration. Both these values describe low-velocity impact resistance of this type of materials and make it possible to compare materials in this respect. The essence of

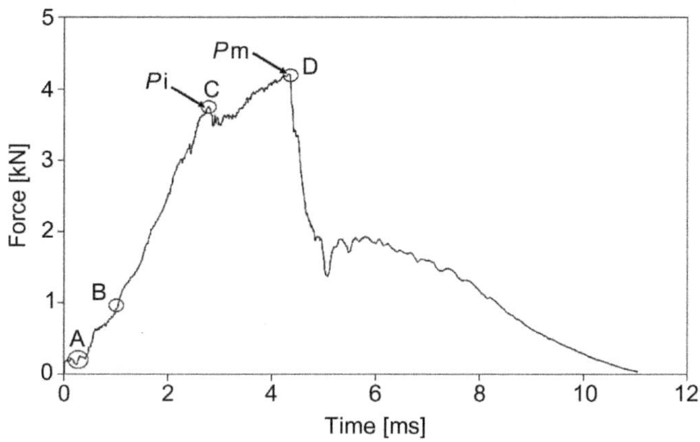

Figure 12.1 Force−time curve of aluminum−carbon fiber laminate after low-velocity impact with 20 J energy.

low-velocity impact resistance of laminates based on the *f*−*t* curve is the curve shape on which characteristic points that describe the impact response of materials in more detail can be specified.

The force−time curve (Fig. 12.1) consists of two parts. The first part comprises the rising part of the curve, corresponding to the force value growth in time. In the second, falling, part, force values decrease steadily in the time function. The rising part of the *f*−*t* curve is the part where the stresses in the material accumulate at the moment of impact until reaching the maximum force P_m (Caprino et al., 2007). The force−time diagram at the initial loading phase is rather steady. However, fluctuations are observable. Fluctuations in the initial period (Fig. 12.1A) are related to the stabilization of the material—indenter—holder system. In this period, there appear low-amplitude vibrations generated mostly from dynamic phenomena, which are related to a sudden plate acceleration. This period represents stabilization of the indenter−plate system. Regular force undulations are caused by vibrations, e.g., of a relatively stiff polymer−fabric composite (Guan and Yang, 2002). What follows is the beginning of the proper phase of an object impact on the material, whose stiffness and membrane effect (Sadighi et al., 2012a; Wu et al., 2007) initiate a stable force growth (Fig. 12.1B). Although the system stabilization phase is an important element of the assessment of a correctly performed experiment, it may be skipped during the assessment of impact resistance of FMLs. The stable force growth phase during the impact process is a phase when the initial quantity and quality assessment of the effect of impact on the material is performed. In the case of an increasing loading, further force growth during the impact is observed, which leads to further material degradation. The point on the *f*−*t* diagram that marks the first significant drop in force (Fig. 12.1C) is specified as impact resistance of laminates (Sohn et al., 2000; Song et al., 2010), and the force value at that point is defined as damage initiation force P_i (Caprino et al., 2004; Shyr and Pan, 2003). At this point, the probability of damage growth increases significantly in multiple ways. The presence of the P_i point on *f–t* diagrams in FML tests is conditioned by sufficiently deep indenter penetration and initial stress relaxation causing degradation. It may be a fiber fracture, a significant delamination growth, or an initiation of the metal layer fracturing. Each of these mechanisms absorbs only part of the energy, hence the possibility of further force increment and indenter penetration (Abrate, 1998). Degradation mechanisms of FMLs resulting from low-velocity dynamic loading have been discussed in more detail in Section 15.4.

Eventually, force reaches critical level (Fig. 12.1D). The culminating point of the force increment is specified as P_m (*maximum force*), which is assumed to be the end of the laminate resistance to further impact force transmission (Caprino et al., 2004; Shyr and Pan, 2003). A sudden drop in force immediately after reaching P_m indicates perforation of the material. Perforation is a result of a massive pileup of stresses within fibers and the immediately following stress relaxation through fiber fracturing. At the same time, a significant portion of the impact force is absorbed (Bienias et al., 2016a; Liaw et al., 2001; Múgica et al., 2014). Laminate perforation entails a loss of laminate resistance to the penetrating indenter. The remaining part of impact energy following a drop in force after reaching P_m is absorbed to enlarge the opening and move the indenter in the case of perforation. In the case of laminates where there is no perforation caused by impact, the diagram, after reaching P_m, is more symmetric, with a mild drop in force (Bienias et al., 2016a, 2016b).

12.1.2.2 Force–displacement curves

Another method of quantitative description of FMLs' low-velocity impact resistance is to plot a curve representing dependencies between changes in force in a displacement function (*f–d*), where force and displacement values contain a significant portion of information. Force values provide analogical information as in the case of the forces assessed on *f–t* curves. Discovering laminate displacement values makes it possible to compare materials with respect to the material macrodeformation kinetics. An example of force–displacement curves of aluminum/carbon-epoxy composite laminates is shown in Fig. 12.2.

Associating force and displacement values on *f–d* diagrams supplements the quantitative description of laminate loading response by indicating force values at which the displacement value changes in a different way than has been determined or predicted. Analysis of the impact force–indentation relationship with known energy at the point of impact allows one to determine the maximum strain point for the material, as well as the amount of energy expended on deformation and further damage extension (Hyla and Lizurek, 2002). Moreover, a frequently measurable parameter describing FML and composite impact resistance is bending stiffness, which is determined on the basis of a relatively rectilinear part of a force increment in the displacement function (Jakubczak et al., 2014b; Liu et al., 2000; Asaee et al., 2015). Bending stiffness had been found to be an important parameter to delamination resistance and perforation resistance. Bending stiffness is a parameter that specifies material stiffness in concentrated force load conditions and is one of the parameters that determines material resistance to this type of loads (Liu et al.,

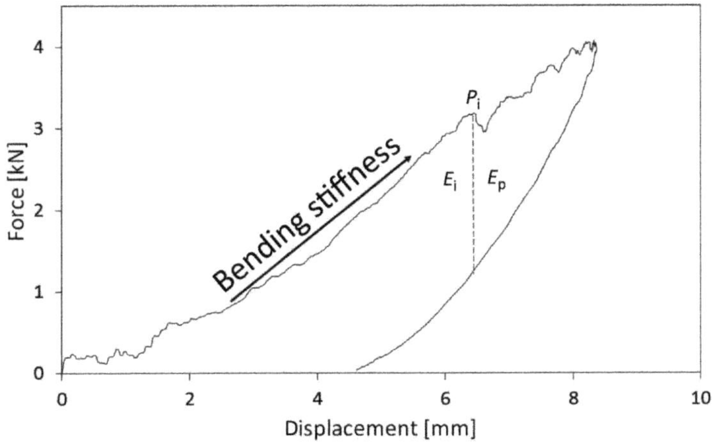

Figure 12.2 Example of force–displacement curves of aluminum/carbon-epoxy composite after low-velocity impact with energy of 15 J.

2000). For instance, Asaee et al. (2015) and Asaee and Taheri (2016) used the bending stiffness parameter to assess the improvement of low-velocity impact resistance of FMLs with a 3D fiberglass composite. Many researchers (Vlot and Gunnink, 2001; Sohn et al., 2000; Atas and Sayman, 2008), apart from conducting direct analyses of *f–d* curves, are performing further calculations based on curves and aimed at creating a quantitative description of the portions of damage initiation and propagation energy. One of the parameters used in the process of composite structure damage assessment resulting from dynamic impact is impact energy E_u and absorbed energy E_a. Energy E_a is defined as the amount of energy absorbed by the composite structure during dynamic impact. The absorbed energy can be determined from the force–displacement curves (*f–d*) registered during dynamic impact (Vlot and Gunnink, 2001; Sohn et al., 2000; Atas and Sayman, 2008). The point of reaching maximum force (P_i) determines the areas of damage initiation energy E_i till the maximum force point is reached, as well as the area of damage propagation energy E_p after reaching the maximum force point. The aggregate energy E_a absorbed by the material during dynamic impact is the sum of the initiation E_i and propagation E_p energy (Sohn et al., 2000). The literature containing classic composite analysis suggests that ply toughness has very little influence on energy absorption during dynamic impact. However, the most vital factor is the fiber stress–strain characteristics. Replacing brittle fibers with, for example, glass ones in composite structures results in reaching much higher energy levels (Beaumont et al., 1974). Brittle and highly durable materials will be characterized by higher initiation energy and lower propagation energy. Carbon fiber reinforced composites can be assigned to this group. On the other hand, more plastic but less durable materials will have a lower initiation energy and higher propagation energy. This may also concern laminates containing glass fibers (Beaumont et al., 1974). This thesis is appropriate also for FMLs.

12.1.2.3 Damage description

Another way to assess the effect of a low-velocity impact on the structure of FMLs is a quantitative assessment of the size of damage understood as a permanent change of the micro-, meso- and macrostructure of individual components of a laminate (metal, fibers, polymer matrix) and the interfaces between them. The ASTM standard (ASTM D 7136/D 7136M, 2007) refers to strength tests after low-velocity impacts and explicitly specifies the method for measuring the size of defects of classic polymer fiber composites. The authors of the available literature (Vlot, 1993; Guo-Cai et al., 2015; Bienias et al., 2016b) on the results of studies conducted on FMLs adopt a similar approach and determine the basic geometric dimensions of defects detected by means of, for example, nondestructive methods or with sectional views (Langdona et al., 2007). The basic measurable features of defects include their span in directions parallel to the sample edges, as well as the longest rectilinear segment within the defect (ASTM D 7136/D 7136M, 2007). Additionally, researchers indicate the area of the damage assessed with nondestructive methods as a significant parameter of a quantitative analysis of the impact damage ratio of FMLs (Jakubczak et al., 2014b; Nakatani et al., 2011). A quantitative damage assessment of laminates, without pointing to any particular degradation mechanisms, is an important element of a comparison of laminates with different structures and different impact conditions. Fig. 12.3 shows an example of the assessment of the effect of impact energy on the damage area and width in the case of Al/CFRP (0/90)/Al and Al/GFRP (0/90)/Al laminates.

Using measurements of a laminate impact-affected zone, it is possible to plot a curve representing dependencies, to compare and to assess the influence of the impact parameters on the postimpact condition of FMLs. The available literature offers some examples of the use of geometric measurements of a laminate damage zone to assess low-velocity impact resistance of FMLs. Many researchers (Caprino et al., 2004; Bienias et al., 2016b) assess the effect of impact energy on the size of damage of FMLs, among other things. Quantitative data have been used to plot a curve representing the dependencies between the effect of impact energy and the size of damage measured with nondestructive testing (NDT) methods. The authors

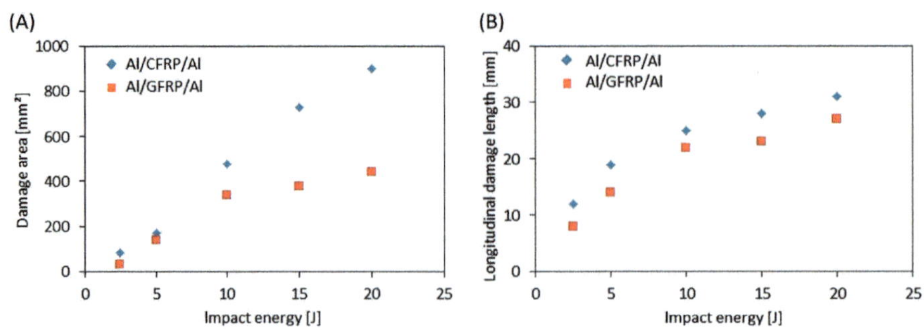

Figure 12.3 The influence of impact energy on the damage area (A) and damage length (B) of selected FMLs.

have obtained an almost linear dependence between the increase of the damage area and the increase of impact energy. Moreover, Sarlin et al. (2014) concluded that the damage area is linearly dependent on the impact energy in the studied energy range (3–30 J). This could facilitate the evaluation of the expected damage under known impact load level of an industrial application. However, some relevant impact parameters, such as the dissipated energy and the residual displacement, appear to be more sensitive than the $F-d$ curve to the main failure initiation phenomena (Caprino et al., 2004; Shyr and Pan, 2003).

Other quantitative parameters discussed in the literature (Vlot, 1993; Guo-Cai et al., 2015; Seyed Yaghoubi et al., 2011; Múgica et al., 2014), which indirectly capture the impact resistance of FMLs, include, for example, measurements of the size of the permanent deformation of a laminate subjected to impact and the length of the bottom metal layer after exceeding the so-called first crack energy (Vlot, 1993; Caprino et al., 2007; Shyr and Pan, 2003; Vo et al., 2012).

12.1.3 Experimental methods for damage assessment

There are some popular methods for damage assessment of FMLs after low-velocity impact. The expected damage modes, such as deformation, metal cracks, fiber cracks, and internal delaminations, make each method somewhat advantageous. Plastic deformation, debonding, delamination, fiber fracture, and matrix cracking, all are identified as energy absorption mechanisms in FMLs (Langdona et al., 2007) which will be described in detail in the following part of the manuscript. In general, macroscopic evaluation, NDT, and cross-sectional analysis are used.

The basic method for the assessment of the effect of low-velocity dynamic loading on the condition of the structure of FMLs is a macroscopic assessment of the contact zone between the indenter and the material, both on the impact side and the opposite side (Nakatani et al., 2011; Guo-Cai et al., 2015; Caprino et al., 2007; Abdullah and Cantwell, 2006). On this basis, it is possible to assess the results of the impact (damage) that were initiated or propagated on the laminate surface. Conducting a macroscopic analysis allows for a qualitative assessment of the condition of the metal layer both on the side affected by the impact and on the unaffected side, i.e., depending on the impact energy—deformation, crack, or perforation. Depending on their needs, after revealing external damage to postimpact FMLs, researchers conduct detailed quantitative and qualitative analyses to assess the depth of the deformation on both the affected and unaffected layer of a laminate, the quantity and direction of cracks propagation, their correlation with the known configuration of the composite fibers, and the direction of metal rolling used to build the laminate. The length of individual cracks is also significant. Diamond and cross-shaped back face damage is observed and varies according to panel thickness (Langdona et al., 2007).

Data collected in this way provide the basis for the assessment of low-velocity impact resistance of FMLs with the use of macroscopically described assessment parameters.

In the assessment of the impact resistance of FMLs, apart from macroscopic tests, methods involving nondestructive tests are also used to assess the damage scale of postimpact laminates. Methods involving nondestructive tests of FMLs are being intensively developed. Constant quality control of structure is required especially in the case of laminates that might be subject to concentrated force impacts. In service conditions, only nondestructive tests make it possible to confirm the suitability of the construction elements for further use. They are also important with respect to the application of damage tolerance principles since they allow for a quantitative assessment of the size of detected damages. The approach selection and methodology for the assessment of the structure condition of postimpact FMLs with nondestructive methods entail a number of problems. What is essential is the method's sensitivity to any potential type of damage to laminates and repeatability of results.

There have been several works published on the subject of nondestructive tests of FMLs (Dragan et al., 2012; Bisle et al., 2006; Sinke, 2003; Ibarra-Castanedo et al., 2011). The authors argue that the structure of skins, whose production may involve this type of materials, should be subject to strict control using nondestructive methods immediately after production and during operation. The problems with nondestructive tests of FMLs arise in connection with their hybrid structure. Components that differ from one another with respect to physical properties reduce the possibility to identify the signals received during a test (Sinke, 2003). On the basis of the research to date, it has been concluded that the most advantageous research method in the case of the GLARE-type structures is the ultrasound method (Seyed Yaghoubi et al., 2011; Bisle et al., 2006). The through transmission method is the most advantageous during the production process, whereas the reflection method—during an in-service inspection. Moreover, it is argued that the phased array technique (PA) ensures the highest accuracy in localizing and dimensioning structural discontinuities of FMLs due to the possibility of generating maps of a test piece as its surface view (C-scan). Even though defect localization using this method is characterized by high accuracy, the dimensioning of discontinuities has an error rate of up to 18% (Dragan et al., 2012). The ultrasound immersion technique in an air-coupled system does not ensure correct results of a nondestructive assessment of FMLs. Other NDT methods used for polymer–fiber composite materials can be used also for diagnosing postimpact FMLs. Although an X-ray-based method has high accuracy and resolution, the cost to results ratio is not satisfactory. Moreover, the size of a CT scanner's chamber limits its test capabilities, especially in the case of aircraft structures. The thermography method, in turn, is advantageous due to test duration and costs, yet small discontinuities located far from the surface may remain undetected (Dragan et al., 2012; Ibarra-Castanedo et al., 2011).

Many researchers, for instance (Bieniaś et al., 2013, 2015b; Bisle et al., 2006; Kim et al., 2009), make use of ultrasound maps in the testing of aircraft coating structures, including GLARE-type materials.

One of the most promising methods for testing postimpact FMLs in fatigue and corrosion conditions is the through transmission phased array (TTPA) method which ensures accurate quantitative measurements of the damage zone. The TTPA method is currently being developed in the Department of Materials Engineering at

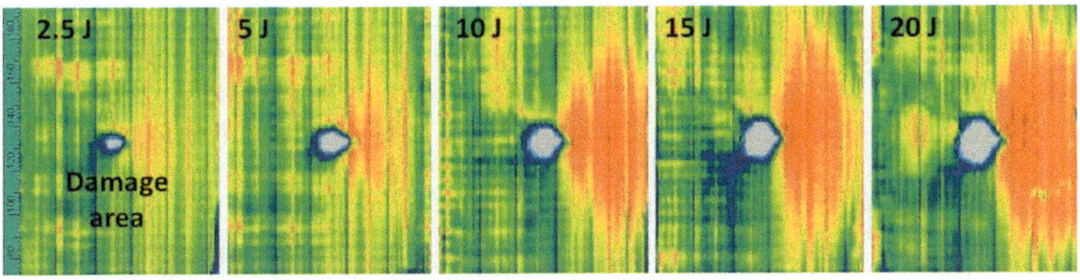

Figure 12.4 Damage area of Ti-FRP laminates after low-velocity impact.

the Lublin University of Technology, Poland, where a prototype TTPA workstation has been designed and constructed. Fig. 12.4 shows examples of an assessment of titanium polymer–fiber–composite laminates subjected to impacts with the energy range of 2.5–15 J.

To sum up, at present, it is argued that only the profile of damage zone could be detected through the ultrasonic C-scan. The mechanical-sectioning technique must be adopted in order to get the details of damage inside the fiber–metal laminates. The drop-weight-induced damage included indentation around impact center, delamination between aluminum and composite, cracks in aluminum layers, and damage in composite layers. More severe damage occurred on the nonimpacted side of FMLs what generate several serious problems (Seyed Yaghoubi et al., 2011).

Cross section of FMLs after low-velocity impact is also used in the assessment of their damage. Many authors were used this method for some quantitative and qualitative analyses. Among others, delamination length observed in cross section is used to represent the degree of delamination failure (Guo-Cai et al., 2015). The damage mechanisms observed in cross section made at the point of impact allows identification of damage mechanisms (Guo-Cai et al., 2015) the assessment of the influence of impact energy level on damage growth and changing the dominant mechanisms (Abdullah and Cantwell, 2006) and correlation the experimental damage with numerical simulations (Fan et al., 2011). Pitting, global displacement, and ring buckling of the front face are also discerned (Langdona et al., 2007).

12.1.4 Failure modes in low-velocity impact damage

12.1.4.1 Matrix cracks and delaminations

Laminate matrix damage is the first damage mode caused by impact, which usually assumes the form of matrix cracking and degradation of connection at fiber–matrix interface. In the case of FMLs, plastic deformation, especially in the aluminum layers, matrix cracking with delaminations as well as the connection loss at the aluminum–composite interface, is normal (Cortés and Cantwell, 2006; Sarlin et al., 2014; Bagnoli et al., 2009; Morinière et al., 2012). Delaminations occurring between particular composite layers and at the metal–composite interface are caused by plastic deformation of the aluminum layers and the presence of high shear stresses (Morinière et al., 2012). Matrix cracks are connected with some incompatibility of properties between the fiber and the matrix. Epoxy resins creating the matrix and composite—metal interface are brittle and they have low resistance to crack propagation, including delaminations (Richardson and Wisheart, 1996; Abrate, 1998; Aymerich and Priolo, 2008). The principal damage of FMLs is located in the lower part of laminates structure consisting of layers (Lawcock et al., 1997; Nakatani et al., 2011). Generally, delaminations are caused by the presence of matrix cracks and interlayer shear stress along the interface, incompatible stiffness among the adjacent layers, layer grouping, as well as by the deformation of the laminate (Abrate, 1998).

There are two types of matrix cracks: flexural cracks and shear cracks (Nakatani et al., 2011; Abrate, 1998). These cracks occur as a result of the presence of very large transverse shear stresses related to pressure force and the impactor—laminate contact area during impact (Richardson and Wisheart, 1996; Abrate, 1998; Aymerich and Priolo, 2008).

Fig. 12.5 shows a diagram of matrix cracks and delaminations propagating in FMLs as a result of low-velocity impacts.

There are three primary material-based causes of extensive delaminations in FMLs due to impacts. They include a lack of resistance of the polymer matrix of the composite filler to the initiation and propagation of cracks due to high brittleness and low strength (R_m ~50 MPa) (Abrate, 1998), bending stiffness misadjustment between adjacent composite layers with different orientation of reinforcing fibers and low mechanical properties of the metal—composite interface (Shyr and Pan, 2003; Morinière et al., 2012; González et al., 2011). Material elasticity, especially the Young's E_1/E_2 modulus ratios of unidirectional composite layers, has a primary influence on the misadjustment of the bending stiffness coefficient between layers with different orientation. A significant difference between transverse and longitudinal modules of material elasticity leads to greater misadjustments of bending stiffness (e.g., for layers with a 0/90 fiber direction) and delamination growth (Abrate, 1998). The influence of the fiber configuration on the initiation and propagation of FML damage resulting from low-velocity dynamic loading will be discussed in more detail in Section 15.5.2.

Fig. 12.6 shows two mechanisms of bending crack growth near the metal—composite interface in FMLs subjected to low-velocity impacts. The mechanism of the transmission of matrix cracks onto metal and the change of the crack propagation direction from perpendicular to parallel with the use of the interface.

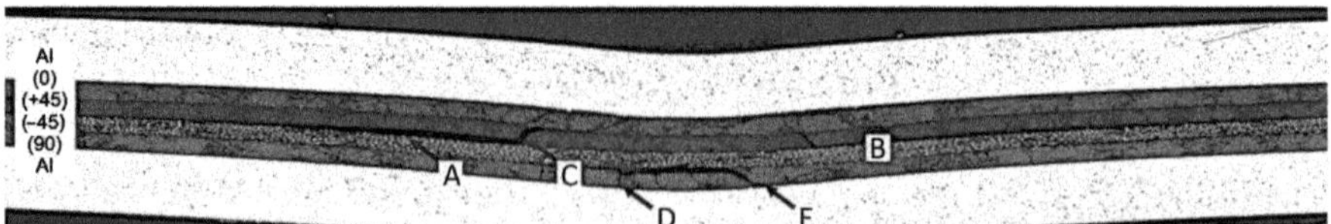

Figure 12.5 Cross section of Al/CFRP laminate subjected to an impact of 2.5 J (A—delaminations, B—ply separations, C, E—transverse cracks, D—flexural cracks).

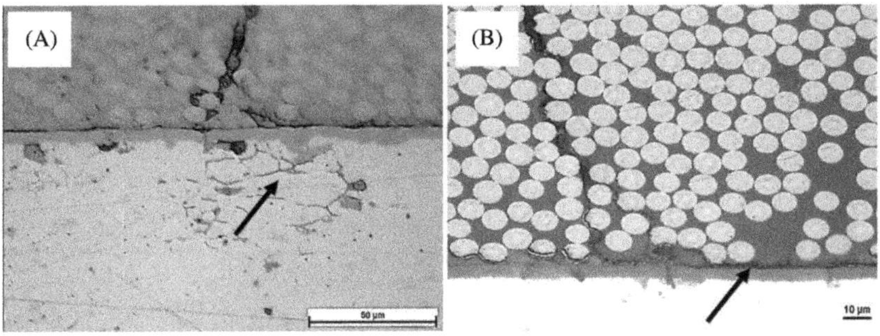

Figure 12.6 Microcracks in the aluminum layer of an FML after a low-energy impact (A) and ply separations at the metal–composite interface (B).

Laminates' resistance to ply separation propagation at the metal–composite interface is significant. It is argued that the metal–composite interface is one of the most unreliable elements in laminates (Ostapiuk et al., 2012). Adhesive properties, the quality of metal–composite connection, and the manufacturing process may directly lead to the occurrence of delaminations and at the same time affect impact resistance of FMLs (Morinière et al., 2012; Pärnänen et al., 2008). By using damage area measurement, it can be discovered that damaged area of GLARE laminates with poor interfacial adhesive bonding was much larger than that of with good bonding. FMLs which had a very good interfacial connection were characterized by much higher load carrying values as well as lower maximum deformations in comparison to laminates with low-quality connections (Morinière et al., 2012; Ardakani et al., 2009).

The size of delaminations in FMLs is also heavily influenced by the impact energy value. In the literature, the relationship between delamination length and impact energy is widely discussed. The delamination length increases at first and decreases afterward. The reason is that delamination is the main damage under low impact energy. The metal cracking and fiber fracture transfer to the main damage as the impact energy increases over some value (Bieniaś et al., 2015b; Guo-Cai et al., 2015).

To sum up, it needs to be noted that the presence of initiating matrix cracks and insufficiently developed ply separations has no significant impact on the laminate stiffness. That is why FMLs can transfer systematically incremental loading in a stable manner. Due to the presence of deformable metal in FMLs, ply separation growth, which is stretched over time, leads to a continuous absorption of the energy of the penetrating indenter, which makes these materials unique with respect to impact resistance, compared to classic composite structures (Bieniaś et al., 2015b).

12.1.4.2 Fiber and metal damage

Regardless of the components of fiber–metal laminates, some of the most important effects of low-velocity dynamic loads include, apart from composite matrix cracks and delaminations, fiber cracks and metal layer cracks which are mainly conditional on the impact energy.

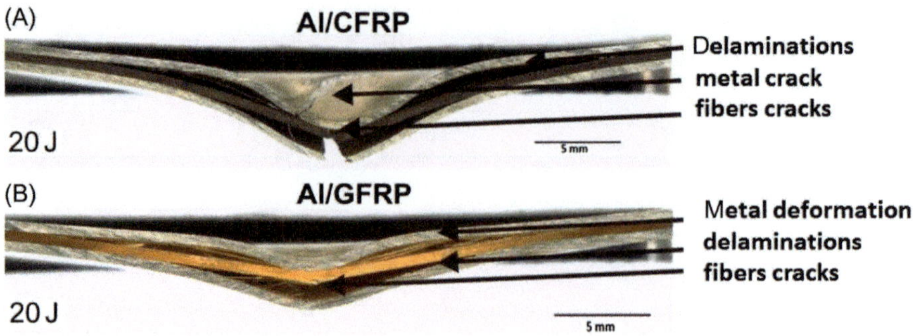

Figure 12.7 Cross section of (A) Al/CFRP and (B) Al/GFRP laminates in area of impact.

Depending on the mechanical properties of the components, the share of individual mechanisms in the damage may vary. High value of Young's modulus in the composite fill in FML makes it prone to fiber cracking and perforation (Fig. 12.7A), whereas lower values of the modulus (e.g., in GLARE laminates) often results in extensive delaminations (Fig. 12.7B).

12.1.4.3 *Fiber damage*

The degradation mechanism in fiber–metal laminates which is often described in literature and occurs especially at higher impact energy is cracking of fibers. Force increase at high impact energy is sufficient for delaminations in relatively stiff composites and on the phase boundaries between metal and composite not to absorb the energy completely. The energy is absorbed in more disastrous damage, such as material perforation, for which high value of energy supplied is considered to be absorbed mainly in the mass fiber cracking (Caprino et al., 2004; Guan and Yang, 2002; Wu et al., 2007; Aymerich and Priolo, 2008; Padaki et al., 2008). Fig. 12.8 shows an example of a yield line in GLARE laminate following a 20 J impact with a 0.5 inch. spherical indenter.

Fiber cracks exhibit highly undefined course of destruction. Fibers are fractured at a different height, which indicates multidimensional cracking, without foreseeable or consistently dispersed sites where the cracking originates. Residual resin is visible in the surroundings of the fibers, in the form of scarps and ribs (Greenhalgh, 2009) (Fig. 12.8). Upon initiation, cracking of fibers leads to the same effects regardless of the reinforcement fibers in laminates. However, glass fiber undergoes mass cracking at much higher impact energy values than carbon fiber. Fiber cracks are triggered by microfiber buckling due to compressive stress in composite. At this stage of indenter penetration, damage originates in the microscale and then propagates to the mesoscale and macroscale. The scale of the damage is strongly dependent on the impact energy (Seltzer et al., 2013). Cracking fibers produce additional loads in microareas, creating pressure points on the surface of the aluminum at the same time, which initializes the cracking of the lower metal layer (Liaw et al., 2001). Many research authors have also noted that cracking fibers in the lower

Figure 12.8 Fiber cracks in Al/GFRP laminate following a 20 J impact.

composite layers determine the direction of cracking propagation in the lower metal layer (Rajkumar et al., 2012b). At the same time, it is believed that the type of fibers in the fiber–metal laminates affects the dominant mechanisms of laminate destruction in the same impact conditions which have been described in Section 15.5.1.

12.1.4.4 Metal damage

One of the more important FML laminate impact-related damage mechanisms is the cracking of metal layers, as it initiates the perforation of the laminate. Laminate perforation is tantamount to total loss of resistance to the indenter penetrating the material. It is assumed that the initiation and the development of cracking in the lower aluminum layer occur when the impact achieves the force value P_i (Fig. 12.9). At this time, local damage sites inside the structure combine each other (Nakatani et al., 2011; Richardson and Wisheart, 1996).

The issue which has been analyzed extensively in the available literature (Caprino et al., 2004; Bienias et al., 2016a; Sadighi et al., 2012a; Moriniere et al., 2013) is the direction of propagation of the bottom aluminum layer cracking as a result of impact causing perforation. Moriniere et al. (2013) noted that the direction of metal cracking for GLARE 3/2 (0/90/90/0) is interpreted as being compatible with sheet rolling direction. In Sadighi et al. (2012a) study, it was summed up that in case of FMLs, in which aluminum dominates in load carrying, cracking of the bottom aluminum layers propagates in the rolling direction, related to the observation in glass fiber laminates. In their own study on carbon fiber laminates, the authors have demonstrated that the rolling direction is of secondary importance in respect of the direction of the crack development in the bottom aluminum layer

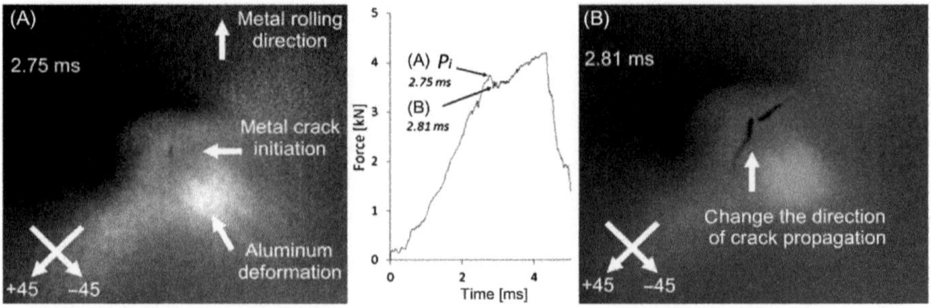

Figure 12.9 FML metal crack initiation (A) and propagation (B).

(Bienias et al., 2016a). The direction of the cracking development in the bottom aluminum layer in carbon fiber laminates is closely linked with the fiber arrangement in the composite layer adjacent to the metal layer, as the authors have shown in other works as well (Bienias et al., 2016a; Jakubczak et al., 2014b). A more detailed description of the problem is presented in Section 15.5.2.

Fractographic examination of the damage in metal layers of aluminum fiber–metal laminates indicates plastic nature of aluminum cracks despite large dynamics of crack development (Fig. 12.10).

Rounded grain areas are dominant, but certain areas (voids, indentations, and surface roughness) point to a more dynamic process than in the case of static loads. There are currently no studies in the available literature which describe damage fractography in fiber–metal laminates and their components following an impact.

12.1.5 Parameters affecting impact damage of FMLs

12.1.5.1 Influence of type of metal and fibers

Metal type

In view of the significant share of fiber cracking in the initiation and propagation of cracks in the bottom metal layer, as well as its effect on the delamination development due to potential deformation of the entire laminate (Pärnänen et al., 2008), the type of metal, particularly its stiffness and resistance to cracking, may prove crucial in improving the resistance of FML laminates to low-velocity dynamic load. Since density is one of the most vital parameters of FML laminates, current studies and application involves FML laminates based on light metal alloys, mainly aluminum (Vlot and Gunnink, 2001).

In FML laminates, 2024-T3 and 7475-T6 aluminum alloys are commonly used due to their durability. In the case of low-velocity dynamic loads, studies have shown lower deflection, earlier initiation of cracks, and bottom delamination of GLARE1 due to the high stiffness and strength of 7475-T6 aluminum alloy (Liu and Liaw, 2010). For GLARE, the application of 7000 grades of aluminum alloys might lead to a less favorable damage resistance than with Al 2024-T3 (Sadighi et al., 2012a).

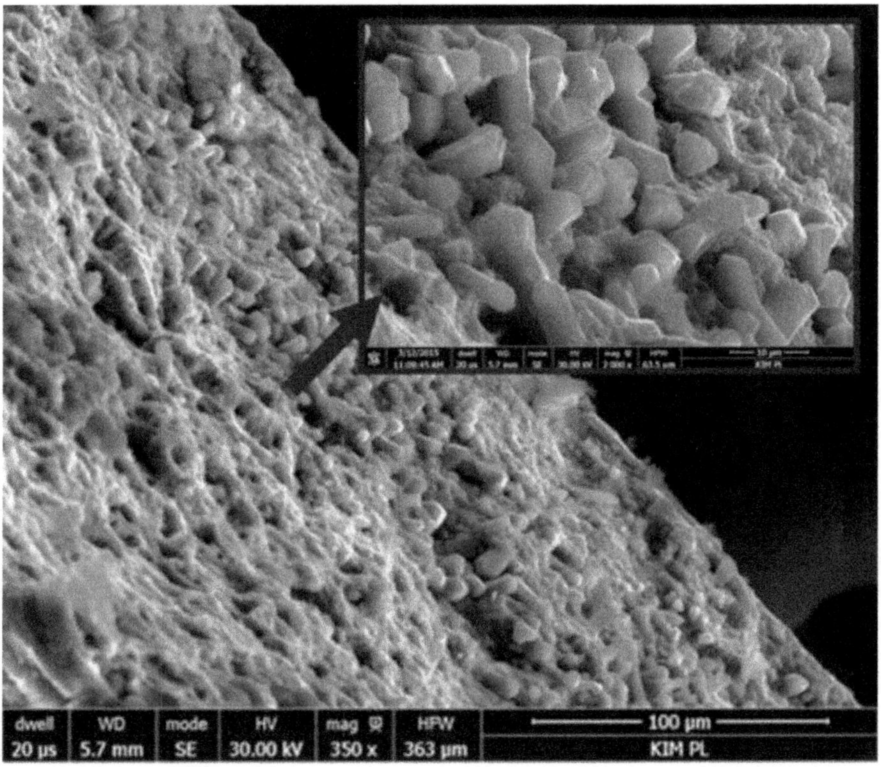

Figure 12.10 Al/GFRP FML aluminum bottom layer crack after 20 J impact energy.

New aluminum alloys for FML from the group of high damage tolerance are alloy 2524/2024. 2524-T3 (Alcoa) is a 2024 derivative requiring stringent control of alloy purity by limiting Fe and Si. Low Fe and Si minimizes the formation of constituents which have been found to exert an adverse effect of fracture toughness, crack initiation, and crack growth resistance (Auffret and Gennai, 2001). Al2024A (Pechiney) is an alloy for highly damage-tolerant fuselages. Cu and Mg contents are optimized to levels necessary to achieve the strength without degrading the damage tolerance properties through potential undissolved, intermetallic phases. Low Fe and Si contents also contribute to decrease the size and number density of the associated phases, responsible for lowering overall material damage tolerance (Auffret and Gennai, 2001).

The third generation Al–Li alloy, 2195, is cost-effective because its high strength provides superior weight savings. Alloy 2097 offers cost savings because it eliminates the need of periodic replacement of 2124-T851 bulkheads due to fatigue cracks (Auffret and Gennai, 2001).

Mg-based FMLs perform significantly better than glass-epoxy/aluminum FML and offer a higher specific perforation energy than the comparable aluminum FML. Substitution of aluminum by magnesium, in spite of some apparent advantages like similar perforation energy or lower dent depth, due to considerable damage area and lower specific properties (normalized by weight of the laminate) will not yield improvements (Sadighi et al., 2012a; Sadighi et al., 2012b).

Other, future FMLs can be built with titanium alloys. An advantage of these laminates is their combination of high stiffness, high yield strength, good fatigue, and good impact properties at both room and elevated temperatures. Titanium and carbon fibers can easily be exposed to up to 300°C, which makes this combination perfect for high-temperature applications (de Boer, 2001). There are limited information about this type of FMLs in the literature. One of the most important issues is good adhesion of titanium to composite. In spite of suitable fatigue performance of titanium-based FMLs (Sadighi et al., 2012a; Tarpani et al., 2009) and improvement in the static strength compare to GLARE (Sadighi et al., 2012a), the low ductility of the titanium alloys caused relatively poor impact properties for the FML (Sadighi et al., 2012a; Cortes and Cantwell, 2007), which can be connected with poor titanium–composite interface. However, some new publications show that titanium is a really perspective material for FML, e.g., in terms of impact resistance (Jakubczak et al., 2016).

Other, less common metal types in FMLs such as magnesium and aluminum–lithium alloy were also investigated in the literature (Vlot and Gunnink, 2001; Pärnänen et al., 2008). In the case of impact resistance of magnesium fiber laminates, it has been demonstrated that the cracking energy limits of magnesium-based FML were clearly lower when compared to traditional Al 2024-T3-based GLARE FMLs. However, the specific perforation limit of the magnesium-based FML was at a level equal to the GLARE FMLs. In addition, the damage sizes of magnesium-based FML specimens were higher due to extensive metal cracking and delaminations. For these reasons, it is suspected that fully magnesium-based FMLs do not offer an improvement in impact resistance or impact damage resistance over the traditional Al 2024-T3-based GLARE FMLs in general (Pärnänen et al., 2008).

Fiber type

The mechanism and the size of the damage in fiber–metal laminates following a low-velocity impact are determined by the type of fibers which are used as reinforcement for the composite part of the FML laminate (Caprino et al., 2004; Bienias et al., 2016a; Sadighi et al., 2012a).

Fiber cracking in glass fiber and aramid laminates occurs much less frequently than in carbon laminates (Lawcock et al., 1997; Sadighi et al., 2012a; Nakatani et al., 2011). Due to relatively low Young's modulus in glass fibers (~50 GPa), as well as the strain to failure (~5%), delamination are considered to be the main degradation mechanism in GLARE laminates. However, with an increase in the number of glass composite layers, fiber cracking assumes a more dominant position (Shyr and Pan, 2003). This is also the case in ARALL laminates with aramid fiber (Sadighi et al., 2012a) Cracking of fibers due to bending and, consequently, their tensing, is more dominant in carbon laminates (Vlot, 1996). In terms of two types of failure of FMLs, i.e., "fiber" or "aluminum" critical, CARAL, due to its low strain to failure of carbon fibers (~2%), presents a fiber critical behavior irrespective of its layup (Vlot, 1996).

From the review of available literature, it appears that extensive investigations have been conducted in recent years in order to research the impact behavior mainly for GLARE composites (Vlot and Krull, 1997; Caprino et al., 2004; Fan et al., 2011; Sadighi et al., 2012a; Abdullah and Cantwell, 2006; Liaw et al., 2001; Laliberté et al., 2014). Relatively high strain glass fibers and in consequence extensive delamination allows GLARE laminates to deform and fracture in a more membranous way and it contributes to total energy absorption (Wu et al., 2007). This membranous way of deformation is one of the reasons for low final deflection of GLARE-type laminates. The maximum depth has nearly linear progress character with impact energy increase during impact of GLARE laminates (Caprino et al., 2004). Nonmembranous effect in the case of CARAL-type laminates is caused by lower strain to failure of carbon fibers. In the case of FML laminates with carbon fiber, the most likely damage initiating a catastrophic destruction is cracking of the fibers in the bottom composite layers due to bending stress. Cracking of the metal layer in the direction transverse to the one of fibers from the bottom composite layer can be observed in case of FMLs with carbon fibers, but studies concerning this subject are few and far between (Bienias et al., 2016a; Lawcock et al., 1997; Song et al., 2010).

Cracking of the bottom aluminum layer in CARAL laminates, noted more frequently, may result from fiber breaking, which does not occur in the other laminate types (GLARE and ARALL), where fibers remain consistent throughout the whole contact with the indenter (Sadighi et al., 2012a).

12.1.5.2 The fibers arrangement

Fiber arrangement in the composite part of FML laminates corresponds with the laminate resistance to low-velocity dynamic load (Sadighi et al., 2012a; Seyed Yaghoubi et al., 2012). Moreover, different layer arrangement affects the dominant degradation mechanisms in FML laminates at the same load conditions (Abdullah and Cantwell, 2006). At lower impact energy values, the influence of composite layer arrangement in FML laminates is not relevant as there is no strong interference of the indenter in the structure of the composite part of the laminate. Most of the impact energy is absorbed by the aluminum deformation and cracking of the polymer matrix. With increasing impact energy, the role of fibers becomes more important, including their orientation in space (Shyr and Pan, 2003; Liaw et al., 2001).

Quasi-isotropic laminates offer the highest resistance to dynamic load applied with focused force, expressed with maximum force criteria (Fig. 12.11), crack length, permanent deflection, damage width, and maximum displacement (Vlot and Krull, 1997; Vlot, 1996; Liaw et al., 2001).

The quasi-isotropic prepreg configuration (0/45/-45/90) for FMLs has better impact damage tolerance than cross-ply prepregs (0/90/90/0) and than UD configuration $(0)_4$, which performs the worst, in terms of impact load and crack formation (Liaw et al., 2001; Seyed Yaghoubi et al., 2012).

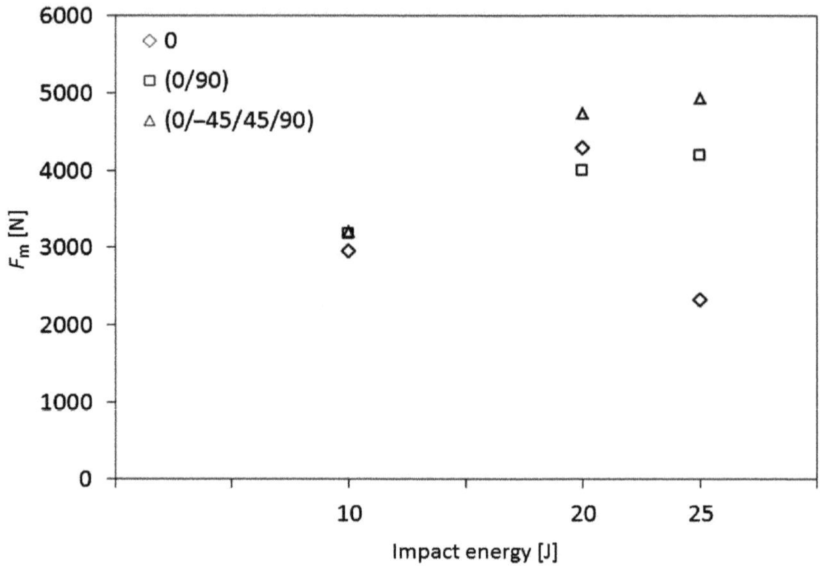

Figure 12.11 Relation between composite layers arrangement in Al/CFRP laminate and the maximum force value.

Literature does not provide a clear explanation for these relations. Prevalence of resistance to impact in quasi-isotropic laminates is considered to be a result of the most even distribution of stress transfer on the surface, which facilitates more dissipated power. Shaping of the multidirectional positioning of each composite layer leads to the creation of interlayer boundaries, which significantly improves the ability of the composite to absorb the energy from the indenter penetrating the material. In a composite laminate, delamination will occur between the adjacent plies with different fiber orientation (Bieniaś et al., 2015b; Abrate, 1998), so for cross-ply configuration it is expected to have more energy dissipation through the delamination process within composite plies. At the same time, initiation and propagation of cracks in metal layers and the final perforation is delayed (Periasamy et al., 2012).

Considering fact that fiber arrangement in FML effects on their impact resistance, it is expected that it also significantly influenced the damage mechanisms (Sadighi et al., 2012a). Fig. 12.12 shows the impacted area and metal cracks of selected FMLs with different layup scheme.

In general, cracks of the bottom metal layer are parallel in GLARE-type FMLs and perpendicular in CARALL laminates type. The reasons of that were explained in Section 15.5.1 In simple terms, in the case of high impact energy, the number of cracks is more or less the same as number of directions of fibers (Zhu and Chai, 2012). For example, (0/90) fiber directions cause two major cracks perpendicular to each other (Pärnänen et al., 2008), (0/ ± 45/90) cause four cracks (Fig. 12.12). In literature can be found information that in UD GLARE, which the majority of fibers are laid up in the 0 direction, a single crack oriented parallel to the major

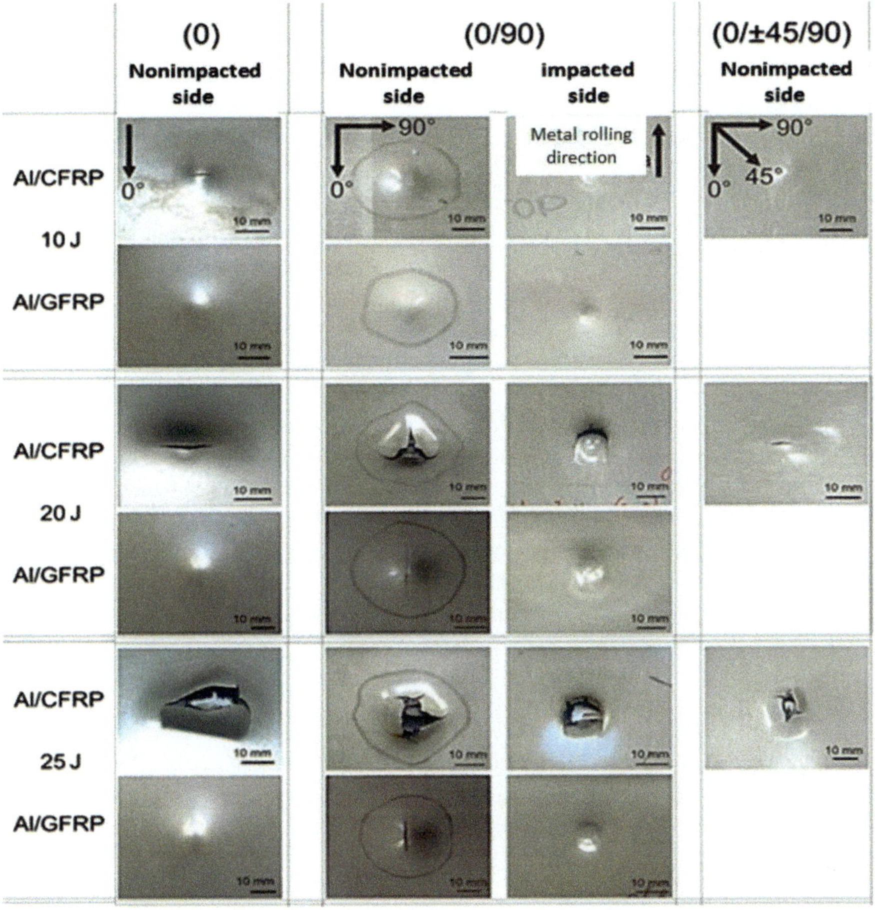

Figure 12.12 The damage area and metal cracks of selected FMLs with different layup scheme.

fiber direction is produced and its length increases with increasing the impact energy while in non-UD GLARE multiple cracks occurred in both direction of 0 and 90 (Sadighi et al., 2012a). It does not correspond with authors (Bienias et al., 2016a) and others' research (Pärnänen et al., 2008). Authors did not observe double crack in case of GLARE. Certainly it requires more detailed research. However, it should be noted that the number of cracks in GLARE laminates depends on impact energy and glass fibers arrangement.

12.1.6 Numerical modeling of low-velocity impact of FMLs

Currently, scientific research studies are also available in the scope of analyses and tests of degradation process in FML structures subjected to low-velocity impact using numerical finite element analyses (FEA) (Fan et al., 2011; Guo-Cai et al., 2015; Sitnikova et al., 2014; Morinière et al., 2014; Chai and Manikandan, 2014). FEA consists in use of mathematical criteria describing the initiation and propagation of damages for matrix and fibers in composite on the basis of actual stresses and mechanical properties examined before. Analysis of literature indicates that there is a need to deepen the knowledge on numerical modeling of FML laminates, especially with comparison to experimental procedures. The present research issue consists in development of an adequate numerical model considering the known mechanisms of FML laminates after impact with consideration of interlayer phenomena. The purpose of numerical simulations completed in many studies is to determine the degree of FML damages in numerical model characterized by means of known damage criteria and to determine their conformity with real research carried out in identical conditions (Fan et al., 2011; Guo-Cai et al., 2015; Bienias et al., 2016b; Morinière et al., 2014). Chai and Manikandan (2014) presented actually review of FML laminates modeling methods and described the most important studies completed up to the present time. The author emphasizes that the modeling of these materials is still in the initial phase of analyses and tests.

In general, there is one most common model of low-velocity impact of FMLs. Model of metal layers is based on ductile criterion (Hooputra et al., 2004) where the occurrence of metal damage is a function of stresses triaxiality and plastic strain rate:

$$\bar{\varepsilon}_{\mathrm{D}}^{\mathrm{pl}}\left(\eta, \dot{\bar{\varepsilon}}^{\mathrm{pl}}\right) \tag{12.1}$$

where $\eta = -p/q$, η is stress triaxiality, p is pressure stress, q is Mises equivalent stress, $\dot{\bar{\varepsilon}}^{pl}$ is equivalent plastic strain rate. This criterion is met if the following condition is met:

$$\omega_{\mathrm{D}} = \int \frac{d\bar{\varepsilon}^{\mathrm{pl}}}{\bar{\varepsilon}_{\mathrm{D}}^{\mathrm{pl}}\left(\eta, \dot{\bar{\varepsilon}}^{\mathrm{pl}}\right)} = 1 \tag{12.2}$$

where ω_{D} is a state variable that increases monotonically with plastic deformation.

Model of composite layers is based on progressive damage analysis.

For undamaged and elastic orthotropic laminates, the stress–strain relationship can be written:

$$\begin{Bmatrix} \sigma_{11} \\ \sigma_{22} \\ \sigma_{33} \\ \tau_{23} \\ \tau_{31} \\ \tau_{12} \end{Bmatrix} = \begin{vmatrix} c_{11} & c_{12} & c_{13} & 0 & 0 & 0 \\ c_{21} & c_{22} & c_{23} & 0 & 0 & 0 \\ c_{31} & c_{32} & c_{33} & 0 & 0 & 0 \\ 0 & 0 & 0 & c_{44} & 0 & 0 \\ 0 & 0 & 0 & 0 & c_{55} & 0 \\ 0 & 0 & 0 & 0 & 0 & c_{66} \end{vmatrix} \begin{Bmatrix} \varepsilon_{11} \\ \varepsilon_{22} \\ \varepsilon_{33} \\ \gamma_{23} \\ \gamma_{31} \\ \gamma_{12} \end{Bmatrix} \tag{12.3}$$

where σ_{ij} are stresses in ij directions, c_{ij} are stiffness coefficients, ε_{ij} are strains, τ_{ij} are shear stress and γ_{ij} are shear strains.

Failure criteria for fiber failure in tension, based on 1980 Hashin-3D tensile criterion (Hashin, 1980). Damage variable $d_{\mathrm{ft}} = 1$, when:

$$\begin{cases} \left(\dfrac{\sigma_{11}}{X_{1T}}\right)^2 + \left(\dfrac{\sigma_{12}}{S_{12}}\right)^2 + \left(\dfrac{\sigma_{13}}{S_{13}}\right)^2 \geq 1 \\ \sigma_{11} > 0 \end{cases} \tag{12.4}$$

Failure criteria for fiber failure in compression, based on Max-stress criterion (Orifici et al., 2008). Damage variable $d_{\mathrm{fc}} = 1$, when:

$$\begin{cases} \dfrac{\sigma_{11}}{X_{1c}} \geq 1 \\ \sigma_{11} < 0 \end{cases} \tag{12.5}$$

Failure criteria for matrix failure in tension, based on literature data (Asaee and Taheri, 2016; Puck and Schürmann, 1998). Damage variable $d_{\mathrm{mt}} = 1$, when:

$$\begin{cases} \left(\dfrac{\sigma_{11}}{X_{1T}}\right)^2 + \left(\dfrac{\sigma_{12}}{S_{12}}\right)^2 + \left(\dfrac{\sigma_{22}^2}{X_{2T}X_{2C}}\right) + \dfrac{\sigma_{22}}{X_{2T}} + \dfrac{\sigma_{22}}{X_{2C}} \geq 1 \\ \sigma_2 + \sigma_3 \geq 0 \end{cases} \tag{12.6}$$

Failure criteria for matrix failure in compression (Vo et al., 2012):

$$\left[\frac{X_{2c}}{2S_{23}}^2 - 1\right]\frac{(\sigma_{22} + \sigma_{33})}{X_{2c}} + \frac{(\sigma_{22}+\sigma_{33})^2}{4S_{23}^2} + \frac{(\sigma_{23}^2 - \sigma_{22}\sigma_{33})}{S_{23}^2} + \frac{\sigma_{12}^2 + \sigma_{13}^2}{S_{12}^2} = 1, \quad d_{\mathrm{mc}} = 1$$
$$\sigma_2 + \sigma_3 < 0 \tag{12.7}$$

where:

X_{1t} denotes tensile failure strength in fiber direction.
X_{1c} denotes compressive failure strength in fiber direction.
X_{2t} denotes tensile failure strength in direction 2 (transverse to fiber direction).
X_{2c} denotes compressive failure stress in direction 2 (transverse to fiber direction).
S_{12} denotes failure shear strength in 1−2 plane.
S_{13} denotes failure shear strength in 1-3 plane.
S_{23} denotes failure shear strength in 2-3 plane.

Sometimes authors (Guo-Cai et al., 2015; Asaee and Taheri, 2016; Vo et al., 2012; Shi et al., 2012; Li et al., 2010; Feng and Aymerich, 2013; Donadon et al.,

2008) use other failure criteria for matrix failure in compression (Puck and Schürmann, 1998) based on damage variable $d_{mc} = 1$, when:

$$\begin{cases} \frac{\sigma_{11}}{X_{1T}}^2 + \frac{\sigma_{12}}{S_{12}}^2 + \frac{\sigma_{22}^2}{X_{2T}X_{2c}} + \frac{\sigma_{22}}{X_{2T}} + \frac{\sigma_{22}}{X_{2c}} \geq 1 \\ \sigma_2 + \sigma_3 < 0 \end{cases} \tag{12.8}$$

It means that d_{mt} is equal d_{mc}.

Regardless of the method, currently published results of numeric calculations in respect of the resistance of FML laminates to low-velocity impact require authors to apply multiple evaluation criteria with the comparison of measurable response parameters of laminates assessed in the experiment and during the damage analysis. Irrespective of the type of the laminates analyzed, most advanced studies present force vs time, impact vs deformation, and impact energy vs the length of the longest occurred delaminations (Fan et al., 2011; Guo-Cai et al., 2015; Bienias et al., 2016b). High conformity between the results of performed experimental tests and the numerical model means that actually used subroutines based on 3D damage criteria and perspective cohesive zone would be successfully applied for modeling of dynamic phenomena.

12.1.7 The future perspectives of low-velocity impact resistance of FML

12.1.7.1 Titanium alloys in FML

Fiber–metal laminates are relatively new materials of numerous modification possibilities. This is one of the reasons why not all their properties are yet known, whereas the data concerning their response to different loads are limited to a narrow group of laminates (e.g., GLARE). A number of benefits attributed to those materials give rise to constant development of new material and technology concepts for future applications.

One of the most intensively developed research directions in the field of fiber–metal laminates is the application of titanium and its alloys in connection with glass, carbon, or aramid fibers. It has been confirmed in preliminary studies that combination of titanium and carbon fibers may have the most significant impact on improving the resistance of fiber–metal laminates to dynamic loads, including the low-velocity impact (Cortes and Cantwell, 2007; Jakubczak et al., 2016; Bernhardt et al., 2007; Bieniaś and Jakubczak, 2012). As written earlier, some authors concluded that the low ductility of the titanium alloys caused relatively poor impact properties for the FML (Cortes and Cantwell, 2007). However, there can be many causes of unsatisfactory results in preliminary studies. Titanium alloys used in FMLs to improve impact resistance require appropriate material selection (including a relatively low ratio of Young's modulus of titanium and composite), proper surface preparation of titanium (Surowska, 2009; Molitor et al.,

2001), and appropriate manufacturing technologies (e.g., autoclaved method). As some authors have demonstrated (Bieniaś and Jakubczak, 2012), titanium–carbon is perspective material to protection against low-velocity impact in aerospace and other applications.

12.1.7.2 CAI of FML

One of the most important features of FMLs is compressive strength after impact. It is known that extensive delaminations between external layers caused buckling as a result of compression. This failure comes as the delamination progresses with the local buckling of the sublaminates produced by impact (Sanchez-Saez et al., 2008; Prichard and Hogg, 2000). Several organizations and companies have published recommendations for the CAI test (NASA (NASA, 1983), Boeing (Boeing, 1988), SACMA (SACMA, 1994), CRAG (Curtis, 1988)), but there is no universal standard (ASTM or ISO) that would state the specimen geometry and the test variables (Sanchez-Saez et al., 2008). ASTM D7137 (2005) recommend minimal thickness of composite plates of 5 mm. Composites are used in range of up to 1 mm. This problem is still unresolved (Sanchez-Saez et al., 2008). For classic composite materials it is known that even BVID (Barely Visible Impact Damage) impact damage caused significant reduction of compressive strength (even ~80%) (Jakubczak et al., 2014a).

Unfortunately, the feature of CAI (Compression After Impact) of FML is actually unknown. Even laminates with the thickness of 5 mm and more do not allow test residual strengths. The reason of that is elastic–plastic metal as an outer layers, which are not stiff. Under compression, they undergo gradual deformation, which does not cause sudden delamination development as a result of stress cumulation (Jakubczak et al., 2014a). This should be provisionally assumed as a benefit compared to conventional composite structures. However, the residual strength of FMLs cannot be tested because of lack of methods for this kind of hybrid materials.

Acknowledgments

Part of this research was financially supported by Structural Funds in the Operational Programme—Innovative Economy (IE OP) and financed from the European Regional Development Fund Project No POIG.0101.02-00-015/08.

References

Abdullah, M.R., Cantwell, W.J., 2006. The impact resistance of polypropylene-based fibre–metal laminates. Comp. Sci. Technol. 66, 1682–1693.

Abrate, S., 1991. Impact on laminated composite materials. Appl. Mech. Rev. 44 (4), 155–190.

Abrate, S., 1998. Impact on composite structures. Cambridge University Press, Cambridge, UK, pp. 135–160. (Chapter 4).

Alderliesten, R.C., Homan, J.J., 2006. Fatigue and damage tolerance issues of Glare in aircraft structures. Int. J. Fatigue. 28, 1116–1123.

Ardakani M.A., A.A. Khatibi, S.A. Ghazavi, 2008. A study on the manufacturing of glass-fiber-reinforced aluminum laminates and the effect of interfacial adhesive bonding on the impact behavior, Proceedings of the XIth International Congress and Exposition, June 2–5, Orlando, FL.

Ardakani M.A., A.A. Khatibi, H. Parsaiyan, An experimental study on the impact resistance of glass-fiber-reinforced aluminum (Glare) laminates, 17th International Conference on Composite Materials, Edinburgh, UK, 2009.

Asaee, Z., Taheri, F., 2016. Experimental and numerical investigation into the influence of stacking sequence on the low-velocity impact response of new 3D FMLs. Comp. Struct. 140, 136–146.

Asaee, Z., Shadlou, S., Taheri, F., 2015. Low-velocity impact response of fiberglass/magnesium FMLs with a new 3D fiberglass fabric. Comp. Struct. 122, 155–165.

ASTM D7137, 2005. Standard Test Method for Compressive Residual Strength Properties of Damaged Polymer Matrix Composite Plates 1, Book of Standards, vol. 15.03.

ASTM D 7136/D 7136M, 2007. Standard test method for measuring the damage resistance of a fibre reinforced polymer matrix composite to a drop-weight impact event, American Society for Testing and Materials.

Atas, C., Sayman, O., 2008. An overall view on impact response of woven fabric composite plates. Comp. Struct. 82, 336–345.

Auffret, A., Gennai, A., 2001. New aluminium alloys for Glare. In: Vlot, A., Gunnink, J.W. (Eds.), Fibre Metal Laminates: An introduction. Kluwer Academic Publishers, Dordrecht—Boston—London, pp. 79–97.

Aymerich, F., Priolo, P., 2008. Characterization of fracture modes in stitched and unstitched cross-ply laminates subjected to low-velocity impact and compression after impact loading. Int. J. Impact. Eng. 35, 591–608.

Bagnoli, F., Bernabei, M., Figueroa-Gordon, D., Irving, P.E., 2009. The response of aluminum/GLARE hybrid materials to impact and to in-plane fatigue. Mater. Sci. Eng. A. 523, 118–124.

Beaumont, P.W.R., Riewald, P.G., Zweben, C., 1974. Methods for improving the impact resistance of composite materials, in foreign object impact damage to composites, ASTM STP 568. Am. Soc. Testing Mater.134–158.

Bełzowski, A., Rechul, Z., Stasieńko, J., 2002. Impact-related damage in glass roving reinforced laminate. Kompozyty (Composites). 2 (5), 394–399.

Bernhardt, S., Ramulu, M., Kobayashi, A.S., 2007. Low-velocity impact response characterization of a hybrid titanium composite laminate. J. Eng. Mater. Technol. 129, 220–226.

Bieniaś, J., Jakubczak, P., 2012. Selected durability properties of hybrid titanium carbon-epoxy composite laminate. Inżynieria Materiałowa. 4, 307–310.

Bieniaś, J., Jakubczak, P., Majerski, K., Ostapiuk, M., Surowska, B., 2013. Methods of ultrasonic testing as an effective way of estimating durability and diagnosing operational capability of composite laminates used in aerospace industry. Eksploatacja i Niezawodność (Maint. Reliab.). 15 (3), 284–289.

Bienias, J., Surowska, B., Jakubczak, P., 2015a. Influence of repeated impact on damage growth in fibre reinforced polymer composites, Eksploat. Niezawodn. Maint. Reliab. 12 (2), 194–198.

Bieniaś, J., Jakubczak, P., Surowska, B., Dragan, K., 2015b. Low-energy impact behaviour and damage characterization of carbon fibre reinforced polymer and aluminium hybrid laminates. Arch. Civil Mechan. Eng. 4 (15), 925–932.

Bienias, J., Surowska, B., Jakubczak, P., 2016a. The comparison of low velocity impact resistance of aluminum/carbon and glass fibres metal laminates. Polymer Comp. 4 (37), 1056–1063.

Bienias, J., Jakubczak, P., Dadej, K., 2016b. Low-velocity impact resistance of aluminium glass laminates—experimental and numerical investigation. Comp. Struct. 152, 339–348.

Bisle, W., Meier, T., Mueller, S., Rueckert, S., 2006. In-service inspection concept for GLARE®—an example for the use of new UT array inspection systems. ECNDT1–9.

Boeing, 1988. Advanced composite compression test, Boeing Specification Support Standard BSS 7260.

Cantwell, W.J., Morton, J., 1991. The impact resistance of composite materials—a review. Composites. 22 (5), 347–362.

Caprino, G., Spatarob, G., Del Luongo, S., 2004. Low-velocity impact behaviour of fibreglass–aluminium laminates. Composites Part A. 35, 605–616.

Caprino, G., Lopresto, V., Iaccarino, P., 2007. A simple mechanistic model to predict the macroscopic response of fibre glass–aluminum laminates under low-velocity impact. Composites Part A. 38, 290–300.

Chai, G.B., Manikandan, P., 2014. Low velocity impact response of fibre–metal laminates—a review. Comp. Struct. 107, 363–381.

Cortés, P., Cantwell, W.J., 2006. The fracture properties of a fibre–metal laminate based on magnesium alloy. Composites Part B. 37 (2–3), 163–170.

Cortes, P., Cantwell, W.J., 2007. The impact properties of high-temperature fiber–metal laminates. J. Comp. Mater. 41, 613–632.

Curtis PT, 1988. CRAG test methods for the measurements of the engineering properties of fibre reinforced plastics, Royal Aerospace Establishment Technical Report TR88012.

Dadej, K., Jakubczak, P., Bieniaś, J., Surowska, B., 2015. The influence of impactor energy and geometry on degree of damage of glass fiber reinforced polymer subjected to low-velocity impact. Comp. Theor. Pract. 2 (15), 163–167.

de Boer, T., 2001. Next generation fibre metal laminates. In: Vlot, A., Gunnink, J.W. (Eds.), Fiber Metal Laminates. Kluwer Academic Publishers, Dordrecht, pp. 39–51.

Donadon, M.V., Iannucci, L., Falzon, B.G., Hodgkinson, J.M., de Almeida, S.F.M., 2008. A progressive failure model for composite laminates subjected to low velocity impact damage. Comp. Struct. 86 (11–12), 1232–1252.

Dragan, K., Bieniaś, J., Leski, A., Czulak, A., Hufenbach, W., 2012. Inspection methods for quality control of fibre metal laminates (FML) in aerospace components. Comp. Theor. Pract. 12 (4), 272–278.

Fan, J., Guan, Z.W., Cantwell, W.J., 2011. Numerical modelling of perforation failure in fibre metal laminates subjected to low velocity impact loading. Comp. Struct. 93, 2430–2436.

Fan, X.L., Sun, Q., Kikuchi, M., 2010. Review of damage tolerant analysis of laminated composites. J. Solid Mech. 2 (3), 275–289.

Feng, D., Aymerich, F., 2013. Damage prediction in composite sandwich panels subjected to low-velocity impact. Composites Part A. 52, 12–22.

González, E.V., Maimí, P., Camanho, P.P., Lopes, C.S., Blanco, N., 2011. Effects of ply clustering in laminated composite plates under low-velocity impact loading. Comp. Sci. Technol. 71, 805–817.

Greenhalgh, E.S., 2009. Failure analysis and fractography of polymer composites. Woodhead Harsh Environment Publishing Limited, Cambridge, UK.

Guan, Z., Yang, Ch, 2002. Low-velocity impact and damage process of composite laminates. J. Comp. Mater. 36, 851–871.

Guo-Cai, Y., Lin-Zhi, W., Li, M., Jian, X., 2015. Low velocity impact of carbon fiber aluminum laminates. Comp. Struct. 119, 757–766.

Hashin, Z., 1980. Failure criteria for unidirectional composites. J. Appl. Mech. 47 (2), 329–334.

Hooputra, H., Gese, H., Dell, H., Werner, H., 2004. A comprehensive failure model for crashworthiness simulation of aluminium extrusions. Int. J. Crashworthiness. 9 (5), 449–464.

Hyla, I., Lizurek, A., 2002. Application of dynamic testing to the analysis of impact strength in laminates composites. Kompozyty (Composites). 5 (2), 374–377.

Ibarra-Castanedo, C., Avdelidis, N.P., Grinzato, E.G., Bison, P.G., Marinetti, S., Plescanu, C.C., et al., 2011. Delamination detection and impact damage assessment of GLARE by active thermography. Int. J. Mater. Product Technol. 41, 5–16.

Jakubczak, P., Bieniaś, J., Dadej, K., Zawiejski, W., 2014a. The issue of residual strength tests of thin fibre metal laminates. Comp. Theor. Pract. 14 (3), 134–138.

Jakubczak, P., Bienias, J., Majerski, K., Ostapiuk, M., Surowska, B., 2014b. The impact behavior of aluminum hybrid laminates. Aircraft Eng. Aerospace Technol. Int. J. 86 (4), 287–294.

Jakubczak, P., Surowska, B., Bieniaś, J., 2016. Evaluation of force–time changes during impact of hybrid laminates made of titanium and fibrous composite. Arch. Metall. Mater. 2A (61), 689–694.

Kim, J.H., Pierron, F., Wisnom, M.R., Avril, S., 2009. Local stiffness reduction in impacted composite plates from full-field measurements. Composites Part A. 40, 1961–1974.

Laliberté J.F., C. Poon, P.V. Straznicky, 2014. Low velocity impact damage in Glare© fibre metal laminates ICCM, http://www.iccm-central.org/Proceedings/ICCM12proceedings/site/papers/pap421.pdf.

Langdona, G.S., Lemanski, S.L., Nurick, G.N., Simmons, M.C., Cantwell, W.J., Schleyer, G.K., 2007. Behaviour of fibre–metal laminates subjected to localised blast loading: Part I—Experimental observations. Int. J. Impact. Eng. 34 (7), 1202–1222.

Lawcock, G.D., Ye, L., Mai, Y.W., Sun, C.T., 1997. Effects of fibre/matrix adhesion on carbon-fibre-reinforced metal laminates—II. Impact behavior. Comp. Sci. Technol. 57, 1621–1628.

Lee J., C. Soutis, C. Kong, 2009. Prediction of compression-after-impact (CAI) strength of CFRP laminated composites, 18th International Conference on Composite Materials, Edinburgh 27–31.

Léonard F., Y. Shi, C. Soutis, P.J. Withers, C. Pinna, 2014. Impact damage characterisation of fibre metal laminates by X-ray computed tomography, Conference on Industrial Computed Tomography (ICT).

Li, Z., Sun, B., Gu, B., 2010. FEM simulation of 3D angle–interlock woven composite under ballistic impact from unit cell approach. Comp. Mater. Sci. 49 (1), 171–183.

Liaw B.M., Y.X. Liu, E.A. Villars, 2001. Impact damage mechanisms in fiber metal laminates, Proceedings of the SEM Annual Conference on Experimental and Applied Mechanics, Portland, Oregon.

Liu, D., Malvern, L.E., 1987. Matrix cracking in impacted glass/epoxy plates. J. Comp. Mater. 21, 594–609.

Liu, D., Raju, B.B., Dang, X., 2000. Impact perforation resistance of laminated and assembled composite plates. Int. J. Impact. Eng. 24, 733–746.

Liu, Y., Liaw, B., 2010. Effects of constituents and lay-up configuration on drop-weight tests of fiber–metal laminates. Appl. Comp. Mater. 17, 43–62.

Liu Y.X., B.M. Liaw, 2004. Drop-weight impact on fiber–metal laminates using various indenters, Proceedings of the SEM X International Congress and Exposition on Experimental and Applied Mechanics, Costa Mesa, CA.

Manikandan, P., Chai, G.B., 2014. A layer-wise behavioral study of metal based interply hybrid composites under low velocity impact load. Comp. Struct. 117, 17–31.

Molitor, P., Barron, V., Young, T., 2001. Surface treatment of titanium for adhesive bonding to polymer composites: a review. Int. J. Adhes. Adhesiv. 21 (2), 129–136.

Morinière, F.D., Alderliesten, R.C., Yarmohammad Tooski, M., Benedictus, R., 2012. Damage evolution in GLARE fibre–metal laminate under repeated low velocity impact tests. Central Eur. J. Eng. 2 (4), 603–611.

Moriniere, F.D., Alderliesten, R.C., Benedictus, R., 2013. Low-velocity impact energy partition in GLARE. Mech. Mater. 66, 59–68.

Morinière, F.D., Alderliesten, R.C., Benedictus, R., 2014. Modelling of impact damage and dynamics in fibre–metal laminates—a review. Int. J. Impact. Eng. 67, 27–38.

Múgica, J.I., Aretxabaleta, L., Ulacia, I., Aurrekoetxea, J., 2014. Impact characterization of thermoformable fibre metal laminates of 2024-T3 aluminium and AZ31B-H24 magnesium based on self-reinforced polypropylene. Composites Part A. 61, 67–75.

Nakatani, H., Kosaka, T., Osaka, K., Sawada, Y., 2011. Damage characterization of titanium/GFRP hybrid laminates subjected to low-velocity impact. Composites Part A. 42, 772–781.

NASA, 1983. Standard test for toughened resin composites. NASA Reference Publication.

Orifici, A.C., Herszberg, I., Thomson, R.S., 2008. Review of methodologies for composite material modelling incorporating failure. Comp. Struct. 86 (1–3), 194–210.

Ostapiuk, M., Surowska, B., Bieniaś, J., 2012. Interface analysis of fiber metal laminates. Comp. Interface. 4 (21), 309–318.

Padaki, N.V., Alagirusamy, R., Deopura, B.L., 2008. Low velocity impact behavior of textile reinforced composites. Review article. Indian J. Fibre Textile Res. 33, 189–202.

Pärnänen, T., Alderliesten, R., Rans, C., Brander, T., Saarela, O., 2008. Applicability of AZ31B-H24 magnesium in fibre metal laminates—an experimental impact research. Comp. Struct. 83, 247–258.

Periasamy, M., Manickam, B., Hariharasubramanian, K., 2012. Impact properties of aluminium—glass fiber reinforced plastics sandwich panels. Mater. Res. 15 (3), 347–354.

Prichard, J.C., Hogg, P.J., 2000. The role of impact damage in post impact compression testing. Composites. 21 (6), 503–511.

Puck, A., Schürmann, H., 1998. Failure analysis of FRP laminates by means of physically based phenomenological models. Comp. Sci. Technol. 58 (7), 1045–1067.

Rajkumar, G.R., Krishna, M., Narasimha Murthy, H.N., Sharma, S.C., Vishnu Mahesh, K.R., 2012a. Experimental investigation of low-velocity repeated impacts on glass fiber metal composites. JMEPEG. 21, 1485–1490.

Rajkumar, G.R., Krishna, M., Narasimha Murthy, H.N., Sharma, S.C., Vishnu Mahesh, K.R., 2012b. Investigation of repeated low velocity impact behaviour of GFRP/aluminium and CFRP/aluminium laminates. Int. J. Soft Comput. Eng. 1 (6), 50–58.

Richardson, M.O.W., Wisheart, M.J., 1996. Review of low-velocity impact properties of composite materials. Composites Part A. 27, 1123–1131.

SACMA, 1994. Recommended Methods SRM 2 94, Suppliers of Advanced Composites Materials Association.

Sadighi, M., Alderliesten, R.C., Benedictus, R., 2012a. Impact resistance of fiber–metal laminates: a review. Int. J. Impact. Eng. 49, 77–90.

Sadighi, M., Parnanen, T., Alderliesten, R.C., Sayeaftabi, M., Benedictus, R., 2012b. Experimental and numerical investigation of metal type and thickness effects on the impact resistance of fiber–metal laminates. Appl. Comp. Mater. 19 (3), 545–559.

Sanchez-Saez, S., Barbero, E., Navarro, C., 2008. Compressive residual strength at low temperatures of composite laminates subjected to low-velocity impacts. Comp. Struct. 85, 226–232.

Sarlin, E., Apostol, M., Lindroos, M., Kuokkala, V.-T., Vuorinen, J., Lepistö, T., et al., 2014. Impact properties of novel corrosion resistant hybrid structures. Comp. Struct. 108, 886–893.

Seltzer, R., Gonzalez, C., Munoz, R., Llorca, J., Blanco-Varela, T., 2013. X-ray microtomography analysis of the damage micromechanisms in 3D woven composites under low-velocity impact. Composites Part A. 45, 49–60.

Seyed Yaghoubi, A., Liu, Y.X., Liaw, B.M., 2011. Drop-weight impact studies of GLARE 5 fiber–metal laminates. Exper. Appl. Mech. 6, 267–279.

Seyed Yaghoubi, A., Liu, Y., Liaw, B., 2012. Stacking sequence and geometrical effects on low-velocity impact behaviors of Glare5 (3/2) fiber–metal laminates. J. Thermoplastic Comp. Mater. 25, 223–247.

Shi, Y., Swait, T., Soutis, C., 2012. Modelling damage evolution in composite laminates subjected to low velocity impact. Comp. Struct. 94 (9), 2902–2913.

Shivakumar, K.N., Elber, W., Illg, W., 1985. Prediction of low velocity impact damage in thin circular laminates. AIAA J. 23 (3), 442–449.

Shyr, T.W., Pan, Y.H., 2003. Impact resistance and damage characteristics of composite laminates. Comp. Struct. 62, 193–203.

Sinke, J., 2003. Some inspection methods for quality control and in-service inspection of GLARE. Appl. Comp. Mater. 10, 277–291.

Sitnikova, E., Guan, Z.W., Schleyer, G.K., Cantwell, W.J., 2014. Modelling of perforation failure in fibre metal laminates subjected to high impulsive blast loading. Int. J. Solid. Struct. 51 (18), 3135–3146.

Sjoblom, P.O., Hartness, J.T., Cordell, T.M., 1988. On low-velocity impact testing of composite materials. J. Comp. Mater. 22, 30–52.

Sohn, M.S., Hu, X.Z., Kimb, J.K., Walker, L., 2000. Impact damage characterization of carbon fibre/epoxy composites with multi-layer reinforcement. Composites Part B. 31, 681–691.

Song, S.H., Byun, Y.S., Ku, T.W., Song, W.J., Kim, J., Kang, B.S., 2010. Experimental and numerical investigation on impact performance of carbon reinforced aluminum laminates. J. Mater. Sci. Technol. 26 (4), 327–332.

Surowska, B., 2009. Metody przygotowania powierzchni stopu tytanu w laminatach metal-kompozyt włóknisty. Inżynieria Materiałowa. 5, 404–407.

Tarpani, J.R., Maluf, O., Gatti, C.A., 2009. Charpy impact toughness of conventional and advanced composite laminates for aircraft construction. Mater. Res. 12, 395–403.

Varatharajoo, R., Abdullah, E.J., Majid, D.L., Romli, F.I., Mohd Rafie, A.S., Ahmad, K.A., 2012. The behaviour of fibre–metal laminates under high velocity impact loading with different stacking sequences of Al alloy. Appl. Mech. Mater. 225, 213–218.

Vlot, A., 1993. Impact properties of fibre metal laminates. Comp. Eng. 3 (10), 911–927.

Vlot, A., 1996. Impact loading on fiber–metal laminates. Int. J. Impact. Eng. 18, 291–307.

Vlot, A., Gunnink, J.W., 2001. Fiber Metal Laminates: An introduction. Kluwer Academic Publishers, Dordrecht.

Vlot A., M. Krull, 1997. Impact damage resistance of various fiber metal laminates, J. Phys. IV France 7, Paris, France.

Vo, T.P., Guan, Z.W., Cantwell, W.J., Schleyer, G.K., 2012. Low-impulse blast behaviour of fibre–metal laminates. Comp. Struct. 94 (3), 954–965.

Vogelesang, L.B., Vlot, A., 2000. Development of fibre metal laminates for advanced aerospace structures. J. Mater. Process. Technol. 103, 1–5.

Wu, G., Yang, J.M., 2005. The mechanical behavior of glare laminates for aircraft structures. JOM. 75, 72–79.

Wu, G., Yang, J.M., Hahn, H.T., 2007. The impact properties and damage tolerance and of bi-directionally reinforced fibre metal laminates. J. Mater. Sci. 42, 948–957.

Zhou, G., Davies, G.A.O., 1995. Impact response of thick glass fibre reinforced polyester laminates. Int. J. Impact. Eng. 16 (3), 357–374.

Zhu, S., Chai, G.B., 2012. Low-velocity impact response of fibre–metal laminates—experimental and finite element analysis. Comp. Sci. Technol. 72, 1793–1802.

Recent progress and perspectives on biofunctionalized CNT hybrid polymer nanocomposites

13

Shadpour Mallakpour[1,2,3] *and Vajiheh Behranvand*[1]
[1]Organic Polymer Chemistry Research Laboratory, Department of Chemistry, Isfahan University of Technology, Islamic Republic of Iran, [2]Nanotechnology and Advanced Materials Institute, Isfahan University of Technology, Islamic Republic of Iran, [3]Center of Excellence in Sensors and Green Chemistry, Department of Chemistry, Isfahan University of Technology, Islamic Republic of Iran

Chapter Outline

13.1 Introduction to carbon nanotubes

One of the important allotropes of carbon which were first discovered by Iijima in 1991 are carbon nanotubes (CNTs) (Souza et al., 2015). CNTs are categorized as single-walled (SWCNTs), double-walled (DWCNTs), or multiwalled (MWCNTs) (Liu et al., 2014). CNTs exhibit extraordinary properties, such as high surface-to-volume ratio, high chemical stability, exceptional mechanical property, and excellent electrical and thermal conductivity which facilitate their applications in catalysts energy conversion and storage, drug and gene delivery, fuel cells, field emission displays, probes, sensors, and so on (Mallakpour and Behranvand, 2015d; Liu et al., 2014). Due to the strong attractive van der Waals attraction between tubes, CNTs tend to aggregate and form bundles thus do not disperse well in the organic matrixes (Mallakpour and

Hybrid Polymer Composite Materials: Properties and Characterisation. DOI: http://dx.doi.org/10.1016/B978-0-08-100787-7.00013-5

Soltanian, 2014b; Mallakpour and Zadehnazari, 2014b). Surface modification not only is necessary to solubilize CNTs but also can provide low toxicity and biocompatibility of CNTs for medical applications (Mallakpour and Behranvand, 2015c). This chapter highlights recent development and our work in functionalization of CNTs with biomolecules, leading to biocompatible CNTs, and their applications in various fields as well as polymer nanocomposites (NCs) preparation.

13.2 Strategies of CNT functionalization

There are two different approaches to disperse CNTs: noncovalent and covalent treatments (Mallakpour and Soltanian, 2014a). Noncovalent functionalization keeps the original nanotube properties intact. In this method, CNTs interact with modifier through various adsorption forces, such as hydrogen bonds, van der Waals and electrostatic force, and p-stacking (Meng et al., 2009). With respect to noncovalent modification, in covalent functionalization, functional −COOH or OH groups can be formed on CNT surface which a broad range of functional groups could be merged onto the surface of CNTs by more adaptable modification approaches (Mansur et al., 2012; Gao et al., 2011).

Conjugating CNTs with proteins, DNA, or carbohydrates is the general biofunctionalization method which creates a new class of bioactive CNTs for the applications such as drug carriers, biosensors, and scaffold materials. Actually, biomolecules provide a biocompatible interface which opens interesting views in biomedicine (Meng et al., 2009).

13.3 CNT embedded polymer NCs

A class of high-performance novel materials, which is receiving significant consideration both in industry and in academia is polymer NCs (Mallakpour and Behranvand, 2014; Choudhary and Gupta, 2011). Because nano-fillers are about 10,000 times finer than a human hair in dimension, ultra-large interfacial area per volume will be created between the polymer matrix and nano-element (Choudhary and Gupta, 2011). Because of special structures of CNTs composed of sp^2-hybridized carbon networks, electrical and tensile strength, these nano-materials are considered promising and ideal filler phases to enhance the electrical, thermal, and mechanical features of polymers (Gatti et al., 2016). To maximize the benefit of CNTs as effective reinforcements in composites, they should be well dispersed in the polymer matrix. In situ polymerization, optimum physical blending, and chemical functionalization are several techniques for dispersion improvement of CNTs in polymer matrixes (Xie et al., 2005).

13.4 Functionalization of CNTs with biomolecules and their applications

In recent times, several studies have been extensively reviewed kinds of CNT functionalization with various molecules (Meng et al., 2009; Saifuddin et al., 2013;

Gao et al., 2012). Our focus here is to update and examine publications about surface functionalization of CNTs with biomaterials, respectively, and their applications. Most of papers which will be studied here were mainly taken from the publications of 2013 and later; however, for a basic description of the structural principles, older publications have also been cited.

13.4.1 Proteins, amino acids, and enzyme-functionalized CNTs

13.4.1.1 Proteins

One of the approaches to immobilize antibodies on CNT surfaces is the use of bacterial proteins like protein A and protein G. These proteins have affinity for the constant fraction "Fc" of immunoglobulins guaranteeing full antigen-binding capacity. Hence, they can be applied as cross-linking elements between the CNT and the immunoglobulins (Villamizar et al., 2009). Villamizar et al. (2009) examined the characteristics of the immobilization process of immunoglobulin G against *Aspergillus* spp. First protein G was adsorbed on top of an SWCNT network then the antibody was adsorbed on it. Actually, a prototype for an ion-sensitive field-effect transistor (FET) based on CNT networks was presented by this group. Huge surface area is the advantage of CNT-based FETs which make them extremely sensitive even to weak adsorption processes.

Carbon-based materials are extensively used in water and gas distillation as well as food drug production and processing. Activated carbon has microscopic pores which are often blocked during adsorption, while CNTs' open structure proposes easy, undisrupted access to reactive sites located on nanotubes' outer surface (Wang et al., 2014). Smith et al. (2014) studied the protein absorption capacity and the mechanisms of adsorption of three carbon-based nano-materials: graphene (G), graphene oxide (GO), and SWCNT in different water chemistries, using lysozyme as a model protein. The results exhibited that GO exhibited the highest adsorption capacity for lysozyme. Results also showed that the adsorption mechanism for GO was mainly electrostatic, while for G and SWCNT was ascribed to the van der Waals forces and some electrostatic interactions. Papper et al. (2014) synthesized the enzyme electrode from glucose oxidase immobilized onto the MWCNT-poly (pyrrole-Con A) protein coating. The resulting electrode was used for the amperometric detection of glucose demonstrating a high sensitivity of 36 mA cm^{-2} mol^{-1} L and a maximum current density of 350 μA cm^{-2}. González et al. (2015) attached perfluorophenyl protein on the nitrogen-doped MWCNTs (CNx-MWCNTs) to generate an efficient and homogeneous coating of Ag-nanoparticles (NPs) on the CNT surface. This protein could act as a reducing agent for $AgNO_3$. High-resolution transmission electron microscopy (HRTEM) images were shown in Fig. 13.1. It is possible to see that in both systems (pristine CNx-MWCNTs or acid-treated CNx-MWCNTs), Ag-NPs with spherical-like shape were anchored on the nanotubes surfaces.

Because of the excellent distribution and small diameter of Ag-NPs of the synthesized systems, they analyzed the CS_2 sensing capabilities of the Ag–CNx-MWCNTs

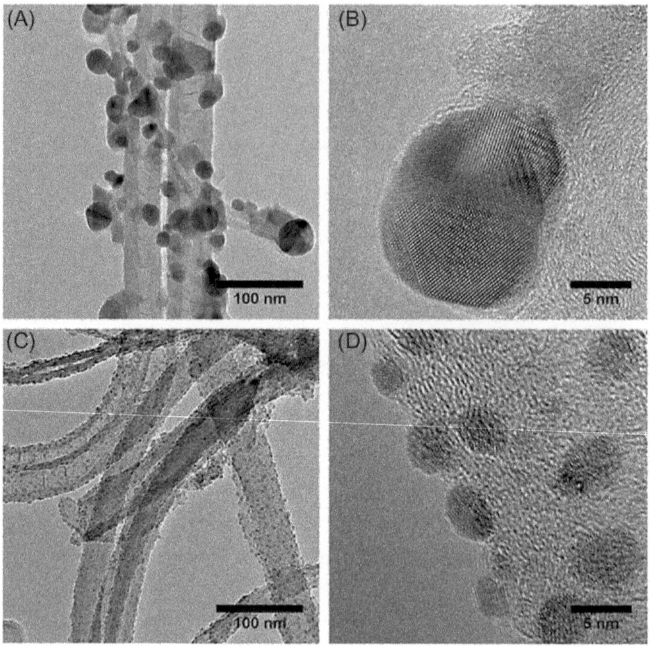

Figure 13.1 TEM images of Ag-NPs deposited on biotinized CNx-MWCNTs. (A and B) correspond to biotinized CNx-MWCNTs. (C and D) correspond to acid-treated CNx-MWCNTs covered with biotin. High-resolution micrograph in (D) reveals that the Ag-NPs are crystallographically ordered, as suggested by X-ray diffraction.
Source: Adapted from González et al. (2015). With kind permission of Elsevier.

hybrid composites. Their results indicated that, in the Ag-NPs biotinized CNx-MWCNTs composite, the biotin molecules not only have passivated the Ag-NPs, improving the homogeneity and decreasing the particle size, but also have covered most of its surface and thus makes difficulty in the interaction between the Ag-NPs and CS_2 gas which causing a considerably reduction in the sensor sensitivity. Sekar et al. (2016) examined the interaction of bovine albumin fraction-V (BSA) with pristine CNTs and their toxicity against *Donax faba*. From the observation, they achieved that the cellular integrity of the whole body tissues treated with pristine SWCNTs has exhibited higher loss in the cellular integrity on comparison with the BSA-conjugated CNTs. These results can show that the adsorption of biomolecules such as the proteins over CNTs surface could have possible influence on decreasing their toxicity profile in nature.

13.4.1.2 Amino acids

Covalent surface functionalization of MWCNTs with different natural amino acids (L-leucine (Leu), L-isoleucine (Ile), *S*-valine (Val), L-alanine (Ala), *S*-methionine

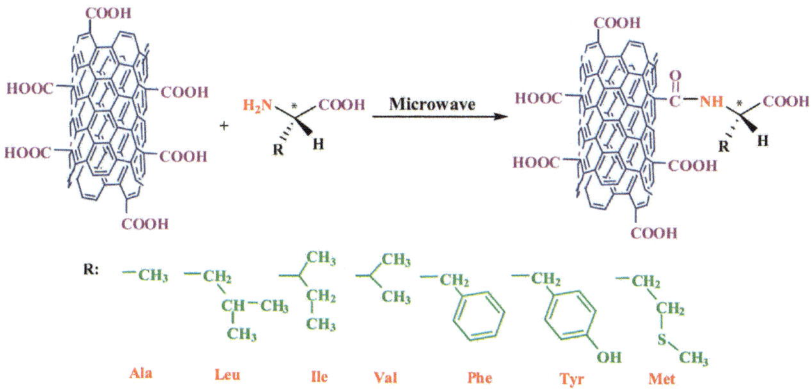

Figure 13.2 The condensation reaction between carboxyl groups on MWCNT and amino groups of amino acid.
Source: Adapted from Mallakpour, S., Zadehnazari, A., 2014a. A facile, efficient, and rapid covalent functionalization of multi-walled carbon nanotubes with natural amino acids under microwave irradiation. Prog. Organic Coat. 77, 679–684. With kind permission of Elsevier.

(Met), L-phenylalanine (Phe), and L-tyrosine (Tyr)) was successfully accomplished under microwave irradiation by Mallakpour and Zadehnazari (2014a) (Fig. 13.2). TEM micrographs of functionalized MWCNT with *S*-valine (MWCNT-Val) showed the formation of helix structures in CNTs which confirmed the functionalization. These structures could be due to the hydrogen bond interaction between the residues of the amino acids and MWCNTs (Fig. 13.3).

Histidine which is an essential amino acid for growth and repair of tissues was selected by Mathavan (2014) for the surface functionalization of MWCNTs due to its high reactivity and wealth of chemistry. They proposed sonication process as relatively simple and effective way for the functionalization. Rahmani and Ketabi (2015) investigated the effect of adsorption of two amino acids (β-alanine and histidine) on the solvation of armchair SWCNTs. According to Monte Carlo simulations, it was found that the binding energy of the interaction of β-alanine with nanotube was larger than histidine. The results of computer simulation in aqueous solution represented that amino acid functionalization increases the intermolecular interactions of CNTs and water. Deborah et al. (2015) prepared and characterized valine-functionalized MWCNTs under ultrasonic irradiations as it was shown in Fig. 13.4. They chose this kind of functionalization because functionalization of CNTs with amino acids makes them soluble and biocompatible. Therefore, they have potential applications in the field of biosensors and targeted drug delivery.

In 2016, Eguílaz et al. investigated the functionalization of MWCNTs with cytochrome c (Cyt c) noncovalently, the direct electron transfer after drop-coating deposition of MWCNTs-Cyt c distribution on glassy carbon electrodes, and the analytical applications for the highly sensitive quantification of hydrogen peroxide. The obtained biosensor showed to be a highly competitive approach to quantify hydrogen peroxide, with excellent performance in different samples.

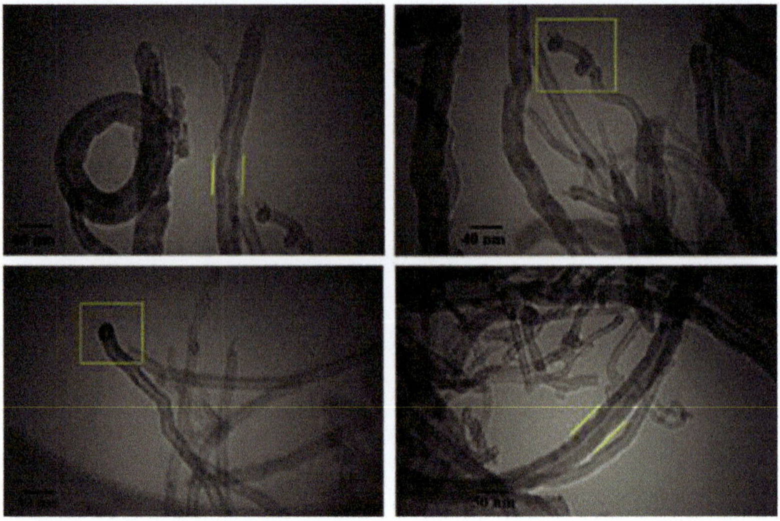

Figure 13.3 TEM images of MWCNT-Val.
Source: Adapted from Mallakpour, S., Zadehnazari, A., 2014a. A facile, efficient, and rapid covalent functionalization of multi-walled carbon nanotubes with natural amino acids under microwave irradiation. Prog. Organic Coat. 77, 679–684. With kind permission of Elsevier.

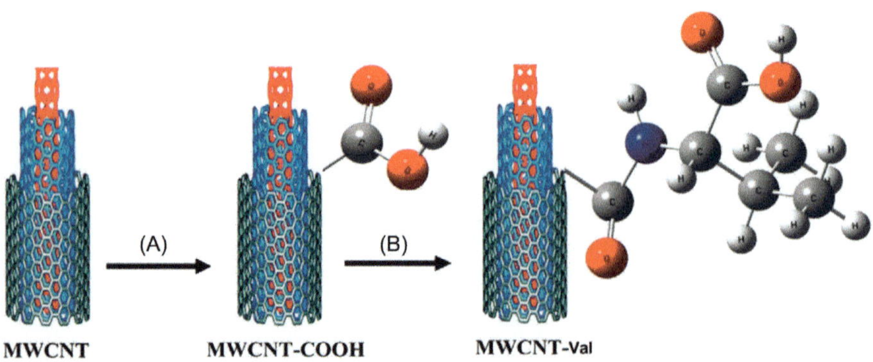

Figure 13.4 Scheme for the synthesis of valine-functionalized MWCNTs. (A) Pristine MWCNTs was mixed with (H_2SO_4/HNO_3 (3:1), sonicated for 3 h and the mixture was added to cold distilled water. The samples were filtered and dried in vacuum at 80°C for 4 h. (B) The oxidized MWCNTs was mixed with valine suspension, sonicated for 1 h at room temperature. The samples were filtered and dried in vacuum at room temperature for 16 h.
Source: Adapted from Benial et al. (2015). With kind permission of Elsevier.

13.4.1.3 Enzymes

Enzyme immobilization is a promising biotechnological application of CNTs especially for fabrication of biosensors and biofuel cells. The conjugation of enzymes with CNTs causes increasing enzyme activity and stability, even when subjected to strongly denaturing environments. The performance of CNT–enzyme complexes is influenced by a combination of the nanotube chemistry and immobilization method. Enzymes can be immobilized over CNTs through covalent and noncovalent approaches which among them the noncovalent route is normally preferred since it is done without chemical additives and conserves the native conformation of the enzyme (Tavares et al., 2015; Feng and Ji, 2011).

Cellulases, enzymes that destroy cellulose, are produced by some bacteria, fungi, and protozoa (Hames and Hooper, 2005). This enzyme was immobilized onto functionalized MWCNTs via physical adsorption method by Mubarak et al. (2014) to yield a stable and ease of separate enzyme. Fourier transform infrared spectroscopy and field emission scanning electron microscopy (FE-SEM) were used to approve the successful immobilization of cellulase enzyme. One of the advantages for enzyme immobilization is that it makes the enzyme possible to be reused because it can be easily recovered or separated from the reaction medium. From Fig. 13.5, it was discovered that as the recycle number of reaction increases,

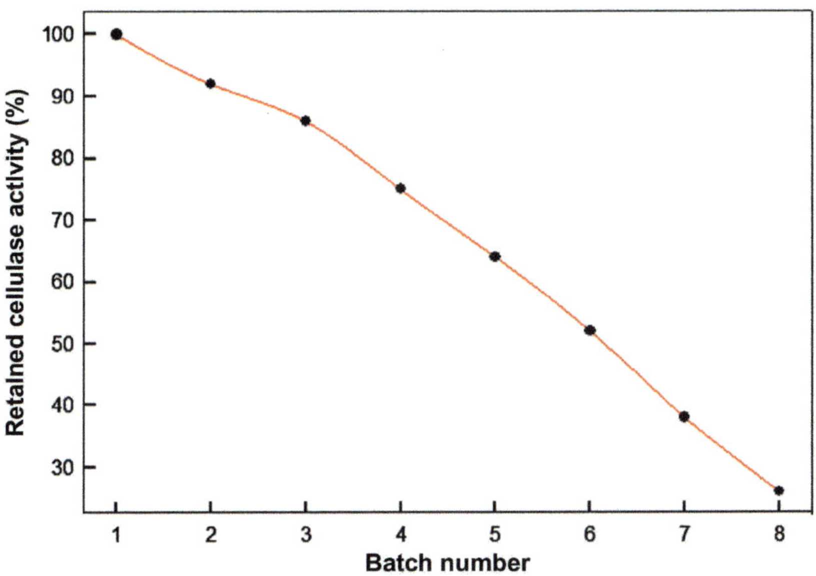

Figure 13.5 Retained cellulase activity at different cycle of reactions.
Source: Adapted from Mubarak, N.M., Wong, J.R., Tan, K.W., Sahu, J.N., Abdullah, E.C., Jayakumar, N.S., et al., 2014. Enzymatic immobilization of cellulase enzyme on functionalized multiwall carbon nanotubes. J. Mol. Catal. B 107, 124–131. With kind permission of Elsevier.

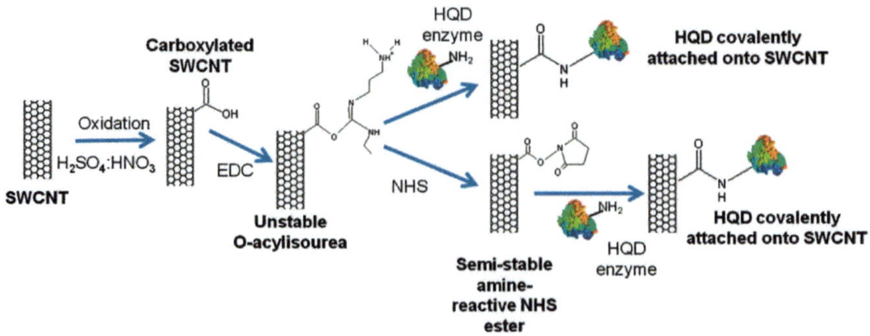

Figure 13.6 Schematic of covalent immobilization of enzyme on SWCNTs functionalized with 1-ethyl-3-(3-dimethylaminopropyl) carbodiimide (EDC) and NHS.
Source: Adapted from Suma et al. (2015). With kind permission of Springer.

there is an increase of the loss of residual activity of the immobilized cellulase enzyme. After six cycles, still 52% of cellulase activity of MWCNT–cellulase composite was retained. This property is advantageous to the industrial applications because of its potential to be easily detached from the end product at the end of the reaction, reuse for multiple times, and permit the improvement of multiple enzyme reaction system.

The degradation of different aromatic hydrocarbon intermediates was studied by Suma et al. (2015) using a recombinant hydroxyquinol 1,2-dioxygenase (CphA-I) enzyme immobilized on SWCNTs. The immobilization was done by physical adsorption and covalent coupling in the absence and presence of *N*-hydroxysuccinimide (NHS). The overall procedure for covalent immobilization of the enzyme on the SWCNTs was shown in Fig. 13.6.

FE-SEM images of the SWCNTs were obtained before and after enzyme immobilization. Unmodified SWCNTs represented a clear weblike network (Fig. 13.7A), while nonuniform structures with aggregates were evident in the images of the enzyme-immobilized SWCNTs by physical adsorption (Fig. 13.7B). It was found from energy-filtering transmission electron microscopy (EF-TEM) that untangling of the SWCNT bundles was not efficiently induced by the enzymes; however, the enzymes interacted strongly with the outer surface of the SWCNT bundles with the creation of networks and resulting precipitation of the assemblies from the dispersion.

Kinetic analysis showed that the immobilized enzyme was almost as effective as the free enzyme for the oxidative decomposition of various aromatic hydrocarbons assessed, and conformational change due to the immobilization procedure was minimal. So they expressed that the immobilized enzymes and immobilization strategy have great potential for treatment of wastewater contaminated with aromatic compounds. Tavares et al. (2015) immobilized MWCNTs by laccase noncovalently. They expressed that the main driving forces involved in the noncovalent immobilization of laccase over MWCNTs were hydrophobic interactions and $\pi-\pi$ stacking.

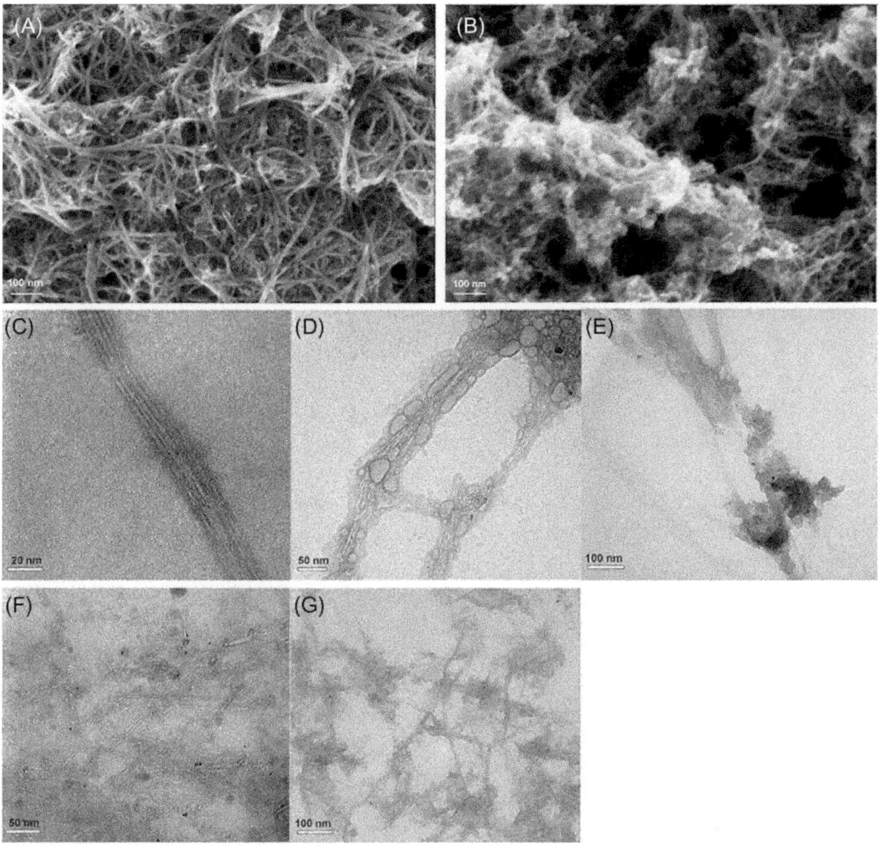

Figure 13.7 FE-SEM images of (A) unmodified SWCNTs and (B) enzymes immobilized on SWCNTs by physical adsorption; EF-TEM images of (C) unmodified SWCNTs and (D and E) enzyme immobilized on SWCNTs by physical adsorption and (F and G) by covalent coupling.
Source: Adapted from Suma et al. (2015). With kind permission of Springer.

TEM analysis showed the presence of globular structured laccase spread over the sidewalls of the CNTs (Fig. 13.8).

Immobilized laccase showed an increased activity in comparison to the free enzyme at a moderate temperature of 50°C. In addition, operational stability of laccase immobilized on MWCNTs was observed over nine cycles, which represented the high capacity of reutilization of this biocatalyst. Mohamad et al. (2015) reported using of carboxylated MWCNTs as a supportive material for the immobilization of *Candida rugosa* lipase (CRL) through physical adsorption procedure to prepare biocatalyst. The resulting CRL-MWCNTs biocatalysts were applied for synthesizing geranyl propionate, an important ester for flavoring agent as well as in fragrances. According to the obtained results, it was concluded that the facile physical adsorption of CRL onto MWCNTs improved the activity and stability of

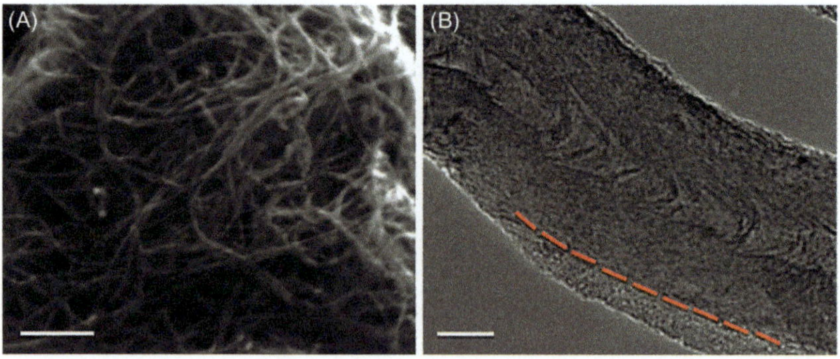

Figure 13.8 SEM (A) and TEM (B) micrographs of laccase immobilized over MWCNTs. The arrows indicate the presence of enzyme over the MWCNTs' surface; dashed lines represent the limit (boundary) between the MWCNTs and the enzyme.
Source: Adapted from Tavares, A.P.M., Silva, C.G., Drazic, G., Silva, A.M.T., Loureiro, J.M., Faria, J.L., 2015. Laccase immobilization over multi-walled carbon nanotubes: kinetic, thermodynamic and stability studies. J. Colloid Interface Sci. 454, 52–60. With kind permission of Elsevier.

CRL as well as served as an alternative method for the synthesis of geranyl propionate. Two approaches for immobilization of lipase on oxidized MWCNTs (o-MWCNTs) were explained by Prlainovic et al. (2016). In addition to immobilization under covalent promoting conditions, a set of experiments without activating agents was also accomplished to examine the influence of functionalization of MWCNTs on the surface hydrophobicity and the amount of nonspecific interactions between lipase and o-MWCNTs. According to the results, the higher activity of the biocatalysts gained under covalent promoting conditions is very helpful for their potential usage on the industrial scale, as well as from the viewpoint of the possibilities of reuse.

13.4.2 Carbohydrate-functionalized CNTs

The biofunctionalization of CNTs using soft matters such as carbohydrates has been the focus which makes them possible to produce a new category of bioactive CNTs for the applications in biosensing and drug delivery (Mallakpour and Behranvand, 2015c). Yang et al. (2015) designed novel chitosan-grafted CNTs (CTS-g-CNTs) as radioactive cesium (Cs^+) remover. The schematic of the formation of CTS-g-CNTs and the research adsorption approach was shown in Fig. 13.9. Chitosan is a biodegradable and biocompatible polymer with a large number of $-OH$ functional groups and amino ($-NH_2$) groups, which easily interacted with CNT (Mallakpour and Madani, 2015b).

The smooth surfaces of the raw-CNTs were observed in SEM image of it. Obvious differences were detected after functionalization with CTS (Fig. 13.10).

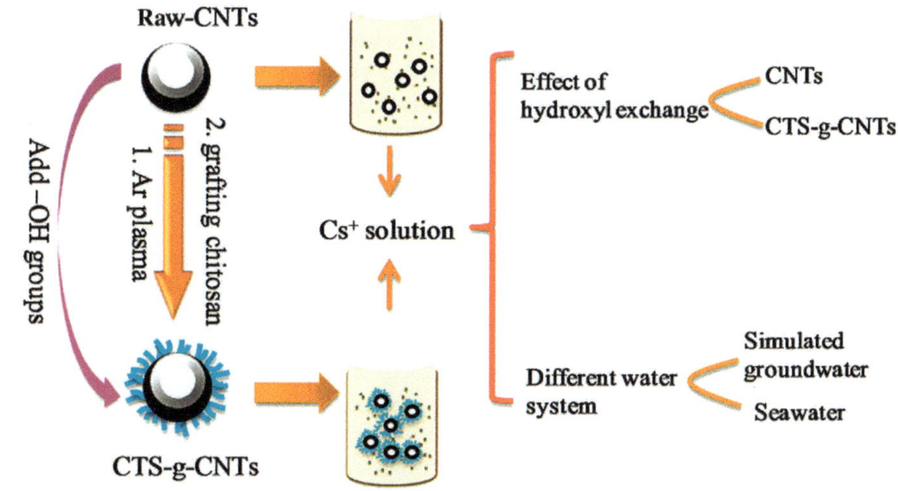

Figure 13.9 Schematic illustration of the designed research approach.
Source: Adapted from Yang et al. (2015). Open access journal.

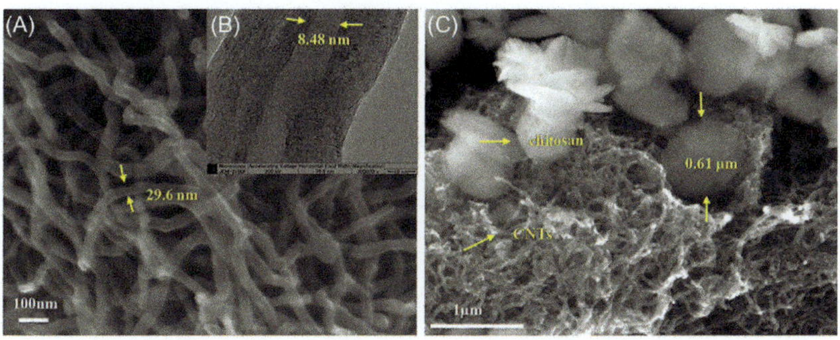

Figure 13.10 SEM (A) and TEM (B) images of raw-CNTs; and SEM image of CTS-g-CNTs (C).
Source: Adapted from Yang et al. (2015). Open access journal.

They examined the effect of –OH functional groups in the Cs^+ adsorption process and compared the adsorption properties of raw-CNTs with CTS-g-CNTs. The results showed direct observational evidence on the influence of –OH functional groups for Cs^+ adsorption. A biosensor based on anti-apolipoprotein B-functionalized CNT-chitosan NC has been designated by Ali et al. (2014) to detect low-density lipoprotein (LDL) molecules. The $-NH_2$ groups present in chitosan were used to create the covalent bond formation with –COOH groups of CNTs. Also, the electrostatic interaction may also be possible through opposite charge on CH (positive) and CNTs (negative). The obtained immunosensor was found to be selective for LDL and exhibited a wide detection range and it was an interesting platform that can be applied for the detection of other lipids such as very LDL and triglyceride. In other work, the production of galactosylated chitosan-grafted oxidized CNTs (O-CNTs-LCH) was reported and their potential to promote hepatic targeting and the tumor therapeutic efficiency of doxorubicin (DOX) was investigated (Qi et al., 2015). Fluorescent images of the tumors in vivo were presented in Fig. 13.11. There was an obvious position of O-CNTs-LCH-DOX in the tumor

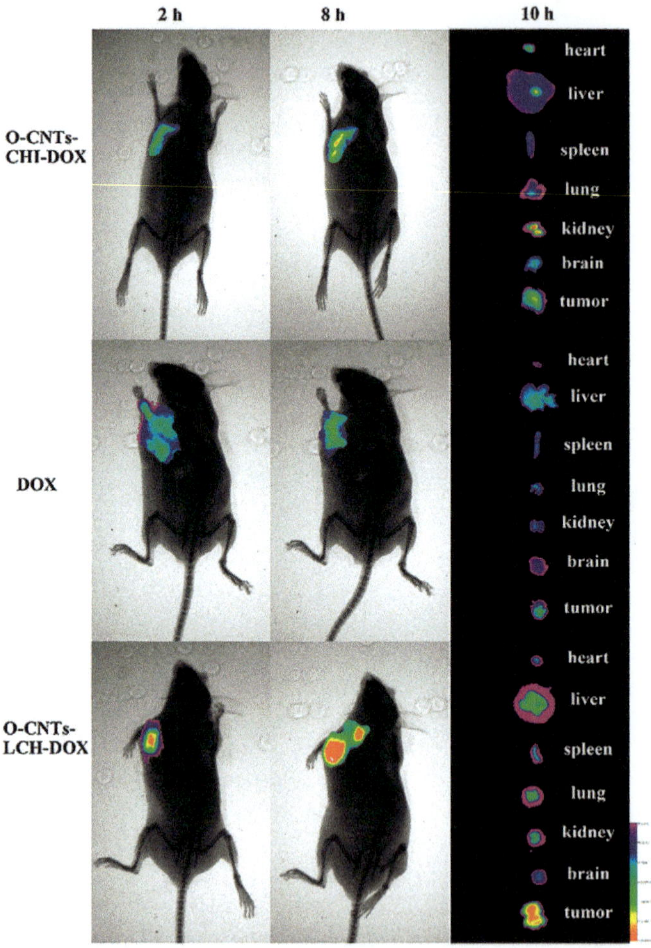

Figure 13.11 The in vivo imaging of H22 tumor-bearing mice after intravenous injection of DOX (control), O-CNTs-CHI-DOX and O-CNTs-LCH-DOX. The images were taken at 2 and 8 h after administration. The tumor-bearing mice were sacrificed 10 h after in vivo injection and representative ex vivo fluorescence images of the major organs (heart, liver, spleen, lung, kidney, and brain) and tumor uptake of DOX ($n = 5$) are shown.
Source: Adapted from Qi, X., Rui, Y., Fan, Y., Chen, H., Ma, N., Wu, Z., 2015. Galactosylated chitosan-grafted multiwall carbon nanotubes for pH-dependent sustained release and hepatic tumor-targeted delivery of doxorubicin in vivo. Colloid. Surface. B, 133, 314–322. With kind permission of Elsevier.

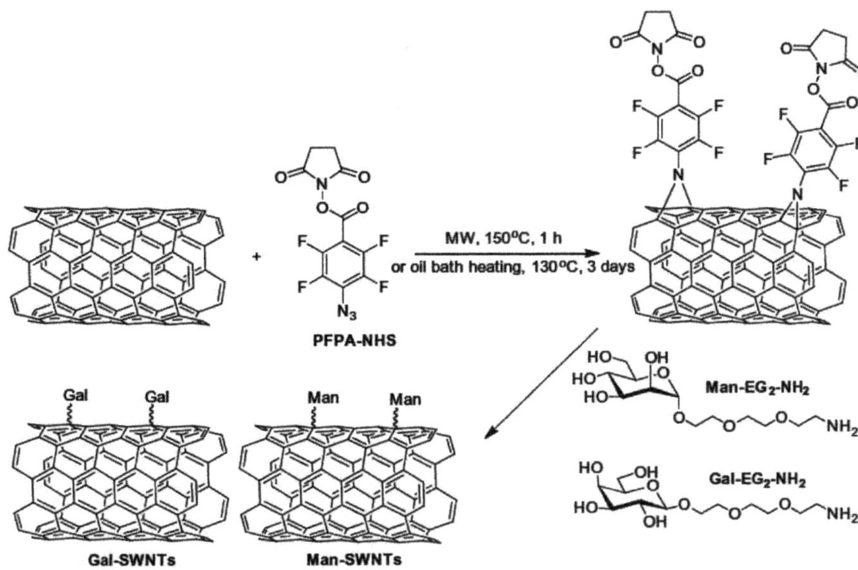

Figure 13.12 Functionalization of SWCNTs using PFPA-NHS and subsequent conjugation with carbohydrates.
Source: Adapted from Kong et al. (2015). With kind permission of Elsevier.

within 2 hours, and an enhanced position was seen 8 hours after injection. In fact, after intravenous administration in mice bearing the H22 tumor, O-CNTs-LCH-DOX showed higher antitumor activity and stronger fluorescent intensity in tumor tissue compared to free DOX.

Kong et al. (2015) used perfluorophenyl azides (PFPA) to conjugate the carbohydrate to SWCNTs through microwave irradiations to produce carbohydrate–SWCNT conjugates for biorecognition. The SWCNTs were functionalized and subsequently conjugated with two carbohydrate derivatives, Man-EG_2-NH_2 and Gal-EG_2-NH as shown in Fig. 13.12. The carbohydrate-functionalized SWCNTs were moreover shown to interact specifically with lectins, resulting in predicted recognition patterns. The method reported here can be readily selected to conjugate a variety of biomaterials easing the use of CNT materials in bioanalytical and biomedical applications.

In our group, utilizing covalent methods or developing new methods to obtain different functionalized CNTs has been well studied. A simple and fast method was developed for the covalent functionalization of MWCNTs with D-sucrose carbohydrate (MWCNT-Suc) (Mallakpour and Behranvand, 2016) (Fig. 13.13). The process is occurred in water under ultrasonic irradiations as safe media and low-cost tool.

The global conformation which could be seen in TEM images of functionalized MWCNTs with sucrose in some areas was due to the presence of many OH groups in the sucrose structure which cause more affinity to hydrogen bonding formation between them with CNTs (Fig. 13.14A, B). We applied this hybrid as a filler for

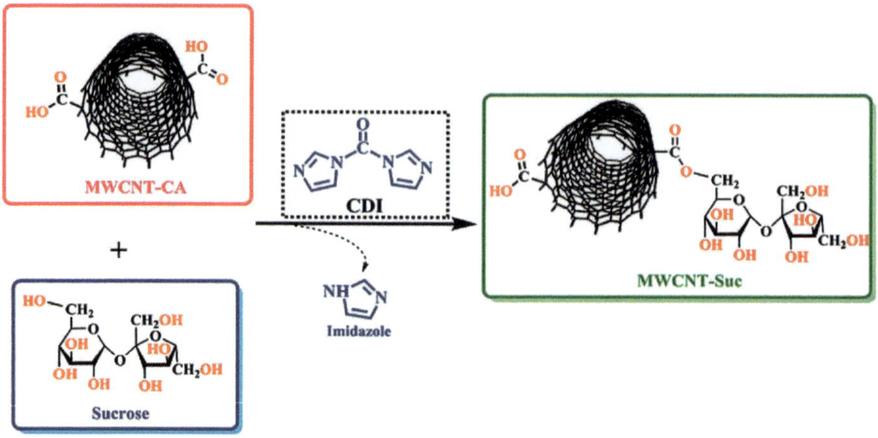

Figure 13.13 Reaction scheme of sucrose-grafted MWCNTs.
Source: Adapted from Mallakpour, S., Behranvand, V., 2016. Chemical adsorption of D-sucrose on MWCNTs for compatibility improvement with alanine-based poly (amide–imide) matrix: morphology examination and thermal stability study. Colloid. Polym. Sci. 294, 239–246. With kind permission of Springer.

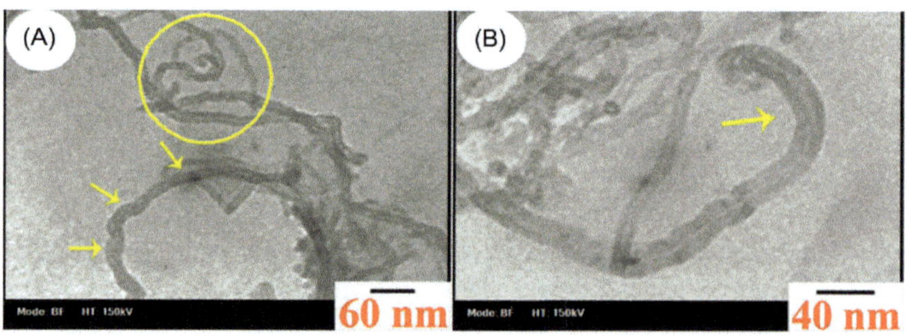

Figure 13.14 TEM images of MWCNT-Suc. (A, B) Arrows in TEM showing the formation of global-like structures in MWCNT-Suc.
Source: Adapted from Mallakpour, S., Behranvand, V., 2016. Chemical adsorption of D-sucrose on MWCNTs for compatibility improvement with alanine-based poly (amide–imide) matrix: morphology examination and thermal stability study. Colloid. Polym. Sci. 294, 239–246. With kind permission of Springer.

the preparation of novel poly(amide–imide) (PAI) NCs which will be examined in the next section.

In other research work, MWCNTs were modified with glucose and fructose carbohydrates to obtain Gl-MWCNTs and Fr-MWCNTs (Mallakpour and Behranvand, 2015a). Sedimentation test, which is an indirect way for the study of successful functionalization, was done for the samples in water. Highly dispersed and stable dark solutions for Gl-MWCNT and Fr-MWCNT in water were obtained. Fr-MWCNT remained stable for a period of about 3 weeks while Gl-MWCNT remained well dispersed in water even after 2 months.

13.4.3 Biopolymer-functionalized CNTs

Among organic material which has been used for the functionalization of CNTs, polymers have received great consideration due to their quite effective dispersity, which could disturb the van der Waals interactions between the walls of CNTs (Gao et al., 2012). Two main approaches have been recognized for the covalent joining of polymers onto the surface of CNTs, namely, the "grafting to" and "grafting from" (or surface-initiated polymerization method). "Grafting from" technique can yield a higher grafting density and chance to grow block copolymers by sequential addition of wanted monomers (Ku et al., 2009; Gatti et al., 2016). We report herein only a few examples related to the functionalization of CNTs with biopolymers and their applications. Poly(vinyl pyrrolidone) (PVP) is a group of biopolymers that has unique high water solubility and biocompatibility (Mallakpour and Behranvand, 2015b). It has been continually used in pharmaceutical formulations since its introduction in the 1940s (Lee, 2005). Popp et al. (2015) described the synthesis of PVP copolymers for wrapping SWCNTs. They decided on a reversible addition-fragmentation chain transfer polymerization method to the synthesis of new PVP copolymers, including new functionality, especially ionic comonomers, to change SWCNT stability and behavior in biological environments. They showed that functionalized PVP copolymers can stabilize SWCNT suspensions in water under biological conditions and maintain strong fluorescence through broad pH changes. Copper NPs deposited MWCNTs with poly(acrylic acid) as dispersant (Cu-PAA/MWCNTs), was prepared by Sheng et al. (2015), and expected to achieve a more effective and well-dispersed disinfection material for water purification. MWCNTs and PAA-modified MWCNTs demonstrated antimicrobial activity toward bacteria in water. Besides, the deposition of Cu NPs greatly improved the antimicrobial ability of MWCNTs and Cu looked to play a more significant role in the antimicrobial action of Cu-PAA/MWCNTs. Zhang et al. (2015) synthesized amino-terminated poly(ethylene glycol) (PEG) using cysteamine hydrochloride as the chain transfer agent (CTA) and PEG monomethyl ether methacylate (PEGMA) as the monomer. Then, amino-terminated polyPEGMA was further connected with polydopamine (PDA)-functionalized CNT (CNT-PDA) through Michael addition reaction (Fig. 13.15).

The obtained PEGylated CNT showed significant enhanced dispersibility in both aqueous and organic solution. Additionally, cell viability investigation suggested that CNT-PDA-PEGMA composites have potential for biomedical applications as they were biocompatible with human alveolar basal epithelial (A549) cells. A thermal-initiation-free radical (polymeric ionic liquids (PIL)) reaction (MWCNT) was involved to create PIL layer on MWCNT. Wang et al. (2015) designed it as an efficient electrode modifier for simultaneous determination of ascorbic acid (AA), DA, and uric acid (UA). The SEM images of PIL-MWCNTs in different

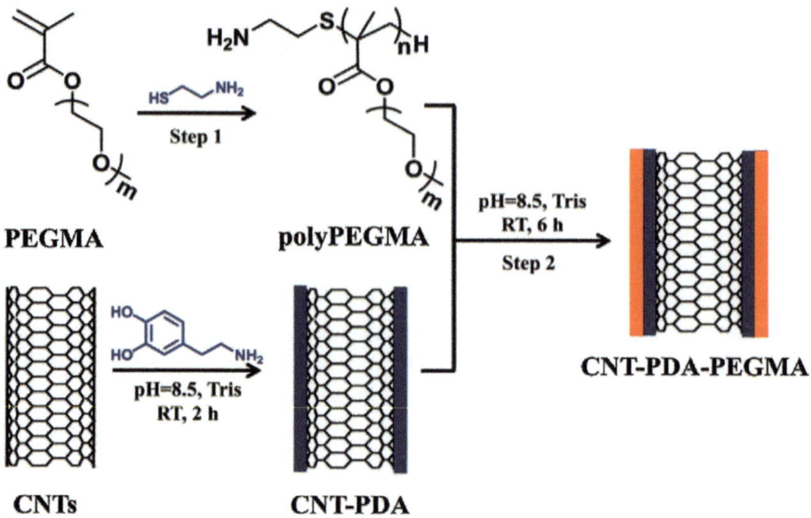

Figure 13.15 Schematic representation for the preparation of CNT-PDA-PEGMA composites via a combination of mussel inspired chemistry and a Michael addition reaction. Step1: Amino-terminated polymers (polyPEGMA) were synthesized via free radical polymerization under a rather mild reaction conditions using cysteamine hydrochloride as CTA and PEGMA as the monomer. Step 2: The PEGMA was further linked with CNT-PDA in deionized water for 6 h at room temperature (pH value is about 8.5).
Source: Adapted from Zhang, X., Zeng, G., Tian, J., Wan, Q., Huang, Q. Wang, K., et al., 2015. PEGylation of carbon nanotubes via mussel inspired chemistry: preparation, characterization and biocompatibility evaluation. Appl. Surface Sci., 351, 425–432. With kind permission of Elsevier.

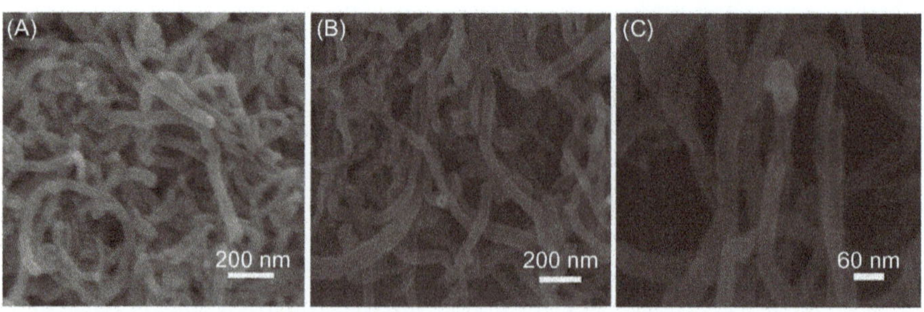

Figure 13.16 SEM images of pristine MWCNTs (A) and PIL-MWCNTs (B and C).
Source: Adapted from Wang et al. (2015). With kind permission of Elsevier.

magnifications showed less compact MWCNTs wrapped with PIL layers and thicker walls compared to the pristine MWCNTs (Fig. 13.16). AA, DA, and UA are known as important molecules for physiological processes in human in which abnormal level of them is usually a symptom of illness. So, simultaneous detection of these biomolecules is important. They examined the possibility of utilizing imidazole-containing PIL-MWCNTs as the active electrode material in determination of AA, DA, and UA for the first time. Simultaneous determination of these biomolecules was attained in a wide concentration range with high selectivity, sensitivity, stability, and good reproducibility.

In another report, hyaluronic acid (HA)-functionalized SWCNTs were designated and selected as a tumor-targeting carrier by Hou et al. (2015). It was applied to deliver magnetic resonance imaging contrast agents (MRI CAs) targeting to the tumor cells specifically. In this system, HA-SWCNTs carrying $GdCl_3$ could considerably decrease the toxicity of Gd^{3+} and increase the circulation time for MRI (Fig. 13.17).

Chelated gadolinium (Gd) compounds are the most widely used in cancer imaging. But these compounds suffer from poor sensitivity and rapid renal clearance, which severely limits the time window for MRI. There were some reports expressing that the internal loading of US tubes with aqueous $GdCl_3$ could significantly improve the effectiveness of its relaxation. Moreover, the SWCNTs with multisites could bind with drugs or molecular targets effectively (Sitharaman et al., 2005; Richard et al., 2008). HA is a naturally occurring linear polysaccharide with

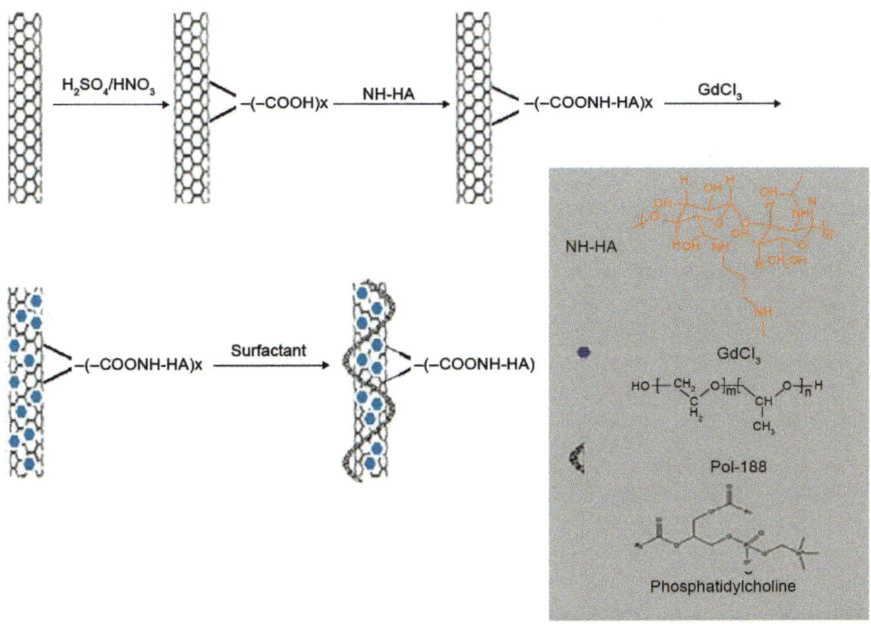

Figure 13.17 Schematic illustration of the formation for the MRI CAs.
Source: Adapted from Hou, L., Zhang, H., Wang, Y., Wang, L., Yang, X., Zhang, Z., 2015. Hyaluronic acid-functionalized single-walled carbon nanotubes as tumor-targeting MRI contrast agent. Int. J. Nanomed. 10, 4507–4520. Open access journal.

negative charge. Cellular uptake was done to estimate the intracellular transport capabilities of HA-SWCNTs for tumor cells and the uptake rank was HA-SWCNTs > SWCNTs due to the presence of HA. The obtained Gd/HA-SWCNTs proposed a low toxicity in clinical practice resulting from the high T_1-relaxivity improvement and the coordination of free Gd with HA-SWCNTs, minimizing Gd release, suggesting that this MRI CA Gd/HA-SWCNTs is promising for high-resolution MR molecular imaging of tumor with high safety.

DNA is a naturally occurring polymer with deoxyribonucleotides as its repeating or monomeric units which plays a central role in biology. Three parts of each monomer unit are a phosphate group, a furanose sugar moiety, and one of the four possible nitrogenous bases: adenine (A), guanine (G), cytosine (C), or thymine (T) (Paul and Bhattacharya, 2010; Hu et al., 2005). DNA functionalization of CNTs has attracted much scientific interest and this hybrid has been applied in solubilization in aqueous media, gene therapy, nucleic acid sensing, and controlled deposition on conducting or semiconducting substrates (Daniel et al., 2007). In 2014, Kang et al. (2014) designed new nanostructured biosensor based on single-stranded DNA (ssDNA)-SWNTs. In this case, they demonstrated the viability of using the negative surface charge of ssDNA-SWNTs to understand layer-by-layer electrostatic self-assembly. The sandwich structure obtained through adsorption of oppositely charged layers was advantageous for steady immobilization of a large amount of glucose oxidase and improvement of stability and anti-interference capability of the biosensor.

13.5 Polymer/biofunctionalized CNT hybrid composites

One of the most interesting applications of CNTs is the polymer/CNT NCs. The combination of CNT properties such as superlative mechanical, thermal, and electronic features makes them an ideal candidate as an innovative filler material for high strength and electronically conductive polymer NCs. However, there are challenges for developing high-performance CNT/polymer NCs: (1) homogeneous distribution of CNTs in the polymer matrix and (2) strong interfacial interactions so as to effect efficient load transfer from the polymer matrix to the CNTs (Liu et al., 2007; Chen et al., 2009). Though intrinsic van der Waals attraction among CNTs tubes, in combination with their high surface area and high aspect ratio, often directs to significant agglomeration, thus avoiding efficient transfer of their unique properties to the polymer matrix. To attain homogeneous dispersion of CNTs in polymer matrix by easy mixing, chemically treated CNTs have been employed (Mallakpour and Zadehnazari, 2013).

In this section, we review different properties, such as mechanical, thermal, and electrical properties of reinforced polymer composites with functionalized CNTs by biomolecules. A new methodology to link covalently MWCNTs with a biodegradable polymer (polylactic acid, PLA) getting the filler remain dispersed in the matrix was reported by Seligra et al. (2013). The reactions scheme for the

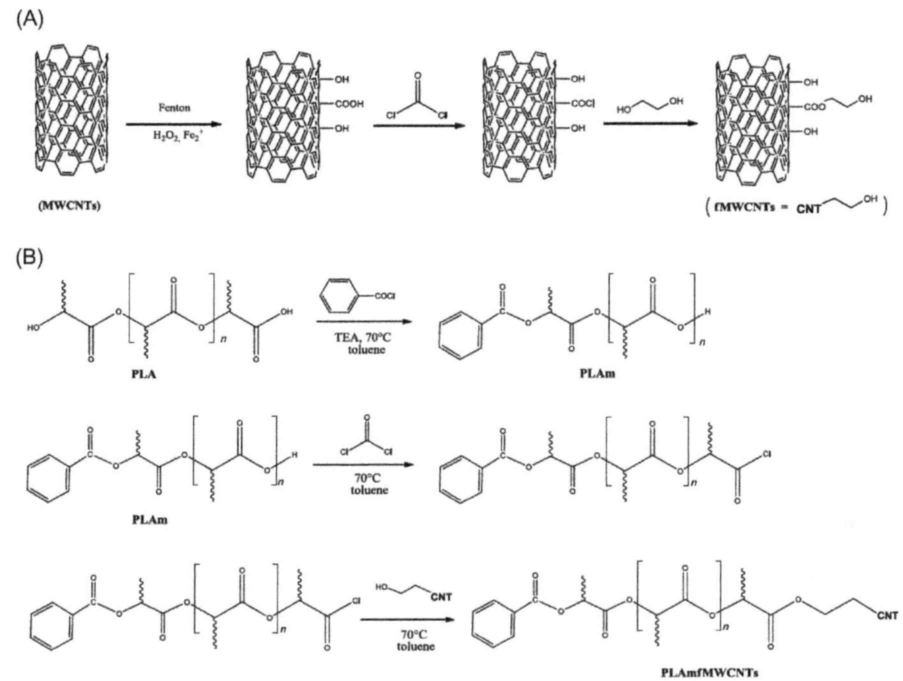

Figure 13.18 Scheme of (A) MWCNTs functionalization (fMWCNT) and (B) synthetic modification of PLA (PLAm) and synthesis of PLAmfMWCNTs.
Source: Adapted from Seligra et al. (2013). With kind permission of Elsevier.

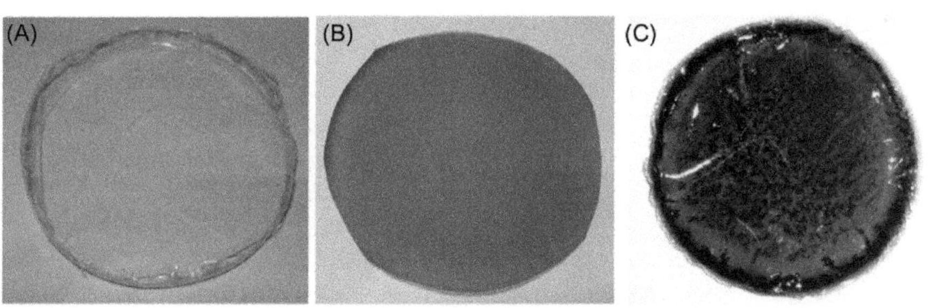

Figure 13.19 Photographs of the conformed films: (A) matrix, (B) NC with fMWCNTs, and (C) NC with pristine MWCNTs.
Source: Adapted from Seligra et al. (2013). With kind permission of Elsevier.

preparation of modified PLA reinforced with fMWCNTs (PLAmfMWCNTs) is shown in Fig. 13.18B.

Fig. 13.19 shows the photographs of different films conformed: matrix (A), NC with fMWCNTs (B) and NC with pristine MWCNTs (C). The observations showed

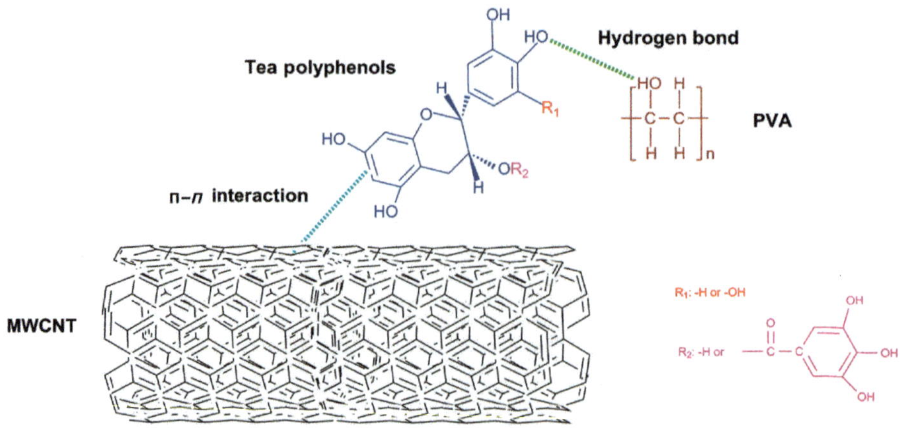

Figure 13.20 Structure of PVA/MWCNT composites.
Source: Adapted from Lu, L., Hou, W., Sun, J., Wang, J., Qin, C., Dai, L., 2014. Preparation of poly(vinyl alcohol) fibers strengthened using multiwalled carbon nanotubes functionalized with tea polyphenols. J. Mater. Sci. 49, 3322–3330. With kind permission of Springer.

that the functionalization of the filler and the modification of the polymer were effective and essential to attain a good dispersion.

Storage modulus and strength at break were enhanced by the addition of fMWCNT without losing deformation properties. They concluded that the obtained novel biodegradable composite material has promising characteristics to be applied in biomedicine and in packing industry. Dai's group functionalized MWCNTs using a green modifier tea polyphenols (TP) noncovalently and then they were incorporated into the poly(vinyl alcohol) (PVA) matrix (Lu et al., 2014). Fig. 13.20 illustrates the expected interactions between the species: π–π interaction between MWCNT and TP, and hydrogen bond between TP and PVA.

The TEM micrograph of PVA/MWCNTs (0.6 wt%) dispersion with TP displayed distinctly improved dispersion of MWCNTs (Fig. 13.21B).

With only a minute quantity of MWCNTs loadings, remarkable increases of tensile strength of the composite fibers were achieved.

Surface functionalization of MWCNTs with biomolecules and their applications in the polymer NCs fabrication have been of interest to our group recently. Of particular interest to our group is how surface functionalization of MWCNTs affects properties of various polymer matrixes.

In this regard, MWCNTs were functionalized covalently with glucose carbohydrate by Mallakpour and Zadehnazari (2013). Glucose-functionalized MWCNTs (MWCNTs-Gl) were applied as reinforcement for the preparation of hydroxylated PAI-based composites. The thermal stability, tensile modulus, and tensile strength of the composites were improved due to the increased interfacial interaction and good dispersion between the PAI chains and MWCNTs-Gl. The influence of MWCNTs-Gl was investigated on other polymer matrixes such as PAI containing

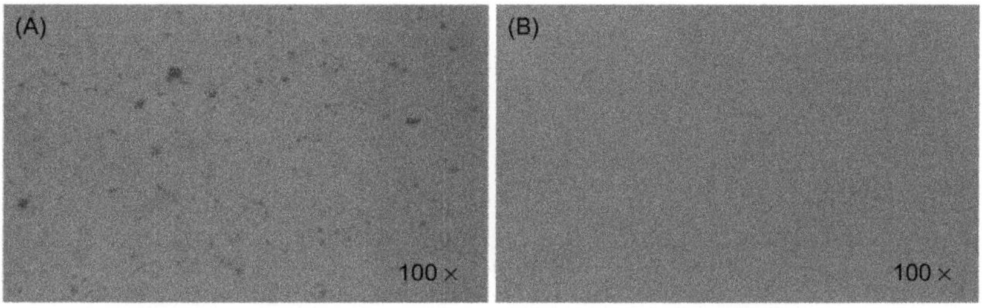

Figure 13.21 Optical micrographs of the PVA/MWCNTs (0.6 wt%) dispersions (A) without TP and (B) with TP.
Source: Adapted from Lu, L., Hou, W., Sun, J., Wang, J., Qin, C., Dai, L., 2014. Preparation of poly(vinyl alcohol) fibers strengthened using multiwalled carbon nanotubes functionalized with tea polyphenols. J. Mater. Sci. 49, 3322–3330. With kind permission of Springer.

N,*N*′-(pyromellitoyl)-bis-*S*-valine and PAI based on *N*-trimellitylimido-*S*-valine (Mallakpour et al., 2015; Abdolmaleki et al., 2015). In all cases, the resulting NCs showed increased thermal stability compared to the pure polymer. Effects of MWCNTs-Gl on the structural, mechanical, and thermal properties of chitosan NC films were studied by our group in 2015 (Mallakpour and Madani, 2015a). The results of thermal properties showed that the chitosan was slightly more thermally stable than the NC films. This could be explained by the fact that the MWCNT-Gl can break the networks created by chitosan chains via intermolecular/intramolecular interactions and cause slight decrease in thermal stability. Though with increasing of the MWCNT-Gl loading in chitosan, the tensile strength and elongation of the NCs were improved.

Sucrose as the most common natural disaccharide was selected for the surface functionalization of MWCNTs by Mallakpour and Behranvand (2016). Nanotube-reinforced PAI NCs were prepared by mixing MWCNT-Suc with alanine amino acid–based PAI matrix under ultrasonic irradiations. Due to the presence of many OH groups in the sucrose structure, there are effective interfacial interactions between nanotubes and PAI backbone (Fig. 13.22).

As can be observed from Fig. 13.23, no agglomerates of nanotubes could be seen in the TEM micrographs of the obtained NCs. On the basis of the limiting oxygen index values, pure polymer and PAI/MWCNT-Suc NCs could be classified as self-extinguishing materials.

In other work, pure MWCNTs were functionalized by L-phenylalanine amino acid (Mallakpour et al., 2014). The effects of modified MWCNT on dispersion in PVA matrix and mechanical, thermal, and morphological properties of the resulting composites were studied. When fMWCNT was embedded into the PVA matrix, the hydroxyl groups of PVA can interact with OH, NH, and carbonyl groups on the MWCNT surface through hydrogen bonding as shown in Fig. 13.24.

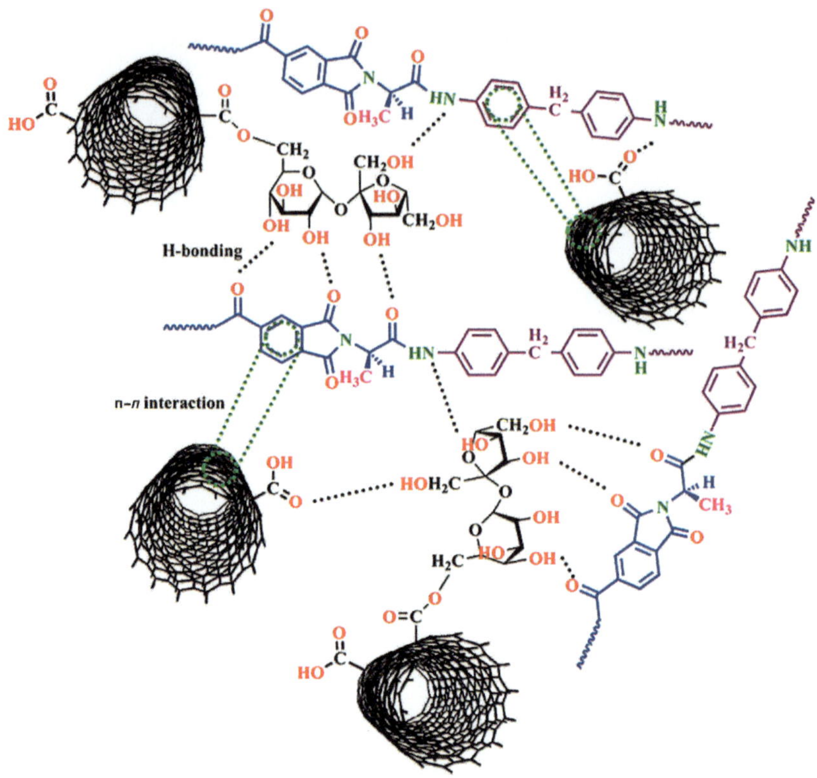

Figure 13.22 H bonding and π−π interactions between the MWCNT-Suc and the PAI chains.
Source: Adapted from Mallakpour et al. (2016). With kind permission of Springer.

The increase in the thermal stability and the significant improvement of mechanical properties of the PVA/fMWCNT composite were resulted due to well dispersion of fMWCNT in PVA matrix in a helix-shaped and nano-sized range as it can be shown in Fig. 13.25.

Mallakpour et al (2015) functionalized carboxylated-MWCNTs with riboflavin (vitamin B2) (RB) under microwave irradiation as an efficient, fast, and simple strategy to improve interfacial interactions and dispersion of CNTs in a poly(ester-imide) (PEI) based on L-phenylalanine linkages. The detailed reaction scheme was demonstrated in Fig. 13.26.

It could be seen from TEM images that the MWCNTs-COOH with a smooth-sided sidewalls were bundled together, but RB-MWCNTs displayed a high surface roughness and had a folded and debundled structure as a result of functionalization (Fig. 13.27).

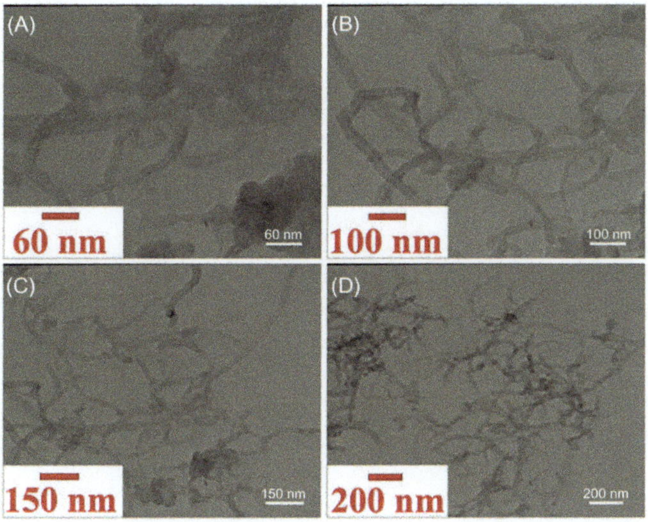

Figure 13.23 TEM images of PAI/MWCNT-Suc NC of 10 wt% at different magnifications: (A) 60, (B) 100, (C) 150 and (D) 200 nm.
Source: Adapted from Mallakpour et al. (2016). With kind permission of Springer.

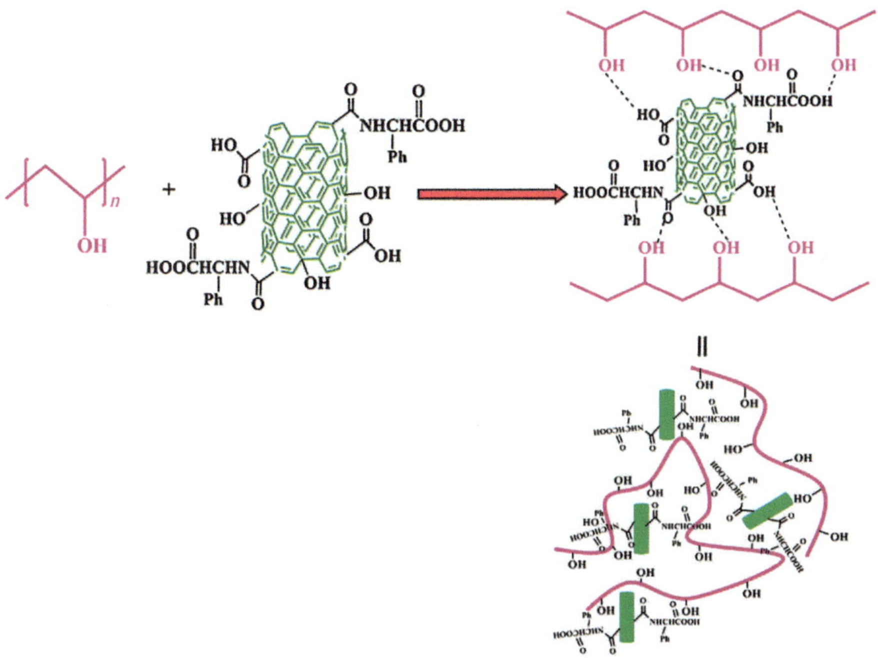

Figure 13.24 Interaction between PVA and fMWCNT.
Source: Adapted from Mallakpour, S. Abdolmaleki, A., Borandeh, S., 2014. L-Phenylalanine amino acid functionalized multi walled carbon nanotube (MWCNT) as a reinforced filler for improving mechanical and morphological properties of poly(vinyl alcohol)/MWCNT composite. Prog. Organ. Coat. 77, 1966–1971. With kind permission of Elsevier.

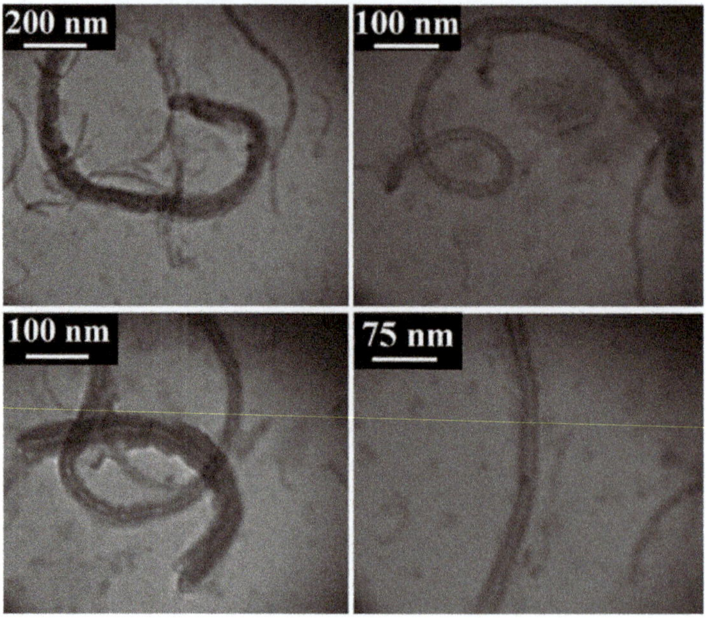

Figure 13.25 TEM photographs of PVA/fMWCNT composite (5 wt%).
Source: Adapted from Mallakpour, S. Abdolmaleki, A., Borandeh, S., 2014. L-Phenylalanine amino acid functionalized multi walled carbon nanotube (MWCNT) as a reinforced filler for improving mechanical and morphological properties of poly(vinyl alcohol)/MWCNT composite. Prog. Organ. Coat. 77, 1966–1971. With kind permission of Elsevier.

Figure 13.26 Chemical attachment of riboflavin to MWCNTs.
Source: Adapted from Mallakpour, S., Soltanian, S., 2015. A facile approach towards functionalization of MWCNTs with vitamin B2 for reinforcing of biodegradable and chiral poly(ester-imide) having L-phenylalanine linkages: morphological and thermal investigations. J. Polym. Res. 22, 183–191. With kind permission of Springer.

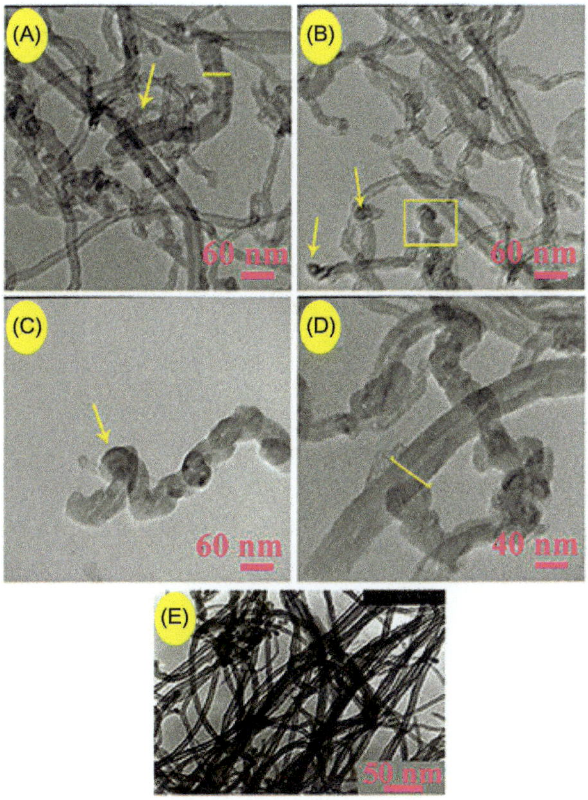

Figure 13.27 TEM images of RB-MWCNTs at different magnifications (A–D) and MWCNTs-COOH (E).
Source: Adapted from Mallakpour, S., Soltanian, S., 2015. A facile approach towards functionalization of MWCNTs with vitamin B2 for reinforcing of biodegradable and chiral poly(ester-imide) having L-phenylalanine linkages: morphological and thermal investigations. J. Polym. Res. 22, 183–191. With kind permission of Springer.

RB-MWCNTs could strongly interact with the PEI via π-stacking and H-bonding as it has been depicted in Fig. 13.28. It was found that the thermal stability of NCs was higher than the pure PEI as a consequence of the presence of thermally stable fillers and efficient interactions of RB-MWCNTs with the PEI matrix.

In 2016, PEI/RB/MWCNT composites containing 4,4′-thiobis(2-tert-butyl-5-methylphenol) were prepared by Mallakpour and Behranvand (2016). According to TEM images of NC with 10 wt% RB/MWCNT, it was observed that the nanotubes were dispersed separately and embedded to the PEI matrix (Fig. 13.29). It was found from thermal properties data that temperature at 10% weight loss was increased from 409°C for pure PEI to 417°C, 420°C, and 424°C for NCs containing 5, 10, and 15 wt% RB/MWCNTs, respectively.

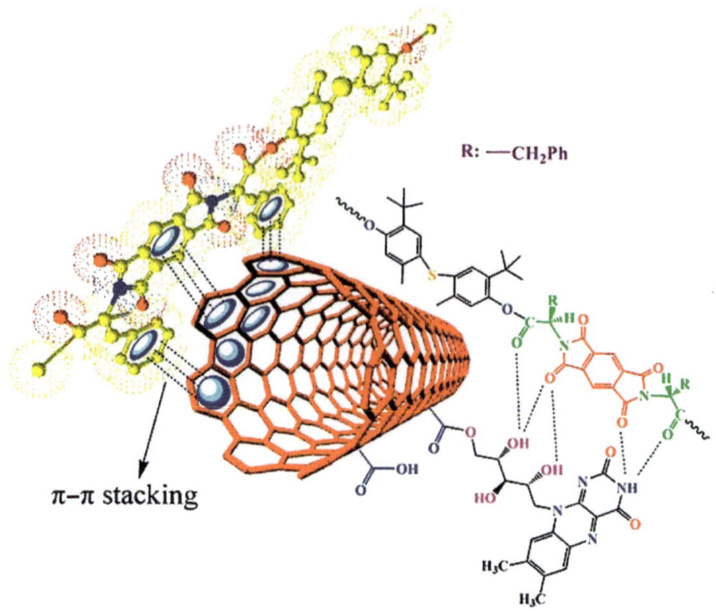

Figure 13.28 Schematic illustration of the possible interaction between the RB-MWCNTs and the PEI matrix.
Source: Adapted from Mallakpour, S., Soltanian, S., 2015. A facile approach towards functionalization of MWCNTs with vitamin B2 for reinforcing of biodegradable and chiral poly(ester-imide) having L-phenylalanine linkages: morphological and thermal investigations. J. Polym. Res. 22, 183–191. With kind permission of Springer.

13.6 Conclusions

For the efficient use of CNTs, an excellent distribution and dispersion is a necessary precondition. The functional groups on the surface of the CNTs and purity as well as mostly their strength of agglomerates affect the dispersability of CNTs in different media. Besides, due to strong van der Waals forces, CNTs tend to agglomerate. Surface biofunctionalization not only is the most effective approach for better dispersion of nanotubes but also has been shown capable of reducing cytotoxicity and improving biocompatibility. In this chapter it could be found that nanotubes modified with biomolecules have practical applications in various fields. These hybrids are of interest for degradation of different aromatic hydrocarbon intermediates, biocatalysis, biosensing, drug delivery, biorecognition, and in sorption processes. Another benefit of application of biofunctionalized CNTs is their nano-reinforcement role for the fabrication of polymer NCs. Some studies have demonstrated that the application of chemical modification to the nanotubes can efficiently improve the distribution of CNTs in polymer matrix and therefore improve the physical properties of the composites. More researches on the preparation and application of biofunctionalized CNT/polymer NCs are still to be carried out in the near future.

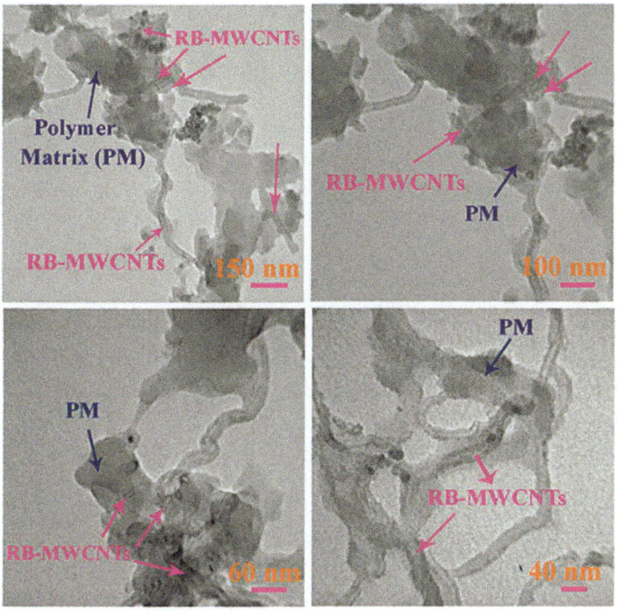

Figure 13.29 TEM micrographs of PEI/RB-MWCNTs NCs (10 wt%).
Source: Adapted from Mallakpour, S., Soltanian, S., 2016. Chemical surface coating of MWCNTs with riboflavin and its application for the production of poly(ester-imide)/ MWCNTs linkages: thermal and morphological properties. J. Appl. Polym. Sci. 1, 1–9. With kind permission of John Wiley & Sons.

Acknowledgments

The authors would like to thank the financial support from the Research Affairs Division, Isfahan University of Technology (IUT), Isfahan, and the National Elite Foundation (NEF).

References

Abdolmaleki, A., Mallakpour, S., Rostami, M., 2015. Surface modification of MWCNTs with glucose and their utilization for the production of environmentally friendly nanocomposites using biodegradable poly (amide–imide) based on *N*-trimellitylimido-*S*-valine matrix. Polym. Adv. Technol. 26, 1141–1147.

Ali, M.A., Singh, N., Srivastava, S., Agrawal, V.V., John, R., Onoda, M., et al., 2014. Chitosan-modified carbon nanotubes-based platform for low-density lipoprotein detection. Appl. Biochem. Biotechnol. 174, 926–935.

Chen, D., Wang, M., Zhang, W.D., Liu, T., 2009. Preparation and characterization of poly (vinylidene fluoride) nanocomposites containing multiwalled carbon nanotubes. J. Appl. Polym. Sci. 113, 644–650.

Choudhary, V. and Gupta, A., 2011. Polymer/carbon nanotube nanocomposites. In: Yellampalli, S. (Ed.), Carbon Nanotubes—Polymer Nanocomposites, pp. 65—91.
Daniel, S., Rao, T.P., Rao, K.S., Rani, S.U., Naidu, G.R.K., Lee, H.Y., et al., 2007. A review of DNA functionalized/grafted carbon nanotubes and their characterization. Sensor. Actuator. B. 122, 672—682.
Deborah, M., Jawahar, A., Mathavan, T., Kumara Dhas, M, Milton Franklin Benial, A., 2015. Spectroscopic studies on sidewall carboxylic acid functionalization of multi-walled carbon nanotubes with valine. Spectrochimica Acta Part A: Mol. Biomol. Spect. 139, 138—144.
D'Souza, O.J., Mascarenhas, R.J., Satpati, A.K., Aiman, L.V., Mekhalif, Z., 2015. Electrocatalytic oxidation of L-tyrosine at carboxylic acid functionalized multi-walled carbon nanotubes modified carbon paste electrode. Ionics. 22, 405—414.
Eguílaz, M., Gutiérrez, A., Rivas, G., 2016. Chemical non-covalent functionalization of multi-walled carbon nanotubes with cytochrome c: enhanced direct electron transfer and analytical applications. Sensor. Actuator. B Chem. 225, 74—80.
Feng, W., Ji, P., 2011. Enzymes immobilized on carbon nanotubes. Biotechnol. Adv. 29, 889—895.
Gao, C., Guo, Z., Liu, J.H., Huang, X.J., 2012. The new age of carbon nanotubes: an updated review of functionalized carbon nanotubes in electrochemical sensors. Nanoscale. 4, 1948—1963.
Gao, Y., Cao, Y., Song, G., Tang, Y., Li, H., 2011. β-Cyclodextrin covalently functionalized single-walled carbon nanotubes: synthesis, characterization and a sensitive biosensor platform. J. Biomater. Nanobiotechnol. 2, 454—460.
Gatti, T., Vicentini, N., Mba, M., Menna, E., 2016. Organic functionalized carbon nanostructures for functional polymer-based nanocomposites. Eur. J. Organ. Chem.1071—1090.
González, V.J., Gracia-Espino, E., Morelos-Gómez, A., López-Urías, F., Terrones, H., Terrones, M., 2015. Biotin molecules on nitrogen-doped carbon nanotubes enhance the uniform anchoring and formation of Ag nanoparticles. Carbon. 8, 51—59.
Hames, B.D., Hooper, N.M., 2005. Biochemistry. School of Biochemistry and Molecular Biology, University of Leeds, Leeds, UK.
Hou, L., Zhang, H., Wang, Y., Wang, L., Yang, X., Zhang, Z., 2015. Hyaluronic acid-functionalized single-walled carbon nanotubes as tumor-targeting MRI contrast agent. Int. J. Nanomed. 10, 4507—4520.
Hu, C., Zhang, Y., Bao, G., Zhang, Y., Liu, M., Wang, Z.L., 2005. DNA functionalized single-walled carbon nanotubes for electrochemical detection. J. Phys. Chem. B. 109, 20072—20076.
Kang, Z., Yan, X., Zhang, Y., Pan, J., Shi, J., Zhang, X., et al., 2014. Single-stranded DNA functionalized single-walled carbon nanotubes for microbiosensors via layer-by-layer electrostatic self-assembly. ACS Appl. Mater. Interface. 6, 3784—3789.
Kong, N., Shimpi, M.R., Park, J.H., Ramström, O., Yan, M., 2015. Carbohydrate conjugation through microwave-assisted functionalization of single-walled carbon nanotubes using perfluorophenyl azides. Carbohyd. Res. 405, 33—38.
Ku, B.C., Kim, D.K., Lee, J.S., Blumstein, A., Kumar, J., Samuelson, L.A., 2009. Synthesis and properties of water soluble single-walled carbon nanotube graft ionic polyacetylene nanocomposites. Polym. Comp.1817—1824.
Lee, J., 2005. Intrinsic adhesion properties of poly(vinyl pyrrolidone) to pharmaceutical materials: humidity effect. Macromolar. Biosci. 5, 1085—1093.
Liu, D., Zhang, Y., Sun, X., Chang, P.R., 2014. Recent advances in bio-sourced polymeric carbohydrate/nanotube composites. J. Appl. Polym. Sci. 40359, 1—13.

Liu, T., Tong, Y., Zhang, W.D., 2007. Preparation and characterization of carbon nanotube/polyetherimide nanocomposite films. Comp. Sci. Technol. 67, 406–412.
Lu, L., Hou, W., Sun, J., Wang, J., Qin, C., Dai, L., 2014. Preparation of poly(vinyl alcohol) fibers strengthened using multiwalled carbon nanotubes functionalized with tea polyphenols. J. Mater. Sci. 49, 3322–3330.
Mallakpour, S., Behranvand, V., 2014. Optical, mechanical, and thermal behavior of poly (vinyl alcohol) composite films embedded with biosafe and optically active poly(amide–imide)-ZnO quantum dot nanocomposite as a novel reinforcement. Colloid Polymer. Sci. 292, 2857–2867.
Mallakpour, S., Behranvand, V., 2015a. Improved solubilization of multiwalled carbon nanotubes (MWCNTs) in water by surface functionalization with D-glucose and D-fructose: properties comparison of functionalized MWCNTs/alanine-based poly(amide–imide) nanocomposites. High Perform. Polym.1–9. Available from: http://dx.doi.org/10.1177/0954008315610395.
Mallakpour, S., Behranvand, V., 2015b. Novel ternary poly(vinyl pyrrolidone)/poly(amide–imide)/ZnO nanocomposite: synthesis, characterization, thermal and optical performance. Prog. Organ. Coat. 86, 18–24.
Mallakpour, S., Behranvand, V., 2015c. Sonochemical production and characterization of D-fructose functionalized MWCNTs/alanine-based poly(amide–imide) nanocomposites. Colloid Polym. Sci. 293, 1817–1826.
Mallakpour, S., Behranvand, V., 2015d. The influence of acid-treated multi-walled carbon nanotubes on the surface morphology and thermal properties of alanine-based poly(amide–imide)/MWCNT nanocomposites system. Colloid Polym. Sci. 293, 333–339.
Mallakpour, S., Behranvand, V., 2016. Chemical adsorption of D-sucrose on MWCNTs for compatibility improvement with alanine-based poly(amide–imide) matrix: morphology examination and thermal stability study. Colloid Polym. Sci. 294, 239–246.
Mallakpour, S., Madani, M., 2015a. Effects of glucose-functionalized multiwalled carbon nanotubes on the structural, mechanical, and thermal properties of chitosan nanocomposite films. J. Appl. Polym. Sci. 42022, 1–9.
Mallakpour, S., Madani, M., 2015b. p-Amino phenol immobilized on multi-walled carbon nanotubes for the preparation of chitosan nanocomposites. J. Comp. Mater.1–9. Available from: http://dx.doi.org/10.1177/0021998315575749.
Mallakpour, S., Soltanian, S., 2014a. Chemical modification of MWCNTs with 5-aminoisophthalic acid and its effects on the thermal and morphological properties of chiral poly(ester-imide)/MWCNT nanocomposites having *N*-trimellitylimido-L-isoleucine moieties. J. Polym. Res. 21, 1–9.
Mallakpour, S., Soltanian, S., 2014b. Studies on preparation and microstructure characterization of novel composites based on functionalized multiwalled carbon nanotubes and chiral poly (ester-imide) containing *S*-valine linkages. Polym. Plast. Technol. Eng. 53, 1583–1589.
Mallakpour, S., Soltanian, S., 2015. A facile approach towards functionalization of MWCNTs with vitamin B2 for reinforcing of biodegradable and chiral poly(ester-imide) having L-phenylalanine linkages: morphological and thermal investigations. J. Polym. Res. 22, 183–191.
Mallakpour, S., Soltanian, S., 2016. Chemical surface coating of MWCNTs with riboflavin and its application for the production of poly(ester-imide)/MWCNTs linkages: thermal and morphological properties. J. Appl. Polym. Sci. 1, 1–9.
Mallakpour, S., Zadehnazari, A., 2013. One-pot synthesis of glucose functionalized multi-walled carbon nanotubes: dispersion in hydroxylated poly(amide–imide) composites and their thermo-mechanical properties. Polymer. 54, 6329–6338.

Mallakpour, S., Zadehnazari, A., 2014a. A facile, efficient, and rapid covalent functionalization of multi-walled carbon nanotubes with natural amino acids under microwave irradiation. Prog. Organ. Coat. 77, 679–684.

Mallakpour, S., Zadehnazari, A., 2014b. The effect of carboxylated multi-walled carbon nanotubes on reinforcement efficiency of thiazole-bearing poly(amide–imide) composites. Des. Monomers Polym. 17, 275–285.

Mallakpour, S., Abdolmaleki, A., Borandeh, S., 2014. L-Phenylalanine amino acid functionalized multi walled carbon nanotube (MWCNT) as a reinforced filler for improving mechanical and morphological properties of poly(vinyl alcohol)/MWCNT composite. Prog. Organ. Coat. 77, 1966–1971.

Mallakpour, S., Abdolmaleki, A., Rostami, M., 2015. Glucose-functionalized multi-walled carbon nanotubes dispersing and hosting nanotubes for poly(amide–imide) bionanocomposites containing *N,N′*-(pyromellitoyl)-bis-*S*-valine. J. Polym. Res. 22, 1–9.

Mansur, H.S., Pereira, M.M., Costa, H.S., Mansur, A.A.P., 2012. Mechanical behavior of nanostructured hybrids based on poly (vinyl alcohol)/bioactive glass reinforced with functionalized carbon nanotubes. J. Nanomater. 2012, 1–9.

Mathavan, T., Kumara Dhas, M., Kanimozhi, C.V., Jothi Rajan, M.A., Umapathy, S., Ramasubbu, A., et al., 2014. Spectroscopic studies on pure and histidine-functionalized multiwalled carbon nanotubes. Spectrosc. Lett. 47, 642–648.

Meng, L., Fu, C., Lu, Q., 2009. Advanced technology for functionalization of carbon nanotubes. Prog. Natur. Sci. 19, 801–810.

Mohamad, N.R., Buang, N.A., Mahat, N.A., Lok, Y.Y., Huyop, F., Aboul-Enein, H.Y., et al., 2015. A facile enzymatic synthesis of geranyl propionate by physically adsorbed *Candida rugosa* lipase onto multi-walled carbon nanotubes. Enzyme Microbiol. Technol. 72, 49–55.

Mubarak, N.M., Wong, J.R., Tan, K.W., Sahu, J.N., Abdullah, E.C., Jayakumar, N.S., et al., 2014. Enzymatic immobilization of cellulase enzyme on functionalized multiwall carbon nanotubes. J. Mol. Catal. B. Enzym. 107, 124–131.

Papper, V., Elouarzaki, K., Gorgy, K., Sukharaharja, A., Cosnier, S., Marks, R.S., 2014. Biofunctionalization of multiwalled carbon nanotubes by electropolymerized poly (pyrrole-concanavalin A) films. Chem. A Eur. J. 20, 13561–13564.

Paul, A., Bhattacharya, B., 2010. DNA functionalized carbon nanotubes for nonbiological applications. Mater. Manufact. Process. 25, 891–908.

Popp, B.V., Miles, D.H., Smith, J.A., Fong, I.M., Pasquali, M., Ball, Z.T., 2015. Stabilization and functionalization of single-walled carbon nanotubes with polyvinylpyrrolidone copolymers for applications in aqueous media. J. Polym. Sci. Part A: Polym. Chem. 63, 337–343.

Prlainovic, N.Z., Bezbradica, D.I., Rogan, J.R., Uskokovic, P.S., Mijin, D.Z., Marinkovicb, A.D., 2016. Surface functionalization of oxidized multi-walled carbon nanotubes: *Candida rugosa* lipase immobilization. Comptes Rendus Chimie. pp. 1–8.

Qi, X., Rui, Y., Fan, Y., Chen, H., Ma, N., Wu, Z., 2015. Galactosylated chitosan-grafted multiwall carbon nanotubes for pH-dependent sustained release and hepatic tumor-targeted delivery of doxorubicin in vivo. Colloid. Surf. B: Biointerface. 133, 314–322.

Rahmani, L., Ketabi, S., 2015. Solvation of alanine and histidine functionalized carbon nanotubes in aqueous media: a Monte Carlo simulation study. J. Mol. Liq. 208, 191–195.

Richard, C., Doan, B.T., Beloeil, J.C., Bessodes, M., Tóth, E., Schermanand, D., 2008. Noncovalent functionalization of carbon nanotubes with amphiphilic Gd^{3+} chelates: toward powerful T_1 and T_2 MRI contrast agents. Nano Lett. 8, 232–236.

Saifuddin, N., Raziah, A.Z., Junizah, A.R., 2013. Carbon nanotubes: a review on structure and their interaction with proteins. J. Chem. 2013, 1–18.

Sekar, G., Vijayakumar, S., Thanigaivel, S., Thomas, J., Mukherjee, A., Chandrasekaran, N., 2016. Multiple spectroscopic studies on the interaction of BSA with pristine CNTs and their toxicity against *Donax faba*. J. Luminescence. 170, 141–149.

Seligra, P.G., Nuevo, F., Lamanna, M., Famá, L., 2013. Covalent grafting of carbon nanotubes to PLA in order to improve compatibility. Comp. Part B. 46, 61–68.

Sheng, L., Huang, S., Sui, M., Zhang, L., She, L., Chen, Y., 2015. Deposition of copper nanoparticles on multiwalled carbon nanotubes modified with poly(acrylic acid) and their antimicrobial application in water treatment. Front. Environ. Sci. Eng. 9, 625–633.

Sitharaman, B., Kissell, K.R., Hartman, K.B., Tran, L.A., Baikalov, A., Rusakova, I., et al., 2005. Superparamagnetic gadonanotubes are high-performance MRI contrast agents. Chem. Commun.3915–3917.

Smith, S.C., Ahmed, F., Gutierrez, K.M., Rodrigues, D.F., 2014. A comparative study of lysozyme adsorption with graphene, graphene oxide, and single-walled carbon nanotubes: potential environmental applications. Chem. Eng. J. 240, 147–154.

Suma, Y., Kang, C.S., Kim, H.S., 2015. Noncovalent and covalent immobilization of oxygenase on single-walled carbon nanotube for enzymatic decomposition of aromatic hydrocarbon intermediates. Environ. Sci. Pollut. Res. 23, 1015–1024.

Tavares, A.P.M., Silva, C.G., Drazic, G., Silva, A.M.T., Loureiro, J.M., Faria, J.L., 2015. Laccase immobilization over multi-walled carbon nanotubes: kinetic, thermodynamic and stability studies. J. Colloid Interface Sci. 454, 52–60.

Villamizar, R.A., Braun, J., Gompf, B., Dressel, M., Rius, F.X., 2009. Morphological and electrical characteristics of biofunctionalized layers on carbon nanotubes. Biosensors Bioelectron. 25, 161–166.

Wang, H., Ma, H., Zheng, W., An, D., Na, C., 2014. Multifunctional and recollectable carbon nanotube ponytails for water purification. ACS Appl. Mater. Interface. 6, 9426–9434.

Wang, X., Zheng, K., Feng, X., Xu, C., Song, W., 2015. Polymeric ionic liquid functionalized MWCNTs as efficient electrochemical interface for biomolecules simultaneous determination. Sensor. Actuator. B. Chem. 219, 361–369.

Xie, X.L., Mai, Y.W., Zhou, X.P., 2005. Dispersion and alignment of carbon nanotubes in polymer matrix: a review. Mater. Sci. Eng. R. 49, 89–112.

Yang, S., Shao, D., Wang, X., Hou, G., Nagatsu, M., Tan, X., et al., 2015. Design of chitosan-grafted carbon nanotubes: evaluation of how the –OH functional group affects Cs^+ adsorption. Marine Drug. 13, 3116–3131.

Zhang, X., Zeng, G., Tian, J., Wan, Q., Huang, Q., Wang, K., et al., 2015. PEGylation of carbon nanotubes via mussel inspired chemistry: preparation, characterization and biocompatibility evaluation. Appl. Surf. Sci. 351, 425–432.

Investigation on morphology, properties, and applications of hybrid poly(vinyl chloride)/metal oxide composites

14

Shadpour Mallakpour[1,2,3] *and Shima Rashidimoghadam*[1]
[1]Organic Polymer Chemistry Research Laboratory, Department of Chemistry, Isfahan University of Technology, Islamic Republic of Iran, [2]Nanotechnology and Advanced Materials Institute, Isfahan University of Technology, Islamic Republic of Iran, [3]Center of Excellence in Sensors and Green Chemistry, Department of Chemistry, Isfahan University of Technology, Islamic Republic of Iran

Chapter Outline

14.1 Introduction

14.1.1 Poly(vinyl chloride)

Poly(vinyl chloride) (PVC) is the third-most extensively fabricated synthetic plastic polymer, after polyethylene and polypropylene (Allsopp and Vianello, 2000). PVC is a thermoplastic produced by polymerization of the vinyl chloride monomer (chloroethene), also it is composed of about 57% chlorine (derived from industrial-grade salt) and 43% of hydrocarbon (derived mostly from oil/gas via ethylene). (Fig. 14.1) (Patrick, 2004).

Hybrid Polymer Composite Materials: Properties and Characterisation. DOI: http://dx.doi.org/10.1016/B978-0-08-100787-7.00014-7

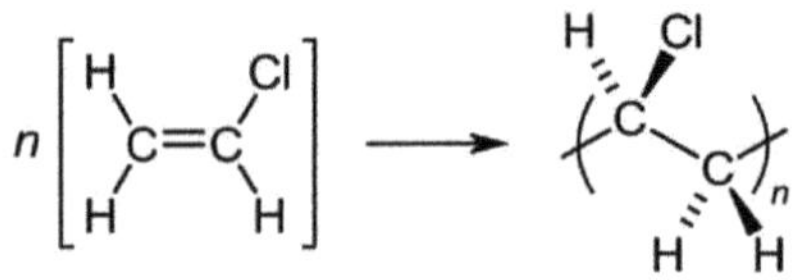

Figure 14.1 Polymerization of the vinyl chloride monomer.

Generally, there are two basic types of PVC available: rigid PVC and plasticized (or flexible) PVC.

Rigid PVC is the most common kind of PVC utilized in the manufacture of pipe, fittings, valves, machining shapes, sheet, and duct. Rigid PVC is ideal for piping and related applications because of its low-cost, high strength to weight ratio, pressure bearing capability, corrosion and chemical resistance, and low friction loss characteristics. PVC can be made softer and more flexible by the addition of plasticizers, the most widely used being phthalates. In this form, it is also used in plumbing, electrical cable insulation, imitation leather, signage, inflatable products, and many applications where it replaces rubber (Titow, 1984).

14.1.2 Metal oxide nanoparticles

Among all the functional materials were synthesized on the nanoscale, metal oxides have attracted particular interest due to their special properties in several scientific and technological areas (Nicolais and Carotenuto, 2014). Metal oxides are formed as a consequence of coordination tendency of metal ions so that oxide ions form coordination sphere around the metal ions and give rise to close packed structure.

There are great interests in the different physical, magnetic, optical, and chemical properties of metal oxides because these are extremely sensitive to change in composition and structure. These functional properties are because of the interplay between factors such as size, shape, morphology, crystal structure, and surface chemistry.

Three important groups of basic properties are influenced by particle size in any material. The first one comprises the structural characteristics, namely the lattice symmetry and cell parameters (Ritu, 2014; Corr, 2012).

Bulk oxides are usually stable systems with well-defined crystallographic structures. In order to achieve mechanical or structural stability, having a low surface free energy is basically needed for a nanoparticle (NP). In consequence of this condition, phases that have a low stability in bulk materials can become very stable in nanostructures. Hence, there are some kinds of materials such as TiO_2, VO_x, Al_2O_3, or MoOx oxides that the mentioned structural phenomenon can be found in them. Moreover, in NPs of NiO, Fe_2O_3, ZrO_2, MoO_3, CeO_2, and Y_2O_3, size-included structural distortions associated with changes in cell parameters have been perceived. The more the particle size decreases, the more number of surface and interface atoms increases and consequently, more stress/strain and concomitant structural perturbations are produced (Ayuub et al., 1998; Garvie and Goss, 1986; Hernandez et al., 2004; Skandan et al., 1992).

The second important effect of size is related to the electronic properties of the oxide. The third group is quantum size or confinement effects. Quantum size effect is unusual properties of extremely small crystals that arise from confinement of electrons to small regions of space in one, two, or three dimensions.

In their bulk state, many oxides have wide band gaps and a low reactivity. When the size of a metal oxide particle is reduced, the number of aromatic orbitals within the bands decreases and discrete energy levels are formed instead of the more or less homogenous energy band in bulk semiconductors. Also, the energy difference or the band gap between the valence band and the conduction band is increased with reducing nanocrystalline size (Selegard, 2013). In the past few decades, they have had some traditional applications. For instance, they were utilized to improve some features of polymeric products such as durability and appearance. Moreover, they have been commonly considered as inert materials. As nanosized particles, these materials are capable of absorption of broadband UV which has been a great benefit only in sunscreen applications. Furthermore, the addition of the nanoparticles may led to increase in some properties of polymeric materials such as stiffness, toughness, and service life, for instance, in applications in which mar resistance is significant. The material characteristics of metal oxide NP/polymer composites, the microstructure and dispersion (sizes and spatial distribution) of NPs must be optimized as well as characterized as a function of different process conditions (Wyckoff, 1964).

14.1.2.1 Synthesis of metal oxide nanoparticles

The first requirement of any novel study of nanoparticulate oxide is the synthesis of the material (Peacock, 2000; Wyckoff, 1964). Considering the involvement of chemical reactions, metal oxide NPs can be prepared using two procedures: chemical and physical methods. Regarding the state of the reaction system, however, these methods can also be divided into liquid phase, solid phase, and gas phase methods. The first class of NP synthesis methods comprises the liquid phase methods, which apply chemical reactions in solvents. This route is particularly attractive due to its simplicity and versatility and leads to colloids, in which the NPs formed can be stabilized against aggregation by surfactants or ligands. These methods involve mainly precipitation, hydrolysis, spray, solvent thermal methods (high temperature and high pressure), solvent evaporation pyrolysis, oxidation reduction (room pressure), emulsion, radiation chemical synthesis, and sol–gel processing.

The second group consists of solid phase methods, which include thermal decomposition, solid-state reactions, and spark discharge, stripping, and milling methods.

The last group involves the gas phase methods. Several gas phase synthetic techniques have been used to prepare NPs. These methods use both inert and reactive atmospheres at a variety of pressures. The gas phase methods include gas phase evaporation methods (resistance heating, high frequency induction heating, plasma heating, electron beam heating, laser heating, electric heating evaporation method, vacuum deposition on the surface of flowing oil, and exploding wire method), chemical vapor reactions (heating heat pipe gas reaction, laser-induced chemical vapor reaction, plasma-enhanced chemical vapor reaction), chemical vapor condensation, and sputtering methods (Kalia, 2015; Kruis et al., 1998).

14.1.3 Metal oxide–polymer hybrid nanocomposites

Since the end of the last century, the discovery of polymer NCs and their ever expanding use in various applications has been the result of continuous developments in polymer science and nanotechnology. Polymers are substances whose molecules have high molar masses and are composed of a large number of repeating units. These materials are encountered in everyday life and are used for many purposes because they are lightweight, easily processed, and allow for design flexibility. Polymer composites, consisting of the polymer phase and an additional component or components (fillers), generally balance performance, mechanical properties, cost, and processability.

The addition of fillers into the polymer matrix to make composites was expected to result in material properties (mechanical, thermal stability and expansion, fire retardant, electrical and barrier properties) not achieved by either phase alone, and often lower cost. This procedure led to an improvement in polymer properties despite the fact that their lightweight and ductile nature were preserved. Hence, the addition of fillers to polymers grew in importance over the years. Based on their chemical family, fillers are divided into two categories: inorganic (e.g., oxides, hydroxides, silicates, metals) and organic (e.g., carbon, natural, and synthetic polymers). Fillers may also be considered as continuous (long fibers or ribbons) or discontinuous (short fibers, flakes, or particulates). The fillers are also classified regarding their shape and size or aspect ratio (ratio of length to diameter for a fiber or the ratio of diameter to thickness for platelets and flakes). These fillers must be added to the polymer matrix in a thoughtful and cost-effective way. Also, additional component or fillers are classified as reinforcing and nonreinforcing fillers.

Reinforcing fillers aid in improving mechanical properties and abrasion resistance whereas nonreinforcing fillers may decrease the cost, modify density, improve barrier properties, or change color (Sperling, 2006; Patel, 2005).

Over the past two decades, polymer composites in which nanosized fillers are distributed homogeneously (known as filler/polymer NCs) have been widely studied because of their unique mechanical, thermal, diffusion barrier, optical, electric, and magnetic properties (Ajayan et al., 2004; Morikawa et al., 1992; Claude et al., 1992; Luther-Davies et al., 1996; Novak et al., 1994; Schmit and Wolter, 1990; Popall et al., 1990; Glaser and Wilkes, 1988; Pope and Mackenzie, 1987; Pope et al., 1989). Because the properties of the nanoscale fillers can be extraordinary, very low content of NPs can produce a high level of reinforcement in polymers (Zare, 2016). Incorporation of nanomaterials in polymers makes a new class of materials which have versatile properties and show potential ability in various fields such as flexible light emitters (Zhang et al., 2015; Lee et al., 2014), magnetic storage (Frey et al., 2009), energy storage (Onlaor et al., 2014), and optical power limiters (Haripadmam et al., 2014).

Polymer NCs are manufactured via various methods that can be categorized into three major routes: melt mixing, in situ polymerization, and solution mixing. The melt mixing process involves heating a polymer and filler mixture under shear, usually in an extruder, above the softening temperature of the polymer. In situ polymerization consists of two stages. First a monomer solution or liquid monomer including the fillers must be prepared and then the mixture needs to be subjected to polymerization. The solution method involves dissolution polymers in common solvents and addition of the fillers followed by evaporation of the solvent in order to achieve the composites (Patel, 2005).

Synthesis of organic/inorganic hybrid systems using different nanoscaled particles such as layered silicates, nanosilica, carbon nanotubes (CNTs), expandable graphite, inorganic NPs, metal oxides, layered titanate, inorganic nanotubes, cellulose nanowhiskers, polyhedral oligomeric silsesquioxanes (POSS), etc., is one of the hottest topics of modern nanotechnology within the last few years (Kar et al., 2014). Among these, the synthesis of polymer/metal oxide NCs has attracted considerable attention because of their potential applications. This study discusses the fabrication of PVC/metal oxide NCs. A limited number of papers have been published on the synthesis of PVC/metal oxide NCs. This chapter will provide a general overview of the techniques and strategies used to prepare these NCs.

14.2 Synthesis of poly(vinyl chloride)/metal oxide hybrid nanocomposites

Numerous experimental methodologies have been developed for synthesizing PVC/metal oxide hybrid NCs, as the hybrid NCs not only inherit the functionalities of NPs but also possess advantages of polymers such as flexibility, film integrity, and conformity (Sangawar and Golchha, 2013). In this paper, we provide an overview of the recent advances in processing and characterization of metal oxide-reinforced PVC composite. Three of the most important and widely used PVC—metal oxide NCs have been presented. In addition, other kinds of the publications have been mentioned briefly in the fourth category.

14.2.1 ZnO/PVC nanocomposite

ZnO is scientifically interesting n-type wide band gap semiconductor. It has room temperature band gap energy of 3.34 eV with high exciton binding energy of 60 meV. It has other advantages such as low cost, nontoxicity, and high stability (Tamgadge et al., 2016). Polymeric ZnO NC materials have been increasingly obtaining researchers' interest because introduction of ZnO filler into polymeric matrices can modify the optical (e.g., shielding from UV and NIR radiation), electrical, and mechanical properties (Mu et al., 2011).

Nano-ZnO/PVC NCs could be extensively applied in coating, rubbers, plastics, fibers, and other applications. Elashmawi et al. studied the effect of the addition of

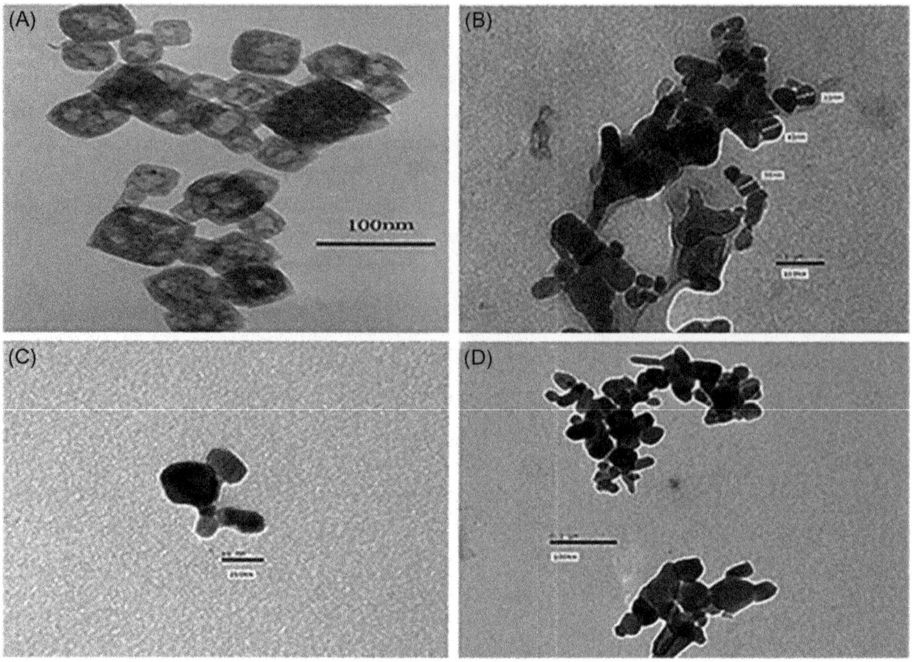

Figure 14.2 TEM of (A) pure ZnO nanoparticle and (B)–(D) ZnO/PVC (20 wt%).
Source: Adapted from Elashmawi et al., 2010, with kind permission of Elsevier.

different concentrations of ZnO NPs on improving the structural, optical, and thermal properties of PVC. ZnO/PVC NCs films of different compositions of ZnO (0, 2.5, 5, 10, 15, and 20 wt% of ZnO nanopowder) have been prepared by the solvent casting method. Fig. 14.2 shows a good dispersion of the ZnO NPs in the PVC polymeric matrix.

The introduction of nano-ZnO into PVC could improve the structural, mechanical, and thermal properties of pure PVC because of its small size, large specific area, quantum effect, and a strong interfacial interaction. Also, the NC films had higher glass transition temperature, specific heat, and thermal stability relative to those of pure PVC because of strong interaction among ZnO NPs and PVC. The prepared samples are stable over 460°C and are chosen in polymer batteries (Elashmawi et al., 2010).

When Sawai and his colleagues found that ZnO powders had antibacterial activities against some bacteria strains in 1995, more and more researchers have embarked on the studies on ZnO NPs as an antibacterial agent (Li et al., 2015). Antibacterial activity of the ZnO NP PVC composite against *Staphylococcus aureus* was studied by Seil et al. The composite is prepared by an ultrasonication reaction between ZnO NPs (at weight percentages of 0%, 2%, 10%, 25%, and 50%) and tetrahydrofuran-dissolved PVC. *S. aureus* cultured in wells containing ZnO NP PVC composites were less active than bacteria cultured on pure polymer samples according to optical density experiments, crystal violet assays, and live/dead staining. Optical density data indicated a significant reduction in total bacteria counts for bacteria cultured in the presence of the composites compared to bacteria cultured in the presence of a pure polymer sample (Fig. 14.3). These composites could be used for biomaterial applications prone to excessive bacterial growth such as orthopedic implants and endotracheal tubes.

There was no significant difference in the reduction of biofilm formation among the different ZnO weight percent composites. SEM images showed that composites with low weight percentages of ZnO NPs had areas of exposed ZnO NPs, but these areas were widely dispersed (Fig. 14.4) (Seil and Webster, 2011).

Incorporation of ZnO NPs of varying sizes (from 10 nm to 200 nm in diameter) and functionalization (including no functionalization to doping with aluminum oxide and functionalizing with a silane coupling agent) into PVC either with or without ultrasonication were studied using a procedure developed by Maschhoff et al. to synthesize ZnO/PVC NCs for numerous medical device applications. Fibroblasts appear to play an important role in wound healing because they not only inhibit bacterial growth but also maintain or promote healthy mammalian cell. Results presented the first evidence of greater fibroblast density after 18 hours of culture on the smallest ZnO NP incorporated PVC samples with dispersion aided by ultrasonication. Specifically, the greatest amount of fibroblast proliferation was

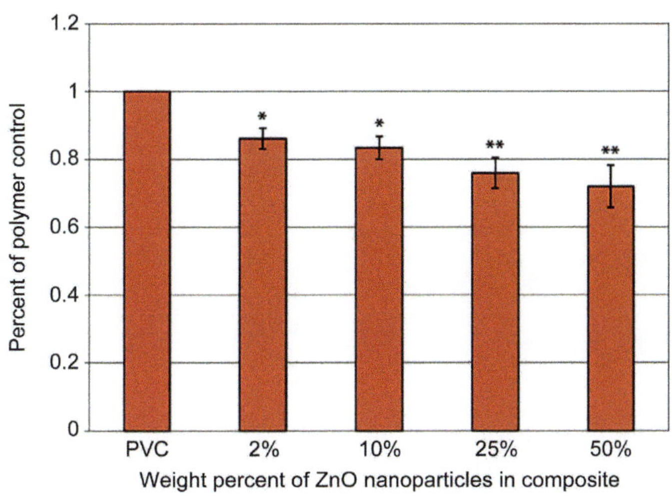

Figure 14.3 Bacteria populations (as a percentage of populations on polymer control) determined by optical density readings of bacteria suspensions cultured in wells containing composite samples at various weight percents for 24 hours. All composite samples showed a significant reduction compared to pure PVC ($P < 0.05$). Composites with weight ratios of 25% and 50% significantly reduced bacteria growth compared to 10% composites ($P < 0.05$). Data = mean ± SEM; $N = 3$.
Source: Adapted from Seil and Webster, 2011, with kind permission of Elsevier.

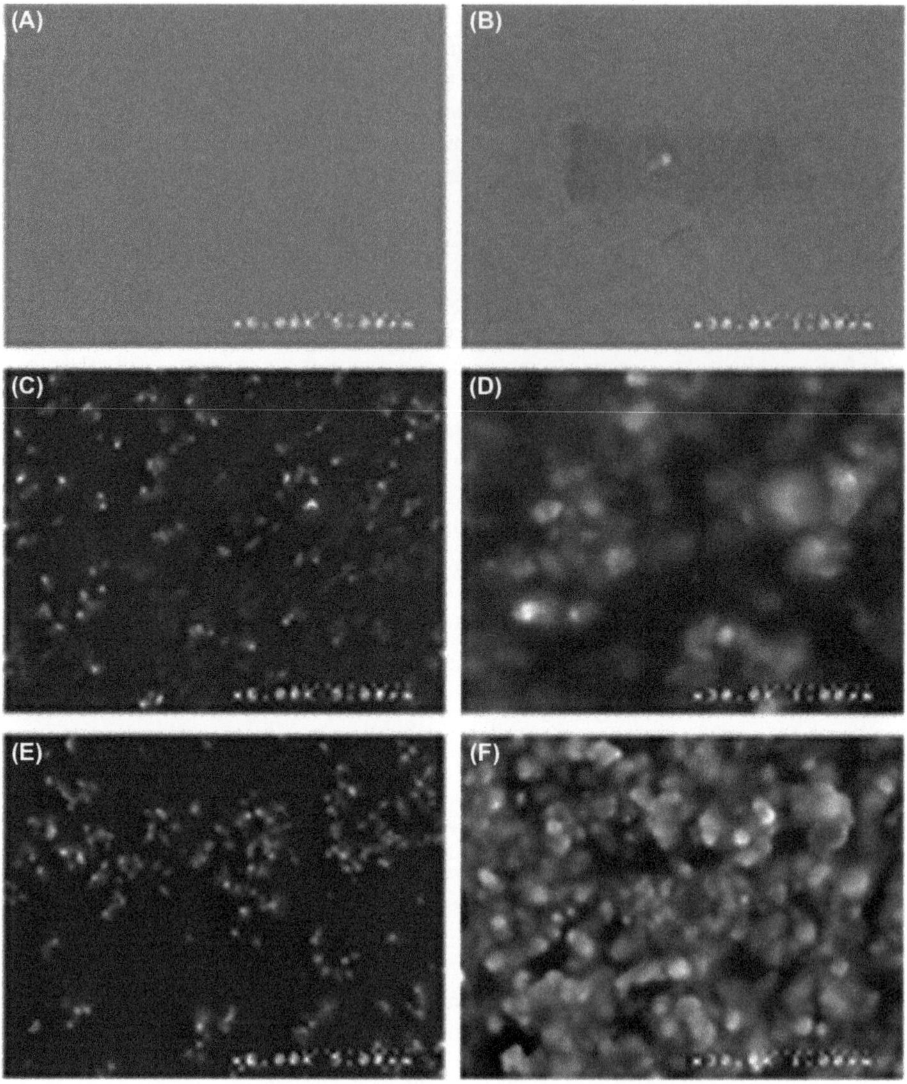

Figure 14.4 Scanning electron micrographs of the ZnO/PVC composites. 100:0 (PVC: ZnO) wt% at 6000 × (A) and 30,000 × (B); 98:2 (PVC:ZnO) wt% at 6000 × (C) and 30,000 × (D); 90:10 (PVC:ZnO) wt% at 6000 × (E) and 30,000 × (F); 75:25 (PVC:ZnO) wt% at 6000 × (G) and 30,000 × (H); 50:50 (PVC:ZnO) wt% at 6000 × (I) and 30,000 × (J). Scale bar for images in left column = 5 lm. Scale bar for images in right column = 1 lm.

Source: Adapted from Seil and Webster, 2011, with kind permission of Elsevier.

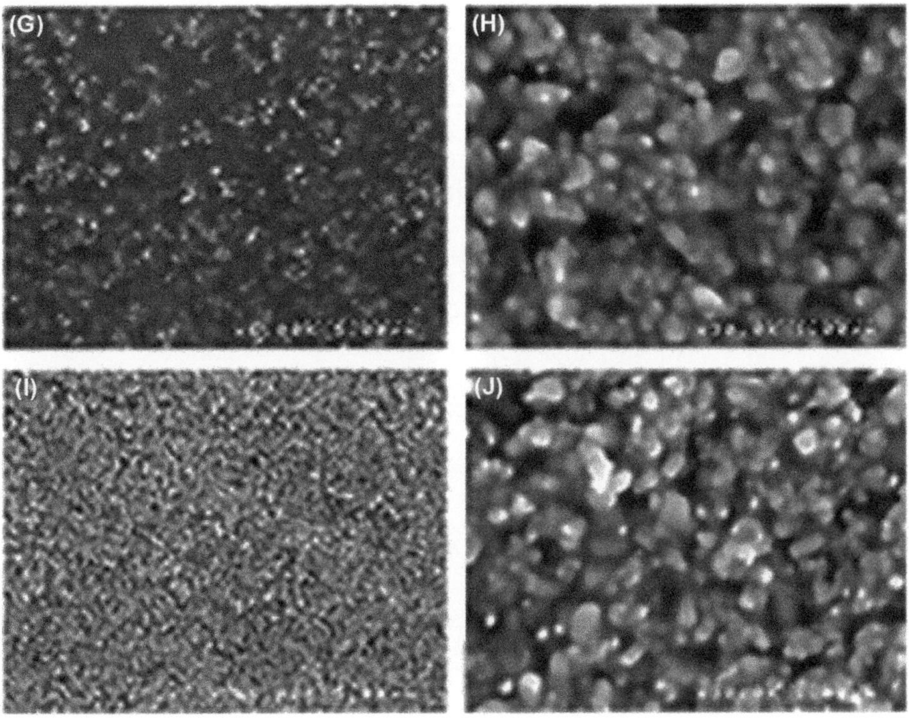

Figure 14.4 (Continued)

measured on ZnO NPs functionalized with a silane coupling agent KH550; this sample exhibited the greatest dispersion of ZnO NPs (Maschhoff et al., 2014).

Al-Taa'y et al. investigated the effects of different concentrations of nano-ZnO (1–20) wt% on the optical properties of PVC films. Nanocomposites were prepared by the casting technique and results revealed that nanosize zinc oxide can effectively dope PVC and enhance its optical properties. The presence of ZnO leads to an increase in the absorption (Fig. 14.5) and to a decrease in the transmission as ZnO concentration increases (Fig. 14.6). Fig. 14.5 demonstrated that the absorption spectra for all films decreased with increasing wavelength, while for the doped films of samples (1%, 5%, 10%, 15%, and 20%), the absorption increased with increasing doping concentration of ZnO. According to Fig. 14.6, the transmittance intensity increases with the increasing of the wavelength, and as the concentration of doped material nano-ZnO increases, the transmittance decreases. The reason for this behavior is that the increases of concentration of ZnO lead to increases of the localized state density which reduces the transmittance values (Al-Taa'y et al., 2014).

Machovsky et al. demonstrated that the antibacterial effect of ZnO-based materials is strongly dependent on their particle morphology. They reported the preparation of three different ZnO-based antibacterial fillers having different

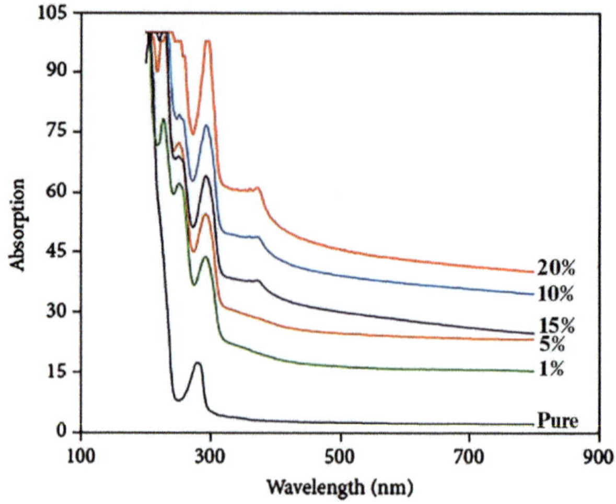

Figure 14.5 Absorption spectra of all PVC samples.
Source: Adapted from Al-Taa'y et al., 2014, open access journal.

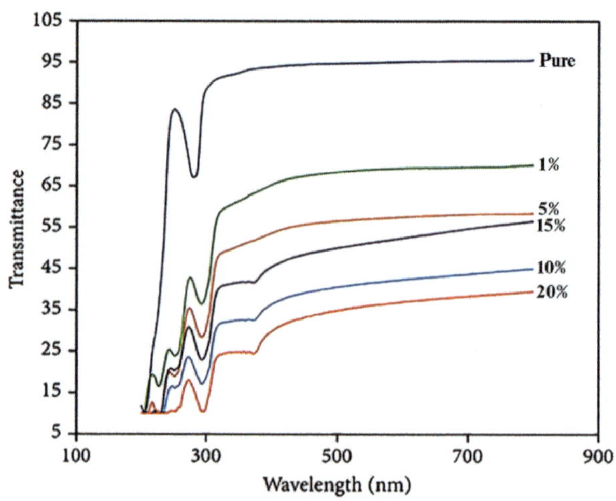

Figure 14.6 Transmission spectra of all PVC samples.
Source: Adapted from Al-Taa'y et al., 2014, open access journal.

morphologies in microscale region by the use of the microwave-assisted synthesis protocol with additional annealing in one case. As antibacterial fillers, the basic zinc hydroxide acetate (ZHA) and two ZnO powders denoted as ZnO 1 and ZnO 2 were prepared. 0.5, 1, 2, and 5 wt% of ZHA, ZnO 1 and ZnO 2 were incorporated into PVC by melt mixing for the synthesis of ZnO/PVC NCs. The medical-grade flexible PVC composite materials with ZnO 2 filler loading higher than 2 wt% showed an excellent antibacterial polymer composite system with a considerable surface antibacterial activity value higher than 6 against *E. coli* and higher than 5 against *S. aureus*. Almost all prepared composite materials were found efficient enough for less demanding applications as they have antibacterial activity at least 2. The studied ZnO-based nanostructured microfillers, namely ZnO 2, have potential in medical plastic industries as additive materials for PVC medical devices (Machovsky et al., 2014).

Mallakpour et al. reported the production of novel PVC/ZnO-CA NC films by ultrasonic irradiation as a green synthesis method with different concentrations of ZnO-CA NPs (4, 8, and 12 wt%) in the PVC matrix. First ZnO NPs were modified with citric acid (CA) to prevent agglomeration and achieve good dispersion of NPs in PVC matrix along with controlling the size (Fig. 14.7).

Fig. 14.7 has demonstrated different interactions between the carboxylate groups of the CA and −OH on the surface of ZnO NPs. PVC/ZnO-CA NC films were prepared by the incorporation of different concentrations of modified ZnO-CA filler into PVC (Fig. 14.8).

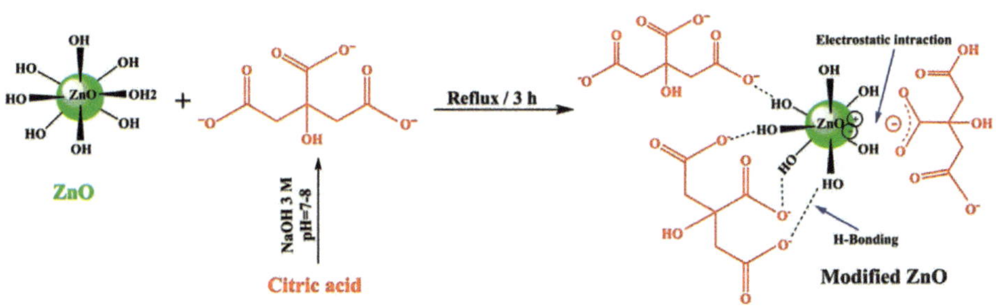

Figure 14.7 Possible forms of interactions between modifier and NPs.
Source: Adapted from Mallakpour and Javadpour, 2015a, with kind permission of Springer.

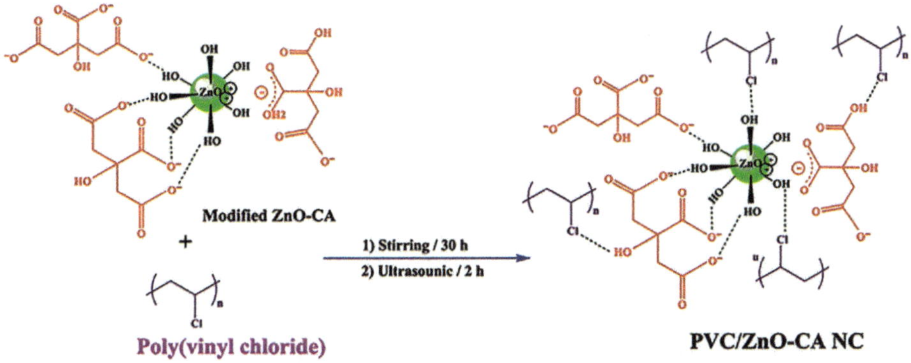

Figure 14.8 Fabrication of PVC/ZnO-CA NC films.
Source: Adapted from Mallakpour and Javadpour, 2015a, with kind permission of Springer.

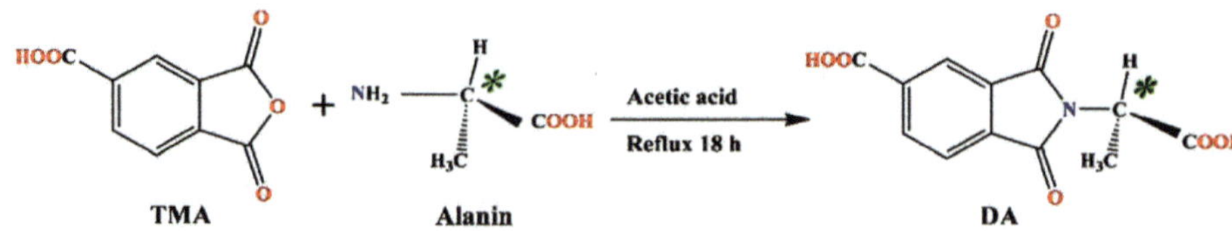

Figure 14.9 Synthesis of DA.
Source: Adapted from Mallakpour and Javadpour, 2015b, with kind permission of Taylor & Francis.

The thermal degradation temperature of PVC matrix was obviously improved by the incorporation of ZnO-CA fillers, and also, the obtained NCs could be classified as self-extinguishing materials on the basis of the LOI values. The presence of ZnO NPs also reclaimed remnant ecofriendly property of the PVC by replacing of HCl harmful gas with Zn salts (Mallakpour and Javadpour, 2015a).

In another work, they utilized ultrasonic irradiation for the surface modification of ZnO NPs by the reaction between −OH group on the surface of NPs and the −COOH groups of the synthesized diacid (DA) coupling agent containing alanine amino acid and the NCs were manufactured by the introduction of modified ZnO into the PVC. Surface modification of ZnO NPs was carried out with an optically active DA containing L-alanine amino acid under ultrasonic irradiation conditions which were reported before by this research group (Fig. 14.9).

The surface modification of ZnO NPs by diacid as a biocompatible coupling agent to prevent aggregation is shown in Fig. 14.10. These ZnO@DA NPs fillers were reinforced into the valuable PVC matrix as filler to fabricate PVC/ZnO@DA NC films (Fig. 14.11). Introduction of modified ZnO NPs into the PVC matrix can improve self-extinguishing parameter in NCs. More than environmental friendly properties of obtained NCs, better degradability of valuable PVC NCs have been expected due to chiral DA containing alanine amino acid linked on reinforced ZnO QDs. Due to countless amount of medical devices made by PVC, the antibacterial activity of biosafe ZnO NPs as well as biocompatible modifier may bestead health care system in novel PVC/ZnO@DA NCs (Mallakpour and Javadpour, 2015b).

Incorporation of ZnO@DA into the PVC matrix also can show new approach for environmentally benign recycling of waste PVC and producing fuel from PVC wastes (Sarker et al., 2012).

Mallakpour et al. also investigated the structural, optical, and mechanical properties of PVC NC films with the addition of different concentrations of the ZnO-PVA NPs. Surface modification of biosafe ZnO NPs was accomplished with the reaction between their −OH groups and the −OH groups of the polyhydroxy PVA chains as a capping agent. ZnO-PVA NPs (4, 8, 12 wt%) were embedded into the PVC matrix via safe ultrasonic irradiation (Fig. 14.12).

TEM analysis was carried out to observe directly the hybrid structure and internal morphology of the obtained PVC/ZnO-PVA NCs as well as the modified ZnO NPs. TEM images of PVA-coated ZnO NPs are presented in Fig. 14.13E and F. Fig. 14.13A−D show TEM images of PVC/ZnO-PVA NC 8 wt% at two different

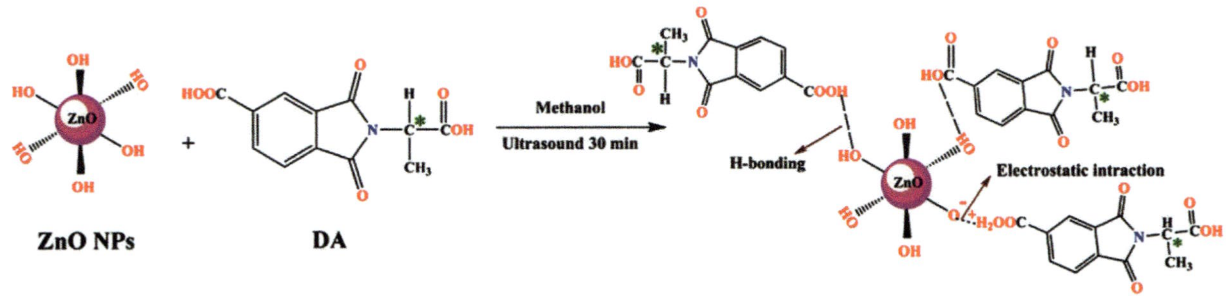

Figure 14.10 Modification of ZnO NPs with DA.
Source: Adapted from Mallakpour and Javadpour, 2015b, with kind permission of Taylor & Francis.

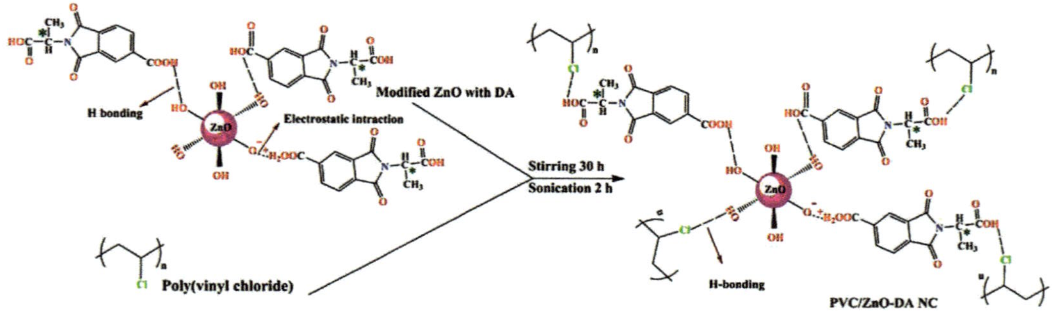

Figure 14.11 Fabrication of PVC/ZnO@DA NCs.
Source: Adapted from Mallakpour and Javadpour, 2015b, with kind permission of Taylor & Francis.

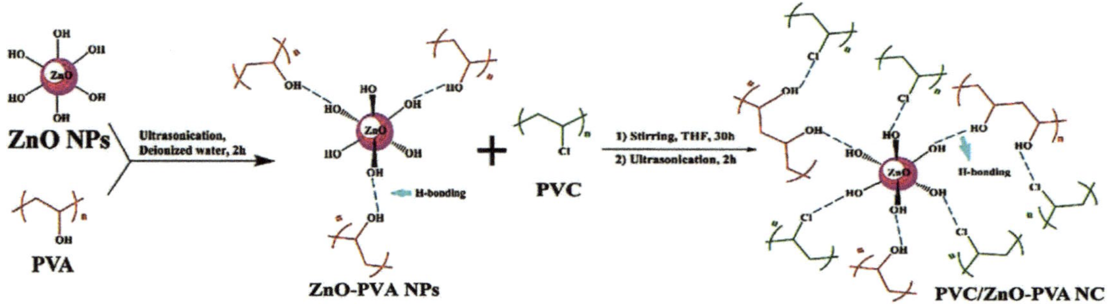

Figure 14.12 Modification of ZnO NPs with PVA and fabrication of PVC/ZnO-PVA NCs.
Source: Adapted from Mallakpour and Javadpour, 2015c, with kind permission of John Wiley & Sons.

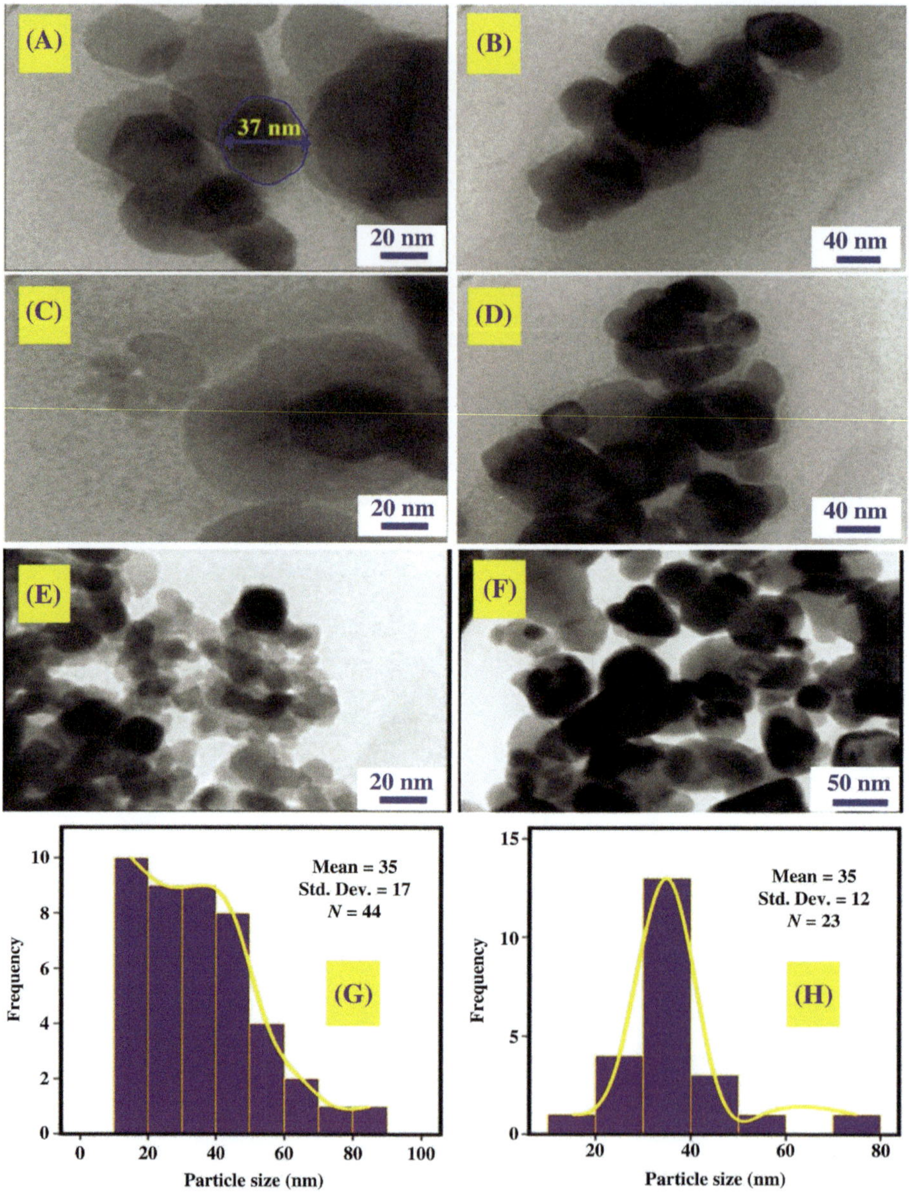

Figure 14.13 TEM images of (E, F) PVA-coated ZnO NPs, (A–D) PVC/ZnO-PVA NC 8 wt % at two different magnifications.
Source: Adapted from Mallakpour and Javadpour, 2015d, with kind permission of John Wiley & Sons.

magnifications. The particles dispersed uniformly in the PVC/ZnO-PVA NC and their sizes were about 37 nm. Self-extinguishing parameter in NCs is improved by the incorporation of modified ZnO NPs into PVC matrix. It was expected to have degradable NCs for decreasing environmental impact of valuable PVC NCs. Therefore, in the future, modified ZnO with bioactive PVA could be useful for the manufacturing of polymer NCs with environmental friendly and antibacterial specifications to apply in medical fields. Also, the better degradability of valuable PVC NCs has been expected due to PVA linked on reinforced ZnO NPs. Owing to great amount of medical devices made by PVC, the antibacterial activity of ZnO NPs may bestead health care system in novel PVC/ZnO-PVA NCs (Mallakpour and Javadpour, 2015c).

Recently, a great effort has been made by the community of researchers for the synthesis and utilization of nanomaterials in the field of medical materials (Geilich and Webster, 2013).

Sedlak et al. described a fast and simple solvothermal microwave-assisted preparation of hierarchical nanostructured mesoscale ZnO. A composite was obtained by mixing these particles with the plasticized medical-grade PVC as a model polymer matrix to develop an antibacterial polymer system for utilization in medical devices.

The filler dispersion in the prepared composites and the morphology of the prepared filler are presented in the SEM images in Fig. 14.14. As can be seen in Fig. 14.14A, the material consists of globular-like particle aggregates with a diameter ranging from 200 nm up to 1 μm. The filler dispersion in the fabricated composites was accomplished at the aggregate level and its distribution in the PVC matrix is shown in the SEM image in Fig. 14.14B, for the sample with $w = 2\%$ of the filler. There are no visible agglomerates of the globular aggregates within the polymer matrix. On the other hand, the aggregates were not disaggregated during the mixing procedure.

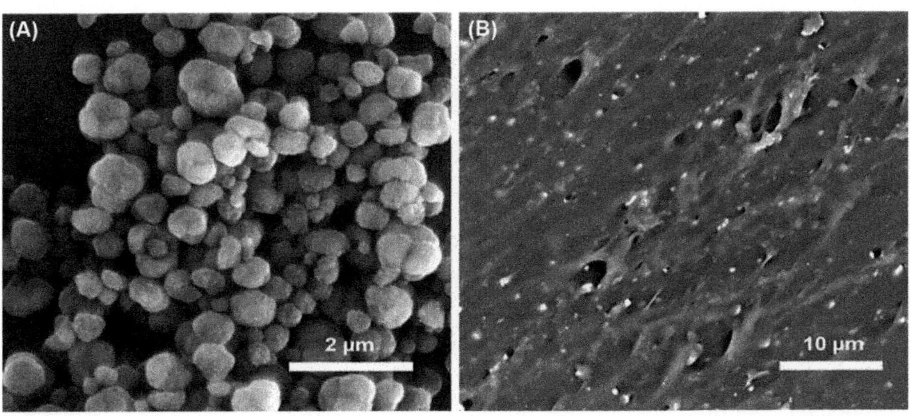

Figure 14.14 SEM images of: (A) prepared ZnO filler and (B) cross-section of the prepared composite material with $w = 2\%$ of ZnO filler.
Source: Adapted from Sedlak et al., 2015, open access journal.

The composite materials showed excellent antibacterial activity values against *E. coli* and satisfactory antibacterial activity values against *S. aureus*. They have an application potential in medicine as the material for medical devices being in the direct contact with the human body, besides many other possible utilizations (Sedlak et al., 2015).

14.2.2 TiO_2/PVC nanocomposite

Titanium dioxide (Titania, TiO_2) has attracted large interest because of its physical and chemical stability, nontoxicity, regular pore structure, and uniquely large specific surface area (Alipour et al., 2016). In the wide field of organic–inorganic composite materials, TiO_2 is a popular filler of organic matrixes. Introduction of TiO_2 filler into the polymeric matrixes results in new and often unique properties, unapproachable for the individual components (Kierys et al., 2013). Introduction of TiO_2 into a polymer can significantly affect electrical, optical, and photocatalytic properties of TiO_2, resulting in NCs with a wide range of applications in future technologies (Ansari and Mohammad, 2011; Irimpan et al., 2008; Kun et al., 2006).

Hasan et al. prepared the TiO_2-PVC NCs by incorporating TiO_2 in PVC followed by solution casting to synthesis of TiO_2-PVC NCs thin films. TiO_2@PVC has higher thermal stability through the incorporation of TiO_2, because of the stabilizing interactions between TiO_2 with PVC. The incorporation of TiO_2 in the PVC matrix would strengthen the mechanical properties of PVC, owing to the good dispersion and strong interfacial adhesion. The mechanical performance of the TiO_2@PVC NCs film was significantly higher than that of pure PVC. These NCs are suitable for a range of practical applications, such as photodegradation and construction, due to its improved optical, thermal, and mechanical performance. This work is the first experimental finding of the strain-induced band gap reduction of TiO_2@PVC NCs, which may open up a novel approach for tuning the band gap of similar anisotropic and/or superlattice materials. The TiO_2@PVC NC film compared with pure TiO_2 showed a lower band gap for the anatase due to the generation of strain within the NCs, which might find interesting applications in visible light-induced photocatalytic activities (Hasan et al., 2015).

It is generally known that environmental stress cause the polymers such as PVC to degrade, as indicated by changes in color, gloss, mechanical properties, etc. (Rabinovitch et al., 1993; Bedia et al., 2003; Raab et al., 1982). Therefore, it is a clear need for the addition of stabilizers and pigments to PVC for its protection from UV irradiation. For most applications, TiO_2 has been used to improve the opacity, appearance, and durability of PVC products (Gardette and Lemaire, 1991; Burn, 1992; Real et al., 2003). Yang et al. demonstrated that TiO_2 particle in PVC composites inhibits photooxidation and chain scission during QUV accelerated weathering. SEM images were investigated to observe the surface morphology of PVC composites with 5 phr and 10 phr of TiO_2 during the accelerated weathering (Fig. 14.15).

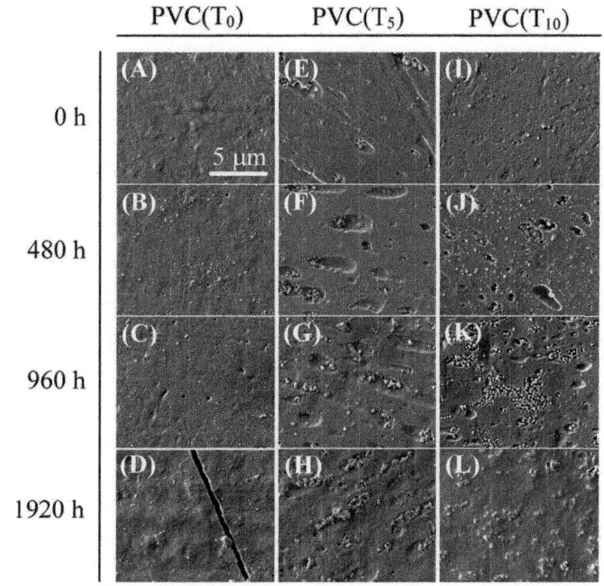

Figure 14.15 SEM images of various composites over 1920 hours of accelerated weathering. (A–D): PVC (T_0), (E–H): PVC (T_5), and (I–L): PVC (T_{10}).
Source: Adapted from Yang et al., 2015, with kind permission of John Wiley & Sons.

It is evident from Fig. 14.15 that the surface morphology of PVC without TiO_2 particle did not exhibit changes up to 960 hours, but exhibited a rough and brittle surface after 1920 hours of QUV accelerated weathering. The weatherability of PVC composites was found to be affected by TiO_2 particle in two distinct and opposing ways, i.e., on one hand, TiO_2 particle acts as a UV absorber to protect the PVC matrix, and on the other hand, TiO_2 particle acts as a UV-activated oxidation catalyst to degrade the PVC composite surface layer (Yang et al., 2015).

Recently, the incorporation of different types of NPs for the modification of filtration membranes has received special attention. Many research groups have devoted effort to improve the membrane performance via different types of TiO_2 because of its suitable properties such as desirable hydrophilicity and fine dispersion in polymeric solution (Lee et al., 2003).

Rabiee et al. used the immersion precipitation phase inversion method to produce emulsion poly(vinyl chloride)/titanium dioxide (EPVC/TiO_2) NCs ultrafiltration membranes with NMP as the solvent and water as the nonsolvent using different TiO_2 contents (0.2, 0.5, 1, 2, and 4 wt% of polymeric solution) for the first time. EDAX analysis showed fine and homogeneous dispersion of TiO_2 NPs in membrane matrix at low concentrations (Fig. 14.16). The NPs have a tendency to aggregate slightly at higher loadings (particularly at 4 wt% TiO_2) that is frequent in other NP-modified membranes. The incorporated TiO_2 NP could improve antifouling properties of the EPVC membrane to be used as an ultrafiltration membrane for the treatment of high fouling wastewaters (Rabiee et al., 2014).

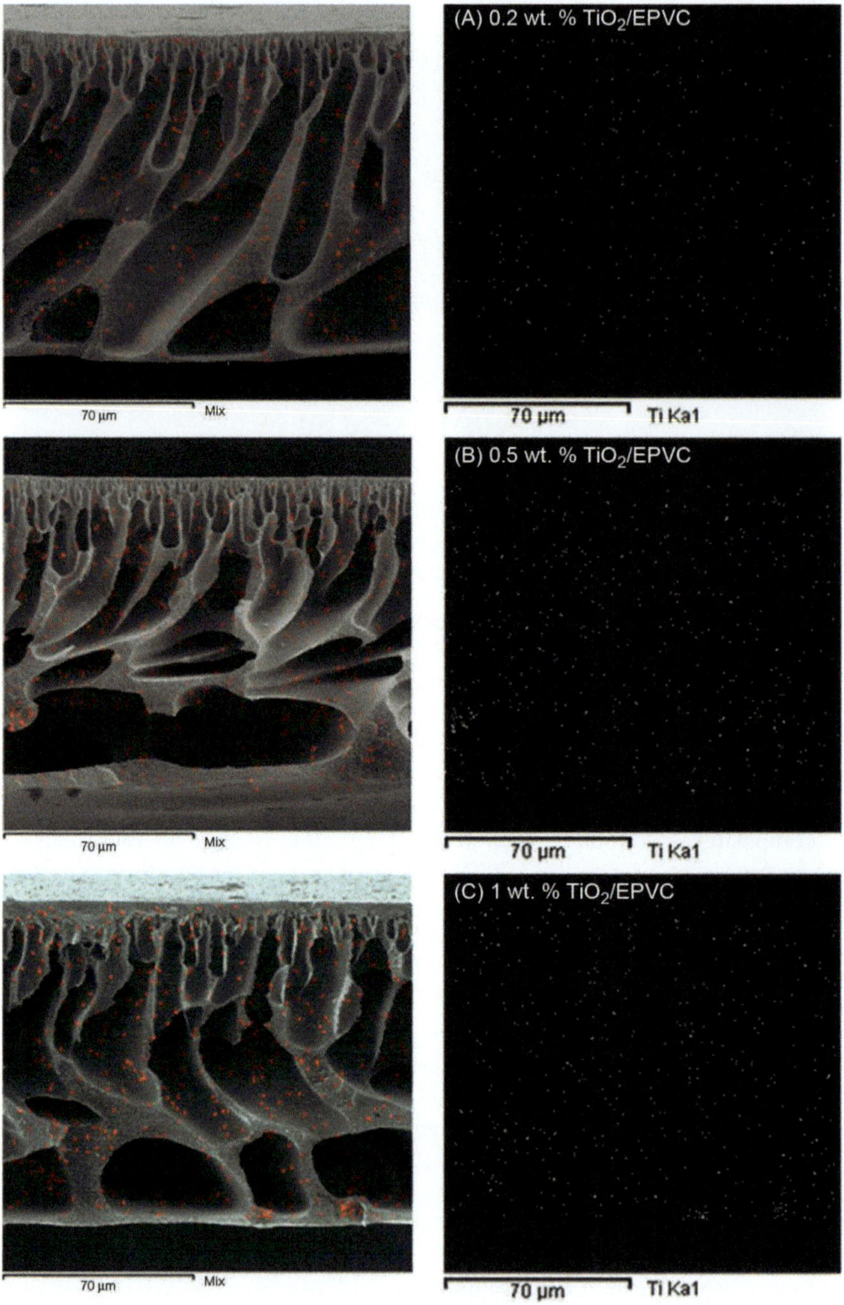

Figure 14.16 EDAX of the cross-section of TiO_2/EPVC nanocomposite membranes: (A) 0.2 wt%, (B) 0.5 wt%, (C) 1 wt%, (D) 2 wt%, and (E) 4 wt%.
Source: Adapted from Rabiee et al., 2014, with kind permission of Elsevier.

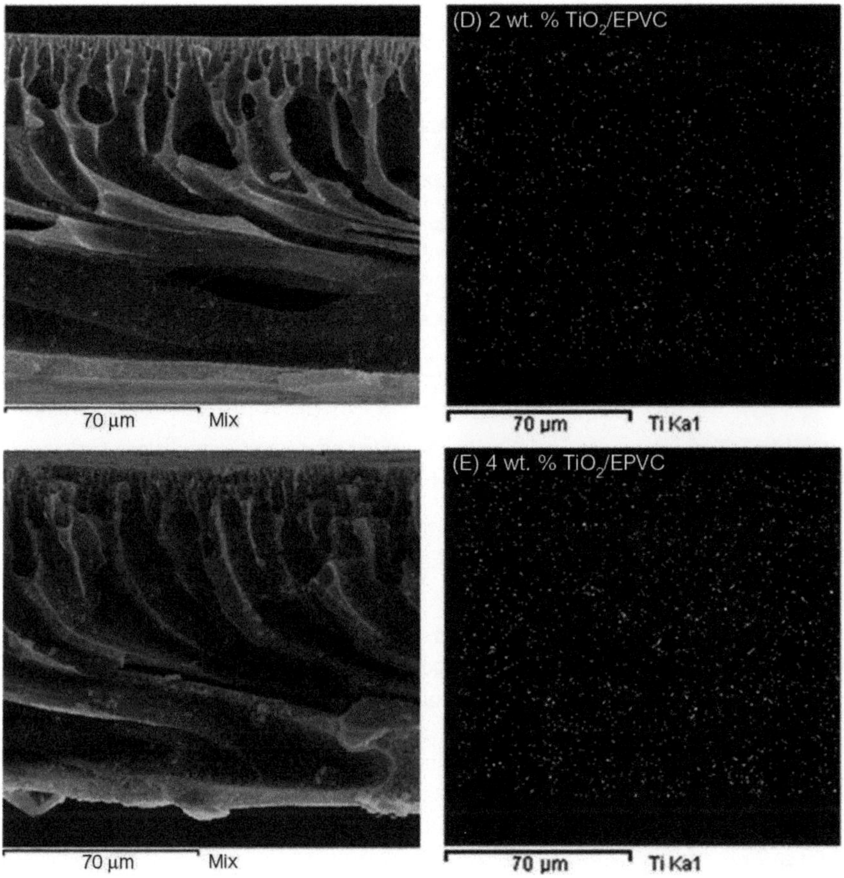

Figure 14.16 (Continued)

The effect of UV and visible lights irradiation on the PVC/TiO_2 NC films containing different amounts of synthesized TiO_2 NPs and commercial rutile powder was studied by Ghaebi Mehmandoust et al. in order to investigate the kinetic parameters of photodegradation over long periods of exposure and evaluate the effect of irradiation intensity as well as wavelength on the rate of photodegradation process. Rigid NCs of UPVC were synthesized by melt blending using the synthesized carbon-coated TiO_2 NPs and commercial powder of TiO_2 (with rutile structure) (Sokhandani et al., 2014). The samples were artificially aged in UV and visible chambers, for 5112 hours with continuous irradiation. Irradiations were executed at room temperature (Fig. 14.17) with a relative humidity range from 40% to 70% during irradiation period.

The degradation kinetics was studied by monitoring the photoinduced weight loss of samples under irradiation of UV and visible lights after 5112 hours (Fig. 14.18). It should be mentioned that with increasing the concentration of TiO_2

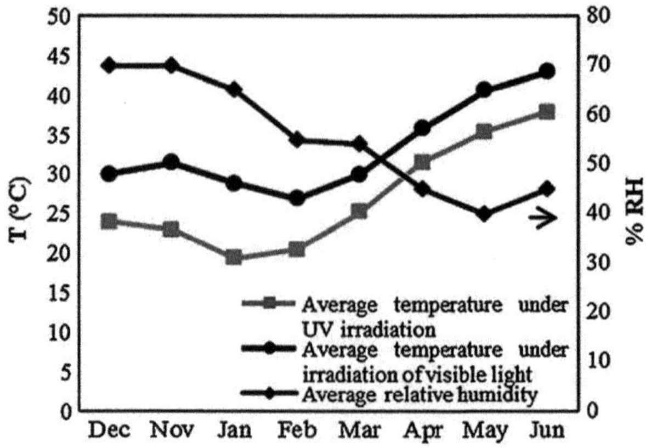

Figure 14.17 Temperature and relative humidity variation during irradiation period. *Source*: Adapted from Ghaebi Mehmandoust et al., 2014, with kind permission of John Wiley & Sons.

NPs in the PVC matrix, the fractional weight loss was decreased, as presented in Fig. 14.18.

The effect of irradiation intensity on the photodegradation process of these NCs was discussed according to the reciprocity law experiment, and it was found that photodegradation occurred in two regimes with respect to irradiation intensity (Ghaebi Mehmandoust et al., 2014).

Mallakpour et al. has reported the surface modification of TiO_2 NPs with citric acid (CA) and then with vitamin C (VC) as bilayer surface modification for better compatibility with PVC matrix. The surface modification process for TiO_2 NPs is shown in Fig. 14.19.

Interfacial interaction between the efficient functional groups of CA and VC and the hydroxyl groups on the surface of TiO_2 NPs can prevent agglomeration of the NPs in the polymer. CA and VC can act as cross-linkers between NPs and the polymeric matrix. Fig. 14.20 shows the possible interactions between polymer and the NPs. The optical properties of the NCs changed because of the presence of the modified TiO_2. The maximum UV absorption of NC films was more than the pure PVC, and its absorption was transferred to higher wavelengths in the visible region. These can be attributed to the existence of the TiO_2 NPs, which can increase UV absorption of the NCs because of its UV-resistant property. Since UV−Vis light has undesirable effects on materials, increases in absorption or reductions in transmission of UV−Vis radiation of NCs are considered as an advantage when these are employed in medical or food packaging (Mallakpour and Jarang, 2015).

14.2.3 Al_2O_3/PVC nanocomposite

Today, ceramics like alumina, silicon carbide, and silicon oxide have been widely used in many engineering applications such as automotive, electronics, space

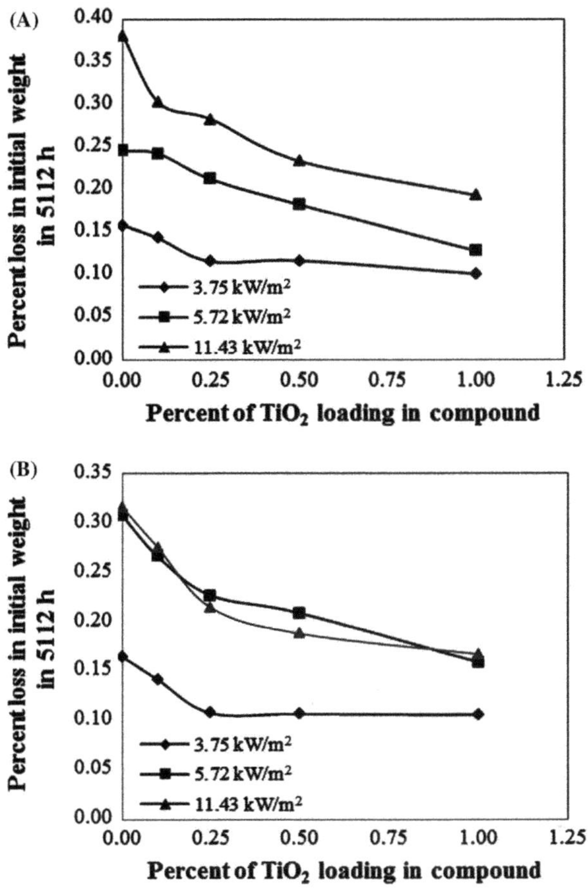

Figure 14.18 Percentage degradation after 5112-hour exposure versus TiO_2 concentration at different intensities under irradiation of (A) ultraviolet and (B) visible light.
Source: Adapted from Ghaebi Mehmandoust et al., 2014, with kind permission of John Wiley & Sons.

shuttle, and ballistic amour. Among these ceramics, alumina or aluminum oxide (Al_2O_3) of various levels of purity is one of the most cost-effective and widely used materials in the family of engineering ceramics because of their several desirable properties such as excellent hardness and wear resistance, good dielectric properties, resistance to strong acid, and alkali attack at elevated temperatures and high thermal conductivity (Kajohnchaiyagual et al., 2014).

Polymer NCs containing Al_2O_3 NPs are a new group of the hybrid materials that demonstrate remarkable improvement in properties with very low nanofillers loading. In these materials, the individual shortcomings of the metal oxides and the polymeric phases can be overcome without compromising the parent properties of either (Sarkar et al., 2012; Kango et al., 2013).

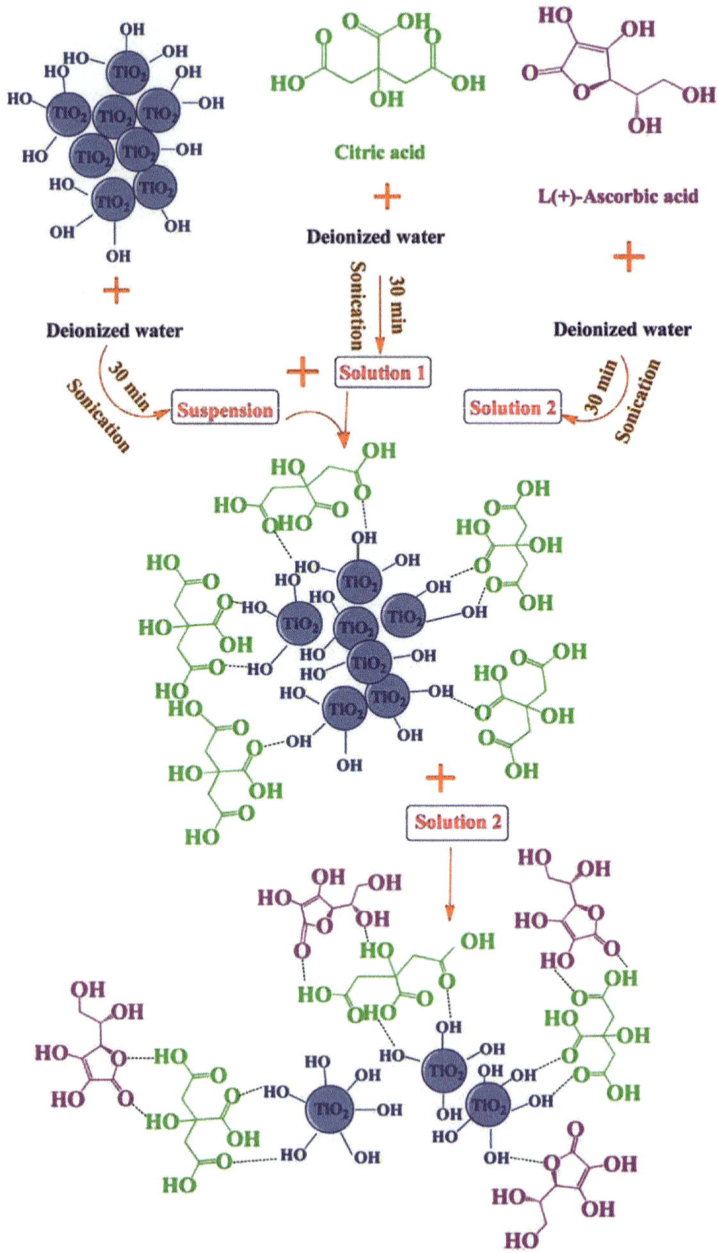

Figure 14.19 Surface modification of TiO_2 NPs with CA and VC.
Source: Adapted from Mallakpour and Jarang, 2015, with kind permission of John Wiley & Sons.

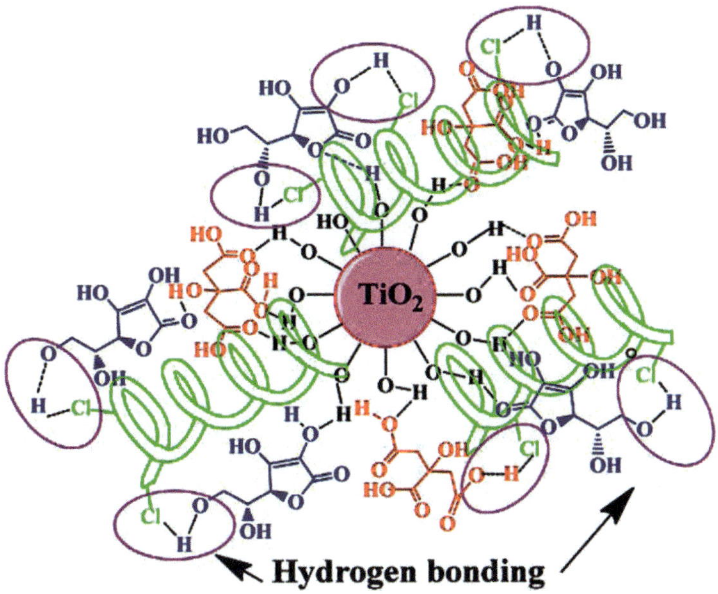

Figure 14.20 Probable interaction between TiO_2-CA-VC NPs and PVC matrix.
Source: Adapted from Mallakpour and Jarang, 2015, with kind permission of John Wiley & Sons.

The effects of various amount of Al_2O_3 nanopowder on the mechanical properties of poly PVC-based films were investigated by Abu-Abdeen. Various amounts of Al_2O_3: 0, 0.5, 1.0, 2.0, and 5.0 wt% were incorporated into PVC matrix for preparing NCs. Loading of PVC with nanopowder Al_2O_3 with concentrations up to 5.0 wt% decreases the elastic modulus, complex viscosity, and storage modulus (Abu-Abdeen, 2012).

Nanocomposites based on polymer (PVC) reinforced with nanofiller (Al_2O_3) were obtained using melt-blending process with plasticizers and heat stabilizer by Nikam et al. Four different concentrations of nano-Al_2O_3, such as 1, 2, 4, and 6 phr, were added to blend system. Nano-Al_2O_3 particles were well dispersed in the matrix at lower concentrations. Also, tensile strength results indicated that the reinforcement in the NCs was more effective than that in the pristine PVC. The increment of Young's modulus of NCs from 1 to 6 phr is about 75%, which is higher than pristine PVC. Dielectric properties of plasticized PVC (p-PVC) NC sheets were carried out in the frequency range from 20 Hz to 1 MHz at various DC bias potentials using the LCR meter impedance analyzer. In the p-PVC NCs, dielectric constant values are increased up to 4 phr whereas with the addition of higher concentration of NPs, the dielectric constant values are decreased for 6 phr content of nano-Al_2O_3. It can be used as a good electrical insulation for different electronic devices (Nikam and Sunita, 2014). Citric acid (CA) and ascorbic acid (AA) were utilized as biocompatible modifiers for the surface modification of Al_2O_3 NPs by

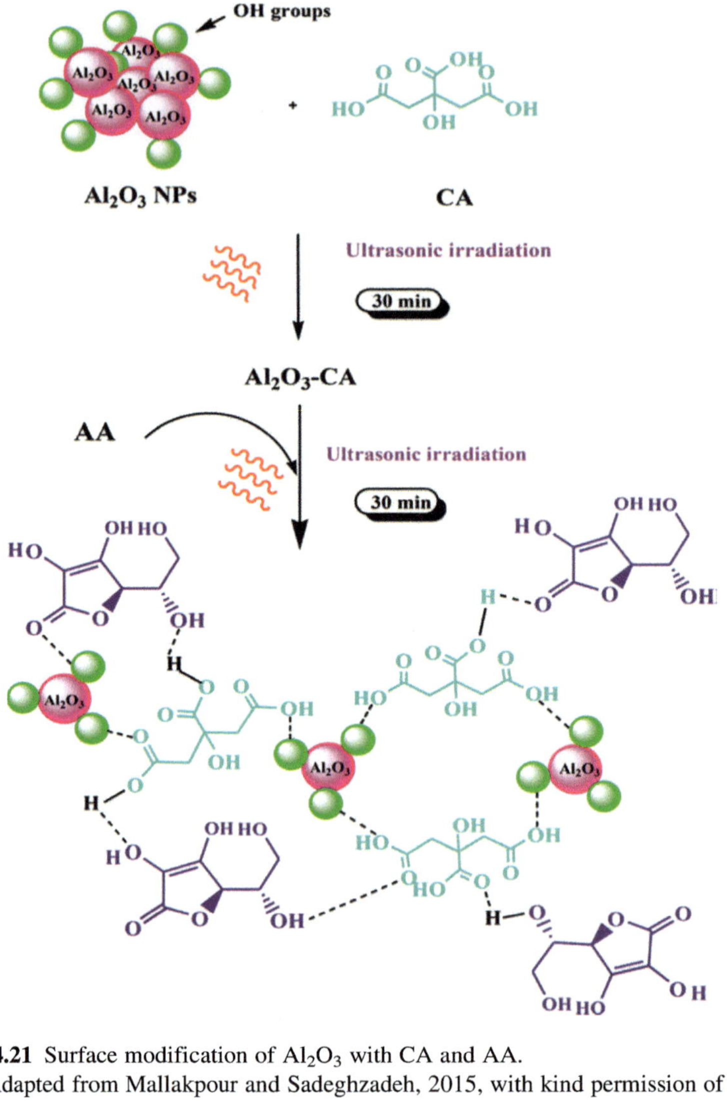

Figure 14.21 Surface modification of Al_2O_3 with CA and AA.
Source: Adapted from Mallakpour and Sadeghzadeh, 2015, with kind permission of John Wiley & Sons.

Mallakpour et al. (Fig. 14.21). It can be seen from Fig. 14.21 that the modified Al_2O_3 has ability to form a homogeneous hybrid film with the PVC matrix because of the organic chains of CA and AA which enabled it to be used as filler. Nanocomposites of PVC and Al_2O_3 were prepared via the introduction of varying amounts of modified Al_2O_3 NPs (4, 8, and 12 wt%) into the PVC matrix using ultrasonication method. Char yield can be applied as a decisive way for estimating the limiting oxygen index (LOI) and its calculation is based on the Van Krevelen Hoftyzer equation obtained from the original thermograms.

$$LOI = 17.5 + 0.4CR$$

where CR is the char yield. The LOI values for PVC/α-Al_2O_3-CA-AA NCs were calculated in the range of 27–29%. According to the results, the LOI values of PVC and PVC/Al_2O_3-CA-AA NCs were higher than 21% (21% is the percentage of oxygen in air). This result confirmed that all NCs could be categorized as self-extinguishing materials. Based on strong or weak interfacial adhesion between chains of the polymer matrix and NPs, tensile strength and elongation could be increased or decreased. Tensile strength for 4 wt% NCs is increased because of strong bonding between modified Al_2O_3, whereas tensile strength for NCs 8 and 12 wt% decreased due to poor bonding between matrix and modified NPs (Mallakpour and Sadeghzadeh, 2015).

14.2.4 Other metal oxides/PVC nanocomposite

The use of a variety of other metal oxides in PVC has been reported. PVC/copper oxide composites with 1% of CuO or Cu_2O were obtained by using melt-blending method in a twin-screw extruder by Rodriguez-llamazares et al. These composite materials inhibit the bacterial adhesion compared to neat PVC. The PVC/Cu_2O composite inhibits to a large extent the bacterial adhesion than the PVC/CuO composite material. Copper oxides alter the membrane of *Escherichia coli* ATCC 25922 and reduce the length of the bacterial cell (Rodriguez-llamazares et al., 2012). Mallakpour et al. reported the synthesis of iron oxide (Fe_3O_4) NPs, the modification of NPs with organic molecules of citric acid (CA) and ascorbic acid (AS), and the incorporation of different amounts of the modified magnetic NPs (4, 8, and 12 wt%) into PVC matrix. Incorporating the NPs improved flame retardancy of the NCs. The modification of NPs with CA and AS was expected to improve biocompatibility of PVC NCs. Furthermore, because of the many kinds of medical equipment made by PVC and the antibacterial activity of capping agents, novel PVC/Fe_3O_4, CA, AS NCs can likely be used in medical devices (Mallakpour and Javadpour, 2015d). Chiscan et al. prepared the PVC/Fe_3O_4 composite nanofibers with the iron oxide NPs completely embedded in the PVC matrix. The transmission loss of PVC/Fe_3O_4 NCs measured in the microwave frequency range of X-band was found to be below -16 dB proving that these materials can be used as electromagnetic radiation protection material (Chiscan et al., 2012). Mallakpour et al. synthesized PVC/α-MnO_2-KH550 NC films embedded with different amounts of the modified α-MnO_2 nanorod (1, 3, and 5 wt%). To compare the properties of pure PVC and its NCs, optical, thermal, mechanical, and adsorptive properties were investigated by different techniques. The experimental results demonstrated that PVC/α-MnO_2-KH550 NCs could be used for the removal of Pb (II) ions from the aqueous solution. In addition, the thermal and mechanical properties of the neat polymer was dramatically increased even at very low modified α-MnO_2 nanorods contents such as 1–5 wt% (Mallakpour et al., 2016). Xiao et al., (2012) prepared the spherical PVC-$Li_4Mn_5O_{12}$ ion-sieve with particle diameter 2.0–3.5 mm where using $Li_4Mn_5O_{12}$ ultrafine powder as ion-sieve precursor, PVC as binder and it was treated with HCl solution to obtain spherical PVC-MnO_2 ion sieve, which well maintains the excellent adsorption behavior of the ultrafine powder with a high Li^+ adsorption capacity of 4.50 mmol/g and also it has a high Li^+ selectivity when recovering Li^+ from the solution containing Li^+, Na^+, K^+, Mg^{2+} (Xiao et al. 2012). PbO was successfully prepared using a sol–gel method and loaded into pure PVC at 1.0 and 2.0 wt% by El Sayed et al. The dielectric constant was increased by adding PbO NPs to PVC due to the resulting interfacial polarization and PVC/PbO NCs films could be used in a variety of dielectric and optical components and devices (El Sayed and Morsi, 2013). Mallakpour et al. reported the incorporation of SiO_2 NPs into the PVC matrix to prepare PVC/SiO_2 NCs. For effective dispersion of SiO_2 NPs, the surface modification of NPs was carried out using citric acid (CA) and L (+)-ascorbic acid (AA) and ultrasonic irradiation technique. Then different percentages of modified NPs (4, 8, and 12 wt%) were introduced in the PVC matrix and PVC/SiO_2–CA–AA NCs were prepared. Microscopic observations showed good dispersity of SiO_2–CA–AA NPs in the polymer matrix. Resulting NC films showed more flexibility than pure PVC (Mallakpour and Naghdi, 2015).

Yu et al. prepared a novel low-cost SiO_2/PVC membrane with different nano-SiO_2 particles loading (0–4 wt%) using the phase inversion process. They found that the membrane with 1.5 wt% nano-SiO_2 addition showed better capabilities against the protein absorption and bacterial attachment, better antifouling performance, and higher flux recovery ratio in filtration of the supernatant liquor which collected from a secondary sedimentation tank in a municipal wastewater plant (Yu et al., 2015).

Chen et al. used ultrasonic oscillations and high energy vibromilling to modify nano-SiO_2 particles with various interfaces and interfacial interactions between the particles and PVC matrix. They demonstrated that direct surface treatment of nano-SiO_2 particles with a silane coupling agent (KH550) is not effective for improving the mechanical properties of PVC/SiO_2 composites. The mechanical properties of this composite with high energy vibromilling modified SiO_2 particles were outstandingly improved (Chen et al., 2006).

A new method of surface chemical modification of nano-SiO_2 was developed by Zhao et al. Hyperbranched poly(amine-ester) (HPAE) was successfully grafted from nano-SiO_2 surface using A_2B monomer N,N-dihydroxyethyl-3-amino methyl propionate by one-step polycondensation. Then the resulting modified nano-SiO_2 was blended with PVC to improve its properties and processing. Compared with pure PVC, PVC/HPAE-g-SiO_2 composites processing improved to a certain extent (Zhao et al., 2008).

Mallakpour et al. combined the PVC with different concentrations (4, 8, and 12 wt%) of ZrO_2 NPs with the purpose of preparing novel NCs. In order to achieve uniform dispersion of the NPs in polymer matrix, double layer modification of ZrO_2 NPs with citric acid (CA) and ascorbic acid (AA) was carried out. According to LOI values, these NCs can be classified as the self-extinguishing materials (Mallakpour and Nezamzadeh Ezhieh, 2015).

Uma reported the ionic conductivity, structural and thermal studies of polymer electrolytes based on PVC and Li_2SO_4 as lithium salt, with and without the presence of different concentrations (5–15 wt%) of dispersed ZrO_2 ceramic powder. It was found that conductivity increases and then dips with the addition of ZrO_2 (Uma et al., 2004).

Chromium oxides (Cr_2O_3) NPs were synthesized using a sol–gel method and mixed with PVC by Hassen et al. The dielectric permittivity and ac conductivity of pure PVC increased with adding Cr_2O_3 due to the formation of conductive three-dimensional networks throughout the NC films and interfacial polarizations. Adding the Cr_2O_3 NPs to PVC films modified their optical properties considerably and these composites could be used in optical devices (Hassen et al., 2014).

El Sayed et al. investigated the influences of CdO NPs on the optical as well as the dielectric properties of PVC. CdO NPs were synthesized by using a sol–gel process and different amounts of the synthesized NPs (0.0, 0.3, 0.7, 1.0, and 1.4 wt%) were added to PVC. It was found that the dielectric constant increased with the addition of CdO NPs to PVC due to the interfacial polarization. Also, the ac conductivity increased with increasing CdO content. These results may reflect the importance applications of these NC films such as optical and/or electrical devices (El Sayed et al., 2014). Linda et al. prepared nano-CdO/ZnO/PVC composite thin films by simple solution cast method, using tetrahydrofuran as solvent. The photocatalytic activity of CdO/ZnO/PVC NCs was studied for the degradation of congo red dye aqueous solution under UV light irradiation and results showed that the CdO/ZnO/PVC NCs significantly enhance the photo catalytic activity towards the degradation of congo red dye (Linda et al., 2015). Zampronio et al. described the synthesis of a new PVC/V_2O_5 hybrid organic–inorganic material. A very important feature from this material is the stable electrochemical response as well as the electrochromic effect in aqueous medium. Another very interesting property, from a morphological point of view, is the flexibility/plasticity of the new material due to the PVC molecules (Zampronio et al., 2003).

Recently, the synthesis of metal oxide/polymer NC is growing rapidly as evidenced from the number of publications in this field.

14.3 Conclusion

Polymeric materials are widely used in industries owing to the ease of fabrication and their ductile and lightweight nature. However, due to their limitation such as low thermal stability and brittleness, much attention has been paid to improve these drawbacks. To overcome these defects, particulate fillers are used to modify the physical and mechanical properties of polymers in many ways. Recently, PVC is one of the most important polymers used these days, since it has many applications in medical equipment, pipes, as well as some machine elements. A small amount of the nano-additives such as metal oxides could improve the overall performances of this polymer. This chapter reviews the recent advances in preparation of metal oxides/PVC NCs, the structure of these NCs, and the resulting rheological, mechanical, and thermal properties. The introduction of nano-ZnO into PVC could improve the structural, mechanical, and thermal properties of pure PVC owing to its small size, large specific area, quantum effect, and a strong interfacial interaction. Up to now, a large variety of methods have been developed for the synthesis of ZnO/PVC NCs which could be widely applied in coating, rubbers, plastics, fibers, and other applications. Also, these NCs have antibacterial properties without the use of pharmaceutical agents; this makes such a material desirable for use in a wide array of medical devices. The combination of TiO_2 with PVC can significantly affect electrical, optical, and photocatalytic properties of TiO_2, resulting in NCs with a wide range of applications in future technologies. The dispersion of alumina in the PVC matrix is achieved by various methods. Al_2O_3/PVC NCs are a new group of the hybrid materials that demonstrate remarkable improvement in properties with very low nanofillers loading. In these materials, the individual shortcomings of the metal oxides and the polymeric phases can be overcome without compromising the parent properties of either. Other metal oxides which have been used as nanofillers in PVC matrix include Cu_2O, Fe_3O_4, MnO_2, PbO, SiO_2, ZrO_2, Cr_2O_3, CdO, and V_2O_5.

Acknowledgments

The authors gratefully acknowledge the financial support provided by Research Affairs Division, Isfahan University of Technology (IUT), Isfahan, and the National Elite Foundation (NEF).

References

Abu-Abdeen, M., 2012. Static and dynamic mechanical properties of poly(vinyl chloride) loaded with aluminum oxide nanopowder. Mater. Des. 33, 523–528.

Ajayan, P.M., Schadler, L.S., Braun, P.V., 2004. Front matter. Nanocomposite Science and Technology. Wiley-VCH Verlag GmbH & Co. KGaA, pp. 1–75.

Alipour, D., Keshtkar, A.R., Moosavian, M.A., 2016. Adsorption of thorium (IV) from simulated radioactive solutions using a novel electrospun PVA/TiO_2/ZnO nanofiber adsorbent functionalized with mercapto groups: study in single and multi-component systems. Appl. Surf. Sci. 366, 19–29.

Allsopp, M.W., Vianello, G., 2000. Poly(vinyl chloride). Ullmann's Encycl. Ind. Chem. 28, 441–468.

Al-Taay, W., Abdul Nabi, M., Yusop, R.M., Yousif, E., Abdullah, B.M., Salimon, J., et al., 2014. Effect of nano ZnO on the optical properties of poly(vinyl chloride) films. Int. J. Polym. Sci. 2014, 6.

Ansari, M.O., Mohammad, F., 2011. Thermal stability, electrical conductivity and ammonia sensing studies on p-toluenesulfonic acid doped polyaniline:titanium dioxide (pTSA/Pani:TiO_2) nanocomposites. Sens. Actuat. B Chem. 157, 122–129.

Ayuub, P., Multani, M., Barma, M., Palkar, V.R., Vijayaghavan, R.J., 1998. Size-induced structural phase transitions and hyperfine properties of microcrystalline Fe_2O_3. Phys. C: Solid State Phys. 21, 2229.

Bedia, E.L., Paglicawan, M.A., Bernas, C.V., Bernardo, S.T., Tosaka, M., Kohjiya, S., 2003. Natural weathering of polypropylene in a tropical zone. J. Appl. Polym. Sci. 87, 931–938.

Burn, L.S., 1992. The influence of titanium dioxide on the durability of uPVC pipe. Polym. Degrad. Stab. 36, 155–167.

Chen, G., Tian, M., Guo, S., 2006. A study on the morphology and mechanical properties of PVC/nano-SiO_2 composites. J. Macromol. Sci. B. 45, 709–725.

Chiscan, O., Dumitru, I., Postolache, P., Tura, V., Stancu, A., 2012. Electrospun PVC/Fe_3O_4 composite nanofibers for microwave absorption applications. Mater. Lett. 68, 251–254.

Claude, C., Garetz, B., Okamoto, Y., Tripathy, S., 1992. The preparation and characterization of organically modified silicates that exhibit nonlinear optical properties. Mater. Lett. 14, 336–342.

Corr, S.A., 2012. Nanoscience: Volume 1: Nanostructures through Chemistry. RSC Publishing.

Elashmawi, I.S., Hakeem, N.A., Marei, L.K., Hanna, F.F., 2010. Structure and performance of ZnO/PVC nanocomposites. Phys. B Condens. Matter. 405, 4163–4169.

El Sayed, A.M., Morsi, W.M., 2013. Dielectric relaxation and optical properties of polyvinyl chloride/lead monoxide nanocomposites. Polym. Compos. 34, 2031–2039.

El Sayed, A.M., El-Sayed, S., Morsi, W.M., Mahrous, S., Hassen, A., 2014. Synthesis, characterization, optical, and dielectric properties of polyvinyl chloride/cadmium oxide nanocomposite films. Polym. Compos. 35, 1842–1851.

Frey, N.A., Peng, S., Cheng, K., Sun, S., 2009. Magnetic nanoparticles: synthesis, functionalization, and applications in bioimaging and magnetic energy storage. Chem. Soc. Rev. 38, 2532–2542.

Garvie, R.C., Goss, M.F., 1986. Intrinsic size dependence of the phase transformation temperature in zirconia microcrystals. J. Mater. Sci. 21, 1253–1257.

Geilich, B.M., Webster, T.J., 2013. Reduced adhesion of *Staphylococcus aureus* to ZnO/PVC nanocomposites. Int. J. Nanomed. 8, 1177–1184.

Ghaebi Mehmandoust, S., Alizadeh, R., Babaluo, A.A., 2014. Kinetic study of the poly(vinyl chloride)/titanium dioxide nanocomposites photodegradation under accelerated ultraviolet and visible light exposure. Polym. Adv. Technol. 25, 799–808.

Glaser, R.H., Wilkes, G.L., 1988. Structure property behavior of polydimethylsiloxane and poly (tetramethylene oxide) modified TEOS based sol–gel materials. Polym. Bull. 19, 51–57.

Haripadmam, P.C., John, H., Philip, R., Gopinath, P., 2014. Switching of absorptive nonlinearity from reverse saturation to saturation in polymer-ZnO nanotop composite films. Appl. Phys. Lett. 105, 221102.

Hasan, M., Banerjee, A.N., Lee, M., 2015. Enhanced thermo-mechanical performance and strain-induced band gap reduction of TiO_2@PVC nanocomposite films. Bull. Mater. Sci. 38, 283–290.

Hassen, A., El-Sayed, S., Morsi, W.M., Sayed, A.M.E., 2014. Preparation, dielectric and optical properties of Cr_2O_3/PVC nanocomposite films. J. Adv. Phys. 4, 571–584.

Hernandez-Alonso, M.D., Belen Hungria, A., Martinez-Arias, A., Coronado, J.M., Carlos Conesa, J., Soria, J., et al., 2004. Confinement effects in quasi-stoichiometric CeO_2 nanoparticles. Phys. Chem. Chem. Phys. 6, 3524–3529.

Irimpan, L., Krishnan, B., Nampoori, V.P.N., Radhakrishnan, P., 2008. Luminescence tuning and enhanced nonlinear optical properties of nanocomposites of ZnO–TiO_2. J. Colloid. Interface. Sci. 324, 99–104.

Gardette, J.L., Lemaire, J., 1991. Photo-oxidation of poly(vinyl chloride): Part 3—Influence of photo-catalytic pigments. Polym. Degrad. Stab. 33, 77–92.

Kajohnchaiyagual, J., Jubsilp, C., Dueramae, I., Rimdusit, S., 2014. Thermal and mechanical properties enhancement obtained in highly filled alumina-polybenzoxazine composites. Polym. Compos. 35, 2269–2279.

Kango, S., Kalia, S., Celli, A., Njuguna, J., Habibi, Y., Kumar, R., 2013. Surface modification of inorganic nanoparticles for development of organic–inorganic nanocomposites—a review. Prog. Polym. Sci. 38, 1232–1261.

Kar, K.K., Pandey, J.K., Reddy, K.R., Mohanty, A.K., Misra, M., 2014. Handbook of Polymernanocomposites. Processing, Performance and Application. Springer, pp. 246–248.

Kierys, A., Zaleski, R., Buda, W., Pikus, S., Dziadosz, M., Goworek, J., 2013. Nanostructured polymer–titanium composites and titanium oxide through polymer swelling in titania precursor. Colloid Polym. Sci. 291, 1463–1470.

Kruis, F.E., Fissan, H., Peled, Aaron, 1998. Synthesis of nanoparticles in the gas phase for electronic, optical and magnetic applications—a review. Aerosol Sci. 29, 511–535.

Kun, R., Mogyorósi, K., Dékány, I., 2006. Synthesis and structural and photocatalytic properties of TiO_2/montmorillonite nanocomposites. Appl. Clay Sci. 32, 99–110.

Lee, B.R., Jung, E.D., Park, J.S., Nam, Y.S., Min, S.H., Kim, B.S., et al., 2014. Highly efficient inverted polymer light-emitting diodes using surface modifications of ZnO layer. Nat. Commun. 5, 4840.

Lee, Y.C., Hong, Y.P., Lee, H.Y., Kim, H., Jung, Y.J., Ko, K.H., et al., 2003. Photocatalysis and hydrophilicity of doped TiO_2 thin films. J. Colloid. Interface Sci. 267, 127–131.

Li, Y.N., Xu, W.M., Zhang, G.Q., 2015. Effect of coupling agent on nano-ZnO modification and antibacterial activity of ZnO/HDPE nanocomposite films. IOP Conf. Ser. Mater. Sci. Eng. 87, 012054.

Linda, T., Muthupoongodi, S., Sahaya Shajan, X., Balakumar, S., 2015. Enhancement of CdO/ZnO/PVC nanocomposites behavior on photo-catalytic degradation of congo-red dye under uv light irradiation. Mater. Sci. Forum. 807, 91–99.

Luther-Davies, B., Samoc, M., Woodruff, M., 1996. Comparison of the linear and nonlinear optical properties of poly(p-phenylenevinylene)/sol-gel composites derived from tetramethoxysilane and methyltrimethoxysilane. Chem. Mater. 8, 2586–2594.

Machovsky, M., Kuritka, I., Bazant, P., Vesela, D., Saha, P., 2014. Antibacterial performance of ZnO-based fillers with mesoscale structured morphology in model medical PVC composites. Mater. Sci. Eng. C. 41, 70–77.

Mallakpour, S., Jarang, N., 2015. Exploration of the role of modified titania nanoparticles with citric acid and vitamin C in improvement of thermal stability, optical property, and mechanical behavior of novel poly(vinyl chloride) nanocomposite films. J. Vinyl Addit. Technol.. Available from: http://dx.doi.org/10.1002/vnl.21526.

Mallakpour, S., Javadpour, M., 2015a. Design and characterization of novel poly(vinyl chloride) nanocomposite films with zinc oxide immobilized with biocompatible citric acid. Colloid Polym. Sci. 293, 2565–2573.

Mallakpour, S., Javadpour, M., 2015b. An efficient preparation and characterization of nanocomposite films based on poly(vinyl chloride) and modified ZnO quantum dot with an optically active diacid containing amino acid as coupling agent. Polymer-Plast. Technol. Eng. 55, 498–509.

Mallakpour, S., Javadpour, M., 2015c. Effective methodology for the production of novel nanocomposite films based on poly(vinyl chloride) and zinc oxide nanoparticles modified with green poly(vinyl alcohol). Polym. Compos.

Mallakpour, S., Javadpour, M., 2015d. Effective strategy for the production of novel magnetite poly(vinyl chloride) nanocomposite films with iron oxide nanoparticles doublecapped through citric acid and vitamin C. J. Vinyl Addit. Technol http://dx.doi.org/10.1002/vnl.21509.

Mallakpour, S., Javadpour, M., 2016. An efficient preparation and characterization of nanocomposite films based on poly(vinyl chloride) and modified ZnO quantum dot with an optically active diacid containing amino acid as coupling agent. Polymer-Plastics Technol. Eng. 55, 498–509.

Mallakpour, S., Naghdi, M., 2015. Application of SiO_2 nanoparticles with double layer coverage consist of citric acid and l(+)-ascorbic acid for the production of poly(vinyl chloride)/SiO_2 nanocomposite films with enhanced optical and thermal properties. Polym. Bull.1–17.

Mallakpour, S., Nezamzadeh Ezhieh, A., 2015. A simple and environmentally friendly method for surface modification of ZrO_2 nanoparticles by biosafe citric acid as well as ascorbic acid (vitamin C) and its application for the preparation of poly(vinyl chloride) nanocomposite films. Polym. Compos.. Available from: http://dx.doi.org/10.1002/pc.23746.

Mallakpour, S., Sadeghzadeh, R., 2015. A benign and simple strategy for surface modification of Al_2O_3 nanoparticles with citric acid and L(+)-ascorbic acid and its application for the preparation of novel poly(vinyl chloride) nanocomposite films. Adv. Polym. Technol.. Available from: http://dx.doi.org/10.1002/adv.21622.

Mallakpour, S., Abdolmaleki, A., Tabebordbar, H., 2016. Production of PVC/a-MnO_2-KH550 nanocomposite films: morphology, thermal, mechanical and Pb (II) adsorption properties. Eur. Polym. J. 78, 141–152.

Maschhoff, P.M., Geilich, B.M., Webster, T.J., 2014. Greater fibroblast proliferation on an ultrasonicated ZnO/PVC nanocomposite material. Int. J. Nanomed. 9, 257–263.

Morikawa, A., Iyoku, Y., Kakimoto, M.-A., Imai, Y., 1992. Preparation of a new class of polyimide-silica hybrid films by sol–gel process. Polym. J. 24, 107–113.

Mu, S., Wu, D., Qi, S., Wu, Z., 2011. Preparation of polyimide/zinc oxide nanocomposite films via an ion-exchange technique and their photoluminescence properties. J. Nanomater. 2011, 10.

Nicolais, L., Carotenuto, G., 2014. Nanocomposites: In Situ Synthesis of Polymer-Embedded Nanostructures. Wiley, United States of America, pp. 165–166.

Nikam, P.N.P., Sunita, M., 2014. Effect of concentration of nano-Al_2O_3 on mechanical and dielectric properties of plasticized poly(vinyl chloride) nanocomposites. Int. J. ChemTech Res. 6, 1757–1759.

Novak, B.M., Ellsworth, M.W., Verrier, C., 1994. Organic–inorganic nanocomposite materials. Mechanical and thermal properties. Polym. Mater. Sci. Eng. 70, 266–267.

Onlaor, K., Thiwawong, T., Tunhoo, B., 2014. Electrical switching and conduction mechanisms of nonvolatile write-once-read-many-times memory devices with ZnO nanoparticles embedded in polyvinylpyrrolidone. Org. Electron. 15, 1254–1262.

Patel, N.S. (2005). Synthesis and Characterization of Polymer Nanocomposites. pp. 1–88.

Patrick, S. (2004). PVC Compounds and Processing. p. 15.

Peacock, A.J., 2000. Handbook of Polyethylene: Structures, Properties and Applications. Marcel Dekker Inc, NewYork.

Popall, M., Meyer, H., Schmidt, H., Schulz, J., 1990. Inorganic–organic composites (ormocers) as structured layers for microelectronics. MRS Online Proc. Librar. Archiv. 180, 995 (7 pages).

Pope, E., Mackenzie, J., 1987. Incorporation of organic dyes in polymer/oxide composites. MRS Bull. 12, 29–31.

Pope, E., Asami, M., Mackenzie, J., 1989. Transparent silica gel–PMMA composites. J. Mater. Res. 4, 1018–1026.

Raab, M., Kotulák, L., Kolařík, J., Pospíšil, J., 1982. The effect of ultraviolet light on the mechanical properties of polyethylene and polypropylene films. J. Appl. Polym. Sci. 27, 2457–2466.

Rabiee, H., Farahani, M.H.D.A., Vatanpour, V., 2014. Preparation and characterization of emulsion poly(vinyl chloride) (EPVC)/TiO_2 nanocomposite ultrafiltration membrane. J. Memb. Sci. 472, 185–193.

Rabinovitch, E.B., Summers, J.W., Northcott, W.E., 1993. Changes in properties of rigid PVC during weathering. J. Vinyl Technol. 15, 214–218.

Real, L.P., Rocha, A.P., Gardette, J.-L., 2003. Artificial accelerated weathering of poly(vinyl chloride) for outdoor applications: the evolution of the mechanical and molecular properties. Polym. Degrad. Stab. 82, 235–243.

RITU (2014). Synthesis and characterization of some nanosized metal mixed metal oxides of transition metals. Doctor of Philosophy, Maharishi Markendeshwar University Mullana, India, pp. 1–107.

Rodri'guez-Llamazares, S., Mondaca, M.A., Badilla, C., Maldonado, A., 2012. PVC/ COPPER oxide composites and their effect on bacterial adherence. J. Chil. Chem. Soc. 57, 1163–1165.

Sangawar, Vijaya S., Golchha, M.C., 2013. Optical properties of ZnO/low density polyethylene nanocomposites. Int. J. Sci. Res. 2.

Sarkar, S., Guibal, E., Quignard, F., SenGupta, A.K., 2012. Polymer-supported metals and metal oxide nanoparticles: synthesis, characterization, and applications. J. Nanopart. Res. 14, 1–24.

Sarker, M.A.M.M.R., 2012. Polyvinyl chloride (PVC) waste plastic treatment using zinc oxide (ZnO) with activated carbon and produced hydrocarbon fuel for petroleum refinery. Int. J. Eng. Sci. 1, 29–41.

Schmidt, H., Wolter, H., 1990. Organically modified ceramics and their applications. J. Non-Crystal. Solid. 121, 428–435.

Sedlák, J., Bažant, P., Klofá, J., Pastorek, M., Kuøitka, I., 2015. Antibacterial composite based on nanostructured ZnO mesoscale particles and a poly(vinyl chloride) matrix. Mater. Technol. 49, 55–59.

Seil, J.T., Webster, T.J., 2011. Reduced *Staphylococcus aureus* proliferation and biofilm formation on zinc oxide nanoparticle PVC composite surfaces. Acta Biomater. 7, 2579–2584.

Selegard, L., 2013. Synthesis, surface modification and characterization of metal oxide nanoparticles. PhD Thesis. Linkoping University, Sweden, pp. 1–58.

Skandan, G., Foster, C.M., Frase, H., Ali, M.N., Parker, J.C., Hahn, H., 1992. Phase characterization and stabilization due to grain size effects of nanostructured Y_2O_3. Nanostruct. Mater. 1, 313–322.

Sokhandani, P., Abdi, M.A., Ghaebi Mehmandoust, S., Babaluo, A.A., Mehdizadeh, R., Rezaei, M., et al., 2014. Effects of carbon-coated titanium dioxide nanoparticles as a photostabilizer on the photodegradation of rigid poly(vinyl chloride). J. Appl. Polym. Sci. 131. Available from: http://dx.doi.org/10.1002/app.40228.

Sperling, L.H., 2006. Introduction to Physical Polymer Science. Wiley-Interscience, USA, pp. 19–20.

Susheel Kalia, Y.H., 2015. Organic–Inorganic Hybrid Nanomaterials. Springer Link, USA.

Tamgadge, Y.S., Sunatkari, A.L., Talwatkar, S.S., Pahurkar, V.G., Muley, G.G., 2016. Passive optical limiting studies of nanostructured Cu doped ZnO–PVA composite thin films. Opt. Mater. 51, 175–184.

Titow, W.V. (1984). PVC Technology.

Uma, T., Mahalingam, T., Stimming, U., 2004. Characterization of PVC–ZrO_2 composite polymer electrolytes. J. Mater. Sci. 39, 2901–2904.

Wyckoff, R.W.G., 1964. Crystal Structures. Wiley, NewYork.

Xiao, G.-P., Tong, K.-F., Sun, S.-Y., Yu, J.-G., 2012. Preparation of spherical PVC-MnO_2 ion-sieve and its lithium adsorption property. Chin. J. Inorg. Chem. 28, 2385–2394.

Yang, T.-C., Noguchi, T., Isshiki, M., Wu, J.-H., 2015. Effect of titanium dioxide particles on the surface morphology and the mechanical properties of PVC composites during QUV accelerated weathering. Polym. Compos. Available from: http://dx.doi.org/10.1002/pc.23537.

Yu, Z., Liu, X., Zhao, F., Liang, X., Tian, Y., 2015. Fabrication of a low-cost nano-SiO_2/PVC composite ultrafiltration membrane and its antifouling performance. J. Appl. Polym. Sci. 132.

Zampronio, E.C., Greggio, D.N., Oliveira, H.P., 2003. Preparation, characterization and properties of PVC/V_2O_5 hybrid organic–inorganic material. J. Non-Crystal. Solid. 332, 249–254.

Zare, Y., 2016. Study on interfacial properties in polymer blend ternary nanocomposites: role of nanofiller content. Comput. Mater. Sci. 111, 334–338.

Zhang, Z., Guo, K., Li, Y., Li, X., Guan, G., Li, H., et al., 2015. A colour-tunable, weavable fibre-shaped polymer light-emitting electrochemical cell. Nat. Photon. 9, 233–238.

Zhao, H., Sun, R., Luo, Y., Li, J., 2008. A novel method of hyperbranched poly(amide-ester) modifying nano-SiO_2 and study of mechanical properties of PVC/nano-SiO_2 composites. Polym. Compos. 29, 1014–1019.

Hybrid optically active polymer/metal oxide composites: Recent advances and challenges

15

Shadpour Mallakpour[1,2,3] and Elham Khadem[1]

[1]Organic Polymer Chemistry Research Laboratory, Department of Chemistry, Isfahan University of Technology, Islamic Republic of Iran, [2]Nanotechnology and Advanced Materials Institute, Isfahan University of Technology, Islamic Republic of Iran, [3]Center of Excellence in Sensors and Green Chemistry, Department of Chemistry, Isfahan University of Technology, Islamic Republic of Iran

Chapter Outline

15.1 Introduction

Chirality is one of the most captivating phenomena that plays essential roles in our daily life, in physicists, chemists, biologists, and in the pharmaceutical industry (Lu et al., 2016; Mallakpour and Zadehnazari, 2011). Optically active polymers as one of the most important classes of high-performance engineering materials have attracted a great deal of interest for their unique characteristics. Chiral

Hybrid Polymer Composite Materials: Properties and Characterisation. DOI: http://dx.doi.org/10.1016/B978-0-08-100787-7.00015-9

macromolecules supply not only the profits of any other polymers but also some distinctive features such as orderly secondary structure, adjustable chiral parameter, and abundant interchain interaction as well. These properties have been caused that polymers were used as a good candidate for the chiral stationary phases for separate polar racemates in chromatographic techniques, construction of chiral media for asymmetric synthesis, chiral liquid crystals in ferroelectrics and nonlinear optical devices and asymmetric catalysis applications (Brown, 2013; Itsuno, 2011; Okamoto and Nakano, 1994). Nonetheless, some inherent shortcomings of the polymer restrict the usage of these polymers for a wide range of applications. Incorporation of nanofiller-like metal oxides into the polymer matrix has been proved as an effective method to overcome this problem. The superior properties of inorganic metal oxide compared to polymers are also reflected in optically active polymer composites without compromising the parent properties of the two entities (Mallakpour and Khadem, 2015; Mohanty et al., 2015; Pomogailo and Kestelman, 2005). Chiral polymer/metal oxide nanocomposites show superior mechanical, thermal, electrical, and flame retardant properties compared to the neat polymer. In a composite, mechanism interaction between the metal oxides and polymers is controlled by many factors, the most significant among which are the origin of molecular forces, degree of perfection of nanoparticle surfaces, internal stress and electrical charge, solvent nature, temperature, and molecular mass of the polymer (Kalia and Haldorai, 2015; Mohanty et al., 2015). Herein, the structure, preparation, and properties of optically active polymers and their composites based on metal oxide are discussed in general along with detailed examples drawn from the scientific literature.

15.2 Synthesis and categories of optically active polymers

Accurate synthesis of optically active polymers is a stimulating subject in polymer chemistry, because natural optically active polymers are constituted by controlling the absolute configurations of the asymmetric centers in each monomer unit of the main chain. Synthetic optically active polymers have become of great interesting and importance because of their unique structure and intriguing optically active in optical resolution, catalysts, and sensors (Kawakami and Tang, 2000; Selegny, 1979). Optical activity is a physical characteristic of chairalic material, resulted from configurational and/or conformational arrangement in the chain of polymer. Configurational is attributed to chiral centers in their main chains or in the side groups of optically active polymers. Instead, conformational shows optical activity derived from a chiral conformation such as a helical structure, even without any chiral centers in the chain (Brown, 2013; Itsuno, 2011; Kawakami and Tang, 2000). In recent years, the synthesis of optically active polymers to regulate the absolute configuration of the main chain has attracted much attention. Therefore, widespread effort on the precise synthesis of new types of optically active polymers that are beneficial as chiral foundations is highly desirable. Based on polymerization mechanism of the polymers, asymmetric polymerization classified the following

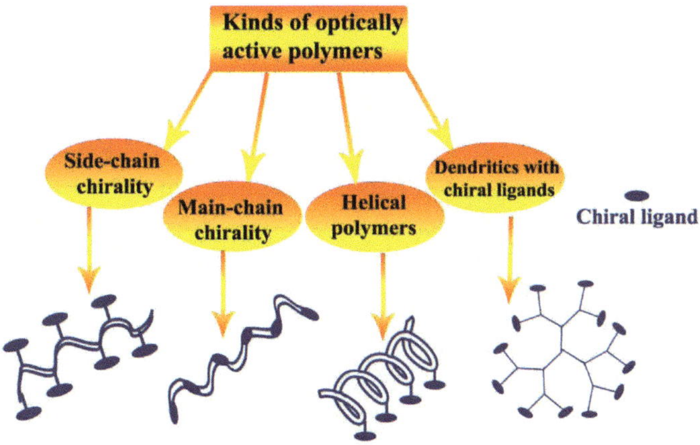

Figure 15.1 Kinds of synthetic chiral polymers.

three categories: (1) asymmetric synthesis polymerization, (2) helix-sense-selective polymerization (HSSP), and (3) enantiomer-selective polymerization. Among these approaches, backbone of synthetic chiral polymers falls into: polymers possessing side-chain chirality, polymers possessing main-chain chirality, dendritic molecules containing chiral ligands, and helical polymers (Fig. 15.1) (Itsuno, 2011; Liu et al., 2014; Yashima et al., 2009). In following, examples of synthesis polymerization to produce optically active polymer and their properties will be discussed.

15.2.1 Optically active polymer based on methacrylate and its derivatives

Radical pocess is one of the methods used for polymerization. In this method, growing free radical is often very active and electrically neutral, which prevents it from interacting with other reagents and makes difficult control during the propagation. Over the past two decades, remarkable progress has been made to stereocontrol during radical polymerization. For example, Xu et al. (2013) applied Lewis acids, α,α'-azobisisobutyronitrile (AIBN), to control stereoregularity during the radical polymerization of (*S*)-(2-hydroxy-1-phenylethyl) methacrylamide ((*S*)-HPEMA) (Fig. 15.2). The results show that in the absence of the Lewis acids, syndiotactic-rich polymers ($r = 84\%$) were produced, whereas in the presence of a catalytic amount of the Lewis acids and depending on the polymerization solvent, the polymerization proceeded in an isotactic-specific manner ($m = 64\%$). They investigated the effect of solvents on the radical polymerization and shown that *n*-butyl alcohol possesses the highest isotactic selectivity.

The investigation of circular dichroism (CD) spectra of novel homopolymeric methacrylates bearing in the side-chain L-lactic acid residue linked to tetraphenylporphyrin chromophore shows a chiral conformations of one prevailing helical handedness, at least for chain sections (Angiolini et al., 2011). Actually, the

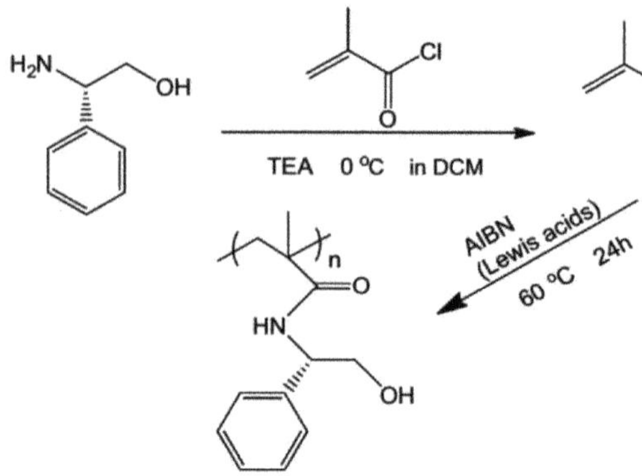

Figure 15.2 Synthetic route of monomer (*S*)-HPEMA and the corresponding polymer PHPEMA. *Source*: Adapted from Xu et al., 2013, with kind permission of Elsevier.

conformational rigidity of the optically active groups interjected between the main chain and the porphyrins moieties of polymer plays a significant role in defining the extent of conformational order in solution. Benelli et al. (2014) synthesized chiral methacrylic copolymers by radical copolymerization of the optically active porphyrin monomers with methyl methacrylate in different molar amounts. Among various classes of photoconductive and photorefractive compounds, polymeric derivatives containing side-chain optically active carbazolyl moieties attract the promising attentions due to their outstanding and unique advantage for using in photoconductive and photorefractive media, electroluminescent devices, as blue emitting materials and in holographic memories. Therefore, Angiolini et al. (2010) prepared polymeric derivatives bearing in the side-chain a 9-phenylcarbazole moieties linked to the (*S*)-2-hydroxy-succinimide, or the (*S*)-3-hydroxy-pyrrolidinyl ring, as chiral moieties covalently linked to the main chain through ester bonds (Fig. 15.3). Novel chiral homopolymeric methacrylates have been produced by AIBN as free radical initiator and THF solvent. The chiroptical properties of the chiral polymers were quantitatively higher than in the corresponding monomers that it was due to conformational dissymmetry of the macromolecules.

Angiolini et al. (2014) investigated photochromism and photoconductivity of synthesized methacrylic polymer containing three distinct functional groups (i.e., azoaromatic, carbazole, and chiral of one single configuration) jointly linked to the side chain. A flexible spacer interposed between the carbazole unit and the main chain is a chiral group of one prevailing configuration which provides a better charge mobility via hopping process. It shows substantial photoconductive and optical activity than that the related monomer, evidencing the existence of conformational dissymmetry in the macromolecules and higher chiral perturbation on the electronic transitions of the azocarbazole chromophore both in the trans- and cis-form. In addition, this polymer due to having the photoreversible variations of optical activity could be used for chiroptical switching.In addition, the polymer with high-optical activity and low-molecular weight monomer showed a defined

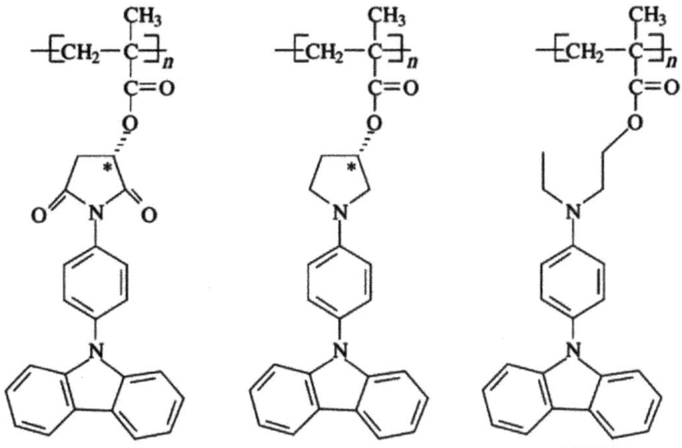

Figure 15.3 Chemical structures of the investigated polymers.
Source: Adapted from Angiolini et al., 2010, with kind permission of Elsevier.

conformational regularity and a prevailing dissymmetry of the macromolecules, which maintained after the photoisomerization of azoaromatic moieties to the cis configuration. Furthermore, the polymers exhibit a photoreversible CD behavior upon irradiation with UV light or by thermal isomerization at room temperature, so, they could be used for chiroptical switching.

In another work they used $(nbd)Rh^+B^-(C_6H_5)_4$ as a catalyst in dry $CHCl_3$ for the preparation of the helical copolymers composed from three substituted acetylene monomers in various feed ratios (Li et al., 2011). In this process, optically active macroporous poly(*N*-isopropylacrylamide) (PNIPAM) hydrogels were polymerized via free radical process of NIPAM, helical copolymer, and cross-linker *N,N′*-methylenebisacrylamide. Moreover, polyethylene glycols with molecular weight of 1000, 1500, and 6000 were used as pore-forming agent to create distinct macroporous architectures and smooth inner walls (Fig. 15.4). The obtained optically active macroporous hydrogels may be applied for instance in drug delivery and biotechnology fields. Based on CD test result and optical rotation measurement, preferential adsorption of the hydrogels towards D-tryptophan is more than L-tryptophan in the corresponding two amino acid enantiomers.

The CD spectrum of the prepared poly[*N*-methacryloyl L-leucine methyl ester] (PMALM) by atom transfer radical polymerization shows a strong negative cotton effect and a weak positive cotton effect at 218 nm and 194 nm, respectively, resulting in $\pi-\pi^*$ and $n-\pi^*$ transition of amide group (Feng et al., 2007). Also, the results expressed that the macromolecule has random coil conformation rather than any secondary structure conformation structure (α-helix, β-sheet, etc.) and value of absolute rotation decreased with the increase of polymer's average number molecular weight (M_n) from 3856 g/mol (PMALM-1) to 11,000 g/mol (PMALM-4).

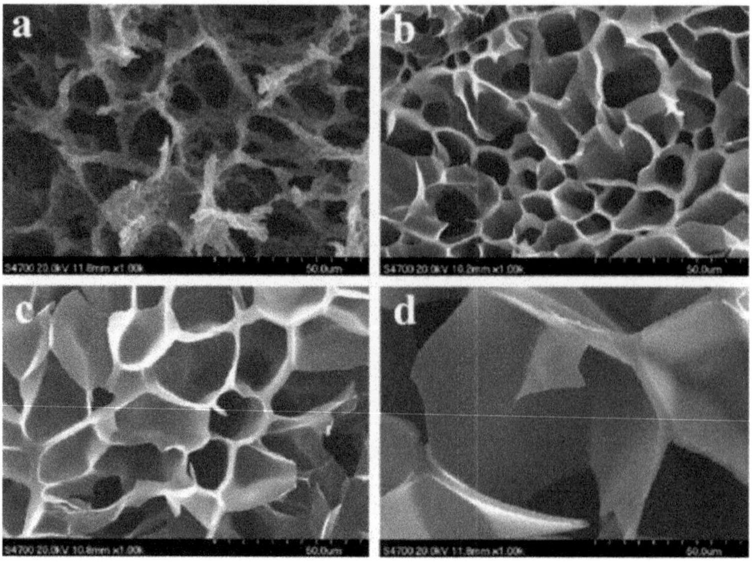

Figure 15.4 SEM micrographs of the hydrogels: (A) PEG-0; (B) PEG-1000; (C) PEG-1500; (D) PEG-6000 (the length bar is 50 lm).
Source: Adapted from Li et al., 2011, with kind permission of Elsevier.

15.2.2 Optically active polymer based on acetylene and its derivatives

Among polymerization approaches, helical polymers are of great interest and the extensive studies have been made because of their helical structures and intriguing optical activity. A helix is one of the simplest and best-organized chiral motifs. The majority of the synthetic helical polymers were constructed by chiral monomers. Nevertheless, limitation of enantiomeric monomers and their high cost restrict the types and numbers of helical polymers. To overcome this problem, achiral monomers can be used instead of chiral monomers. The HSSP techniques were majorly applied in solution polymerization systems for preparing optically active polymers from achiral monomers forming predominantly one-handed screw sense (Liu et al., 2009; Yashima et al., 2009). Liu et al. used HSSP method to directly produce optically active helical polymers by using achiral monomers (Fig. 15.5) (Liu et al., 2014). They first prepared three chiral emulsifiers with different alkyl chain length and different amino acid groups, and then achiral acetylene monomer underwent HSSP with catalyst $[(nbd)RhCl]_2$ to form chiral micelles. According to integration ratio, Raman spectra between the cis and trans C=C and C–C peaks in the main chains of L-poly1 and and D-poly1, the cis contents of poly1 were respectively calculated as 89.0% and 89.1%.

Recently, Hung et al. reported first protocol for constructing optically active helical polymer particles through helix-sense-selective precipitation polymerization (HSSPP) (Huang et al., 2015). In this reaction, achiral acetylenic monomer with

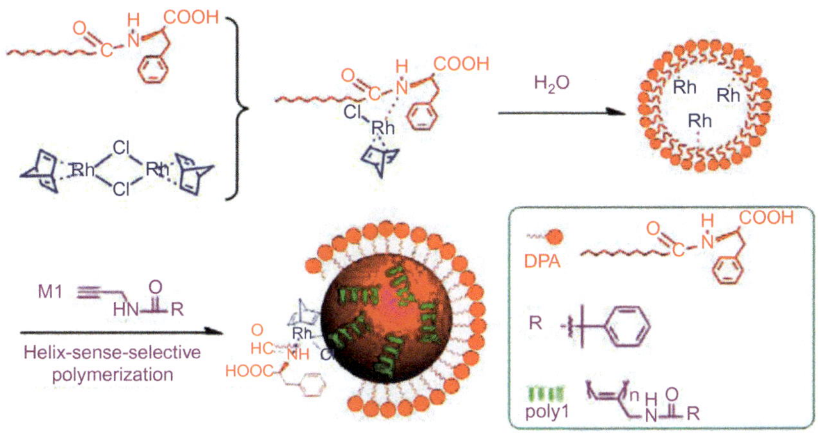

Figure 15.5 Schematic representation of HSSP of achiral acetylene in chiral micelles.
Source: Adapted from Liu et al., 2014, with kind permission of Elsevier.

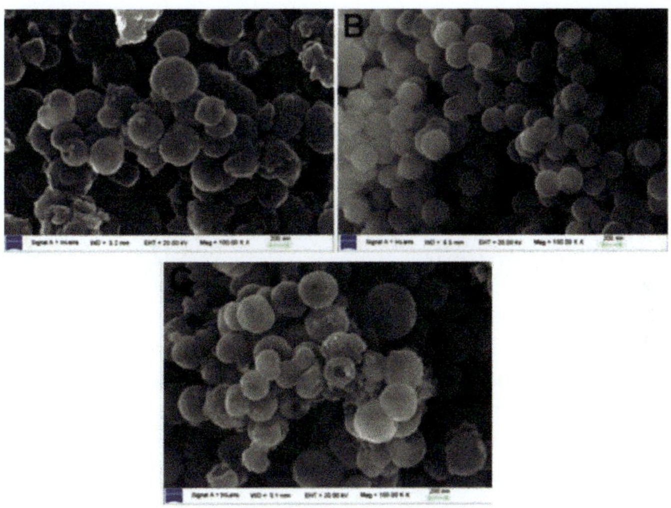

Figure 15.6 SEM images of particles prepared in THF/*n*-heptane with Boc-D-alanine as chiral additive. THF/*n*-heptane: (A) 1/9; (B) 1/10; and (C) 1/11 (mL/mL).
Source: Adapted from Huang et al., 2015, with kind permission of American Chemical Society.

bulky adamantyl group smoothly was underwent HSSPP in the presence of catalyst $(nbd)Rh^+B^-(C_6H_5)_4$ and Boc-L- or Boc-D alanine as chiral additive. SEM images (Fig. 15.6) show that the polymer particles were in average diameter approximately 300 nm with regular spherical shape.

On the basis of the investigations, a mechanism for the occurrence of HSSPPs and the formation of stable helical structures was tentatively proposed by taking Boc-D-alanine in Fig. 15.7. Hydrogen bonds formed between each chiral additive molecule and the amide groups in the polymer chains played essential roles for controlling the screw

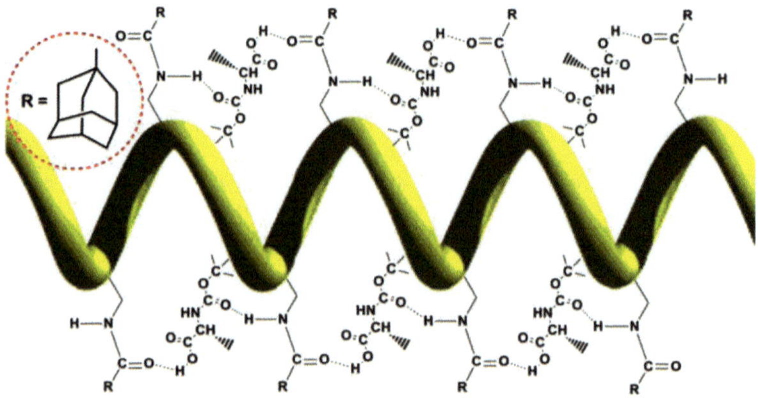

Figure 15.7 Proposed mechanism of HSSPP and the possible architecture of the polymer chains.
Source: Adapted from Huang et al., 2015, with kind permission of American Chemical Society.

sense of the helical polymer chains. This novel polymer can be used as chiral stationary phase for HPLC, enantioselective recognition probes, chiral catalysts, etc.

Yao et al. prepared optically active polymer in effect of catalytic polymerization of chiral monomer *N*-propargyl abietamide (M1) in presence of (nbd)$Rh^{+}B^{-}(C_6H_5)_4$ which was an unstable helice in tetrahydrofuran (Yao et al., 2013). However, copolymerization of M1 and the achiral N-propargylamide monomer (M2) were performed in a same way in dry solvent ($CHCl_3$, CH_2Cl_2, or THF) and caused to the synthesis of helical optically active copolymers and characterized by CD studies and UV–vis spectroscopy. According to UV−vis absorption, with increasing M1 content (from 14% to 45%) in copolymer, the maximum absorption of the peak gradually decreased from 2.2 to 0.8 (M^{-1} cm^{-1}) and copolymer chain lost its ordered helical conformations. CD spectra of copolymer in range of 350−430 nm exhibited that the CD signal progressively intensified by increasing of the M1 content from 14% to 32%; however, an additional increase in M1 content led to a decrease in CD intensity (Fig. 15.8).

The existence of π-conjugated structure in polymers can induce helical configuration with a predominantly one-handed screw sense through the introduction of appropriate chiral substituents into the side chain or through the HSSP of achiral monomers. The results of polarimetric, circular dichroism, and UV−vis spectroscopic analyses on polymerization of salicylidene Schiff-base-containing polyacetylenes indicated that the formed polymers possess a helical structures with a predominantly one-handed screw sense and its optical rotations were 5−20 times larger than corresponding monomers (Zhang et al., 2015).

15.2.3 Optically active polymer based on amino acids and amino alcohols

Amino acids are the most common chiral resources that have been widely used in the synthesis of optically active materials for biomedical application (Mallakpour

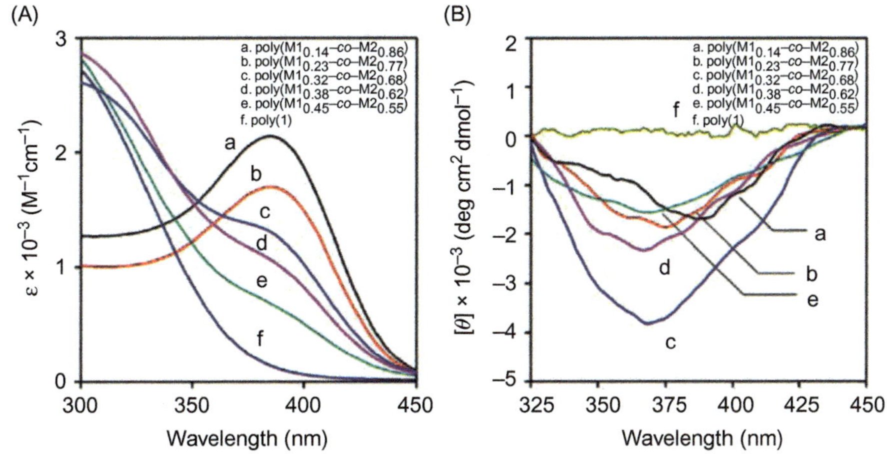

Figure 15.8 (A) UV–vis and (B) CD spectra of the copolymers (measured in THF, $c = 0.3$ mM).
Source: Adapted from Yao et al., 2013, with kind permission of Elsevier.

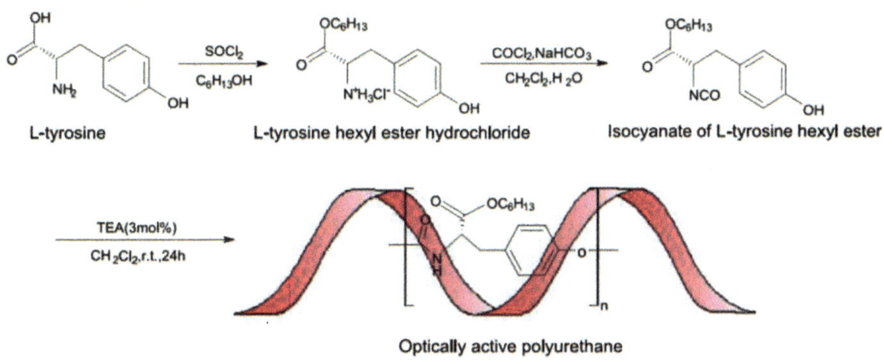

Figure 15.9 Synthetic procedure of isocyanate-phenol and polyurethane.
Source: Adapted from Yang et al., 2011a, with kind permission of Elsevier.

and Dinari, 2011). Moreover, amino acid-based chiral polymers due to secondary interactions such as hydrogen bonds and van der Waals forces could have induced a helical configuration with the ability to form higher ordered structures that show improved properties (Qu et al., 2009). Yang et al. reported that the crystallinity and thermal decomposition temperature of a couple of enantimorphs of polyurethanes (LPU and DPU) synthesized by the self-polyaddition of the isocyanate-phenols derived from chiral and racemic tyrosine (Fig. 15.9) were higher than that of the racemic polyurethane (RPU) (Yang et al., 2011a). These results can be related to the more regular secondary structure which facilitate the constitution of a large number of interchain hydrogen bonds. Optical rotations of LPU and DPU were estimated as +124.5° and −120°.

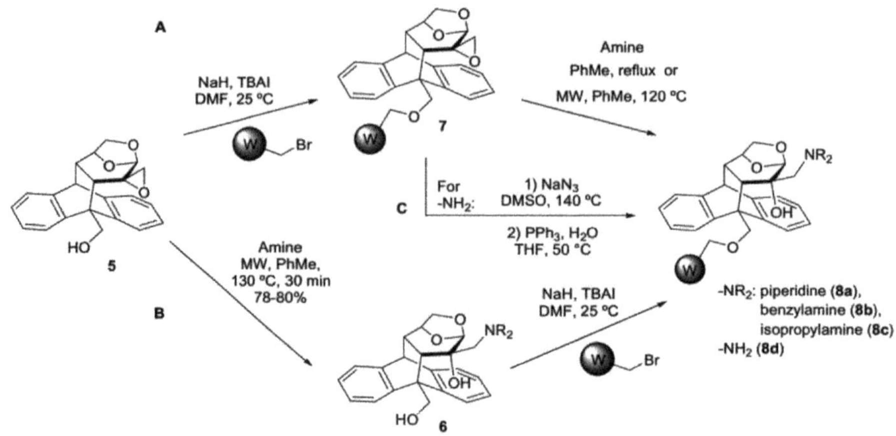

Figure 15.10 Synthesis of polymer-supported amino alcohols 8a–d.
Source: Adapted from Botta et al., 2016, with kind permission of Elsevier.

Reaction of addition of dialkylzinc reagents to aldehydes shows an enantioselective transformation performed under flow conditions with the help of a polymer-supported chiral catalyst. For the first time, in the early 1990s Frechet and Itsuno used polymers functionalized with some amino alcohol derivatives as enantioselective catalysts for adding diethyl zinc to aldehydes (Itsuno et al., 1990). Chiral amino alcohols are among the most important family of chiral ligands that could be employed as chiral inducers for this transformation and asymmetric catalysis. Recently, Botta et al. reported the use of generated resin-bound 1,2-amino alcohols as asymmetric ligands for the addition of $ZnEt_2$ to aromatic aldehydes (Fig. 15.10) (Botta et al., 2016). It is worth mentioning that the moderate enantioselectivity observed for this system under flow conditions than the enantioselectivity obtained for analogous 1,2-amino alcohols in solution.

15.2.4 Optically active polymer based on binaphthol

The conjugated polymer incorporating optically active binaphthyl moiety in the main-chain backbone could adopt a rigid and regular structure in the polymer main-chain backbone. These ordered chiral binaphthyl-based polymers could lead to the properties of fluorescence sensors with good fluorescence quantum efficiency due to the extended π-electronic structure between the chiral repeating unit and the conjugated linker unit. In this regard, Huang and his coworkers reported the synthesis of four chiral polybinaphthyls incorporating 2,2′-bipyridyl moieties in the polymer main-chain backbone by Pd-catalyzed Heck reaction or Wittig–Horner reaction (Huang et al., 2009). Based on the great differences of specific rotation values and CD spectroscopy, they show that each of polymers may adopt a zigzag chain or a helical configuration. Chen et al. investigated synthesis and polymerization of (*S*)-3-vinyl-2,2′ bis (methoxymethoxy)-1,1′-binaphthyl using *n*-BuLi as an initiator (Chen et al., 2014). The results show that the polymer kept a prevailing helicity of backbone in solution and this conformation of polymer was stable and insensitive to temperature variation.

15.2.5 Optically active polymer based on amide and its derivatives

Asymmetric synthesis polymerization is a reaction that produces polymers with main-chain configurational chirality starting from chiral and achiral monomers (Nakano and Okamoto, 2000; Okamoto and Nakano, 1994). Microwave heating is a simple approach to accelerate polycondensation reaction of optically active diacids with a thiazole-bearing diamine in a medium consisting of molten tetrabutylammonium bromide as a green molten salt medium and triphenyl phosphite as the homogenizer to prepare optically active poly(amide-imide) (Fig. 15.11) (Mallakpour and Zadehnazari, 2014). Thermogravimetric analysis (TGA) of the polymers demonstrated most of these PAIs have a high thermal stability with initial decomposition temperatures being in the range of 365−389°C. Also, limiting oxygen index (LOI) calculated from Van Krevelen equation shows PAIs can be applied as self-extinguishing polymers for engineering plastics.

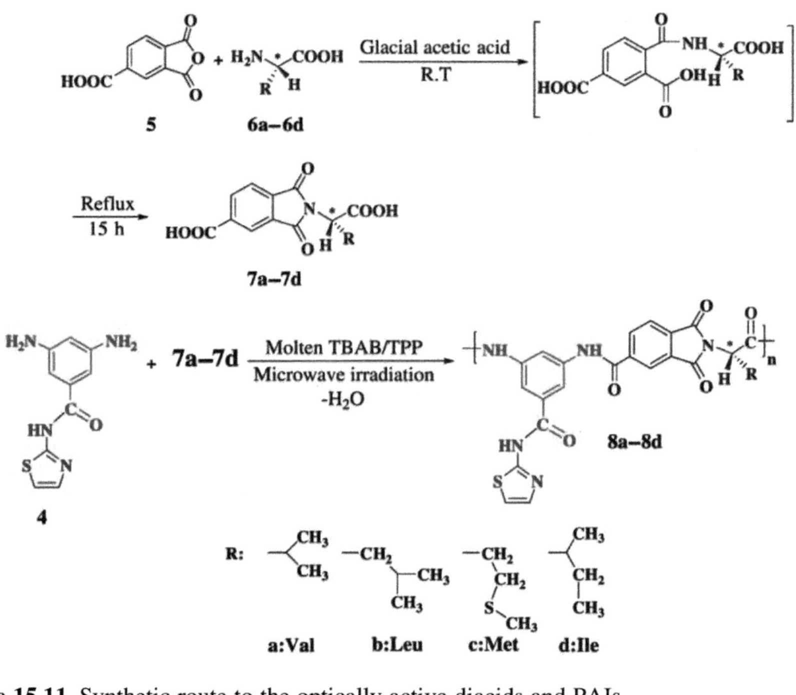

Figure 15.11 Synthetic route to the optically active diacids and PAIs.
Source: Adapted from Mallakpour and Zadehnazari, 2014, with kind permission of Elsevier.

Mallakpour et al. (2012) synthesized biodegradable poly(ester-imide)s (PEI) by polycondensation reaction of *N,N′*-(pyromellitoyl)-bisdimethyl ester tyrosine as a diphenolic monomer with two chiral trimellitic anhydride-derived diacid monomers containing S-valine and L-methionine under dissolving tosyl chloride in a mixed solvent of pyridine and *N,N*-dimethylformamide. The results obtained from wheat seedling growth in the soil buried with synthetic polymers demonstrated that PEIs were nontoxic and showed possibility of phytoremediation in polymer-contaminated soils.

15.2.6 Other optically active polymers

Ohkawa et al. investigated chiroptical properties of the polymer constructed from a monomer ((*S*)-configuration at stereogenic center) in a cholesteric liquid crystal (CLC) medium consisting of chiral molecules in (R)-configuration (Ohkawa et al., 2013). They showed that the polymer from the monomer with (R)-configuration prepared in the CLC with (R)-configuration exhibits weak Cotton effect compared to polymer with configuration (S). This result is applicable to intermolecular interaction between the monomer and the CLC medium in the polymerization process. Also, they displayed that polymerization of achiral monomers in CLCs produces various pairs of chiroptically active polymers without chiral centers or axial chirality in the primary structure. Whereas, the polymerization of the chiral substituent shows chiroptical activity polymers with a main chain twisted in a predominant direction (secondary structure) (Ohkawa et al., 2013; Ohta et al., 2007). In other work, Liu et al. investigated the steric effects of chiral azobenzene derivative with chiral (K)-menthyl moiety on the induction of CLCs and the sensitivity of the photoisomerizable azobenzene derivatives (Liu and Yang, 2006). The results showed that the obtained chiral polymers are led to induce the helical structure of the CLC phases and stabilize the liquid crystals.

Cataldo and coworkers prepared poly-β-pinene (pBp) resins by radiation-induced polymerization of pure b(+)-pinene and b(+)-pinene and their scalemic mixtures under vacuum at 1181 kGy (Cataldo et al., 2011). The polymer resins synthesized were accurately studied by optical rotatory dispersion (ORD) spectroscopy and by FT-IR spectroscopy in comparison to pB(+)p and pB(-)p obtained by thermal processing under the action of a chemical free radical initiator. According to results observed by ORD in the spectral range between 375 nm and 625 nm, all pBp resins synthesized showed a curious asymmetry in the ORD of pB(+)p versus the ORD of pB(−)p. The results of FT-IR showed that discrimination between the resins produced with radiation and those produced with a free radical initiator is easy. In addition, TGA analysis showed that pB(+)p processes lower thermal stability compared with the pB(−)p.

Kanbayashi et al. reported a new synthetic approach for the preparation of optically active polymer by a combination of asymmetric allylic amidation catalyzed by planar-chiral ruthenium (Cp′Ru) complexes (Fig. 15.12) (Kanbayashi et al., 2015). The results showed that the Cp′Ru allows construction of a polymer with high regio- and enantioselectivities. Ring closing metathesis (RCM) as a

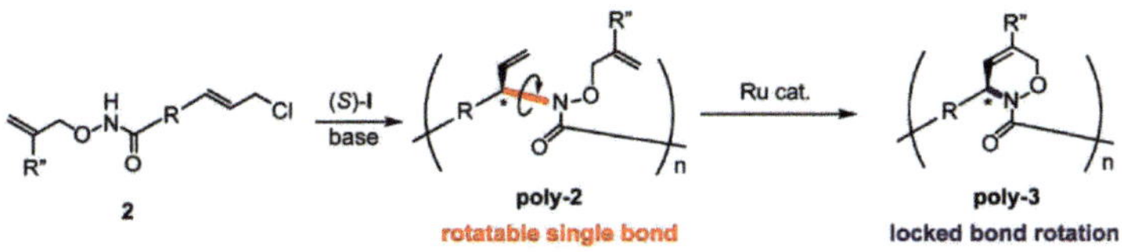

Figure 15.12 Combination of asymmetric allylic substitution and RCM.
Source: Adapted from Kanbayashi et al., 2015, with kind permission of American Chemical Society.

candidate for transformation of the terminal olefin which restricts freely rotation single bonds around an asymmetric center and constructs a useful chiral foundation. Therefore, they have designed bifunctional monomers bearing allylic chloride and *N*-alkoxyamide possessing an olefinic moiety for RCM reactions in each monomer unit.

Iida et al. prepared novel optically active polymer involving of riboflavin units as the main chain (poly-1) (Iida et al., 2012). The riboflavin residues of poly-1H were converted to 5-ethyl-4a-hydroxyriboflavins by a hydroxylation/dehydroxylation reaction through the corresponding 5-ethylriboflaviniums (poly-2)..The results of absorption, CD, and NMR spectra revealed that cationic poly-2 self-assembled can be formed a supramolecularly twisted, duplex-like helical structure with excess one-handedness through intermolecular face-to-face π-stacking of the riboflavinium units. Additionally, the poly-2 enantioselectively catalyzed the asymmetric oxidation of sulfides in the presence of H_2O_2 through the production of oxidatively active 4a-hydroperoxyriboflavins (giving poly-2OOH) with enantioselectivity of up to 60% ee, which is higher than reaction catalyzed by the monomeric counterpart.

15.3 Characterization of optically active polymers

To evaluate the architecture of the main chain formed by asymmetric polymerization, FT-IR (examine the chemical bands and functional group on the polymer), optical rotation (rotation of the plane of linearly polarized light on passing through the sample), ellipticity (almost never measured directly), single crystal X-ray crystallography (when crystals can be grown), and circular dichroism (CD; differential absorption of left and right circularly polarized light) were conducted. In this context, the last two techniques provide the most information, but in the case of most macromolecules, X-ray crystallography is not possible due to the lack of appropriate crystals. Thus the most suitable technique for the analysis of optically active polymers is CD spectroscopy, which permits the direct analysis of chiral backbone physical and electronic structures (Fujiki and Koe, 2000; Kaneko et al., 2005). Several review papers are available to summarize the research results mostly with focuses on the characterization of optically active polymers (Huang et al., 2009; Lu et al., 2016; Zahmatkesh and Yazdanpanah, 2012).

15.4 Optically active polymer/metal oxide nanocomposite

In the past half century, optically active polymer matrix-based nanocomposites have generated many attentions in the recent literatures. Although, a great deal of effort has been made to preparation of optically active polymers, low thermal stability, poor mechanical strength, and intractable processability of polymer are still detrimental to their potential technological application. Thus further hybridization of the polymer with filler-like inorganic metal oxides is truly required. Because the large surface area at a rather low nanofiller concentration creates a large interphase and interfacial interactions in nanocomposites which presents various unexpected performance compared to large-scale particulate reinforcement (Zare, 2016a,b). Incorporation of metal oxide nanoparticle offers the potential of combining the best parts of the distinctive properties of the two entities in a single hybrid. However, when the polymer and metal oxide nanoparticles are combined together, only inhomogeneous mixtures can be obtained due to the strong self-aggregation and phase separation of both the incompatible components. This phenomenon produce many defects and stress concentrations in nanocomposites, and inversely deteriorate the properties of nanocomposites (Bu et al., 2014b; Zare, 2016c). Since a few strategies in regard to hybridization and decrease of aggregation/agglomeration have been presented such as particle coating by a capping agent, application of coupling agent or compatibilizer, and charging the filler surface to separate them via electrostatic revulsions. Also, observing optimal parameters in hybridization process can induce to an effective decreasing of aggregates. For instance in melt mixing by extruder, ideal screw speed and feeding rate create much stress to melt materials which break the particle aggregates. The effects of reinforcement of nanoparticles on the characteristic properties of polymer hybrids have been largely discussed in the previous reports (Kalia and Haldorai, 2015; Mohanty et al., 2015).

15.4.1 Polyurethane/metal oxide nanocomposites

Polyurethane is one of the most versatile polymer materials of which the properties can be easily tailored through structure design and polymer chemistry techniques for using in a variety of industrial parts (Chen et al., 2008; Mallakpour et al., 2010). Yang et al. prepared the multilayered OPU/TiO_2/MnO_2 nanorods by coating of TiO_2 on MnO_2 with the thickness around 15 nm and then amount of 0.24 g/(g inorganics) of optically active polyurethane (PU) was grafted of TiO_2/MnO_2 (Yang et al., 2014). TEM images (Fig. 15.13) showed that smooth surface of MnO_2 after deposition of titania generated a heterogeneous outside with the core–shell structure. After adding of PU, the surface of crystalline titania was tightly wrapped with amphorous substance and constructed a unique multilayered core–shell architecture.

The result of the infrared emissivity values of the samples at wavelength of 8–14 μm showed that the infrared emissivity values of MnO_2 decreased after the wrapping indicating a potential for practical application. In chiral polyurethane/CdS–SiO_2

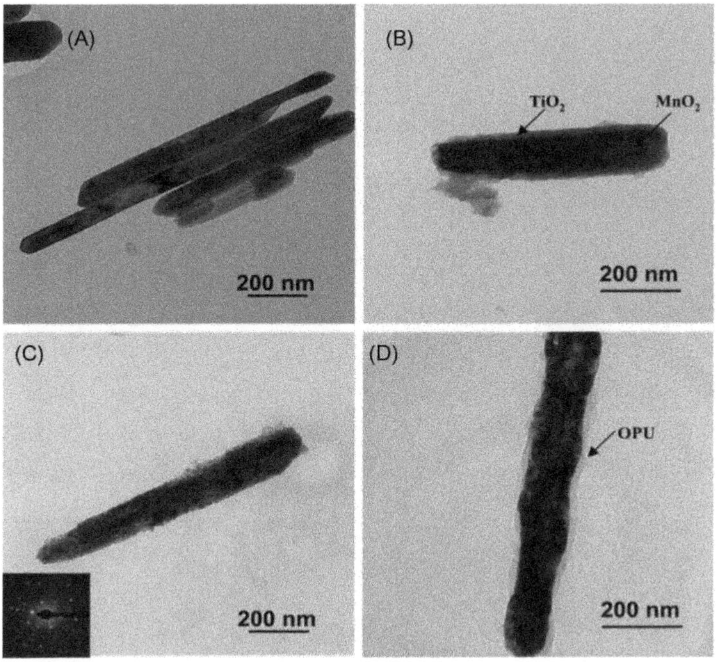

Figure 15.13 TEM micrographs of (A) MnO_2, (B) TiO_2 (amorphous)/MnO_2, (C) TiO_2 (anatase)/MnO_2 (after the calcination at 500°C), and (D) OPU/TiO_2 (anatase)/MnO_2.
Source: Adapted from Yang et al., 2014, with kind permission of Elsevier.

nanocomposites prepared under ultrasonic irradiation revealed that the nanocomposites possessed much lower infrared values compared with those of the neat polymers and nanoparticles, respectively (Chen et al., 2008). This occurrence can be attributed to the intense reflectivity against thermal radiation, variation of the vibration mode of molecules, atoms, or pendant groups at the interface of the composite, and the helical conformation of OPU. Chen et al. synthesized optically active PU from chiral 1,10-binaphthyl (BINOL) and 2,4-toluene diisocyanate (TDI) by the simple hydrogen transfer addition reaction (Chen et al., 2007). The presence of stronger CD signals with positive and negative Cotton effect in CD spectrum confirmed chirality of polymer. TGA studies manifested improvement of thermal stability of the optically active PU with existence of TiO_2 nanoparticles. Yang et al. studied the effect of dispersion of indium tin oxide (ITO) modified with KH550 on morphology and properties of optically active polyurethane (LPU) and racemic polyurethane (RPU) based on tyrosine (Yang et al., 2012). In TEM photograph (Fig. 15.14), ITO exhibits globular shape and some of them accumulate into larger aggregates. After the graft with polyurethanes, RPU@ITO and LPU@ITO nanocomposites (Fig. 15.14B and C) show a clear layer of amorphous surroundings with the thickness of about 10 nm. It is remarkable that the dispersion degree of ITO in LPU is more than RPU through the ultrasonic dispersion. It contributes to

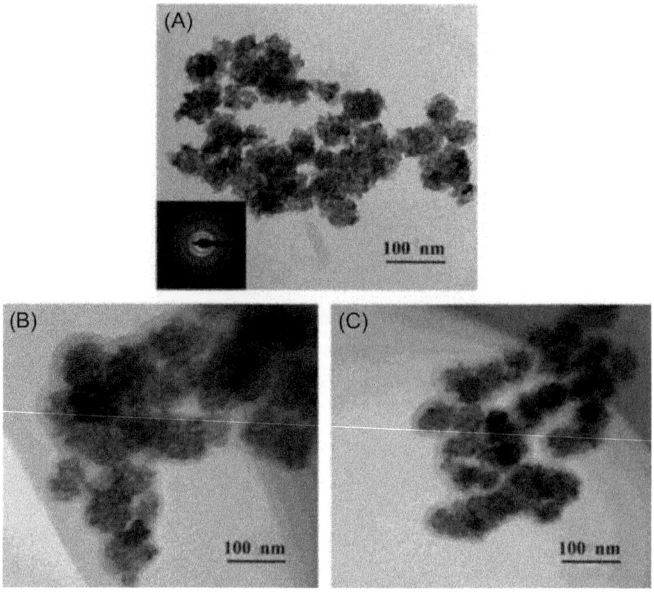

Figure 15.14 TEM micrographs of (A) ITO, (B) RPU@ITO, and (C) LPU@ITO composites.
Source: Adapted from Yang et al., 2012, with kind permission of Elsevier.

the uniform wrapping and orderly secondary structure of LPU which provides the advantage of less steric hindrance.

Based on TGA result, amounts of polyurethanes grafted onto the ITO are about 0.26 g/g inorganics for RPU@ITO and 0.32 g/g inorganics for LPU@ITO. This variation in weight loss may result from the different chain conformations. In addition, LPU@ITO due to the regular secondary structure of organics would easily lead to the formation of intermolecular interactions and exhibit an infrared emissivity value lower than RPU@ITO. The similar results were also observed in grafted optically active polyurethane on the surface of TiO_2/SiO_2. They showed that the infrared emissivity value of SiO_2 from 0.782 is decreased to 0.553 in the multilayered core–shell composite microspheres (Yang et al., 2011b).

Waterborne polyurethane (WPU) is a newly developed PU system prepared by the incorporation of ionic groups into the polymer structure to be dispersed well in the water phase (Zhang et al., 2012). Chen et al. (2016) examined mechanical properties, electrical conductivity, and magnetic properties of WPU nanocomposites with addition of various amount of 3-(triethoxysilyl) propyl isocyanate (IPTS)-treated Fe_3O_4 nanoparticles by in situ polymerization technique. The WPU/Fe_3O_4-IPTS 8 wt% showed high electrical conductivity (8.43×10^{-4} S/cm) compared to that of pure WPU (1.53×10^{-13} S/cm). Also, the nanocomposites showed superior magnetic properties (14.22 emu/cm^3) with a low Fe_3O_4-IPTS content of 8.0 wt%. The excellent properties were mainly attributed to the good dispersion of Fe_3O_4-IPTS in WPU through strong covalent bonding. The resulting nanocomposite with excellent physical performance and superior magnetic properties can effectively reduce electromagnetic reflections of particular vehicles or devices and provide a promising route to prepare environmentally friendly microwave absorption materials.

15.4.2 Poly(amide-imide)/metal oxide nanocomposites

Poly(amide-imide) is one of the most thermoplastic polymers with good solubility in polar amide-type solvent. They can present both the advantages of polyimide and polyamide such as high mechanical properties and thermal resistant. One of the approaches to enhance certain physical properties and chemical stability of PAI is incorporating of nanoparticles into the polymer matrix. Various nanoparticles, such as metal oxide, clay, CNT, silica, and graphene, have been embedded within PAI, which in this chapter investigated only effect of metal oxides. In this context, Mallakpour's research group is one of the vanguard in synthesis of optically active PAI derived from L-amino acid-based diacid and diamine and its nanocomposites which followed this approach to apply treatment of metal oxides for improving physicochemical properties of optically active PAIs. The solution casting method has been the most used method to prepare the PAI nanocomposites based on Al_2O_3, TiO_2, ZnO, ZrO_2, SiO_2, CuO modified with silane coupling agent, and diacids based on hiral *N,N′*-(pyromellitoyl)-bis-L-amino acid and hiral *N*-trimellitylimido-L-amino acid (Mallakpour and Hatami, 2013; Mallakpour and Khadem, 2014b; Mallakpour and Madani, 2012). For example, they prepared optically active PAI by combining of hiral *N,N′*-(pyromellitoyl)-bis-L-phenylalanine diacid with 2-(3,5-diaminophenyl) benzimidazole under green condition using molten tetrabutylammonium bromide (TBAB). Then the Al_2O_3 NPs were treated with 3-aminopropyltriethoxylsilane and incorporated into PAI matrix by ultrasonic irradiation. Homogeneous dispersion of alumina nanoparticles in polymer was shown in Fig. 15.15 (Mallakpour and Dinari, 2013).

The mechanical properties showed that the maximum ultimate strength (85.6 MPa) was found to 10 wt% of Al_2O_3 NPs into PAI. Nevertheless, if the fillers exceed 15 wt%, a 28% decrease is observed in the failure strain. The thermal decomposition of PAI and its nanocomposites showed that the initial decomposition temperature of the pure PAI was about 310°C while it shifted toward higher temperatures by increasing the percentage of SiO_2 NPs (Mallakpour and Marefatpour, 2014). Similar results have also been reported by the incorporation of various weight percentages of ZrO_2 modified with 3-aminopropyltriethoxylsilane (Mallakpour and Zeraatpisheh, 2014). TiO_2 NPs were modified with bioactive *N,N′*-(pyromellitoyl)-bis-L-valine (Mallakpour and Aalizadeh, 2014) and Al_2O_3 modified with *N*-trimellitylimido-L-phenylalanine (Mallakpour and Khadem, 2014a), into PAI via ultrasound assistant technique. Inorganic NPs encapsulated into the polymer matrix as a higher thermal insulator and a mass transport barrier to the unstable products produced can increase thermal stability of nanocomposite. Also, modified nanoparticles by hydrogen bonding reduce mobility of the PAI

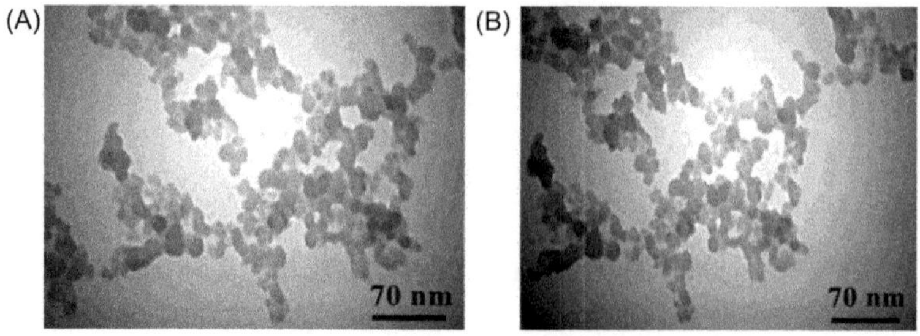

Figure 15.15 TEM images of (A) PAI/Al_2O_3 NC (5 wt%) and (B) PAI/Al_2O_3 NC (15 wt%). *Source*: Adapted from Mallakpour and Dinari, 2013, with kind permission of Elsevier.

chains and produce tortuous path in the matrix prevented diffusion of the volatile pyrolysis products in the composites. The LOI values of PAI and PAI/metal oxide nanocomposites were calculated from char yield and showed that aromatic material due to thermal resistant can be classified as self-extinguishing materials and tend not to be burned (Mallakpour and Khadem, 2014b). Occasionally, the addition of functionalized NPs has deleterious effect on the thermal degradation of the PAI matrix. This denoted by the oxide, causing a reduction in the thermal stability of the polymer in some compositions. Also, aggregation of metal oxide and weak interaction between filler and matrix can assist fast thermal decomposition of PAI in nanocomposite (Mallakpour et al., 2014a; Mallakpour and Sirous, 2015).

Mallakpour and Hattami synthesized the optically active PAI from the hydroxyl substituted diamine with synthetic diacid chloride by polycondensation reaction in DMAc in the presence of propylene oxide (PO). Then they investigated the effect incorporating of TiO_2 NPs treated by 3-aminopropyltriethoxysilane (KH550) on water contact angle (CA) of PAI (Fig. 15.16). The results showed the nano-TiO_2 improved the hydrophilic properties by reducing CAs from about 78.5 to 45.6 with increasing TiO_2 content to 25 wt% (Mallakpour and Hatami, 2012).

TEM images of ZnO NPs embedded in PAI matrix were shown in Fig. 15.12. In line with this goal, a novel coupling agent containing bromine elements with flame retardancy potential for the surface modification of ZnO nanoparticles was synthesized from the reaction between tetrabromophthalic anhydride and p-aminophenol. In Fig. 15.17 the black spots of ZnO NPs with hexagonal shapes and average size 33 nm were dispersed in the polymer matrix with gray background (Mallakpour and Behranvand, 2014).

Optical properties of PAI were compared with PAI nanocomposites reinforced with metal oxide. The presence of modified ZrO_2 NPs in optically active PAI displayed better UV absorption than the pure PAI in the range of UV-A (315–380 nm) which it may be due to the light scattering and nanofiller UV absorption (Mallakpour and Mani, 2014). UV–vis absorption spectra of PAI/TiO_2 nanocomposites showed UV absorption peak maximum at 307 and 319 nm. It is attributed to the charge transfers of the chromophoric unit of the PAI structure and that of Ti–O–Ti segment (Rafiee and Zare, 2015).

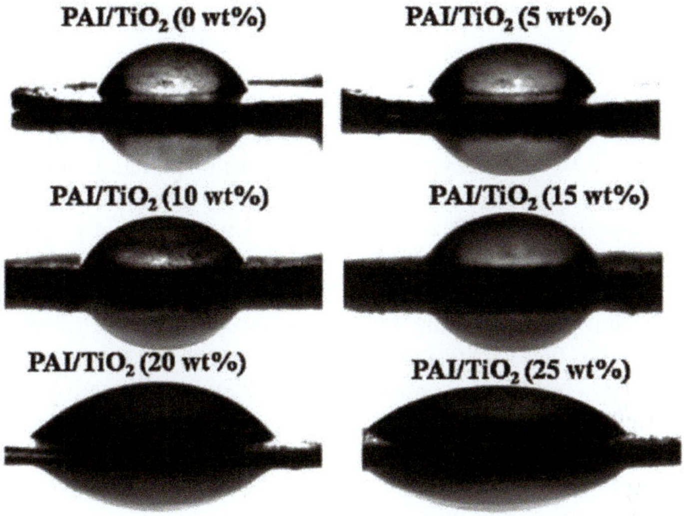

Figure 15.16 CA measurements for different contents of nano-TiO_2 in NCs.
Source: Adapted from Mallakpour and Hatami, 2012, with kind permission of Elsevier.

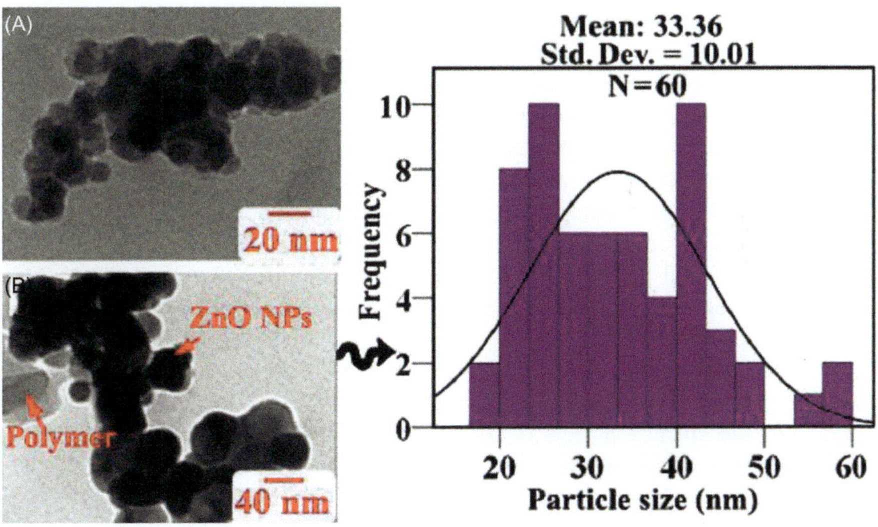

Figure 15.17 TEM images of PAI/ZnO-CA NC 8 wt% at different magnifications and its histogram.
Source: Adapted from Mallakpour and Behranvand, 2014, with kind permission of Springer.

Mallakpour et al. observed two absorption peaks at 264 and 345 nm for PAI. While, the slight blue shift in 264 nm and red shift in 345 nm were observed by the addition of ZnO nanoparticle into PAI matrix (Mallakpour et al., 2014b). The UV shielding ability attributes to the scattering and/or absorbance of nanoparticles. These changes may confirm the interaction between PAI and modified metal oxides.

Poly(amide-ester-imide) is one of the derivative of PAI. Abdolmaleki et al. reported the synthesis of nanostructured poly(amide-ester-imide) (PAEI) based on two different amino acids, prepared via direct polycondensation of biodegradable *N,N′*-bis[2-(methyl-3-(4-hydroxyphenyl)propanoate)]isophthaldiamide and *N,N′*-(pyromellitoyl)-bis-L-phenylalanine diacid (Abdolmaleki et al., 2012). Then PAEI/ZnO bionanocomposites (BNCs) were prepared via interaction of pure PAEI and ZnO nanoparticles modified with two different silane coupling agents. Based on the UV/vis absorption, by increasing of NP contents the maximum absorption peak of polymer is shifted toward the maximum absorption peak of modified ZnO NP (blue shift). So, these materials can be used as a UV shielding materials. In another work, TiO_2 NPs were coated by chiral diacid monomer, *N*-trimellitylimido-L-leucine, and incorporated within optically active PAEI matrix (Mallakpour and Khani, 2013). The thermal performance of the PAEI/TiO_2 nanocomposites was superior to that of the pure PAEI. For example, with a TiO_2 loading of only 15%, the char yield increased by 31%.

15.4.3 Polyacetylene/metal oxide nanocomposite

Bu et al. (2014b) investigated the infrared emissivity of optically active substituted polyacetylene (SPA)@WO_3 nanorod hybrids fabricated via a "grafting to" strategy which the preparation process is shown in Fig. 15.18. According to TGA, amount of SPA grafted onto WO_3 is about 0.073 g/(g hybrids). Based on the infrared emissivity measured at the wavelength of 8–14 μm, the infrared emissivity value of SPA@WO_3 nanohybrids (0.527) was much lower than that of bare WO_3 nanorods (0.885) and pristine SPA (0.632). The enhancement in lowering infrared emissivity could be attributed to the helical structures of polymer chains and intensified interfacial interactions between organics and inorganics which lead to the formation of massive intermolecular interactions, and reduce the index of hydrogen deficiency.

They also produced SiO_2/TiO_2/SPA multilayered core–shell composite nanospheres (Bu et al., 2014a). In this process amount of 0.115 g/(g composite) of SPA grafted onto the surface inorganic sphere through effective covalent bonds. TEM and SEM images demonstrated that the SPA molecules have been successfully attached to the surface of the inorganic SiO_2/TiO_2 particles. In TEM images (Fig. 15.19), the presence of a light appearance of around the spherical shape of the SiO_2/TiO_2 particles attributed to SPA. The multilayered core–shell structure not only creates the hybrid with excellent thermal stability, reinforced processability, resistance to oxidation and degradation, but also endows hybrid with the lower infrared emissivity than both the optically active SPA and inorganic SiO_2/TiO_2 core–shell nanoparticles.

Prasad and coworkers studied electrical conductivity of the ionic polyacetylenes and nanocomposites based on poly (*N*-octadecyl-2-ethynyl pyridinium bromide) (PNO) or poly (*N*-methyl-2-ethynyl pyridinium iodide) (PNMe) and $NbWO_6$

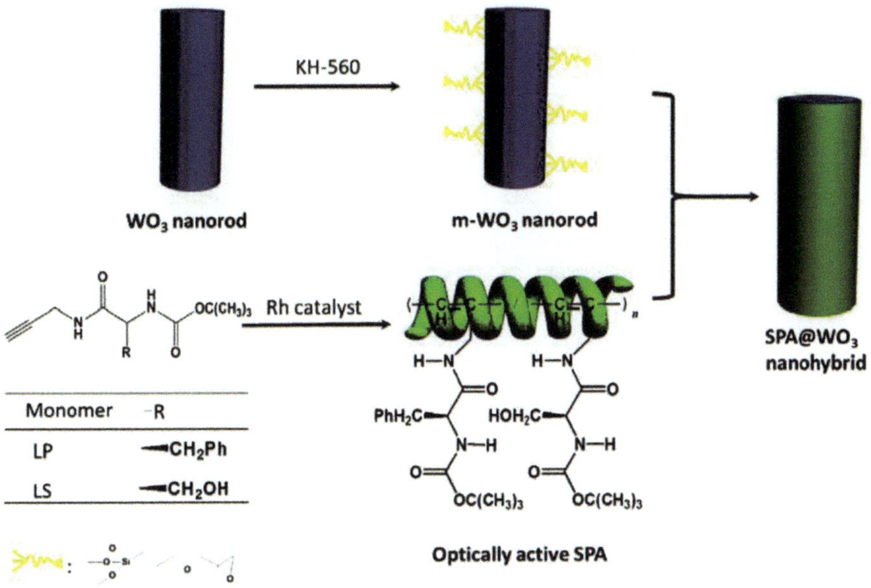

Figure 15.18 Schematic illustration for the fabrication of the SPA@WO_3 nanohybrid.
Source: Adapted from Bu et al., 2014b, with kind permission of Elsevier.

nanosheets (Prasad et al., 2006). Thus first they prepared nanocomposites by exfoliation–reflocculation method under overnight sonication. They reported that electrical conductivity of PNO (10^{-3} S/cm) and PNMe (10^{-4} S/cm) is much higher than the conductivity of the PNO:$NbWO_6$ (10^{-5} S/cm) and PNMe:$NbWO_6$ (10^{-6} S/cm). This occurrence can be attributed to the spatial confinement of the highly conducting polymer inside the nanosheets, which limits the mobility of the charge carriers in effect of separation of the polymer chains by the low conductive $NbWO_6$ nanosheet layers.

15.4.4 Polyaniline/metal oxide nanocomposite

Among the conducting polymers, polyaniline (PAni) has attracted considerable attention of researchers due to its unique properties such as low cost, simple preparation, excellent thermal and environmental stabilities (Bandgar et al., 2014). Main imperfection of PAni is its poor solubility in organic solvents that it causes to low processability and further limits its application. Many approaches have been used to overcome this drawback such as incorporation functionalized organic acids as dopant, protonation with organic acids, grafting with copolymers or preparing it using emulsion polymerization in the presence of surfactants (Patil et al., 2015). Karim and coworkers prepared sulfonated polyaniline–titanium dioxide (SPAni–TiO_2) hybrid composites by combination of aniline and orthoanilinic acid comonomers, a

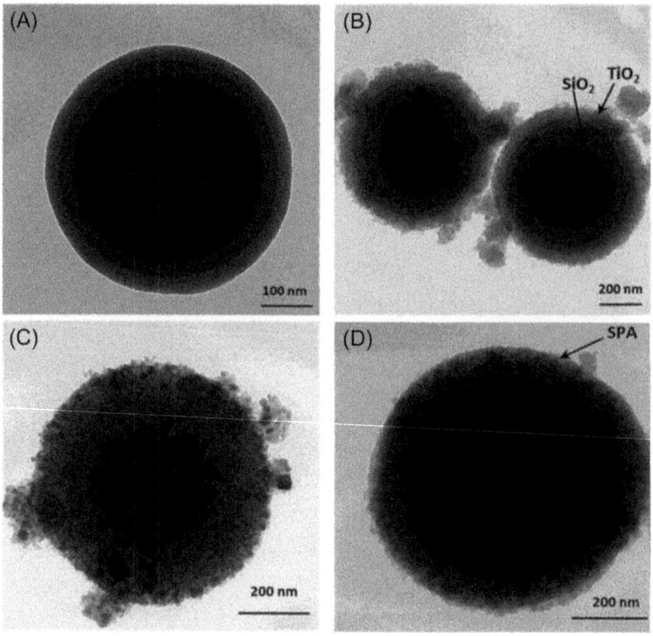

Figure 15.19 TEM images of (A) SiO_2, (B) SiO_2/(amorphous)TiO_2, (C) SiO_2/(anatase)TiO_2, and (D) SiO_2/(anatase)TiO_2/SPA nanosphere.
Source: Adapted from Bu et al., 2014a, with kind permission of Elsevier.

free-radical oxidant and titania precursor in aqueous solution under irradiation of UV rays (Karim et al., 2009). In this process, UV radiolysis not only used in the initial hydrolysis and conversion of titanium tetraisopropoxide to TiO_2, but also it acts as a mild photocatalyst to oxidize aniline and OA to sulfonated polyaniline. Two electronic bands of the pristine SPAni at 331 and 633 nm were attributed to the $\pi-\pi^*$ and $n-\pi^*$ transitions of benzene rings and quinoid exciton bands. The encapsulation of TiO_2 in the polymer matrix led to the blue-shift of the SPAni peaks to 329 and 602 in UV−vis spectrum. TEM image of the nanocomposite showed that nanoparticles with a uniform size distribution encapsulated in the polymer matrix (Fig. 15.20).

Farrokhzadeh et al. used cheap and simple approach and with minimum environmental contamination for the preparation of poly [(7)-2-(sec-butyl) aniline]/silica-supported perchloric acid composites under solid-state condition (Farrokhzadeh and Modarresi-Alam, 2016). They employed poly[(7)-2-sec butylaniline] base (PSBA) and the silica-supported perchloric acid (SSPA) as dopant solid acid in solid state. The conductivity of samples was $\approx$0.07 S/cm in agreement with the percent of doping obtained of the XPS analysis and enhances with increase of doping degree. Also, substituted polyaniline showed lower conductivity than original polyaniline. It is due to the presence of some nonplaner conformations in chain that decrease the conjugation along the backbone. Bandgar et al. reported PAni/Fe_2O_3 NC (50%) thin film sensor that can detect NH_3 in the concentration range from

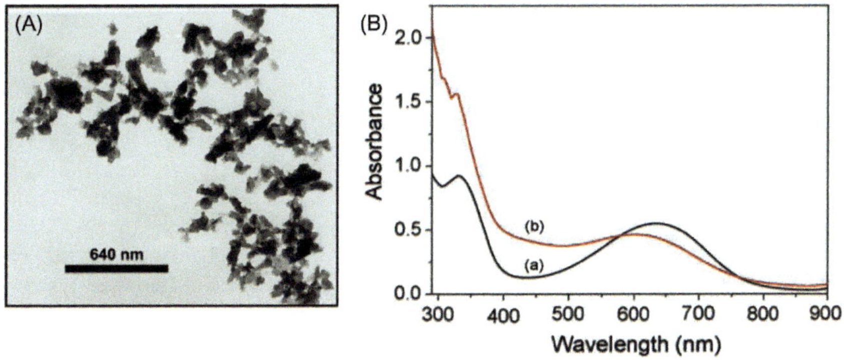

Figure 15.20 (A) TEM images of SPAni–TiO_2 nanocomposites synthesized by the one-pot system of UV-cured polymerization method, (B) UV–visible spectra of bulk PAni powders (A) and SPAni–TiO_2 nanocomposites (B) in DMF solutions synthesized by the one-pot system of UV-cured polymerization method.
Source: Adapted from Karim et al., 2009, with kind permission of Elsevier.

50 to 100 ppm with fast response time (29 s) and highly reproducible response curves (Bandgar et al., 2015).

15.5 Conclusion

Optically active polymers are certainly essential in a variety of potentials, including materials science, chemical and biological sensing, enantioselective separations, asymmetric catalysis, and chiral stationary phases. Synthetic chiral polymers can be produced by several strategies like asymmetric synthesis. Polymerization of optically active monomers simply produces chiral polymers in the main-chain or side-chain chiral polymers. Helical polymers are the other form of optically active polymers, synthesized by asymmetric polymerization. Combination of metal oxides into the chiral polymer matrix acts as a fundamental approach to solve the conventional drawbacks of the polymer. In this chapter, we provided an overview of the research in synthesis and characterization of optically active polymer as well as the effects of nanoparticles on the properties of nanocomposites prepared with various optical active polymers such as polyurethane, polyacethylene, poly(amide-imide), and polyaniline and their derivatives were discussed.

Acknowledgments

We are grateful to the support of the Research Affairs Division of Isfahan University of Technology (IUT) and National Elite Foundation (NEF).

References

Abdolmaleki, A., Mallakpour, S., Borandeh, S., 2012. The use of novel biodegradable, optically active and nanostructured poly (amide-ester-imide) as a polymer matrix for preparation of modified ZnO based bionanocomposites. Mater. Res. Bull. 47, 1123–1129.

Angiolini, L., Giorgini, L., Li, H., Golemme, A., Mauriello, F., Termine, R., 2010. Synthesis, characterization and photoconductive properties of optically active methacrylic polymers bearing side-chain 9-phenylcarbazole moieties. Polymer. (Guildf). 51, 368–377.

Angiolini, L., Benelli, T., Giorgini, L., 2011. Novel optically active methacrylic polymers containing side-chain porphyrin moieties for chiral recognition. Polymer. (Guildf). 52, 2747–2756.

Angiolini, L., Benelli, T., Giorgini, L., Golemme, A., Mazzocchetti, L., Termine, R., 2014. Effect of a chiral substituent on the photochromic and photoconductive properties of a methacrylic polymer bearing side chain azocarbazole moieties. Dye. Pigment. 102, 53–62.

Bandgar, D., Navale, S., Vanalkar, S., Kim, J., Harale, N., Patil, P., et al., 2014. Synthesis, structural, morphological, compositional and electrical transport properties of polyaniline/α-Fe_2O_3 hybrid nanocomposites. Synth. Met. 195, 350–358.

Bandgar, D., Navale, S., Mane, A., Gupta, S., Aswal, D., Patil, V., 2015. Ammonia sensing properties of polyaniline/α-Fe_2O_3 hybrid nanocomposites. Synth. Met. 204, 1–9.

Benelli, T., Angiolini, L., Caretti, D., Lanzi, M., Mazzocchetti, L., Salatelli, E., et al., 2014. Effect of composition on the properties of novel optically active methacrylic copolymers containing side-chain zinc–porphyrin chromophores suitable to chiral recognition. Dye. Pigment. 106, 143–153.

Botta, M.C., Biava, H.D., Spanevello, R.A., Mata, E.G., Suárez, A.G., 2016. Development of polymer-supported chiral aminoalcohols derived from biomass and their application to asymmetric alkylation. Tetrahedron. Lett. 57, 2186–2189.

Brown, C., 2013. Chirality in Drug Design and Synthesis. Academic Press.

Bu, X., Zhou, Y., He, M., Chen, Z., Zhang, T., 2014a. Optically active SiO_2/TiO_2/polyacetylene multilayered nanospheres: preparation, characterization, and application for low infrared emissivity. Appl. Surf. Sci. 288, 444–451.

Bu, X., Zhou, Y., He, M., Chen, Z., Zhang, T., Wang, Y., 2014b. Fabrication and characterization of optically active polyacetylene@WO_3 nanorodhybrids with low infrared emissivity. Mater. Lett. 120, 239–242.

Cataldo, F., Lilla, E., Ursini, O., 2011. Radiation-induced polymerization of β (+)-pinene and synthesis of optically active β (+)/β (−) pinene polymers and copolymers. Radiat. Phys. Chem. 80, 723–730.

Chen, J., Zhou, Y., Nan, Q., Ye, X., Sun, Y., Zhang, F., et al., 2007. Preparation and properties of optically active polyurethane/TiO_2 nanocomposites derived from optically pure 1, 1′-binaphthyl. Eur. Polym. J. 43, 4151–4159.

Chen, J., Zhou, Y.-M., Nan, Q.-L., Ye, X.-Y., Sun, Y.-Q., Wang, Z.-Q., et al., 2008. Synthesis and characterization of polyurethane/CdS–SiO_2 nanocomposites via ultrasonic process. Appl. Surf. Sci. 255, 2244–2250.

Chen, S., Zhang, S., Jin, T., Zhao, G., 2016. Synthesis and characterization of novel covalently linked waterborne polyurethane/Fe_3O_4 nanocomposite films with superior magnetic, conductive properties and high latex storage stability. Chem. Eng. J. 286, 249–258.

Chen, Y.-H., Zhang, J., Yang, L.-W., Yang, N.-F., Yang, Z.-S., 2014. Optically active helical vinyl polymers via helix-sense-selective anionic polymerization of (*S*)-3-vinyl-2, 2′-bis (methoxymethoxy)-1, 1′-binaphthyl. Polymer. (Guildf). 55, 3666–3671.

Farrokhzadeh, A., Modarresi-Alam, A.R., 2016. Complete doping in solid-state by silica-supported perchloric acid as dopant solid acid: Synthesis and characterization of the novel chiral composite of poly [(±)-2-(sec-butyl) aniline]. J. Solid State Chem. 237, 258–268.

Feng, L., Zhao, F.B., Liu, Z.L., Sun, J.P., Hu, J.W., 2007. Synthesis of optically active polymers bearing amino acid moieties by atom transfer radical polymerization. Chin. Chem. Lett. 18, 875–878.

Fujiki, M., Koe, J.R., 2000. Optically active silicon-containing polymers. Silicon-Containing Polymers. Springer, pp. 643–665.

Huang, H., Yuan, Y., Deng, J., 2015. Helix-sense-selective precipitation polymerization of achiral monomer for preparing optically active helical polymer particles. Macromolecules. 48, 3406–3413.

Huang, X., Xu, Y., Miao, Q., Zong, L., Hu, H., Cheng, Y., 2009. Synthesis and properties of chiral helical polymers based on optically active polybinaphthyls. Polymer. (Guildf). 50, 2793–2805.

Iida, H., Iwahana, S., Mizoguchi, T., Yashima, E., 2012. Main-chain optically active riboflavin polymer for asymmetric catalysis and its vapochromic behavior. J. Am. Chem. Soc. 134, 15103–15113.

Itsuno, S., 2011. Polymeric Chiral Catalyst Design and Chiral Polymer Synthesis. John Wiley & Sons.

Itsuno, S., Sakurai, Y., Ito, K., Maruyama, T., Nakahama, S., Fréchet, J., 1990. New solid-phase catalysts for asymmetric synthesis: cross-linked polymers containing a chiral Schiff base-zinc complex. J. Org. Chem. 55, 304–310.

Kalia, S., Haldorai, Y., 2015. Organic–Inorganic Hybrid Nanomaterials. Springer.

Kanbayashi, N., Okamura, T.-A., Onitsuka, K., 2015. New synthetic approach for optically active polymer bearing chiral cyclic architecture: combination of asymmetric allylic amidation and ring-closing metathesis reaction. Macromolecules. 48, 8437–8444.

Kaneko, T., Umeda, Y., Yamamoto, T., Teraguchi, M., Aoki, T., 2005. Assignment of helical sense for poly (phenylacetylene) bearing achiral galvinoxyl chromophore synthesized by helix-sense-selective polymerization. Macromolecules. 38, 9420–9426.

Karim, M.R., Lim, K.T., Lee, M.S., Kim, K., Yeum, J.H., 2009. Sulfonated polyaniline–titanium dioxide nanocomposites synthesized by one-pot UV-curable polymerization method. Synth. Met. 159, 209–213.

Kawakami, Y., Tang, H., 2000. Recent progress in the synthesis of conformationally optically active polymers. Design. Monomer Polym. 3, 1–16.

Li, L., Du, X., Deng, J., Yang, W., 2011. Synthesis of optically active macroporous poly (*N*-isopropylacrylamide) hydrogels with helical poly (*N*-propargylamide) for chiral recognition of amino acids. React. Funct. Polym. 71, 972–979.

Liu, D., Li, Y., Deng, J., Yang, W., 2014. Helix-sense-selective polymerization of achiral substituted acetylene in chiral micelles for preparing optically active polymer nanoparticles: effects of chiral emulsifiers. Polymer. (Guildf). 55, 840–847.

Liu, J., Lam, J.W., Tang, B.Z., 2009. Acetylenic polymers: syntheses, structures, and functions. Chem. Rev. 109, 5799–5867.

Liu, J.-H., Yang, P.-C., 2006. Synthesis and characterization of novel monomers and polymers containing chiral (−)-menthyl groups. Polymer. (Guildf). 47, 4925–4935.

Lu, H., Mack, J., Nyokong, T., Kobayashi, N., Shen, Z., 2016. Optically active BODIPYs. Coord. Chem. Rev. 318, 1–15.

Mallakpour, S., Aalizadeh, R., 2014. Application of TiO_2 nanoparticles modified with bioactive diacid in manufacturing of polymer nanocomposites containing 4, 4′-sulfonyl dianiline and *N,N′*-(pyromellitoyl)-bis-L-valine diacid. Synth. React. Inorganic Metal-Organic Nano-Metal Chem. 44, 1450–1456.

Mallakpour, S., Behranvand, V., 2014. Surface treatment of nano ZnO using 3, 4, 5, 6-tetrabromo-*N*-(4-hydroxy-phenyl)-phthalamic acid as novel coupling agent for the preparation of poly (amide-imide)/ZnO nanocomposites. Colloid Polym. Sci. 292, 2275–2283.

Mallakpour, S., Dinari, M., 2011. Progress in synthetic polymers based on natural amino acids. J. Macromol. Sci. A. 48, 644–679.

Mallakpour, S., Dinari, M., 2013. Investigating the nanostructure and thermal properties of chiral poly (amide-imide)/Al_2O_3 compatibilized with 3-aminopropyltriethoxysilane. Mater. Res. Bull. 48, 3865–3872.

Mallakpour, S., Hatami, M., 2012. Production and evaluation of the surface properties of chiral poly (amide-imide)/TiO_2 nanocomposites containing L-phenylalanine units. Prog. Org. Coat. 74, 564–571.

Mallakpour, S., Hatami, M., 2013. Optically active poly (amide-imide)/zinc oxide hybrid nanocomposites based on hydroxyphenyl benzamide segments: compatibility by using 3-methacryloxypropyltrimethoxysilane coupling agent. Polym. Sci. B. 55, 643–650.

Mallakpour, S., Khadem, E., 2014a. A green route for the synthesis of novel optically active poly (amide-imide) nanocomposites containing *N*-trimellitylimido-L-phenylalanine segments and modified alumina nanoparticles. High Perform. Polym. 0954008313516820.

Mallakpour, S., Khadem, E., 2014b. Reinforcement of poly (amide-imide) containing *N*-trimellitylimido-L-phenylalanine by using nano α-Al_2O_3 surface-coupled with bromo-flame retardant under ultrasonic irradiation technique. J. Mol. Struct. 1075, 196–203.

Mallakpour, S., Khadem, E., 2015. Recent development in the synthesis of polymer nanocomposites based on nano-alumina. Prog. Polym. Sci. 51, 74–93.

Mallakpour, S., Khani, M., 2013. Construction and characterization of poly (amide-ester-imide) nanocomposites containing *N*-trimellitylimido-L-leucine toughened with a combination of bioactive surface-grafted TiO_2. Prog. Org. Coat. 76, 1608–1615.

Mallakpour, S., Madani, M., 2012. A facile route for the preparation of novel optically active poly (amide-imide)/functionalized zinc oxide nanocomposites containing pyromellitoyl-bis-L-phenylalanine moieties. Polym. Bull. 68, 1201–1214.

Mallakpour, S., Mani, L., 2014. Improvement of the Interactions between modified ZrO_2 and poly (amide-imide) matrix by using unique biosafe diacid as a monomer and coupling agent. Polymer-Plastics Technol. Eng. 53, 1574–1582.

Mallakpour, S., Marefatpour, F., 2014. Novel chiral poly (amide-imide)/surface modified SiO_2 nanocomposites based on *N*-trimellitylimido-L-methionine: synthesis and a morphological study. Prog. Org. Coat. 77, 1271–1276.

Mallakpour, S., Sirous, F., 2015. Surface coating of α-Al_2O_3 nanoparticles with poly (vinyl alcohol) as biocompatible coupling agent for improving properties of bio-active poly (amide-imide) based nanocomposites having L-phenylalanine linkages. Prog. Org. Coat. 85, 138–145.

Mallakpour, S., Zadehnazari, A., 2011. Advances in synthetic optically active condensation polymers—a review. Express Polym. Lett. 5, 142–181.

Mallakpour, S., Zadehnazari, A., 2014. Molten salt-supported polycondensation of optically active diacid monomers with an aromatic thiazole-bearing diamine using microwave irradiation. J. Adv. Res. 5, 311–318.

Mallakpour, S., Zeraatpisheh, F., 2014. Novel flame retardant zirconia-reinforced nanocomposites containing chlorinated poly (amide-imide): synthesis and morphology probe. J. Exp. Nanosci. 9, 1035–1050.

Mallakpour, S., Tirgir, F., Sabzalian, M.R., 2010. Novel biobased polyurethanes synthesized from nontoxic phenolic diol containing L-tyrosine moiety under green media. J. Polym. Environ. 18, 685–695.

Mallakpour, S., Zeraatpisheh, F., Sabzalian, M.R., 2012. Construction, characterization and biological activity of chiral and thermally stable nanostructured poly (ester-imide) s as tyrosine-containing pseudo-poly (amino acid) s. J. Polym. Environ. 20, 117–123.

Mallakpour, S., Abdolmaleki, A., Rostami, M., 2014a. Morphological and thermal properties of poly (amide-imide)/ZnO nanocomposites derived from 4, 4′-methylenebis (3-chloro-2, 6-diethyl trimellitimidobenzene) and 3, 5-diamino-*N*-(4-hydroxyphenyl) benzamide. Polymer-Plastics Technol. Eng. 53, 1615–1624.

Mallakpour, S., Madani, M., Roshandel, S., 2014b. Applications of ultrasound for modification of zinc oxide and fabrication of optically active poly (amide-imide)/zinc oxide bionanocomposites. Design. Monomer Polym. 17, 364–371.

Mohanty, S., Nayak, S.K., Kaith, B., Kalia, S., 2015. Polymer Nanocomposites Based on Inorganic and Organic Nanomaterials. John Wiley & Sons.

Nakano, T., Okamoto, Y., 2000. Asymmetric polymerization of methacrylates. Macromol. Rapid Commun. 21, 603–612.

Ohkawa, S., Yang, F., Kawabata, K., Goto, H., 2013. Synthesis of polymers from chiral monomers in chiral liquid crystals. J. Appl. Polym. Sci. 128, 3586–3591.

Ohta, R., Togashi, F., Goto, H., 2007. Synthesis of chiral conjugated polymers bearing azobenzene moieties using cholesterics. Macromolecules. 14, 5228–5230.

Okamoto, Y., Nakano, T., 1994. Asymmetric polymerization. Chem. Rev. 94, 349–372.

Patil, U., Ramgir, N.S., Karmakar, N., Bhogale, A., Debnath, A., Aswal, D., et al., 2015. Room temperature ammonia sensor based on copper nanoparticle intercalated polyaniline nanocomposite thin films. Appl. Surf. Sci. 339, 69–74.

Pomogailo, A.D., Kestelman, V.N., 2005. Nanoparticle modifying action on polymers. Metallopolymer Nanocomposites 427–458.

Prasad, G., Takei, T., Arimoto, K., Yonesaki, Y., Kumada, N., Kinomura, N., 2006. Nanocomposites based on exfoliated $NbWO_6$ nanosheets and ionic polyacetylenes. Solid State Ion. 177, 197–201.

Qu, J., Sanda, F., Masuda, T., 2009. Synthesis and chiroptical properties of amino acid-derived poly (methylpropargyl ester) s. Eur. Polym. J. 45, 448–454.

Rafiee, Z., Zare, E., 2015. Optically active poly (amide-imide)/TiO_2 nanocomposites containing amino acid moieties: synthesis and properties. Amino Acids. 47, 2293–2302.

Selegny, E., 1979. Optically Active Polymers. Springer.

Xu, X., Feng, S., Zhu, Y., Li, H., Shen, X., Zhang, C., et al., 2013. Stereospecific radical polymerization of optically active (*S*)-*N*-(2-hydroxy-1-phenylethyl) methacrylamide catalyzed by Lewis acids. Eur. Polym. J. 49, 3673–3680.

Yang, Y., Zhou, Y., Ge, J., Wang, Y., Chen, X., 2011a. Synthesis, characterization and infrared emissivity property of optically active polyurethane derived from tyrosine. Polymer. (Guildf). 52, 3745–3751.

Yang, Y., Zhou, Y., Ge, J., Wang, Y., Zhu, Y., 2011b. Preparation, characterization, and infrared emissivity property of optically active polyurethane/TiO_2/SiO_2 multilayered microspheres. J. Solid State Chem. 184, 2617–2622.

Yang, Y., Zhou, Y., Ge, J., Yang, X., 2012. Optically active polyurethane@ indium tin oxide nanocomposite: preparation, characterization and study of infrared emissivity. Mater. Res. Bull. 47, 2264–2269.

Yang, Y., Zhou, Y., Wang, T., 2014. Preparation of optically active polyurethane/TiO_2/MnO_2 multilayered nanorods for low infrared emissivity. Mater. Lett. 133, 269–273.

Yao, F., Zhang, D., Zhang, C., Yang, W., Deng, J., 2013. Preparation and application of abietic acid-derived optically active helical polymers and their chiral hydrogels. Bioresour. Technol. 129, 58–64.

Yashima, E., Maeda, K., Iida, H., Furusho, Y., Nagai, K., 2009. Helical polymers: synthesis, structures, and functions. Chem. Rev. 109, 6102–6211.

Zahmatkesh, S., Yazdanpanah, A., 2012. Microwave-assisted synthesis and characterization of L-lysine-derived optically active poly (hydrazide-imide)s. Arab. J. Chem (In press). Available from: http://dx.doi.org/10.1016/j.arabjc.2012.06.004.

Zare, Y., 2016a. Modeling the strength and thickness of the interphase in polymer nanocomposite reinforced with spherical nanoparticles by a coupling methodology. J. Colloid. Interface Sci. 465, 342–346.

Zare, Y., 2016b. Modeling the yield strength of polymer nanocomposites based upon nanoparticle agglomeration and polymer–filler interphase. J. Colloid. Interface. Sci. 467, 165–169.

Zare, Y., 2016c. Study of nanoparticles aggregation/agglomeration in polymer particulate nanocomposites by mechanical properties. Compos. Part A: Appl. Sci. Manuf. 84, 158–164.

Zhang, C., Zhang, Y., Zhang, H., Hu, Y., Zhang, X., Masuda, T., 2015. Synthesis and ion responsiveness of optically active polyacetylenes containing salicylidene Schiff-base moieties. Reactive Funct. Polym. 87, 46–52.

Zhang, Q., Shi, Y., Zhan, X., Chen, F., 2012. In situ miniemulsion polymerization for waterborne polyurethanes: kinetics and modeling of interfacial hydrolysis of isocyanate. Colloid Surf. A: Physicochem. Eng. Asp. 393, 17–26.

Index

Note: Page numbers followed by "*f*" and "*t*" refer to figures and tables, respectively.

D

G

Q

R

S

CPI Antony Rowe
Chippenham, UK
2017-06-13 18:22

Contents

List of Tables and Figures

Introduction

Criminal law is a means of social control and governance which regulates human behavior and interaction and serves certain purposes such as retribution, deterrence, incapacitation, rehabilitation and restoration. It also benefits society by maintaining order, resolving disputes, protecting individual rights and property, providing for a smooth functioning of society and safeguarding civil liberties. This area of law includes a body of rules and statutes that define conduct prohibited by the government, it establishes criminal response for the commission of such acts (substantive law), and in a broad sense, it describes the process through which criminal laws are enforced (adjective law). If substantive criminal law, as the normative section of any criminal justice system, is defective, then justice cannot be retrieved and recovered by procedural criminal law and through fair trial. In accordance with criminological and sociological findings, this issue gains in significance when criminal law relates to juveniles.

Juvenile delinquents and children in conflict with the law have always been of special interest for authorities, policy-makers, professionals, and researchers to find out how they should be protected by the law until they are deemed capable of interacting as adults in their society. Particular attention to children and juveniles should be extended when they get into the criminal realm and where their behaviors violate the provisions of the Penal Code. In this case, they are no longer viewed as other ordinary children and juveniles but treated as a delinquent or deviant. Therefore, society should take measures as well as help such children and juveniles find a way out of the vicious cycle they got stuck in so as to start a purposeful life. According to research in this field, they should be treated differently from adults by a special section of the criminal justice system called the "juvenile justice system". The most important step in establishing a juvenile justice system is to enact a special code that contains both substantive and procedural rules for juvenile delinquents, in line with the best interests of the child and in accordance with the requirements of due process. When talking about juvenile and child criminal law, what first comes to mind are the criminal proceedings and juvenile courts in the portion of the law called adjective law, while the special procedural law for children and juveniles can achieve its objectives when specific substantive law is developed. Therefore, differentiated substantive and adjective criminal law should be considered and enacted by a legislature as the first stage.

The term "juvenile delinquent" refers to any person below a certain age who has come into contact with the justice system as a result or being suspected of committing a crime. Juvenile criminal law and the juvenile justice system – which is generally based on an age index (childhood, adolescence) and the result of clinical criminology studies and research – can be considered as specialized fields of criminal law and, because of their features, categorized as technical criminal law. This field of criminology has originated in the positivist school and the Sixth Congress of Criminal Anthropology held by its followers in Turin, Italy in 1906. In terms of delinquency, this congress recommended that the judge must be able to apply a variety of actions and measures depending on the individual-social circumstances of each child, and applying any corrective measures toward a delinquent should be preceded by medical-psychological examinations and an evaluation of their family and ancestry history. Therefore, according to the pioneers of criminology, children should be viewed as complete human beings independent of adults, as their personality, mind, and body are still forming and growing, which is why they are more vulnerable to delinquency, victimization, and deviance; therefore specific social-legal actions and measures are required in handling them.[1] These criminological views were gradually welcomed by legal doctrines and penal statutes, so legislation both in delinquency and victimization of children as well as special rules and measures were developed for the delinquents and child victims in criminal law. Specialized courts and judges, the flexibility of procedures, a diversity of measures and responses that should be educational, helpful, supportive, protective, and regulatory, independence and specialty centres and institutions where the decisions of juvenile courts are executed, as well as participation of people and NGOs in different stages of the process of addressing delinquency are among the most important features of juvenile criminal law. A differentiated treatment of juvenile delinquents grew to upmost importance and held a special position with the advent of four theories and schools called "social reactions criminology" (labeling theory, structural/organizational theory, radical theory, and victimology) in the 1960s. These four theories and schools questioned the intervention of the criminal system in addressing and suppressing crimes and called it bothe a cause of crime and a "cause of recidivism". Accordingly, some mechanisms were implemented for decriminalization, depenalization, diversion, and supporting victims. So the emphasis was more on them being used in the treatment of children as well as of adolescent delinquents and victims. This new way of thinking rapidly found its manifestation in the domestic criminal law of countries as well as regional and international programs, activities, and agreements.[2]

Juvenile criminal law differs significantly from country to country. However, many similarities across the boundaries of domestic criminal law are also an issue in every legal system, such as the violation of legal norms and interests, the criteria for criminal responsibility and its limitations, as well as the determination of sanctions and sentencing. Despite the fact that the persistent peculiar parochialism of criminal law – which is deeply intertwined with the history of criminal law itself and basically whose importance exists in every legal system –, it is notable that the aspiration of

1 For further reading about the specific social-legal actions and measures for juvenile delinquents, see, *Abachi* 1384/2005 and *Abachi* 1387/2008.

2 *Nadjafi Abrandabadi* 1388/2009, p. 9.

evolution and subsequently the harmonization of criminal law is stronger today, especially in the field of juvenile delinquency. As a system of norms, rules, procedures, and institutions, law is a social phenomenon which is mutually related to other social phenomena. This means that in the process of the regulation and control of phenomena, law both affects and is influenced by social relations and phenomena. Furthermore, the fact that criminal law can involve the life and death of a human being makes its concern for humanity even more crucial, which is why it must be founded on social realities and humanity emphasizes the need for the protection of human rights. As societies change, the laws that govern behavior should be changed, too. Evolving and improving value systems normally results in new laws and regulations supporting modern beliefs and thoughts. Although a certain stability is essential for the enforcement of rules, these must change occasionally, which is why the importance of comparative criminal law at the national and international level has increased enormously in the present age of evolution[3] and globalization.[4]

At an international level, the United Nations has developed certain rules and regulations on how to deal with those children and adolescents who are perpetrators of crime. In fact, since 1959, many documents have been issued by the United Nations in the form of declarations, resolutions, and treaties about the standards and norms in crime prevention and criminal justice in general as well as juvenile justice in particular. The United Nations differentiated criminal policies toward children and juveniles and stated that they can be considered as a desirable criminal thinking pattern for the treatment of juvenile delinquents. Obviously, these documents (both binding and non-binding instruments), do not have the same legal validity, but at least, as international references of juvenile criminal law for the member states, they can play an essential role in harmonizing policies in the world. Therefore, a scientific overview of these instruments seems to be necessary for legislative, judicial, and executive uses at a national level. Despite the fact that most (even developing) countries have established a "juvenile justice system" in response to criminological findings and in accordance with international standards, Iran is one among those that have not paid thorough attention to the conception of child rights as well as sociological and criminological findings in the formulation of juvenile criminal law rules. However, over the past few years, there have been attempts to re-establish a juvenile justice system based on differentiated criminal policy, although it should be noted that the structure of the juvenile justice system in Iran is distinct. Iran is a state governed by the rule of *Shari'a* (Islamic law). It is worth mentioning that the ideology and thus basis of Iran's criminal law, in general, and juvenile criminal law, in particular, is *Shari'a*, and as a civil and continental system the legislations, Islamic law, and the judicial precedent are considered, respectively, as the main criminal law sources in

3 For further reading about the evolution of criminal law and the importance of comparative studies, see *Dubber* 2005; *Eser* 2017; *Chamelin et al.* 2012; *Hazel* 2008; *Ghassemi* 2015.

4 For further reading about the globalization of the juvenile justice system, see *Khajeh Noori* 1396/2017.

Iran. Therefore, the structure of Iran's juvenile criminal law is a combination of Islamic jurisprudence, international obligations, and customary law. Because of the clashes and many other problems which have led to inconsistencies in the legal system, as well as due to the demands of society, the necessity for reforming criminal law based on modern models is of particular importance in Iran. The German juvenile criminal law and juvenile justice system are, by contrast, the product of a centuries-long project of scientific findings undertaken by scholars who traditionally have defined the science of criminal law in contradistinction to the markedly unscientific business of criminal justice policy. With its democratic system, Germany is a state governed by the rule of law. The objective of German criminal law is the "protection of legal interests", and its ideology is based on "doctrine". The constitution, legislations, and case law, respectively, are considered the main legal sources in Germany as a continental system.[5]

In recent decades, respect for basic child rights and the importance of sociological and criminological research have become major themes in international discourse and politics, initiating a strong movement toward the recognition and implementation of common minimum standards of juvenile justice.[6] Therefore, analyzing German juvenile criminal law and the United Nations' standards in this field can be useful for reforming the structure of Iran's juvenile criminal law, which has lost its bilateral relationship with the mentioned realities and is furthermore suffering from freezing and hardening. For this reason, this study investigates, historically and comparatively, the criminal law governing juvenile delinquency in Iran, Germany, and the United Nations on the one hand and specifically examines and analyses public attitudes of elites toward juvenile justice in Iran on the other hand. In the theoretical part of this study (*Chapter 1*), the historical evolution of criminal law concerning juvenile delinquents from ancient times will be discussed for Iran and Germany, as well as for its establishment of the United Nations. It is necessary to emphasize that when investigating the history of juvenile criminal law, the experiences of society with the way children are treated may appear to be in conflict with the law and juvenile delinquents. This also plays an important role in the new policies to be put into effect in society. In *Chapter 2*, the comparative study of criminal law concerning juvenile delinquents in Iran, Germany, and the United Nations will be analyzed. The main objective in this chapter is to investigate and evaluate current juvenile criminal law and to grasp the legal reality in Iran, Germany, and the United Nations in this regard. In this section, a further aim is to illustrate to what extent criminal laws and practices relating to juvenile delinquents in Iran and Germany are in accordance with the recognized standards of the United Nations and its human rights sources. In the empirical part (*Chapter 3*), the attitude of elites toward juvenile justice in Iran will

---

5 For further reading about criminal law sources in Germany, see *Bohlander* 2009, pp. 7–16.

6 For further reading about reforming juvenile justice, see *Junger Tas et al,* 2009.

be studied. In order to evaluate how elites consider issues concerning juvenile delinquents in Iran, two target groups were sampled. The main criteria for differentiating these target groups were their potential influence on the revision and enactment of laws as well as the recognition of juvenile justice as a differentiated and separate system in the Iranian criminal justice system. The first group comprised those who work in the judicial system, including criminal trial judges (judges of lower, appellate, and high courts), prosecutors and investigators, social workers, and defense lawyers (totaling 130 respondents). Due to their duties, they are the most influential group in the formation of criminal policy and, subsequently, juvenile justice in the legal system. The second group comprised those who do not work in the justice system but are influential in policy formulation and law-making processes, including law faculty members, Islamic lawyers and master's degree students in criminal law and criminology (totaling 92 respondents). The empirical study was conducted in three provinces in Iran (Tehran, Gilan, and Qum) so as to evaluate the various regional attitudes of elites.

Chapter 1

The Historical Study of Criminal Law Governing Juvenile Delinquency in Iran, Germany and the United Nations

In order to demonstrate the importance of studying the legal history and historical evolution of law, we should focus on the nature of law as well as the need for law, which lies in the history of humanity. In early times laws were not needed because there were few conflicts. However, gradually, different desires, needs, and necessities in society caused conflicts and laws became necessary as a means of social control, either to alleviate conflicts or to settle them in a manner most advantageous to the respective group.[7]

It can be said that law, like any other social phenomenon, has gone through a historical evolution. Both the content and the format of this phenomenon, of any scope and capacity, are products of a historical process which was born in the context of human societies and was influenced by various social, cultural, economic, and even geographical factors. The diversity of legal systems and the ongoing development of their rules and institutions cast no doubt whatsoever on the historic nature of this phenomenon. A review of law history from a philosophical perspective can play a decisive role in the formulation of general theories on law, in the same way it was effective in the emergence of historic fulfillment and realism schools in the philosophy of law. In addition, from a sociological perspective, constrictive effects can be observed in the processing of sociological theories on general law, as the sociology of law is one of the highly-valued branches of sociology.[8]

Furthermore, understanding the history of law is not only of scientific value but also plays an undeniable role in social management and planning and can help the judicial organization recognize the needs of society, as well as meeting expectations and requirements. There is a significant difference between the historical perspective on law and reviewing the history of law. This is an epistemological difference which refers to the boundary between science and philosophy. But law is also considered to be a historical phenomenon in which its whole identity is based on the requirements of any one period of time. By contrast, in the review of law history, such judgments do not exist, and only the process of formation, its evolution and interaction with other social and environmental factors over time are studied and analyzed,

---

7 *Chamelin et al.* 2012, p. 1.

8 *Aramesh* 1380/2001, p. 72; *Rasekh* 1381/2002, p. 70.

which may ultimately provide a rich source of material for the historical investigation. The benefits of understanding the history of law firstly include theoretical ones such as understanding the legal nature as such, its possibility for comparative studies, and enriching training; secondly, they encompass practical benefits such as developments in legislation, justice realization, and the evolution of international law.[9]

The mutual relationship between law and human collective life has been accepted by the natural and the historical-social schools of thought. From the perspective of the natural school, which is influenced by stoic philosophy and the ideas of *Thomas Aquinas*, general laws already exist in nature and human-related laws should be established in accordance with these basic natural rules. Thus, according to the natural school of law, it may be stated that the above-mentioned norms and rules are independent of the realities and relations determined by them, but this relationship should be realized through the comprehensively mentioned laws and in accordance with the respective environmental conditions. Therefore, according to natural law, the system of human laws is the product of a mutual relationship between social laws and realities.

From the perspective of the historical-social school, which influenced by the ideas of many theorists like *Hume* and *Montesquieu*, there is no preexisting norm or rule which can form and organize the relationship between humans. These norms and rules are merely the result of interaction between human demands and desires on the one hand and the natural-humanistic environment on the other. Accordingly, it is evident that this school does not consider any independent existence for legal rules and norms, nor does it, consequently, claim that the realities of human collective life merely follow the rules in a one-way fashion. Furthermore, the belief in a bilateral relation and interaction between the law and these realities consists of integral components of this school of thought. Therefore, in any layer, social phenomena do not come into existence out of the blue. This means that all levels and forms of human collective life are based on a long and tortuous process. Phenomena and realities have evolved over time, and we are now witnessing a more highly developed form of them that cannot be correctly understood without paying sufficient attention and consideration to their evolution. Hence, it can be claimed that law has a historical and flowing nature, and since it is exposed to the realities of human collective life with its mutual relationships, another reality logically exists; the law has changed and evolved along the developments of these realities.[10]

The development of law over the course of time is called the history of law. As a system of norms, rules, procedures, and institutions, it is a social phenomenon which could be influenced by developments just like any other social phenomenon and, as a result, has its own history.[11] As a science, law empowers humans by teaching them argument and reasoning, rather than making them open-minded and

---

9 *Rasekh* 1381/2002, pp. 71–72.

10 *Rasekh* 1381/2002, p. 73.

11 *Rasekh* 1381/2002, pp. 73.

liberal. However, the history of law, in addition to empowering humans, has directed our minds toward liberalism and open-mindedness by providing a wider perspective for lawyers to observe the emergence, evolution, and disappearance of legal rules and to realize that the first mission of law is to be in accordance with social demands and requirements.[12]

Our ideas about juvenile delinquency and juvenile justice are inevitably shaped by the time period in which we live. These ideas emerge at certain points in history, and they reflect the historical conditions of the time. Ideas are elaborated into philosophies, in the sense of their being rational, coherent, and organized ways of understanding and interpreting the world. The philosophies then form the basis for laws that define juvenile delinquency, direct the juvenile justice system, and determine juvenile justice policies. What we define as juvenile delinquency and juvenile justice can only be understood by studying these ideas in the context of history, philosophy, and law. Therefore, the main reason for studying the past is to better understand how we have arrived at our present state of being and where we might be heading in the future. This is true for all aspects of social life, but, in particular, for juvenile justice.[13]

To this purpose, the historical evolution of juvenile criminal law in Iran, Germany, and the United Nations will be discussed in this 331 B.C., *Alexander the Great* conquered and defeated the *Achaemenid Empire*. Afterwards, his successors, the section.

1.1 Historical Evolution of Juvenile Criminal Law in Iran

As an introduction to the main discussion, it is necessary to outline historical documents in *Iran*.[14] The history of *Iranian Ariyans*[15] is divided into four parts[16]:

---

12 For further reading, see *Katoziyan 1393/2014*.

13 *Bernard et al* 2010, p. 6.

14 It should be noted that the term Greater Iran/Persia, which has its source in the history of the *Achaemenid Empire* in *Persis*, is not limited to the modern state of *Iran*. It is worth mentioning that Persia was an ancient kingdom within Greater Iran, and the term Persia was used by classical Greek authors in ancient times. However, after the Arab conquests, Greater Iran lost many of its territories due to changes of boundaries. Later, the term Iran was recognized as the official legal name of the region by *Reza Shah*, the king of the *Pahlavi Dynasty*, on both the national and the international level in 1935. See *Pirnia et al.* 1387/2008, pp. 21–31. The term Iran is a modern pronunciation of *Aireya* or *Ariya* in the *Avesta* language, referring to Iranian descent and meaning honorable, decent, and noble. See *Mohammadi et al.* 1383/2004, p. 99.

15 According to *Avesta*, the oldest Iranian piece of writing and the holy book of Zoroastrianism, the origin of the term Iranian was *Airyamonjo*, which was changed into *Ayronovic* in the *Pahlavi* language, which is spoken in a very cold region where the *Ariyans* lived before emigrating. See *Mohammadi et al.* 1383/2004, p. 99. See also *Sunderland* 1968, pp. 394–409. At first, dozens of tiny kingdoms occupied the region. Eventually two major powers emerged: the *Medes* and the *Persians*. In time, a remarkable ruler would lead *Persia* to dominate the *Medes* and to found a huge empire.

16 *Pirnia* 1362/1983, p. 258.

1) In the ancient era, which began in the late 8th century B.C. and ended in the middle of the 7th century A.D., Iran's history can be divided into various periods as follows (the periods a) and b) complement each other):

   a) The *Medes*[17] period: The *Medes* (612–549 B.C.) were ancient Iranian people who lived in *Media*, a region located in northwestern Iran. This period is known for its promotion of *Iranian Ariyans* in the old east.

   b) The *Achaemenid*[18] period: The *Achaemenid Empire* (550–330 B.C.) was founded by *Cyrus the Great* and was one of the largest empires in ancient history. This period is known as a period of *Iranian Ariyan* domination over the old east.

The *Macedonian* and *Seleucid*[19] periods: In *Seleucids*, ruled from 312 to 63 B.C,. following the division of the *Macedonian Empire*. This period is known as one of the superiority of Greeks and Macedonians in *Persia* and the old east.

   c) The *Parthian*[20] period: After the defeat of *Alexander the Great's* successors,[21] the *Seleucids*, the *Parthian Empire* ruled from 247 B.C. to 228 A.D. This period marks a fascinating era of Persian history because it revived the greatness of the *Achaemenid Empire*.

   d) The *Sassanid*[22] period: The *Sassanid Empire* (224 to 651 A.D.) was the last Iranian empire before the rise of Islam that followed the *Arab* conquest of *Persia* in the 7th century.[23] The Sassanid Iran influenced Roman civilization. This influence continued to the early Islamic world.

2) The middle era refers to the period between the first half of the first century A.D. and the beginning of the *Safavid* dynasty. This period can be divided into two sub-periods:

   a) before the *Mongol*[24] conquest;

   b) after the *Mongol* conquest, which is considered to be the second and greatest shock for *Persia* in the 12th century, after the arrival of the Arabs and until the establishment of the *Safavid* Empire in the 15th century, which can be considered as the most significant dynasty in *Persia* after the *Arab* conquest.

---

17 For further reading, see *Pirnia* 1362/1983, pp. 205–740; *Diakonoff* 1968, pp. 36–149.

18 For further reading, see *Pirnia* 1362/1983, pp. 205–740.

19 For further reading, see *Steans* 1912; *Lane* 1994; *Pirnia* 1362/1983, pp. 1020–1059; *Moors* 1968, pp. 420–502; and *Davis* 1912

20 For further reading, see *Pirnia et al.* 1387/2008, pp. 115–140.

21 For further reading about *Alexander the Great*, see *Fox*, 1994.

22 For further reading, see *Pirnia et al.* 1387/2008, see also *Yasami* 1361/1982.

23 For further reading, see *Spencer* 1975, pp. 1–56.

24 *Pirnia et al.* 1387/2008, pp. 523–570; see also *Schwartz* (ed.) 1994, pp. 201–217.

3) The third era ranged from the *Safavid* Empire until the Constitutional Revolution[25] that was the first and most important event of its kind in the Middle East; it opened the path to cataclysmic change and heralded the dawn of the modern era in the 19th century (the forcible secularization era).

4) The fourth era lasted from the Constitutional Movement to the present time. It is considered as the contemporary era. The secularization of society was one of the most important events in this time, which started with the *Pahlavi* dynasty (1925–1979) and ended with the Islamic Revolution[26] in 1979 (ending the forcible secularization era).

5) The long era in ancient times, which spanned approximately 14 centuries, can be described in the following two ways:[27]

    a) Consistent with inscriptions of the ancient kings of *Persia*, the findings in excavations at historic sites in Iran, as well as of other monuments of the old places, directly or indirectly provide some information on the history of old Iran, as well as religious books of the Iranian Arians, monuments and old coins, writings of foreign historians, and books by some authors of the Islamic era.

    b) Consistent with a variety of stories which have been handed down from generation to generation since the ancient period, were compiled during the *Sassanid* period, and have become the subject of much writing, the most famous being *Ferdowsi's Shahnameh*.[28]

It is clear that the documents of the first type are superior to the second type of documents in terms of their accuracy and realization. Since the stories mentioned in the second item have been transferred from generation to generation across centuries, they might have been gradually distorted, so they lack the required accuracy and realization. With respect to the documents of the first type, it should be noted that orientalists have not managed to read *Iranian*, *Assyrian*, *Babylonian*, and *Egyptian*, as well as other kinds of inscriptions, and there were no excavations in historical places. The sole source of our information on the ancient times of Iran is the writings of *Greek*, *Roman*, *Egyptian*, *Jewish*, and *Armenian* historians, as well as some authors of the Islamic ages. It must be acknowledged that we do not know everything about the history of ancient Iran, because *Greeks*, *Romans*, and others often recorded only those events which related to their own empires, or events which had occurred

---

25 For further reading, see *Pasha Saleh* 1348/1969, p. 56; *Pirnia* 1362/1983, p. 90.

26 For further reading, see *Pasha Saleh* 1348/1969, p. 35; *Pirnia* 1362/1983, p. 95.

27 *Pirnia* 1362/1983, p. 70.

28 The *Shahnameh* is a long epic poem written by the famous Persian poet *Ferdowsi* between 977 and 1010 C.E. It is the national epic of *Greater Iran*, consisting of some 60,000 verses.

in the western parts of the region. Therefore, many important events and incidents which occurred in the north or east of ancient *Persia* have fallen into oblivion.[29]

In the first part of this chapter, historical developments of child criminal law will be discussed in four separate sections from the ancient period to present times. In the first part, issues regarding ancient *Persia* and the *Sassanid* era will be discussed, while in the second part, the rules and regulations relating to children in the Islamic era will be reviewed. In order to follow the methods of dealing with juvenile justice after the Constitutional Revolution of 1906 and the Islamic Revolution of 1979, respectively, these developments will be studied in the third and fourth parts in this study. For the purpose of studying the historical evolution of juvenile delinquent criminal law and the changes made in reaction to child delinquency in the history of Iran, the following categorization will be considered:

1) ancient *Persia*;
2) from the birth of *Persia* to the end of the *Sassanid* empire (the *Zoroastrian* era), the Islamic era;
3) from the Arab conquest of *Persia* to the Constitutional Revolution (the *Islamic* era), the contemporary era being divided into two sections;
4) from the Constitutional Revolution to the Islamic Revolution (the *forcible secularization* era) and from the Islamic Revolution to date (stopping the *forcible secularization* era).

1.1.1 The ancient era

1.1.1.1 Legal system

In each era, the issue of crime, punishment, and treatment of an offender depended on the environment and the social ethics of society. In each period, what the community perceived as contrary to the moral order, and/or what it observed as being harmful to the social order, was considered to be a crime and merited punishment; these interactions in the social system gradually evolved as a tradition, then changed into a custom, and finally became law. Furthermore, in the following periods, with any social change, certain principles which were in compliance with social ethics and the requirements of the society were confirmed and accepted by society; and whatever was not in accordance with new trends was removed, with new rules and procedures being created and established instead. The evolution of penal law and dealing with the criminal was governed by a different order, slowly causing a transition of penal rules from retaliation to rehabilitation.[30] Dealing with young criminals

29 *Pirnia et al.* 1387/2008, pp. 144–146.

30 *Vahedi Navaei* 1380/2001, p. 76.

was not excluded from these social rules, and this shift could be observed in almost the same way in all communities. Criminal law dealt with juveniles and adults in a similar manner.[31]

Due to its vastness, persistence, and historical origin, Iran's territory has seen many ups and downs and has been nourished by various cultures as well as different thoughts and religions.[32] In the *Medes* period (612–549 B.C.), *Pishdadian* were the first families among the *Persians* to introduce the first rules and legislations; the kings of this dynasty, such as *Houshang*, *Tahmures*, *JamshidFereydun*, and *Prince Avrakhshiyeh* are considered the first legislators. Trial in ancient *Persia* was based on religious principles, and was more reliant on the *Avesta*, the holy book of the *Zoroastrian* religion and its rules, in the *Achaemenid* period.[33] In ancient *Persia*, rights and rules were exclusively subject to the king's will and power of the army; no right was respected against this right, and no tradition was of any value without conforming to the king's order and judgment.[34] His decisions and orders were considered to be revelations and inspirations sent down to him by *Ahura Mazda*, the sole God in *Zoroastrianism*. Thus, the empire's laws were seen as divine providence, and any disobedience was considered as disobedience to God's will and command. The judiciary power was also in the king's hands.[35] It can be said that in ancient *Persia*, law, the judiciary, justice, and in general rituals and the formalities of trials originated from the teachings of the *Zoroastrian* religion as well as from commands of the king. In other words, the sources of law in ancient *Persia* were the *Avesta* and its interpretations, the king's commands, the judicial verdicts, customs and habits, and theories of law scholars (doctrines).[36] Historical books regarding the sources of law in ancient *Persia* include:[37]

- *Vendidad:* This is a collection of texts within the greater compendium of the *Avesta* which is not considered a liturgical manual;
- *Matikan Hezardadsetan:* The title literally means "a thousand chapters or laws". It was a compilation of the social, moral, civil, and criminal laws as well as the judicial code of the time;
- *Dadestan Denid:* This work consists of an introduction and 92 religious, social, ethical, legal, philosophical, and cosmological questions and answers – a kind of ancient FAQ, so to speak.

---

31 *Gharavi* 1359/1980, p. 249.

32 *Pirnia et al.* 1387/2008, p. 13.

33 *Karami* 1380/2001, pp. 35–36. For further reading, see *Aram* 1367/1988, pp. 418–419.

34 *Vahedi Navaei* 1380/2001, p. 24.

35 *Karami* 1380/2001, pp. 35–36.

36 *Oshidari* 1371/1992, p. 41; *Vahedi Navaei* 1380/2001, pp. 24–25.

37 *Karami* 1380/2001, p. 37.

After the *Achaemenid* period and the associated events, many chapters of the original *Avesta* were lost, although nearly one third of it was recompiled in the *Sassanid* period. The *Avesta* includes 21 *Nasks* (books). Most of the *Datik's Nasks*, which discuss judicial organization and in particular criminal law, have been lost, while other parts, such as procedure, testimony, the condition of witnesses, crimes, and specific criminals still exist. In some of the 30 topics, the crimes committed by children, such as murder, assault, battery, and crimes against morality, as well as crimes committed by slaves have been discussed.[38]

In ancient *Persia*, according to the remaining texts of the *Avesta*, it was believed that punishment should be based on justice. For some crimes, with the consent of the victim, the penalty was replaced by giving the victim compensation in the form of cash or merchandise. In some cases, if the criminal had a daughter or sister, it would be arranged that she would marry the victim or, if the criminal had certain professional knowledge, he would teach it to the victim. The punishment did not exceed the offender's guilt, and in some cases a rehabilitation response plan was applied. The criminals were trained to serve the community and to become more optimistic about the future. In the *Sassanid* period, particularly because of the king's interest in the ancient religion of *Zoroastrianism*, the judicial administrations were defined by the *Avestan* commands. According to *Matikan Hezardatestan*, there were two types of courts in this era, religious and institutional courts. Judgment over claims rather related to religious affairs were mostly based on orders and rules of the *Avesta*. According to a propaganda letter written by *Tansar* who was a *Zoroastrian* high priest, and although the categorization and principles of *Avesta* were conserved, the conviction rates, the intensity of punishment, and their enforcement changed considerably. Punishment was divided into two categories: for crimes against persons and for crimes against property.[39] The penalty for any crime was determined in the *Avestan* texts, and the judge could not exceed the prescribed limit of the punishment. However, *Denkard*, a book which contains descriptions and summaries of the great *Avesta*, generally presumes that the judge could utilize the minimum and maximum qualities of the penalties (according to the proportionality principle). In *Tansar's* letter, lenience is shown to criminals who are first-time offenders, and financial penalties are given instead of corporal punishments; however, in the case of reoffending, a more severe punishment was implemented.[40]

Historical studies show that in ancient times, especially in some periods, justice and people's rights were considered as a significant subject. However, as mentioned above, the *Zoroastrian* rule was the main governing source to enforce law and order, and the king's commands were to be executed. In any region where the inhabitants had special rules and customs, these were also respected and judicial matters were

---

38 *Gharavi* 1359/1980, p. 257.

39 *Gharavi* 1359/1980, pp. 259–260.

40 *Gharavi* 1359/1980, p. 268.

resolved according to them.[41] It can be said that protecting the political interests of the state and the ruling elite, maintaining *Zoroastrianism* as the official religion of the kingdom, and protecting society against criminals were the most important functions of *Sassanid* criminal law. In substantive criminal law, crimes were classified into various categories and sub-categories, based on which penalties were imposed for individual delicts. In adjective criminal law, especially in proceedings against heretics and members of religious minorities, *Zoroastrian* clergymen played a decisive role. The death penalty was executed by crucifixion, beheading with the sword, burning, stoning, and other crucial methods of execution. Prisons served only as preventive detention, and prisoners could be released on bail.[42]

Furthermore, some main goals were met through the implementation of punishments: the overall administration of justice, the reduction of crime rates, and utilitarian purposes for some offenders.[43]

1.1.1.2 Socio-legal status of children and juveniles

From the formation of the borders of *Persia* up until the end of the *Sassanid* era, *Persian* children were nurtured based on a uniform and consistent culture. Nevertheless, it should be noted there are few documents about *Persian* children's lives before the birth of *Zoroaster*, the ancient prophet and the *Zoroastrian* religion (28 to 551 B.C.).[44] In places where other beliefs and religions dominated the *Persian* society, the destiny of children was strongly influenced by clergymen and wizards. The *Excerpts Zat Seperm*, which is comprised of three books about the *Zoroastrian* rules, indicates that people at the time really believed what their priests told them and did not do anything without consulting them. When a child was born, thepriests predicted the fate of the infant in adulthood. In each case, the order was valid, and parents would adopt their ideas and put them into practice, even if the life of their child was at risk.[45] A great fear from clergymen was rooted in the hearts and minds of people at the time and, as a result, they followed them blindly.[46]

---

41 *Pirniaet al.* 1387/2008., p. 148.

42 *János* 2007, p. 347.

43 *Vahedi Navaei* 1380/2001, pp. 43–44.

44 For further reading, see *Jackson* 1899, p. 68.

45 The mythical story of *Zal* in *Shahnameh* is actually a story about the children of this land whose lives and destinies were in the hands of wizards and priests. Witches and clergymen would send children to death based on unnatural omens, and nobody could stand up against them. For example, *Sam* was an Iranian hero who dreamed a child inheriting the heroic temper from his father. After a while, he had a white-haired son who was named *Zal*. It was a shame for *Sam* to have such a child. Clergymen and wizards believed that having such a child was ominous and indicative of an imminent consequence for his/her family and descent. Based on this belief, *Sam* left his son to the hands of destiny at the *Alborz* Mountain, the place of the *Phoenix*. See *Mohammadi et al.* 1383/2004, p. 117.

46 *Mohammadi et al.* 1383/2004, p. 116.

With the rise of *Zoroastrianism* and its acceptance in a dark period of the *Achaemenid* era, a child's life gradually became more humane. *Zoroaster* called on people to follow their own wisdom and ways of thinking. However, this change was only restricted to free children. In the same period, children of the main and original ethnic groups of *Persia*, such as *Medes*, *Parthians*, and *Scythian*, were considered to be superior to children of other groups. Each of these ethnic groups had their own aristocratic advantages and benefits, which determined the lifestyle of children. For example, children of noblemen had a better status than those of farmers.[47]

When studying the socio-cultural organization of *Persia* in this era, two groups can be identified: free children and slave children. Free children were divided into two sub-categories: those of superior and those of inferior ethnic groups. For instance, during the domination of *Medes*, the *Ariyan* and *Parthian* ethnic groups considered themselves to be superior and had special privileges. Inferior ethnic groups were political affiliates of superior ones. However, the either ethnic groups included nobles, wizards and priests, soldiers, and peasants. Stagnation in *Persia's* social organization had caused stagnation in child destiny, too. A farmer's child, who thus came from the "dregs" of society, could never change his social position and background, unless by the order of the king, which was highly unlikely to happen. Thus, the destiny of children was based on their family's social class more than anything else. Revealing documents have been found in the *Elamite* texts in *Persepolis*, the ceremonial capital of the *Achaemenid* empire, about the status and fate of the children who were enslaved in wars, or those who were born and raised in slave families. These documents, which can be dated to the late 5th century A.D., mentioned that commodities such as sheep, grain, groceries, or cash payments and silver were given to the *Kurtashas*, slaves who worked as clerks in the royal economy system. They were mainly slave children who were captured and employed in wars and received special training to work in the royal family's systems, or the children of *Kurtashas* who carried out this task from generation to generation. Some of the captured children were sent as slaves to the houses of noblemen and great figures. A document that has recently been found indicates conflicts and hostile treatment toward slave children, a practice that has repeatedly occurred throughout *Persian* history. Throughout all these periods, *Persian* rulers tried to become increasingly aware of the culture and ideas of their subordinate ethnic groups. In this regard, *Herodotus* states that *Hovakheshtar*, the *Medes* king, valued a group of *Scythian* warriors who were seeking refuge and asked them to teach their mother language to the *Medes* children.[48]

A *Persian* child's life in the ancient times should be studied in the context of the dominant and rigid social and class structures of that time. For instance, only nobles and courtiers of the king received tribute from other ethnic groups. Depending on

47 *Mohammadi et al.* 1383/2004, p. 118.

48 *Mohammadi et al.* 1383/2004, p. 118.

how the country was organized, this tribute affected life in the lower classes. In this era, the priests were also repressive forces that had their own courts and temples. Furthermore, religious opponents such as the *Manichaeists*, *Mazdakians*, *Christians*, and *Buddhists* who had more followers were enslaved by priests in temples and in the lands around them. As a result, during this period of history, a child's life depended on the position of their family in society. If the family were slaves, their had no choice but to also become slaves, and if their family had been enslaved in war, they were treated as slaves, too.[49]

The importance of child education and a family's responsibility for the child's behavior in ancient *Persia* has been discussed by most philosophers, historians, and thinkers of that era. *Plato* illustrated how the Persian princes and children were educated, so as to set an example to the Greek. In the *Cyropaedia*, *Xenophon* admired the care given to children when describing the education of *Cyrus the Great* and was of the opinion that the grandeur of *Persians* in that era was due to the training and valuable teachings that were given to children – especially in terms of moral and religious teachings – and their habituation to honesty, truthfulness, respect for the law, tolerance, courage, and endurance in the face of hardships and difficulties. *Strabo* also cited singularities about the ancient *Persians* and the education of their children.[50] This training style and the importance of adolescents' education were reflected in the will, decrees, and orders of *Cyrus* and *Dariush the Great* (both *Achaemenied* kings).[51]

It is worth mentioning that *Zoroastrians* believed that if a person passes away without having a child, there would be nobody left to see them cross the "Bridge on Judgment Day". Therefore, those who did not have children selected a child from a close or distant relative or acquaintances by paying them money and including them in their wills, in the form of property. This was known as bridge-building in *Zoroastrianism*.[52]

The legal situation of children was different from adults in ancient times. In a section of *Gonbasr Nygad Nask*, there is reference to the first period in a child's life when they are innocent, to the extent of their criminal responsibility, and guilt in childhood and adolescence. Since a child was considered to achieve a relative sense

---

49 *Mohammadi et al.* 1383/2004, pp. 118–119.

50 *Gharavi* 1359/1980, p. 261; *Vahedi Navaei* 1380/2001, p. 65.

51 *Vahedi Navaei* 1380/2001, p. 66.

52 With respect to the gender differences between children in ancient *Persia*, it should be noted that in civil law, the legacy of the deceased was only inherited to their elder son. If he accepted the legacy, he was also responsible for all debts of the deceased, unless these exceeded the amount of the legacy. In some cases where the deceased had no other heir, his/her daughter received the legacy, and she was obliged to transfer the legacy to her first male child as representing the son of her father. However, the marriage portion of the deceased's wife had to be paid first. See *Aramesh* 1380/2001, pp. 73.

of responsibility by a certain (discerning) age, their parents or guardian had to give them the necessary training and education, and whenever a child committed any crime due to lack of training on the part of the parents or guardian, the adults were held responsible. Also, *Denkard* discusses the level of the penalty for a child's crime. Regarding the laws related to assault and battery, the beginning of the discerning age was considered to be 8 years old. There was no difference between boys and girls in this regard.[53]

Chapter 38 in *Denkard* includes subjects about the role of a man of 15 years or older and elaborates on how important it is for him to learn about laws. Adults were obliged to inform these young men about the laws, and the child was also obliged to be diligent by seeking knowledge about the laws through his/her questions. In fact, criminals *were* divided into two categories: minors and adults. Minors were classified into two further sub-categories, those under 8 years old (who were exempted from criminal liability and punishment) and children between 8 and 16 years old (who had a diminished criminal responsibility). According to various narratives, punishment for minors between 8 and 16 years of age was milder.[54]

Children's exemption from criminal responsibility is also mentioned in *Denkard*; if someone complains about a minor to the court and requests reparation and compensation, the judge should ask the defendant's age before he investigates the case, because minors are not subjected to law. The father or, in some cases, the supervisor or guardian of the minor is responsible for the crimes committed by the child. They are responsible for compensation payment, and they are also liable to severe punishment if necessary. In addition, according to the *Gnbasr Nygad Nask*, whenever a father fails to train his child in correct behavior from 8 years of age, he will be responsible for the child's felony. In this regard, para. 10 of *Ztamystan* is about paternal responsibility when an 8-year-old son commits a murder in front of his/her father's eyes, and it has been discussed as to whether or not the father had the opportunity to stop the killing.[55]

Thus, child crimes were considered differently from adult crimes. In *Denkard*, for cases of murder and other crimes committed by minors as well as the punishment and degree of parents' responsibility were taken into consideration.[56] At this time in history, crimes such as robbery, banditry, murder, and crimes against public morality were associated with physical punishment, which could not be applied to children.

53 *Vahedi Navaei* 1380/2001, p.2.

54 *Vahedi Navaei* 1380/2001, pp. 63–65, cited in *Katebi* 1346/1967; p. 34; *Parvaresh* 1336/1957, p. 140; *Yazdgerdi* 1354/1975, p. 68; *Gharavi* 1359/1980, p. 269. Some researchers believe that the criminal responsibility of children between 7 and 15 years of age had been diminished at that time. For further reading, see *Vahedi Navaei* 1380/2001, pp. 63–65.

55 *Vahedi Navaei* 1380/2001, p. 67; *Gharavi* 1359/1980, p. 257.

56 *Vahedi Navaei* 1380/2001, p. 65.

If a child committed a crime that would result in damages, their guardian or supervisor had the responsibility to compensate damages and losses to the victim, as mentioned above.[57]

Therefore, in ancient Persia and during the *Zoroastrian* era, children and juveniles were treated differently from adults. According to the *Avesta*, the minimum age of children's criminal responsibility was 8 years of age and the maximum was 16 years of age, without any discrimination between boys and girls; punishment for minors between these ages was lighter.[58]

Disregarding the fact that the *Avesta* penal system dates back a thousand years, it is still thought of as more than just a series of primitive thoughts in a penal system. By reviewing the *Avesta's* rules and other legal sources, it is understood that some goals which are still present in the modern legal systems already existed at the time with respect to certain crimes and for juvenile delinquents.[59]

1.1.2 The Islamic era

1.1.2.1 Legal system

The Arab conquest led to the end of the *Sasanid* empire and to the decline of the *Zoroastrian* religion in Iran in the 7th century.[60] However, it can be said that in some cases, regulations, laws, and rights, especially in *Shi'a* jurisprudence, have been affected by the *Avestan* teachings. However, the Muslims' public affairs resolution was introduced based on the *Quran's* teachings and commandments, as the main source and handling reference for Muslims throughout the centuries.[61] According to the *Ja'fari* school,[62] sources of Islamic law include the *Quran* (Holy Book), *traditions* (*Sunnah/Hadith*), the *consensus of the jurists* (*Ijma*), and *reason (Aql/Aghl)*.

The *Quran*, which was recorded by the Prophet. companions shortly after his death in 632 A.D., is the most important source of law for all Muslims. It specifies the moral, philosophical, social, political, and economic basis on which a society should be constructed. The *Quran* contains many behavioral rules expected of Muslims – its legal rules are to be followed in absolute terms. Since it states relatively few explicitly defined rules (only about 600 lines in the *Quran* deal with specific legal matters), three other sources have contributed immensely to the corpus of *Shi'a* law:

57 *Gharavi* 1359/1980, p. 258.

58 *Vahedi Navaei* 1380/2001, p. 63.

59 *Vahedi Navaei* 1380/2001, p. 63.

60 For further reading, see *Zarrinkub* 1975, pp. 1–57; *Pourshariati* 1387/2008, p. 40.

61 *Gharavi* 1359/1980, p. 280.

62 Twelver *Shi'ism* has been particularly influenced by the opinions of the sixth *Imam*, *Abu Abdullah Ja'far bin Mohammad al-Sadeq*, and hence its legal school is known as the *Ja'fari* school of jurisprudence. *Entessar* 1988, p. 94.

- *Traditions* is a second important source of law and is commonly defined as the customs, statements, deeds and sayings, as well as silent assertions of the Prophet *Mohammad* and the *Imams*, which have been collected and codified by people who were close to them and are viewed as reliable sources by the *Ulama*. Along with the *Quran*, the *Hadith* collections constitute the most important sources of the *Shi'a* legal system.
- The third source of *Shi'a* law is *consensus* (*Ijma*), which refers to the means by which the opinions of the *Imams* can be brought to light. Specifically, it refers to the consensus or unanimity of the views on a particular legal issue by *Shi'a* scholars who lived during the time of the *Imams*. Since several of the prominent *Shi'a* scholars were close companions of the *Imams*, the *Shi'a* doctrine places heavy emphasis on the opinions of these early Islamicists. Muslim jurists provide many verses of the *Quran* that legitimize *Ijma* as a source of legislation.
- The fourth source of *Shi'a* law is *reason* or *Aql/Aghl*. *Shi'a* jurists maintain that if a solution to a problem cannot be found in the primary sources, *reason* should be given free reign to deduce a proper response. This process, in which rational efforts are made by the jurist to reach an appropriate ruling, is called *Ijtihad*. In fact, according to the *Shi'a* principle of rule of correlation, it is possible to infer from reasoning a coherent body of religious law. It is important to note, however, that both *consensus* and *reason* (*Ijma* and *Aql/Aghl*) are considered by some *Shi'a* legists as secondary sources of *Shi'a* law. For them, the *Quran* and *traditions* remain the indisputable primary sources.[63]

The four legal sources noted above were the main legal sources for many centuries, with little change, although their interpretation depended on the different Islamic sects which followed in the Islamic era. However, events like the *Mongol* invasion of *Persia* in the 12th century, and the implementation of *Yassa*, a secret written code of law created by *Genghis Khan*, stopped the use of these regulations for a while to some extent in some parts of the Islamic world, but it can certainly be said that this had no impact on Islamic rules and was not established as a source of legal authority. Through the eradication of the reign of aggressive ethnics, the *Mongol* rules and regulations were forgotten and no lasting effect of *Yassa* remains in the current Islamic laws.[64]

Before the rule of the *Safavids* in the Islamic era and the establishment of *Shi'a* Islam as the official religion, Iran's population was mostly *Sunni* of the *Shafi'i* and *Hanafi* legal rites; thus there were relatively few *Shi'a* scholars or authorities in the religious hierarchy of the Islamic religious sciences (*Ulama*) in Iran. Before the *Safavid* period, there was a centrality of Islamic countries ruled by the *Caliphate*, a form of Islamic government, and thus the *Caliph* who believed to be a successor to the

63 *Lotfi* 1384/2005, pp. 150–184; *Sokhansanj* 1380/2001, pp. 20–22; *Entessar* 1988, p. 94.

64 *Gharavi* 1359/1980, p. 282. For further reading, see *Morgan* 1994.

Prophet *Mohammad*. Therefore, all matters were handled in accordance with Islamic law. Religious, conventional, and political laws were not kept apart, and all matters were implemented in accordance with the *Shari'a*. In all cases, the *Caliphs*, governors, and judges were forced to act based on *Shari'a* rulings and orders in accordance with the *Quran* and the *traditions*. Since the *Shi'a* gradually penetrated Iran, the vast majority of Iranians turned to this religion. However, it should be noted that Iran was part of the Islamic countries and territories under *Bani Abbas*, who was *Sunni*, which is why Iranian Muslims had to follow the rules, ideas, and principles of *Sunni* jurisprudence. These clashes led to the *Shi'a* sect settling their affairs, transactions, and other matters such as testaments, worship, marriage, and divorce by referring to the *Shi'a* jurists and scholars. This conflict ended in a division into religious and conventional affairs (*Urf*).[65] From that time onwards, the basis of *Shi'a* jurisprudence strengthened, and its ordinances and rules were developed and codified as a distinct set from other Islamic sects.[66]

However, religion and state were officially united during the *Safavid* era and the *Shi'a* sect was announced as the official religion of Iran, so this division into religious and conventional affairs which existed from the previous government's hegemony persisted. In religious affairs, *fatwas* (religious orders) were issued by the authorities and judges, then executed by court officials. The adjudicator (*Sadrol-Sodor*) was the highest religious authority; he was chosen by the king and remained in this position long-term, often for life.[67] Handling civil complaints was the responsibility of the bureaucrat (*Divan-Beigi*); he also dealt with religious complaints under the adjudicator's provision. Judicial affairs in the provinces were headed by the highest provincial religious authority. The *Shari'a* courts resolved anything that was associated with religion, such as wills, inheritance, divorce, *Qisas*, *Hadd*, and *Diye* based on the judge's personal understanding of religious texts, they also tried affairs related to matters of provincial governors and agents, while political crimes were resolved in accordance with customary rules and regulations.[68]

Therefore, it can be said that criminal law was based on Islamic law before the Constitutional Revolution, and the courts were divided into religious and civic courts, which continued until the *Qajar* dynasty, the last dynasty before the Constitutional Revolution. By the time of the dominion of *Nassereddin Shah* (1848–1896) and the expansion of Iran's relations with Western countries, the social life in Iran gradually changed and the need arose for drafting new regulations in order to address public disputes, which resulted in the creation of new organizations.[69]

---

65 For further reading, see *Markham* 1874, pp. 102–103.

66 Many books were written based on *Hadith* by *Shi'a* Islamic scholars in this era, but "*Alkafie*", "*Man la yahzaratol faqih*", "*Tahzib al-Ahkam*", and "*Al-estebsar*" constituted the main sources of *Shi'a* religious affairs and the law at that time.

67 *Gharavi* 1359/1980, p. 283.

68 *Ghassemi* 2013, p. 45.

69 *Gharavi* 1359/1980, p. 284.

In the 18th century, modern visions and modernity casted their shadows on all cultural, social, and political aspects in Iran. The phenomenon of modernity extended to the realm of governance, law, technology, literature, art, and so on. In this way, modernity affected the old native customs and beliefs and started to change the prevailing values. Hence, with the expansion of a dominant Western culture in developing countries, including Iran, modernity rose as a normative value against the longstanding norms and values of these societies. Various social, technological, and cultural developments increasingly deepened the contrast between traditional and new approaches.[70] In Iran, the Prime Minister at the time, *Mirza Ali Khan Aminoddole*, who was familiar with the thoughts of European scholars and especially with the rationalism and individualism schools, and who had learned many things from philosophical personalities such as *Voltaire* and *Montesquieu*, was commissioned to invite the lords of the state for assessing its economic-political status. He also invited some Belgian experts to correct the administrative and financial system of the Iranian state. Then he established the Council of Education and provided substantial support to philantropists *Mirza Hasan Rashidiye* and *Mirza Mahmoud Khan Ehteshamosaltane* for the establishment of new schools. Classic theories were reflected gradually in the law of the 18th century, especially in the Penal Code of 1810, which was a prominent example for Napoleonic law.[71] Iranian liberal intellectuals such as *Mirza Ali Khan Aminoddole*, *Ehteshamosaltane*, *Mirza Malcom Khan*, and *Mirza Yusuf Khan Mostasharodolle Tabrizi*, who were inspired by the thoughts of European intellectuals, demanded new laws, freedom in every aspect of life, and release from tyranny. They called for equal rights for all people, regardless of their political ideology and religious beliefs, and emphasized the development of law for the establishment of freedom and equality.[72]

Mirza Malcom Khan Nazemoddolleh, who was a politician, journalist, and liberal intellectual in the dominion of *Nassereddin Shah*, as well as under the influence of the positivism school of thought by *Voltaire* and *Montesquieu*, published some articles in Farsi on topics such as liberty, equality, law, justice, and liberalism.[73] It has been asserted that he was the first person to use the term "law" in its Western sense of modernity in the culture of Iranian legal language. In his writings, he claimed that these terms are fully consistent with *Shari'a*. A careful and extensive review of his articles reveals that "law" was of great importance to him.[74]

Specifically in the legal system, one of the most famous efforts to control the judicial and penal powers in late 19th century Iran was the reform agenda of *Amirkabir* who

70 *Zamirian et al.* 1381/2002, p. 70.

71 *Ghassemi* 2013, p. 46; *Gharavi* 1359/1980, p. 284.

72 *Zamirian et al.* 1381/2002, p. 73.

73 One of the most important articles of *Mirza Malcom Khan* was titled "*Occult Manual*", in which – inspired by the liberalism school – wisdom, individual liberty, and conditional governance were discussed and the concept of law was comprehensively described.

74 *Zamirian et al.* 1381/2002, p. 75.

was the high councilor of *Nassereddin Shah* in the *Qajar* dynasty. He tried to bring together all judicial practices under unified state laws, to this purpose: he restricted the competence of *Shari'a* courts to personal status cases such as family inheritance and divorce. Issuing the ordinance to local governors to ban torture and disproportionate punishments, asking them to prove an accused's guilt before punishing, and considering the principle of proportionality according to *Shari'a* and customary law were the other efforts *Amirkabir* undertook in the two years before his death in 1850.[75]

In this era, in order to comply with Western countries, the "Ministry of Justice" was established for the first time in 1885, and certain laws and regulations were introduced before the Constitutional Revolution. Rules for the "Grand Court" were written in 1862,[76] and the introduction of procedural rules for judicial administration was also issued in 1857 by the police chief of Tehran. Furthermore, so as to control and monitor the penal practices of local governors, in 1870, *Nasereddin Shah* ordered local governors to send their case files to the *Shah*, with a description of the respective criminal event and the evidence, including *Shari'a* or customary types of evidence. He also banned governors from executing any punishment without his permission. In 1878, the first law booklet, known as the "Conte Law Booklet", was written by *Comte de Montfort* (*Conte de Monte Forte*),[77] the police chief of Tehran.[78] It was the first law of Iran which defined crimes and punishments in a modern form with a special categorization before the Constitutional Revolution.[79] Disregarding the fact that these endeavors were considered as significant events in the late 19th century, it is worth mentioning that there were no more major changes until the Constitutional Revolution.

Following the changing atmosphere in Iran, *Mirza Yusuf Khan* published an article in Paris in 1870, titled "One Word Article", which was the translation of the French Constitution. He compared its themes with the verses of the holy *Quran*. Although the first edition of this article received high praise, he was arrested and imprisoned by the authorities of *Nassereddin Shah* soon afterwards. This article can be considered as one

75 *Ghassemi* 2013, p. 57.

76 *Gharavi* 1359/1980, p. 286.

77 The *Conte de Monte Forte* was originally born to a French family in 1839 and lived in Sicily after immigrating to Italy. After graduating, he entered the Royal Guards of Naples and Sicily and served in the Austrian empire. In 1869, the *Conte de Monte Forte* ended his military service and was appointed by the Austrian empire to serve in Iran according to *Nasereddin Shah's* request. *Rayejiyan Asli* 2009, p. 690.

78 *Ghassemi* 2013, p. 58. Before the Constitutional Revolution, the journeys of King *Nasereddin Shah Qajar* to some European countries prompted him to reform the judicial system; his chancellors, including *Amirkabir* and *Mirza Hossein Khan Sepahsalar*, tried to apply some reforms, but they did not succeed. On his second journey to Europe, *Nasereddin Shah* asked the Austrian Emperor to dispatch four military counselors to Iran. One of these counselors, the *Conte de Monte Forte*, established a modern police force and wrote a law booklet, including an introduction on police duties and 58 articles about crimes and punishments. See *Fazli* 2018, p. 76.

79 *Rayejiyan Asli* 2009, p. 687; *Ghassemi* 2013, p. 58.

of the most important ones in the contemporary history of Iran in terms of raising and publishing the concept of modernization. *Mirza Yusuf Khan* wrote that "the price of all pains of our society can be summarized in one word, and that word is law". He believed that the long-lasting traditions originating in *Shari'a* were the reasons for the backwardness of Iranian society. The traditional religious scholars and their followers were appalled by this idea. Furthermore, *Mirza Malcolm Khan* also published his ideas and views on criminal law subjects in a magazine called "*Ghanoon*" (Law). He believed that a constitution determines the political structure of a society and its government. He also published another article on criminal law. Here, he mentioned the progress of Western civilization and claimed that it is inextricably interlinked with Iranian civilization, so the latter's achievements, such as liberty, equality, promotion, justice, and law should be adapted uncontestably. In other articles, he wrote about rights, individual freedom, and the necessity of law, and he frequently sharply criticized the government for violating these principles.[80] These publication contributed to the Constitutional Revolution, which will be discussed later.

1.1.2.2 Socio-legal status of children and juveniles

In the Islamic era, and especially in the *Safavid* period, the social situation of children was based on *Shi'a* Islam.[81] As attention has been paid to child rights[82] and the guarantee of a child's physical and spiritual health has been emphasized and repeated in Islamic teachings and sources, it is clear that child rights should be respected in the Islamic society. Accordingly, the family is recognized as the basic social unit,[83] which is why parents have many duties in terms of training and disciplining children.[84]

---

80 The development and dissemination of the ideas of Enlightenment and modernity in Iran embarrassed some scholars, such as *Sheikh Fazlullah Nuri*, and their followers, so these also declared their ideas against modern intellectuals. In his article "*Tazkerat Al-Ghafel and Ershad Al-Jahel*", in which he rejected the ideas of Enlightenment and pointed out their non-compliance with *Shari'a*, *Sheikh Fazlullah Nuri* says: "Enacting the law in general is sometimes incompatible with Islam." This is the mission of prophets, and all prophets were chosen to fulfill this mission. Each prophet endorsed some of the provisions of the former prophet and changed some of the others. Finally, the Prophet *Muhammad* was sent to complete the religion of *Allah* and to provide the perfect divine law. The duty of a *Hojjat al-Islam* is to extract the divine laws from the four components of *Quran* – news, tradition, consensus, and wisdom – and to explain them to the public. They should be careful not to apply analogies and *Istihasn* in the deduction of divine laws." *Zamirian et al.* 1381/2002, p. 77.

81 *Gharavi* 1359/1980, p. 287; *Mohammadi et al.* 1383/2004, p. 223.

82 *Mosavi Bojnordi* 1396/2017, p. 43.

83 The *Quranic* statements about children are mainly concerned with infanticide, adoption, breast-feeding, and orphan children. These statements were of a normative-ethical significance for later Muslim jurists who formed the foundations of Islamic legislation. In 19 verses, the *Quran* prohibits harsh and oppressive treatment of orphaned children while urging kindness and justice toward them.

84 *Mousavi Bojnordi* 1387/2009, p. 45.

In the Islamic era in Iran, the father was responsible for his children and his family – not only for his children's primary education, but also to compensate damages caused by a child to others. The educational and correctional efforts that were applied to children were more likely approached by two principles: a child does not commit a sin against themselves, and he/she does not inflict harm or incur loss to others.[85]

In this period, the education of children was very important and essential. With the exception of children of nobles and clergymen, who were educated at home, rich parents sent their children to schools to study verious subjects, especially religious studies.[86] With regard to the situation of children in the *Safavid* era, *Jean Chardin* writes that it may seem unlikely for Iranian families to have been able to raise their children well, but in fact, the children of both nobles and ordinary families had a good upbringing and grew up very well indeed.[87] At this time, there were many schools, and all children – whichever social class they came from – could go to school, although children from poor families usually helped their parents at home and on the farm. Later, strict rules were enforced for older girls that they must not be seen in public except in case of an emergency and only with a *hijab* – especially if they were from the families of nobles and clergymen. In addition, some girl schools were established for educating girls, where only female teachers were employed.[88]

In the *Shi'a* legal sources, matters relating to minors are mentioned in the sections regarding adults, but their rules were different. It should be noted that in Islamic jurisprudence, children do not have the right of ownership of property until they reach the legal age. Nevertheless, after reaching maturity, they have an obligation to compensate for damages or crimes they may have caused during childhood – e.g. a child has to pay compensation for any unpaid debts due to inflicted harm or damage previously caused when he/she reaches maturity.[89] Based on the *Hadiths*, the principle of children not being responsible until they reach maturity was accepted. Therefore, applying adult penalties to children was not recognized. Accordingly, the rules governing adults could not be applied to children either. For instance, if a child assaulted someone, he/she was not to be beaten or punished in any other manner. In *Shi'a* Islam, a child is considered to be a person who has not reached the age of maturity. If a child committed a crime, he/she could not be considered to be responsible and only correctional measures could be applied. If a child killed another person, this was considered unintentional killing, and due to the lack of the child's capacity for committing a crime, retaliation was not applied, while blood money could

---

85 *Mohammadi et al.* 1383/2004, p. 225.

86 *Mohammadi et al.* 1383/2004, pp. 225–226.

87 *Yaghmaeei* 1372/1993, p. 416.

88 *Rahami* 1380/2002, p. 171; *Cipriani* 2009, pp. 77–84; *Mosavi Bojnordi* 1396/2017, p. 49.

89 *Rahami* 1380/2002, p. 171; *Cipriani* 2009, pp. 77–84; *Mosavi Bojnordi* 1396/2017, p. 49.

be paid by their parents as compensation.[90] It is worth mentioning that a child's exemption from criminal responsibility does not mean that the child was not expected to follow religious orders.[91]

Regarding the definition of a "child", Muslim scholars and jurists have divided the different stages of childhood into three periods:[92]

- *From birth to 7 years old:* in early childhood, a child is not able to understand the wrongfulness of their actions. The majority of jurists and scholars agree on this age and believe that a child has no penal or criminal responsibility in this period and cannot be punished at all, despite their civil responsibilities.
- *From 7 years old to the onset of puberty:* There is an assumption that in this period, a child is able to understand the wrongfulness of their actions. It is now considered as the one who can generally distinguish between benefit and harm. It is notable that jurists and scholars do not have a consensus about the beginning of puberty.
- *From puberty onward:* in this period, a person has full responsibility for their actions, and if they commit a crime, the punishment is effectuated after proving the intellectual development and maturity of the offender.

In order to know the criteria for children's responsibility in criminal matters, it should be noted that there is no special rule regarding the minimum age for criminal responsibility in the *Quran*, but according to tradition as well as the teachings of *Shi'a* leaders, girls under 9 years and boys under 15 years are exempted from criminal responsibility. Although a child is defined as a person who has not reached the age of puberty, intellectual maturity is a factor for criminal responsibility according to the opinion of some Islamic jurists.[93] In Islamic jurisprudence, the beginning of puberty is the end of childhood, meaning that puberty defines the boundary between childhood and adolescence. In a particular sense, a child is thus defined as a person who has not reached the religious puberty age.

1.1.3 Contemporary era

In this section, the legal system as well as the laws and regulations concerning child and juvenile delinquents are divided into two periods: from the Constitutional Revolution of 1906 to the Islamic Revolution of 1979; and from the Islamic Revolution 1979 to date.

---

90 *Mosavi Bojnordi* 91396/2017, p. 51; *Rahami* 1380/2002, p. 173.

91 *Rahami* 1380/2002, p. 174; *Mosavi Bojnordi* 1396/2017, p. 52.

92 *Mosavi Bojnordi* 1396/2017, p. 60; *Rahami* 1380/2002, p. 174.

93 *Nobahar* 1378/2010, p. 202.

1.1.3.1 From the Constitutional Revolution of 1906 to the Islamic Revolution of 1979

1.1.3.1.1 Legal system

In Iran, the 20th century began with the Constitutional Revolution of 1906. In fact, the unstable condition of the country, the efforts of social groups inside Iran and the role of foreign powers led to the Constitutional Revolution, which is considered as the most important event in Middle East at the end of the *Qajar* dynasty.[94] It can be said that the Constitutional Revolution divided Iran's legal history into two periods: before and after the Constitutional Revolution. This date marks the beginning of a new chapter in the Iranian legal system. In fact, the origin of the current laws and regulations as well as the basis of the modern judicial organizations date back to the Constitutional Revolution.[95] Following this event, *Mozafaradin Shah Qajar* signed the first Constitution of Iran and established the national assembly (*Majlis*). Due to certain events, the first national assembly did not function properly,[96] and only on a few occasions could the representatives gather and discuss legal drafts. However, during the initial round, some important laws were passed, including the supplementary parts of the Constitution which were mostly about the duties and position of the national assembly as well as laws concerning the relationship between *Shah*, nation, and the national assembly. Efforts at the centralization and secularization of Iran's legal structures began in 1910 with the establishment of the Central Public Prosecutor's Office. In the second assembly of 1912, the first temporary Criminal Trials Principles Act was drafted, based on the French laws by *Moshiroddole Pirnia*, *Mudarres* (an influential clergyman), and *Monsieur Adolph Prani*. The draft was

---

94 *Ghassemi* 2013, p. 56. It should be noted that the end of the absolutistic monarchy, approval of the Constitution, the rule of law, laying the foundations of an independent judiciary, and a parliament with legislative powers were the most important changes after this revolution; see *Entessar* 1988, p. 97.

95 *Gharavi* 1359/1980, p. 290. The modifications made to the Constitution in 1907, however, recognized the long-standing religious influence in the country's legal system by providing for a two-tiered structure in the new legal system. That is, a distinction was made between the secular law (*urf*) and religious, Islamic law (*Shari'a*). The Muslim clerics were allowed to maintain exclusive control of family law, such as the laws regulating marriage, divorce, succession, and the like. In general, much to the dismay of the secular constitutionalists, the clergy continued to exercise de facto jurisdiction in most areas of civil law. The clergy's power to exercise near-monopolistic control in civil cases was primarily due to the absence of any effective centralized governmental control over the country's nascent legal system. Efforts at centralization and secularization of Iran's legal structures began with the establishment of the Central Public Prosecutor's Office in 1910 and were formalized a year later, when the Ministry of Justice was reorganized to perform its constitutional mandate. However, the first serious and sustained attempt to promulgate secular laws began when *Reza Shah*, the founder of the *Pahlavi* dynasty, assumed power in 1925. *Entessar* 1988, p. 98.

96 The national assembly underwent many tensions and struggles during the first years of its existence. The first assembly was bombarded by *Mohammd Ali Shah*, who was successor of *Mozafardin Shah*. The second assembly was canceled by the state, and the third assembly was closed because of the First World War and the immigration of some of the representatives. See *Ghassemi* 2013, p. 32.

provisionally signed by a board of state ministers[97] and approved by *Mudarres* thereby implying that it was not in contradiction to *Shari'a* rules.[98] The establishment of a "parquet" (prosecution authority, including prosecutor, prosecutor assistants, and interrogators) was one of the main characteristics of the French criminal procedure system reflected in this act.[99] Police officers went as constables under the order of parquet authorities to help them with detecting crimes, investigations, and service of summons. The structure of the criminal courts (appeal courts and the Supreme Court) was given in a supplement of the Constitution. The most important matter in the development of the system of criminal procedure after the Constitutional Revolution was that canonical courts were gradually liquidated due to the shortage of *Mojtahids* and Permitted Judges (*Mazoon*) who were not *Mojtahids* (but were authorized by the king), the adjudication authority for almost all cases.[100]

During the 1920s, Iran faced many social, political, and economic problems, some of which were caused by the First World War. Increasing insecurity and disorder led to much public dissatisfaction; under these conditions, the *Qazaq* army (the army of the *Qajar* dynasty) invaded Tehran and conducted a coup d'état in 1920. *Reza Shah* (*Sardar Sepah*), commander-in-chief of this army, was appointed as prime minister in 1923 by *Ahmad Shah*, the last *Shah* of the *Qajar* Dynasty. In 1925, when *Ahmad Shah* tarried in Europe, the national assembly announced the collapse of the *Qajar* regime and declared *Reza Shah* as the provisional holder of power. The process of reform and modernization was thus ceded to the *Pahlavi* dynasty. Various aspects of life underwent modernization, even by force where deemed necessary. *Reza Shah* excluded clerics from the arena of public policy-making, except those who supported the modernization project.[101] In the social sphere, a range of laws were passed so as to institutionalize the new (modern) way of life, including changes on how people had to dress. The Act of National Clothes was passed in 1923, whose subject was just state officials. The Act on the Male Uniform, known as *Pahlavi* attire, was approved in 1929, according to which long garments and turbans were to be replaced with Western clothing such as shirts, trousers, and hats. One of the most notorious modernizing initiatives of *Reza Shah* was his ordinance on the "*Hijab* (*Chador*, body-long garment for women) Uncovering" in 1936. It started with female teachers and the wives of high state officials. Police officers were to enforce the ordinance, although its implementation among all citizens was impeded for various reasons.[102]

97 Since the national assembly was often closed during this period of time, many of the laws were passed and announced by a board of state ministers.

98 *Ghassemi* 2013, pp. 21–23.

99 The temporary criminal trials principles Act was subject to many amendments before the Islamic Revolution.

100 *Fazli* 2018, p. 76.

101 *Ghassemi* 2013, p. 35.

102 *Ghassemi* 2013, pp. 24–25.

Attempts at a further secularization of the Iranian legal system continued under the rule of *Reza Shah's* son, *Mohammad Reza Shah*, the last of the *Pahlavi* monarchs.[103] In 1954, the modernizing engine started to run faster. The oil price began to rise and escalated Iran's economic transactions with other countries. All kinds of technologies were imported. Expansion of state and private industries brought about new forms of production and reformed the patronage connections. This trend was accelerated by a land reform policy in the late 1960s, which was called "White Revolution of Shah and Nation" and sought to exterminate the traditional feudalistic relationship between grant land owners and their agrarian tenants. The land reform considerably increased the number of migrants from rural to urban areas.[104] In fact, urbanization in modern Iran only began after the land reform in the 1960s.[105]

After the "White Revolution of *Shah* and Nation", clerics and the *Shari'a* were increasingly excluded from the social, political, and cultural agenda of the state. In the penal code of 1974, unlike in former codes, there was no reference to Islam and *Shari'a* any more. Besides penal reforms, many other reforms were also initiated and developed during this period. The enactment of new (Western-oriented) laws[106] clearly meant that "social and legal problems" and their "solutions" were not going to be in accordance with *Shari'a*. Thus, clerics and their legal/judicial apparatus were increasingly losing their power to regulate the public sphere. The modern art of "legal/penal governance" was to reshape and reconstruct the so far traditional/Islamic construction of "social problems" and hence to redefine the legal/penal configuration.[107] It can be said that the pace of penal modernization from the mid-1950s to the late 1970s became considerably faster than the early stages of modernization during *Qajar* and the first *Pahlavi*.[108]

On the judicial level, *Mohammad Reza Shah* sought to professionalize the appointment of judges by making it difficult for a cleric to function as a sitting judge in a court of law. For example, a law passed on December 27, 1946 required that judges hold a degree from the Faculty of Law at Tehran University (or an equivalent foreign university). Since the *Ulama* had received their legal training in religious seminaries rather than at the secular Tehran University, they were legally prevented from sitting as new judges in a court of law. The law also stipulated that the current judges who were not graduates of Tehran University Law Faculty or its foreign equivalent must

---

103 *Entessar* 1988, p. 99.

104 In 1956, only 31% of the population was living in cities, while in 1976, two years before the Islamic Revolution, 51% of the population were doing so. *Ghassemi* 2013, p. 33.

105 *Ghassemi* 2013, p. 25–26.

106 Such as the "Family Protection Act" in 1974, which reformulated the Islamic way of family relations between wife and husband. *Ghassemi* 2013, p. 33. In fact, in this period, many laws following the models of the French legal system were approved, based on the Roman-German legal system/civil law. Therefore, the penal codes were separated from Islamic law.

107 *Ghassemi* 2013, pp. 43–44.

108 *Ghassemi* 2013, p. 26.

pass an examination in both Iranian and international law in order to remain employed as judges by the Ministry of Justice. The net effect of the provisions of the 1946 law was the disbarment of the overwhelming majority of the *Ulama* and their removal from cherished judicial positions.[109]

The institutionalization of such a modern art of governance would be likely to lead to the birth and development of new kinds of knowledge, institutions, and agents, none of which were prescribed in *Shari'a*, and therefore result in the complete elimination of clerics from the social, political, cultural, and, more importantly, economic arena.[110]

It is worth mentioning that after World War II and the transformation of societies toward more industrialized countries (and the consequential social changes) – especially the sudden expansion of cities and urban development due to the continuous migration from rural areas to cities –, the risk of crime and specifically of juvenile delinquency increasing was gradually felt and required new methods of crime prevention based on rehabilitation and education, benefiting from the good experiences of European countries in their implementation of educational and rehabilitative mechanisms for juveniles and children. Therefore, the establishment of the Juvenile Courts Act was approved in 1959, thus establishing a juvenile justice system with a modern framework in Iran for the first time. The recognition of a gradual criminal responsibility, the establishment of special courts, as well as a differentiated sentencing system and procedures were the most important innovations in this regard. However, especially due to the lack of juvenile correctional centers as the main obstacle, the implementation of the Juvenile Courts Act was postponed to 1966,[111] and only four years later, in 1970, the Juvenile Correctional Center Regulation was adapted.[112]

However, before approval of the Juvenile Courts Act, a specific criterion had been considered to address juvenile delinquency in the General Penal Codes. Nevertheless, the correctional methods and proceeding manner in the criminal courts and prisons could not effectively educate and edify children and prevent reoffending.

In this section, all laws and regulations enacted after the Constitutional Revolution will be discussed in detail.

1.1.3.1.2 Laws and regulations concerning child and juvenile delinquents

a) Penal Code of 1926

In the penal code of 1926, the first Iranian general penal code approved by parliament (a translation of the French Penal Code of 1810), children were divided by age into

---

109 *Entessar* 1988, p. 98.

110 *Ghassemi* 2013, p. 46.

111 *Shambayati* 1389/2010, p. 97; *Gharavi* 1359/1980, p. 93; *Mohammadi et al.* 1383/2004, p. 289.

112 Juvenile correctional center regulation is still in effect.

three categories, which determined the types of decisions and punishment: children below 12 years old, children between 12 and 15 years old, and children between 15 and 18 years old.[113]

Children younger than 12 years were considered as immature and irresponsible, and if they committed a crime, there was no punishment for them, so they were left to their parents again with the only requirement to be educate, trained, discipline, and taught good morals (article 34); but regarding children included in the second category, in case of a criminal commitment including both misdemeanors and criminal offenses, there is lash punishment of between 10 and 50 lashes (article 35). Regarding the third category of children, the legislator exerts more force: if these children are guilty of misdemeanors, their sentences will not be less than half the minimum and not more than half the maximum of the sentences that the General Penal Code intends for adults.

Despite the fact that the legislator enacted milder punishments in determining criminal reaction against children and adolescents in conflict with the law as well as juvenile delinquents – in other words: despite recognizing their diminished criminal responsibility –, it should be noted that criminal responses and punishments to child and juvenile crime were still repressive and retributive.[114] In fact, the purpose of the measures was not aimed at educating children and could not be effective in preventing recidivism. Thus, dealing with juvenile delinquents required new and effective ways synchronous with developments in the social situation of Iran. Based on the new criminal policy, the establishment of the Juvenile Court Act was approved in December 1959, from which the developments with regard to trial methods for children and the formation of educational facilities originated.[115] However, in the course of Islamizing the Iranian society after the Islamic Revolution of 1969, this act was abolished and to date, no special acts have replaced it.

b) Juvenile Detention Regulation in Tehran Provisional Prison, 1943

According to the Juvenile Detention Regulation,

- a place of detention for all suspects under 18 years of age was to be kept separate from adult detention centers (article 1).
- If the accused persons were illiterate, a teacher, who was to be determined by the prosecutor office and the police department, had to teach them *Persian* literacy; and if they were literate, other materials were to be taught to them (article 2).

---

113 *Gharavi* 1359/1980, p. 80; *Danesh* 1374/1995, p. 89.

114 *Moazzenadegan* 1389/2010, p. 68; *Gharavi* 1359/1980, p. 95.

115 *Gharavi* 1359/1980, p. 87; *Shambayati* 1389/2010, p. 105.

- A court clergyman had to preach to the accused twice a week and to give them advice on how to get back onto the right path (article 3).
- In juvenile detention centers, juveniles 12 to 15 years old were segregated from those 15 to 18 years old. If an accused became 18 while in custody, he/she was immediately transferred to an adult detention center (article 4).
- The prosecutor had a prison officer investigate the status of the juvenile detention centers once a week (article 5),
- and in addition, he personally visited it once a month and prepared a comprehensive report on the status of the juveniles in aspects such as health and ethical principles, which he then sent to the Ministry of Justice through the relevant legal department (article 6).
- The legal department examined the aforementioned report and with their own comments sent it on to the Minister of Justice (article 7).
- In this regulation, it was stated that as soon as the means were given to implement it elsewhere, the juvenile and adult detention centers should be segregated in other cities, too, and the arrangements specified in these regulations with regard to juvenile detainees should be enforced (article 8).
- Regarding the implementation of the regulation, it is notable that the Tehran prosecutor and police chief appointed the agents to implement this regulation (article 9).[116]

c) Establishment of the Juvenile Courts Act, 1959

In accordance with the Juvenile Courts Act, children and juveniles were divided into three categories:

1) According to article 17, the first category encompassed children between the ages of 6 and 12. If they committed a crime, the following decisions could be taken: delivering the children to their parents or guardian who had to sign a written commitment vowing to educate the child. In cases where the parents or guardian were not qualified to do so or where it was not possible for them to take responsibility for the education of the child, the children would be sent to correction and rehabilitation centers for one to six months.
2) The second category concerned children between the ages of 12 and 15. As was stated in article 18, three kinds of responses could be applied to these children, including delivering them to their parents or guardian who would sign a written commitment to educate the child; blame and advice by the judge and sending them to correction and rehabilitation centers for three months to one year.

116 For further reading, see *Gharavi* 1359/1980, p. 90.

3) Article 18 stipulated the rules for the third category, i.e. mature children aged between 15 and 18. Accordingly, in addition to the responses applicable for the second category, if a particular type of criminal behavior was proven, the judge was obliged to send the child to correction and rehabilitation centers for six months up to five years.

In summary, it can be said that the key features of this act – in the field of protecting these children and juvenile delinquents in conflict with the law – included the following measures:

- the establishment of a special independent court for children (article 1);
- the selection of specialized and experienced judges (article 2);
- in the case of criminal charges, the requirement of the presence of two assessors (teacher, tutor, etc.) in the court (article 33);
- determination of a minimum age of criminal responsibility (according to article 4, children aged under 6 years were completely exempt from criminal responsibility), grading the criminal responsibility of children into three age groups of 6–12, 12–15, and 15–18 (articles 17 and 18);
- the establishment of correction and training centers (article 22);
- prosecution, investigation, and trial by a judge (article 6);
- making decisions and adjudicating according to the personal and social history of the child as well as the views of experts, social workers, and psychologists (article 7);
- emphasizing the specific rule that a trial is not to be held in public in juvenile proceedings (article 9);
- the presence of a defense lawyer in criminal offenses (article 11);
- opting for milder and different reactions and sentences compared to the penalties prescribed for adults (articles 17 and 18);
- the possibility of appeal or a reduction of the sentence duration of children and juveniles kept in correction and training centers, or replacing their sentence with a milder reaction based on the reports and good behavior (article 21);
- the obligation of children to learn and be trained in different careers in the correction centers (article 24);
- and the obligation of the judge to visit the correction centers every three months and to monitor progress (article 25).[117]

At the time, this act was the first comprehensive and progressive law to deal with children and juvenile delinquency. However, because of the delay when adopting the executive regulation of this law, it was not enforced in practice. According to

117 *Gharavi* 1359/1980, p. 76.

article 1, dealing with juvenile delinquency was to be done based on the Public Penal Code in districts where no juvenile correctional centers had been established, but their shortage was the main obstacle for the implementation of the law. For this purpose, some effort was necessary. In fact, with the changes in the administrative organization of the judiciary, the Social Services Department of the Justice Ministry was established, which soon approved the establishment of juvenile correctional centers in Tehran, provincial capitals, and other cities. The budget for constructing these centers was provided by the Prisoners Support Council in the judiciary, and then land was bought for the projects. In compliance with the requirements, a piece of land was allocated to this project in Tehran, too. Despite these arrangements and the urgency expressed by the justice department to implement this act, it was postponed to 1966.[118] Consequently, the first juvenile court in Tehran was not established before December 9th 1966. Due to the multitude of referred cases, other branches were later added.

d) Juvenile Correctional Center Regulation, 1970

As has been discussed above, because of various adverse reasons – especially due to the shortage of juvenile correctional centers –, the implementation of the Juvenile Courts Act was postponed to 1966. Four years later, in 1970, the Juvenile Correctional Center Regulation was finally adopted, the idea having been borrowed from Western countries.

In accordance with this regulation, which is still in effect, the juvenile correctional center is a reformatory and educational center to reform and educate children who are sent there (article 1). A juvenile correctional center has three sectors: provisional detention; education and reform; and the prison sector. These sectors are separated from each other, and in each sector the children are categorized according to their age, records, crimes committed, and degrees of their trainability. In addition, the detention sector for girls is to be segregated from the one for boys (article 2). The provisional detention sector is to be allocated to children who are still awaiting court decisions. Those who are accused of serious crimes and are sent to this sector of the center will remain in custody until the final court decision. In order to prevent collusion or because of other legal reasons, the children can be kept separate in solitary confinement until further orders from the court are received (article 3). During provisional detention of a child in custody, the family, social background, and moral characteristics of the child are supposed to be investigated by social workers; furthermore, the physical and mental status as well as the level of education and any professional talent of the child should be examined by a specialist; a set of the collected records is evaluated in a council in the presence of the center manager and the experts who have examined the child. In the final report, which includes the personal

118 *Moazzenadegan* 1389/2010, p. 70; *Gharavi* 1359/1980, p. 96; *Danesh* 1374/1995, p. 83; *Shambayati* 1389/2010, p. 208.

characteristics of the detainee, the result of the examinations conducted by social workers and specialists as well as the proposed measures are prepared in two copies, one of which is to be sent to the juvenile court, while the other copy is kept in the center (article 4). In the educational and reformatory as well as in the prison sector, those children who are sentenced by the court should be kept in custody. In each of the mentioned sectors, children are first enrolled in the educational and training programs according to their talents, age, and degree of knowledge. Vocational training programs are to be implemented in such a way that they do not clash with the children's regular education (arts. 5, 6, 7). Instructions for the implementation of facilities in the prison sector (such as professional education, ethics programs, eating, sleeping, and so on) are identical with those at the educational and reformatory sector. However, according to the individual child's special situation and type of crime committed, the center manager can divide the detainees into specific groups in terms of their ethical and professional education, their eating or sleeping habits, and the enforcement of educational and reformatory measures, and forbid each group from getting into contact with one another (article 8).

It can be said that the key features of this act in the field of protecting children and juvenile delinquents in conflict with the law include the following:

- taking into account the three separate sectors (article 2) and
- rules for each sector (arts. 4, 5, 6);
- a categorization of children according to their age, records, crime committed, and degrees of their trainability and implementing a special program which suits them (arts. 2, 3);
- instructions for implementation of the facilities in each sector (arts. 7, 8, 9, 10);
- the presence of social workers, psychologists, psychiatrics, doctors, and other related specialists as personnel in the centers (article 11);
- determining the duties of the head of the center and other personnel in line with the rights of the children in mind who are sent to the correctional centers (arts. 12–40).

It is worth mentioning that the first correctional center was established in Tehran in 1970. In the following years, other correctional centers were established with fewer facilities in Mashhad and Ahwaz; plans to establish such centers in Tabriz, Isfahan, Shiraz, Hamadan, and a few other cities were also adopted. It is notable that after some time, these centers lost their main character (rehabilitation based on the differentiated decisions by a court), so their purpose remained abortive. This is why the centers were reformed again with new policies and structures in 1979.

e) Penal code of 1974

Under the influence of new institutions in global criminal justice and new developments in the law schools of thought in this area, Iranian legislature decided to revise the penal code of 1924, which resulted in the penal code of 1974.[119] In fact, "this code was inspired by the "New Social Defense" movement which had been developed after World War II by *Marc Ancel* who illustrated a moderate account of "Defense Social" proposed by *Gramatica*".[120]

In line with the establishment of the Juvenile Courts Act, the penal code of 1974 classified children into two categories: children between 6 and 12 years old and children between 12 and 18 years old. If children of the first category committed a crime, the response was to deliver them to their parents or guardian and/or to public or private agencies or firms which were responsible for retaining and educating them for a period of one to six months. In cases where the parents or guardian were not qualified or able to take responsibility for the education of the child, the children were sent to correction and rehabilitation centers to be educated for one to six months. Regarding children between 12 and 18 years old, the court attempted to adopt one of the following decisions: delivering the child to their parents or guardian for discipline and expecting them to be committed to educating and caring for the child with a high standard of ethics in mind; blame and advice by the trial judge; three months to one year detention in a juvenile correctional center; six months to five years detention in a juvenile correctional center if the child was more than 15 years of age and guilty of murder; and at least two years in a juvenile correctional center if the crime was punishable by the death penalty or life imprisonment (arts. 33, 34, and 35). According to note 2 of article 33, final decisions by a court could be revised once.

1.1.3.2 From the Islamic Revolution 1979 until now

1.1.3.2.1 Legal system

As has been explained, under the *Pahlavi* monarchy, attempts at a further secularization continued for all aspects of life, including the criminal justice system.[121] The secularizations of society and legal reforms in the absence of democracy were major factors in the convergence of popular, nationalist, leftist, and Islamist opposition to Pahlavi rule, which led to the 1979 Iranian Revolution as the second revolution in the twentieth century. Islamist elements gained the upper hand in the new Islamic Republic. In order to reestablish *Shari'a* as the source of law and the clergy as its official interpreters, they set about undoing the secularization of the legal system.[122]

Therefore, the foundations of the Iranian criminal justice system were changed after the 1979 Revolution with the abolition of the Juvenile Courts Act 1959 for the goal of Islamizing society. In fact, according to article 4 of the Iranian Constitution, all laws and regulations (including civil, penal, financial, economic, administrative, cultural, military, and political laws and regulations) must be based on Islamic criteria.

---

119 *Gharavi* 1359/1980, p. 97.

120 *Ghassemi* 2013, p. 35.

121 *Entessar* 1988, p. 100.

122 *Mir Hosseini* 2010, p. 327; see also *Amin* 1386/2007 and *Pasha Saleh* 1348/1969, pp. 325–370.

This article shall apply generally to all articles of the Constitution as well as other laws and regulations. It shall be decided by the jurists of the Guardian Council whether or not such laws and regulations conform to article 93 of the Constitution.

With the victory of the Iranian Revolution in 1979 and the eventual monopolization of political processes by the Shi'a clerics, the Iranian Constitution of 1906–1907 was replaced by a new Constitution in which Shi'ism was declared as the sole source of law in the country.[123]

Therefore, the main sources of law were considered the *Quran*, *tradition*, *consensus*, and *reason*. Accordingly, crime and punishment were classified into *Hadd*,[124] *Qisas*,[125] *Diye*,[126] and *Ta'zir*,[127] which was completely different from the previous classification in the justice system, which had been based on the model of Western European legal systems.

The new classification of crime and punishment based on *Shari'a* formed the basis of the new criminal justice system. The abolition of the Juvenile Courts Act of 1959 led to a legal vacuum in both substantive and procedural criminal law on the one hand, and caused some conflicts between the legislative, judicial, and executive criminal policies on the other hand. This vacuum constituted the most controversial situation within the Iranian legal system. In fact, in the previous Islamic Penal Codes (1982, 1991, and 1996), only a very limited number of articles dealt with children who were in conflict with the law.[128] According to these Codes, the age of criminal responsibility was 9 years for girls and 15 years for boys in accordance with the lunar

---

123 *Entessar* 1988, p. 100.

124 *Hadd* penalties include whipping; death penalty and execution by stoning or hanging (gallows); deportation; amputation; life imprisonment and defect of limb that are applicable for crimes such as adultery, sodomy, lesbianism, pimping, false accusation of another person of sodomy and adultery (*Qazf*), consumption of intoxicating liquors, terrorism (*Moharebe*), and a particular kind of theft. It should be mentioned that the punishments are very severe, but proving *Hadd* crimes is very particular and difficult (in that it requires four confessions or four witnesses and knowledge by the judge). Since most crimes in *Hadd* are considered to be victimless, the punishment can (under some conditions) be removed in cases of the offender's repentance.

125 *Qisas* is an Islamic punishment for murder and other intentional equal crimes against the physical integrity of persons. *Qisas* includes both life and limbs retaliation and should be implemented with the consent of the victim's family. By mediation, it can be changed to *Diye* (blood money, see below), and the punishments can be removed in case of forgiveness on the part of the victim and their family.

126 *Diye* is a pecuniary penalty (blood money), the quantity of which has been determined in *Shari'a*. This form of punishment is settled upon for negligent homicice or an injury or defect of limb caused by accident.

127 All offenses against persons, property and public security (except offenses in *Hadd*, *Qisas*, and *Diye*) are *Ta'zir* crimes. *Ta'zir* penalties include imprisonment, community-based punishment, disciplinary, security, and corrective measures, fines, whipping (the number of lashes must be less than the number of lashes stipulated for Islamic punishment in *Hadd*), and execution for some specific crimes such as illicit drug trafficking.

128 For further reading, see *Norbaha* 2005, pp. 296–299.

year.[129] This criterion was based on the age of maturity in *Shari'a*. The age limit was thus set very low, and there was no transitional period between irresponsibility and full criminal responsibility. In these Penal Codes, the age of criminal liability was determined strictly and without giving any age ranges. In accordance with this age determination, either the natural persons in criminal law were children (below the ages mentioned above) and thus without any criminal responsibility, or they were adults (above the ages mentioned above) and thus with full criminal responsibility. In criminal law, however, and as far as the age of criminal responsibility was concerned, natural persons were divided into three groups: natural persons who have no criminal responsibility (irresponsibility); natural persons who have diminished criminal liability; and natural persons who have full criminal responsibility. Those within the second group were treated separately by the Penal Code and subsequently within the justice system. It is notable that the gradual transition and diminished criminal responsibility for children were ignored in the penal codes approved after the Islamic Revolution, and that only some rules were taken into account about the persons who had no criminal responsibility. Therefore, according to the previous penal codes, mature girls and boys under 18 could receive punishments for adults, especially as far as fixed punishments were concerned. However, due to their discrimination in the age of criminal responsibility, girls could receive penal punishments at a younger age than boys. On the other hand, it should be noted that in the Iranian legal system, there are some severe penal sanctions in fixed punishments. According to the pervious penal codes, these punishments could be applied by judges. This situation was a very controversial subject matter as well as incompatible with social, physiological, and psychological realities, so that it required further changes and reforms.

However, it is noteworthy that at that time, some attempts were made by judges to apply rehabilitation mechanisms under the influence of social and cultural rules as well as international obligations. In order to understand why the judicial system tried to apply other policies – besides the importance of fulfilling international obligations –, the positive impact of the Juvenile Courts Act 1959 on the judicial system should not be ignored. In fact, it can be said that suitable laws and regulations can create a lasting judicial culture even after their abolishment. In other words, the positive message of this abolished act has remained in the judicial system.

1.1.3.2.2 Laws and regulations concerning child and juvenile delinquents

a) Bill for the Establishment of Public Courts, 1979

In accordance with the Bill for the Establishment of Public Courts of 1979, the criminal courts were divided into two public courts: Grade 1 and Grade 2 Courts. According to the note of article 12 in the Act for the Establishment of Public Courts, juvenile crimes are to be tried in a criminal court as mentioned in the Establishment of Juvenile Courts Act. This note led to the liquidation of independent juvenile

---

129 8 years and 9 months for girls and 14 years and 7 months for boys according to the solar calendar.

courts. However, it had already been established when dealing with juvenile crimes in public courts according to the Act of Establishment of Juvenile Courts, but this resulted in conflict and clashes. In this regard, the Judicial Precedent approved by the General Board of the Supreme Court in May 1985 ended these clashes, as well as the use of specific procedures in dealing with juvenile delinquents, which was in accordance with article 12 of the Act for the Establishment of Public Courts. Therefore, this specific procedure was withdrawn from the judicial system of Iran, and procedural supportive regulations for children in conflict with the law, extracted from the Act for the Establishment of Juvenile Courts, were completely abolished. Thus, it can be said that in the absence of procedural supportive regulations for children and juveniles, a throwback took place to the Public Penal Code (1925).[130]

b) Islamic Penal Code 1982 and its amendment in 1991

By approving the Islamic Penal Code of 1982, the Act for the Establishment of Juvenile Courts (1959) was abolished. In addition to providing age groups for criminal responsibility, the new code proposed educational-rehabilitative measures. In terms of determining the reaction to crimes committed by children and juveniles, article 26 of this code states that in cases where children and juveniles commit a crime, they are exempt from criminal responsibility, and if necessary, their family or correction centers are responsible for educating them as determined in the court's decision. In note 1 of this article, a child is defined as a person who has not attained the legal age of maturity. In accordance with note 2, when a minor commits a murder, injury, or battery, their *Aqele/Aghele* (the closest male relative) shall be liable for compensation; but when a minor is responsible for the loss of property of other persons, the child shall be personally liable for compensation and its payment shall be the liability of his/her parents or guardian. Note 3 states that when, in the case of serious crimes, corporal punishment is considered necessary for the correction of the child who committed the crime, it shall be carried out in such a way that it may not cause any serious harm incurring the liability of blood money.

Later, with the approval of the Islamic Penal Code 1982, the legislator made amendments to some articles, including article 29, in 1991. Accordingly, article 26 was changed into article 49 which states: by committing a crime, children are exempted from criminal responsibility, and their correction is the responsibility of their parents or guardian or, if the court decides so, of a correctional center. According to note 1 of this article, a "child" is defined as a person who has not reached the level of maturity as stipulated by Islamic jurisprudence (*Shari'a*). note 2 of this article was changed into article 50. In addition, note 1 was approved without changes, and note 3 was amended to note 2 as follows: when corporal punishment is considered necessary for the correction of the child who committed the crime, it shall be carried out to the extent necessary and in the best interests of the child.

130 For further reading, see *Gharavi* 1359/1980, p. 91; *Moazzenzadeghan* 1389/2010, p. 73.

According to arts. 112 and 113 of the Islamic Penal Code 1991 (which are related to article 49), a flogging sentence of up to 74 lashes is issued for children and juveniles in cases of sodomy, while article 49 of the Code sets children and juveniles innocent of criminal responsibility.

It is notable that these penal codes were silent on what the age of criminal responsibility and of maturity is, and in practice, judges would rather refer to the Civil Code. In this code, the age of maturity is 9 years for girls and 15 years for boys according to the lunar calendar. This rule was used by judges for both discretionary and fixed punishments. Therefore, the gradual transition to full criminal responsibility was not accepted in these codes, and there was no rule which applied to children who had diminished criminal responsibility. Furthermore, there was discrimination between boys and girls.[131]

Therefore, according to article 49, the judge had wide discretion in determining one of the options, such as adopting correctional-educational measures by entrusting the child to their parents or legal guardian, sending them to correction and training centers for any duration they deemed necessary, or corporal punishment. Obviously, this method was unsuitable to support children and juveniles who were in conflict with the law. Hence, judges in most cases – especially concerning *Ta'zir* crimes based on the authority – tried to issue correctional-educational verdicts instead of repressive and punitive punishments. This occurred by adopting principled criminal-judicial policies based on the child rights and under the influence of the abolished Juvenile Courts Act 1959.

c) Ratification of the Convention on the Rights of the Child in 1994

Some attempts were made to fill the legal vacuum in the field of juvenile justice. To this purpose, the [UN] Convention on the Rights of the Child (UNCRC) was signed by the representative of the Government of the Islamic Republic of Iran on 5 September 1991 and ratified in March 1994 by the Islamic Consultative Assembly, with general reservations. On 15 June 1994, the Islamic Republic of Iran acceded to the Convention, and it became binding from 12 August 1994.[132]

It is notable that the right not to apply any provisions or articles of the Convention that may be in contradiction to domestic laws and Islamic standards was reserved by Iran, which constituted an obstacle to the protection of children. The broad generality of the reservation implied that, in principle, Iran could comply with the Convention without modifying or strengthening existing national laws and policies that affect children. However, with respect to the commitment to ensure the survival, development, protection and participation of children, the government decided to review and re-examine this reservation in order to amend it accordingly.

131 For further reading, see *Norbaha* 2005, pp. 296–299.

132 *Abachi* 2015, p. 827.

At that time, in domestic law, the most critical points referred to the age of criminal responsibility (which was set very low) and the possibility to impose severe punishments in accordance with the fixed penalties (*Hadd*, *Qisas*, and *Diye*), which needed to change. Subsequently, with particular regard to the best interests of children, some supportive rules were adopted. Therefore, in line with international obligations as well as social realities and (what is more important) influenced by judicial criminal policies for dealing with juvenile delinquents, some changes were carried out. In fact, the legislative climate changed in favor of children in conflict with the law and toward the re-establishment of an explicit juvenile justice system.

d) Regulations for the Establishment of the General and Revolutionary Courts Act, 1994

The parliament of the Islamic Republic of Iran approved Iran's accession to the UN Convention on the Rights of the Child (1989) in March 1994 when all procedural supportive regulations for children and juveniles were already outdated. In an effort to address this gap, according to a paragraph in article 8 of the Regulations for the Establishment of General and Revolutionary Courts, the legislator decreed that the Head of the Judiciary has the authority to select and appoint each of the judges, based on experience and skills, to any jurisdiction and for any hearings in criminal, civil, family, juvenile, marine, medical, and other cases. Furthermore, the Head of the Judiciary at the time used this legal authority to issue the order of establishing special courts for juveniles as specific branches of general courts. However, it should be noted that these branches did not have specific procedures to deal with crimes of children and juveniles and with the general provisions of the Criminal Procedure Code, so these branches were obliged to follow them and were not effective in supporting children and juveniles when addressing their crimes and charges.[133] This is the reason why the new Criminal Procedure Code, approved in 1999, had a special chapter for dealing with juvenile delinquents in criminal proceedings.

e) Criminal Procedure Code for Public and Revolutionary Courts, 1999

After the Islamic Revolution, the first differentiated legislative criminal policy concerning children and juvenile delinquency was applied in procedural law, as there were no clashes between *Shari'a*, customary law, and international obligations. Therefore, the Criminal Procedure Code for Public and Revolutionary Courts with a special chapter for children and juvenile delinquents was approved in 1999.

Articles 219–231 of this code were related to juvenile delinquency hearings and trials. Accordingly, in each jurisdiction and when necessary, one or more branches of the general court were assigned to proceed with hearings of all crimes related to children (article 219). However, assigning some branches of public courts to child

133 *Moazzenzadeghan* 1389/2010, p. 69; *Khaleghi* 1388/2009, p. 21; *Shamloo* 1390/2011, p. 59.

crimes did not hinder the referral of other cases (article 231). Therefore, professionalism in dealing with child crimes could not be claimed by these branches. In fact, in jurisdictions where such courts had not been established, the general court, in accordance with the code, would provide hearings for all crimes committed by children and juveniles (note 2 of article 219).

In this code, a child was defined as a person who had not reached religious maturity (note 1 of article 219). All crimes committed by persons under 18 years of age were also processed in the juvenile courts, but according to general rules (note of article 220). In other words, with an allegiance to the religious maturity age (9 lunar years for girls and 15 lunar years for boys) in substantive criminal law on the one hand, and in compliance with the age determination in the Convention on the Rights of the Child (18 solar years for both girls and boys) in procedural law on the other hand, the legislator considered some prerequisites for children and juveniles in the trial stages, which was a positive point superficially, yet negative in essence, as severe punishments were applicable for persons under 18 years of age under substantive criminal law.[134]

Other features of arts. 219–231 in this code were extracted from the Act for the Establishment of Juvenile Courts. In line with international standards and according to the differentiated rules for child and juvenile delinquents, the court was bound to inform the parents or legal guardians about a child's rights and proceedings. If the parents or legal guardians neither appeared in court nor hired an attorney, the court assigned a defense lawyer for the child (article 220). In cases of child crimes, the initial process, such as prosecution and investigation, was conducted by the judge (article 221). In children crime cases, the initial process, such as the pursuit of the case and the investigation, was conducted by the judge or the examining magistrate upon request of the judge, by observing the note to article 43.[135] The court itself performed all the actions which were the responsibilities of the judicial officers by law (article 221). Where there was a need to investigate the mental condition of the child (or the parents or legal guardians, the rest of the family and the social conditions), the court could conduct the investigation by any means deemed appropriate and/or could ask for a specialist's opinion if necessary (article 222). The child was summoned through their parents or legal guardians for conducting the investigation and hearing (article 223). If there was no need to keep the child in custody throughout the investigation, other decisions were made in order to maintain accessibility to the child and to prevent them from escaping or hiding based on the type of crime they committed. These included requiring the parents or legal guardian or any other person to present the child to the court when asked to do so if it was deemed necessary

134 However, since this age determination was not accepted in substantive criminal law, it led to conflict and clashes in practice.

135 Article 43: except in cases related to unchaste behavior, judges and examining magistrates can, with proper training, refer to the witnesses, or collection of information, or proof and evidence, or any other action necessary to investigate the crime. These actions are considered to be judicial circumstantial evidence.

to detain the child in order to further the investigation and to prevent collusion due to the importance of the crime; if the child did not have a parent or guardian, or if the parent or guardian did not accept to enter into a recognizance or a bail warrant, the accused child was temporarily accommodated in a correctional center (note 1 of article 224). If there is no correctional center in the jurisdiction of that District Court, the court would decide where to accommodate the child (note 1 of article 224). Children and juvenile delinquency hearings and trials were not done in open sessions – rather only the parents, legal guardians, attorneys, witnesses, and those who had knowledge of the crime, as well as a representative from the correctional center, could be present if deemed necessary by the court. Publishing the trial process through group media, filming, and photography, as well as disclosing the identity and any specification of the accused child was forbidden. Any violator to this rule could be sentenced to legal punishment as stipulated in the Islamic Penal Code (article 225). In the interest of the child, part of the process could be performed in the absence of the accused, and the sentence was considered as if it were given in the child's presence (article 226). The court processed the petition of a private plaintiff according to the regulations and issued the sentence accordingly. There was no need for the child to be present at the trial (article 227). If one or more children had been involved in committing a crime, with the assistance of one or more accomplices, the crime the children committed was processed in child courts (article 228). These could review their previously made decisions to decrease the duration of detention to one fourth of the sentence based on the reports received from the correctional center, which was based on the child's condition, manner, and behavior (article 229). An appeal of a sentence issued by the child court was carried out according to the Islamic Penal Code (article 230).

f) Executive regulation for the prisons, security, and reform measures organization, 2005

The first section of the executive regulations for prisons, security, and the organization of reform measures deals with terminology. Article 19 in this section defines the juvenile correctional center as "a center to reform and educate children under 18 years of age who are detained in this center". Therefore, these centers have been recognized by this regulation; however, there are no other special rules for juvenile correctional centers in this regulation.

The general regulations on the reception, diagnosis, and categorization of prisoners, their daily program in prison, working and learning, rehabilitative programs, and disciplinary punishments are found in the second section. The third part includes some rules on the relations of prisoners with the outside world: receiving visits, post, prisoners going out on day trips as a reward; transporting and releasing prisoners. The conditions for preparing prisoners for release are explained in the last section.[136]

136 For further reading, see *Ghassemi* 2013, pp. 34–42.

g) *Juvenile Delinquent Bill, 2005*

After approving the Criminal Procedure Codes in 1999, further national attempts were made under the UN Convention on the Rights of the Child and on the basis of criminological, psychological, and educational findings to protect children's rights in the justice system. One of the most important efforts was drafting the hearing for the bill on offenses by children and juveniles (Juvenile Delinquents Bill, with five chapters and 55 articles)[137] by the judiciary, which was submitted and approved in the cabinet and handed to parliament in March 2005. However, after innumerous discussions over a long period of time (about eight years), the bill was divided into two sections, that is, substantive and procedural criminal law. The articles relating to substantive criminal law (including the definition of criminal responsibility as well as of juvenile sentencing systems) were merged and added to the new Islamic Penal Code, while the articles relating to procedural criminal law were merged and added to the Procedural Criminal Code. These codes were enacted in 2013 and 2014, respectively. Consequently, it is apparent that the Iranian legal system will continue to be devoid of a special act for juvenile delinquents. However, the legislature has provided some new rules regarding the age of criminal responsibility, the juvenile sentencing system, as well as special police, court and differentiated procedures in these codes, which means that a juvenile justice system has been re-established in Iran. When the Juvenile Delinquents Bill was submitted to Parliament, the legal and political environment for the re-establishment of a juvenile justice system was perceived to be positive.[138] Nevertheless, due to a range of obstacles and probably some other priorities – which appeared later on the political agenda –, it became impossible to approve the bill as a specific act. It was finally was decided to add and merge the content of the bill into those of the Islamic Penal Code and the Criminal Procedure Code which were then under discussion in parliament. Since it was very clear that these bills, being the main and general bills in the legal system, had to be revised and approved, this was a good opportunity to approve the content of other special bills, such as the Juvenile Delinquents Bill, by merging them into these main bills. But although the general political climate was overall positive at that time, the content of the bill was not entirely considered with the needs of juvenile delinquents nor in line with international standards.[139]

---

137 For further reading, see *Abachi* 2015.

138 At that time, two important national conferences were held by the Center of Family Study at Shahid Beheshti University and the Judicial Science Faculty in Tehran. See *Nadjafi Abrandabadi & Jamshidi* 1388/2009; *Jamshidi* 1384/2003, pp. 61–102; see also Journal of Family Studies 1388/2009, Vol. 5 (conference proceeding).

139 In this regard, the research department of parliament states that "the Juvenile Delinquents Bill (www.majlis.ir/mhtml/modules.php) includes new defendable regulations and advantages. Approval of this bill will play the greatest role in organizing the children's criminal code system and pushing it in the direction of scientific criteria, and it can be effective in preventing juvenile delinquency in the future (reoffending)". Hence, the necessity of its adoption is known

h) Penal Code of 2013

In thc new Penal Code of 2013 – due to the clashes between Islamic and customary law on the one hand and international obligations on the other hand –, reforms and changes were made in two different sections and based on the category of punishments in the Iranian legal system. This category, which is influenced and shaped by *Shari'a*, includes discretionary and fixed penalties. Discretionary or *Ta'zir* punishments are applicable for crimes prohibited by Islam, and their types and quantities have not been determined in Islam. In contrast, fixed punishments (including *Hadd*, *Qisas*, and *Diye*) are penalties applicable for crimes prohibited by *Shari'a*, the types and quantities of which have been determined in Islam. It is notable that many crimes and punishments have been categorized in the discretionary or *Ta'zir* section. According to new findings in criminology and penology and (what is more important) in line with international standards, they can be changed by reforming the laws, which means that most of the changes were made in this section. Therefore, as far as *Ta'zir* crimes committed by juveniles are concerned, the new rules and innovations will be applied. In the category of fixed penalties, however, the legislature could extend the rights of juveniles neither in compliance with international instruments nor according to the needs of juveniles in society. Nevertheless, some solutions were introduced to solve problems related to fixed penalties.

Innovations in the rules relating to discretionary punishments include increasing the age of criminal responsibility and the elimination of discrimination between boys and girls. Accordingly, the minimum age for criminal responsibility is now 9 years and the maximum age is 18 years (according to the solar calendar), without any discrimination between girls and boys and in consideration of a gradual transition, which is considered to be the most important innovation in this code. It also provides a differentiated response and sentencing system for *Ta'zir* crimes, while punishment in line with the best interest of child legislation is another innovation. The most important innovation, relating to fixed punishments, is to introduce the notion of intellectual development into the rules governing fixed punishments, which will be explained in the next chapter.

i) Criminal Procedure Code of 2014 and its amendments in 2015

In accordance with the new Criminal Procedure Code 2014, which was enforced in June 2015, a new structure was established prescribing specialized competences for juveniles and a differentiated procedure. The most important innovations are: the

to experts. However, some critical remarks and reformatory proposals are also applicable in this regard: According to *Christian Volksmann*, former representative of the United Nations Children Fund (UNICEF) in Iran, domestic and international support is essential to achieve juvenile justice reform in Iran, which would consequently lead to the necessary legal changes in national law. He believes the current environment is favorable for such activities, as the Iranian authorities themselves have initiated the harmonization of the national and international legal standards with the Act for the Formation of Juvenile Courts.

establishment of a juvenile police unit, prosecution office, and court as well as the expansion of the network of juvenile correctional centers. In accordance with the new Criminal Procedure Code, the rights of juveniles should be respected based on the requirements of due process in criminal procedure. As already explained, the rights of juveniles in criminal proceedings had already been mentioned in the previous Criminal Procedure Code of 1999; however, they have been amended in the new Code. The differentiated procedure is another innovation in this Code, which will be explained in more detail in the next *chapter*.

1.2 Historical Development of Children's Criminal Law in Germany

It is important to emphasize that Germany was not unified as a nation-state before 1871, although German history stretches back to antiquity.[140] The actual historical record begins in Germany in the so-called Migration Period, when Greek and Roman writers first described the inhabitants of the region. In the late 400s C.E., and after the collapse of Roman authority[141] in the western provinces, one of these Germanic peoples, the Franks, gradually brought the various Germanic tribes under their authority and christianized the remaining pagans under the rule of Roman emperors and in the 800s C.E.[142] thus establishing the foundations of the Holy Roman Empire. This empire, a decentralized imperial institution that unified the Germanic duchies into a loose confederation, would dominate political life in central Europe until modern times. After *Charlemagne*, the Carolingian Empire dissolved, fragmenting into several diverse kingdoms. The eastern portion of the Frankish realm, where the German language was spoken, became a distinct kingdom under a Carolingian ruler known as Louis the German. The creation of a Germanic state in central Europe, the

140 It should be noted that "Germany has experienced centuries of immigration, confrontation, and negotiation. The arrangement of constituent parts of Germany has changed repeatedly, with individual territories joining together or breaking apart. Thus, Germany's boundaries, and what it means to be German, have always been unstable and have evolved continually throughout the region's long and troubled history. Situated along ancient migration routes, the area known today as Germany has become home to an endless stream of migrants since prehistoric times. In fact, paleontologists have recovered traces of early hominid habitation in Germany going back almost 50 million years." *Coy* 2011, p. xiii.

141 It has been stated that "the Germanic tribes played an important role in creating the hybrid civilization that prevailed in medieval Europe after the collapse of Roman imperial administration". *Coy* 2011, p. xiii.

142 Politics played a key role in spreading Christianity. "By 600, the Church, with the help of Frankish rulers, had converted many Germanic peoples. These new converts had settled in Rome's former lands. Missionaries also spread Christianity. These religious travelers often risked their lives to bring religious beliefs to other lands. During the 300s and 400s, they worked among the Germanic and Celtic groups that bordered the Roman Empire. In southern Europe, the fear of coastal attacks by Muslims also spurred many people to become Christians in the 600s." *Beck et al.* 1994, p. 99.

ruler of which had inherited the crown of the Holy Roman Empire, was an important milestone in the development of German identity.[143]

During the medieval and early modern eras, the imperial structure provided the fragmented German territories with an administrative and legal framework, while preserving the liberty of hundreds of individual German princes and towns. The unity of the Holy Roman Empire was shattered, however, during the 16th century, with *Martin Luther's* protest[144] against the Catholic Church. This dramatic event, a turning point in German and global history, split Germany and eventually all of Europe into rival confessional – Protestant and Catholic – camps. The nadir of this religious strife was reached in the 17th century, when the Thirty Years' War plunged central Europe into decades of warfare and misery. In the wake of this ruinous struggle, the territorial monarchies of Europe waged their own wars. These wars on German soil, plaguing her with rapacious, mercenary armies, left the empire devastated, with large areas depopulated. The Thirty Years' War greatly diminished the power and prestige of the Holy Roman Emperor and destroyed the delicate balance within the empire between imperial authority and territorial autonomy. Thus, it no longer functioned as a confederation of principalities, but rather became the arena for struggles for dominance by a few powerful princes. In this competition, the centralized, militarized state of Prussia emerged as the dominant power in northern Germany, with Austria, the ancestral home of the Habsburg emperors, being the most powerful principality in the south. These developments would have dire consequences for Germany in the modern era, as German nationalism took on autocratic and militaristic overtones.[145]

The modern period was ushered into Germany by Napoleon's revolutionary armies, who helped facilitate the collapse of the venerable Holy Roman Empire and the foundation of an ambitious German nation-state in its place. In 1806, during the Napoleonic Wars, the French dissolved the Holy Roman Empire and spread revolutionary, nationalist ideals within Germany.[146] After the defeat of Napoleon, the German states were joined in the Austrian-led German Confederation, a loose alliance established by the

143 *Buse* 2005, p. 254.

144 *Martin Luther* challenged the Pope's authority, which marked the beginning of the Protestant reformation. "He received support from the central and northern regionally-based princes. Protestants appealed to the aristocracy of the German nation, underscoring the importance of independent territories within the empire." *Buse* 2005, p. 254. *Luther* reversed his stance by describing Roman law as "heathen wisdom", this being the very law that deprived many peasants of their livehood. *Boes* 2013, p. 29.

145 *Fulbrook* 2015, p. 2.

146 In fact, under the impact of the Napoleonic war, "a fundamental reorganization of the domestic and external affairs of the German states was begun. In 1806 the Holy Roman Empire was abolished. Legal, social and economic reforms were introduced, either as a direct result of Napoleonic rule or in a form of defensive modernization". *Fulbrook* 2015, p. 2.

Congress of Vienna in 1815 in a futile attempt to restore the old political order in central Europe. The rise of Prussia, an aggressive military powerhouse, led to the declaration of a German empire in 1871, a unified Germany ruled by a Prussian monarch. Under Prussian leadership, Germany underwent rapid industrialization, and militant German nationalism flourished in the late 19th century. The reckless ambition of the Prussian ruling house and Germany's delayed unification prompted the new German nation to demand inclusion in the major powers of Europe. These pressures would contribute to the outbreak of World War I which proved a disaster for Germany, and the peace afterward a disaster for the world.[147] According to the punitive Treaty of Versailles, Germany was forced to accept a series of crushing and humiliating terms. The trauma of war ending in defeat fostered a combination of political discontent and economic depression in the country that crippled the fledgling postwar regime, the ill-fated Weimar Republic, Germany's first democratic system.[148]

It is notable that Germany's first experiment with a democratic government proved abortive, ending in tragedy in 1933. Amid political and economic chaos, the "National Socialist German Workers' Party", a radical fascist group, took control after their fanatical leader, *Adolf Hitler*, was appointed chancellor. Under autocratic Nazi leadership, Germany descended into a nightmare of totalitarian dictatorship and racist oppression of its Jewish citizens. In the end, *Hitler's* "Thousand-Year Reich" lasted only little more than a decade but brought the Germans' catastrophic defeat in World War II.[149]

After the war, the victorious Allies partitioned the smoldering wreckage of Germany, a division that was ossified during the Cold War. The United States, the British, and French zones of occupation were combined to establish the Federal Republic of Germany, or West Germany, in 1949, with the German Democratic Republic (East Germany) forming in the Soviet zone. While West Germany returned to democratic government and joined both the North Atlantic Treaty Organization (NATO) and the European Union, East Germany became part of the Soviet-controlled Warsaw Pact, its isolation from the West symbolized by the Berlin Wall. In 1989, in the course of the peaceful *Wende* revolution, the East German government collapsed and the Wall came down. In 1990, East Germany was reunited with West Germany, forming today's unified federal democracy. Having emerged from the trauma of its past, today's Germany is a federal parliamentary republic.[150]

147 "The First World War certainly produced some innovations, not least in criminal law, but they consisted more of changes in the style of legislative technique than long-term changes of normative content." *Vormbaum* 2014, p. 13.

148 *Coy* 2011, pp. xiv–xv.

149 *Coy* 2011, pp. xv–xvi; see also *Biesanz* 1941, pp. 554–559.

150 *Coy* 2011, pp. xv–xvi.

1.2.1 The early era

The legal system and the socio-legal status of child and juvenile delinquents in the Early and High Middle Ages will be separately discussed in this *section.*

1.2.1.1 The Early Middle Ages

1.2.1.1.1 Legal system

It is notable that the current German criminal law and legal system has its roots in Roman law,[151] and the accounts of German criminal law generally consider the time of the Germanic tribes as their starting point. This is the period of the great migration, which ended in the 6th century.[152] The first records of Germanic law arise from the observations of Roman authors, notably *Caesar* and the Germania of *Tacitus*, otherwise there is little evidence. Germanic culture was spoken rather than written, as were language and customary law. The oldest report written in German is alleged to be the "Abrogan", a Germanic-Latin dictionary from A.D. 770. The type of law employed by the tribes was a form of simple unwritten customary law, which was commonly established in a democratic manner by a general open-air assembly of tribes approving as the occasion demanded, customary rules. Legal proceedings were based on the same assembly described as either a *Thing*, *Ding*, or a *Gau* depending on the particular tribe and matter at hand, but also other formal organizations of justice seem to have taken place. The laws were based on collective laws as the originator of rules rather than individual rights or laws based on a particular area or property. Courts were established only when needed, and the leader of the tribe, whatever title they possessed, was usually the spokesman who expressed the findings of the assembly. Developments to this form of dispute resolution included setting aside a particular day or days on a regular basis to resolve disputes, for example at the time of full or new moon. However, this is not to say that there was a universal pattern throughout the Germanic tribes for dispute resolution, and many variations existed with varying levels of democratic participation.[153]

In the Early Middle Ages, the operative unit of society was kindred, the clan, or extended family. Clans mainly lived together in ordered, manageable, and rural communities. According to the traditions and customs of the Germanic tribes, which were pretty much the same, when the member of one clan harmed a member of another, the aggrieved person's relatives sought retribution in what is now termed a vendetta or blood feud. The retribution was not sought specifically from the person at fault,

---

151 "The first compilation of Roman law was contained in the Twelve Tables, which were published during the Roman Republic in 450 B.C. The Twelve Tables provided the basic legal Code of Rome for almost a thousand years. The end of the Western Roman Empire in 476 A.D. following the incursion of Germanic peoples from the east led to a revival of Roman jurisprudence from the classical period (the first and second centuries A.D.) in the eastern part of the Empire." *Fisher* 2009, p. 1.

152 *Robbers* 2012, p. 108.

153 *Foster et al.* 2010, p. 9.

but from any member of his kindred based on the collective manner in which people operated. A process of "tit for tat" could go on for years, with men being killed long after the original basis of complaint had been forgotten. The urban control of bishops and the territorial control exercised by dukes and counts were the main characteristics of society in the Early Middle Ages. There were court days on which all citizens could bring forward their disputes, which were arbitrated by an alderman and jury. Furthermore, there were arbitration boards where the opposing parties could choose a referee who tried to find a solution for the dispute. But due to the fact that the people lived together in close-knit groups, mainly in clans, they continued to deal with most problems within the community. Criminality was one of the issues that were generally solved within the clan and for which no court day or arbitration board was needed. Beyond that, the feud system still played a more important role in the punishment of criminal offenders than the official court days. These feuds were a problem for the ruling councils of the towns, because possessions were destroyed and numerous people killed. Attempts to restrict the feud system by legal means failed over and over again.[154] It should be noted that the "Frankish period" – from the founding of the Merovingian Kingdom to the division of the empire (end of the 5th century to the middle of the 9th century) – brought a strengthening of the central penal authority and the anchoring of criminal law in written sources. The continuing influence of Roman law together with Christianity favored the development of public punishment in the place of the private feud. The laws of some of the tribes were recorded in Latin, for example the Salic law (*lex Salica*) in about 500 A.D.[155]

1.2.1.1.2 Socio-legal status of child and juvenile delinquents

In both extended and stem families, children had to contribute to the homesteads or work on the family farm from early childhood. They did not have separate areas of life from adults because they were considered "small adults". Sons and girls were raised differently. Sons were to become self-disciplined, emotionally controlled, conformist, morally upstanding, and the boundaries between themselves and others were not defined. Girls were raised to be willing and able to take on the serious moral, emotional, managerial responsibilities of organizing a household and raising children, but without having the desire to aspire beyond their household's walls.[156] Education was limited to a few wealthy children of townspeople and noblemen and was primarily a matter for the home or monastery until the beginning of the High Middle Ages when the first schools appeared in more populous areas. Peasants had no incentive to teach their children to read and write – it was rare that they were literate themselves. The characteristics of an age-specific period of development to adulthood cannot be identified; this did not play an important role. Accordingly,

---

154 *Wiese* 2009, pp. 125–126.

155 *Robbers* 2012, p. 108.

156 *Meyers* 1999, p. 17.

there was no special treatment for young offenders. As a result, most children did not have a period dedicated to their education and started helping their parents once they were strong enough to do so. From that point onwards, they led the same life their parents did.[157] Furthermore, the father, as the head of the family, was held responsible for catering to the well-being of the family in general and the wrongdoings committed by members of his household, and he was at risk of forfeiture for the wrongful acts. He held the exclusive right to punish the child, which could extend to killing it if this was thought necessary for the father to fulfill his duty to ensure discipline and order in his household. This authority over the child would end if children found an independent situation by marrying or leaving the father's household.[158]

In the small communities of the Early Middle Ages, child delinquency was not very common. A child's offence was regarded as carelessness on the part of its custodian. Public punishment of children did not exist – the young delinquent was returned to his/her clan for further education and supervision. The predominant response of the clan was castigation of the child by either his/her custodian or by the victim of the offence. The death penalty or corporal punishments were rare for children. If a child offended a member of another clan, both clans met and preferred reconciliation. Owing to the age of the young offender, the clans often agreed upon penal mitigation or exclusion.[159]

1.2.1.2 The High Middle Ages

1.2.1.2.1 Legal system

In the High Middle Ages, the situation changed. In fact, the rise of urban communes marked the beginning of this era. Outside the cities, the power of the central government was greatly reduced. Consequently, government authority and responsibility for military organization, taxation, and law and order was delegated to provincial and local lords, who supported themselves directly from the proceeds of the territories over which they held military, political, and judicial power. This marked the establishment of the feudal system. The High Middle Ages saw the growth of centralized power and new national identities, as strong rulers sought to eliminate competition (and potential threat to their rule) from powerful feudal nobles.[160]

---

157 *Wiese* 2009, p. 127.

158 "Later, the state took a more interventionary approach to family affairs with the aims of ensuring loyalty to the state and religious education. This entailed the state assuming the authority to punish for wrongful acts, leaving the father merely with the right to discipline the children. As the state took over the role of punishment the principle was formed that below a certain age, around the age of reaching puberty, punishment should be milder than that which an adult would receive." *Crofts* 2004, p. 402.

159 *Wiese* 2009, p. 128.

160 *Wiese* 2009, p. 128.

At that time, the power to inflict punishment passed to the territories, towns and petty princes. This strengthened the tendency toward public punishment. However, it also led to the fragmentation of the criminal law and its increasing harshness, indeed cruelty, with the imposition of corporal forms of punishment and torture. The feudal system bloomed in knighthood, but outside fortified strongholds physical danger always loomed. Particularly the towns, with their interest in trade, reacted to these threats with decisive harshness, resorting to torture and summary procedures. Private records of regional customary law, such as the Saxon Mirror (*Sachsenspiegel*) written by *Erike von Repgow* in 1230 and the Swabian Mirror (*Schwabenspiegel*) of 1275, provided a basis for a systematic approach to criminal law. The Third Reich and the Church attempted to discourage the custom of conducting feuds and to create safe periods and areas with the campaigns for the "peace of the lands" and the "peace of God". However, these attempts only achieved partial success with the enteral peace of 1495 right at the end of the mediaeval period. The reception of Roman law starting in the 12th century was of central importance as much for public criminal law as it was for other aspects of law. The detailed and systematic Roman law was conducive to the creation of a class of secular lawyers and was applied in the ecclesiastical courts in Germany from the 13th century onwards. On the one hand, this development was in line with the legitimation of the Holy Roman Empire and the successor of ancient Rome, while on the other hand it promoted the creation and administration of the territorial principles.[161]

The attempts to achieve a uniform system of criminal law through the empire culminated in the *Constitutio Criminalis Carolina* (CCC) of Emperor *Charles V* (*Karl V*). This was promulgated in 1532 and was the first criminal code applicable to all of Germany. It continued the most important provisions concerning crime and punishment and was mainly concerned with matters of procedures. This code already gives expression to the idea of punishment according to the rule of law. It takes into account the principle that the accused must have a guilty state of mind. As far as the procedure aspect was concerned, the use of torture was limited to certain situations defined by statute, and the use of an inquisitorial procedure was mandated, i.e. the court was required to undertake comprehensive investigation into the facts of the case. Further refinement of procedural law took place with coming into force at the imperial police ordinances of 1530, 1548, and 1557. The *Constitutio Criminalis Carolina* was based to a large extent on the *Constitutio Criminalis Bambergensis* produced by *Johann von Schwarzenberg* (1465–1528). In some German territories it was applicable, at least as a subsidiary source of law, until the 1870 criminal code of the North German federation came into force.[162]

In this period, special provisions regarding young offenders are rarely found.

---

161 *Robbers* 2012, p. 109.

162 *Robbers* 2012, p. 110.

1.2.1.2.2 Socio-legal status of child and juvenile delinquents

It is notable that during the High Middle Ages, it was acknowledged that children do not turn directly into adults, so a stage of development called "youth" was also recognized, although this recognition did not lead to special treatment of children or juveniles who had committed offences in the criminal law dimension; but children, juveniles, and other young persons were not punished as severely as adults.[163] In other words, there was no concept of subjecting the young offender to totally different methods of punishment; there was rather merely an exclusion from the more brutal forms of physical punishment. The Saxon Mirror and the Swabian Mirror are examples of codes which determined that up to the age of 12 (Saxon Mirror II 65 I 1) and 14 (Swabian Mirror Ldr. L 177) respectively, a child was not to be sentenced to death or maimed. Some codes in the Middle Ages did not simply provide for milder punishment of the young. The code of the Lübeck law (*Codex zum Lübischen Recht*, 1180–1400) made the degree of punishment dependent on an assessment of understanding of the young person. Accordingly, a child under 12 years of age was generally to be punished more mildly than in the Saxon or the Swabian Mirror. However, they could be punished as an adult if the child possessed an understanding of the wrongfulness of the act. One traditional test of criminal capacity was for the judge to hold out their hands to the child, with a penny in one hand and an apple in the other. If the child reached out for the apple, they had proven their childhood. However, if the child reached for the penny, it followed that the child already knew the value of money and could thus be punished in the same way as an adult.[164]

Later, at the end of the High Middle Ages, due to the partial break-up of the self-education and self-controlling of social communities, child and juvenile delinquency started to increase, and so did subsequently public concern about this matter, as far as public intervention and sanctions were demanded. Criminal law became more and more an issue for the authorities, and consequently, offences and conflicts were solved to a lesser extent by the clans themselves. Public sanctions for young offenders, such as "caning on skin and hair", "whipping with birches", clipping, and executions were introduced. These sanctions were intended to be deterrents and differed only marginally from those imposed on adults. From that time onwards, young offenders in Germany were dealt with by formal criminal proceedings, being prosecuted and convicted in criminal courts to receive criminal sanctions. This notion can be seen as the predecessor of the justice model wherein the German juvenile justice system is rooted. A more marked differentiation between the phases of childhood, youth, and adulthood began in the early Renaissance.[165]

163 *Wiese* 2009, p. 129.

164 *Crofts* 2004, p. 402.

165 *Wiese* 2009, p. 130.

In 1532, the *Constitutio Criminalis Carolina* (named after *Charles V*) came into effect. It was the first statute dealing with the culprit's guilt and was made a requirement for determining someone's liability. Furthermore, it was the first statute to provide for special rules regarding young persons.[166] This statute provided that in general, the young were to receive milder punishment than adults (article 179), but did not fix any age limits except in relation to theft. Article 164 of the *Constitutio Criminalis Carolina* provided that thieves under 14 years of age were not to be punished by death. However, they could be punished on "goods, body or life" if their age was close to 14 and the value of the stolen goods was of a large amount or if there were aggravating circumstances such that malice could make up for the lack of age. The only concession these Middle Age codes made to childhood in determining punishment was to exclude harsher forms of punishment for those under a certain age. In the cases of the *Codex zum Lübischen Recht* and the *Constitutio Criminalis Carolina*, this exclusion depended on certain circumstances of there being no act of malice. It was not, however, until the 18th century that these tests for malice became the determinants of whether the child would be punished at all, rather than determining the degree of punishment. The principle that unless the offender was capable of guilt they were not to be punished became anchored in the later codes.[167]

It should be noted that during the Renaissance, awareness of a special life phase between childhood and adulthood increased, and consequently, the treatment of young persons changed fundamentally. It was also recognized that children and young persons are vulnerable and, because of that, malleable through education during their development. Thus, education and pedagogy became very important issues.[168]

1.2.2 The codification (modern) era

1.2.2.1 Legal system

Further change occurred in the Age of Reason (Age of Enlightenment) in the 17th century: The irrationality and superstition of the Middle Ages finally gave way to an optimistic view of the world according to logical and mathematical principles. The Age of Reason brought about a gradual departure from the traditional principles of Roman law and a return to those of natural law.[169] In this era, criminal law was increasingly included in the area of philosophy. From this followed the implementation

---

166 *Wiese* 2009, p. 130.

167 *Crofts* 2004, p. 404.

168 *Wiese* 2009, p. 131.

169 *Fisher* 2009, p. 2. "Natural justice, from the German point of view, should be seen as a kind of safety valve in a legal system tending toward a positivistic approach, as far as the usual primacy of the written law is concerned. It is difficult to place natural justice firmly into a hierarchy of laws, as it applies in different shapes and forms at any level of the German legal system. It permeates the law as a guiding principle of interpretation. It would, however, not

of important principles (which could be attributed to the justice approach) into substantive criminal law in Germany. One of those innovations was that judges had to confinc themselves to the law and could no longer consider customary law, ancient customs, traditions, or religious approaches anymore. Another innovation was that punishment should be proportionate to the offence.[170]

In fact, from the 17th century onwards, the academic study of criminal law became a significant influence in the creation of a uniform criminal law for all of Germany. This academic activity was inspired primarily by Italian and Spanish jurists as well as the canon law. Perhaps the most important writers were *Benedict Carpzov* (1596–1666), *Johann Samuel Friedrich Boehmer* (1704–1772) and *Christian Thomasius* (1655–1728). The latter was active in opposing the prosecution of people accused of being witches and, together with *Samuel Pufendorf* (1632–1694) and *Christian Wolff* (1979–1754), embodies the start of the Age of Enlightenment in Germany. In this period, academics began to consider what the purpose of punishment is and attempted to organize criminal law on the basis of reason. In this way, the criminal law found a basis in the common good, legality, and the right of the individual. Like all state laws, criminal law became secularized to a significant extent in this period. In other words, it largely lost its religious and theological connections. Detention became the standard for punishment. *Jeremy Bentham* and *Cesre Beccaria* achieved significant influence in Germany. On his accession to the Prussian throne in 1740, *Friedrich II* (*Friedrich the Great*) abolished the use of torture. A systematic approach, the embodiment of the law in statutes, the general applicability of law and its accessibility to everyone as well as especially humanness and justice were key ideas of the Enlightenment. These ideas became important for criminal law with the codification movement in the 18th century and were embodied in the Prussian General Code of 1794, the *Constitutio Criminalis Theresiana* of 1768 in the Austrian territories, and the Codex *Juris Criminalis Bavarici* of 1751 for Bavaria. Against this background, the French Code of 1810 and the Bavarian Criminal Code of 1813, which was created by *Paul Johann Anselm von Feuerbach* (1755–1833), formed the basis of German criminal law in the 19th century. Its development was typified by codification in many German principalities, but at the same time academic writers kept alive the idea of a common criminal law throughout the German-speaking lands. The Prussian criminal code of 1851 formed the basis for the criminal code of the North German federation of 1870. This code became applicable throughout the newly founded German Empire as the Imperial Criminal Code of 1871. Despite numerous amendments and the complete restructuring of the general provisions in 1975, this code is still applicable today in its 1998 version. The impact of European Union law on the criminal law is still relatively small, but it is growing. The numerous reforms of the criminal law were initiated by the Weimar Republic. In

be unfair to say that the principle of natural justice has the force of influencing the application even of the highest-ranking legal rules on the constitutional level. Looking at it that way, one can make the statement that it represents the top tier in the hierarchy of laws." *Bohlander* 2009, p. 13.

170 *Wiese* 2009, p. 132.

particular, it introduced the prison system and special rules for dealing with youthful offenders. The criminal law also suffered from the perverting influence of the National Socialists. The regime introduced the idea of convicting people of new offences by analogy with existing statutory crimes and introduced castration as a punishment for sexual offences. These innovations were abolished by the Allies after the Second World War. Alterations of the law during this period which were not tainted by National Socialist ideology and therefore not abolished after the war were the introduction in 1933 of provisions for securing, reforming, and treating more leniently persons with reduced criminal capacity, as well as the introduction in 1943 of special provisions and the criminal law for juvenile offenders.[171]

1.2.2.2 Laws and regulations concerning child and juvenile delinquents

1.2.2.2.1 The German state codes before unification

It should be noted that in the 18th century and under the influence of the movement away from physical punishment to imprisonment, in the Bavarian Codex *Juris Bavaraci* of 1751 and the more influential General Law for the Prussian States of 1794, imprisonment was to take the place of traditional physical punishments and a direct link was made between mental capacity and punishment. The Prussian law stated in article 18 that everything which affects the ability of a person to think and act freely also affects the degree of punishment. Specifically, with regard to the young, this meant that minors (those under 14 years of age [article 25 I 1]) could be disciplined to prevent further wrongs, but were not to be punished according to the full strength of the law (article 17 II 20). Although not mentioned in the code, a significant development was that the Prussian courts adopted the practice of not applying any sanctions at all to children under 7 years of age. This shows the acceptance of the principle (which became anchored in later codes) that there should be a fixed age under which a child was thought too young to ever be punished.[172]

According to the General Law for the Prussian States, the death penalty and imprisonment were expressly excluded for all offenses for children under the age of 14. Later an amendment was passed which provided that children under 14 could be punished with imprisonment if malice supplied the lack of age. Minors who committed an act without malice were subjected to milder punishment in order to be disciplined (article 17); however, some Prussian provinces interpreted "disciplined" to mean physical punishment. Dissatisfaction with the Prussian law led to a number of draft criminal codes being drawn up in the early part of the 19th century.[173]

These drafts reveal a significant change in the attitude toward young offenders. Milder punishment was to be reserved only for those children who were found to be

---

171 *Robbers* 2012, pp. 110–111; see also *Overland et al.* 1968, pp. 35-64; and *Krehl* 2003, pp. 421–431.

172 *Crofts* 2004, p. 404.

173 *Crofts* 2004, pp. 404–405.

criminally responsible, while those lacking the capacity to form a guilty mind were not to be punished. Gradual responsibility was also recognized. In line with the practice of the Prussian courts, most of the drafts provided that there should be a minimum age under which all children were deemed to be incapable of criminal capacity and thus should never be punished at all. Beyond this minimum age, there would be milder punishment up to an age level where adult punishment would be applied without exception. The proposed age level under which a child would be regarded as being absolutely criminally incapable ranged from 10 to 14, with 14 being the most common suggestion. Under this age level, children were not to be punished, not even more mildly. Instead, they were to be left to home discipline, subject to the supervision and involvement of the state. In support of the age of 14 as the age of complete criminal incapacity were the arguments that most people left school and took confirmation at this age. Therefore, it was thought that until the child had completed their religious education, they were not yet fully aware of the extent of their obligations and so could not be held responsible for any wrongdoing. Social conditions were also recognized as affecting a child's understanding, and concern was expressed that poor nourishment and care among the lower classes was hindering their physical and mental development. A significant change proposed in the drafts was pushing out the special protection provided to the young beyond this minimum age level until the age of 16 – in recognition of the circumstance that until this age, some young persons may not yet be developed enough to form a guilty mind. The early drafts did not, however, propose a test to establish whether those within this age range were capable of guilt; it was rather thought that milder punishment would adequately allow for any reduced criminal capacity. An alternative proposal to fixing a minimum age under which there would be a blanket exclusion of punishment and an age range beyond this of blanket milder punishment was to formulate of a test of criminal capacity. This test would make allowance for the fact that children develop at different levels and would ensure that punishment was related to the ability of the offender to form a guilty mind. This was the approach taken by the French in the Penal Code of 1810, and it was the model supported by the Rhine provinces in the Prussian parliament. The Penal Code of 1810 did not have a minimum age level under which punishment was entirely excluded. It simply provided in article 66 that a youth under the age of 16 was to be acquitted unless they possessed discernment (the ability to discern).[174]

In the early 19th century, efforts to realize the special situation of children, juveniles, and young persons increased in Germany. In 1851, the German legislation implemented section 42 of the Prussian Criminal Code, under which the capacity to understand a wrongdoing was essential for the punishment of a person under 16 years. If young persons possessed this capability, they were punished like adults. If they

174 *Crofts* 2004, pp. 404–405.

lacked it, they were to be exonerated or sent to a reformatory.[175] In line with the perception that earlier Prussian law had not been much of an advancement over Middle Age codes, the final draft code of 1850, which was passed one year later as the Prussian Criminal Code, incorporated a proposal similar to the French Penal Code. It therefore did not include a lower age limit of absolute criminal incapacity as had been suggested in all the other drafts.[176]

The Prussian Criminal Code from 1851 translated article 66 virtually literally into article 42 and required that unless the youth under the age of 16 had the ability to discern, they were to be acquitted. This required an individual examination in each case of the ability of the offender to understand the punishment ability of the particular wrong committed. It was considered not to be enough that the youth could discern between good and bad but rather that they understood it was their duty not to commit this act, and that in so doing they were opening themselves up to a more severe and serious punishment than discipline at home. If the offender was found not to have the ability to discern the wrongfulness of the act it was to be included in the judgment of the court whether they were to be passed on to their parents or taken to a reformatory. Until the age of 16, the young person was only to be subjected to milder punishment if it was ascertained that they had acted with the ability to discern (article 43). In contrast to the Prussian Criminal Code of 1851 and the very similar Bavarian Criminal Code of 1861, most other German state codes provided for an age limit of absolute criminal incapacity under which there was to be no punishment at all. This age level varied between 8 and 14, with 12 appearing to have been the most common age. Despite these codes having been deviated from the Prussian Criminal Code (1851), the Bavarian Criminal Code (1861), and the Penal Code (1810) in providing for an age level of absolute criminal incapacity, the influence of the latter can still be seen. It lies in the fact that most of these codes also had age periods of conditional criminal responsibility beyond the minimum age where the liability of the youth to punishment depended on whether they had the ability to discern. These codes similarly provided that if a youth was found not to be criminally capable, they were not to be punished, but instead could be left to be disciplined at home or sent to the police for some reform or educational measure to be ordered (for instance article 78 Criminal Code of Baden (1845). Only if the youth was found to have been criminally capable could they be punished, but the punishment applied was to be milder than adult punishment[177] (for example article 79 Criminal Code of Baden 1845, which also detailed a scale of how punishment was to be made milder).[178]

---

175 *Wiese* 2009, p. 133.

176 *Crofts* 2004, p. 405.

177 *Crofts* 2004, pp. 405–406.

178 For further reading on laws and regulations concerning child and juvenile delinquents, see *Laubenthal* 2015, pp. 11–31.

1.2.2.2.2 The German criminal codes after unification

a) *Imperial Criminal Code of 1871*

It is notable that at the end of the 18th and throughout the 19th centuries, due to the changes in society which appeared as a consequence of industrialization, modernity, and urbanization, an acknowledgement of a youth phase developed, which was completely dedicated to the education of young persons and therefore formed a clear contrast between childhood and adulthood. Much employment during the industrial revolution required school education and professional training. A special system of youth policy and social control was set up, including youth-specific institutions of education and socialization, such as different types of schools, a Youth Welfare Department, and Youth Aid. Soon after the establishment of this youth-related system, new concepts were demanded of dealing with young persons who offended or were in need of care and protection. The rise of the social work professions and their motivation to support former convicts and young persons at risk resulted in a more welfare-oriented approach concerning the treatment of young offenders. Pedagogues and criminologists criticized the previously existing, mostly repressive sentences of the justice system, which they mistrusted and regarded as outdated. Thus, a new way of dealing with young persons in trouble was sought, which provided for appropriate intervention possibilities if problems should arise with the education and socialization of young persons. Attempts to create a special criminal law for juveniles were made at the end of the 19th century.[179]

Furthermore, there were political moves toward German unification. Until these political movements, different state criminal codes were applied, with varying approaches to the criminal responsibility of the young. A new uniform code was therefore drafted for the German Empire, which led to passing the Imperial Criminal Code of 1871.[180] This code was influenced by the Prussian Penal Code of 1851, which itself had been inspired by the Bavarian Penal Code of 1813 as prepared by *Feuerbach*.[181] The Imperial Criminal Code deviated from the approach of the two largest states and introduced an age limit of absolute criminal incapacity. It was a response to the Prussian and Bavarian codes being seen as defective in failing to acknowledge that there is a point in time in a person's life in which their criminal capacity is always lacking. This feeling of dissatisfaction was fueled in Prussia by the prosecution of a 6-year-old for arson. Such an absolute age level would also ensure that the law would be applied consistently regarding children. The age under which a child could never be found criminally responsible and punished was set to 12 years in the Imperial Criminal Code of 1871. This was in line with the majority of state codes and was the level recommended by the Prussian Royal Scientific Deputation for Medicine, which

179 *Wiese* 2009, pp. 133–134.

180 *Crofts* 2004, p. 407.

181 *Ghassemi* 2013, p. 38.

was commissioned to report on the criminal capacity of young delinquents. For those aged 12 but not yet 16, the Code replaced the test of the ability to discern the wrongfulness of the act with the question of whether the youth had the understanding necessary to recognize the punishment ability of the act (article 54). For those who were deemed to have this understanding, the appropriate response was still seen in adult punishment being applied in a milder form; otherwise, they were to be acquitted and could then be handed over to their parents or sent to a reformatory (article 56) and subjected to forced education.[182]

It should be mentioned that the "modern school of thought" in criminal justice theory has called for the conversion of classical, retaliatory criminal justice into a specialized prevention-oriented criminal justice since the 1880s.[183] On the other hand, in the late 19th century, debate about the adequacy of the existing responses to youth crime intensified. This was fueled by the publication of criminal statistics, which commenced in 1882. In the following years, these statistics revealed that crime rates among the young were starkly increasing, which directed attention to the causes of youth crime. In 1883, the influential criminal law professor *Franz von Liszt* gave his pioneering lecture, known as the *Marburger Programm*, in which he called for a fundamental rethinking of the purpose of punishment.[184] In the lecture he stated that if a juvenile and even an adult commits a crime, the probability that they will commit yet another crime is lower if we let them go than if we punish them.[185] Since children were no longer regarded as merely small adults but were seen as being in a distinct phase of their development, where they were especially susceptible to influence, they received the special attention of reformers. Up to this time, the main concern had been to set age levels of criminal responsibility such that only those capable of guilt would be punished. From this time onward, the educational measures by reformers to combat criminal influences came into account.[186]

Gradually, with the social changes, crime was seen as a consequence of the dire social circumstances of the young, of increasing poverty and deficient education, often due to broken families, mothers having to work and children being born illegitimately. Industrialization was therefore seen as destroying traditional control mechanisms which had previously helped prevent children from falling into crime. In the eyes of the reformers, the answer to the crime problem was for the state to provide

182 This did not necessarily improve the situation of the young offender, since punishment, which was a direct response to the crime committed, was replaced by indeterminate custody designed to overcome the educational deficit that had led to the crime. It has also been criticized that forced education was used as a way to disseminate middle-class values and to extend social control while diminishing the rights of working-class children. *Crofts* 2004, p. 407.

183 *Kaiser* 1994, p. 321.

184 *Crofts* 2004, p. 408.

185 *Kaiser* 1994, p. 322.

186 *Crofts* 2004, p. 408.

education to cure the young offender from their criminal tendencies.[187] Imprisonment was seen as counterproductive to the aim of educating the young. The concern was expressed that when the child returns to school after having been in prison, they will either be regarded by their school companions as a hero and their moral degeneracy will be spread in all directions, or they will be mocked, despised, and avoided, and might learn to hate society and all its moral principles. The general motto of the time became: where education is necessary, there is no space for imprisonment! Efforts were therefore made to increase the age of absolute criminal incapacity, in line with the view that children who were still at school (i.e. under 14) had no place in prison. As the recognition grew that poor social conditions and neglect were contributing to crime, so did the concern to step in earlier in cases where children were so neglected or morally degenerated that a criminal career was to be expected. This idea of intervening in the lives of neglected youths to prevent them becoming offenders was not new – it had formed early on in the 19th century, but now became a more widespread concern to provide education for offenders and the neglected. Houses of rescue were set up, where it was hoped that work and strict discipline would teach children an ordered and law-abiding lifestyle. The desirability of a shift in focus from the criminal capacity of the young and the possibility of punishment to the question of whether educational measures could help the young offender was discussed at the assembly of the International Criminalist Association in Bern in 1890, at the International Penitentiary Congress in Petersburg in the same year, and at the second assembly of the German section of the International Criminalist Association in Halle in 1891. It was recommended at these assemblies that the question of criminal responsibility be replaced with that of the suitability of educational measures. It was also proposed that the age of absolute criminal incapacity, and thus the age under which punishment would be excluded, be raised to 14.[188] Another intensely criticized issue at that time was the practice of dealing with juveniles in the adults' courts and imprisoning adults and juveniles in the same prisons. Experts demanded that proceedings against juveniles should be held in special juvenile courts, and that young persons should be placed in probation homes, reformatories, or at least in special juvenile prisons.[189] The basic idea behind institutionalizing a separate domain

187 *Aschrott*, one of the leading criminal law reform campaigners, saw the increasing crime rate as a signal that the traditional measures of criminal law were powerless in fighting youth crime. He also argued that through timely and energetic intervention, the seed and tendency of antisocial behavior in the endangered youth should be repressed and a good seed planted; a task which was not considered to be achieved through imprisonment. *Crofts* 2004, p. 409.

188 Naturally, there was opposition to increasing the absolute age level of criminal incapacity and moving away from punishment to education. Some felt that excluding the threat of punishment would contribute to the increasing crime rate, and they therefore opposed any increase in the age level. These opposing attitudes, on the one side, asserted that education and not punishment was necessary for the young and, on the other side, postulated that punishment had to be available, which continues to mark the discussion of youth criminal law up to the present day. *Crofts* 2004, p. 409.

189 *Wiese* 2009, p. 136.

for juvenile justice was a juvenile sentencing policy in keeping with the educational concept that encourages the judge to act as an educator and requires a special qualification or training as well.[190]

It is important to emphasize that the first explicit youth criminal court chambers were established in Frankfurt, Cologne, and Berlin in 1908. In 1912, the first separate prison for juvenile delinquents was built. Furthermore, police administrators advised the police officers to be more sensitive and careful when interviewing a child or young person who was charged with an offense. With the enactment of the German Civil Code (*Bürgerliches Gesetzbuch*, BGB) in 1900, section 1666 was introduced. It provided for welfare, education, and supervision and directed that the Guardianship Court could impose the required measures when the mental or physical well-being of a child or young person was endangered due to neglect or because of the culpable moral conduct of their legal guardian. The Guardianship Court could withdraw the right of custody if the parents misused their legal guardianship or if the child or young person committed a criminal offense. It can be concluded that the treatment of young offenders in Germany was purely based on a justice-model-orientation until the early 20th century. Young offenders were tried in criminal courts and received formal sentences. Welfare-oriented and diversionary ideas, introduced by criminologists or practitioners, were not yet implemented in substantive law.[191]

b) *Drafts of the German Criminal Codes of 1909, 1911, 1919*

In 1909, increasing the age of absolute criminal incapacity to the 14th birthday (article 68) and setting the period of conditional criminal responsibility to 14 to 18 (article 12) was suggested in a pre-draft of the German Criminal Code. Interestingly, this draft no longer included the requirement of an individual assessment of the abilities of the young person. Instead, article 69 provided either for milder punishment or for education: If the act appeared to be mainly the result of poor education or if it was otherwise to be assumed that educational measures are necessary to accustom the perpetrator to a law-abiding life, the court could order him, alongside or instead of punishment, to state-supervised education; the court might also order detention in an educational institution or reformatory. This draft clearly shows that the belief in the importance of education over punishment was gaining ground. Indeed, this was the first official draft code in Germany to incorporate the concept of replacing or supplementing punishment with educational measures. Up to this time, forced education was only available for those young people who were not found to be criminally responsible, while the only option for those found criminally responsible was punishment, albeit in a milder form.[192]

---

190 *Kaiser* 1994, p. 324.

191 *Wiese* 2009, pp. 136–137.

192 *Crofts* 2004, p. 410. This draft would have meant that for the first time in history, the judge would have the discretion to not order punishment but an educational measure. It therefore

In 1911, Professor *von Liszt* and others presented an opposing draft. This draft also suggested the age of absolute criminal incapacity to be set at 14 (article 15). However, by including a test of criminal responsibility (article 16), this was to ensure that there could be no punishment without a finding of criminal responsibility. The priority of education over punishment was made even clearer in this draft in article 17, which provided that even when there was a finding of criminal responsibility, there was to be no punishment if the crime was the result of deficient education. In such cases, educational measures were to be applied instead of (and not alongside) punishment. With the advent of World War I, the reform of the law concerning young offenders took a back seat. During the economic crisis which followed the war, there was a large rise in youth crime and in the rate of youth imprisonment, which was taken as evidence of the need to continue the reform efforts. In 1919, a draft Criminal Code was presented which contained a segment entitled "Children and Young Persons" (15th part article 129 to article 137). This segment, which is reminiscent of the 1911 draft, revealed the growing acceptance of the reformers' aims of putting education before punishment. It proposed an age of absolute criminal incapacity of 14 years and provided in article 130 for a test of criminal responsibility. Accordingly, a young person aged 14 but not yet 18 would not be held to be capable of guilt if they committed an act due to slower development or lack of mental or moral maturity such that they were not capable of understanding the unlawfulness of the act or determining their will according to this understanding. Compared to earlier tests of criminal responsibility, this test would introduce a new requirement that the youth was not only morally mature but also mature of will. The priority of education was also made clear in that even when a young person was found criminally responsible, the judge was required to examine whether educational measures were necessary to accustom the young person to a law-abiding lifestyle. If such measures were regarded as sufficient, the judge was to refrain from punishing the offender. Furthermore, even when punishment was to be applied, article 136 laid down the basic principle that it was to be designed to foster the young person's education and required that the young were detained separately from adults.[193]

This draft also did not reach the statute book, which meant that young offenders were still being dealt with under the provisions of the Imperial Criminal Code of 1871. This was generally regarded as an unfortunate situation because according to that Code, the only option for dealing with those between 12 and 18 who were found criminally responsible was milder punishment. There was no legal provision for educational measures. Again, the high crime rate was used to support opposing

---

provided for a shift from the question of whether the child was developed enough to form a guilty mind to that of whether educational measures could be a sufficient response. This was the approach recommended at the earlier International Criminalist Association assemblies. Despite the educational focus, this draft concerned some of the reformers because punishment was no longer to be dependent on criminal responsibility.

193 *Crofts* 2004, p. 412.

views on the relationship between educational measures and punishment. The crime rate was seen as evidence that education rather than punishment was needed to steer the young away from crime.[194] By World War I, the progressive development of a special treatment for young offenders was disrupted. Only after World War I was the idea of specific legislation successfully pursued by opting for the "dualistic" approach of welfare and justice. This dualistic approach, which distinguishes between criminal sentences on the one hand and welfare measures on the other, has remained in force until today. In 1922, the Youth Welfare Act[195] was passed, dealing with children and young persons who were in need of care and protection. It was a classic legal statute providing intervention along the lines of the *Parens Patriae* doctrine. Under this new Act, the state was able to replace parents who were considered unable or unwilling to fulfill their educational duties. The educative measures available under this Act were, for example: supervisory directives, care orders to improve the educational abilities of parents, and placement in foster families or in residential care.[196]

c) *Youth Court Act of 1923*

As was explained above, it can be said that the system of specific social control for minors in Germany goes back to the beginning of 20th century. In 1908, the first courts started developing chambers that specialized in issues concerning young delinquents.[197] Thus, one year after enacting the Youth Welfare Act dealing with young persons in need of care, the Youth Court Act, dealing with juvenile offenders who had committed a delinquent act, was enacted in 1923.[198] This Act was the first

---

194 *Crofts* 2004, pp. 412–413. The view that punishment was harmful to the young led the left-wing parties to push for the age level of absolute criminal incapacity to be raised to 16 or even 18. In fact, the Social Democratic Party passed a resolution at their 1906 party congress that the minimum age of criminal responsibility should be 16 years. This would have completely removed the young from the criminal justice system and would have allowed offenders to be provided with educational measures in the same welfare system as those who were neglected. On the other hand, there was the view that the crime rate provided evidence that there could be no relinquishing of the threat of punishment especially for hardened young criminals.

195 The Youth Welfare Act dealt solely with neglected children, dashing the reformers' hopes for a single welfare system for offenders and the neglected. As a reform of the whole criminal code was still not foreseeable at this time, the regulations concerning young offenders were also separated from those of the general criminal law, making the way for a separate Youth Court Act. *Crofts* 2004, p. 412.

196 *Wiese* 2009, p. 138.

197 The conflict between what is generally referred to as welfare and criminal law principles appears to have featured in at least some of the changes that have taken place in the latter part of the twentieth century. Criminal law, however, incorporates a number of quite independent concepts. When contrasted with welfare, criminal law can mean a greater focus on the offense rather than the offender, more focus on due process issues, proportional responses to offending, or on punishment as a justification for intervention. *Doob et al.* 2004, p. 12.

198 *Dünkel* 2005, p. 4.

significant result of rehabilitative thinking in modern German criminal legislation[199] as well as the first Act dealing only with young criminal offenders who had committed a delinquent act as defined by the German General Penal Code. As the legislators had been pressed for time, the Youth Court Act was not as revolutionary as it had been intended to be. A totally welfare-oriented model of juvenile justice did not fit the German mentality, which remained intent upon keeping the penal option to deal with young offenders. Thus, the new Act was rather a compromise: a mixed system of juvenile justice, combining elements of educative measures with legal guarantees and a procedural approach. This, in fact, reflected the characteristics of the justice model. The punishments and ranges of punishment of the German General Penal Code, however, remained applicable. By enacting the Youth Court Act, the legislator did not create a distinctive juvenile penal law. Punishable crimes were – and still are – the same ones as defined for adults in the General Penal Code. What is known as status offenses did not form an element of the Youth Court Act. Instead, the Youth Court Act consisted of a specific system of responses and sanctions applicable only to young offenders, and of some specific procedural rules for the juvenile court and its proceedings.[200]

The 1923 Youth Court Law increased the minimum age of criminal responsibility from 12 to 14. A young person was now described as a boy or girl from the age of 14 but under the age of 18. Thus, the threshold of unlimited criminal responsibility at 18 years of age remained untouched. The preceding regulation concerning acquittal of a young person under the age of 18, who lacked the capacity to understand, was modified by the new Act. Thus, pursuant to section 3 Youth Court Act, not only intellectual and mental but also moral maturity was now a precondition for being able to impose a punishment against young persons. In addition, besides the condition of capacity to understand the wrongfulness of an action, it had to be proven that young people possessed the ability to affect and control their own actions (sections 2 and 3).[201]

At the beginning of the 20th century, prison sentences began to be regarded as unsuitable to meet the educational needs of and to rehabilitate young offenders. Criminologists emphasized that incarceration has damaging effects when applied to young offenders. Thus, in 1923, for the first time in German criminal law history, judges were given the option to suspend a prison sentence and to replace it with probation. This regulation, however, only applied to juveniles. Additionally, the judges could also refrain from punishment, i.e. without imposing a sentence at all, if the case just dealt with a petty offense. This provision expressed the legislator's

---

199 *Albrecht* 2002a, p. 444.

200 *Wiese* 2009, p. 139.

201 *Wiese* 2009, p. 139.

new aim regarding juvenile justice: the reform of the offender instead of the traditional compensation of guilt through punishment (section 9(4)).[202] In this Act, the range of criminal sanctions was expanded to include so-called educational measures besides prison sentences. Sentencing rules were adjusted to the goal of education. The Youth Court Law explicitly stated that prison sentences were to be used only if educational measures were assessed to be insufficient to respond to juvenile offenders' needs. A dogma was introduced that still prevails today: a youth prison sentence represents the course of last resort to implement education within the juvenile justice system. Special procedural rules for juvenile offenders restricted the principle of legality, allowed for expediency in prosecution decisions, and prohibited public juvenile trials.[203]

In 1923, the German legislature suspended the applicability of the principle of mandatory prosecution to the prosecution of juvenile crime. As a result, it was initially in the area of juvenile crime, rather than in low-level crime, where prosecutors enjoyed the widest latitude in their case disposition practices.[204] In comparison with the general penal law for adults, the legislator of 1923 for the first time "opened the floor" for educational measures instead of punishment, particularly of imprisonment. Also opened was the possibility to abandon the otherwise strictly applied principle of obligatory prosecution (principle of legality).[205]

The Youth Court Act thus was a forerunner of the notion of giving the prosecutor a discretion whether and how to prosecute or dismiss a case because of the petty nature of the offense or educational measures taken by other institutions or persons (sections 45, 47).[206] The aim was to avoid stigmatization, social discrimination, and unnecessary criminal proceedings in light of a prevention of reoffending (section 32, Youth Court Act – the possibility of abandoning the principle of legality, which is part of the Basic Law and accordingly a strictly applied principle in adult criminal law). Summing this up, the Youth Court Act of 1923 implemented rules which were more than ever geared toward the protection and education of the young offender instead of toward punishment. Nevertheless, the Youth Court Act of 1923 did not become a welfare model or a completely education-oriented Act. Instead, it remained a part of criminal law. This situation was reflected by the new sharp partition of juvenile welfare legislation on the one hand and juvenile criminal justice on the other. Under the new legislation, young offenders were clearly separated from otherwise endangered children and young persons.[207]

---

202 *Wiese* 2009, p. 139.

203 *Albrecht* 2004, p. 447.

204 *Boyne* 2014, p. 192.

205 *Dünkel* 2005, p. 5.

206 *Dünkel* 2005, p. 5.

207 *Wiese* 2009, p. 142. As a result, many saw the new Act as an advancement, but not as the reform they had been hoping for. Rather, it was regarded as a compromise, as part of a reform

d) Youth Court Act 1943

It is crucial to mention that academics still differ when interpreting the development of the juvenile justice system during the Third Reich. Some authors believe that it developed continuously between 1933 and 1945, or that it was merely good fortune that the Nazis did not destroy the work and achievements of many years. Other authors, however, argue that the juvenile justice system and its provisions were exploited for National Socialist ideals. These different points of view result from different perspectives. Adherents of the first theory consider only the new measures and sanctions introduced under the Reich Youth Court Act, whereas those of the other theory also look at the socio-political background and the intentions behind the new provisions. In fact, the reforms of the National Socialist system were ambivalent insofar as they introduced innovative educational ideas that had already been previously discussed in the Weimar Republic, but at the same time established a totally different meaning and use of the educational principle by defining "education" as education through (not instead of) punishment. Consequently, the educational principle of the former Youth Court Act of 1923 was undermined by its racist interpretation and use. Indeed, the Nazis kept most of the provisions of the Youth Court Act of 1923, but at the same time implemented special regulations which ensured that mitigation and educational provisions did not apply to Jewish and foreign juveniles anymore.[208]

Therefore, it can be said that a partial setback occurred in 1943.[209] Under the influence of German fascism, an amendment concerning juvenile felons introduced the possibility of the transfer of juvenile offenders aged 16[210] and older to adult criminal courts with adult criminal penalties, including the death penalty.[211] The minimum age of criminal responsibility was lowered to 12 for serious crimes (section 3(2)). In accordance with the adherents who stated that the progressive development of the juvenile justice system continued during the Third Reich, it has to be emphasized that the Reich Youth Court Act introduced some of the former ideas of the Juvenile court movement, such as renaming the term "imprisonment" as "youth penalty" and

procedure not yet completed. The left-wing parties continued to call for the age level of absolute criminal incapacity to be raised to 16 or 18 because they saw the school-leaving period as the most unsuitable age conceivable for the start of criminal responsibility. This duality of a criminal justice system with an educational emphasis for young offenders separate from the welfare system for the neglected – which was introduced with the Youth Welfare Act of 1922 and the Youth Court Act of 1923 – has persisted to the present day. *Crofts* 2004, p. 413.

208 *Wiese* 2009, p. 143.

209 This law, in force for three years, was abolished immediately after World War II as part of the general cleansing of fascist ideology from the legal system. *Albrecht* 2004, p. 447.

210 Furthermore, a special decree dealing with the treatment of dangerous juvenile criminals was implemented which allowed the possibility of applying adult criminal law sanctions including death penaltyto juvenile felonies over the age of 16 years. *Albrecht* 2004, p. 447; and see *Biesanz* 1941, pp. 554-559.

211 *Albrecht* 2004, p. 446.

therewith making it independent from the range of prison penalties provided for adults; and establishing a separate and uniform set of minimum and maximum youth prison sentences.[212]

A third category of juvenile sanctions called “disciplinary measures” was provided in the Youth Court Act Amendment of 1943. Short-term detention (labeled as “arrest”) not exceeding four weeks is one of the examples of disciplinary measures. But short-term imprisonment for juveniles was abolished completely, and the minimum prison sentence was set to three months. Prison sentences in the Youth Court Law were renamed “youth imprisonment”. However, the change in names was not merely a formal act, but made youth imprisonment independent from the range of (prison) penalties provided for adults. A separate and uniform set of minimum and maximum youth prison sentences, justified by the offender orientation of juvenile criminal sanctions, was thus established. Suspended youth prison sentences were abolished.[213] Additionally, the power of the executive authority, especially of the police, increased more and more. Preventive and repressive methods became more extreme. The police were able to impose measures without even referring to the court. This procedure led to enormous injustice, and many persons – including children and juveniles – were taken to concentration camps.[214]

e) Youth Court Act of 1953

I should be noted that the general situation in Germany after the German defeat in 1945 was characterized by a breakdown of law and order. Germany was partitioned into four zones of occupation, which were coordinated by an Allied Control Council. The original division was made between the USA, the Soviet Union and the United Kingdom. *Josef Stalin* agreed to also give France a zone, but it had to be cut out from the US or British zones. The US, British, and French zones joined in 1949 to become the Federal Republic of Germany (FRG), while the Soviet zone, situated in the eastern part of Germany, became the German Democratic Republic (GDR).[215]

As a result of this political development, the juvenile justice system developed differently in the two German republics. Straight after founding the GDR, a youth justice system was established, which was quite similar to the one that had existed under the Youth Court Act of 1923. This system remained effective until 1968. After its abolition, the adult criminal law of the GDR with its regular sentences and ranges of punishment became applicable – except for some penal mitigation – for juvenile offenders (young persons between 14 and 18 years of age). The adolescents were thus treated like adults and did not receive any mitigation at all.[216]

212 *Wiese* 2009, p. 143–144.

213 *Albrecht* 2004, p. 447.

214 *Wiese* 2009, p. 145.

215 *Wiese* 2009, p. 146.

216 *Wiese* 2009, pp. 146–147.

In the FRG, the Allied Control Council abolished all National Socialist elements from juvenile justice legislation. Apart from that, the Youth Court Act initially remained in force because it was not regarded as being a typical National Socialist body of thought. On 4 August 1953, the FRG enacted a new Youth Court Act.[217] The amendments brought important changes in authority to sentence young adults as juveniles and to place juvenile offenders under probation supervision on suspension of a youth prison sentence.[218] In fact, this new Act reintroduced the suspension of imposing the youth penalty (section 27 Youth Court Act) on probation because it was acknowledged that imprisonment generally entails many adverse influences and does not advance the young offender's rehabilitation. In connection with this development, probation services and supervision were established.[219] Finally, the minimum youth prison sentence was set to six months. More important was the placement of young adults under the Youth Court Law regime.[220] Under the new Act of 1953, all young people up to the age of twenty-one were transferred to the jurisdiction of the youth court. This provision introduced a flexible system to deal with young adult offenders and allowed the judge to choose an appropriate sentence from either the Youth Court Act or the German General Penal Code after having considered the personality and maturity of the offender.[221] By enacting the Youth Court Act of 1953, the legislators finally implemented nearly all the ideas of the juvenile court movement which had been revolutionary visions only at the beginning of the 20th century.[222]

During the 1960s and 1970s, the juvenile justice system once more became the focus of criticism and discussion, followed by several proposals for reform.[223] A principled debate took place on whether to place juvenile delinquents completely under the regime of welfare laws and welfare administration, thereby practically abolishing juvenile criminal law. In particular, the Labor Welfare, an organization associated with the labor union, in 1970 proposed the Extended Juvenile Welfare Law that would have treated juvenile criminals mainly as juveniles in need of care and supervision, and subjected them entirely to the regime of juvenile welfare law.[224] The

---

217 *Wiese* 2009, p. 147.

218 *Albrecht* 2004, p. 446.

219 *Wiese* 2009, p. 14; *Würtenberger* 1954, p. 32.

220 *Albrecht* 2004, p. 447.

221 The broadening of the juvenile justice scope was based on the belief that adolescents still differ from adults psychologically, physiologically, and in their social status due to stress and conflicts arising during the transitional development period from childhood to adulthood. This period is rarely completely concluded when a young person turns 18. Instead, transitional periods have been extended and entrance into the adult world has been made more difficult. *Wiese* 2009, p. 147.

222 *Wiese* 2009, p. 148.

223 *Wiese* 2009, p. 148.

224 Raising the minimum age of criminal responsibility further, to 16, was discussed in the 1970s in West Germany in connection with a proposed reform of the Youth Welfare Act. *Schäfer* 2004a, p. 46.

Ministry of Youth, Family, and Health contributed a draft "Juvenile Law" guided by the idea of restricting punitive sanctions to those aged 16 and 17 while using welfare measures for offenders aged 14 and 15.[225]

It was during this period that Germany came closest to abolishing juvenile criminal law and replacing it with a "Youth Welfare Act" or a "Law for Youth in Conflicts". However, the reform dynamic, driven by the student movement as well as new liberal and welfare-oriented policies implemented by the Social Democratic liberal government in office since 1969, faded without leaving significant traces. One reason was strong resistance by criminal justice professionals, which made a compromise impossible.[226] As a consequence, the Youth Welfare Act of 1922 remained in force alongside the Youth Court Act and provided measures for dealing with children and young persons in need of care and protection. Those measures were similar to, or even the same as, the educative measures stipulated in the Youth Court Act, including supervisory directives, care orders, orders to improve the educational abilities of parents, and placement in a foster family or in residential care.[227]

f) Youth Court Act of 1990

In 1990, the Youth Welfare Act was replaced by a modern law based on social welfare. It established juvenile welfare boards to help young persons and their families and to offer assistance instead of being agents of intervention. The juvenile justice legislation had remained unchanged for almost 40 years, but the juvenile justice system had experienced major changes since the 1970s. This had happened without any legislative amendment and had therefore been called "reform through practice". The slogan of the 1980s had been "internal reform" of the juvenile justice system. This label signaled two considerations: On the one hand, the reform should not touch the basic structures of the Youth Court Act; and on the other hand, the term indicated that the 1980s reform of youth criminal law was planned and implemented essentially through its practice. Since 1953, the practitioners working in the field of juvenile justice had adjusted to the social and political conditions relating to young persons and their lives. They had gained experience and insight from several years of work in this specific area, from the results of criminological research projects and the increasing influence of educational and psychological findings. Accordingly, the administration of juvenile justice had changed significantly since the early 1980s and, to a considerable extent, turned away from repressive and institutional sanctions, tending and more and more toward helping and supporting non-institutional measures as well as informal sanctions.[228]

Finally, following the developments initiated by practice, a "partial reform" proceeded in 1990 when the legislator enacted the Youth Court Act (Amending Act to

225 *Albrecht* 2004, p. 447.

226 *Albrecht* 2004, p. 447.

227 *Wiese* 2009, p. 149.

228 *Wiese* 2009, pp. 149–150.

the Youth Court Act). Its purpose was to implement the educational approach of the Youth Court Act to a greater extent.[229] This amendment brought fewer changes than many had expected, despite a vocal debate on asserted needs to amend both juvenile criminal law and child welfare law.[230] The changes only included some consolidations of diversion, decarceration, and depenalization policies that had been changing juvenile court practice since the late 1970s.[231] In fact, longitudinal research had proven that as many as 70% of young offenders are one-time offenders. This suggested that prevention efforts should rely heavily on non-prosecution and diversion rather than on youth penalties and other punitive orders. An implementation of diversion and community-based orders, such as community services, social training courses, restitution, conflict mediation, and reconciliation was therefore aspired.[232] Therefore, it can be said that the 1990 Youth Court Amendment Act extended a paradigm shift away from prison and toward community-based sanctions and social-work-based measures that had been under way since the end of the 19th century.

In the new Act of 1990, the last symbol of purely rehabilitative detention – the indeterminate youth prison sentence – was abolished; in practice, however, it has never played a numerically significant role. Decarceration was expanded by introducing social training courses, supervision and support orders, and victim-offender mediation (and restitution) in place of short-term confinement, as well as by expanding the scope of existing alternatives such as community services. Finally, pre-trial detention was restricted, allowing detention prior to trial for juveniles aged under 16 only if foster care is not available. In order to implement this policy, the juvenile court aide was directed to provide a report containing mandatory information on the need for detention and the unavailability of alternative sanctions.[233]

Furthermore, in spite of considerable criticism concerning youth detention and the youth penalty imposed because of "harmful inclinations" (section 17 Youth Court Act), the first Youth Court Act reformed these topics very cautiously. Indeed, the debate concerning these topics was taken into consideration as the German Bundestag, parallel to the enactment of the first Youth Court Act, adopted a resolution in which the Federal Government was invoked to present a draft of a second Amendment Act of the Youth Court Act no later than 1 October 1992.[234] In 1992, a demand was voiced at the youth court convention for youth penalty – the strongest form of intervention, resulting in the offender being sent to a youth prison – to

229 *Wiese* 2009, pp. 149–150.

230 These debates were triggered by media coverage of seemingly uncontrollable child career criminals, juvenile immigrant offenders, and youth violence (particularly hate violence). *Albrecht* 2004, p. 446.

231 *Albrecht* 2004, p. 448.

232 *Wiese* 2009, p. 150.

233 *Albrecht* 2004, p. 448.

234 *Wiese* 2009, p. 151.

be applicable only to young people aged 16 or older. Such efforts have been unsuccessful, however, and the minimum age of criminal responsibility in Germany has thus remained unchanged for almost half a century. This has not prevented some politicians from demanding that it should be reduced to 12 years, a proposal clearly rejected by social scientists and those responsible for formulating and implementing youth policies. Approximately 14 years is now the European average age of criminal responsibility.[235]

g) *Ratification of the United Nations Convention on the Rights of the Child in 1992*

The United Nations Convention on the Rights of the Child was signed by the Federal Republic of Germany on 26 January 1990 and came into force in Germany on 5 April 1992 after the instrument of ratification had been deposited. However, at that time, Germany made interpretative statements which show that Germany views the Convention as a welcome development in international law that will hopefully improve the situation of children worldwide, and that Germany will play its part, in keeping with article 3 para. 2 of the Convention, by drafting legislation to live up to the spirit of the Convention and to ensure the well-being of the child. At that time, a few reservations had also been made but were withdrawn in 2010. It is worth mentioning that more than 90 nationally active organizations have merged in the National Coalition for the Implementation of the United Nations Convention on the Rights of the Child in Germany. For instance, in October 2000, the National Coalition organized the third German Day of Rights of the Child, with the main focus on the rights of children and young people in case of deprivation of liberty. The aim was to focus on the statutory regulations and everyday practice in Germany against the background of the United Nations Convention on the Rights of the Child – in particular articles 37 and 40.[236]

h) *Ratification of the European Convention on the Exercise of Children's Rights in 2002*

The European Union's (EU) approach toward children's rights protection has evolved throughout the years, influenced by the pace of European integration, the development of the concept of European citizenship, demographic changes, and European Union enlargement. Until the 1990s, the issue of children's rights was less relevant for the European Union agenda, and the European Union had a rather limited competence in the field of fundamental rights. For the past decade, children's rights have gradually taken on increased importance thanks to new competences acquired by the EU in this field. The EU's approach toward the protection

---

235 *Schäfer* 2004a, p. 46.

236 *Holthusen* 2004, p. 133.

of children's rights[237] has evolved through the years from an economic integration approach toward the development of the concept of a shared region of freedom, security, and justice.[238]

The European Convention on the Exercise of Children's Rights, which was ratified by Germany, is a major instrument containing provisions aimed at protecting the best interests of children and promoting their rights, in particular in family proceedings before judicial authorities. It provides a number of measures so as to allow children to exercise their procedural rights.[239] The Convention aims at protecting the best interests of children. It has set up a Standing Committee which shall keep under review problems relating to this Convention. The Convention provides for measures which aim at promoting the rights of children before judicial authorities. These, or a person appointed to act before a judicial authority on behalf of a child, have a number of duties designed to facilitate children's exercise of their rights. Children should be allowed to do so (for example the right to be informed and the right to express their views) either themselves or through other persons or bodies. Each party is required to specify at least three categories of family proceedings which this Convention is to be applied to. This EU legal instrument will also facilitate the implementation by parties of the United Nations Convention on the Rights of the Child.[240]

i) Youth Court Act of 2007

According to article 74 para. 1 no. 1 of German Basic Law, the Constitutional Court acknowledged the legislature to provide adequate control of judicial protection by providing the second Youth Court Act in the determined deadline in 2007. On 13/12/2007, the federal legislator finally passed the second Youth Court Act, which

237 The European Union has a long-standing tradition of promoting human rights. In 1992, the "Maastricht Treaty" introduced the obligation for the European Union to respect fundamental rights in its policies and actions.

238 Furthermore, the Convention on the Protection of Children against Sexual Exploitation and Sexual Abuse, the European Convention on the Repatriation of Minors, as well as various conventions regarding family life, in particular the adoption and custody of children, have been adopted under the Council of Europe's leadership to promote and protect children's well-being. Finally, the Guidelines of the Committee of Ministers of the Council of Europe on child-friendly justice seeks to enhance children's access to and treatment in justice. Furthermore, to the Council of Europe Third Summit of Heads of State, the program "Building a Europe for and with children" has been established so as to ensure the promotion of children's rights and their protection from violence. The program has set a strategy aimed at promoting the mainstreaming of children's rights and focusing on specific objectives, including children's access to justice, the eradication of all forms of violence against them, the participation of children and their influence in society, as well as a special focus on vulnerable children, such as children with disabilities or without parental care. The Convention for the Protection of Human Rights and Fundamental Freedoms, most commonly known as the European Convention on Human Rights (ECHR), is the Council of Europe's cornerstone fundamental rights instrument. EU Framework of Law for Children's Rights 2012, pp. 7–13.

239 EU Framework of Law for Children's Rights 2012, pp. 7–13.

240 See www.coe.int/en/web/conventions/full–list/–/conventions/treaty/160.

entered into force on 01/01/2008.[241] In accordance with this Act, for the first time the education principle was emphasized. Paragraph 1 of article 2 states: the application of criminal law relating to young people is above all to counter renewed criminal offenses on the part of a youth or young adult: In order to achieve this goal, the legal consequences, and with respect for the parental right of upbringing also the procedure, shall be orientated primarily in line with the educational concept. Therefore, in accordance with this statutory basis, the education principle as an objective definition should apply in the procedural and substantive juvenile criminal law and in line with the criminological, educational, youth psychological, and other professional findings. The explanatory paper emphasizes that the application of juvenile sanctions is to be based on empirical evidence and the principles of "what works", which is in line with the Recommendation of the Council of Europe (2003) 20 on "New Ways of Dealing with Juvenile Delinquency and the Role of Juvenile Justice".[242] Another significant change can be seen in the sections about the enforcement of juvenile detention and custodial sanctions.[243] In this regard, article 91 states: (1) a convicted person who has reached the age of 18 and who is not suitable for execution of the youth penalty, the youth penalty may be executed in accordance with the provisions relating to the execution of sentences applicable to adults, instead of in accordance with the provisions for youth custody. If the convict has reached the age of 24, youth custody should be executed in accordance with the provisions relating to the execution of sentences applicable to adults; (2) the enforcement officer shall decide on exceptions from the youth penalty.[244]

j) Federal States' Youth Prison Acts from 2008 onward

The Prison Act was passed by the federal legislature in 1976 and came into force in 1977. The Federal Prison Act fixed reintegration as the goal to be pursued by the prison administration. Section 202 states: (1) the provisions on execution of youth imprisonment shall apply to the execution of sentences of imprisonment imposed on juveniles and young adults in accordance with the Criminal Code of the German Democratic Republic, and the provisions on the execution of youth detention shall apply to the execution of youth custody; (2) the provisions of the Prison Act on the execution of imprisonment shall otherwise apply to the execution of a sentence of imprisonment that has become final and binding in accordance with the Criminal Code of the German Democratic Republic and to detention.[245]

---

241 *Laubenthal et al.* 2015, p. 26.

242 *Dünkel* 2011, p. 551.

243 The Federal Government (consisting of Christian Democrats and Social Democrats) decided in July 2007 to go one step further and to introduce post-sentence preventive detention also for young offenders sentenced by Youth Courts. The draft law did not even require new facts, because the dangerousness of the offender is established by the sentencing court in the first place. The federal legislature was supposed to vote on this proposal in the course of 2008. *Boetticher et al.* 2004, p. 367.

244 *Laubenthal et al.* 2015, p. 26.

245 *Boetticher et al.* 2004, p. 8.

The Federal Prison Act did not apply to juvenile correctional facilities. Attempts to design special legislation for youth prisons have failed in the past. As a consequence, these prisons were run on a rather meager legal basis provided by two norms of the Youth Court Act. This situation was seen by many legal experts as violating the Federal Constitution. In May 2006, the Federal Constitutional Court ruled that an adequate legal basis for running youth prisons had to be created until the end of 2007.[246] The legal situation for young prisoners changed at the beginning of 2008. Before, only a few general legal provisions existed in the Youth Court Act and in the Prison Act for adult prisoners. There had not been a differentiated legal framework covering the legal rights and duties of young prisoners. The Federal Constitutional Court outlawed this missing primary legislation as being unconstitutional, as in Germany any restriction of fundamental human rights has to be based on regulations in law. Administrative rules are deemed an insufficient basis. The Federal Constitutional Court obliged the legislators of the Federal States to pass primary legislation before the end of 2007. In September 2006, a general reform of the legislative competences came into force, transferring them for prison legislation to the Federal States. The new State laws in the Federal States vary to some extent and express different political orientations on what is to be seen as being the most promising concepts of rehabilitation. Nevertheless, there is a strong consensus that the organization of youth prisons (even more than in adult prisons) must be oriented toward rehabilitation and education.[247]

Furthermore, the prevailing opinion is that youth prisoners shall be accommodated in small living groups and individual cells during the night. All youth prisons should also provide a variety of school and vocational training programs, special (social) therapeutic units, and a system of progressive preparation for release (including leaves of absence, early release schemes and continuous care and aftercare). Although the competence of youth prison legislation has been transferred to the Federal States, legislation concerning prisoners' complaints, rights, and procedures are still regulated by Federal Law. The reform law of December 2008 brought major improvements, guaranteeing juvenile and young adult inmates an oral hearing as well as regular legal advice when complaining to the court.[248]

k) Youth Court Act of 2012

It is notable that in 2012, the German parliament introduced the so-called "warning-shot detention". This implied a toughening up of the juvenile justice system, since detention (of up to four weeks) could now be imposed in addition to a suspended juvenile imprisonment sentence or to a conditional sentence/probation. The aim is to use short but intensive educative supervision in order to make it less likely that the young person will reoffend during their suspended juvenile imprisonment sentence; it should show young offenders what they can expect if they reoffend by subjecting

246 *Boetticher et al.* 2004, p. 8.

247 *Dünkel* 2011, pp. 596–597.

248 *Dünkel* 2011, pp. 596–597.

them to a short period of detention. Furthermore, this option should prevent the young convict from misinterpreting a suspended imprisonment sentence or a conditional sentence/probation as a "neutral" response.[249] However, the legal prerequisites are so narrowly formulated that almost no cases can be found with this combination of sentences. Another, more punitive approach can be seen in increasing the maximum penalty for 18- to 20-year-old young adults in very serious murder cases from 10 to 15 years. The recent reform history demonstrates that the German legislature is maintaining a generally moderate approach to juvenile delinquency, handing out more punitive sanctions only at the extreme "edges" of the sanctions system for seriously violent offenders.[250]

1.3 Historical Study of Criminal Law Governing Juvenile Delinquency in the United Nations

After World War I – and its negative effects throughout the world –, it was realized that people generally and children specifically should be protected at the international level. Therefore, with the establishment of the League of Nations at the end of World War I, the "Declaration of the Rights of the Child" was adopted and ratified by this intergovernmental organization in 1924. This declaration, known as "Geneva Declaration", is considered to be the first international human rights document in history to specifically address children's rights and needs on the one hand, and the responsibility which should be taken toward children by adults on the other. Because of its numerous shortcomings, the League of Nations was dissolved in 1946. However, World War II also led to a reinvention of the League of Nations, insofar as the United Nations (UN) was built on the foundations of the earlier organization and formally came into existence on 24 October 1945, when the Charter of the United Nations was ratified by a majority of countries. This unique organization of independent countries was formed with the main objectives of world peace and social progress.[251] The philosophy of human rights had begun with the Age of Enlightenment and emerged stronger then ever after World War II; to this reason, numerous human rights documents were adopted by the United Nations.[252] The notion of children's rights has wrought a historical sea change in how children are perceived and treated as members of and participants in societies.[253] Therefore, the historical evolution of children and juvenile criminal law in the United Nations' instruments will be discussed in this section in detail.

---

249 *Persson* 2017, p. 126

250 *Dünkel* 2011, p. 5.

251 For further reading on international documents to be discussed in this section, see *Buereh* 1998, pp. 10–62; see also www.ohchr.org/; www.humanium.org/; www.encyclopedia.com/.

252 *Buereh* 1998, pp. 10–62.

253 *Wall* 2008, p. 523. For further reading about children's rights, see *Malekian et al.* 2012; *Abachi* 2017; and *Malekian et al.* 2013.

1.3.1 Validity of the United Nations instruments concerning juvenile delinquents

At the international level, the United Nations has developed a range of rules and regulations about how to deal with children and adolescents who are perpetrators of crime. In fact, since 1959 the United Nations has issued many documents in the form of declarations, resolutions, and treaties about the standards and norms on crime prevention and criminal justice in general and on juvenile justice in particular. The United Nations differentiated Criminal Policy toward children and juveniles can be considered as a desirable criminal pattern for the treatment of juvenile delinquents. Obviously, these documents do not have the same legal validity as national legislation, but, as international references of criminal law of juveniles for the Member States, they can at least play an essential role in harmonizing policies throughout the world. Therefore, a scientific overview of these instruments seems to be necessary for legislative, judicial, and executive uses at the national level. In general, these United Nations documents and instruments can be divided into two main categories: binding and nonbinding instruments which will be discussed in this part.

1.3.1.1 Binding instruments

The binding instruments and documents are categorized in the hard law instruments which are legally binding at the international level. In contrast to soft law, hard or firm law gives State members actual binding responsibilities as well as rights.[254] In international law, hard law includes self-executing international agreements such as treaties, conventions, covenants; Security Council resolutions; as well as customary international law.[255] Regarding children in conflict with the law and juvenile delinquents, the "Convention on the Rights of the Child" and the "Human Rights Covenant" are the instruments which are legally binding for States that are parties to them. This means that the State parties are obliged under international human rights law to respect, protect, and fulfill their provisions as well as to report on the ways in which national legislation, policy, and practice reflect this. Each binding instrument has established a committee of experts to monitor the implementation of the provisions by the States parties.[256] Therefore, for instance, when countries ratify a convention, they are obligated to change their national laws in accordance with the contents and articles of the binding instrument which has been signed. They should also report to the respective committee about the actions which have been taken at the national level. It should be noted that the mechanisms for ratification and implementation of these kinds of documents differ at the national level. Some states are dualist in their

---

254 *Hanjani* 1385/2006, p. 73; *Bigdeli* 1375/1996, p. 32; *Habibzade et al.* 1391/2012, pp. 2–20.

255 *Pollack* 2010, pp. 707–712.

256 Protecting Children's Rights in Criminal Justice Systems 2013, p. 12.

approach to international law, meaning that treaties must be transformed into domestic legislation (e.g. by way of an act of the domestic legislature) before their provisions will be considered part of the domestic legal system. The approach of other states is monistic: international treaties automatically become part of domestic law by the time a state accedes or ratifies the instrument.[257] It is necessary to emphasize that some international institutions and issue areas approach the theoretical ideal of hard legalization, but most of international law is soft in distinctive ways.[258]

1.3.1.2 Nonbinding instruments

In addition to the binding documents, other instruments regularly approved by the United Nations and guiding the Member States are merely advisory and not legally binding. In international law, these instruments and documents are categorized in the subjects of soft law that refers to quasi-legal instruments which do not have any legally binding character. Although the concept of soft law has existed for many years, scholars have not reached consensus on why states should use soft law or even whether "soft law" is a coherent analytic category. In part, this confusion reflects a deep diversity in both the types of international agreements and the strategic situations that produce them.[259]

However, what is very important to emphasize here is that soft law can contribute to creating international norms and standards. Most resolutions and declarations approved by the United Nations General Assembly are nonbinding, even if the approval is unanimous. The General Assembly is the body where all Member States of the United Nations come together and discuss international issues. Therefore, the General Assembly resolutions could be considered a reflection of global public opinion; to this reason, these instruments are very important. The countries that vote in favor of a resolution have only an ethical responsibility in obeying that resolution. However, this highlights the key issues in the interaction between the General Assembly resolutions and international customary law.[260]

In addition, some elements such as statements, recommendations, declarations, standard minimum rules, basic principles, general comments, proclamations, guidelines, codes of conduct, codes of ethics, and so on are considered to be soft law. However, it is worthy to say that some soft law instruments, such as declarations, could be termed "basic soft law". This is because they are usually considered to be the starting point for drafting hard law instruments.[261] Human rights standards that are concerned with criminal justice law and policy reform are also formed and enshrined in

---

257 *Steinerte et al.* 2009, p. 12.

258 *Abbott et al.* 2000, p. 420.

259 *Guzman* 2010, p. 171.

260 *Flowers* 2016; http://hrlibrary.umn.edu/edumat/hreduseries/hereandnow/Part-1/short-history.htm [20/05/2021].

261 *Flowers* 2016; http://hrlibrary.umn.edu/edumat/hreduseries/hereandnow/Part-1/short-history.htm [20/05/2021]; see also www.humanium.org/ [20/05/2021].

United Nations soft law instruments. These complement the hard law instruments, have significant moral force and provide useful as well as practical guidance. As has been mentioned above, the value of the soft law instruments rests on their recognition and acceptance by a large number of States.[262]

1.3.2 Categorization of the United Nations instruments concerning juvenile delinquents

In this part, the United Nations instruments concerning juvenile delinquents will be categorized and discussed on three levels: basic soft law, hard law, and soft law.

1.3.2.1 Basic soft law

1.3.2.1.1 Universal Declaration of Human Rights of 1948

After World War II, the Universal Declaration of Human Rights (UDHR) was adopted early in the life of the United Nations, by a General Assembly Resolution in 1948. With the adoption of the Universal Declaration of Human Rights, the United Nations as an international community affirmed the values to struggle against fascism and Nazism. The Universal Declaration of Human Rights was a milestone instrument in the history of human rights, as it was drafted by representatives with different legal and cultural backgrounds from all regions of the world. However, as a General Assembly resolution, this document is not a legally binding instrument, but it has achieved the status of customary international law because people all over the world regard it as a common standard of achievement for all people and all nations. Its principles have been incorporated into the constitutions of most United Nations Member States, which shows to what extent the influence of the Universal Declaration of Human Rights has been substantial. It contains 30 articles relating to those rights and freedoms which are regarded as being every person's birthright.[263]

According to the Universal Declaration of Human Rights, all human beings are born free and equal in dignity. Although children are seldom mentioned in this text (articles 25(2) and 26), it is nevertheless a significant document, and its impact on all human beings, including children, is what makes the Declaration so important. As a matter of fact, children's rights are based on human rights. As has been mentioned above,[264] the document sets out fundamental rights and has also influenced and been the basis for the adoption of numerous other human rights instruments, standards, and guidelines.[265]

---

262 Protecting Children's Rights in Criminal Justice Systems 2013, p. 12.

263 *Steinerte et al.* 2009, p. 11.

264 Steinerte et al. 2009, p. 11; Flowers 2016, http://hrlibrary.umn.edu/edumat/hreduseries/hereandnow/Part-1/short-history.htm [20/05/2021]; see also www.humanium.org/ [20/05/2021].

265 Protecting Children's Rights in Criminal Justice Systems 2013, p. 12.

1.3.2.1.2 Universal Declaration on the Rights of the Child 1959

In fact, the advancement of human rights by the Universal Declaration of Human Rights of 1948 revealed the need for drafting a specific corresponding universal instrument for children. Therefore, for the purpose of expanding children's human rights, in particular, the Declaration of the Rights of the Child was drafted and adopted unanimously by all Member States at the United Nations General Assembly in 1959. This basic soft law instrument is considered the first major international consensus on the fundamental and essential principles of children's rights. Although the principles of the Geneva Declaration were influential, they cannot be described exactly as binding. It consists of a preamble and ten principles. The principles of this declaration have played a significant part in the promotion of the rights of children globally as well as in shaping various forms of international cooperation in this sphere. According to the 1924 declaration, the child was seen as an object of international law.By 1959, however, children are beginning to emerge no longer as passive recipients but as subjects of international recognition as being able to enjoy the benefits of specific rights and freedoms.[266] According to the Declaration of the Rights of the Child, the child is recognized, universally, as a human being who must be able to develop physically, mentally, socially, morally, and spiritually, with freedom and dignity. However, there is no age determination of children in this instrument. Neither the 1924 Geneva Declaration nor the 1959 Declaration of the Rights of the Child define at what age childhood starts and when it ends.[267]

1.3.2.2 Hard law

1.3.2.2.1 Covenants

In order to establish the general enforcement mechanisms of the Universal Declaration of Human Rights, the United Nations Commission on Human Rights proceeded to draft the International Covenant on Economic, Social and Cultural Rights (ICESC) as well as the International Covenant on Civil and Political Rights (ICCPR) and its optional Protocol in 1966. Of the two covenants, the former was the first to enter into force.

a) International Covenant on Economic, Social and Cultural Rights of 1966

The International Covenant on Economic, Social and Cultural Rights was adopted by the United Nations General Assembly in 1966, and entered into force in 1976. It applies to all men and women and therefore should be applied to children, too. The Covenant specifically refers to children in articles 10 and 12. In article 100, the States

266 *Buereh* 1998, p. 11.

267 *Flowers* 2016, p. 6.

recognize the family as the natural and fundamental group unit of society and therefore accords the widest possible protection and assistance to the family. This article contains a broad ambit of protection. Article 12 is concerned with the right of everyone to the highest attainable standard of physical and mental health and incorporates a specific provision concerning children. The Covenant establishes a reporting procedure on the measures the States parties have adopted and the progress made in achieving the observance of the rights contained in the Covenant (article 16). A State party is under a duty to the maximum of its available resources to implement progressively the rights enshrined in the International Covenant on Economic, Social and Cultural Rights. This is in contrast to the manner of implementation of the rights guaranteed in the International Covenant on Civil and Political Rights, which places states under a duty to implement the rights immediately, regardless of the States parties' resources. One of the principal reasons for this difference in implementation obligations is the assumption that measures implementing economic, social, and cultural rights will have far greater resource implications than measures implementing civil and political rights. The Covenant elevates the status of children by making them eligible to special measures of protection and assistance.[268]

b) International Covenant on Civil and Political Rights of 1966

The International Covenant on Civil and Political Rights was also adopted by the United Nations General Assembly in 1966, and entered into force in 1976. This important instrument contains general provisions which children are entitled to benefit from, as well as certain specific provisions on safeguards for children in the administration of justice and as members of a family unit. Therefore, it should be noted that, for the first time, the specific provisions regulating the administration of juvenile justice were enshrined in the international Covenant on Civil and Political Rights 1966. This aspect of child protection in the criminal justice system had been ignored in both the 1924 and the 1959 declarations[269] on the rights of the child. Regarding juvenile delinquents, some specific provisions regulating the administration of juvenile justice can be reviewed. Article 14(1) states: any judgment rendered in a criminal case or in a suit at law shall be made public except where the interest of juvenile persons requires otherwise or the proceedings concern matrimonial disputes or the guardianship of children. According to article 14(4), criminal proceedings should take account of age and the desirability of promoting their rehabilitation; the penal system should segregate juvenile offenders from adults and accord them a treatment appropriate to their age and legal status (article 10(3)). Article 24 of the International

---

268 *Buereh* 1998, p. 20; United Nations Office of the High Commissioner for Human Rights and International Bar Association 2003, p. 39.

269 Italy attempted to include a specific provision on the administration of juvenile justice in the 1959 Declaration on the Rights of the Child, but this had failed.

Covenant on Civil and Political Rights is specifically devoted to children. It stipulates that every child shall have – without any discrimination as to race, colour, sex, language, religion, national or social origin, property or birth – the right to such measures of protection as are required by their status as a minor, on the part of their family, society, and the State.[270]

1.3.2.2.2 Conventions

a) Convention on the Rights of the Child of 1989

One of the most important instruments categorized in the hard law instruments is the Convention on the Rights of the Child.[271] It was adopted in the General Assembly of the United Nations, through its Resolution 44/25 of 20 November 1989. The Convention was opened for signature in New York, by acclamation (without voting), on 26 January 1990 and, in accordance with article 49 thereof, entered into force on 2 September 1990, the 30th day after the date of depositing the 20th instrument of ratification or accession with the Secretary-General of the United Nations.[272] The year 1989 is symbolic because it marks the 30th anniversary of the Children Rights Declaration and the 10th anniversary of the International Year of Children. The text of the Convention of 1989 goes further than the Declaration of 1959: it includes new rights and recognizes children as subjects to rights. The Convention on the Rights of the Child affirms and describes the fundamental human rights of all children below the age of 18, and the governments that have ratified it have legally agreed to fulfill its provisions.[273] It is also the most comprehensive and well-established international standard for children's rights and provides the framework for the actions of the United Nations Children's Fund (UNICEF), which is the United Nations children's agency. The Convention on the Rights of the Child is the most widely ratified Convention, having been ratified by 190 out of 192 Member States – Only the United States of America and Somalia have signed but not ratified it. A wide range of children's rights are set out in the Convention, which obliges the ratifying States to protect them. States must report on their implementation of the rights at regular intervals to the Committee on the Rights of the Child, which subsequently delivers "Concluding Observations" on the situation in the said country. It should be noted, however, that the Convention on the Rights of the Child

---

270 *Bureh* 1998, p. 170; *Flowers* 2016, p. 13.

271 The Convention has three optional protocols entitled: the Optional Protocol to the Convention on the Rights of the Child on the involvement of children in armed conflict, 2002; the Optional Protocol on the Convention on the Rights of the Child on the sale of children, child prostitution and child pornography, 2002; the Optional Protocol to the Convention on the Rights of the Child on a communications procedure, 2014.

272 *Detrick* 1995, pp. 446–447.

273 For further reading, see *Cohen et al.* 1991, pp. 60–65.

holds no sanctioning powers and the Committee has to rely on persuasion and admonishment rather than enforcement.[274] Also, regarding the realm of the Convention on the Rights of the Child, article 41 provides that nothing in the present Convention shall affect any provisions which are more conducive to the realization of the rights of the child and may be contained in (a) the law of a State party or (b) international law in force for that State.

Articles 37 and 40 specifically address the juvenile justice system. According to these articles, no child shall be subjected to torture or other cruel, inhuman, or degrading treatment or punishment. No child shall be deprived of their liberty unlawfully or arbitrarily. The arrest, detention, or imprisonment of a child shall be in conformity with the law and shall be used only as a measure of last resort and for the shortest appropriate period of time. Every child deprived of liberty shall be treated with humanity and respect for the inherent dignity of the human person, as well as in a manner which takes into account the needs of persons of their age. In particular, every child deprived of liberty shall be separated from adults unless it is considered in the child's best interest not to do so. Every child deprived of their liberty shall have the right to prompt access to legal and other appropriate assistance, as well as the right to challenge the legality of the deprivation of their liberty before a court or other competent, independent, and impartial authority, and to a prompt decision on any such action (article 37a, b, c, d).[275]

States parties recognize the right of every child alleged as, accused of, or recognized as having infringed the penal law to be treated in a manner consistent with the promotion of the child's sense of dignity and worth, which reinforces their respect for the human rights and fundamental freedoms of others, and which takes into account the child's age and the desirability of promoting their reintegration as well as the child's assuming a constructive role in society (article 40 (1)). States parties shall seek to promote the establishment of laws, procedures, authorities, and institutions specifically applicable to children alleged as, accused of, or recognized as having infringed the penal law, and, in particular, they shall establish a minimum age below which children shall be presumed not to have the capacity to infringe the penal law. Whenever appropriate and desirable, measures for dealing with such children are to be implemented without resorting to judicial proceedings, providing that human rights and legal safeguards are fully respected (article 40(3) a, b). The Convention on the Rights of the Child is binding under international law and carries a clear obligation for nation states to ensure its full implementation.[276]

---

274 Protecting Children's Rights in Criminal Justice Systems 2013, pp. 11–12.

275 *Woll* 2000, pp. 71–72.

276 At the most basic level, the state party reporting process to the Committee on the Rights of the Child requires governments, at least once every five years, to collect some information on the advancement of children's rights in their country. Used in an optimal manner, however, the

b) Other related conventions

Children's rights are also generally protected under the other United Nations human rights instruments, which include the International Labour Organization conventions; the Convention against Torture and Other Cruel, Inhuman or Degrading Treatment or Punishment of 1984 and its Optional Protocol of 2002; the International Convention on the Elimination of all Forms of Racial Discrimination of 1969; the Convention on the Elimination of all Forms of Discrimination against Women of 1981; the International Convention on the Protection of the Rights of all Migrant Workers and Members of their Families of 2003; and the Convention on the Rights of Persons with Disabilities and its Optional Protocol of 2006.[277]

1.3.2.3 Soft law

Resolutions, general comments, and other instruments regarding soft law will be discussed in this section.

1.3.2.3.1 Resolutions

a) United Nations Standard Minimum Rules for the Administration of Juvenile Justice of 1985

It should be noted that with the adoption of the International Covenant on Civil and Political Rights in 1966, an increasing number of States began to develop separate juvenile justice system, so on the international level, the need became apparent for a complete framework they could utilize for guidance in establishing and operating their own national juvenile justice systems. In 1980, the sixth United Nations Congress on the Prevention of Crime and Treatment of Offenders called for the preparation of minimum rules regulating the administration of juvenile justice. In 1985, appropriately on Human Rights Day, the General Assembly adopted the United Nations Standard Minimum Rules for the Administration of Juvenile Justice, known as the Beijing Rules. As intended, they provide a framework within which a national juvenile justice system should operate, as well as for States a model of a fair and humane response to juveniles who may find themselves in conflict with the law. The Beijing Rules, which are divided to six parts, cover the whole range of juvenile justice processes: general principles, investigation and prosecution, adjudication and deposition, non-institutional treatment, research and planning, policy formulation, and

---

reporting process can create an opportunity for governments, NGOs (non-governmental organizations), children and young adults, as well as other members of civil society (including the media) to engage in a spirited discussion about priorities, successes and challenges, and to create agendas for change. Indeed, the reporting process could be used as a catalyst for domestic review, debate, and policy change. *Woll* 2000, pp. 71–72.

277 For further reading, see www.ohchr.org/ [20/05/2021].

evaluation.[278] There are several fundamental principles in this nonbinding instrument according to the best interest of child,[279] including: fair humane treatment based on rehabilition objectives, use of diversion; taking the children's opinion into consideration; decarceration, detention as a last resort and for a minimum necessary period; deprivation of liberty only for extremely serious cases; abolition of capital and corporal punishment; institutionalization as last resort; professionalization.

b) United Nations Rules for the Protection of Juveniles Deprived of their Liberty of 1990

The United Nations Rules for the Protection of Juveniles Deprived of their Liberty, known as the Havana Rules, were adopted in 1990. They are a set of principles that apply to every juvenile deprived of their liberty in any facility or institution and promote their development and well-being. They also operate to ensure that juvenile detainees, and offenders, are given fair treatment and receive consideration for their age. The standards for material conditions of detention, the protection of juveniles whilst in detention, as well as educational, vocational, and work programmers, are the subjects discussed in this resolution.[280] The United Nations Rules for the Protection of Children Deprived of their Liberty define the deprivation of liberty as any form of detention or imprisonment or of placement of a person in a public or private custodial setting from which this person is not permitted to leave at will, by order of any judicial, administrative, or other public authority. The Havana Rules cover different establishments corresponding to the closed character such as arrest cells, detention centers, prisons, and other closed institutions such as educational institutions, rehabilitation centers, psychiatric institutions, establishments for the disabled, reception centers for illegal immigrants, etc.[281]

c) United Nations Guidelines for the Prevention of Juvenile Delinquency of 1990

The United Nations Guidelines for the Prevention of Juvenile Delinquency, known as the Riyadh Rules, were adopted in 1990. They stress the need for and importance of progressive delinquency prevention policies. These Guidelines have three main characteristics: a) they are very comprehensive; b) they promote a proactive approach to prevention; c) they consider children to be fully-fledged participants in society.[282] In other words, social and situational crime prevention strategies and policies are the most important subjects in this resolution. However, it should be noted

---

278 *Buereh* 1998, p. 173.

279 For further reading about the best interest of the child, see *Zermaten* 2010, pp. 483–499.

280 Protecting Children's Rights in Criminal Justice Systems 2013, p. 12.

281 *Cappelaere et al.* 2005, pp. 29–31; *Tobin* 2002, pp. 211–213.

282 *Verhellen et al.* 1996, p. 59.

that prevention polices after committing crime and within juvenile justice systems are also mentioned here. Avoidance of criminalizing and penalizing a child for behavior that does not cause serious damage to their development or harm to others; avoidance of labeling a youth a deviant or delinquent as this contributes to negative patterns of behavior; and avoidance of harsh disciplinary measures, particularly corporal punishment, are the most important prevention policies to be taken by juvenile justice systems.[283]

d) Other related resolutions

Children's rights are also generally protected under other UN resolutions and non-binding rules, such as: the Standard Minimum Rules for the Treatment of Prisoners of 1955; the United Nations Standard Minimum Rules for Non-custodial Measures of 1990 (Tokyo Rules); the United Nations Basic Principles on the Use of Restorative Justice Programs in Criminal Matters of 2002; the United Nations Rules for the Treatment of Women Prisoners and Non-custodial Measures for Women Offenders of 2010 (Bangkok Rules); and the United Nations Principles and Guidelines on Access to Legal Aid in Criminal Justice Systems of 2012.

1.3.2.3.2 General comments

The implementation of the United Nations Convention on the Rights of the Child (1989) by its States parties is monitored by a body of independent experts who work in the Committee on the Rights of the Child. It regularly adopts general comments based on specific articles, provisions, and themes of the Convention to assist the States parties in fulfilling their obligations.[284] General comments and other mechanisms available to the treaty bodies' periodic reporting and the complaints systems are the two most common mechanisms for ensuring and monitoring State compliance with obligations imposed by the United Nations' human rights treaties. However, other devices have been employed by various treaty bodies, including, among others, general comments and inquiries. In essence, general comments present the views of the treaty body about the rights contained in the relevant treaty and constitute the most authoritative interpretation of the substantive rights contained in the treaty. They thus represent another important source of information for States parties in guiding their understanding of the obligations they have undertaken.[285]

a) Children's rights in juvenile justice 2007 and 2019

The Committee on the Rights of the Child issued general comment no. 10 in 2007, which was titled "Children's Rights in Juvenile Justice", providing the interpretation

283 *Verhellen et al.* 1996, p. 61.

284 *Verhellen et al.* 1996, p. 58.

285 *Steinerte et al.* 2009, p. 20. Interagency Panel on Juvenile Justice (IPJJ) (n.d.); www.ipjj.org/en/juvenile-justice/overview/ [20/05/2021].

of the Convention on the Rights of the Child provisions for children in conflict with the law. The general comment elaborates on articles 37 and 40 of the Convention on the Rights of the Child pertaining to children's rights in juvenile justice while also taking into account the general principles enshrined in articles 2, 3, 6, and 12, as well as all other relevant articles of the Convention, such as articles 4 and 39. It also highlights shortcomings in the development and implementation of juvenile justice policies by States parties and provides guidance to States in developing comprehensive juvenile justice policies that comply with the Convention and relevant international standards. The general comment explores several aspects of juvenile justice policies, such as the prevention of juvenile delinquency; the diversion from judicial proceedings; the minimum age of criminal responsibility; the guarantee to a fair trial; the prohibition of the death penalty and life imprisonment without parole; and the deprivation of liberty. It should be noted that the objectives of the general comments are to encourage States

- to implement a comprehensive policy on juvenile justice with an emphasis on the prevention of juvenile delinquency and with assistance from the Interagency Panel on Juvenile Justice (IPJJ);
- to provide guidance and recommendations to States for the content of such a policy, which aims at preventing juvenile delinquency, and to implement alternative measures from judicial proceedings;
- to promote the integration into a national and comprehensive juvenile justice policy of other international standards, in particular the United Nations Standard Minimum Rules for the Administration of Juvenile Justice (Beijing Rules), the United Nations Rules for the Protection of Juveniles Deprived of their Liberty (Havana Rules), and the United Nations Guidelines for the Prevention of Juvenile Delinquency (Riyadh Rules).[286]

In 2019, general comment no. 24 replaced general comment no. 10 (2007) on children's rights in juvenile justice. This reflects the developments that have occurred since 2007 as a result of the promulgation of international and regional standards, the Committee's jurisprudence, new knowledge about child and adolescent development, and evidence for effective practices, including those relating to restorative justice. It also reflects concerns such as the trends relating to the minimum age of criminal responsibility and the persistent use of deprivation of liberty. This general comment covers specific issues, such as those relating to children recruited and used by non-State armed groups, including those designated as terrorist groups, and children in customary, indigenous, or other non-State justice systems (general comment no. 24, para. 6). The objectives and scope of general comment no. 24 are:

---

286 Interagency Panel on Juvenile Justice (IPJJ) (n.d.); www.ipjj.org/en/juvenile-justice/overview/ [20/05/2021].

- to provide a contemporary consideration of the relevant articles and principles in the Convention on the Rights of the Child, and to guide States toward a holistic implementation of child justice systems that promote and protect children's rights;
- to reiterate the importance of prevention and early intervention, as well as of protecting children's rights at all stages of the system;
- to promote key strategies for reducing the especially harmful effects of contact with the criminal justice system, in line with increased knowledge about children's development, in particular:
    - setting an appropriate minimum age of criminal responsibility and ensuring the appropriate treatment of children on either side of that age;
    - scaling up the diversion of children away from formal justice processes and to effective programs;
    - expanding the use of non-custodial measures to ensure that detention of children is a measure of last resort;
    - ending the use of corporal punishment, capital punishment, and life sentences;
    - for the few situations where deprivation of liberty is justified as a last resort, ensuring that its application is for older children only, is strictly time-limited, and is subject to regular review;
- to promote the strengthening of systems through improved organization, capacity-building, data collection, evaluation, and research; and
- to provide guidance on new developments in the field, in particular the recruitment and use of children by non-State armed groups, including those designated as terrorist groups, and children coming into contact with customary, indigenous, and non-State justice systems.

b) Related general comments

Other general comments should also be considered in this section. They include: general comment no. 1, 2001: The Aim of Education; general comment no. 2, 2002: The Role of Independent National Human Rights Institutions in the Protection and Promotion of the Rights of the Child; general comment no. 3, 2003: HIV/AIDS and the Rights of the Child; general comment no. 4, 2003: Adolescent Health and Development in the Context of the Convention on the Rights of the Child; general comment no. 5, 2003: General Measures of Implementation of the Convention on the Rights of the Child; general comment no. 6, 2005: Treatment of Unaccompanied and Separated Children outside their Country of Origin; general comment no. 7, 2006: Implementing Child Rights in Early Childhood; general comment no. 8, 2006: The Right of the Child to Protection from Corporal Punishment and other Cruel or Degrading Forms of Punishment; general comment no. 9, 2007: The Rights of Children with Disabilities; general comment no. 10, 2007: Children's Rights in Juvenile Justice; general comment no. 11, 2009: Indigenous Children and their Rights under the Convention; general comment no. 12, 2009: The Right of the Child to Be Heard; general comment no. 13, 2011: The Right the Child to Freedom from all Forms of Violence; general comment no. 14, 2013: The Right of the Child to Have his or her Best Interests Taken as a Primary Consideration; general comment no. 15, 2013: The Right of the Child to the Enjoyment of the Highest Attainable Standard of Health; general comment no. 16, 2013: State Obligations Regarding the Impact of the Business Sector on Children's Rights; general comment no. 17, 2013: The Right of the Child to Rest, Leisure, Play, Recreational Activities, Cultural Life and the Arts; general comment no. 18, 2014: Joint General Comment of the Committee on the Elimination of Discrimination Against Women and the Committee on the Rights of the Child on Harmful Practice; general comment no. 19, 2016: Public Budgeting for the Realization of Children's Rights; general comment no. 20, 2016: Implementation of the Rights of the Child during Adolescence; general comment no. 21, 2017: Children in Street Situations; general comment no. 22, 2017: The Human Rights of Children in the Context of International Migration; general comment no. 23, 2017: The Human Rights of Children in the Context of International Migration in Countries of Origin, Transit, Destination and Return; and general Comment no. 25 (draft), 2020: Children's Rights in relation to the Digital Environment.

1.3.2.3.3 Other soft law instruments

Fact sheets, general recommendations, statements, recommendations, declarations, standard minimum rules, basic principles, proclamations, guidelines, codes of conduct, and codes of ethics are considered as other soft law instruments. These kinds of instruments are considered to be semi-legal and issued for guiding the Member States. Therefore, they are only advisory and not legally binding.

1.3.3 International obligations for applying children's rights

When speaking about obligations on the international level,[287] this means that the binding treaties should be considered and discussed. As mentioned above, the most

---

287 It should be noted that the national governments are the primary agents for the realization of children rights. It is important to assure that human rights are clearly articulated in national law. When states undertake obligations to honor particular human rights by becoming parties to international human rights agreements, they are expected to elaborate their understandings of those obligations by spelling them out in their own national law. This national law should not only spell out the definitions but also the means for its own implementation, which then becomes the major instrument through which the human rights are realized. In principle, internationally recognized rights can be realized within nations even if they are not articulated in national law. However, they are much more likely to be realized when the law regarding these rights at the national level is well-crafted. 1) the individual's rights, 2) the government's obligations that correspond to those rights, 3) the government's response mechanisms (including both identification of the agencies responsible for fulfilling the obligations and the procedures through which they are to operate), and finally 4) the accountability mechanisms which assure that the obligations are honored are the four core components of national human rights

important treaty about children's rights is the United Nations Convention on the Rights of the Child of 1989. But even many years after the Convention came into force, children's issues continue to remain at the forefront of social, political, and legal struggles in both the international and domestic arenas. In principle, the international community has an important role to play in setting out widely accepted guidelines, but once general principles are set out in international human rights agreements, the primary locus of human rights work is within the States. Therefore, after taking obligations by becoming parties to international human rights agreements, it is expected from States to elaborate the understandings of those obligations by setting out national mechanisms.[288] Despite the international efforts to recognize children's rights, the outcomes on the national level have not necessarily had the expected impact. The lack of political will, the absence of systematic engagement within the academia, and ignoring the operationalization of children's rights by States can be considered as the most important elements in this field.[289] Therefore, the progress made in terms of children's rights and subsequently in juvenile justice has been very mixed. Despite ratification of the Convention on the Rights of the Child, not all States parties have adopted the standards for establishing juvenile justice systems that are specialized, comprehensive, and separate from adult criminal justice systems, with the primary aims of rehabilitation and social reintegration. It should be noted that while some improvement has been made in recent years, the Committee on the Rights of the Child continually keeps revealing gaps in national juvenile justice legislation and implementation.[290]

After ratifying the Convention on the Rights of the Child, States parties take on obligations under international law to implement its contents. Implementation in this context addresses the process whereby States parties take action to ensure the realization of Convention rights for all children within a jurisdiction. The core obligation to be implemented is set out in article 4, which requires States parties to undertake all appropriate legislative, administrative, and other measures for the implementation of the rights recognized in the present Convention on the Rights of the Child. With regard to economic, social, and cultural rights, States parties shall undertake such measures to the maximum extent of their available resources. In general, international human rights treaties do not specify how States parties are to give effect to their obligations at the domestic level, but they do require them to take all appropriate measures. As such, it is up to the States to determine how best to implement their international treaty obligations, subject to the satisfaction of those obligations in

---

law. Implementation of the latter three parts of the law help assure realization of the rights stated in the first part. *Kent* 1987, pp. 439–440.

288 *Kent* 1987, pp. 439–440.

289 *Carvalho* 2008, pp. 885–586.

290 Interagency Panel on Juvenile Justice (IPJJ); www.ipjj.org/en/juvenile-justice/overview/ [20/05/2021].

practice. According to the United Nations Committee on Economic, Social and Cultural Rights, each State party must decide for themselves which means are the most appropriate with respect to each right under the given circumstances.[291]

In this section, the obligations of Iran and Germany will be discussed with respect to children rights and juvenile justice standards under international human rights law (binding treaties).

1.3.3.1 International obligation of Iran

Some core international human rights treaties were approved by the representatives of Iran at the United Nations. These treaties include: the International Convention on the Elimination of All Forms of Racial Discrimination of 1969 (signature 1967/ratification 1968); the International Covenant on Civil and Political Rights of 1976 (signature 1968/ratification 1975); the International Covenant on Economic, Social and Cultural Rights of 1976 (signature 1968/ratification 1975); the Convention on the Rights of the Child of 1989 (signature 1991/ratification 1994); the Optional Protocol to the Convention on the Rights of the Child on the Sale of Children, Child Prostitution and Child Pornography of 2002 (ratification 2007); the Optional Protocol to the Convention on the Rights of the Child on the Involvement of Children in Armed Conflict of 2002 (signature 2010/ratification: in process);[292] the Convention on the Rights of Persons with Disabilities of 2008 (signature 2007/ratification 2009).[293]

All the legally binding treaties above, in general, and the Convention on the Rights of the Child, in particular, should be considered for protecting juvenile delinquents and children in conflict with the law in Iran. As already mentioned, the Convention on the Rights of the Child as a specific hard law instrument was signed by the representative of the Government of the Islamic Republic of Iran on 5 September 1991 and ratified in March 1994 by the Islamic Consultative Assembly (Parliament), with general reservations. On 15 June 1994, the Islamic Republic of Iran acceded to the Convention on the Rights of the Child, and the Convention became binding from 12 August 1994.[294] It is notable that the right not to apply any provisions or articles of the convention that may be in contravention of domestic laws and Islamic standards was reserved by Iran, which constitutes an obstacle to the protection of children.[295]

---

291 *Lundy et al.* 2013, p. 18, and *Ledogar* 1993, pp. 377–391, see also *Leblanc* 1996, pp. 357–381.

292 The bill was passed to the Iranian Parliament in 2016.

293 For further reading, see http://indicators.ohchr.org/ [20/05/2021].

294 *Abachi* 2015, p. 827.

295 Prior to ratification, Iran signed the Convention on the Rights of the Child in 1991. At that time, Iran submitted a similar reservation, except that this one read contrary to the Islamic *Shari'a*, rather than the 1994 formulation, which focuses on provisions that are incompatible with Islamic laws and the internal legislation in effect. The change in the wording of Iran's reservation upon signature and ratification seems to reflect the understanding that Islamic laws in Iran are subject to a specific interpretation. In fact, according to article 4 of the Iranian

In principle, due to the broad generality of the reservation, Iran could comply with the Convention on the Rights of the Child without modifying or strengthening existing national laws and policies that affect children. Iran has committed itself to implementing the Convention's provisions within its domestic law by way of ratifying the treaty; the Convention's provisions may thus be understood as part of domestic law, as States must integrate the obligations imposed by these treaties within their national legal framework. According to the Iranian Civil Code, treaty provisions which have been, in accordance with the Constitutional Law, concluded between the Iranian Government and other governments shall have the force of law.[296]

Iran recognizes that it is bound by the Convention on the Rights of the Child as part of its domestic law, which also implies that the courts should encourage implementation of international human rights law within domestic decisions, although the blanket reservation to the Convention also implies that all rights derived from it are subject to Islamic law as enforced in Iran. Therefore, Iran's reservation would allow the State to apply the entirety of the Convention in a flexible manner and subject to internal legislation. According to article 51(2) of the Convention on the Rights of the Child, a reservation incompatible with its object and purpose shall not be permitted.[297]

This provision reflects one of the core principles of the Vienna Convention on the Law of Treaties (VCLT) that deals with reservations to treaties: when States ratify a treaty, they are allowed to formulate a reservation so as to express their differences with specific provisions. These reservations, however, must not be incompatible with the object and purpose of the treaty. If the object and purpose of the Convention is to ensure greater rights and protection to children, then Iran's blanket reservation would appear to be invalid on the basis of its generality.[298] However, it should be

---

Constitution, all civil, penal, financial, economic, administrative, cultural, military, political, and other laws and regulations must be based on Islamic criteria. Article 4 further stipulates that this provision applies absolutely and generally to all articles of the Constitution as well as to all other laws and regulations, and that the Islamist jurists of the Guardian Council shall judge compliance with this principle. The principle of conformity with "Islamic criteria" is explicitly invoked in other specific provisions of the Iranian Constitution as well. In article 20, for instance, the Constitution states that all citizens of the country, both men and women, equally enjoy the protection of the law as well as all human, political, economic, social, and cultural rights – in conformity with Islamic criteria. *Birnbaum et al.* 2014, p. 14.

296 *Birnbaum et al.* 2014, pp. 14–17.

297 *Birnbaum et al.* 2014, p. 17.

298 In its 2000 and 2005 Concluding Observations on Iran's first and second periodic reports under the Convention on the Rights of the Child, respectively, the Committee on the Rights of the Child encouraged Iran to withdraw or narrow its reservation, criticizing the "broad and imprecise nature of the State party's general reservation [which] potentially negates many of the Convention's provisions and raises concern as to its compatibility with the object and purpose of the Convention". This concern was echoed by the United Nations General Assembly, which in Resolution A/68/184 (2014) called upon Iran to effectively implement its human rights obligations, to withdraw overly general or imprecise reservations to human rights treaties or any

noted that with respect to the commitment to ensure the survival, development, protection and participation of children, Iran has decided to review and re-examine this reservation in order to amend it accordingly.[299] Furthermore, on 16 April 2012, the president's cabinet established the National Body on the Convention on the Rights of the Child (NBCRC), which is generally charged with promotion of the Convention on the Rights of the Child. The establishment of a national body to monitor implementation of the Convention is thus a positive step. However, this national body is not independent since it is overseen by the Ministry of Justice. Its duties include evaluating respect for the rights of children at the national level, reviewing draft legislation and regulations as well as government reports on child rights, monitoring reported violations of the rights of the child, and conducting public outreach related to child rights. However, the powers of the national body, as described in interviews by its leadership, are primarily advisory and geared toward coordination between government agencies.[300]

1.3.3.2 International obligation of Germany

With two exceptions, all core international human rights treaties [301] have been approved by the representative of Germany in the United Nations. These treaties include: the International Convention on the Elimination of all Forms of Racial Discrimination of 1969 (signature 1967/ratification 1969); the International Covenant on Civil and Political Rights of 1976 (signature 1968/ratification 1973); the Optional Protocol to the International Covenant on Civil and Political Rights of 1976 (signature 1966/ratification 1993); the Second Optional Protocol to the International Covenant on Civil and Political Rights, Aiming at the Abolition of the Death Penalty of 1991 (signature 1990/ratification 1992); the International Covenant on Economic, Social and Cultural Rights of 1976 (signature 1968/ratification 1973); the Convention on the Elimination of all Forms of Discrimination against Women of 1981 (signature 1980/ratification 1985); the Optional Protocol to the Convention on the Elimination of All Forms of Discrimination against Women of 2000 (signature 1999/ratification 2002); the Convention against Torture and other Cruel, Inhuman or Degrading Treatment or Punishment of 1987 (signature 1986/ratification 1990); the Optional Protocol to the Convention against Torture and other Cruel, Inhuman or

---

reservations that could be considered incompatible with the object and purpose of the treaty, and to consider acting upon the concluding observations of relevant treaty bodies. *Birnbaum et al.* 2014, pp. 14–17.

299 *Abachi* 2015, p. 827.

300 Rights of the Child in Iran, March 2015, joint alternative report by civil society organizations on the implementation of the Convention on the Rights of the Child by the Islamic Republic of Iran, Submitted within the framework of the United Nations Committee on the Rights of the Child's review of the 3rd and 4th periodic reports of the Islamic Republic of Iran.

301 The international Convention on the Protection of the Rights of all Migrant Workers and their Families, 2003; and also the Optional Protocol to the the International Covenant on Economic, Social and Cultural Right, 2013 have not been ratified.

Degrading Treatment or Punishment of 2006 (signature 2006/ratification 2008); the Convention on the Rights of the Child of 1989 (signature 1990/ratification 1992); the Optional Protocol to the Convention on the Rights of the Child on the Sale of Children, Child Prostitution and Child Pornography of 2002 (signature 2000/ratification 2009); the Optional Protocol to the Convention on the Rights of the Child on the Involvement of Children in Armed Conflict of 2002 (signature 2000/ratification 2004); the Optional Protocol to the Convention on the Rights of the Child on a Communications Procedure of 2014 (signature 2012/ratification 2013); the International Convention for the Protection of all Persons from Enforced Disappearance of 2010 (signature 2007/ratification 2009); the Convention on the Rights of Persons with Disabilities of 2008 (signature 2007/ratification 2009); and the Optional Protocol to the Convention on the Rights of Persons with Disabilities of 2008 (signature 2007/ratification 2009).[302]

All the legally binding treaties above, in general, and the Convention on the Rights of the Child, in particular, should be considered for protecting juvenile delinquents and children in conflict with the law in Germany. As already mentioned, Germany signed the Convention on the Rights of the Child on 26 January 1990, which came into force on 5 April 1992, after the instrument of ratification had been deposited.[303] However, at that time, Germany made interpretative statements and reservations which show that it views the Convention as a welcome development in international law that will hopefully improve the situation of children worldwide, and that Germany will play its part, in keeping with article 3 para. 2 of the Convention, by drafting legislation to live up to its spirit and to ensure the well-being of the child.[304] It was also made clear that the Convention is not considered as being self-executing by making the following sweeping statement: The Federal Republic of Germany also declares that domestically the Convention does not apply directly. It establishes state obligations under international law that the Federal Republic of Germany fulfills in accordance with its national laws, which conform to the Convention.[305] In addition to denying any domestic effects of the Convention, Germany made specific reservations or interpretative statements to several articles of the Convention, such as those relating to child custody.[306] Moreover, Germany made exceptions to the Convention's guarantees of legal representation and appellate proceedings by stating that these guarantees might not be provided in some cases for minor infringements.

---

302 For further reading, see http://indicators.ohchr.org/ [20/05/2021].

303 *Holthusen* 2004, p. 133.

304 As an example of such legislation, Germany promised to reform parental custody.

305 This statement may contain an implied reference to the fundamental rights guaranteed by the German Constitution, in particular its article guaranteeing the family and the welfare of children, as well as its articles 1 and 2 guaranteeing human dignity and personal freedom.

306 For instance, concerning article 18(1), dealing with the common responsibility of the parents to raise the child, Germany declared that parental custody has to be awarded on a case-by-case basis, taking into consideration the situation of the parents, and declared furthermore that this

Germany made an even more serious reservation in favor of its immigration laws by stating: "Nothing in the Convention may be interpreted as implying that unlawful entry by an alien into the territory of the Federal Republic of Germany or his unlawful stay there is permitted; nor may any provision be interpreted to mean that it restricts the right of the Federal Republic of Germany to pass laws and regulations concerning the entry of aliens and the conditions of their stay or to make a distinction between nationals and aliens. It should be noted that public interest groups, particularly those advocating a higher level of rights for child immigrants, urged Germany to withdraw the reservations.[307] The German Parliament also recommended to withdraw the reservations toward the Convention on the Rights of the Child on 26 September 2001. Since then, many parliamentary decisions repeated this recommendation. The withdrawal of reservations entered numerous party programs and government declarations. However, consecutive governments usually explained their failure to withdraw the reservations by pointing to the resistance of regional governments within Germany and their competences within the federal political system. On 26 March 2010, the representation of all regional governments finally gave up its blockade and decided to welcome a withdrawal of the reservations.[308]

---

article does not derogate German law concerning legal representation of minors, custody for children born in wedlock, as well as family and inheritance law for children born out of wedlock.

307 Children's Rights: The Law Library of Congress, Global Legal Research Center 2007, pp. 86–96; www.law.gov [20/05/2021].

308 For further reading, see *Lundy et al.* 2013, pp. 42–49; see also www.proasyl.de [20/05/2021].

Chapter 2

Comparative Study of Current Criminal Law Governing Juvenile Delinquency in Iran, Germany and the United Nations

Historically, criminal law has been closely associated with sovereignty, which can be seen in the state's virtually unlimited discretion to discipline members of the society. Contrary to a democracy, state and society in an autocratic regime – because here they are fortified as basic entities – form the foundation of criminal law and its culture; human rights are delegated to an insignificant status, and sociological or criminological findings are also ignored.[309] Iran is one of the countries which have not paid as much attention as necessary to the conception of child rights as well as sociological and criminological findings in the formulation of juvenile criminal law rules. However, over the past few years, attempts have been made to re-establish a juvenile justice system based on differentiated criminal policy.

The German juvenile criminal law and juvenile justice system is, by contrast, the product of a centuries-long project of scientific discovery undertaken by scholars who have traditionally defined the science of criminal law in contradistinction to the markedly unscientific business of criminal justice policy. It should be noted that at an international level, the United Nations has developed some rules and regulations about how to deal with children and adolescents who are perpetrators of crime. In fact, since 1959, the United Nations has issued many documents in the form of declarations, resolutions, and treaties setting standards and norms for crime prevention and criminal justice in general and for juvenile justice in particular. They differentiated criminal policy toward children and juveniles in the hope of bringing into effect a desirable criminal pattern for the treatment of juvenile delinquents. In order to study the current criminal law applicable to juvenile delinquents in Iran and Germany, the international instruments and standards as well as the content of Iranian and German legal sources will be comparatively reviewed in this chapter.

Traditionally, comparative law is defined as a comparison of one country's law with that of another and more generally is the study of differences and similarities between the laws of different countries. It essentially includes the description and analysis of

309 For further reading about the evolution of criminal law and the importance of comparative studies, see *Dubber* 2005; *Ghassemi* 2015; *Chamelin et al.* 2012; Hazel 2008.

foreign criminal laws. Many distinguished comparative lawyers have insisted on the virtues of comparative law as a means of expanding knowledge generally and of achieving a better understanding of law. On a purely cognitive level, comparative law should therefore contribute to a better understanding of one's own national law through the contrasts and greater range of information it provides. As a practical consequence, it should be a primary instrument of domestic law reform through legislation, and indeed, the need for comparative research is a constant theme on the agenda of law reform agencies and ministries of justice. It has been said, however, that comparative law is also a means of evaluating the efficiency of different models as well as of recognizing the resistance of particular national traditions (path dependency) in the "evolution" of efficiency. Thus, it is necessary to bear in mind the different political, socio-economic and cultural contexts within which systems and policies operate.[310]

A larger pragmatic objective seems to be the regional or international harmonization of law in the worldwide process of the development of international and transnational law. Comparative criminal law has attracted little attention, at least compared to other types of law. Apart from basic research, it is application-oriented research which serves the interpretation of applicable law or at least the determination of the requirements for its application. In this function, comparative criminal law can assist in the process of finding new legislative solutions and of evaluating them by drawing on foreign or international law.[311]

When it comes to basic research concerning criminal law, the comparison of different systems is an important method. A comparison might inspire a researcher to learn more about and to rethink his or her own culture. A crucial advantage when employing a comparative approach is that one is forced to remove oneself from one's own legal system. It is a tool that allows to break free from the national constraints that bind almost every national lawyer – namely, the authoritativeness of one's own legal system and its sources. This approach creates a certain distance from the familiar system and thereby sharpens one's own perception. This is not only a benefit of comparative law but also an important precondition for it.[312]

---

310 *Hazel* 2008, p. 15. It is necessary to bear in mind the different political, socio-economic and cultural contexts within which systems and policies operate. It would be wrong to assume that one country's policies and practices could necessarily be transferred to another country with the same level of success. In relation to youth justice, it would be dangerous to imagine that we are about to see the possibility of universally agreed ideas of "what works in youth justice": Juvenile and youth justice may be becoming more globalized through the impact of neo–liberalism, policy transfer, and international conventions; but at the same time it is becoming more localized through national, regional, and local enclaves of difference, coalition, and resistance.

311 *Persson* 2017, pp. 36–40.

312 *Persson* 2017, pp. 36–40.

The characteristic and objective of juvenile criminal law and justice, the principles and measures concerning juvenile delinquents, the criminal responsibility of juvenile delinquents, delinquency and sanctions, as well as special institutions, authorities, and proceedings in justice system in the mentioned dimensions will be comparatively reviewed in this chapter.

2.1 Juvenile Criminal Law

Juvenile criminal law/justice includes a differentiated body of rules and statutes that define prohibited conduct and establishes a criminal response for the commission of such acts by juvenile delinquents (substantive law) on the one hand; in a broad sense, it describes the process through which criminal laws are enforced for juveniles (adjective law) on the other hand. In fact, the center of any criminal justice system consists of a set of officially prescribed rules for the behavior of persons within the system's jurisdiction. If these fundamental substantive prohibitions are defective, justice cannot be retrieved through the most meticulous observance of procedural due process or even by the most enlightened system of correction.[313] This issue can become more crucial if the prescribed rules relate to juvenile delinquents.

Regarding the term juvenile justice system, it should be noted that its definition is broader than juvenile criminal law. From a legal perspective, a juvenile justice system consists of legal norms, procedural aspects, institutions, and the particular agents within institutions concerning juveniles. Such a juvenile justice system is a juvenile criminal system for upholding social control, minimizing juvenile crimes, and imposing legal consequences for criminal offences by those who have not reached the age of criminal capacity. In this system, young persons' lack of maturity implies less culpability and also a higher level of vulnerability and formability. Furthemore, it has an impact on issues of deterrence and proportionality which are discussed in the criminal justice system. In the English version of general comment no. 24 (2019), the term "Child Justice System" was used instead of "Juvenile Justice". According to this general comment, a child justice system includes: the legislation, norms and standards, procedures, mechanisms, and provisions specifically applicable to – as well as institutions and bodies set up to deal with – children considered as offenders.[314] In this section, all aspects about juvenile criminal law will be reviewed.

2.1.1 Characteristics, objectives and application of juvenile criminal law

In this section, the characteristics and objectives of juvenile criminal law/justice will be studied in three dimensions: United Nations, Iran, and Germany.

---

313 *Irving et al.* 1980, p. 1.

314 *Persson* 2017, pp. 20–22, see also *Smith* 2005, and *Soler* 2009, pp. 483–541.

2.1.1.1 United Nations

2.1.1.1.1 Characteristics of juvenile criminal law

According to article 1.4 of the Beijing Rules, juvenile justice shall be conceived as an integral part of the national development process of each country, within a comprehensive framework of social justice for all juveniles. Thus, at the same time, it contributes to the protection of the young and the maintenance of a peaceful order in society. The specific characteristics should be considered when establishing a juvenile justice system. In this regard, article 40(3) of the Convention on the Rights of the Child provides that States parties shall seek to promote the establishment of laws, procedures, authorities, and institutions specifically applicable to children alleged as, accused of, or recognized as having infringed the penal law. In terms of promoting and applying juvenile criminal law, an effort should be made to develop a set of rules and regulations in each national jurisdiction unit so as to address the crimes of juvenile offenders (Beijing Rules 2, 3). In fact, national governments and legislatures should take responsibility for the realization of children's rights and juvenile justice standards by providing special provisions, procedures, and institutions. Enacting particular laws as a first and most important step clarifies the individual's rights, the government's obligations that correspond to those rights, the government's response mechanisms (including both identification of the agencies responsible for fulfilling the obligations and the procedures through which they are to operate), and also the accountability mechanisms which assure that the obligations are honored. Such a differentiated system should be based on the core elements, including: special provisions; protection of children and juveniles in conflict with the law as vulnerable persons; participation of juveniles; prevention of crime (recidivism); professionalization; diversion; decriminalization; depenalization; decarceration and deinstitutionalization in line with the best interest of the child.

2.1.1.1.2 Objectives of juvenile criminal law

Article 40(1) of the Convention on the Rights of the Child refers to the right of every child to be treated in a manner consistent with the promotion of the child's sense of dignity and worthiness, reinforcing the child's respect for human rights and fundamental freedoms of others as well as taking into account their and the desirability of promoting their reintegration as well as their assumtion of a constructive role in society. Paragraph 1 of article 40 of the Convention upholds the positive rehabilitative aims of the juvenile justice system, which have been underlined by the Committee in the wider context of the best interests of the child. The paragraph links to the provisions in paras. 3 and 4 of article 40, which stress the importance of excluding younger children from criminal responsibility, of avoiding judicial proceedings, and of developing a variety of dispositions, in particular including alternatives to institutional care. It also echoes the aims of education set out in article 29, which include development of respect for human rights as well as the preparation of the child for a responsible life in a free society. In relation to criminal justice systems, the International Covenant on Civil and Political Rights requires: in the case of juvenile persons, the procedure shall be such as will take account of their age and the desirability of promoting their rehabilitation (article 14(4)).[315]

Therefore, according to the spirit and provisions of the Convention on the Rights of the Child and other United Nations instruments, the objectives of the juvenile justice system should be based on well-being and proportionality (justice). In accordance with article 5.1 of the Beijing rules, the juvenile justice system shall emphasize the well-being of the juvenile and ensure that any reaction to juvenile offenders is always in proportion to the circumstances of both the offender and the offense. The most important objectives of juvenile justice should be the promotion of the juvenile's well-being. This is the main focus of those legal systems in which juvenile offenders are dealt with by family courts or administrative authorities, but the well-being of the juvenile should also be emphasized in legal systems that follow the criminal court model, thus contributing to the avoidance of merely punitive sanctions. The second objective is "the principle of proportionality". This principle is well-known as an instrument for curbing punitive sanctions, mostly expressed in terms of just desert in relation to the gravity of the offense. The response to young offenders should be based on the consideration not only of the gravity of their offense but also of personal circumstances. The individual circumstances of the offender (for example social status, family situation, the harm caused by the offense, or other factors affecting personal circumstances) should influence the proportionality of the reaction (for example by having regard to the offender's endeavor to indemnify the victim or to their willingness to turn to a wholesome and useful life). By the same token, reactions aiming at ensuring the welfare of the young offender may go beyond necessity and therefore infringe upon the fundamental rights of the

---

315 It is worth mentioning that the importance of human rights principles is emphasized, and it is necessary to recognize that the performance of the juvenile justice system and the effectiveness of other community responses to juvenile crime cannot simply be measured in terms of their compliance with those principles. Such an evaluation must also encompass the main objectives of the criminal justice system and the guarantees it offers (or fails to offer) to offenders, children, and adults. In its general comment no. 10 (2007), the Committee on the Rights of the Child has acknowledged that the preservation of public safety is a legitimate aim of any justice system. That aim dictates, at least partly, the choice of criteria used to evaluate a juvenile justice system or some of its programmatic components. The question is thus how best to achieve the public safety objectives of the justice system while still protecting the rights of the child and respecting the principles of juvenile justice enshrined in the Convention on the Rights of the Child. As the criminal justice process applies to both adult and child offenders, its public safety objectives are at the center of the public's expectation of that process. To ignore that basic expectation when designing new responses to juvenile crime means to risk depriving such initiatives of the public support required for their success and may even encourage unsupportive or hostile public attitudes toward rights-based reforms in general. This, too, is crucially relevant to the elaboration of criteria for measuring the success of reforms and the overall performance of a juvenile justice system. See *Grahn Farley* 2002, pp. 297–335, *Shaygan*, pp. 350-321; UNOCD 2010, p. 6.

young individual, as has been observed in some juvenile justice systems. Here, too, the proportionality of the reaction to the circumstances of both the offender and the offense, including the victim, should be safeguarded. In essence, rule 5 calls for no less and no more than a fair reaction in any given case of juvenile delinquency and crime. The issues combined in the rule may help stimulate development in both regards: new and innovative types of reactions are as desirable as precautions against any undue widening of the net of formal social control over juveniles. According to the objectives explained above, the model of the juvenile justice system should be based on the welfare and justice model.[316]

2.1.1.1.3 Application of juvenile criminal law

According to general comment no. 24, the child/juvenile justice system should apply to all children above the minimum age of criminal responsibility but below the age of 18 years at the time of committing the offense. The Committee recommends that States parties that limit the applicability of their child justice system to children under the age of 16 years (or lower), or that allow by way of exception that certain children are treated as adult offenders (for example, because of the offense category), change their laws to ensure a full, non-discriminatory application of their child justice system to all persons below the age of 18 years at the time of the offense (see also general comment no. 20, para. 88). Child justice systems should also extend protection to children who were below the age of 18 at the time of committing the offense but who turn 18 during the trial or sentencing process. The Committee commends States parties to allow the application of the child justice system to persons aged 18 and older, whether as a general rule or by way of exception. This approach is in line with developmental and neuroscience evidence which shows that brain development continues into the early twenties (general comment no. 24, paras. 29–32).

2.1.1.2 Iran

Regarding the characteristics of current juvenile criminal law and justice in Iran, it should be mentioned that the rules governing juvenile delinquency are not laid down in a specific juvenile justice act, although specific rules can be found in the Penal Code and the Criminal Procedure Code, which are considered as general codes applicable to both adults and juveniles. These codes provide some speific provisions regarding the age of criminal responsibility, sanctions, prosecution, adjudication,

---

316 Within a juvenile justice system, justice and welfare are considered to be competing systems which have a duty to cooperate with one another. This tension is expressed not only by several different scholars but also in the commentary of the Beijing Rules: the main difficulty in formulating guidelines for the adjudication of young persons stems from the fact that there are unresolved conflicts of a philosophical nature, such as: a) rehabilitation versus just desert; b) assistance versus repression and punishment; c) reaction according to the singular merits of an individual case versus reaction according to the protection of society in general; d) general deterrence versus individual incapacitation. *Persson* 2017, p. 16. See also, *Herz* 2012.

and sentencing. In the new Penal Code of 2013, due to the clashes between Islamic and customary law on the one hand and international obligations on the other hand, reforms and changes in this regard were made in the two different sections and based on the category of punishments in the Iranian legal system. This category, which is influenced and shaped by *Shari'a*, includes discretionary and fixed penalties. The rules governing discretionary crimes and punishments (*Ta'zir*) are flexible, while the rules governing fixed punishment (*Hadd*, *Qisas*, and *Diye*) are stable, according to *Shari'a*.

Thus, it can be said that there is no integrated system to deal with juvenile delinquents in Iran. However, it is notable that Iran has made progress in considering juvenile rights and establishing a juvenile justice system in line with international standards and obligations. Therefore, the new codes are generally an improvement over the previous ones in this respect, because the approach used in the new codes for *Ta'zir* crimes and punishments is based on corrective as well as rehabilitation objectives and measures within the justice system. Nevertheless, in the fixed punishments, the rights of juveniles have not yet been completely respected. In this section, the legislature was completely influenced by the conservative view of *Shari'a*. However, in adjective criminal law, according to the Criminal Procedure Code of 2014, special institutions are allocated for trials of juveniles, with a special process being in line for the best interest of children and juveniles. It can be said that in procedural criminal law – and to the contrary of substantive criminal law –, there is no a major clash with *Shari'a*. Furthermore, there is no specific objective dealing with juvenile delinquents arising from the ideology of Iran's criminal law.

Iran is a state largely governed by the rule of *Shari'a*. It is worth mentioning that the ideology of Iran's criminal law, in general, and its juvenile criminal law, in particular, is *Shari'a*, and as a civil/continental system, the legislations, the Islamic law, and also the judicial precedent are considered, respectively, as the main criminal law sources in Iran.[317]

2.1.1.3 Germany

In line with international instruments and standards, there is an integrated system for dealing with juvenile delinquents in the German justice system. The Youth Court Act, as a specific act, comprises substantive and procedural criminal law applicable to juvenile delinquents. It can be said that a separate system with special institutions and procedures has been established for juvenile delinquents and those juveniles who are considered as persons in need handled by the Youth Welfare Act.[318] It should be emphasized that a separate juvenile justice system can only function in conjunction with an infrastructure for educational dispositions. In Germany, such a structure has

---

317 For further reading about the legal sources in Iran, see *Abbasi* 1391/2012, pp. 247–260.

318 *Albrecht* 2004, pp. 450–454.

been implemented through a two-track system of welfare and justice laws[319] (the dualistic approach): the Child and Youth Welfare Act for children in need of care and the Juvenile Justice Act for children who have committed an offense.[320] These laws demand the implementation of a network of private and state welfare as well as justice agencies.[321]

The German youth justice system is characterized by a high level of stability in terms of procedures and sentencing. This is despite the rhetoric in political debates sometimes being dominated by penal populism. For example, violent subway crimes in late 2007 led to the Christian Democratic Party calling for increased juvenile sanctions, followed swiftly by a backlash from criminal justice practitioners and an election loss for this party in early 2008.[322]

Nevertheless, the Youth Court Law focused on educational and rehabilitative needs of juvenile offenders. German juvenile justice, however, was never dominated by a social welfare model and never adopted more than marginal traces of welfare and social support. Throughout the 20th century, the idea prevailed that punishment and education could be reconciled within the framework of juvenile justice. Strict separation between juvenile offenders on the one hand and children and juveniles in need

---

319 It is notable that this dualistic approach causes problems, specially in relation to the rule of law, which demands proportionality, predictability and legal certainty, transparency, and equality. These legal principles are vague and general and seldom give clear and explicit answers to legal questions of law. Instead, they offer a framework for balancing opposing interests. The educational guiding principle introduces an additional legal principle to the framework of the juvenile criminal justice system. It implies a strong focus on the individual instead of the offense. This in turn gives the juvenile court broad discretion and can lead to conflict with the justice principles. *Persson* 2017, p. 90.

320 The basic characteristics and philosophy of the German Juvenile Welfare Act include the right to support and education; and to being protected by the child and youth welfare agencies established at the local community level with the priority of private non-profit organizations which must be accredited by the regional (i.e. federal state-level) youth welfare departments of the ministries of social affairs, and also co-operation of welfare and justice agencies. Furthermore, the basic characteristics and philosophy of the German Juvenile Justice Act include the following priorities: of diversion and minimum intervention; of mediation and restorative justice; of educational community sanctions; of youth imprisonment to be as short as possible and only as a last resort; of no transfer of juveniles to adult courts; of the jurisdiction of a juvenile court. *Dünkel* 2018, pp. 4, 12–16.

321 The agencies organized on the basis of the Juvenile Welfare Act are the community youth welfare departments and the youth services in youth court proceedings which have a double task: to fulfill purely welfare-oriented tasks (family aid, protection of children in need of care); to support the juvenile prosecutor and court by delivering personal and family background information for the trial. They are also partly responsible for the execution of educational measures (mediation, social training, etc.). The youth services in youth court proceedings are also responsible for avoiding unnecessary pre-trial detention. Their personnel are social workers or social pedagogues with at least three years of university education. The personnel of private institutions in most cases have the same professional education; sometimes they also have teachers, psychologists, and social workers with special training, e.g. as mediators, at their disposal. There are special courses for further professional specialization, e.g. as a mediator, provided by the federal probation service. *Dünkel* 2018, pp. 4, 12–16.

322 *Matthews et al.* 2018, p. 68.

of care and education on the other can be observed throughout the history of juvenile laws. Voices were repeatedly raised in favor of a unified welfare approach toward delinquent children and juveniles, but juvenile criminal law never deviated far from general criminal law. It remained bound to the rule of law with respect to the criminal offense statutes that triggered juvenile criminal-justice-based intervention, as well as to a compromise between punishment and education.[323]

Youth criminal law under the Youth Court Law targets education and rehabilitation of the young offender. Although a juvenile is held legally responsible for a crime (if the mens rea is proven), the primary goals are education and rehabilitation. In fact, the objective of juvenile criminal law and justice is to transform the young perpetrator into a law-abiding citizen. German juvenile criminal law focuses on the offender and his or her rehabilitative needs, while the adult criminal system concentrates mainly on the offense itself. Therefore, the German juvenile criminal justice system is sometimes called "offender criminal law" or "educative criminal law". The focus on the individual and education still involves gaining insight into the wrongdoing itself, but it should also avoid stigmatizing young people as criminals. This rests on an understanding that young offenders differ from adult perpetrators psychologically, physiologically, and in their social status during the transitional developmental period from childhood to adulthood. As a consequence, the principle of education prevails over general prevention and arguments of just desert.[324]

It should be emphasized that the adoption of the principle of education and disregard for general prevention, general deterrence, and seriousness of the offense as guides in sentencing does not mean that the principle of proportionality must not be applied. It is derived from the German Constitution and requires that the educational measure chosen be sufficient to reach the goal that is sought, be the least severe measure among several possibilities equally suited to attain that goal, and be proportional to the goal to be achieved. This principle sets an upper limit to legal intervention in juvenile criminal cases, thus boiling down educational needs and their possible impact to what is seen to be proportional, from the viewpoint of the seriousness of the offense. Thus, while any disposition must be proportional to the offense, it must also be guided by the goal of education.[325]

Germany is a state governed by the rule of law. It should be noted that the ideology of German criminal law, in general, as well as that of juvenile criminal law, in particular, is doctrine, and as a civil/continental system, the constitution, the legislations, and also the case law are respectively considered as the main criminal law sources in Germany.[326]

---

323 *Albrecht* 2004, pp. 450–454

324 *Persson* 2017, p. 87; and *Albrecht* 2004, pp. 450–454; and see also, *Holthusen & Schäfer* 2004, pp. 11–23.

325 *Albrecht* 2004, pp. 453–454, see also *Viehmann* 2009.

326 For further reading about the criminal law sources in Germany, see *Bohlander* 2009, pp. 7–16.

2.1.2 Fundamental rights of delinquents in juvenile criminal law

In this part, the fundamental rights of delinquents in juvenile criminal law/justice will be discussed for Iran, Germany, and the United Nations.

2.1.2.1 United Nations

2.1.2.1.1 Right to life, survival, and development

Article 6 of the Convention on the Rights of the Child requires States parties to recognize that every child has the inherent right to life and to ensure to the maximum extent possible the survival and development of the child. This applies to children at all stages of their life and in all settings, including those in the juvenile justice system. Regarding juvenile delinquents in protecting their right to life and addressing the abolition of capital punishment for persons under 18 years of age should be emphasized.[327] In addition to the inherent right to life, article 37a of the Convention provides explicitly that neither capital punishment nor life imprisonment without possibility of release shall be imposed for offenses committed by persons below the age of 18. The prohibition of imposing the death penalty continues to apply when an adult is tried for an offense that was committed when the person was still under the age of 18. The domestic legislation of all States should contain clear provisions setting out these two prohibitions.[328]

Ensuring the right to a harmonious development of children who are subject to the juvenile justice system, particularly for those who are deprived of liberty, is more challenging. Development in the context of the Convention on the Rights of the Child is a holistic concept: it is about children being provided with optimal conditions that will enable them to access all their rights during childhood. Children who are detained by the juvenile justice system or the adult criminal justice system are often not provided with optimal conditions and may be housed in poor facilities, with little educational or vocational opportunities, and, of course, are separated from their families, often with little contact. Such children find it difficult to obtain education, accommodation, or employment when released, and they also experience difficulties when reintegrating into their family and the community. However, legislation providing for the right to development is often, by itself, not sufficient: it should cover the manner in which institutions are to be operated and monitored, as this is the key to ensuring the child's right to development and survival. The Havana Rules contain minimum standards on provision and conditions in custodial institutions.[329]

---

327 *Hamilton* 2011, pp. 25–27; *Hodgson* 1994, p. 369.

328 Despite the clear statement contained in the Convention on the Rights of the Child, Iran, Pakistan, Sudan, Saudi Arabia, and Yemen, all of whom have ratified the Convention, have continued to execute juveniles.

329 *Hamilton* 2011, pp. 27–29.

2.1.2.1.2 Right to dignity

This principle reflects the fundamental human right enshrined in article 1 of the Universal Declaration of Human Rights (UDHR), which stipulates that all human beings are born free and equal in dignity and rights. This very important principle has also been mentioned in other human rights instruments. However, there is no consensus on the definition of human dignity, but a basic minimum content of its meaning can be discerned: that each human being possesses an intrinsic worthiness that should be respected; that some forms of conduct are inconsistent with respect for this intrinsic worthiness; and that the state exists for the sake of the individual, and not vice versa. However, the fault lines here lie in the disagreement on what that intrinsic worthiness consists of, which forms of treatment are inconsistent with that worthiness, and what the implications are for the role of the state.[330]

Article 40(1) of Convention on the Rights of the Child (and other instruments concerning children's rights) provides a set of fundamental principles for the treatment to be accorded to children in conflict with the law: treatment that is consistent with the child's sense of dignity and worthiness; treatment that reinforces the child's respect for the human rights and freedoms of others; and treatment that takes into account the child's age and promotes their reintegration and their assumption of a constructive role in society. Furthermore, international instruments stipulate that respect for the dignity of the child requires that all forms of violence in the treatment of children in conflict with the law must be prohibited and prevented. The Committee on the Rights of the Child acknowledges that the preservation of public safety is a legitimate aim of the justice system. However, it is of the opinion that this aim is best served by full respect for and implementation of the leading and overarching principles of juvenile justice as enshrined in the Convention on the Rights of the Child (general comments no. 10, p. 13).

2.1.2.1.3 Right to equality and non-discrimination

When explaining non-discrimination, we must emphasize the principle of equality, which is a deep-rooted principle in human morality and has quickly been recognized legally as one of the fundamental principles of modern democracies. It occupies pride of place in most written constitutions, and numerous countries now have anti-discrimination legislation, whether against all or some specific forms of discrimination, such as racial or sexual discrimination.[331] Article 2 of the Convention on the Rights of the Child prohibits discrimination of any kind, irrespective of the child's or their parents' or legal guardians' race, color, sex, language, religion, political or other opinion, national, ethnic, or social origin, property, disability, birth, or other status. Therefore, States parties shall take all appropriate measures to ensure that the child is protected against all forms of discrimination or punishment on the basis of the

---

330 *McCrudden* 2008, p. 723.

331 *Besson* 2005, p. 434.

status, activities, expressed opinions, or beliefs of the child's parents, legal guardians, or family members. Although the right to non-discrimination is a well-established human rights principle in international law, less attention has been paid to the particular ways in which children experience discrimination. The Convention on the Rights of the Child was created to recognize children's inherent entitlement to fundamental rights and freedoms. The principle of non-discrimination is crucial for achieving this given that discrimination is one of the main reasons why children's rights are violated.[332]

Furthermore, the Committee on the Rights of the Child has specifically identified a number of other grounds of discrimination in its Concluding Observations, including discrimination on the basis of sexual orientation and against children affected by HIV/AIDS. The Committee has established that States must take proactive measures where necessary to guarantee the principle of non-discrimination, including law and policy reform, education and awareness-raising, and monitoring, among others. The Committee has also said that special measures may sometimes be needed to promote the rights of certain groups of children.[333] Such positive measures may, however, exceptionally need to be of a permanent nature, such as interpretation services for linguistic minorities and special consideration for members of ethnic minority groups, or for girls who are alleged to have committed a criminal offense.[334] For example, special healthcare and hygiene needs for girls whilst in detention, equal access for all children to education and vocational programs, or the use of an interpreter and providing translated information for children with a different language are of primary importance. At every stage of a child's contact with the justice system, as well as when forming legislation and policy, thought must be given to the situation of the child. It must be ensured that all children have their rights upheld and are offered appropriate services without discrimination. Each situation must be handled with sensitivity and an understanding of the issues that a child, or a group of children, may face due to their sex, age, race, disability, etc.[335]

2.1.2.1.4 Right to best interest

Disregarding the fact that there is no standard definition of "best interests of the child", it means considering the child before a decision affecting his/her life is made. This is a principle that has established itself through all matters and legislation affecting the well-being of the child.[336] The term "generally" refers to the deliberation courts undertake when deciding what type of services, actions, and orders will best serve a child, as well as who is best suited to take care of him/her. "Best interests"

332 Child Rights Information Network 2009, p. 6.

333 Child Rights Information Network 2009, p. 6.

334 *Hamilton* 2011, pp. 25–27.

335 Protecting Children's Rights in Criminal Justice Systems 2013, p. 15.

336 *Dausab* 2009, p. 145.

determinations are generally made by considering a number of factors related to the child's as well as the parents' or caregivers' circumstances and capacity to the parents, with the child's ultimate safety and well-being being the paramount concern.[337]

Article 3 of the Convention on the Rights of the Child requires that in all actions concerning children – whether undertaken by courts of law, administrative authorities, or legislative bodies –, the best interests of the child shall be a primary consideration. Regarding juvenile delinquents, the Committee on the Rights of the Child in general comment no. 10 notes that the protection of the best interests of the child means, for instance, that the traditional objectives of criminal justice, such as repression/retribution, must give way to rehabilitation and restorative justice objectives when dealing with child offenders. This can be achieved in concert with attention to effective public safety. Therefore, it would seem that the "best interests" test requires that the juvenile justice system should not use repressive or retributive measures but promote the reintegration of children who have offended. Furthermore, Juvenile justice legislation should contain a clear section or article stating that in making any decision in relation to an individual child, the principle of best interests applies. Also, the legislation should state that the principle of best interests applies to all aspects and operations of the juvenile justice system. The principle of best interests therefore also applies to all decisions taken in the juvenile justice system. It is applicable from the child's first contact with a law enforcement authority right through to sentencing and post-sentence services, particularly after-care services for those who have been deprived of their liberty. In determining the child's best interests, the personality, wishes, circumstances, family situation- and well-being of child should be considered by the decision-makers. It is recommended that these hear the views of the child as they will be key in determining the best interests. These must be a primary consideration in the decision-making process. Other interests at stake include public safety and the rights of the defendant(s), which should also be considered.[338]

2.1.2.1.5 Right to be heard

According to article 12 of the Convention on the Rights of the Child, all children who are capable of forming their own views have the right to express them freely in all matters that affect them, and these views should be given due weight in accordance with the age and maturity of the child. This article is explicit about children's rights to express their views and to have an opportunity to be heard in any judicial and administrative proceedings affecting them. It will focus on the responsibility of courts and, in particular, that of the judicial officer, to facilitate children's participation.[339] In general comment no. 12, the Committee on the Rights of the Child has

337 Child Welfare Information Gateway 2012, p. 2. See also *Zermatten* 2010.

338 *Hamilton* 2011, pp. 23–25.

339 *Cashmore et al.* 2007, p. 43

placed great emphasis on the right of the child to be heard at every stage of the juvenile justice process, and it takes the view that the voices of children involved in the juvenile justice system are increasingly becoming a powerful force for improvements and reform as well as for the fulfillment of their rights. Article 40 also reflects the important principle that part of the right to be heard is the one to choose when to be heard and what to say.[340]

Children have the right to take an active role in legal proceedings but must not be compelled to give testimony or to confess guilt. They have the right not to exercise their right to be heard. This is also enshrined in article 13 of the Convention on the Rights of the Child, which protects the right to freedom of expression. Therefore, it is important that the right to be heard is clearly set out in legislation. The criminal justice system as a whole can be very intimidating, and police officers, prosecutors, and judges should be trained to interact with children in ways that enable these to express themselves. It is difficult to expect that a child could be heard without the assistance of parents and/or a legal representative. In addition, if the child does not speak the same language as the lawyer or defendant, this will be a clear barrier to communication. Generally, the language used in court is complex and difficult to follow for a lay person, particularly a child. In all these instances, the child's rights under article 12 of the Convention could potentially be violated. The right to express his/her views is crucial when the child charged with or being tried for a criminal offense appears before a court. The ability of the child to express his/her views has been treated as being synonymous with their ability to defend themselves.[341]

In accordance with the above principle of non-discrimination, children with disabilities and those speaking foreign languages should be helped participate in the same way as all other children. Legislation must contain specific provisions covering the right to be heard both pre- and at trial.[342] In sum, children must be able to express their views, opinions, and concerns, as well as to actively participate throughout the judicial process in accordance with their best interests and through a representative where necessary. In order to ensure this right, a child must receive adequate information about the process, the choices they have, and possible consequences of these choices.[343]

2.1.3 General principles in juvenile criminal law

In this part, the general principles in juvenile criminal law/justice will be discussed for Iran, Germany, and the United Nations.

340 *Hamilton* 2011, pp. 29–33.

341 *Hamilton* 2011, pp. 29–33.

342 *Hamilton* 2011, pp. 29–33.

343 Protecting Children's Rights in Criminal Justice Systems 2013, p. 15.

2.1.3.1 United Nations

2.1.3.1.1 The principle of presumption of innocence

This principle of the presumption of innocence is – especially in the different stages of a trial and when there is inadequate evidence – considered to be a common legal heritage of all developed countries. It has received special consideration and was welcomed by scholars, lawyers, and legislators, especially after World War II; it has also been the subject of regional and international declarations and treaties.[344] The principle of innocence has been considered to be one of the basic and most important principles in the United Nations' instruments concerning children and juveniles. Article 40(b) i of the Convention on the Rights of the Child provides that to be presumed innocent until proven guilty according to the law should be respected as a trial guarantee. As a basic procedural safeguard, this principle should be applied for juveniles who have been arrested or are waiting for trial (Beijing Rules, article 7-1; and Havana Rules, article 17). In the General Comments concerning the rights of children in the criminal justice system, the United Nations Committee on the Rights of the Child acknowledges that the presumption of innocence is fundamental to the protection of the human rights of children in conflict with the law. It means that the burden of proof of the charge(s) brought against the child is on the prosecution. The child alleged as or accused of having infringed the penal law has the benefit of doubt and is only guilty as charged if these charges have been proven beyond reasonable doubt. The child has the right to be treated in accordance with this presumption, and it is the duty of all public authorities or others involved to refrain from prejudging the outcome of the trial. States parties should provide information about child development to ensure that this presumption of innocence is respected in practice. Due to a lack of understanding of the process, immaturity, fear, or other reasons, the child may behave in a suspicious manner, but the authorities must not assume that the child is guilty without proof of guilt beyond any reasonable doubt (general comment no. 10, p. 4; and general comment no. 24, para. 43).

2.1.3.1.2 The principle of legality

The principle of legality is a fundamental principle and defense in criminal law prosecution according to which no crime or punishment can exist without a legal ground, which is in fact a guarantee of human liberty. The principle of legality can protect individuals from state abuse and unjust interference; it ensures the fairness and transparency of the judicial authority, therefore it is considered as a core value in criminal proceedings.[345] Regarding juvenile delinquents, article 4(2) of the Convention on the Rights of the Child states: "no child shall be alleged as, be accused of, or recognized as having infringed the penal law by reason of acts or omissions that were not prohibited by national or international law at the time they were committed." There are

344 *Ashori* 1381/2002, p. 133; *Khazani* 1372/1993, p. 125.

345 *Crisan* 2010, p. 2.

two possible interpretations of this article: first, it is the law that recognizes a child accused or guilty; and second, the punishment applied to the child should be according to the law that already existed before committing the crime; an act cannot be considered criminal based on the rules that may be enacted in the future. The first interpretation of this article refers to the principle of legality of crimes and punishments, according to which a criminal law should match a corresponding law in national or international rules. The second interpretation refers to the principle of non-retroactivity of criminal law. This means that criminal behaviors should be investigated according to the existing rules and regulations at that time and not according to future ones.

According to article 40(3) of the Convention on the Rights of the Child, States parties shall seek to promote the establishment of laws, procedures, authorities, and institutions specifically applicable to children alleged as, accused of, or recognized as having infringed the penal law. Accordingly, the principle of legality of crimes and punishments should be considered, because it encourages the States parties to enact laws for accused and criminal children and then addresses the crimes of them based on the existing rules and regulations.

Also, according to article 37(b) of the Convention on the Rights of the Child, "no child shall be deprived of his or her liberty unlawfully or arbitrarily". This is indicative of the principle of legality of punishments, which means that the judge shall determine punishments based on the existing rules and regulations.

2.1.3.1.3 The principle of non-retroactivity of criminal law

Basically, rules and regulations are enacted for future acts and cannot be applied to previous acts. The principle of non-retroactivity of criminal law is the essential result of the principle of legality of crimes and punishments. It is composed of general and regulatory aspects. Generally, a judge cannot extend a new law that recognizes an act (which has not been considered a crime in the past or stipulates a more severe punishment that has not existed in the past) to the committed act before the new law is, investigated or is waiting to be, investigated.[346] The principle of non-retroactivity of criminal law has been adopted in the international instruments relating to juvenile offenders, and it has been decreed that "no child shall be alleged as, be accused of, or recognized as having infringed the penal law by reason of acts or omissions that were not prohibited by national or international law at the time they were committed". However, it seems that the rules governing the principle of non-retroactivity of criminal law and its expectations[347] also govern the international instruments, because they are to the benefit of children and juveniles. The United Nations Committee on the Rights of the Child also acknowledges that no child can be charged with,

---

346 *Noorbaha* 1379/20001384/2005, pp. 149–150.

347 For instance, the principle of non-retroactivity of criminal law is applied to cases where the new law is milder. In other words, the new, milder law is immediately applicable.

or sentenced under the penal law for, acts or omissions which at the time they were committed were not prohibited under national or international law.[348] According to general comment no. 24 (2019) para. 42, no child shall be held guilty of any criminal offense that did not constitute a criminal offense (under national or international law) at the time it was committed. States parties that expand their criminal law provisions to prevent and combat terrorism should ensure that those changes do not result in the retroactive or unintended punishment of children. No child should be punished with a heavier penalty than the one applicable at the time of the offense, but if a change of law after the offense provides for a lighter penalty, the child should benefit.

2.1.3.1.4 The principle of equality

The principle of equality expresses that the same and equal penalty shall be determined for offenders who have committed similar crimes in similar circumstances.[349] Penalties in the past were different depending on the situation of personalities of individuals. However, the principle of equality of penalties is more theoretical than practical, because people have different standings and judges treat individuals differently based on their own particular personality: strict judges punish those disobedient to lawful order very severely, while mild judges overlook violations of the law as much as possible. The fate of two criminals who have committed the same crime is thus dramatically different in two branches of the court. If the judges are allowed to apply penalties based on their own personal taste, the principle of equality is likely to be severely damaged. Therefore, this principle can be restricted by different conditions and social standing of criminals, moods of judges, public opinion, and many other issues.[350]

There are many rules on this principle in international instruments. Article 2(1) of the Convention on the Rights of the Child urges that States parties shall respect and ensure the rights set forth in the present Convention to each child within their jurisdiction without discrimination of any kind, irrespective of the child's or their parents' or legal guardians' race, color, sex, language, religion, political or other opinion, national, ethnic, or social origin, property, disability, birth, or other status. Article 2(2) of Convention also emphasizes that States parties shall take all appropriate

348 In the light of the fact that many States parties have recently strengthened and/or expanded their criminal law provisions to prevent and combat terrorism, the Committee recommends that they ensure that these changes do not result in retroactive or unintended punishment of children. The Committee also wishes to remind States parties that the rule that no heavier penalty shall be imposed than the one that was applicable at the time when the criminal offense was committed – as expressed in article 15 of the International Covenant on Civil and Political Rights – is (in light of article 41 of Convention on the Rights of the Child) applicable to children in the States parties to the International Covenant on Civil and Political Rights. No child shall be punished with a heavier penalty than the one applicable at the time of his/her infringement of the penal law. But if a change of law after the act provides for a lighter penalty, the child should benefit from this change (general comment no. 10 p. 41).

349 *Saneie* 1379/2000, p. 141.

350 *Noorbaha* 1384/2005, pp. 402–403.

measures to ensure that the child is protected against all forms of discrimination or punishment on the basis of the status, activities, expressed opinions, or beliefs of the child's parents, legal guardians, or family members. In fact, equality in the Convention's rights of the child also encompasses equality in sentencing and imposing punishments. The Beijing Rules state in article 2-1 that the principle of equality shall be applied to juvenile offenders impartially, without distinction of any kind, for example as to race, color, sex, language, religion, political or other opinions, national or social origin, property, birth, or other status. This indicates non-discrimination between children and their equality before the law and execution of punishments. Also, according to general comment no. 24 (2019), safeguards against discrimination are needed from the earliest contact with the criminal justice system and throughout the trial, and discrimination against any group of children requires active redress. In particular, gender-sensitive attention should be paid to girls and to children in general who are discriminated against on the basis of their sexual orientation or gender identity. Accommodation should be made for children with disabilities, which may include their physical access to the court and other buildings, support for children with psychosocial disabilities, assistance with communication and the reading of documents, and procedural adjustments for testimony.

2.1.3.1.5 The principle of proportionality

International standards stipulate that in all cases, the final disposition of a case must be proportionate both to the circumstances and the gravity of the offense, to the circumstances and needs of the juvenile, as well as to the needs of society. Therefore, it cannot be delivered until the background and circumstances have been investigated in which the juvenile is living or the conditions under which the offense has been committed. This assessment should take place as soon as possible after the child has been arrested.[351] The juvenile justice system shall emphasize the well-being of the juvenile and ensure that any reaction to juvenile offenders is always in proportion to the circumstances of both the offenders and the offense. The principle of proportionality is well-known as an instrument for curbing punitive sanctions, mostly expressed in terms of just deserts in relation to the gravity of the offense. The response to young offenders should be based on the consideration not only of the gravity of the offense but also of personal circumstances. The individual circumstances of the offender (for example social status, family situation, the harm caused by the offense, or other factors affecting personal circumstances) should influence the proportionality of the reactions (for example, by having regard to the offender's endeavor to indemnify the victim or to their willingness to turn to a wholesome and useful life) (Beijing Rules 5.1). In this regard, the Beijing Rules (17.1-a) also state that the reaction taken shall always be in proportion not only to the circumstances and the gravity of the offense but also to the circumstances and needs of the juvenile, as well as to the needs of society.

351 Protecting Children's Rights in Criminal Justice Systems 2013, p. 78.

2.1.3.1.6 The principle of child-oriented interpretation

In the interpretation of criminal law, it is assumed that this interpretation is strict and restricted to the explicit appearance of law. This is a corollary of the principle of legality of crimes, because if the judge has the power to punish offenders on the items not explicitly mentioned in law with their own broad interpretation, the action which has not directly been under the protection of law seems necessary to be punished. For this purpose, in cases where the rules of committed acts have not been explicitly forecasted in criminal law, the judge is required to ensure that the accused is acquitted and released. Since the principle of legality of crimes and its logical consequences are inspired by liberalistic thoughts and feelings in order to protect and support the rights of the accused, it should not be used against the accused. Accordingly, all criminal laws should not be spontaneously interpreted in a strict and rigid way, but the principle is that lenient laws toward the accused be interpreted broadly and logically.[352]

However, a comparison should not resort to an interpretation of criminal issues. This means that the penalties predicted by the law for criminal acts cannot be extended to other acts based on personal desire, even if there are many similarities between criminal and non-criminal acts. Strict interpretation does not mean literal interpretation – it means that if it is found that a mistake has been made in the development of law and that the legislator's intention is contrary to what is mentioned in the law text, the judge must issue their verdict based on the real purpose and intention of the legislator.[353] Regarding juvenile delinquents, international instruments have explained child-oriented interpretation or interpretation to the benefit of children and stipulate that a child-oriented approach should be followed for the interpretation of the rules on juvenile delinquents (Riyadh Rules, article 3). According to the provisions contained in international instruments, it can be inferred that the rules should be interpreted in a way which is in line with the interests and benefits of children.

2.1.3.1.7 The principle of participation

All international instruments on juvenile offenders have directly or indirectly mentioned the principle of participation. Especially in general comments concerning the rights of children in the criminal justice system, the United Nations Committee on the Rights of the Child acknowledges that effective participation of children in all stages of the procedure guarantees a fair and just trial. Article 4 of the Beijing Rules states that proceedings shall be conducted in an atmosphere of understanding, which shall allow the juvenile to participate therein and to express themselves freely (general comment no.10, article 46)· The child's right to be heard is also in line with the principle of participation of children. In this regard, the Convention on the Rights of the Child acknowledges that States parties shall assure to the child

352 *Saneie* 1379/2000, pp. 114–121.

353 *Saneie* 1379/2000, pp. 114–121.

who is capable of forming their own views the right to express those views freely in all matters affecting the child, their views being given due weight in accordance with their age and maturity.[354]

For this purpose, the child shall in particular be provided the opportunity to be heard in any judicial and administrative proceedings affecting them – either directly or through a representative or an appropriate body, in a manner consistent with the procedural rules of national law (Convention on the Rights of the Child, article 12). In diversion cases, including mediation, the child shall be provided the opportunity to express voluntary consent. In addition, the opportunity to visit a consultant shall be given to the child for determining the appropriate method of diversion (general comment no. 10, article 28).

In order to effectively participate in the proceedings, any child should be quickly informed of their charges, proceedings, and the decision taken in their own or an understandable language. Proceedings shall be conducted in an atmosphere that allows the juvenile to participate therein and to express themseves freely (general comment no. 10, article 60). According to a generally accepted principle, the court or any other authority addressing the crimes shall continue the proceedings behind closed doors. Exceptions to this principle should be limited and explicitly specified in the internal rules of any country in order to ensure the highest and most excellent interests of the child (general comment no. 10, article 68).

In sum, the right to participation, as formulated in article 12 of the Convention on the Rights of the Child, includes three main components: the right to express one's own views; the right to do so freely; and the right to have those views given due weight. Accompanying this right is the right to receive information. The right to participation was defined by the United Nations Committee as a symbol of the change process in the perception of the rights of the child expressed in the Convention.[355]

2.1.3.2 Iran and Germany

At first glance, it might seem that the guiding principles provided in the international instruments and standards concerning juvenile delinquents are protected by the Iranian Constitution. But if we pinpoint the Constitution Law as well as other legal sources, we can see that some of the principles stipulated in the international instruments are violated. Therefore, the principle of legality, for example, which has been protected by the Constitution, is violated by another article in this law, because the *Shari'a* itself is introduced as a second legal source after the legislations of Parliament (article 167 of the Constitution Law); or, as another example, the principle of equality is protected in the Iranian Constitution (article 19), but there are some discriminations regarding the age of criminal responsibility or the execution of punishments in the Penal Code (articles 91and 147). Furthermore, where there are no

354 *Morag* 2014, pp. 539.

355 *Morag* 2014, pp. 539.

clashes with Iranian laws and regulations, the concepts of the protected principles are defined in a very different way due to them being linked to *Shari'a*. For instance, the basis of the principle of proportionality, child-oriented interpretation, and the participation of children in *Shari'a* are completely different form the basis for these fields in modern theories. In line with the international instruments and standards, these principles are protected by the German Constitutional Law as well as other legal sources. They are also applied in practice by the juvenile justice authorities.[356]

2.1.4 Specific measures in juvenile criminal law

In this part, the specific measures in juvenile criminal law/justice will be discussed for Iran, Germany, and the United Nations.

2.1.4.1 United Nations

2.1.4.1.1 Diminished criminal responsibility

Someone who commits a criminal act is not necessarily found guilty. In addition to the general conditions necessary to punish a perpetrator, there should be further characteristics. In addition to criminal intent, the criminal act could be attributed to the person who actually commits it. The general account of criminal responsibility is based on the relationship between the performed act and the character of the agent performing it. An act can be attributed to a defendant, it is argued, only if that act is connected to the agent's character in a proper way. For this reason, criminal law recognizes some people as being irresponsible and does not find them criminally liable even if they committed a criminal act – some of them because of diminished criminal responsibility. Obviously, the physical and mental structure of the perpetrator should have reached perfection, and their mental faculties should be intact to be found criminally responsible.[357] Therefore, it can be concluded that responsibility refers to perfect physical and mental growth that is not disordered due to an internal factor. Different criminal schools of thought have submitted different views to this issue. All these theories about responsibility reflect the fact that some perpetrators should be exempt from criminal responsibility, with children and juveniles being those who should not be held responsible if committing a crime. There is a variety of ways in which age interacts with criminal responsibility. Legal systems tend to use two distinct senses of this age. One concerns the capacity of a child to commit a crime; the other relates to the exemption of a child from full criminal responsibility and the adult system.[358] Thus, determining the age of criminal responsibility based on the best interests of the child is very important in every legal system.

In criminal law, however, and as far as the age of criminal responsibility is concerned, natural persons are divided into three groups: those who have no criminal

---

356 For further reading, see *Abbasi* 1390/2011, pp. 248–259; *Bahadori* 1391/2012, pp. 13–17.

357 *Saneie* 1379/2000, p. 93; *Bojnordi* 1379/2000, p. 30.

358 *Gerry* 2005, p. 494.

responsibility (irresponsibility); those who have diminished criminal liability; and, finally, those who have full criminal responsibility. Persons in the second group are treated separately by the Penal Code and subsequently within the justice system. Diminished criminal responsibility leads to gradual liability and transition from irresponsibility to full criminal responsibility, depending on which is in line with the best interest of the child, and it is shaped by considering minimum and maximum ages for criminal responsibility. This model is recommended by the United Nations' instruments and standards (Convention on the Rights of the Child, article 40.3, general comments no. 10 and 24), which will be discussed in detail in the next parts.

Furthermore, in consideration of the protection of the best interests of the child, the minimum age of criminal responsibility should be determined based on the gradual transition model (with a given minimum and a maximum age) in order to recognize diminished responsibility without any discrimination between boys and girls.

2.1.4.1.2 Decriminalization

In a narrow sense, decriminalization means to remove the label or title of "criminal" from an act or behavior which is described as criminal by law which also leads to the removal of punishment from a specific act or behavior. In other words, decriminalization involves the processes through which the related authority in the criminal system imposes the sanctions, as a reaction toward some behaviors. Decriminalization may be achieved through legislation or a method in which the judiciary system interprets the law.[359] Generally, it can be concluded that decriminalization should be applied to acts which are not serious and do not inflict serious injury to oneself or others. Furthermore, the status offenses (such as truancy, disobedience, public drunkenness, and so on) which are not considered as crimes if the perpetrator is an adult should also be decriminalized. The United Nations' instruments on human rights have allocated some discussion to this issue. For instance, article 5 of the Riyadh Rules stipulates that in order to prevent further stigmatization, victimization, and criminalization of young persons, legislation should be enacted to ensure that any conduct not considered an offense or not penalized if committed by an adult is not considered an offense and thus not penalized if committed by a young person (see also general comments no. 10 and 24).

In addition to the above-mentioned items, decriminalization in juvenile crimes can be investigated from another point of view. Some acts turn a child into a delinquent-victim with victimization being more dominant. In many legal systems, such behaviors are addressed in the field of victimization rather than criminal acts. Sexual offenses and drug abuse by children are examples of such behaviors. In such cases, rather than punishing children, supportive measures should be applied for them and they should be treated as victims instead of offenders.

---

359 *Nadjafi Abrandabadi et al.* 1397/2018, p. 76.

2.1.4.1.3 Depenalization

Depenalization means partial decriminalization, in which an intervention and social policy replaces punishment, but the act or omission is still considered a crime. Depenalization may be realized in various ways. It may be the result of the measure reached by the legislator converting a serious crime into a less serious one, in which the offender is only subjected to disciplinary punishment, or which may be the result of efforts of judicial authorities such as the courts. For instance, a judge can make a crime less severe by providing less severe descriptions about it during the investigation. Depenalization may even originate from a victim who refuses to complain or give their consent, and based on this consideration, some legal systems have subjected the criminal prosecution to the formal complaint of the victim. It can be concluded that, as a whole, the concept of depenalization involves all forms of decentralization of the penal system.[360]

International instruments emphasize that explicitly punitive approaches are not appropriate. In juvenile cases, such considerations should always be outweighed by the interest of safeguarding the well-being and the future of the young person. Probation should be granted to the greatest possible extent via suspended or conditional sentences, board orders, and other dispositions. Capital punishment should be entirely eliminated. The use of alternatives should be applied to institutionalize to the maximum extent possible, and thus, deprivation of liberty can only be applied as a last resort and for a short period (Beijing Rules, articles 17.1, 17.2; Convention on the Rights of the Child, article 37, general comment no. 24, p. 79–81).

2.1.4.1.4 Decarceration

One of the most fundamental of human rights is individual liberty, which is legally violated by incarceration. Governments have a duty to justify the use of incarceration as being necessary to achieve an important societal objective for which there are no less restrictive means for the objective to be achieved. The loss of liberty that results from imprisonment and correction centers is then inevitable and impinges several other human rights as well.[361] Detention has a profoundly negative impact on young people's mental and physical well-being, their education, and their employment. The process of incarcerating them will reduce their future earnings as well as their ability to remain in the workforce, and could turn formerly detained youth into less stable employees. Most importantly, for a variety of reasons, there is viable and significant research which suggests that the experience of detention may make it more likely that youth will continue to engage in delinquent behavior, and that the detention experience may increase the odds that they will recidivate, further compromising public safety.[362]

---

360 *Mehra* 1385/2006, p. 308; *Nadjafi Abrandabadi & Hashem Beigi* 1397/2018, pp. 105–106.

361 UN 2007, p. 16.

362 *Holman et al.* 2013, pp. 1–2.

Regarding depenalization policies, international instruments and standards (Convention on the Rights of the Child, Beijing Rules, and general comments no. 10 and 24) provide that any restrictions on the personal liberty of the juvenile should be imposed only after careful consideration and should be limited to a possible minimum. Subsequently, deprivation of personal liberty should not be imposed unless the juvenile is adjudicated of a serious act involving violence against another person or of persistence in committing other serious offenses, and unless there is no other appropriate response. The well-being of the juvenile shall be the guiding factor in the consideration of their case.

2.1.4.1.5 Diversion

According to the theory of labeling and stigmatization, informal intervention is the best way to deal with juvenile delinquents. In fact, under diversion strategies, authorities focus on dealing with people in other ways, especially children and juveniles who could be processed through the criminal justice system. Criminal justice systems typically should process only a small proportion of criminal law offenses, because if the system investigates, prosecutes, trials, and convicts all offenders, its various parts, including prisons, would soon be unable to cope with the sheer numbers of delinquents. As a result, police and prosecutors, who introduce offenders into the system, have to exercise a degree of discretion in deciding whom to take action against and whom to ignore.[363]

In this regard, international instruments and standards propose to deal with juvenile offenders without resorting to formal trial, wherever appropriate and by the competent authority (see Beijing Rules, article 11). To this reason, the police, the prosecution, or other agencies dealing with juvenile cases should be empowered to dispose of such cases, at their discretion, without recourse to formal hearings under certain conditions which will be reviewed in the next sections. In accordance with general comment no. 24 (2019) para. 8, diversion is defined as any measures for referring children away from the judicial system, at any time prior to or during the relevant proceedings. Also, restorative justice is recommended – as any process in which the victim, the offender, and/or any other individual or community member affected by a crime actively participates in the resolution of matters arising from the crime, often with the help of a fair and impartial third party. Examples of restorative processes include mediation, conferencing, conciliation, and sentencing circles. The Committee on the Rights of the Child emphasizes that the competent authorities – in most States the public prosecutor – should continuously explore the possibilities of avoiding a court process or conviction, through diversion and other measures. In other words, diversion options should be offered from the earliest point of contact, before a trial commences, and be available throughout the proceedings. In the process of

363 UN 2007, p. 16.

offering diversion, the child's human rights and legal safeguards should be fully respected, bearing in mind that the nature and duration of diversion measures may be demanding and that legal or other appropriate assistance is therefore necessary. Diversion should be presented to the child as a way to suspend the formal court process, which will be terminated if the diversion programme is carried out in a satisfactory manner (general comment no. 24 para. 72).

2.1.4.1.6 Due process

Due process can be interpreted as the right to be treated fairly, efficiently, and effectively by the judicial administration. The rights to due process place limitations on laws and legal proceedings, in order to guarantee fundamental fairness and justice. Due process is interpreted as the rules administered through courts of justice in accordance with established and sanctioned legal principles and procedures, and with safeguards for the protection of individual rights. The rules applicable to the administration of justice are extensive and refer to fair trial, presumption of innocence, as well as independence and impartiality of the tribunal.[364]

In this regard, the Beijing Rules provide that the proceedings shall be conducive to the best interests of the juvenile and shall be conducted in an atmosphere of understanding, which shall allow the juvenile to participate therein and to express themselves freely; the minimum standards that are applied almost universally for any criminal defendant under the procedure known as "due process of law" shall be followed. In accordance with due process, a "fair and just trial" includes such basic safeguards as the presumption of innocence, the right to be notified of the charges, the right to remain silent, the right to counsel, the right to the presence of a parent or guardian, the right to appeal to a higher authority the presentation and examination of witnesses, the common legal defenses, the right to have the last word in a hearing, and the right to appeal, which will be studied in the next sections (see articles 7.1 and 14.2 of the Beijing Rules, article 40(2) of the Convention on the Rights of the Child). General comment no. 24 (2019) in section D emphasizes guarantees for a fair trial and explains about no retroactive application of child justice, presumption of innocence, right to be heard, effective participation in the proceedings, prompt and direct information of the charge, legal or other appropriate assistance, decisions without delay and with the involvement of parents or guardians, freedom from compulsory self-incrimination, presence and examination of witnesses, right of review or appeal, free assistance by an interpreter, and full respect of privacy. The general comment also elaborates on an emerging view that trials of civilians by military tribunals and state security courts contravene the non-derogable right to a fair trial by a competent, independent, and impartial court. This is an even more concerning breach of rights for children, who should always be dealt with in specialized child

---

364 UN 2007, p. 16, see also *Slobogin* 2006, pp. 955–990.

justice systems. The Committee has raised concerns about this in several concluding observations (para. 96).

2.1.4.2 Iran and Germany

In Iran, these specific measures have – according to the international instruments and standards as well as in line with the modern theories – been considered by the legislator and are applied in *Ta'zir* crimes and punishments. All the above-mentioned specific measures and policies have also been applied in line with global standards. In the next sections, all these measures and policies as well as their advantages and disadvantages in Iran and Germany will be discussed.

2.2 Juvenile Delinquents and Criminal Responsibility

2.2.1 Juvenile delinquents

It should be noted that the term "juvenile delinquent" is an ambiguous concept of many different intermingled criteria such as: age, maturity, intellectual development, nature of offenses, as well as, more importantly, responsibility for crimes. In its broadest sense, a juvenile delinquent is a non-adult person who violates the formal norms in society. In a narrower sense, a juvenile delinquent is a person designated as a non-adult who has committed a crime that would make him/her subject to the juvenile court. In the criminal justice system, an underage person who has been adjudicated based on their criminal responsibility is considered a juvenile delinquent.[365] Therefore, in this context, being held liable for having committed a crime is a core element, which will be explained in detail in this section. It should be noted that there are several terms to address non-adult persons in different legal systems, such as "children", "juveniles", "young adult persons", or "youth", and also some terms for addressing non-adult persons who violate the law, such as "children in conflict with the law", "juvenile delinquents", "juvenile offenders", "young adult offenders", and "young offenders". Nevertheless, the terms "child", "juvenile", "young adult", and "juvenile offender" have been used for non-adult persons who have a criminal responsibility, but there is no definite determination of a age for criminal responsibility in the international instruments.

The term "juvenile delinquent" has not been defined in international documents; Instead, the term "juvenile offender" is commonly used. In order to know who is a juvenile delinquent at the international level, we should refer to the Beijing Rules that are directly related to the administration of juvenile justice.[366] According to this advisory instrument, a juvenile offender is a child or young person who is alleged or

365 *Dussich* 1989, pp. 70–71.

366 *Abachi* 1388/2009, p. 36.

has been found to have committed an offense (article 2.2-c). Furthermore, a juvenile is defined as a child or young person who, under the respective legal system, may be dealt with for an offense in a manner which is different from an adult (article 2.2-a); in this respect, according to article 1 of the Convention on the Rights of the Child, a child means every human being below the age of 18 years, unless maturity is attained earlier under the law applicable to the child.[367] It is notable that article 2 of the Convention on the Rights of the Child emphasizes that the right set for children should be respected without discrimination. Therefore, based on the best interests of the child and without any discrimination, the age determination is the most important element for defining the term "juvenile delinquent".[368]

Accordingly, being a delinquent is indicative of the special mental and age conditions of the perpetrator, and because of this feature, children are not considered offenders if they commit an act against the law but cannot be treated as delinquents in the realm of criminal law. In this regard, the United Nations' documents have put more emphasis on this issue and describe labeling as considerably harmful to children. In fact, juveniles are more susceptible to stigmatization. Criminological studies on labeling processes have provided evidence on a variety of adverse effects caused by the permanent notoriety of young people as delinquents or criminals. According to recent research, any kind of labeling of offenders, even just using the word "delinquent", will have adverse consequences for the child. According to the rules of the United Nations, labeling a child as "delinquent" itself indicates their shortcomings and never refers to their conditions before delinquency, even if they have been truly rehabilitated. However, there is controversy over labeling methods, which is the result of social reactions and can be effective in increasing criminal behavior and perceptions of self-criminalization.[369]

---

367 This article focuses on all aspects of children's rights in general and of juvenile justice systems in particular.

368 Another similar term – "children in conflict with the law" – refers to anyone under 18 years of age who comes into contact with the justice system as a result of being suspected or accused of committing an offense. Most children in conflict with the law have committed petty crimes or minor offenses such as vagrancy, truancy, begging, or alcohol use. Some of these delinquencies are known as "status offenses" and are not considered criminal when committed by adults. In addition, some children who engage in criminal behavior have been used or coerced by adults. Too often, prejudice related to race, ethnicity, or social and economic status may get a child into conflict with the law even when no crime has been committed, or result in harsh treatment by law enforcement officials. See *Maxim et al.* 2008.

369 *Abachi* 1388/2009, p. 53. Studies on child delinquents have revealed that labeling under terms such as "criminal", "delinquent", and "law violator" happens within families rather than criminal justice institutes (some sort of labeling happens in these institutes through their special procedures). On the other hand, some studies on this issue indicate that juvenile delinquency is rather the result of labeling by others who consider the child as a lawbreaker. However, according to the results of both of these studies, labeling by the family and other people in society plays a more decisive role in increasing delinquency among juveniles than labeling by criminal justice institutes.

A person often decides to commit a crime by themselves, although sometimes two or more people collaborate with each other in doing so.[370] Generally, all of them are legally answerable by criminal law and considered liable as main offenders or perpetrators on the one hand – or mere accomplices, instigators, aiders, and abettors on the other. However, in most legal systems, the liability of accomplices, instigators, aiders, and abettors is derived from the liability of the main perpetrators, which is called "derivative responsibility".[371] Since they are vulnerable persons in society, juvenile delinquents can play a role in committing a crime as accomplices, instigators, aiders, and abettors; therefore, according to the United Nations' instruments, special laws must be enacted and enforced in order to prevent children and juveniles from participating in the criminal activities of adults. In particular, the use of children and juveniles in the production and trafficking of illicit drugs should be banned, and necessary measures must be taken in order to prevent children from participation in such crimes (article 33 of the Convention on the Rights of the Child; article 53 of the Havana Rules).

With regard to the participation of children and juveniles in criminal activities of adults, it was emphasized that governments should adopt effective measures to prevent the abuse of children and juveniles who have run away from home and school, those who are rebellious and delinquent, and street children who might participate in criminal activities of adults; criminal activities of adults in which children may primarily participate in include drug trafficking, sexual activities, pornography, robbery, and murder. Therefore, good strategies should be applied to develop measures for such children who are vulnerable of being abused in criminal activities of adults. The main aim of adopting such measures is to reintegrate these children and juveniles into society. For this purpose, governments should consider and apply appropriate punishments for adults who force and stimulate children to engage in criminal activities. They should pay special attention to the fact that child engagement in criminal activities of adults is a serious crime. This way of thinking would lead to bringing into effect severe punishments for such crimes. It is notable that adult offenders abuse children and juveniles in illegal activities because they know that children have a diminished criminal responsibility.[372] Due to their specific physical and psychological conditions as well as under the influence of external factors, children and juveniles sometimes commit acts that are illegal and illegitimate, but they are not found to be criminals or delinquents. Special attention should be paid to the group of children and juveniles who are considered and defined as victim-delinquents, or even as victims and not as offenders. In fact, they should be cared for in rehabilitative and health care centers so as to prevent their re-victimization and recidivism.

In this regard, there is evidence that criminals and offenders are often former victims as opposed to ordinary individuals. There are many examples of victims who commit

---

370 *Norbaha* 1389/2010, p. 216.

371 *Vogel* 2009, p. 152.

372 *Gregory* 1995, p. 781.

illegal acts or lead a deviant or criminal life. Both the Convention on the Rights of the Child (articles 19 and 34) and the Riyadh Rules (articles 25 and 53) have clarified the victimization of children in these crimes and state that the adults who use children for such services should be subjected to severe criminal penalties. According to article 39 of the Convention on the Rights of the Child and article 46(b) of the Riyadh Rules, appropriate rehabilitation measures and programs should be developed and executed for such children who have been victimized. One of these measures is to keep children in special institutions for rehabilitation and treatment as well as to prevent their future delinquency. Due to the vulnerability of children, as stipulated in international conventions, special attention should be paid to child victims, and they should be given rehabilitative health care.[373]

It is noteworthy to say that the vulnerability of children and adolescents, their involvement in certain crimes, and identifying them as victims in specific crimes such as sexual abuse and exploitation is often a difficult task due to the sensitive nature of the committed acts. Although it is difficult to detect whether a child is a delinquent or a victim in sexual exploitation or pornography, most legal systems assume such a child to be a victim or at least an offender-victim. Perhaps this is due to the fact that children suffer the highest level of mental, psychological, and physical injury in these crimes and should be given rehabilitative care. With regard to other crimes – such as financial crimes or organized robberies –, although the child may sometimes be a victim (i.e. forced by adults to commit this type of crime), he/she is basically considered a criminal (however, the child may be only an abettor to the crime and not the perpetrator). According to the criminalization basis of a criminal act, this is the difference in the type of crime, in terms of whether it has been committed against people, theft of property, or disturbing the public order, and the extent of material or spiritual damage. Although finding children guilty and proclaiming them an offender in the latter case is not a barrier to rehabilitative and treatment programs and punishment may vary depending on the severity of the crime, any punishment for these crimes should be rehabilitation-oriented.[374]

So as to study the legal status of juvenile delinquents in the dimensions of this research, criminal responsibility will be explained in full, and the common terms in the legal sources will be defined for both the international (United Nations) and the national levels (Iran, Germany).

The concept of juvenile delinquency is linked to criminal responsibility, which are both tied to age determination on the one hand and the notion of maturity, intellectual development, and an ability of understanding on the other. These elements will be studied in the international instruments as well as in the applicable laws of Iran and Germany in the following parts. But before explaining the elements mentioned above, the different types of juvenile delinquents will be discussed.

---

373 *Abachi* 1388/2009, p. 54.

374 *Abachi* 1388/2009, p. 55–57.

2.2.2 Elements of criminal responsibility concerning juvenile delinquents

In this part, the elements of criminal responsibility concerning juvenile delinquents will be first studied for United Nations and then for Iran and Germany.

2.2.2.1 United Nations

2.2.2.1.1 Age and gradual criminal responsibility

As already explained, diminished criminal responsibility is recommended based on considering minimum and maximum ages[375] for criminal responsibility by the United Nations' instruments. The juveniles and children who have diminished responsibility are assumed to be criminally liable only if they have the required maturity in that regard. This kind of age determination is called gradual criminal responsibility, which is recommended in line with the best interest of the child by the international instruments and standards.[376] Regarding the minimum age of criminal responsibility, article 40.3 of the Convention on the Rights of the Child provides that States parties shall seek to promote the establishment of laws, in particular a minimum age below which children shall be presumed not to have the capacity to infringe the penal law. The Beijing Rules document also emphasizes that the beginning of the age of criminal responsibility shall not be fixed at too low an age level (article 4.1).

According to general comment no. 24 (2019) section C, children who are below the minimum age of criminal responsibility at the time of committing an offense cannot be held responsible in criminal law proceedings. Children at or above the minimum age at the time of committing an offense but younger than 18 years can be formally charged and subjected to child justice procedures, in full compliance with the Convention. The Committee reminds States parties that the relevant age is the one at the time of committing the offense. Under article 40(3) of the Convention, States parties are required to establish a minimum age of criminal responsibility, but the article

---

375 Two ways to approach age in the legal context include: "the status approach, which is based on biological age; and the competence approach, which is based on the maturity of the young offender and takes into account particular characteristics of the individual. Identifying a boundary between childhood and adulthood becomes important because a person's legal identity as either a child or an adult may confer a series of legal entitlements, duties, responsibilities, and accountability for actions. In relation to criminal law, these approaches seek to establish a point in a child's life after which they may be held accountable for their actions if they cause harm to others or the society generally. However, the disadvantages of each approach are rather obvious: a status approach might be too static, not paying enough attention to the individual and therefore conflicting with, for example, the United Nations' Convention on the Rights of the Child. On the other hand, a competence approach, with its individualized shape, may clash with the rule of law, specifically with predictability and transparency"; *Persson* 2017, p. 22.

376 As a positive improvement at the national level, the Committee on the Rights of the Child notes that some States parties allow for the application of the rules and regulations of juvenile justice to persons aged 18 and older, usually until the age of 21, either as a general rule or by way of exception.

does not specify this age. More than fifty States parties have raised the minimum age following their ratification of the Convention, with the most common minimum age of criminal responsibility now being 14 years internationally. Nevertheless, reports submitted by States parties indicate that some States retain an unacceptably low minimum age of criminal responsibility.

The general comment also emphasizes that several States parties apply two minimum ages of criminal responsibility (for example 7 and 14 years), under the presumption that a child who is at or above the lower age but below the higher age lacks criminal responsibility unless sufficient maturity is demonstrated. Initially devised as a protective system, it has not proven to become one in practice. Although there is some support for the idea of an individualized assessment of criminal responsibility, the Committee has observed that this leaves much to the discretion of the courts and results in discriminatory practices. States parties are urged to set one appropriate minimum age and to ensure that such legal reform does not result in a retrogressive position.

2.2.2.1.2 Maturity, intellectual development, ability of understanding, and diminished criminal responsibility

In addition to the age determination and in accordance with the United Nations' instruments and standards, findings on emotional, mental, and intellectual maturity should be considered for the notion of criminal liability. This approach would mean to consider whether a child can live up to the moral and psychological components of criminal responsibility; that is, whether a child, by virtue of her or his individual discernment and understanding, can be held responsible for essentially anti-social behavior.[377] The international instruments emphasize that there is a close relationship between the notion of responsibility for delinquent or criminal behavior and other social rights and responsibilities, such as marital status, civil majority, etc. They recommend to agree on a reasonable lowest-age limit that is applicable internationally (article 4 of the Beijing Rules).

Children in conflict with the law who at the time of committing the crime are at or above the lower minimum age but below the higher minimum age are assumed to be criminally responsible only if they have the required maturity in that regard. This means that the age determination should be assessed by the notion of maturity. This assessment should be left to the court, with the requirement of involving a psychological expert, and in practice results in the use of the lower minimum age in cases of serious crimes. Documented evidence in the fields of child development and neuroscience indicates that maturity and the capacity for abstract reasoning is still evolving in children aged 12 to 13 years due to their frontal cortex still developing. Therefore, they are unlikely to understand the impact of their actions or to comprehend criminal proceedings.

377 Protecting Children's Rights in Criminal Justice Systems 2013, p. 1.

They are also affected by their entry into adolescence. As the Committee notes in its general comment no. 20 (2016) on the implementation of the rights of the child during adolescence, adolescence is a unique defining stage of human development characterized by rapid brain development, which affects risk-taking, certain kinds of decision-making, and the ability to control impulses. States parties are encouraged to take note of recent scientific findings and to increase their minimum age accordingly, to at least 14 years. Moreover, developmental and neuroscience evidence indicates that adolescent brains continue to mature even beyond the teenage years, still affecting certain kinds of decision-making at a later stage. Therefore, the Committee commends States parties that have a higher minimum age, for instance 15 or 16 years of age, and urges them not to reduce the minimum age of criminal responsibility under any circumstances, in accordance with article 41 of the Convention. The Committee recognizes that although setting a minimum age of criminal responsibility at a reasonably high level is important, an effective approach also depends on how each State deals with children above and below that age. The Committee will continue to scrutinize this in reviews of State party reports. Children below the minimum age of criminal responsibility are to be provided with assistance and services according to their needs by the appropriate authorities, and they should not be viewed as children who have committed criminal offenses. If there is no proof of age and if it cannot be established that the child is below or above the minimum age of criminal responsibility, he/she is to be given the benefit of doubt and is not to be held criminally responsible.

The Committee is concerned about practices that permit the use of a lower minimum age of criminal responsibility in cases where, for example, the child is accused of committing a serious offense. Such practices are usually created so as to respond to public pressure and are not based on a rational understanding of children's development. The Committee strongly recommends that States parties abolish such approaches and set one standardized age below which children cannot be held responsible in criminal law, without exception (general comment no. 24 section C).

2.2.2.1.3 Irresponsibility

In the international instruments, there are some rules on age determination and irresponsibility of children. Accordingly, children who commit an offense at an age below that minimum cannot be held responsible in a penal law procedure. Even very young children do have the capacity to infringe the penal law, but if they commit an offense while below the minimum age, the irrefutable assumption is that they cannot be formally charged and held responsible in a penal law procedure. International standards provide special protective measures for these children to be taken if necessary, in line with the best interest of the child (general comments no. 10 para. 31). Therefore, the establishment of a minimum age below which children shall be presumed not to have the capacity to infringe the penal law is necessary.

Even if there is no proof of age and it cannot be established that the child is at or above the minimum age of criminal responsibility, he/she is to be given the benefit of the doubt and shall not be held criminally responsible. Every child must be provided with a birth certificate free of charge whenever he/she needs it to prove his/her age. Children with developmental delays or neurodevelopmental disorders or disabilities (for example autism spectrum disorders, fetal alcohol spectrum disorders, or acquired brain injuries) should not be in the child justice system at all, even if they have reached or passed the minimum age of criminal responsibility. If not automatically excluded, such children should be individually assessed (general comment no. 24 paras. 24, 25, 28).

A child who does not have a birth certificate should promptly be provided with one and free of charge by the state whenever it is required to prove his/her age. If there is no proof of age through a birth certificate, the authority should accept any documentation that can do so, such as a notification of birth, extracts from birth registries, baptismal or equivalent documents, or school reports. These documents should be considered genuine unless there is proof to the contrary. Authorities should allow for interviews with or testimony by parents regarding the child's age, or for permitting affirmations to be filed by teachers, religious or community leaders who know the age of the child. Only if these measures prove unsuccessful may there be an assessment of the child's physical and psychological development, conducted by specialist pediatrician or other professionals skilled in evaluating different aspects of development. Such assessments should be carried out in a prompt, child- and gender-sensitive as well as culturally appropriate manner, including interviews with children and parents or caregivers in a language the child understands. States should refrain from using only medical methods based on, inter alia, bone and dental analysis, which is often inaccurate due to wide margins of error and can also be traumatic for the child. The least invasive method of assessment should be applied. In the case of inconclusive evidence, the child or young person is to have the benefit of the doubt (general comment no. 24 paras. 33, 34).

2.2.2.2 Iran

2.2.2.2.1 Criminal responsibility in discretionary punishments

In Iran, discretionary or *Ta'zir* punishments are applicable for crimes prohibited by Islam, and their types and quantities have not been determined in Islam (article 18 of the Penal Code 2013).[378] All offenses against persons, property, and public security (except offenses in *Hadd*, *Qisas*, and *Diye*) are *Ta'zir* crimes. It is notable that many crimes and punishments have been categorized in the discretionary or *Ta'zir* section in accordance with new findings in criminology and penology – and, what is more important, in line with international standards they can be changed by reforming the

378 The English translation of some sections of this code is available at www.iranhrdc.org [20/05/2021].

respective laws, which means that most of the innovations and changes can be made in this section. Therefore, as far as the *Ta'zir* crimes committed by juveniles are concerned, the new rules and innovations are applicable in line with the United Nations standards.

a) Age determination and gradual criminal responsibility

As explained in the above discussions, in consideration of protecting the best interests of the child, the age of criminal responsibility should be determined based on the gradual transition model (with a given minimum and a maximum age) in order to recognize diminished responsibility and without any discrimination between boys and girls. In the Penal Code 2013, the legislature tried to find a good solution in accordance with the needs of children and in line with international conventions and standards. For discretionary or *Ta'zir* punishments, the minimum age of criminal responsibility is 9 years and the maximum age is 18 years (according to the solar calendar),[379] without any discrimination between girls and boys and in consideration of gradual transition, which is considered to be the most important innovation in this code (articles 88 and 89 of the Penal Code 2013). In *Ta'zir* crimes, children and juveniles are divided into two sections: children between 9 and 15 years of age and children between 15 and 18 years of age. Article 88 stipulates that in cases of children and juveniles whose age at the time of committing a *Ta'zir* crime is between 9 and 15 years (according to the solar calendar), the court shall make disciplinary measures and based on rehabilitative objectives. For the second group, article 89 states that for juveniles whose age at the time of committing a *Ta'zir* crime is between 15 and 18 years, punishments such as deprivation of liberty for a short time, paying a fine, and public services are impelied.

As can be seen in this section, the Iranian legislature is influenced by international standards, not only as regards the content but also in terms of the literature. The designations "child" and "juvenile" are used instead of "mature" and "immature", which are the terms used for fixed punishments (*Hadd*, *Qisas*, and *Diye*) in *Shari'a*, although the term "child" is only used for persons between 9 and 15 years of age. A gradual criminal responsibility has been accepted without any discrimination between girls and boys and in line with international standards.

b) Maturity and diminished criminal responsibility

It should be mentioned that in the articles related to criminal responsibility of children and juveniles in *Ta'zir* crimes (articles 88 and 89 of the Penal Code), no condition is stipulated by the legislature concerning the notion of maturity. It seems that the age determination and the categories for children and juveniles were more important than emphasizing the notion of maturity for the legislature – although this

379 The main Iranian calendar is a solar calendar; however, besides this calendar, the lunar calendar is also in use in Islam.

does not mean that maturity has been ignored. Regarding this very important condition, as a general rule, article 140 of the Penal Code provides that in cases of *Hadd*, *Qisas*, and *Ta'zir*, criminal responsibility shall be established only when the person is sane, mature, and free at the time of committing the offense. Therefore, regarding children and juveniles, their maturity should be established for all crimes. The Iranian legislature has defined the term of "maturity" along the age determination. In this regard, article 147 of the Penal Code states that the age of maturity for girls and boys a full 9 and 15 lunar years, respectively.[380] This age determination is based on their physical and sexual maturity rather than their moral and mental development.

As becomes clear from the content of the above-mentioned articles, there are clashes between the articles related to children and juveniles. In fact, the legislature could not draw a clear line between the rules governing discretionary and fixed punishment. This is because the rules governing fixed punishment and influenced by *Shari'a* are considered a main criterion for legislature according to the Iranian Constitution, which has been explained in the previous chapter. However, in practice, there is no mechanism to assess maturity as a main element for bearing criminal responsibility, while, as already mentioned, the main criterion for assessing maturity is the age determination, which is not a correct assessment. In other words, if a child or juvenile between the ages of 9 and 18 (as stipulated in articles 88 and 89) commits a *Ta'zir* crime, and if it is recognized that he/she commits the crime with a free will while being sane, this person is assumed to be a mature person and considered liable, provided that the mens rea is proven (articles 144–145 of the Penal Code). It is worth mentioning again that the new interpretation of *Shari'a*[381] shows that the age determination is not sufficient for assessing maturity, and the notions of the child's ability of understanding and knowing that something is forbidden are those elements which should be considered for assessing his/her maturity and subsequently for bearing criminal responsibility in accordance with *Shari'a*.

c) Irresponsibility

Children under the age of 9 years who commit *Ta'zir* crimes are not criminally responsible. They are assumed to be immature children. In this regard, there is no discrimination between girls and boys. Nevertheless, according to article 148 of the Penal Code, if children (boys or girls) under 9 years of age commit crimes, disciplinary measures can be applied in accordance with the provisions in *Ta'zir* crimes.

---

380 Despite this stipulation in the Penal Code, some Iranian *Shi'a* clerics consider the age of maturity for girls to be higher. *Ayatollah Yousef Sanei* and *Ayatollah Mosavi Bojnordi*, for example, set the age of maturity for girls at 13 instead of 9 years. But the Penal Code has followed the *fatwa* of the majority of conservative clerics, who deem 9 years to be the age of maturity for girls.

381 For further reading about the new interpretation of *Shari'a* concerning assessing maturity, see *Nobahar* 1391/2012, pp. 168–183.

2.2.2.2.2 Criminal responsibility in fixed punishments

In general, fixed punishments, including *Hadd*, *Qisas*, and *Diye*, are penalties applicable for crimes prohibited by *Shari'a*. The types and quantities of these punishments on the one hand and the special rules governing these responses on the other have been determined in Islam. Therefore, the rules governing fixed punishments are stable, and any changes in very exceptional cases should be done by a religious authority (*faqih*) by issuing a scholarly opinion on the matter of Islamic law (*fatwa*).

a) Age determination and non-gradual criminal responsibility

In the new Penal Code of 2013, in the fixed penalties (*Hadd*, *Qisas*, *Diye*), the age that is set for criminal responsibility is based on *Shari'a*, which is very different from the model used in discretionary punishments. In fact, in this section, no changes or innovations have been made as far as the age of liability is concerned. In other words, immature persons have no criminal responsibility (article 146 of the Penal Code 2013), and the age of maturity is defined as 9 lunar years for girls and 15 lunar years for boys (article 147 of the Penal Code 2013). In this section, there is discrimination between girls and boys, and the gradual transition to full responsibility has not been accepted by legislature. Here, the legislature has been completely influenced by the conservative view of *Shari'a*, not only as regards the content but also in terms of the considered literature and calendar. In *Hadd*, *Qisas*, and *Diye*, the designations "mature" and "immature" are used instead of "child" and "juvenile", and the calendar used for evaluating the age of responsibility is the Islamic calendar.

This method of age determination as introduced by the legislature is due to the obstacles existing in the Iranian Constitution, which provides that all laws should be in accordance with *Shari'a* (article 4 of the Constitution). Therefore, in the category of fixed penalties, the legislature could neither extend the rights of juveniles in compliance with international instruments nor do so according to the needs of juveniles in society. Nevertheless, some solutions were introduced to solve problems relating to the fixed penalties. Introducing the notion of intellectual development to these rules is the most important solution to be discussed.

b) Maturity, intellectual development, and diminished criminal responsibility

In the fixed punishments, all girls above 9 years and all boys above 15 years (according to the lunar calendar)[382] are considered mature persons with full criminal responsibility; they are treated as adults and thus according to the general rules in the Penal Code. This age determination is based on the non-gradual model, which is a very controversial subject matter, since it violates international obligations.

382 8 years and 9 months for girls and 14 years and 7 months for boys according to the solar calendar.

In order to solve the problem in the fixed punishments, for the first time, the legislature has introduced the notion of intellectual development into the Penal Code 2013,[383] as a key element regarding the sentencing of juvenile delinquents in *Hadd*[384] and *Qisas*. article 91 of the Penal Code stipulates that in cases of offenses punishable by *Hadd* or *Qisas*, if mature persons under 18 years do not realize the nature of the committed crime or its prohibition, or if there is uncertainty about their full intellectual development, according to their age, they shall be sentenced in accordance with the punishments prescribed in the *Ta'zir* section. According to the note of this article, the court may ask the opinion of forensic medicine or resort to any other method that is deemed appropriate in order to establish the delinquent's full mental development. Thus, in *Hadd* crimes such as adultery or homosexuality and also in *Qisas* for crimes such as murder, the new code gives judges the discretion to decide whether or not the mature person below 18 years has understood the nature of the crime. The most important point here concerns the notion of maturity, which is mentioned in article 91.

When comparing articles 147 and 91 of the Penal Code in particular, one realizes that the notion of maturity determined by age is based on the person's physical and sexual development, while his/her intellectual development is determined based on his/her moral and mental maturity. In this regard, as it will discussed in the following sections, there are differences between Iranian and German criminal law. Furthermore, it should be noted that the concept of maturity and intellectual development in the Iranian Penal Code are influenced by *Shari'a* while in the German Penal Code these notions are explained and described by Doctrine which is a scientific method.

c) *Irresponsibility*

In the fixed punishments, girls under the age of 9 and boys under the age of 15 are assumed as immature children. These children are exempt from criminal responsibility, and if they commit crimes, only disciplinary measures as described in the *Ta'zir* section are applied (articles 146–148 of the Penal Code). As already mentioned above, the crimes punishable by fixed penalties are considered as serious crimes, with discrimination between boys and girls in the age determination in the fixed punishments.

---

383 For a discussion about the definition of intellectual development or growth in *Shari'a*, see *Rahami* 2002, pp. 167–198; see also *Hashemi* 2005, pp. 247–297.

384 *Hadd* crimes are proven by confession, witnesses, and the knowledge of the judge, which are considered the specific and traditional evidences prescribed in *Shari'a*. Therefore, proving *Hadd* crimes is very difficult, and due to their different nature, the requirement of criminal responsibility for *Hadd* crimes is also different. In fact, in cases of offenses punishable by *Hadd*, the offender shall be liable only if – in addition to having knowledge, intention, and meeting the requirements for criminal responsibility – he/she is aware of the prohibition of the conduct committed under *Shari'a* rules. Furthermore, in the cases of offenses punishable by *Hadd*, if the accused claims that he/she, at the time of committing the offense, did not have the knowledge or intention to do so, or if he/she claims that one of the obstacles to criminal liability exists, in case there is the likelihood of veracity of the claim, or if he/she claims that his/her confession has been made under threat or fear of torture, the claim shall be accepted without resorting to testimonies and oaths (articles 217 and 218 of the Penal Code).

In fact, girls can receive adult punishments earlier than boys if they commit crimes. For example, if a 14-year-old girl kills someone, she is assumed as an adult and punished by execution if her intellectual development is not established; on the other hand, a 14-year-old boy with the same condition is assumed as an immature child, with no need to establish his intellectual development, and he can only be treated according to article 88. Accordingly, the response is cautioning and warning or taking a written promise not to commit an offense again – or deprivation of liberty from three months to one year will be applied (note 2 of article 88 of the Penal Code).

2.2.3.3 Germany

In Germany, there is an integrated rule for addressing juvenile delinquents without discrimination. The minimum age of criminal responsibility is formed based on gradual criminal responsibility. Furthermore, notions of maturity, intellectual development, and the ability to understand and control oneself are considered as key elements to bear or incur criminal responsibility.

2.2.3.3.1 Age and gradual criminal responsibility

There are two main classifications of age determination in the Youth Court Act[385] (sections 3 and 1.2) based on the criteria of childhood and adulthood. Persons who are between the ages of 14 and 18 years, and those who are older than 18 but not yet 21. Youth who, at the time of committing the criminal act, have reached the age of 14 but not yet 18 years, are described as being conditionally capable of guilt, based on the criteria of childhood. In this regard, section 3 of the Youth Court Act provides for an individual assessment of whether the young person can be found to be criminally responsible. Only when this is the case can the measures of the Youth Court Act be applied. If the young person is found not to be criminally responsible at the main trial, then he/she is to be acquitted and the youth court judge can, according to sections 2 and 3 of the Youth Court Act, order the same measures as the family court judge in order to educate the young person. These measures may, however, only be applied on the same basis as the ones the family court judge can apply.[386]

The rules governing criminal responsibility are completely different for the second group. A young adult who, at the time of committing the criminal act, has reached the age of 18 but not yet 21 years is, based on the criteria of adulthood, assumed to be fully criminally responsible, and there is no requirement of a special assessment of their capacity for guilt. For the purpose of criminal responsibility, they are regarded as adults. Nonetheless, it is recognized that such young adults are often not yet fully developed and should not be treated in the same way as adults. Thus young adults are to be dealt with in the youth court instead of the adult criminal court, and

385 The English version of the Youth Court Act is available at www.gesetze–im–internet.de/englisch_jgg/.

386 *Crofts* 2002, p. 56.

certain measures of youth criminal law can be applied if, according to section 105 of the Youth Court Act, 1) the overall assessment of the perpetrator's personality, taking account of his/her living environment, demonstrates that at the time of the act he/she was still equivalent to a youth in terms of his moral and intellectual development, or 2) the type, circumstances, and motives of the act indicate that it constituted youth misconduct.[387] This age determination, which is based on the gradual model, is applied for all youth and young persons without discrimination between girls and boys and governs all crimes committed by these persons. However, using the age associated with 'normal' development as a measure for judging the individual case is especially inappropriate. By recognizing this, the Federal Supreme Court has held that an age-corresponding development is not proof of the existence of moral and mental maturity, as required by section 3 of the Youth Court Act. As has been mentioned, besides age determination, other elements such as moral and intellectual development as well as the ability to understand and control should be considered for bearing criminal responsibility, which will be explained in detail below.

2.2.3.3.2 Moral and mental development and maturity

As stipulated in section 3 of the Youth Court Act, if, at the time of the criminal act, the young person has reached a level of moral and intellectual development/maturity sufficient to enable them to understand the wrongfulness of the act and to conduct themselves in accordance with such understanding, criminal liability should be incurred. The maturity of the young offender is judged according to their moral and mental development. It is notable that the maturity criterion is not concerned with the level of development of the young person in general, but only in relation to its effects on their ability to understand the wrongfulness of the act and to act according to this understanding in the concrete situation.[388] Moral maturity and development concerns the ability to evaluate one's behavior according to moral standards, which requirea that the young person has internalized the socio-ethical content of norms and has an anchored feeling for what is right or wrong. Furthermore, the young person must have grasped the social meaning and value of the legal demands placed on them and have integrated this in their person, such that they understand the seriousness of the norms. Only if the norms of society have a real value for the young person can they be guided by them and can be rightly blamed if they do not.[389]

---

387 *Crofts* 2002, p. 56.

388 It is interesting to note that in the Youth Court Act of 1923 reference was made to the "mental and moral" development, which was changed in the Acts of 1943 and 1953 to "moral and mental" development. This, it has been claimed, shows that the legislator wanted to emphasize the importance of the ability of the young person to ethically evaluate his/her action. *Crofts* 2002, p. 60.

389 Despite the emphasis on the moral development of the young person, it is generally accepted that the recognition of the act having been immoral is not sufficient to establish the required understanding, because legal and moral demands are not always the same. What is important

Regarding the mental or intellectual development, it should be noted that this requires a certain degree of intellectual maturity and refers to the cognitive abilities of the juvenile. This concerns the ability to distinguish between right and wrong and to rationally grasp the wrongfulness of one's action. Mental development, however, is not solely based on the intellectual abilities of the young person, but also touches the growth of one's willpower and the stability of their emotions. The type of factors used to assess the mental or intellectual development of the juvenile are: school knowledge, general experience, understanding of concepts and imagination, ability to remember things, flexibility of thought, perceptiveness, abstract-logical reasoning, grasp of connections, the ability to plan and foresee, etc.[390]

2.2.3.3.3 Ability to understand and control

Following the assessment of the moral and intellectual maturity of young persons described above, the next stage should focus on the question of whether the young person was able to understand the wrongfulness of the act and act accordingly. With respect to the ability of understanding, it should be clear to the young person that the act has caused harm to the legal interests of others or of the community, and that such behavior is forbidden by the state. The connection between the ability to understand and the moral development of the young person is thought to make clear that he/she must have been in a position to understand and take seriously the social content of the prohibition as well as of the legal interests protected in that way. He/she must at least have begun to understand the earnestness and necessity of the prohibition for social co-existence.[391] However, if the young person understood the wrongfulness of the act, he/she must also have been able to control his/her actions according to this understanding. Concern here thus touches a volitional element. Even if a

---

is that the young person is able to recognize the social and legal relevance of the act. Moral norms, however, lack the features that are essential to legal norms, i.e., they are not concrete, binding, and mutual. In order to understand an act as wrongful, the juvenile must be able to understand that he/she has an obligation toward society to obey such norms, rather than just having an obligation to another person such as a parent, relative, or friend. Therefore, the maturity required to be criminally responsible has often been termed "social maturity". *Crofts* 2002, p. 60.

390 *Crofts* 2002, p. 60.

391 In this field, the young person needs not know that his/her act violates a specific norm of the criminal law; it is not necessary that he/she understands that this act stands under the threat of punishment. A decision of the German Federal Supreme Court simply stated that the requirements of section 3 of the Youth Court Act are always fulfilled when the young person is conscious that he/she is doing something forbidden (BGH, Herlan [1959] GA 47). In fact, children tend to evaluate all prohibitions as carrying the same weight. For instance, a child may be unable to differentiate between the prohibition on playing in the courtyard and that on stealing. An adult, however, knows that there are differences between the weight of these prohibitions. As such, for the required understanding, the young person must have the ability to distinguish between those types of acts which are disapproved by adults for more or less obvious reasons (e.g. breaching disciplinary rules, not behaving properly, harming oneself, etc.) and such acts which – even despite being approved by individual adults (e.g. by negative examples, instigators) – violate the fundamental order and trust of communal living. *Crofts* 2002, pp. 60–61.

young person knows that an act is wrong, he/she may not yet be mature enough to resist committing it. For young people, the stimulus to commit a certain act may be of such elementary force that the motivation not to commit the act, connected with the knowledge that the act is wrong, is easily overridden.[392]

To what extent a youth can determine and control his/her actions depends on how developed and how strong the young person's will power and ability to resist are; it also depends on which factors, such as temptations and fears, are in play in each individual case. In order to be able to control his/her actions, the young person must have sufficient experience in dealing with frustration and stressful situations, such that he/she is mature enough to behave reasonably. In situations of conflict, the young person may not be mature enough to see other "non-delinquent" ways out of the circumstances which he/she finds himself in. It must be remembered that factors which may seem irrelevant to adults can have an overbearing importance to a young person. The latter's life takes place within the norms, values, and interests of their peers, which may be very different from those of adults. It may then be unrealistic to expect the child to act in the same way an adult would.[393]

2.2.3.3.4 Irresponsibility

Section 19 of the General Penal Code stipulates that persons who have not attained the age of 14 at the time of the commission of the offense shall be deemed to act without guilt. Accordingly, they are not capable of guilt, and regardless of how mature a child may be, the Youth Court Act does not apply and the child cannot be dealt with by the youth court. There can therefore be no criminal proceedings at all against children. The fact that children under the age of 14 are assumed to be incapable of guilt does not, however, mean that no consequences may follow from their criminal acts. Such acts may be seen as an indication that the child is not receiving proper education and that his/her welfare is endangered. If the youth prosecution officer thinks this is the case, he/she should bring to the attention of the parents or guardians and their child that they have a claim to help with education according to sections 27–35 of the Social Code Book VIII, and he/she may also inform the youth welfare

---

392 *Crofts* 2002, pp. 60–61.

393 Furthermore, young people are especially vulnerable to external influence, such as from older people, peer group pressure, or aggressive advertisements, and there may also be cases where a young person commits an act out of fear or admiration of an authoritative figure. A youth may, for example, fear punishment by his/her parents more than the comparatively remote punishment by the state if he/she does not commit an offense they require him/her to do. It is also often noted that during the age of puberty, the young person may have difficulty controlling awakening sexual desires, which may lead to sudden impulses leaving the young person without the power of reflection or restraint. The effect of such factors in establishing criminal responsibility will be examined in detail in the following section. Here they suffice to show that only having regard to the ability of the young person to understand the wrongfulness of a criminal act would not be sufficient to establish whether he/she was criminally responsible. *Crofts* 2002, pp. 60–61.

office.[394] Furthermore, with respect to children up to age 13 and juveniles aged 14 to 17 who are found to be in need of care and education, it should be noted that they are handled by youth welfare departments set up under the Youth Welfare Law. In general, youth problem behavior does not justify juvenile court proceedings; therefore, a criminal offense committed by a child may be used as an indicator for his/her need of care and education.[395]

2.3 Juvenile Delinquency and Sanctions

According to criminologists, any act or omission that is dangerous to the majority of the people in a society is called a delinquency or crime.[396] From a different point of view, sociologists believe that any act which violates the strong and specific states of a collective conscience is criminal.[397]

From the perspective of psychologists, in turn, delinquency is a bio-psycho-social phenomenon. It is sometimes misbehavior, but not all individuals who misbehave are seen as delinquents. The quality of behavior and the range of acts of social deviance are factors for judging whether a person is a criminal or rather mentally ill. On the other hand, some acts of misbehavior are illegal and some just are unethical, but not all illegal behaviors are also unethical. The difference between illegal and unethical behavior is very important and should be considered in defining crime and delinquency. Literally, delinquency means to leave a legal duty or to commit a wrong action which is not necessarily a crime. However, the term of delinquency is often used in relation to crime and especially to the crimes committed by children and juveniles. In its conventional and non-legal sense, child delinquency involves behaviors such as school absenteeism, truancy, public drunkenness, and disobedience (status offenses), which are not considered criminal if an adult commits these acts. Therefore, it can be concluded that delinquency is a series of minor offenses, and, since children typically commit minor offenses, the term “delinquency” is used for them,[398] whereas legally, crime or delinquency is defined as an act or omission that is punishable by law.

---

394 Guideline no. 2 in section 1 of the Youth Court Act also requires that when a child is reported to have committed a crime, the youth prosecution officer should consider whether to refer the case to the family court judge, who may apply civil care measures if necessary. Forced measures are, however, normally only applied if there is an actual neglect or danger of neglect of the child and if the parents or guardians are not able or not willing to avert this neglect or danger (section 1666 I Civil Code).

395 *Albrecht* 2004, p. 68. It is worth mentioning that public youth welfare is generally understood to be a last resort, with private youth welfare having priority over public interventions.

396 *Danesh* 1374/1995, p. 45.

397 *Ardebili* 138/2014, p. 10.

398 *Nadjafi Abrandabadi & Hashem Beigi* 1397/2018, pp. 9–10.

The legal definition of crime and delinquency is the most important definition applied in the justice system, and the same definition is considered for juvenile delinquents. Only by considering the legal definition can the discussion of sanctions be proposed and designed. Therefore, punishment or penal sanction can only be imposed for offenses against legal rules and applicable by the justice system according to legislation and with justification.[399] However, a very important question is raised here: are there any differences between juvenile and adult crimes and punishments from the point of view of the legal definition in the criminal code? If so, what are the criteria and justifications? In order to answer this question, we should distinguish between legal systems. In some systems, there are different criteria and justifications for criminalization and penalization in the juvenile justice system, as status offenses are criminalized for children and juveniles but not for adults. In fact, criminalizing status offenses is justified on the basis of crime prevention and the protection of children and juveniles. This justification is criticized by many scholars due to its negative effects on children's life in society. In addition, it is contrary to human rights standards.

In fact, according to stigmatization, when a person is described and labeled in a certain way by others, as a result of this social pressure, they try to change the perception of their own behavior so as to adapt to this label; the theory of labeling comes into play here. Therefore, criminalization should be limited especially for children and juveniles, because juvenile delinquents who are subjected to the procedures and convictions of the justice system are officially labeled as criminals or delinquents, so they are expected to behave criminally according to the theory of labeling, which would put their social status at risk.[400] In contrast, according to this theory, as far as children and juveniles are concerned, the notion of delinquency is a result of the special mental and age conditions of the perpetrator, and because of this feature, children are not thought of as criminals if they commit an act against the law that is not severe. Therefore, not only should status offenses be decriminalized but also other crimes which are common in adults and children; in line with the best interest of the child, a policy of decriminalization should thus be taken.[401]

With respect to decriminalization, and in relation to children and juveniles, it is even argued that there is a need for a separate catalogue of offenses, because the normative commands and prohibitions of the adult world are often not compatible with the experiences and norms understood by juveniles. This is because they have limited experience and often cannot foresee all the consequences connected to their actions, nor can they always understand the need for protection that may be the reason for

---

399 For further reading, see *McPherson* 1967, pp. 21–24.

400 *Nadjafi Abrandabadi & Hashem Beigi* 1397/2018, p. 79.

401 *Siegel et al.* 1997, p. 18.

the norm. Thus, the norms of the General Penal Code do not at all or at least not adequately cover youthful forms of interaction, and a linear application of all these norms to the young could lead to unjust results. The juvenile often acts on an impulse, i.e. spontaneously with little thought for the consequences of their actions, and many acts may be committed out of playfulness or adventure rather than with a criminal intention, such as stealing the clothes of an opponent football team or letting air out of car tyres.[402]

In light of such considerations, *Ostendorf* has suggested that there should be a rethinking of the definitions of offenses in relation to the child, juvenile, or young person in order to achieve a decriminalization of certain types of behavior. He discusses three groups of offenses which should be decriminalized for young offenders: 1) offenses which are generally not understood by a child, juvenile, or young person; 2) offenses which do not fit the child, juvenile, or young person; 3) offenses which need not apply to the child, juvenile, or young person.[403] However, other scholars believe that the arguments for a new catalogue of offenses may be persuasive to a certain degree, but they do not totally convince them of the need for a separate catalogue of offenses for minors.

Most of the points made above can be suitably addressed by a proper assessment of moral and mental maturity, which requires an individual examination of a young person's criminal responsibility. Accordingly, if the offense were one that a youngster had not understood to be wrong, or if, for example, the youth was not able to control his/her actions due to peer pressure, then he/she should not be found criminally responsible. Similarly, if the offense was minor, diversion may be a more appropriate response.[404]

In addition, a more important point is perhaps that criminal law aims at making clear what sort of behavior is wrongful and should not be committed, and insofar it has an educational purpose. It would therefore be misleading for a young person if an act which was not an offense under the age of 18 suddenly became one upon reaching that age. Indeed, children, juveniles, and young people must be prepared for the full responsibility of adulthood, although with care; this also includes understanding what is a punishable act. Therefore, according to the latter view about the justification of criminalization, it should be emphasized that the only offenses which should be decriminalized are those which come under the heading of "status offenses".[405]

---

402 *Croft* 2008, p. 132.

403 *Croft* 2008, pp. 132–133.

404 *Croft* 2008, p. 133.

405 *Croft* 2008, p. 133.

2.3.1 United Nations

Juvenile delinquency, sanctions, and responses, as well as finally depenalization and decarceration policies of the United Nations will each be discussed in this section.

2.3.1.1 Juvenile delinquency

2.3.1.1.1 Delinquency and crime

As already explained, according to scientific criminology (which deals with the causes of crime and the treatment of offenders), delinquency is basically referred to as any anti-social behavior or tension affecting society. In line with the criminological definition of crime and delinquency, the United Nations' instruments provide that delinquency, in any sense of the word, is affected by various factors such as poverty, family problems, inheritance, and so on, and they recommend accordingly that effective measures be taken to prevent the notion of crime and delinquency. However, international documents recommend social and situational crime prevention policies and measures to curb anti-social behaviors in society (article 5 of the Riyadh Rules). Furthermore, so as to define anti-social acts prohibited by law, as far as juveniles who should be treated in the justice system are concerned, the United Nations has not used the term "delinquency" and, instead uses and defines the term "offense", acknowledging that an offense is any behavior (act or omission) that is punishable by law under the respective legal system (article 2.2 of the Beijing Rules). This definition is derived from the legal concept of crime. Therefore, only anti-social behavior and misconduct, which are punishable by the Penal Codes, can be treated in the justice system, as well as other anti-social behaviors which are not protected by Penal Codes and should be handled by institutions of the social system. Furthermore, in line with the legality principle of crimes and punishments, article 40a of Convention on the Rights of the Child provides that no child shall be alleged as, be accused of, or recognized as having infringed the penal law by reason of acts or omissions that were not prohibited by national or international law at the time they were committed.

There are several definitions for delinquency and crime, as well as various interpretations. According to criminologists, any act or omission that is dangerous to the majority of people in society is called a "crime". Sociologists believe that any act which violates the strong and specific states of a collective conscience is criminal. From the perspective of psychologists, delinquency is a bio-psycho-social phenomenon. According to a United Nations report, delinquency, in any sense of the word, is affected by a range of factors, such as poverty, family problems, inheritance, and so on. The latter definition of delinquency is based on these factors. Delinquency is sometimes an act of misbehavior, but not all individuals who misbehave are delinquents. The quality of behavior and the range of social deviance are factors for judging whether a person is a criminal or mentally ill. Some misbehaviors are illegal and some are unethical, but not all illegal behaviors are unethical. The difference between illegal and unethical behavior is very important and should be considered when defining delinquency. Literally, delinquency means to leave a legal duty or to commit a wrong action which is not necessarily a crime. However, this term is often used in relation to crime and especially those committed by children and adolescents. In its conventional and non-legal sense, child delinquency involves behaviors such as school absenteeism and escaping parental control which are not considered criminal in adults. Therefore, it can be concluded that delinquency is a series of minor offenses, and since children typically commit only minor offenses, the term "delinquency" is used for children and the term "criminal" is applied for adults who commit a wrongdoing. The United Nations has not defined the term "delinquency" and, instead, has used the term "crime", acknowledging that a crime includes any act (or omission) that is punishable legally in a particular system of law.

Legally, there are two definitions of crime. According to the first definition, an act is considered a crime if the law finds it to be a crime. Based on the second definition, a crime is an act or omission that is punishable by law. The first definition of crime is simple but vague and does not add anything to our knowledge of crime. By contrast, the second definition is more complete and precise, and the United Nations has used the same definition for juvenile crimes. As we know, crime has been defined in criminal law, because the legislator is responsible for determining the act or omission which disrupts the social order and causes social damage. However, the legislator cannot ignore the different views and comments of experts about various phenomena, especially crimes, in its definition of crime.[406]

In criminology, which deals with the causes of crime and the treatment of offenders, crime is basically referred to as all anti-social behavior or tensions affecting society, whether their causes are psychological or social. However, due to legal and social restrictions, criminology has no choice but to use the legal definition of crime in most cases. In addition to the general criminal rules governing juvenile offenders, the United Nations has made attempts to describe special principles and rules for children. Following the criminological studies and given the specific situation of adolescents, these principles have been included and mentioned in the rules for juvenile offenders.[407]

2.3.1.1.2 Status offenses

A status offense is a non-criminal act that is considered a violation of law only because of a youth's status as a minor.[408] Typical status offenses include truancy, running away from home, violating curfews, underage use of alcohol, and general non-

406 For further reading, see *Abachi* 1388/2009, p. 40.

407 *Abachi* 1388/2009, pp. 39–42.

408 A status offender is a juvenile charged with or adjudicated for conduct that would not, under the law of the jurisdiction in which the offense was committed, be a crime if committed by an adult.

governability. Status-offending behavior is often a sign of underlying personal, familial, community-based, and systemic issues, similar to the risk factors that underlie general offenses. Sometimes these underlying issues contribute to delinquency later in life, putting youths at a higher risk for drug use, victimization, engagement in risky behavior, and an overall increased potential for physical and mental health issues, including addictions. Some evidence supports the notion that less serious forms of delinquency often precede the onset of more serious delinquent acts. However, the precursor to the delinquency view of status offending does not take into account the normal experimentation of childhood and adolescence or the diverse developmental pathways that can lead to serious delinquency. Children and adolescents commonly experiment with behaviors that are not considered positive or pro-social, such as lying, playing truant, or defying one's parents. Such experimentation allows youths to discover the negative consequences of their behaviors and to learn from their mistakes. Most who engage in status and other minor offenses never progress to more serious behaviors.[409]

There are numerous possible causes of status offense behaviors. These non-criminal behaviors are often caused by poor family functioning or dynamics, school problems, youth characteristics, or community problems. For example, research indicates that risk factors for potential truancy include domestic violence, academic problems, substance abuse, lack of parental involvement in education, and chronic health problems. Research also indicates that many youths who run away were physically or sexually abused at home in the year prior to their runaway episode. Family dysfunction and drug use in the company of the child are also risk factors for youth running away. In its general comment no. 10 on "Children's Rights in Juvenile Justice", the Committee states that criminal codes quite often include provisions criminalizing behavioral problems of children, such as vagrancy, truancy, running away, and other acts, which are often the result of psychological or socio-economic problems. Such acts (also known as status offenses) are not considered to be offenses if committed by adults. The Committee recommends that States should abolish status offenses in order to establish equal treatment under the law for children and adults; such behavior should be dealt with through the implementation of child-protective measures, including effective support for parents and/or other caregivers and measures which address root causes (general comments no. 10, para. 4). The Committee refers to article 56 of the Riyadh Rules: in order to prevent further stigmatization, victimization, and criminalization of young persons, legislation should be enacted to ensure that any conduct not considered an offense or not penalized if committed by an adult is neither considered an offense and neither penalized if committed by a young person.[410]

Therefore, criminalization of status offenses is prohibited by the United Nations' standards. Although it is quite common for criminal codes to contain provisions

409 *Abachi* 1388/2009, p. 42.

410 United Nations Children's Fund 2007, p. 407.

criminalizing behavioral problems of children, in order to protect children and juveniles, the Beijing Rules state that the provisions of administering justice for juveniles should be applied not only to juvenile offenders but also to those juveniles who may be proceeded against for any specific behavior that would not be punishable if committed by an adult; these are called "status offenses", such as truancy, school and family disobedience, public drunkenness, etc. (article 3 of the Beijing Rules). In this regard, the Committee on the Rights of the Child emphasizes that it is particularly a matter of concern that girls and street children are often victims of status offenses and recommends that the States parties abolish the provisions on status offenses in order to disestablish unequal treatment under the law for children and adults. In this regard, the Committee also adds that behaviors such as vagrancy, roaming the streets, or runaways should be dealt with through the implementation of child-protective measures, including effective support for parents and/or other caregivers and measures which address the root causes of these behaviors (general comment no. 10, paras. 8–9). More complex cases of child crimes indicate that their criminal actions such as truancy, alcoholism, drug abuse, sexual exploitation, and violent crimes are in direct relationship with social changes.

Furthermore, in relation to decriminalization, article 5 of the Riyadh Rules provides that the need for and importance of progressive delinquency prevention policies, as well as the systematic study and the elaboration of measures, should be recognized. These should avoid criminalizing and penalizing a child for behavior that does not cause serious damage to their development or harm to others, with consideration that youthful behavior or conduct that does not conform to overall social norms and values is often part of the maturation and growth process and tends to disappear over time in most individuals during their transition to adulthood on the one hand; there should also be an awareness that labeling a young person as "deviant", "delinquent", or "pre-delinquent" often in itself contributes to the development of a consistent pattern of undesirable behavior by young persons on the other (article 5e, 5f of the Riyadh Rules).

2.3.1.2 Sanctions and responses

2.3.1.2.1 Principles concerning sanctions and responses

Making a decision about a juvenile delinquent who has been identified by the relevant and competent authorities and whose case is under investigation is a very crucial and important process, because children and juveniles have special moods; their mental status and the laws for determining sanctions must be enacted and enforced in a way that provides their reintegration. Hence, (national or international) legislators provide special rights and facilities for children and juveniles (compared to adults), due to their specific physical and mental conditions. Accordingly, the sanctions should be guided by the principle of proportionality as the most important principle for responses to juvenile delinquency. The reaction taken should always be in proportion not only to the circumstances and the gravity of the offense, but also to the circumstances and needs of the juvenile, as well as to the needs of society (article 17-1a of the Beijing Rules).

a) Taking reaction according to the circumstances prevailing at the time of the crime

The crime itself is one of the factors that should be taken into account, because these conditions sometimes affect the motivation for committing a crime, the selection of its type, its severity, and so on. Age and gender status as well as economic, social, cultural, and educational conditions can be effective in committing a crime. Street children, minorities, immigrants, refugees, children in bottlenecks, and so on are to a higher degree at risk, depending on their age and gender, so they are sometimes in conflict with the law. Hence, these groups of people are of special importance (articles 15, 16-1, 3-1, 13 of the Beijing Rules).

b) Taking reaction according to the severity of crime

This principle is known as a tool to control severe and harsh penalties and is usually mentioned in conjunction with the term "fair punishment", i.e. as being proportionate to the gravity of the crime. The reaction to juvenile offenders should not only be based on individual situations such social and family status, damages resulting from crimes, and other factors that influence a person's position, but should also take into account the severity of the crime. As we know, public interest lies in a less frequent occurrence of crimes. Accordingly, the obstacles that prevent people from committing crimes should be as strong as the extent to which these crimes are against public interest. Therefore, a balance should be established between crimes and penalties. In this regard, *Montesquieu* says that penalties must be uniformed. However, illegal acts that cause more damage to society should be avoided and prevented more strongly than smaller actions that cause less damage.[411] Serious crime corrupts the public morals on a higher level; thus, they deserve more severe reactions; however, the reaction should never be more severe than the act, and there should be a balance between them. Nevertheless, it has been noted that the principles of proportionality and uniformity are usually linked with the theories of retributive and just desert[412] if they are considered alone, while lenient reactions appear with other elements. Furthermore, we should distinguish between the concept of retroactive and prospective proportionality.[413] Retroactive proportionality, which is internal to the penal system, considers the measure of blameworthiness, while prospective proportionality emphasizes the goal that is to be achieved, and in relation to juvenile delinquents, the legislator should rather focus on the latter in the legislating stage, and judges or other authorities should do so in the sentencing stage.

---

411 *Ardebili* 1374/1995, pp. 104–105.

412 *Jareborg* 2005, p. 529

413 *Frase* 2005, p. 75.

c) Taking reaction according to the needs of juveniles and society

This important issue should be taken into account in the stage of both crime detection and preliminary investigation and in decision-making when a crime is committed by a child or juvenile. In this regard, article 3 of the Convention on the Rights of the Child indicates that in all actions concerning children – whether undertaken by public or private social welfare institutions, courts of law, administrative authorities, or legislative bodies –, the best interest of the child shall be a primary consideration. The main issue emerges when the interests of the child are in opposition to the interests of society. According to the Convention on the Rights of the Child and other international rules related to juvenile delinquents, the interests of children must then be considered first. Regarding the principle of proportionality, the Committee on the Rights of the Child emphasizes that the reaction to an offense should always be in proportion not only to the circumstances and gravity of the offense, but also to the age, lesser culpability, circumstances, and needs of the child, as well as to the various and particularly long-term needs of society (general comment no. 10).

Article 17-1 of the Beijing Rules emphasizes that the conflict between some approaches for determining sanctions – which is more pronounced in juvenile than in adult cases, such as rehabilitation versus just desert; assistance versus repression and punishment; reaction according to the singular merits of an individual case versus reaction according to the protection of society in general; general versus individual deterrence – should be solved by States parties. The Beijing Rules also imply that strictly punitive approaches are not appropriate. While in adult cases (and possibly also in cases of severe offenses by juveniles), just desert and retributive sanctions might be considered to have some merit, in juvenile cases, such considerations should always be outweighed by the interest of safeguarding the well-being and future of the young person (article 17-1 of the Beijing Rules, commentary).

2.3.1.2.2 Types of sanctions and responses

In this section, prohibited and applicable sanctions according to the international instrunments will be discussed.

a) Prohibited sanctions and responses

aa) Death penalty

The international instruments prohibit imposing capital and corporal punishment on children and juveniles (article 6, paragraph 5, and article 7 of the ICCPR; article 37a of the Convention on the Rights of the Child). Article 37a of the Convention on the Rights of the Child reflects the customary international law prohibition of imposing the death penalty for a crime committed by a person who is under 18 years of age. Some States parties assume that the rule only applies to persons who are below 18 years at the time of execution. Others defer the execution until the person has reached the age of 18. The Committee reiterates that the explicit and decisive criterion is the age at the time of committing the offense. If there is no reliable and conclusive proof that the person was below 18 at the time the crime was committed, he/she should have the benefit of the doubt and the death penalty cannot be imposed. The Committee calls upon the few States parties that have not yet abolished the imposition of the death penalty for all offenses committed by persons below the age of 18 years to do so urgently and without exception. Any death penalty imposed on a person who was below 18 years at the time of committing the offense should be commuted to a sanction that is in full conformity with the Convention (general comment no. 24, paras. 78, 79).

bb) Life imprisonment without parole

In accordance with the international instruments, life imprisonment without the possibility of release is prohibited for children and juveniles (article 37a of the Convention on the Rights of the Child). Furthermore, incarceration should be avoided in the case of juveniles unless there is no other appropriate response (Convention on the Rights of the Child; Havana Rules; and Beijing Rules). General comment no. 24 (2019) emphasizes that no child who was below the age of 18 at the time he/she committed the offense should be sentenced to life imprisonment without the possibility of release or parole. The period to be served before consideration of parole should be substantially shorter than the one for adults, it should be realistic, and the possibility of parole should be regularly reconsidered.

The Committee reminds States parties sentencing children to life imprisonment with the possibility of release or parole that when applying this sanction, they should strive for realizing the aims of article 40(1) of the Convention. This means, inter alia, that a child sentenced to life imprisonment should receive education, treatment, and care aiming at his/her release, reintegration, and the ability to assume a constructive role in society. This also requires a regular review of the child's development and progress in order to decide on his/her possible release. Life imprisonment makes it very difficult, if not impossible, to achieve the aims of reintegration. The Committee notes the 2015 report in which the Special Rapporteur on torture and other cruel, inhuman or degrading treatment or punishment finds that life imprisonment and lengthy sentences, such as consecutive sentencing, are grossly disproportionate and therefore cruel, inhuman, or degrading when imposed on a child (A/HRC/28/68, para. 74). The Committee strongly recommends that States parties abolish all forms of life imprisonment, including indeterminate sentences, for all offenses committed by persons who were below the age of 18 at the time committing the offense (general comment no. 24, para. 81).

cc) Applicable sanctions and responses

Article 18.1 of the Beijing Rules provides that a large variety of disposition measures shall be made available to the competent authority, allowing for flexibility so as to avoid institutionalization to the greatest extent possible. Such measures, some of which may be combined, include: a) care, guidance, and supervision orders; b) probation; c) community service orders; d) financial penalties, compensation, and restitution; e) intermediate treatment and other treatment orders; f) orders to participate in group counselling and similar activities; g) orders concerning foster care, living communities, or other educational settings; h) other relevant orders.

2.3.1.2.3 Depenalization and decarceration policies

The Beijing Rules encourage the use of alternatives to institutionalization to the maximum extent possible, bearing in mind the need to respond to the specific requirements of the young. Thus, full use should be made of the range of existing alternative sanctions, and new alternative sanctions should be developed, bearing the public safety in mind. Probation should be granted to the greatest possible extent via suspended sentences, conditional sentences, board orders, and other dispositions. By considering articles 40(1), 40(4) and 37b of the Convention on the Rights of the Child, the Committee explains that children in conflict with the law, including child recidivists, have the right to be treated in ways that promote their reintegration and their assuming a constructive role in society. The arrest, detention, or imprisonment of a child may be used only as a measure of last resort. Therefore, it is necessary – as part of a comprehensive policy for juvenile justice – to develop and implement a wide range of measures so as to ensure that children are dealt with in a manner appropriate to their well-being and proportionate to both their circumstances and the offense committed. These should include care, guidance and supervision, counselling, probation, foster care, educational and training programs, as well as other alternatives to institutional care (general comment no. 10, para. 23).

2.3.2 Iran

According to article 2 of the Penal Code, crime – a general term applicable for both adults and juveniles – is defined as any behavior (including acts or omissions) for which punishment is provided by law. Therefore, any juvenile delinquency and misconduct can only be punished under this article and the provisions of general law. On the other hand, status offenses have not been criminalized in the Penal Code. By considering the definition of crime, it should be noted that no difference is made between crimes committed by adults and those committed by children and juveniles; however, the sanctions applicable for children and juveniles are different. Also, some privileges are introduced by legislature for children and juveniles to enjoy. For instance, regarding the aggravating circumstances of offenses committed by adults, the legislature has emphasized that the provisions regarding reoffending and its severe responses shall not be applicable in the cases of offenses committed by children and juveniles (article 138 of the Penal Code).

In order to explain juvenile delinquency in the Iranian Penal Code, we should overview the exclusive rules concerning each category of crime. Under the Penal Code, crimes and punishments are divided into four categories: *Hadd*, *Qisas*, *Diye*, and *Ta'zir* (article 14 of the Penal Code). This classification can be subdivided in two sections: as already mentioned, the Iranian penal system is shaped and also influenced by two different systems of crimes and punishments, i.e. discretionary and fixed crimes and punishments. Therefore, juvenile delinquency and the sanctioning system concerning juveniles will be explained according to the criteria of discretionary and fixed punishments. In *Taz'ir* crimes and punishments, the differentiated policy for children has been accepted by the legislator, while there is no differentiated policy in the fixed punishments – unless establishing intellectual maturity, which has already been explained above.

2.3.2.1 Crimes, differentiated measures and sanctions for juveniles in Ta'zir

Ta'zir as a discretionary sanction is a punishment which does not fall under the categories of *Hadd*, *Qisas*, or *Diye* and is determined by law for the commission of prohibited acts under *Shari'a* or for violations of state rules. It is notable that many crimes against a person, property, and public security have been placed in the discretionary or *Ta'zir* section, most of which are categorized as customary law. The type, amount, conditions of execution, as well as mitigation, suspension, cancellation, and other relevant rules of *Ta'zir* crimes are determined by law. When making decisions in *Ta'zir* crimes, in accordance with legal rules, the court considers the following issues: a) the offender's motivation as well as his/her mental and psychological conditions when committing the crime; b) the method of committing the crime, the extent of a breach of duty and its harmful consequences; c) the conduct of the offender after committing the crime; d) the offender's personal, family, and social background, as well as the effect of the *Ta'zir* punishment on him/her (article 18 of the Penal Code). Therefore, *Ta'zir* punishment is considered as a flexible instrument. As a general rule, *Ta'zir* penalties (for adults) include imprisonment, community-based punishment, disciplinary, security and corrective measures, fines, whipping (the number of lashes must be less than the number of lashes stipulated for Islamic punishment in *Hadd*), and execution for some specific crimes such as illicit drug trafficking.

In accordance with the new Penal Code of 2013 and with regard to the discretionary or *Ta'zir* punishments, children are divided into two groups: those between the ages of 9 and 15 years, and those between the ages of 15 and 18 years. In this code, for each group, some differentiated responses and sanctions have been determined, as can be seen below.

2.3.2.1.1 Children and juveniles between the ages of 9 and 15 years

If children and juveniles between the ages of 9 and 15 years (article 88 of the Penal Code) have committed *Ta'zir* crimes, the judge can order disciplinary, corrective, and security measures to be taken, such as handing over the children to their parents or natural and legal guardians, or handing them over to other natural or legal persons with promise of correction. In consideration of the best interests of the child, the aforementioned persons can be asked to take some measures, such as referring the child to a social worker or psychologist, sending him/her to an educational or cultural institute, rehabilitating and treating his/her addiction under the supervision of a doctor, banning him/her from visiting specific persons or places, or advising the child by the judge. In respect of these measures, the juvenile court, taking into account the investigations made and also the reports of social workers about the condition of the juvenile and his/her behavior, can review its decision as many times as the best interests of the juvenile require (article 88 note 3 of the Penal Code). Cautioning and warning or taking a written promise not to commit an offense again is another applicable response to children and juveniles in this group – or, as a last resort, sending the child to a correctional center for a period from three months to one year in the case of *Ta'zir* crimes of the first to fifth degree,[414] which are considered to be the most severe *Ta'zir* crimes. The two latter sanctions can only be applied if the child is between 12 and 15 years old. It should be emphasized here that the sanctions imposed most often for this group of children are disciplinary, corrective, and security measures without deprivation of liberty, which is in accordance with the recognized standards.

2.3.2.1.2 Juveniles between the ages of 15 and 18 years

If children and juveniles between the ages of 15 and 18 years (article 89 of the Penal Code) commit *Ta'zir* crimes, the judge can order deprivation of liberty by sending the child to a correctional center or sentencing him/her to a fine or to unpaid public service. Depending on the degree of crimes in this section, the degree of punishment can vary. It should be noted that in the case of severe crimes punishable in

---

414 According to article 19 of the Penal Code of 2013, *Ta'zir* punishments are divided into eight degrees:

- 1st degree: imprisonment for over 25 years, fine of more than one billion Rials;
- 2nd degree: imprisonment from 15 to 25 years, fine from 550 million to one billion Rials;
- 3rd degree: imprisonment from 10 to 15 years, fine from 360 million to 550 million Rials;
- 4th degree: imprisonment from 5 to 10 years, fine from 180 million to 360 million Rials;
- 5th degree: imprisonment from 2 to 5 years, fine from 80 million to one 180 million Rials, deprivation of social rights from 5 to 15 years;
- 6th degree: imprisonment from 6 months to 2 years, fine from 20 million to 80 million Rials, whipping (from 31 to 74 lashes and up to 99 lashes) in indecent crimes, deprivation of social rights from 6 months to 5 years, publication of the final judgment in the media;
- 7th degree: imprisonment from 91 days to 6 months, fine from 10 million to 20 million Rials, whipping (from 11 to 30 lashes), deprivation of social rights up to 6 months;
- 8th degree: imprisonment of up to 3 months, fine of up to 10 million Rials, whipping (up to 10 lashes).

law by a *Ta'zir* punishment of the first to third degree, deprivation of liberty is limited to a maximum of five years, and for crimes punishable in the fourth degree, it is limited to a maximum of three years. In these *Ta'zir* crimes, deprivation of liberty by detention is the only criminal justice sanction or punishment applicable to children in this group. However, for petty and minor crimes punishable in law by a *Ta'zir* punishment of the fifth degree, if deprivation of liberty based on the discretion of the judge is necessary, it shall not exceed one year. It should be mentioned that the fine cannot exceed 40 million Rials for crimes punishable in law by a *Ta'zir* punishment of the fifth degree, whereas in the case of a crime punishable by a *Ta'zir* punishment of the sixth degree, it is limited to a maximum of 10 million Rials, and a fine of up to one million Rials can only be applied in the case of petty and minor crimes punishable by the seventh and eighth degrees of *Ta'zir* punishments. It should be noted that in the seventh and eighth degrees of *Ta'zir* punishments, a fine is the only penalty applicable to children in this group. It is furthermore notable that the sanction of unpaid public service can only be imposed in the fifth and sixth degrees of *Ta'zir* punishments. In the case of offenses punishable in law by a *Ta'zir* punishment of the fifth degree, the legislature prescribes 180 to 720 hours of unpaid public service. In the sixth degree of *Ta'zir* punishments, this is limited to 180 hours. In any case, unpaid public service shall not be provided for more than four hours a day.

Under certain conditions, alternative punishments can be applied for this group of children. Taking into account the accused person's condition and the crime committed, instead of sending the child to a correctional center or sentencing him/her to the payment of a fine, the court, at its discretion, can order house arrest during specific hours or detention in a correctional center only during the weekend for three months to five years. However, it is notable that the most important change in *Ta'zir* crimes is the abolition of the death penalty for persons under 18 years, which is a considerable innovation in the Iranian justice system.

As we have seen above, the legislature has reconsidered some correctional measures for each group in the *Ta'zir* section. In the discretionary punishments, if the mitigating factors are applicable, the court can reduce the punishment up to half of the minimum penalty provided in the Code and replace the correctional measures for juveniles by other measures. The reduction of the detention duration to one fourth is possible once, in the case of which the right to apply for conditional release should also be preserved. In any *Ta'zir* crimes committed by children and juveniles, the court can defer the conviction or suspend the execution of the punishment altogether. In order to protect the rights of juveniles, the aggravation rules for recidivism cannot be applied to them. Furthermore, criminal convictions of children and young offenders shall have no effect on their criminal records, which is a very important supportive measure for juvenile delinquents.

2.3.2.2 Crimes and common sanctions for adults and juveniles in Hadd, Qisas, and Diye

In the new Penal Code of 2013, in the fixed penalties (*Hadd*, *Qisas*, *Diye*), the age that is set for criminal responsibility based on *Shari'a* is very different from the model used in the discretionary punishments. In fact, in this section, immature persons have no criminal responsibility, and the age of maturity is defined as 9 lunar years for girls and 15 lunar years for boys. Actually, in fixed punishments, the gradual transition to full responsibility has not been accepted by legislature. Therefore, girls above 9 years and boys above 15 years are considered mature persons and treated as adults with full criminal responsibility.

However, in the case of offenses punishable by *Hadd* or *Qisas*, if mature people under 18 years do not realize the nature of the crime they committed or its prohibition, or if there is uncertainty about the full mental development of these mature people, according to their age, they are sentenced to the punishments for children between the ages of 9 and 15 years as prescribed in the *Ta'zir* section. Nevertheless, we should note that if a boy between 12 and 15 years who has not become mature commits any of the offenses punishable by *Hadd* or *Qisas*, he shall be sentenced to one of the measures provided by the rules for *Ta'zir* crimes and punishments. Likewise, for immature children (girls under 9 years and boys under 15 years) who commit crimes punishable by *Ta'zir*, *Hadd*, *Qisas* or *Diye*, only security and correctional measures shall be applied in accordance with the provisions of this Code. In addition, the provisions about *Diye* (blood money) are the same as for adults. In order to know about sanctions for fixed punishment, they should be defined. As already mentioned, the fixed system which is affected by *Shari'a* is stable and includes *Hadd*, *Qisas*, and *Diye*:[415]

- *Hadd* crimes punishable by *Hadd* punishments are prescribed by *Shari'a* (article 15 of the Penal Code). *Hadd* crimes are also considered as crimes against the rights of God, with an exclusive rule. Crimes punishable by *Hadd* are very limited yet important due to the severe punishments applicable for them. *Hadd* crimes include: adultery; sodomy; lesbianism; pimping; false accusation of another person of sodomy and adultery; consumption of intoxicating liquors; insulting the Prophet Mohammad; waging war against God; rebellion and corruption on Earth; and a kind of theft. It is notable that some of the *Hadd* crimes such as adultery, sodomy, lesbianism, and consumption of intoxicating liquors are considered as victimless crimes, which are more likely to be committed by juveniles. The *Hadd* punishments for the above-mentioned crimes include the death penalty, amputation of limbs, and whipping. Proving *Hadd* crimes in the court is very difficult, although the punishment can be removed by repentance of the perpetrator.

---

415 For further reading on criminal sanctions in Iran, see *Ghassemi* 2009, pp. 158–180.

- *Qisas* is an Islamic punishment for murder and other intentional crimes against the physical integrity of a person (article 16 of the Penal Code). *Qisas* includes both life and limbs retaliation. These are considerd as crimes against the rights of persons, and therefore should be implemented with the consent of the victim's family. By mediation, *Qisas* can be changed to *Diye* (blood money), and in case of forgiveness on the part of the victim and the victim's family, the punishments in this section can be removed.
- *Diye* is a pecuniary penalty (blood money), the quantity of which has been determined in Islam. This form of punishment is settled upon for unintentional murder, injury, or defect of limb caused by an accident (article 15 of the Penal Code).

2.3.2.3 Depenalization and decarceration policies

In the Penal Code of 2013, the depenalization and decarceration policies have been extended not only to children and juveniles but also to adults. Hence, the new depenalization policies, such as deferment of sentence, have been introduced by the legislature on the one hand, while the previous policies were emphasized on the other hand. However, the new measures can be applied for *Ta'zir* crimes and punishments, while in the fixed punishments there are specific policies which are completely different from them.

2.3.2.3.1 Discretionary punishments: *Ta'zir*

a) Exemption from punishment

According to article 39 of the Penal Code, when the mitigating factor is recognized in *Ta'zir* crimes of the seventh and eighth degree, and if the court finds the accused guilty but believes that the offender will be corrected even without execution of the punishment – provided that they have no effective criminal record, the complainant has forgiven the offender, and the losses have already been compensated for (or appropriate measures are taken to compensate the loss) –, the court may decide to exempt the offender from punishment.

b) Deferred sentence

In all *Ta'zir* crimes, after the accused is found guilty, the court – subject to the following conditions and considering the juvenile's personal, family, and social conditions and background, as well as the circumstances that resulted in committing the offense – may postpone the deliverance of the judgment from six to two years under the following conditions: (a) existence of mitigating factors; (b) foreseeable correction of the offender; (c) compensation of, or taking appropriate measures to compensate, the loss (articles 94 and 40 of the Penal Code).

Deferred sentences can have two forms: simple or supervised. In a simple deferment of sentence, the juvenile shall promise in writing that in the period determined by the court, they will not commit any crime, and it is believed from their good behavior that they will not commit any crime in the future either. In a supervised deferment of sentence, in addition to the conditions mentioned for the simple deferment, the offender promises to comply with and execute the orders and measures set by the court during the deferment period. The court cannot issue the warrant of deferment of sentence in absentia. In supervised deferment of sentence, the following measures shall be taken:

a) Punctual attendance at the time and place determined by the judicial authority or the supervisory social worker.
b) Providing the required information and documents in order to facilitate the supervision of the social worker over the compliance of the convict with their obligations.
c) Declaring any change of job, residence, or relocation within 15 days and providing a report to the social worker.
d) Application to the judicial authority for permission to travel abroad.

These measures can be accompanied by other, supportive measures at the court's discretion, such as referral of the juvenile to supporting organizations. In a supervised deferment of sentence, while considering the offense committed, the characteristics of the juvenile, and the conditions of their life, the court can ask the juvenile to carry out one or more of the following orders during the period of deferment, provided that this will not significantly or fundamentally disrupt their own or their family's life:

a) Learning or holding a specific job or career.
b) Residence or non-residence in a specific place.
c) Treatment of an illness or rehabilitation for an addiction.
d) Refraining from operating any or some motor vehicles.
e) Refraining from professional activities relating to the offense committed or by using the means of the offense.
f) Refraining from contacting and associating with accomplices and abettors to the offense or other people such as the victim of the offense at the discretion of the court.
g) Attending a special program for training and learning basic skills of life, or participating in training, ethical, religious, educational, or physical education classes.

After the period of deferment of sentence ends, considering the level of the juvenile's compliance with the court orders as well as reports of the social worker, and taking into account the general conditions of the juvenile delinquent, the court can either sentence the juvenile to punishment or grant them an exemption. However, it should be noted that if the offender commits a *Hadd* or *Qisas* crime or an intentional crime punishable by *Diye* or *Ta'zir* of up to the seventh degree during the period of deferment, the court will cancel its warrant and deliver the judgment of conviction. In the case of non-compliance with the court orders, the court (only once) can either add to the period of delivery up to half the time determined in the warrant, or deliver the judgment of conviction. In this case, once the warrant of deferment of sentence is cancelled and the juvenile is convicted, a suspension of the execution of punishment is forbidden.

c) Differentiated alternative to punishments for juveniles

With regard to the following punishments applicable for *Ta'zir* crimes, by considering the personal conditions of accused juveniles (boys or girls) aged between 15 and 18 and the crime committed, the court, at its discretion, can – instead of sentencing them to detention or a fine – order house arrest at specific times determined by the court or detention/deprivation of liberty in a Correction and Rehabilitation Center at the weekend for three months to five years. The punishments (detention/deprivation of liberty and fines) which can be replaced by the new and differentiated alternatives include:

a) Detention and deprivation of liberty in a Correction and Rehabilitation Center from two to five years in the case of offenses punishable by law with a *Ta'zir* punishment of the first to third degree.

b) Detention and deprivation of liberty in a Correction and Rehabilitation Center from one to three years in the case of offenses punishable by law with a *Ta'zir* punishment of the fourth degree.

c) Detention and deprivation of liberty in a Correction and Rehabilitation Center from three months to one year or a fine of ten million to forty million Rials or providing 180 to 720 hours of unpaid public service in the case of offenses punishable by law with a *Ta'zir* punishment of the fifth degree (article 89 (note 3) of the Penal Code).

d) Mitigation and replacement of punishment

Article 93 of the Penal Code stipulates that if the court recognizes mitigating factors, in the case of *Ta'zir* crimes committed by children and juveniles, it can reduce the punishments by up to half of the minimum punishment provided and replace security and correctional measures for children and young people with another measure. As a general rule, according to article 38 of the Penal Code, mitigating factors are:

a) forgiveness by the complainant or private claimant;

b) effective cooperation of the accused in recognition of accomplices and accessories to the offense and in finding the proceeds of the offense; or discovering the property and goods that resulted from, or the means used in commission of the offense;

c) specific circumstances under the influence of which the accused has committed the offense, such as: provocative behavior and goading of the victim or honorable motive for committing the offense;
d) statement of the accused prior to prosecution, or their effective confession during investigation and prosecution;
e) regret, good reputation, or specific condition of the accused such as their age or an illness;
f) efforts by the accused to reduce the effects of the offense and their measures to compensate the loss or damage resulting from it;
g) when the loss or damage imposed to the victim or the consequences of the offense are slight;
h) slight contribution of an accomplice or accessory to the offense in its commission.

Note 1 provides that the court must stipulate the mitigating factors in its judgment, and note 2 states that if the same mitigating factors as mentioned in this article are provided in specific articles, the court may not mitigate the punishment again for the same mitigating factors.

e) Repentance

In the case of *Ta'zir* offenses of the sixth, seventh, or eighth degree – which are considered to be less serious crimes –, if the juvenile repents and his/her correction is recognized by the judge, the punishment shall be removed (i.e., it shall not be given). In other, more serious *Ta'zir* crimes, if the juvenile repents, the court can apply the provisions relating to mitigation of the punishment (article 115 of the Penal Code). Repentance is a special mechanism which is more often used to remove the fixed punishment and is exceptionally applied for some *Ta'zir* crimes. It is notable that the rules governing repentance are introduced by *Shari'a*, so it is a bit different from regret, which can be assessed by customary law. As already explained, due to the dual approach in the Iranian criminal justice system – which leads to classifying crimes and punishments into two categories and introducing a separate rule for each of them –, there are some similar institutions or mechanisms in place which are different in essence; due to the obstacles to integrate them, the legislature has to keep all of them, and sometimes, this approach leads to confusion and misunderstanding.

f) Suspension of execution of punishment

In the case of all *Ta'zir* crimes committed by young people, the court can suspend execution of all or part of the punishment from one to five years, subject to the same requirements provided for a postponement of a deferment of sentence. Also, after executing one third of the punishment, the public prosecutor or judge in charge of the execution of criminal sentences can ask the court to suspend execution of the punishment. Furthermore, after completing one third of the punishment, subject to legal requirements, the convict can request suspension through the Public Prosecutor or Prosecutor in charge of the execution of criminal sentences. Subject to the same provisions provided for a deferment of sentence, a suspension of execution of punishment can take two forms: simple or supervised. The suspension of execution of punishment shall be issued by the court within or after the judgment of conviction. If in custody, juveniles whose execution of punishment has been completely suspended shall be released immediately.

However, it is notable that if a convict whose punishment has been suspended without a reasonable excuse does not comply with the court orders during the period of suspension, the court can – upon the proposal of the Public Prosecutor or the judge in charge of execution of punishment – add one to two years to the period of suspension or cancel the suspended sentence on the first occasion. Non-compliance with the court orders for the second time shall result in a cancellation of the suspended sentence and the execution of punishment. Furthermore, if the juvenile does not commit any intentional offense punishable by *Hadd*, *Qisas*, *Diye*, or *Ta'zir* of up to the seventh degree, then the suspended punishment shall be ineffective (articles 94, 46, 48, 49, 50, 54 of the Penal Code).

g) Semi-liberty

Article 57 of the Penal Code stipulates: in *Ta'zir* imprisonments of the fifth to seventh degree, subject to forgiveness of the complainant, pledging an appropriate guarantee, and promising to pursue a vocational, professional, or educational activity, or contributing in continuity of the family life, or treating an addiction or illness which are effective in the process of rehabilitation (of the convict), or compensation of the victim, the court of final judgment can put the convict (with their consent) under the regime of half-release. Additionally, the convict can request the order of half-release while serving his/her sentence, provided that they meet the legal requirements and the court is obliged to consider the request.

h) Reduction of the time of deprivation of liberty

A reduction of the time of deprivation of liberty is a differentiated policy applicable to children and juveniles. Accordingly, the court can review its decision once according to the reports received about the condition of the child or youth and their behavior in the Correction and Rehabilitation Center, and it may reduce the detention term by up to one third or replace it with handing the child or youth over to their natural or legal guardians. The court's decision to review the original decision shall be made if the child or youth has spent at least one fifth of the detention term in a Correction and Rehabilitation Center. The court's decision in these cases is deemed final; however, this shall not prevent the offender from enjoying conditional release and other mitigations prescribed in the law if their requirements are met (article 94 of the Penal Code).

i) Conditional release

In *Ta'zir* prison sentences, upon the proposal of the public prosecutor or the judge in charge of the enforcement of judgment, and subject to the following conditions, the court can issue the order of conditional release for juveniles after one third of the sentence has been served:

a) The convict constantly shows good behavior whilst serving their sentence.
b) From the conditions and behavior of the convict, it is predicted that they will not commit any further offenses after being released.
c) The court can confirm that the convict has compensated, or has arranged to pay the loss or damage contained in the judgment or agreed upon by the private claimant.
d) The convict has not previously used conditional release.

After being reported by the warden of the prison, lapse of the above-mentioned periods, as well the conditions mentioned in paragraphs (a) and (b) of this article, shall be approved by the judge in charge of the enforcement of judgments. This judge is obliged to consider the prescribed periods as well as the condition of the prisoner and checks whether the requirements are met, in which case he/she must submit a proposal for conditional release to the court. The period of conditional release shall be the same as the remaining duration of the sentence; however, the court can also change its duration. In any event, it cannot be less than one year and more than five years – unless where the remaining sentence is less than one year, in which case the period of conditional release shall be the same as the remaining duration of the sentence. Considering the circumstances under which the crime has been committed, as well as the convict's psychological conditions and character, the court can require him/her to comply with the same orders provided in the deferment of sentence during the conditional release term. In its judgment, the court shall state and inform the juvenile convict about the said requirements and the consequences of non-compliance with them, as well as the consequences of committing a new offense. However, it should be mentioned that if the juvenile convict, without a reasonable excuse, does not comply with the orders of the court during the conditional release term, one to two years shall be added to the conditional release term on the first occasion. If non-compliance with orders occurs repeatedly or in the case of committing an intentional offense punishable by *Hadd*, *Qisas*, *Diye*, or *Ta'zir* of up to the seventh degree, then (in addition to the punishment for the new crime) the remaining duration of the original sentence shall be executed; otherwise, their release shall become unconditional. In *Ta'zir* crimes of the fifth to eighth degree, which are considered less serious crimes, subject to the same conditions provided for a supervised deferment of sentence, the court can, with the convict's consent, put them under the supervision of electronic systems inside a specific area. If required, the court can put the convict subject to supervised orders or orders provided for a supervised deferment of sentence (articles 58–62 of the Penal Code).

2.3.2.3.2 Fixed punishments: *Hadd*, *Qisas*, *Diye*

a) Repentance in Hadd

Hadd crimes are proven by confession, witnesses, and knowledge of the judge; furthermore, as already mentioned, proving *Hadd* crimes is very difficult in court. As crimes against the right of God, *Hadd* crimes are stable. Article 219 of the Penal Code stipulates that "the court cannot change the conditions, type, and amount of *Hadd* punishments or reduce, replace, or remove a *Hadd* punishment. Such punishments can only be removed, reduced, or replaced through repentance and pardon under the conditions prescribed in this Code." In the case of offenses punishable by *Hadd* – with the exception of false accusation of another person of sodomy and adultery (*Qazf*) or waging war against God (*Moharebeh*) –, if the accused repents any time before the commission of the offense is proven, and if his/her regret and correction is certain in the eyes of the judge, the *Hadd* punishment shall not be given. In addition, if the above-mentioned offenses (except for *Qazf*) are proven by confession, and if the offender repents (even after the commission of the offense is proven) the court can, through the Head of Judiciary, apply for pardon of the offender by the Leader. It should be noted that contrary to other *Hadd* crimes, false accusation of another person of sodomy and adultery (*Qazf*) or waging war against God (*Moharebeh*) is not considered as a victimless crime, which is why there are some exceptions for these crimes. *Diye*, *Qisas*, and *Hadd* punishments of *Qazf* and *Moharebeh* shall not be removed by repentance.

b) Forgiveness in Qisas and Diye

In *Qisas* and *Diye*, which are crimes against personal rights, the depenalization policy is different. In *Qisas*, recourse to mediation can also replace retaliation to *Diye* and imprisonment. Therefore, if someone commits a murder, the punishment applicable is retaliation (*Qisas*), but with the consent of the victim's immediate family, *Qisas* can be forgiven or replaced by *Diye*. It is notable that imprisonment here is a *Ta'zir* punishment – which is applied for the sake of public security –, and the rule governing this punishment is different and cannot be removed with the consent of victim's immediate family. *Diye* or blood money as a financial penalty in fixed punishments, like *Qisas*, is considered a personal right and can be forgiven by the victim's family.

2.3.3 Germany

2.3.3.1 Juvenile delinquency

Disregarding the fact that the Youth Court Act (as a specific act) is separate from the Criminal Code (as a general code), it is notable that it does not contain its own catalogue of crime definitions.[416] In fact, under section 1 of the German Criminal Code,

416 *Crofts* 2008, p. 73.

the Youth Court Act shall apply if a youth or young adult engages in misconduct punishable under the provisions of general law. The term “misconduct” refers to offenses or crimes that are subject to full criminal punishment under either the Criminal Code or a subsidiary penal law, e.g. tax legislation, economic, or military defense laws. For this reason, administrative offenses that are prosecuted under the Law on Administrative Offenses and merely result in fines do not fall under the Youth Court Act.[417] Therefore, as a general rule, there is no difference between juveniles' and adults' crimes with regard to the definition of crimes – and if there is sufficient reason to believe that a juvenile has committed a criminal offense, the procedure can be started. Hence, the age of the perpetrator at the time of the offense is key to the definition of juvenile crime.[418]

Nevertheless, the same basic rules must be followed for adults and juveniles when establishing criminal responsibility, and the differences lie only in the type and range of sanctions that can be imposed in the juvenile justice system. It is notable that the status offenses have not been recognized by the legislature – not only in the Youth Court Act but also in the Criminal Code.[419] However, some instances of these behaviors, which are not protected by the criminal justice system, *are* handled by social justice system. In fact, the penal law does not include any status offenses like drinking in public or truancy. Those actions are defined as administrative offenses and are subject to a special Administrative Offenses Act, which can result in an administrative fine.[420]

Unlike in some other European legal systems, a system of antisocial behavior orders has not been established in Germany. If juveniles show antisocial behavior that does not result in criminal offenses, the Child Welfare System does have to prove if the juvenile and/or their parents need and want support. All support measures are principally voluntary. Only if the child is in extreme danger, the youth welfare agencies can ask the family court to take him/her out of their family or to apply special support measures to the parents or other responsible persons.[421] However, it is worth mentioning that there is debate about a special catalogue of crimes for juveniles due to distinctive aspects of the lives and lifestyles of young people.[422]

---

417 *Kaiser* 1994, p. 324.

418 *Albrecht* 2004, p. 452.

419 *Albrecht* 2004, p. 452.

420 *Pruin* 2012, p. 2.

421 *Pruin* 2012, p. 2.

422 For example, offenses such as fraud or forgery and certain aggravating circumstances make sense in judgments on adult behavior, but, it is said, they may be too complicated to be fully understood by a 14- or -15year-old. Similarly, some juvenile behaviors, such as taking away items belonging to opposing soccer fans, are said to be triggered by expressive motivations and not to reflect the instrumental considerations at the center of offenses such as robbery or theft. *Albrecht* 2004, p. 452.

Furthermore, according to section 1 of the Youth Court Act, youth is anyone who, at the time of the act, has reached the age of 14 but not yet 18 years, and a young adult is anyone who, at the time of the act, has reached the age of 18 but not yet 21 years. Thus, the age of the perpetrator, at the time of the crime, can be an essential factor for a differentiated policy in the criminal justice system. It is notable that section 4 of the Youth Court Act emphasizes that the provisions of general criminal law shall be applied both to classify an unlawful act by a youth as a serious or less serious criminal offense and in assessing when the act shall be barred by statute.

Nevertheless, there are some privileges regarding juveniles and youth. For example, in case of a youth committing several offenses, the Youth Court Act considers a differentiated policy, where only one set of supervisory or disciplinary measures or a single youth penalty shall be imposed by the judge.[423] If one sentence is passed simultaneously for a combination of offenses, where youth criminal law would apply to some offenses and general criminal law would apply to the others, youth criminal law shall be applied to all of them if the main focus lies on those offenses which should be assessed under youth criminal law. If that is not the case, general criminal law shall apply to all of them (section 32).

2.3.3.2 Sanctions and measures

In the case of criminal offenses, the interventions of the Youth Court Act are characterized by the principle of subsidiarity or minimum intervention. Accordingly, penal action should only be taken if absolutely necessary. Furthermore, sanctions must be limited by the principle of proportionality, as already explained. The legislative reform of the Youth Court Act in 1990, passed in the same year as the Youth Welfare Act, underlines the principle of youth court sanctions as a last resort (*ultima ratio*). Therefore, priority is given to diversion (which will be discussed in the next section) and where the youth courts do impose sanctions, primacy is given to educational or disciplinary measures instead of youth imprisonment.[424]

---

423 Insofar they are provided for in section 8 of the Youth Court Act, different types of supervisory and disciplinary measures may be ordered in combination, or they may be combined with youth penalty. The statutory maximum limits applicable to youth detention and penalty may not be exceeded. Furthermore, if the youth's guilt has already been finally established in relation to some of the criminal offenses or a supervisory or disciplinary measure, or to a youth penalty determined though not yet completely implemented, served or otherwise disposed of, account shall be taken of the judgment, and similarly only measures or a youth penalty shall be imposed. The judge shall have discretion to give credit for periods of youth detention already served when imposing a youth penalty. If it is conducive to the purposes of supervision, the judge may refrain from including in the new decision offenses for which a conviction has been obtained. In so doing, he/she may declare supervisory or disciplinary measures if he/she imposes a youth penalty.

424 *Dünkel* 2016, pp. 4–5. For further reading on the sentencing system in Germany, see *Albrecht* 2013, pp. 293–330; *Albrecht* 2002b; *Subramanian* 2013, and *Jehle* 2015.

2.3.3.2.1 Main sanctions and measures

A range of sanctions are applicable to youth and young adults, if committing crime, which differ in forms but are similar in the objectives that should be followed. In fact, all sanctions shall be implied in line with an educational concept (article 2 of the Youth Court Act). According to section 5, supervisory measures as the first priority should be considered by the judge, and in case these do not suffice, disciplinary measures or youth penalty may be imposed. It is notable that theses sanctions are dispensed with if placement in a psychiatric hospital or institution for withdrawal treatment renders punishment by the judge dispensable.

a) Supervisory measures

Section 5 I of the Youth Court Act stipulates that supervisory measures may be ordered in response to a criminal offense committed by a youth. Thus, these measures have priority, albeit they are sufficient at the discretion of the judge. Under sections 9 and 10 of the Act, there are two supervisory measures: (a) the issuing of instructions; (b) an order to avail oneself of supervisory assistance. With regard to the first measure, in particular, the judge may instruct the youth to:[425]

1) comply with instructions relating to his/her place of residence;
2) live with a family or in residential accommodation;
3) accept a training place or employment;
4) perform certain work tasks;
5) submit him-/herself to the care and supervision of a specific person (care assistant);
6) attend a social skills training course;
7) attempt to achieve a settlement with the aggrieved person – settlement between offender and victim;
8) avoid contact with certain persons or frequenting places providing public hospitality or entertainment; or
9) attend a road-traffic training course. The judge may also require the youth to undergo specialist rehabilitative or addiction withdrawal treatment.[426]

It should be noted that, according to section 11 of the Act, the duration of instruction, which should be determined, may not exceed two years. However, in the case of instruction no. 5, the duration should not exceed one year; and in the case of instruction no. 6, it should not exceed six months. The judge may amend instructions, lift them, or prior to expiry extend their duration to no more than three years if this is

425 Instructions shall be directions and prohibitions by which the youth can conduct his/her life and which are intended to promote and guarantee his/her education. Instructions must not place unreasonable demands on the way the youth conducts his/her life.

426 With the consent of the parent or guardian and the legal representative, and if the youth is more than 16 years of age, such condition should be imposed only with his/her consent.

conducive to the purposes of supervision. Furthermore, if the youth culpably fails to comply with these instructions, youth detention may be imposed if he/she has previously been cautioned as to the consequences of culpable non-compliance. It seems that the imposition of short-term detention on top of an educative measure almost inevitably leads to the actual legal consequence exceeding the level of what is proportionate; however, this is authorized by the law.[427] The period of youth detention imposed in such cases may not exceed a total duration of four weeks if there is a conviction. The judge shall dispense with enforcement of youth detention if the youth complies with the instruction after the detention has been imposed.

It can also be a supervisory assistance under section 12 of the Youth Court Act[428] –in the shape of a social worker within the meaning of section 30 of the Eighth Book of the Social Code, in a day-and-night-time institution, or in another form of supervised accommodation within the meaning of section 34 of the Eighth Book of the Social Code.[429]

b) Disciplinary measures

Under sections 13–16 of the Youth Court Act, disciplinary measures shall be applied if youth penalty is not indicated, but the youth has to be made acutely aware that they must assume responsibility for their wrongdoing. Disciplinary measures include:

a) reprimands, with the purpose of making absolutely clear to the youth the wrongfulness of their actions; or
b) imposition of conditions, in which
    - the judge can require the youth to make good, to the best of their ability, for damage caused as a result of the offense; to apologize personally to the aggrieved person; to perform certain tasks; to pay a sum of money to a charitable organization; or

427 *Persson* 2017, p. 120.

428 It is notable that according to section 12 of the Youth Court Act, "the juvenile court seeks to correct aspects of the young offender's personality which obstruct his/her becoming a law-abiding citizen by imposing orders or enforced juvenile care. Orders imposed by the juvenile court may include community service for juveniles, participation in social training courses, victim-offender mediation, or traffic education, and attendance at vocational training. The maximum number of hours for community service for juveniles is not prescribed but limited by the principle of proportionality. The German Constitutional Court has clarified that community service for juveniles is not in conflict with the prohibition of compulsory labor since it is only a short, selective commitment and does not degrade the person the way compulsory labor would do under a totalitarian regime. Educative support and residential care originate from youth welfare law and are supervised by a social worker according to section 12 of the Youth Court Act"; *Persson* 2017, p. 120. For further reading, see *Schaffstein et al.* 2014, pp. 121–122.

429 In fact, after hearing the youth welfare office, the judge may, under the conditions set out in the Eighth Book of the Social Code, require the youth to avail themselves of supervisory assistance.

- the judge should order payment of a sum of money only if the youth has engaged in just minor misconduct and it is to be assumed that they will pay the sum from money which they are allowed to personally dispose of, or the proceeds the youth has gained from their offense or the payment they received for committing the criminal offense is to be withdrawn from them; or the judge may subsequently vary conditions or dispense with compliance with them either in full or in part where this is conducive to the purposes of supervision;[430]

c) youth detention,[431] which means detention of the youth during leisure time, which shall be imposed during the youth's weekly leisure time and is to be counted as one or two periods of leisure time; or

d) short-term detention,[432] which shall be imposed in lieu of detention during leisure time if an uninterrupted period of execution appears to be expedient given the purpose of the supervision, and neither the youth's education and training nor their employment are adversely affected;[433] or

e) long-term detention, which shall be at least one week and not more than four weeks in duration. It shall be counted in entire days or weeks.[434] What is very important to note here is that no type of disciplinary measure shall carry the same legal consequences as a criminal sentence.

c) Youth penalty or imprisonment

It can be said that the only criminal penalty applicable for juveniles in the juvenile justice system is youth imprisonment, which is only a sanction of last resort in line with the international rules.[435] A verdict of youth imprisonment is registered in the

---

430 Section 11, subsection 3 of the Youth Court Act shall apply mutatis mutandis where the youth culpably fails to comply with conditions. Where youth detention has been enforced, the judge may declare conditions to have been met either in full or in part. In so doing, no unreasonable demands may be made of the youth.

431 "Youth detention" shall mean detention of the youth during leisure time, short-term or long-term detention.

432 It is notable that in 2012, the German parliament introduced so-called "warning-shot detention". This implied a toughening up of the juvenile justice system since detention (of up to four weeks) could now be imposed in addition to a suspended juvenile imprisonment sentence or to a conditional sentence/probation. *Persson* 2017, p. 126.

433 A two-day period of short-term detention shall be deemed equivalent to one leisure period.

434 Considered as a "sharp shock treatment", detention is imprisonment for up to four weeks. It means placement in a special unit separated from the juvenile prison and is subject to ongoing review. According to section 90 of the Youth Court Act, short-term detention should feature an educational design and should, in addition to getting the young person to reflect and having a deterrent effect, help the young offender address the personal problems which contributed to the criminal conduct. *Persson* 2017, p. 122. For further reading, see *Schaffstein et al.* 2014, p. 162.

435 *Dünkel* 2016, p. 7.

adult criminal record once a juvenile reaches age 18. All other juvenile measures are recorded only in juvenile records, which are accessible only to the prosecution services and the court system. All aspects of access to information on prior records are regulated in the Law on the Federal Register of Judicial Information, which is guided by the goal of reducing the stigma of prior criminal convictions.[436]

Under section 17 of the Youth Court Act, "youth penalty" shall mean deprivation of liberty in a facility for its execution. The judge shall impose a youth penalty if, as a result of the harmful inclinations demonstrated by the juvenile during the act, supervisory or disciplinary measures are not sufficient for the purpose of supervision or if such a penalty is necessary given the seriousness of the juvenile's guilt. Also, under section 18 of the Youth Court Act, the minimum duration of a youth penalty shall be six months and its maximum shall be five years, in case of particularly serious murder. If the act constitutes a serious criminal offense for which general criminal law prescribes a maximum sentence of more than ten years' deprivation of liberty, the maximum duration of youth penalty shall be ten years, in the case of particularly serious murder. It is notable that the statutory range of penalties under general criminal law shall not apply. Furthermore, youth penalty shall be calculated such as to make it possible to achieve the desired supervisory aim. Youth imprisonment sentences of up to two years can be suspended – which is a similar sanction as probation – in cases of a favorable prognosis; in all cases the probation service gets involved. The period of probationary supervision is one to two years, and the period of probation lasts for a total of two to three years.[437]

Regarding the main sanctions and measures, it should be mentioned that the combination of measures and youth penalty can only be implied under section 8 of the Youth Court Act. Accordingly, supervisory measures and disciplinary measures, as well as several supervisory measures or several disciplinary measures, may be ordered in combination. Youth detention may not be combined with an order to provide supervisory assistance. The judge may only impose instructions and conditions and order supervision by a social worker in combination with a youth penalty. Where the youth is subject to probationary supervision, any concurrent order for supervision by a social worker shall be suspended until expiry of the probationary period. In addition to supervisory measures, disciplinary measures, and youth penalty, the judge may order the imposition of those incidental penalties and consequences admissible under the Youth Court Act.[438]

---

436 *Albrecht* 2004, p. 473.

437 *Dünkel* 2016, p. 7. According to researches, in Germany, the recidivism rate for four years after short-term detention is considerably higher (at 70%) than that for juvenile imprisonment with probation (60%). *Persson* 2017, p. 122.

438 Under section 6 of the Youth Court Act, the court may not hand down a decision entailing loss of the capacity to hold public office or to attain public electoral rights, or the right to elect or vote in public matters. The court may not order a public announcement of the conviction. There shall be no loss of the capacity to hold public office and to attain public electoral rights (section 45, subsection 1 of the German Criminal Code).

According to researches and statistics, the most common legal consequences for young offenders in Germany are disciplinary measures in the form of warnings, community service for juveniles, or fines, followed by educational measures, which may themselves take the form of community service for other juveniles or participation in a social training course.[439]

2.3.3.2.2 Other measures with reformative and preventive goals

Under section 7 1of the Youth Court Act, placement in a psychiatric hospital or an institution for withdrawal treatment, supervision of conduct or withdrawal of the permission to drive may be ordered as measures of reform and prevention within the meaning of general criminal law. However, it is notable that under some conditions the court may subsequently order placement in preventive detention. Furthermore, according to section 7 2, if – after being sentenced to at least seven years' youth penalty because of a serious criminal offense

- against life, physical integrity, or sexual self-determination, or
- because of a robbery causing death (article 251 of the Criminal Code), theft and use of force to retain stolen goods (article 252 of the Criminal Code), blackmail and use of force, threats against life or limb (article 255 of the Criminal Code),

by means of which the victim suffered severe mental or physical damage or was exposed to such a danger – facts become known prior to the end of the execution of this youth penalty indicating a considerable danger for the public emanating from the convict, the court may subsequently order placement in preventive detention if it emerges in the overall evaluation of the convict, of his/her offense or offenses, and additionally of his/her development during the execution of the youth penalty that he/she is highly likely to commit criminal offenses of the above nature again.

The regular deadline for reviewing as to whether the further enforcement of placement in preventive detention is to be suspended on probation (section 67e of the Criminal Code) shall be one year in cases falling under subsections 2 and 3.

2.3.3.3 Depenalization and decarceration policies

2.3.3.3.1 Suspension of sentence

Under the Youth Court Act, if it is expected that the youth will regard the sentence itself as a warning, (while missing out on the experience of serving the sentence) will gain from the supervisory influence of the probation and henceforth conduct themselves in a law-abiding manner, enforcement of punishment shall be suspended on probation, provided that the imposition of the youth penalty does not exceed one

---

439 *Persson* 2017, p. 122.

year. Account shall be taken of the youth's personality, his/her prior life, the circumstances in which he/she acted, his/her conduct after the act, his/her living environment, and the effects the suspension of sentence can be expected to have on him/her. By considering the youth conditions, the judge shall also suspend probation enforcement of a longer period of youth penalty not exceeding two years if enforcement is not indicated on grounds relating to the youth's personal development. Furthermore, a suspension of sentence cannot be limited to part of the youth penalty and shall neither be excluded because of credit given for periods of remand detention or other deprivation of liberty (section 21). The judge shall fix the duration of the probationary period. It may not exceed three years nor be of less than two years[440] (section 22). During the probationary period, the supervisory influence on the youth's conduct should be exercised by the issuance of instructions. The judge may also impose conditions on the youth and also make or revoke such orders subsequently. In the case that the youth gives assurances concerning his/her future conduct or offers to provide services apt to make amends for the wrong he/she has done, imposing instructions and conditions shall be refrained from if it is expected that the youth will comply with his/her assurances or offers (section 23).

The youth shall be placed under the supervision and guidance of a full-time probation officer for a maximum of two years during the probationary period. The judge may also place the youth under the supervision of a volunteer. The probation officer[441] shall monitor the fulfillment of instructions, conditions, assurances, and offers and provide the youth with help and guidance. The probation officer should promote the youth's supervision and, wherever possible, work together on a basis of trust with the youth's parent or guardian or his/her legal representative. In the exercise of his/her office, the probation officer shall have rights of access to the youth. He/she may require the youth's parent or guardian, legal representative, school, or the person providing him/her with training to provide information about the youth's conduct (section 24).

If the youth commits a criminal offense during the probationary period and thereby demonstrates that the expectation on which the suspension was based has not been fulfilled, or if he/she seriously or persistently violates instructions or persistently evades the probation officer's supervision and guidance and thereby gives cause for concern that he/she will commit further criminal offenses, or if he/she seriously or

440 The probationary period shall commence on the day the decision to suspend the youth penalty enters into force. It may subsequently be shortened to one year or, prior to its expiry, be extended to a maximum of four years. However, in the cases designated in section 21, subsection 2, the probationary period may be shortened to no less than two years.

441 The probation officer shall be appointed by the judge. The judge may issue instructions for the performance of his/her tasks in accordance with section 24, subsection 3. The probation officer shall report, at intervals fixed by the judge, on the manner in which the youth conducts himself/herself. He/she shall inform the judge of serious or persistent violations of instructions, conditions, assurances, and offers (section 25).

persistently violates conditions, the probationary suspension of the youth penalty shall be revoked by the judge. However, the judge shall refrain from this if that is sufficient for further instructions to be issued or conditions to be imposed, or to extend the suspension or supervision period to a maximum of four years, or to place the youth under the supervision of a probation officer once more prior to expiry of the probation period. No reimbursement shall be effected for services rendered by the youth in compliance with instructions, conditions, assurances, or offers (section 23). However, if the judge revokes suspension, he/she may give credit against the youth penalty for services rendered by the youth in compliance with conditions or corresponding offers (section 26). If the judge does not revoke the suspension, he/she shall release the offender from serving the youth penalty upon expiry of the suspension period (section 27).

2.3.3.3.2 Suspension of imposition of youth penalty (probation)

Under section 27 of the Youth Court Act, if, after exhausting all forms of investigation, there can be no certainty as to whether, while committing the criminal offense, the youth's harmful inclinations were demonstrated to such an extent as to necessitate the imposition of youth penalty, the judge may issue a finding as to the youth's guilt while suspending the decision to impose a youth penalty for a probationary period to be fixed by the judge. It should be mentioned that the probationary period may not exceed two years, nor be of less than one year's duration, and it shall commence on the day the judgment establishing the youth's guilt enters into force. It may subsequently be shortened to one year or, prior to its expiry, be extended to a maximum of two years (section 28). The youth shall be placed under the supervision and guidance of a probation officer for all or part of the probationary period (section 29). If – primarily from the youth's poor conduct during the probationary period – the offense censured in the court's verdict is a result of the youth's harmful inclinations demonstrated to an extent requiring the imposition of youth penalty, the judge shall order the imposition of the very penalty he would have handed down at the time of the verdict had a certain assessment of the youth's harmful inclinations been possible (section 30).[442]

2.4 Juvenile Delinquents and Proceedings

When a child behaves in conflict to the law, the controversial issue in the juvenile criminal justice system is how to treat such a child. According to the labeling theory, international instruments emphasize on informal intervention in the field of juvenile delinquency. Under this theory, being delinquent is indicative of the special mental

---

442 If the conditions in this section do not obtain upon expiry of the probationary period, the sentence shall be considered to have already been spent.

and age conditions of the perpetrator, and because of this feature, children and juveniles are not considered as offenders (guilty) if they commit an act against the law, and they can be treated as delinquents in the realm of criminal law. From the viewpoint of this theory, delinquency does not require formal intervention and punishment but educational, corrective, and rehabilitative measures. Hence if a child or juvenile commits a crime which deserves punishment, they should be called criminals or delinquents, and according to the theory, this would put their social status at risk. Therefore, the best intervention is informal. However, in some cases a formal intervention should be considered, provided that the differentiated policies in the justice system apply to juvenile delinquents. Thus, intervention can be either formal or informal, each of which has its own features and criteria to be discussed in this section.

2.4.1 United Nations

2.4.1.1 Formal institutions and proceedings

Formal intervention and assessing to justice is one of the human rights measures that has been strongly emphasized by all international human rights instruments. Despite all criticism of formal intervention, especially in the case of juvenile delinquency, it cannot be ignored, yet requires a special mechanism which should also be considered as a last resort.

According to the international instruments and standards, juvenile justice agencies are an integral part of the national development process of each country. Under the comprehensive framework of social justice for all juveniles, these agencies play an important role in society for the protection of juveniles and the establishment of a peaceful order. Therefore, the juvenile justice services and agencies should be developed and coordinated so as to improve and preserve the strengths of authorities and employees who provide such services, such as procedures and methods with respect to treating children and juveniles. In terms of considering the diverse needs of juvenile offenders, of the protection of fundamental rights, as well as full and fair provision of the requirements of society, institutes and councils should be established for the administration of justice agencies. In addition, the supply of services in the field of managing and administering juvenile justice agencies should be regularly and nationally planned and executed (Beijing Rules).

According to article 40-3 of Convention on the Rights of the Child, States parties shall seek to promote the establishment of laws, procedures, authorities, and institutions specifically applicable to children alleged as, accused of, or recognized as having infringed the penal law. In this field, the Committee on the Rights of the Child states that for a child in conflict with the law to be dealt with in a way that promotes reintegration requires that all actions should support the child in becoming a full, constructive member of his/her society (general comments no. 10, p. 29). The formal institutions, authorities, and procedures will be reviewed in this section.

2.4.1.1.1 Pre-trial stage

a) Specialized police units for juveniles

The United Nations' instruments for juvenile delinquents have paid particular attention to the role of special youth police in the criminal process. Accordingly, regarding specialization within the police force and in order to best fulfill their functions, article 12 of the Beijing Rules stipulates that police officers who have frequently or exclusively dealt with juveniles, or who are primarily engaged in the prevention of juvenile crime, shall be specifically instructed and trained. In large cities, special police units should be established for that purpose. As police are the first point of contact with the juvenile justice system, it is most important that they act in an informed and appropriate manner. Furthermore, it should be noted that increasing juvenile crime has been associated with the growth of large cities; therefore specialized police units would be indispensable – not only in the interest of implementing specific principles in the justice system, but more generally for improving the prevention and control of juvenile crime as well as the handling of juvenile offenders.

In the Second Congress of the United Nations on Crime Prevention and Criminal Justice in 1960,[443] it was empathically emphasized that the police forces shall take special measures for the prevention of juvenile delinquency. [444] Albeit, such measures should not turn the police into a social organization or cause interference with social and educational organizations. It was also stated that all possible measures should be taken in consideration of and respect for human rights, adding that a special police force for the prevention of juvenile delinquency should take into account the needs and requirements of each country. It was provided that in the instance of a first crime committed by a juvenile, his/her records shall be checked and filed after fingerprinting. However, it is worth mentioning that he role of police and especially the use of female police officers for dealing with female delinquents has been emphasized in other United Nations Congresses on Crime Prevention and Criminal Justice.[445] According to the international instruments, proactive and reactive policing should be considered in the case of juvenile delinquency.

---

443 The Second Congress made recommendations on special police services for the prevention of juvenile delinquency, with particular emphasis on developing the greatest possible cooperation between the police, various specialized national agencies, and the general public regarding measures to prevent juvenile delinquency.

444 Contemporary studies of policing have emphasized the importance in police work of duties unrelated to crime detection and the arrest of offenders. The policing of young people exemplifies this aspect of police work. This can be illustrated by tracing the historical and ideological context within which police remedies for juvenile delinquency were developed throughout the 19th and 20th centuries. *Lee* 1998, p. 1.

445 *Loeber et al.* 2003, p. 32.

In the framework of the justice system in general, and in crime detection and juvenile arrest specifically, the role of the police is essentially reactive. This is the responsibility of the judicial police, who should be present at the scene of the crime as soon as possible after notification of the crime, take measures to maintain the evidence, trace materials of the crime, and prevent the escape of the suspects. Then the proceedings should be declared to an independent prosecutor's office.[446] However, as already explained, in addition to these items, proactive policing plays an important role in crime prevention. In fact, the police are responsible for maintaining peace and order in society, for crime detection, arresting offenders, and crime prevention. Sometimes, prevention, control, and even rehabilitation measures overlap, and often one effective method can be used to address several aspects of delinquency.

Once a crime is committed, it should first be specified whether or not it is the responsibility of the police to intervene. The way the police treat the perpetrator is very important in juvenile delinquency cases. Hence, the legislator has given broader powers and authority to the police handling juveniles. According to the cases mentioned in the international instruments, the police should observe the principles assigned by the law when dealing with juvenile delinquency. These principles include giving warnings to children, diversion, and redirection, according to the rules. The most important development of the past quarter century has been the creation of various measures for juvenile offenders, especially police warnings and alerts. However, it has taken a long time for all those in the police forces to become aware of the issue that referral of children and juveniles to the court for addressing their crime was not suitable, and that an informal warning to the child or reporting their predicament to the family (in the form of an informal settlement) were more appropriate ways to handle such a situation.[447]

b) Prosecutor's office for juvenile offenders

In the justice system, the prosecutor, as the representative of the government, is a person who is responsible for protecting the state's interests. In the case of juvenile delinquency, the prosecutor does their duty by issuing an indictment against the accused who is either a child or a juvenile. However, differentiated policies in the justice system also require other duties which should be carried out by the prosecutor. Accordingly, in addition to the above-mentioned duties, the legal powers, and the authority of the prosecutor in the juvenile justice system were included, as well as participation in programs for changing and reforming the criminal justice system in favor of children and juveniles, discussions on legal issues, and playing an active role in society to further the prevention of delinquency and the protection of children's rights.[448] Although it is believed that the role of the prosecutor in juvenile

---

446 *Akhondi* 1368, p. 38.

447 *Ashoori* 1376/1998, p. 68; *Khaleghi* 1388/2009, p. 23.

448 *Seigel et al.* 1989, p. 544.

courts is to control the policies of these courts and not to restrict them, the duties and opinions of the prosecutor are different from those of the juvenile court's staff. Therefore, these multiple duties have led to criticism in some legal systems.

In fact, there are two approaches to the establishment of a prosecutor's office besides the juvenile court. Some believe that with respect to the objectives and policies envisaged under juvenile criminal law, there is no need for a prosecutor's office (and thus a prosecutor), because their primary task is criminal prosecution, conviction, and enforcement of penalties, while juvenile criminal law is based on other foundations such as the rehabilitation of delinquents and the prevention of reoffending. Others argue that all duties of the prosecutor's office in the general rules of criminal law can also be applied to juvenile delinquency cases, but with different priorities and orientations. It is obvious that many of the measures that are appropriate for the juvenile justice system can be taken under the supervision of the prosecutor's offices. Some authors also believe that the philosophy behind this measure is facilitation of addressing the juvenile crimes, considering the presumption of lack of criminal responsibility for children and juveniles.[449] In any case, it seems that the establishment of a special prosecutor's office is the basis for most legal systems in the world.

In most legal systems, whenever a child or young person under age (in terms of having criminal responsibility) commits a serious crime, the prosecutor shall be immediately informed and issues an order for investigation and prosecution. After completion of a preliminary investigation, the case will be referred to the competent authorities through the prosecutor's office. In general, as far as juvenile delinquency cases are concerned, duties of the prosecutor include investigation, cooperation with the police when investigating the facts, preparing and reviewing the indictment to the court, making decisions on the duration of pre-trial detention (preventive detention), justifying the child or juvenile before referral to the court, discussion and exchange of views on the lawyer's arguments, explaining the condition of the case, and so on.[450] Whenever the circumstances of a crime require a more meticulous investigation, the preliminary investigations are conducted by the investigators under the supervision of the prosecutor. Therefore, in line with the best interest of the child, prosecutors and investigators should be trained in order to perform their duty in the best way possible. The United Nations Committee on the Rights of the Child also emphasizes that in formal investigations by a competent authority – which usually begin at the prosecutor's office –, the principles of a fair trial should be observed. Under investigation, no child shall be subjected to torture or other cruel, inhuman, or degrading treatment or punishment, and generally the imposition of any corporal punishment for children is forbidden. After crime detection and the preliminary investigation, the child may need to be detained before trial but for the shortest possible time.

---

449 *Goldoost* 1382/2003, p. 96.

450 *Danesh* 1374/1995, p. 130; *Seigel et al.* 1989, pp. 545–547.

c) Juvenile rights in pre-trial proceedings

According to international hard law and soft law, including the Convention on the Rights of the Child (article 37), the Beijing Rules, the United Nations Model Law on Juvenile Justice and Related Commentary, as well as the United Nations General Comments on Children's Rights in juvenile justice systems, children's rights should be observed in pre-trial proceedings. Every child who is apprehended or arrested shall be informed immediately of the reason for such arrest and of his/her rights, in a manner which is consistent with the age of the child and his/her level of understanding. The law enforcement officer concerned shall immediately notify the child's parents or legal guardian of the arrest, providing information on why and where the child is being held. Use of force and of instruments of restraint by the police is prohibited during arrest and in police custody, except in very special circumstances. Every child who is arrested shall have the right to have his/her parents or legal guardian present at the place where he/she is being held in custody, and to be questioned about the alleged offense only in the presence of his/her parents or legal guardian. A child who is arrested shall be provided with free legal aid, be permitted to consult with his/her legal representative before being questioned by the police or the prosecutor, and be questioned about the alleged offense only in the presence of his/her legal representative. A child who is apprehended or arrested shall have free assistance by an interpreter if he/she cannot understand or speak the language used.

Where the child is brought to a police station after being apprehended or arrested, or if he/she is apprehended or arrested at the police station after attending voluntarily, the details of the child shall be registered by the law enforcement officer concerned. The child or juvenile shall not be placed in a locked cell or secure area, unless he/she poses a danger to him-/herself or others. Only law enforcement officers (including police officers, investigating officers, and prosecutors) who have received specialized training in working with children shall question a child about an alleged offense. No child shall be subjected to coercive questioning or interrogation and be compelled to confess or acknowledge guilt. A child who is apprehended or arrested shall be searched only by a police or investigating officer of the same gender as that of the child. Before the child is searched, he/she shall be informed of the purpose of the search and the grounds on which the relevant authority for the search has been given. Where the law enforcement officer wishes to conduct an intimate search, a warrant shall be obtained from the juvenile court. This rule should be observed in sampling the child or juvenile.

Where a law enforcement officer has charged a child or juvenile with an offense, he/she shall be released without delay and handed over to his/her parents or legal guardian on the condition that he/she may have to return to the police station or appear before a competent court on a specified date. In exceptional cases, the juvenile court may order pre-trial detention if the child is being investigated for a serious offense or if he/she is a persistent offender, and such detention is a measure of last resort; it is necessary where the child poses a danger to him-/herself or to other persons, to prevent interference with a witness or other obstruction of the course of justice, or to ensure that the child does not avoid further judicial proceedings. A child held in pre-trial detention shall be entitled to all rights and guarantees under international instruments.

Children and juvenile delinquents have the right to appeal against the order of detention and to access a lawyer free of charge so as to enable them to prepare for their case or to appeal against detention. Where a child is held in pre-trial detention, his/her deprivation of liberty should be reviewed by the juvenile court every two weeks. The child shall be present in court at each review of the pre-trial detention. He/she shall be legally represented at the review. The juvenile court shall order continued detention only where the prosecution satisfies the court that it is necessary. Where the criteria have ceased to apply, the child shall be released immediately. The juvenile court shall ensure that the child is being detained in suitable conditions and not together with adults or convicted children. Where the conditions of detention are not suitable, the child shall be released or moved to a suitable alternative place.[451]

2.4.1.1.2 Trial stage

a) Juvenile Court

The main objectives for the establishment of a juvenile delinquents' court include the necessity of determining the causes of delinquency, studying the personality of the child, adoption of specific measures for rehabilitation and educating them, as well as the prevention of further crimes. In trials for adult criminals, the judicial authorities do their best to collect evidence and proof, to investigate the elements of the crime, and to prove or deny the charges, and non-legal experts are rarely used. By contrast, in juvenile courts, in collaboration with physicians, psychologists, psychiatrists, sociologists, and social workers, the judge tries to identify the personality traits of the child and adopts the required rehabilitative, educational, and treatment measures. Juvenile courts, with their own agencies, are different from judicial agencies for adults and specifically address the crimes of children and juveniles. In a juvenile court, the judge is not the only one who makes decisions about the child; the comments and views of non-legal experts also play a decisive role in determining cases.[452]

With the presence of psychologists, physicians, social scientists, and other experts in juvenile courts, and due to the specific character of the judges in juvenile courts, in addition to the law, they should have a rudimentary knowledge about social science,

---

451 UNODC 2013, pp. 19–27.

452 *Danesh* 1374/1995, p. 65; *Shambayati* 1389/2010, p. 39.

sociology, psychology, and so on. It can be concluded that all members of a juvenile court – from the judge to experts – aim at helping the juvenile delinquents and at rehabilitating them. This distinguishes a juvenile court from an ordinary court of adult hearings and also indicates the excellent and noble goals of such courts.[453]

A juvenile court is basically headed by a judge or other competent authority. The judge is the central axis and is responsible for a variety of duties, which have become more prominent in recent years. Given the specific agencies of juvenile courts, the path to appointing judges for juvenile courts is different from the one through which judges are appointed to ordinary adult courts. Due to the greater vulnerability of children and juveniles and the special spirit of this stratum of society, the international and national rules recommend special judges with specific characteristics to be chosen for this highly sensitive and important mission. In addition to judicial competence and mastery, having knowledge about social and cultural affairs as well as psychological issues is one of the primary conditions for a person if they are to be appointed as a juvenile court judge.

For international cooperation to improve the social, economic, and psychological status of children and juveniles, the general assembly of the International Association of Children's Court Judges (AIJE) in Geneva in 1970 approved the necessity of specialized juvenile courts judges and emphasized the offer of courses on juvenile delinquency in law schools. On a summit in Oxford in 1974, the second branch of the AIJE recommended:[454]

1) Social workers shall participate in special courses for training and work as apprentices in juvenile courts for a while after passing the course;
2) the task of the juvenile court judge is to determine the type of educational measures that are expected to be executed by probation officers, and to make the necessary changes to them;
3) in the case of conditional release, the judge of a juvenile court should strive to somehow preserve the parental guardianship of the child and the emotional relations of his/her family;
4) even where the intervention of the juvenile court is restricted to parole or educational guardianship of the child, the decisions made by the judge should be based on medical, social, and psychological studies on upbringing methods of the child.[455]

---

453 *Danesh* 1374/1995, p. 87.

454 It has been a long time since the Modern Social Defense School has pointed out and recommended that such professionals and experts should be involved in courts. This is of greater importance for juvenile courts because of the special conditions of children and juveniles.

455 *Danesh* 1374/1995, p. 87–89. For further reading about the cycle of the juvenile justice system, see *Bernard et al.* 2010, pp. 380–381.

b) *Juvenile rights in trial proceedings*

According to the Convention on the Rights of the Child (article 37), the Beijing Rules, the United Nations Model Law on Juvenile Justice and Related Commentary, and the United Nations General Comments on Children's Rights in juvenile justice, children's rights should be observed in trial proceedings. Every child is entitled to a fair trial before a juvenile court, and to have the matter tried without delay. Before commencing proceedings, the juvenile court shall ensure that the child understands the nature of the charges and the matters that must be established before he/she can be found guilty. Where the child does not understand the language used in trial, the juvenile court shall appoint an interpreter for the child suitable to his/her needs. Handcuffs or other restraints are not permitted while the child is in court or in transit to or from the court, unless he/she poses a danger to him-/herself or to others. The parents or legal guardian shall have the right to be present at the trial unless their presence is considered not to be in the best interests of the child.[456]

Every child shall be entitled to free legal aid during the trial. Where a child appears in court and is not legally represented, the court shall direct that he/she be legally represented free of charge. The child shall have the right to dismiss his/her legal representative and to appoint a replacement. Every foreign child has the right to consular assistance during his/her trial. A child who cannot understand or speak the language used by the court, or if the child is deaf or has no speech or a speech impairment or other disability that reduces his/her ability to communicate, he/she shall be provided free and without charge with an interpreter qualified to assist him/her.[457]

The general rule that a trial is to be held in public shall not apply to court proceedings involving a child or to an appeal against any conviction or order of the court within the proceedings. No person may be present at a sitting of the juvenile court unless his/her presence is necessary to the proceedings of the court, or unless the juvenile court has given permission for him/her to attend. Information that may lead to the identification of the child shall not be published in any oral, written, visual, or virtual form. No child shall be tried in absentia. The child shall be entitled to hear all the evidence in the case and to remain in court at all times unless the court decides that the child's presence is detrimental to his/her best interests. The child shall have the right to examine or have examined the witnesses against him/her as well as to obtain their attendance and examination on his/her behalf under the same conditions as the witnesses against him/her. Every child shall have the right during his/her trial not to be compelled to give testimony, and to refuse to answer to incriminating questions. In deciding whether to admit a confession made by a child, the juvenile court shall take into account his/her age, development, the

456 UNODC 2013, pp. 27–35.

457 UNODC 2013, pp. 27–35.

length of the interrogation, the child's understanding, his/her fear of unknown consequences or of a suggested possibility of detention, any inducements offered, the presence of parents or the legal guardian, and whether the child has a legal representative present.[458]

Every convicted child shall have the right to have the decision and any measures imposed reviewed by the appeals court at the juvenile court, and to be informed, in a language that he/she can understand, of the right to appeal. Every child shall be entitled to free legal representation when making an appeal. The juvenile court shall have the power to discontinue the proceedings at any time and to discharge the child prior to conviction. When a child appears before the court for the first time after being charged with an offense, the juvenile court shall ascertain whether the police have considered the use of diversionary measures rather than proceeding to trial. If the juvenile court considers that the police or prosecutor has not given any or sufficient consideration to the use of alternative measures to judicial proceedings during the pre-trial phase, they may apply such measures. The juvenile court may at any stage of the proceedings refer the child to assessment in order to determine whether he/she is in need of protective services. The purpose of a sentence imposed on a child by the juvenile court shall be to promote his/her rehabilitation and reintegration into society.[459]

In all cases, the juvenile court shall obtain a social enquiry (pre-sentence) report before passing a sentence or pronouncing a decision on a child. Where a child is convicted of a criminal offense, the juvenile court shall consider – with regard to the circumstances of the case – alternatives to detention. Deprivation of liberty as a sentence shall be imposed only after careful consideration, only as a measure of last resort, and only for the shortest appropriate period of time. The juvenile court shall not impose a custodial sentence on a child unless he/she is convicted of a serious offense and unless there is no other appropriate response. The child shall serve the custodial sentence in a detention facility as close as possible to the area where his/her parents or legal guardian reside. No child shall be subjected to capital punishment, life imprisonment, torture, or other cruel, inhuman, or degrading treatment or punishment, to corporal punishment or forced labor as a sentence.[460]

2.4.1.1.3 After trial stage/post-trial

a) Juvenile correctional centers

A juvenile correctional center is considered to be a rehabilitative residential facility for juveniles sentenced to deprivation of liberty. Under article 11(b) of the Havana Rules, deprivation of liberty means any form of detention or imprisonment or placement of a person in a public or private custodial setting, from which this person is

458 UNODC 2013, pp. 27–35.

459 UNODC 2013, pp. 27–35.

460 UNODC 2013, pp. 27–35.

not permitted to leave at will or by order of any judicial, administrative, or other public authority. Educational institutions, rehabilitation centers, and similar institutions can be examples according to the definition in the Havana Rules.[461]

Article 37(c) of the Convention on the Rights of the Child requires that every child deprived of liberty shall be treated with humanity and respect for the inherent dignity of the human person. Also, article 12 of the Havana Rules provides that deprivation of liberty should be effected in conditions and circumstances which ensure respect for the human rights of juveniles. Bearing in mind these provisions as well as the fact that detention facilities serve as indicators for the level of implementation of the rule of law, international law imposes an obligation upon States to secure a detained child's human dignity and worth regarding both his/her treatment in and the conditions of detention. The international framework on child justice (and particularly the Beijing and the Havana Rules) provide a very detailed set of standards on what this treatment should include. The Beijing Rules set out fundamental guarantees for children under the age of 18 who are deprived of their liberty, while the Havana Rules – which apply to all children in a public or private custodial setting from which a person is not permitted to leave at will – by order of any judicial, administrative, or other public authority detail the treatment children should be provided with, as well as the conditions of their detention. The standards were developed with a view to counteracting the detrimental effects of all types of detention as well as to fostering integration into society and are reinforced by the Beijing Rules.[462]

A child can be admitted to a detention facility only by court order, which includes pre-trial detention as well as post-trial deprivation of liberty as a sentence. Separation from adults, between age groups, and by type of offense shall be observed. Girls and children with special needs shall be protected. The juveniles also have the following rights: right of access to healthcare services, physical environment, accommodation and nutrition, education and vocational training, and work opportunities. Furthermore, the requirements for exercise and recreational activities shall be facilitated, as the Havana Rules state that every child has the right to a suitable amount of time for daily free exercise. This should take place in the open air whenever weather permits. During this time, appropriate recreational and physical training should be provided. Freedom of religion, conscience, and thought, contact with family and the outside world, complaints, and requests are the other children's rights to be considered by the authorities. Use of force and/or physical restraint should be prohibited, any search shall be done by an intimate person, and an inspection should be carried out regularly and by an independent system.[463]

---

461 *Cappelare* 2005, pp. 30–31.

462 UNODC 2013, pp. 110–135.

463 UNODC 2013, pp. 135–139.

b) Juvenile after-care centers

After-care centers should be established to assist children and juveniles with their reintegration after release. Some are released without any arrangements for housing, employment, education, counselling, or support, which may result in them reoffending. This is despite the Havana Rules requiring states to ensure that all children leaving detention benefit from supportive arrangements (articles 38 and 59). Making such arrangements can require considerable coordination and planning. That planning process should start on the first day of the sentence, and a plan should be elaborated focusing on the reintegration and rehabilitation of the child into society. Prior to a child's release from the detention facility, its management needs to prepare accordingly and take, as early as possible, all necessary measures. Access programs should be elaborated that prepare children for their return to their communities, with full attention given to their emotional and physical needs, family relationships, housing, schooling and employment possibilities, as well as their socio-economic status.[464]

Unless adequate support is given to the child following a period of detention, there is a high risk that any rehabilitative progress that has been made during the child's detention will be lost. The commentaries to the Beijing Rules (article 97) and the Havana Rules (article 80) emphasize the importance of care and support following a period of detention. When it comes to the actual day of the release, the child detention facility has a positive obligation to ensure that the child is handed over to his/her parents or legal guardian. Where these are not available to collect the child, the management of the detention facility has to put in place mechanisms to safely deliver the child to his/her home. The parents or legal guardian must sign a form presented by the officer concerned upon delivery of the child stating that the child has been handed over. The State has the duty to put in place legislation to facilitate the child's reintegration and rehabilitation after his/her release from the detention facility. The Beijing Rules (article 29.1) provide that the State should also provide half-way houses, educational homes, daytime training centers and other such appropriate arrangements that may assist juveniles in their proper reintegration into society.[465]

2.4.1.2 Informal institutions and proceedings

As already explained, dealing with juvenile delinquents without resorting to formal judicial authorities and institutions is strongly emphasized in the international instruments and standards. This is due to the need to prevent for the delinquent the adverse consequences of exposure to formal criminal authorities.[466] This kind of intervention

---

464 UNODC 2013, pp. 135–139.

465 UNODC 2013, pp. 135–139.

466 According to recent research, labeling to any title, even "delinquent", will have adverse consequences for the child, which is in contrast to the first theory. According to the rules of the United Nations, labeling the child as a "delinquent" itself indicates his/her fault and never

is called informal intervention or diversion. The concept of diversion usually refers to the use of alternatives to the classical and traditional measures by the prosecution authorities in dealing with criminals and delinquents. The desired effect of such a method is to remove the suspect from the cycle of criminal procedure.[467]

In this regard, general comment no. 24 (2019) emphasizes that the decision to bring a child into the justice system does not mean that the child must go through a formal court process. The Committee on the Rights of the Child emphasizes that the competent authorities – in most States the public prosecutor – should continuously explore the possibilities of avoiding a court process or conviction, through diversion and other measures. In other words, diversion options should be offered from the earliest point of contact, before a trial commences, and should be available throughout the proceedings. In the process of offering diversion, the child's human rights and legal safeguards should be fully respected, bearing in mind that the nature and duration of diversion measures may be demanding, and that legal or other appropriate assistance is therefore necessary. Diversion should be presented to the child as a way to suspend the formal court process, which will be terminated if the diversion program is carried out in a satisfactory manner (para. 72). The general comment also explains that a wide range of experience exists with the use and implementation of non-custodial measures, including restorative justice[468] measures. States parties should benefit from this experience, they should develop and implement such measures by adjusting them to their own cultures and traditions. Measures amounting to forced labor, to torture or inhuman and degrading treatment are to be explicitly prohibited and penalized (para. 74).

In other words, the purpose of alternative measures for judicial proceedings (diversionary measures) is to avoid instituting judicial proceedings against a child in conflict with the law or to suspend them, as well as to influence the proper development

---

refers to his/her conditions before delinquency, even if he/she is properly rehabilitated. However, there is a discussion about labeling methods, which is the result of social reactions and can be effective in increasing criminal behaviors and self-criminalization thoughts. Studies on child delinquents have revealed that labeling under titles such as "criminal", "delinquent", and "law violator" is done more often by family members than criminal justice institutes (some sort of labeling happens in these institutes due to their special procedures). On the other hand, some studies on this issue indicate that juvenile delinquency is rather the result of labeling by others who consider the child as a lawbreaker, rather than coming from the child's family. However, according to the results of both of these studies, labeling by the family and other people in society plays a more decisive role in increasing delinquency among juveniles than labeling in the criminal justice institutes. *Abachi* 1388/2009, p. 57.

467 *Nadjafi Abrandabadi & Hashem Beigi* 1397/2018, p. 95.

468 According to para. 8 of general comment no. 24, "restorative justice" is defined as any process in which the victim, the offender, and/or any other individual or community member affected by a crime actively participate in the resolution of matters arising from the crime, often with the help of a fair and impartial third party. Examples of restorative processes include mediation, conferencing, conciliation, and sentencing circles.

of the child and enhancing his/her personal responsibility in order to promote his/her reintegration and assuming a constructive role in society. According to the international instruments and standards, whenever it is appropriate and desirable, the competent authority (including police, prosecutor) dealing with the child's criminal case shall consider whether alternative measures to judicial proceedings would better serve his/her reintegration and protection, the rights of the victim, crime prevention and/or the protection of society, rather than judicial proceedings. They shall also take into account the seriousness of the offense, the age of the child, the circumstances of the case, and any previous offending behavior when considering alternative measures to judicial proceedings. According to article 14-1 of the Beijing Rules, alternative measures to judicial proceedings shall be imposed on a child only where there is compelling evidence that he/she has committed the alleged offense, and if the child freely and voluntarily admits responsibility and, where appropriate, his/her parents or legal guardian. Where the child has no parents or legal guardian, if they cannot be found, if the child is estranged from his/her parents or legal guardian, or if there is a conflict of interest between the parents or the legal guardian and the child, the juvenile court shall appoint a curator (guardian) ad litem who can consent to alternative measures to judicial proceedings. Prior to consenting, the child and the parent or legal guardian shall be provided with adequate and specific information on the nature, content, and duration of the diversionary measures as well as on the consequences of failure to complete the diversionary measures; furthermore, the child shall be given the opportunity to seek legal aid and to discuss the appropriateness and desirability of the diversionary measures offered. Alternative measures to judicial proceedings may include (a) restorative justice programs, such as victim-offender mediation, family or group conferencing, and social work intervention; (b) verbal warning; (c) formal written warning; and (d) counseling for the child and family. The competent authority shall have the power to make a restorative justice order. The purpose of such an order shall be to enable the child to make a reparation for his/her offense to the victim, the community and/or to society. Such an order may require the child to 1) acknowledge responsibility for the offense and to show an understanding of its impact on the victim; 2) repair the harm caused to the victim, the community and/or society; 3) make an apology to the victim; and 4) undertake such activities as may be agreed with the victim or the community. Where a child breaches a condition attached to a diversionary measure, the competent authority may decide to resume proceedings against the child, taking into consideration the part of the measure already completed by the child before sentencing. Admission of responsibility for the alleged offense, made for the purpose of applying an alternative measure to judicial proceedings, shall not be used against the child in court.[469]

469 UNODC 2013, pp. 15–19.

2.4.2 Iran

In this section, the formal and informal institutions and proceedings will be studied for the Iran dimension.[470]

2.4.2.1 Formal institutions and proceedings

2.4.2.1.1 Pre-trial stage

In the Iranian justice system, criminal proceedings slightly differ from those in the juvenile justice system, yet there is common ground. In accordance with the Criminal Procedure Code, the rights of juveniles should be respected based on the requirements of due process in the criminal procedure (article 7). In line with the international instruments, a specialized police unit handles cases of juveniles under 18 years as the first line of intervention at the police station (article 31).[471] Immediately after arresting the accused, the police have an obligation to inform the accused of his/her rights as dictated by law, including the option to consult a lawyer. The police should also record the beginning and ending time of each interrogation, its duration, as well as the duration of pauses between interrogations. They should then ask the accused to sign the recording document. The accused can ask for the presence of a defense lawyer, but not for more than one hour[472] (articles 28 to 64).

Preliminary investigations into crimes committed by persons aged 15 to 18 – except for crimes of adultery, homosexuality, against public decency, and minor crimes which are directly dealt with by a court[473] – will be conducted in this special prosecutor's office. Preliminary investigations into crimes committed by persons aged under 15 are directly conducted by a juvenile court which is responsible for all duties assigned to law enforcement officers and the prosecutor's office by law. In evident crimes, if the perpetrator is a child or juvenile, the law enforcement officers are required to preserve all tools, equipment, proofs, signs, and causes of the crime, but

---

470 For further reading about proceedings and formal and informal institutions in Iran; see *Asadi* 1382/2003, pp. 103–123; *Khaleghi* 1388/2009; *Ashori* 1381/2002; and *Khazani,* 1377/1998.

471 *Abachi* 2015, pp. 837–838.

472 With cases related to organized crime, drug crimes, and crimes against the security of the country, the arrested accused cannot meet with a lawyer for one week.

473 Any prosecution and investigation of adultery, sodomy, and other crimes against public morals is forbidden and questioning the individuals is not allowed, unless the offense has been publicly visible or if there is a private plaintiff. In this case, prosecution and investigation are allowed only in the scope of a complaint or conditions observable by the judge. Where there is no private plaintiff for adultery, sodomy, and other crimes against public morals and if the accused is primarily intended to plead guilty, the judge will advise him/her to not reveal the crime and to restrain from confession. The judge is obliged to remind witnesses of the consequences of a testimony lacking the legal requirements. In these crimes, if the victim is incapacitated, his/her parents or legal guardian have the right to file a complaint (articles 306 and 102).

they are not allowed to conduct a preliminary investigation on the child. After arresting the child, the law enforcement officers are obliged to immediately refer him/her to the prosecutor's office or juvenile court. Expiration of working hours and holidays does not prevent the referral the child to the prosecutor's office or juvenile court (article 285). The personal and social circumstances of the juvenile and his/her family for *Ta'zir* crimes (first to sixth degree) as well as other serious crimes should be evaluated and recorded by the prosecutor's office (article 285).

During the preliminary investigation, the judicial authority can, if necessary, entrust the child and juvenile to their parents or legal guardian – or to any other natural or legal person, in the case of absence, lack of accessibility, or refusal of parents or legal guardian. These individuals are obliged to introduce the child or juvenile to the court whenever necessary. However, children and juveniles aged 15 to 18 are obliged to personally introduce themselves to the court. If necessary, obtaining surety or collateral is possible only from defendants aged over 15. In the case of failure to introduce surety or to provide collateral – or in case of an offense mentioned in article 237 of the Criminal Procedure Code –, the prosecutor's office or the court can, by observing article 238 of the Code,[474] issue the writ (court order/warrant) of a temporary referral of the juvenile to correction and training centers. The writ of temporary referral is subjected to all conditions of temporary detention mentioned in the Criminal Procedure Code (article 287). In crimes that are addressed in the jurisdiction of a Criminal Court grade one, crimes that are subjected to blood money or the value of more than one fifth thereof, and *Ta'zir* crimes of grades six and higher, the juvenile prosecutor's office shall declare to the parents or legal guardian of the child to hire a public defense lawyer. In case no public defense lawyer is hired, or if the lawyer is absent without a reasonable excuse, the judicial authority shall hire an attorney for the child (article 415). In the pre-trial stage, secret investigation, attendance of the defense lawyer in the preliminary investigation, and participation of the parents or legal guardians are the most important elements in the Criminal Procedure Code.

It is notable that besides and in the same location of the juvenile court, a branch of the General and Revolutionary Court should be established as the juvenile prosecutor's office, led by a deputy prosecutor and attended by one or more investigators (article 285). The prosecutors for the specialized prosecutor's offices are appointed by the Head of Judiciary among the judges, who have at least five years of judicial service and the required competency considering their age and other aspects such as being married, having passed training courses, and preferably having a child (article 409).

---

474 Article 237 is about the temporary detention warrant, which is only applicable in serious crimes with severe punishments, while article 238 stipulates the conditions for issuing a temporary detention warrant.

2.4.2.1.2 Trial stage

Under the Criminal Procedure Code, criminal courts are classified into criminal courts of grades 1 and 2, revolutionary courts, juvenile courts and military courts. The juvenile court is held with the presence of a judge and two assessors who only provide consultations (article 294). In every district court, and if required, one or more branches of juvenile courts should be established. As long as the local juvenile court has not been established, all crimes committed by children and juveniles, except for those mentioned in article 315 of this law, will be addressed in a branch of a driminal court grade 1 or the court running the same duties (article 298). In accordance with article 315 of the Criminal Procedure Code, if children or juveniles commit a serious crime[475] that is within the jurisdiction of a criminal court grade 1 or a revolutionary court, their crime will be addressed and tried in a criminal court grade 1 specialized for children and juveniles. The accused (child or juvenile) enjoys all rights that are usually applied in juvenile courts. In each district, the required number of branches of a criminal court grade 1 for the investigation of juvenile crimes should be established to address the crimes mentioned in this article. However, it is notable that assigning some branches of a criminal court grade 1 to children's crimes does not prevent the referral of other cases. Observing the condition mentioned in this law, the presence of assessors is necessary for addressing and dealing with the crimes of children and juveniles in branches of a criminal court grade 1 that is specialized for juvenile crimes (article 315). Judges of juvenile courts are appointed by the Head of Judiciary among those judges with at least five years of judicial service who have the required a competency considering their age and other aspects such as being married, having passed training courses, and preferably having child (article 409).

Whenever the judge of a juvenile court fails to continue his/her duty for any legal reason, the Chief of Justice Department can personally continue the trial or appoint another competent judge to continue the process of proceedings (article 411). The Chief of Justice Department or the President of the court in the capital of each province is also the President of juvenile courts (article 410). Assessors of juvenile courts shall be selected among the experts of educational sciences, psychology, criminology, and social working, as well as academics and educators, either working or retired, who are familiar with psychological and educational issues of children and juveniles. For the selection of assessors, the head of each local jurisdiction proposes at least eight men and women who fulfill the conditions stipulated in this article to the Chief of Provincial Justice Department for each branch. The Chief of Provincial Justice Department appoints four of them to this position for two years, and re-election of them is not forbidden (article 410).

475 The crimes mentioned in article 315 include sexual offenses.

Therefore, as a specialized court with a special judge and two assessors, the juvenile court deals with children as well as with persons under 18 solar years of age (personal jurisdiction of the juvenile court) charged with all offenses except those which carry the death penalty, life imprisonment, imprisonment of more than five years, limb *Qisas* or *Diye* up to one third or more than the total blood money in intentional crimes, offenses punishable by a *Ta'zir* penalty of the first to fourth degree, and finally press and political crimes (subject matter jurisdiction of the juvenile court) which should be tried in criminal courts grade 1 (articles 302, 304, 306, 340). Under article 304 (note 1), children are defined as persons who have not yet reached religious maturity, and according to the Civil Code (note 1, article 1210), as well as the Penal Code (article 147), the age determination for this group is 9 and 15 lunar years, respectively, for girls and boys. Therefore, two separate groups are protected by the Criminal Procedure Code: children who have not reached religious maturity; and persons who are mature according to *Shari'a* but are under 18 years of age (article 304). This age determination has been influenced by *Shari'a* on the one hand and is combined with the international instrument standards on the other hand. Nevertheless, if the age of the accused exceeds 18 years pending trial, he/she will be addressed and tried in the juvenile court; if this happens before trial, he/she should be tried and addressed in a competent criminal court, albeit enjoying all benefits and privileges applicable in the juvenile court (article 304).

As already mentioned, preliminary investigation into crimes committed by persons aged under 15, as well as adultery, homosexuality, crimes against public decency, and minor crimes are directly conducted by the juvenile court, which is responsible for all duties assigned to law officers and the prosecutor's office by law (article 285). Whenever one or more children or juveniles commit a crime with the participation or abetment of adult persons, or if they abet adults in committing a crime, only the crimes of the children or juveniles will be addressed in the juvenile court. In crimes can only be committed by two or more persons and one of the defendants is within the jurisdiction of the juvenile court, the crimes of all defendants will be addressed there. In these cases, the principles governing the addressing of offenses of persons aged over 18 are a function of general rules (article 312). The juvenile court sets the time of trial and declares it to the child's parents or legal guardian, his/her defender, the prosecutor, and the complainant. If the age of the defendant reaches 18 or more during the proceedings, the time of trial will be declared to the accused or his/her defense lawyer.

In *Ta'zir* crimes of grades six, seven, and eight, as well as in *Ta'zir* crimes punishable by penalties other than imprisonment, whenever the accused, his/her parents or legal guardian, and the defense lawyer are present and call for trial, and once the conditions of the trial are provided, the court starts the proceeding and makes the final decision without setting an appointment (article 412). According to article 91 of the Penal Code, the evaluation of the intellectual development of child and juvenile should be done by the judge in *Hadd* and *Qisas*. In fact, in the cases of offenses punishable by *Hadd* or *Qisas*, if a mature person under 18 years does not realize the nature of the crime committed or its prohibition, or if there is uncertainty about their full mental and intellectual development, in accordance with their age, they shall be sentenced to the punishments prescribed in the *Ta'zir* crimes. The court may ask the opinion of forensic medicine or resort to any other method that seems appropriate in order to establish the offender's full intellectual development. Furthermore, parents, legal guardian, defense lawyer, complainant, individuals whose statement has been elicited in the preliminary investigation stage, witnesses, and social workers attend in the juvenile court, while the presence of other persons in the trial sessions is permitted only with the consent of the court (article 413). Whenever the interests are affected of children who have not yet reached the required religious maturity, the whole or part of the trial may be held in the absence of the child. However, the court's verdict is considered in-person (article 414). In cases of crimes that are addressed in the jurisdiction of a Criminal Court grade 1, those that are subjected to blood money or the value of more than one fifth thereof, and in *Ta'zir* crimes of grades six and higher, the juvenile prosecutor's office or court shall declare to the parents or legal guardian of the child to hire a public defense lawyer. In case no public defense lawyer is hired for the child or if the lawyer is absent without a reasonable excuse, the judicial authority shall hire an attorney for the child. In *Ta'zir* crimes of grades seven and eight, the child's parent or legal guardian can defend him/her or hire a defense lawyer. The child can also defend him-/herself (article 415). A petition of loss caused by a crime which has been submitted to the juvenile court will be addressed according to regulations, and the proper verdict will be issued in this regard. During addressing/proceeding the claim of loss, the child is not required to be present in court, unless his/her expressions are necessary for the issuance of the final decision (article 416). Verdicts and decisions of the juvenile court will be declared to the complainant, the defendants as well as the convicts, the parents or legal guardian of the defendants as well as the convicts, and their lawyers (article 417).

Verdicts and decisions of a juvenile court can be appealed in all cases. It is notable that the appeal of decisions and judgments of the juvenile court can be submitted to the office of the juvenile court, to the office of appeal at a provincial court, or to the management of a correction and training center if the child is held there (articles 445–446). Apart from the juvenile delinquent and his/her attorney, the juvenile court as well as the parents and legal guardian of the child also have the right of appeal (article 447). It is also worth mentioning that the conditions for appeal before the final verdict are the same for both juveniles and adults.

2.4.2.1.3 After the trial stage

Regarding juvenile cases, the respective judge who issues the sentence is not responsible for its enforcement. However, contrary to the rules applicable to adults, he/she should be responsible for the execution of the sentence in line with the best interest of the child. The most important sentences here are the custodial ones, which are enforced in correction centers. Correction centers are places established by the Prisons, Security and Reforms Measures Organization in the capital of each province as well as in other cities by recognition of the Head of the Judiciary, for holding and upbringing children and juveniles who have been convicted to deprivation of liberty. The first section of the executive regulations for Prisons, Security and Reform Measures Organization of 2005 deals with the terminology, with article 19 defining the juvenile correctional center as a center for reforming and educating children under 18 years who are held there. Therefore, these centers have been recognized by this regulation; however, there is no other special rule for juvenile correctional centers in this regulation.

The general regulations on the reception, diagnosis, and categorization of prisoners, their daily program of life in prison, working and learning, rehabilitative programs, and disciplinary punishments are found in the second section. The third part encompasses rules on the relations of the prisoners with the outside world. Receiving visits and mail, prisoners' short outings as a reward, transporting and freeing prisoners, and the conditions of preparing them for release are explained in the last section. In fact, most human rights standards have been considered in this regulation. Furthermore, the Juvenile Correctional Center Regulation of 1970 and other related international instruments, such as the United Nations Rules for the Protection of Juveniles Deprived of their Liberty of 1990 (Havana Rules), are named as examples.

Juvenile correctional centers have three sectors which consist of the provisional detention sector, the educational and reformatory sector, and the prison. These sectors are separated from each other, and in each sector children are categorized according to their age, records, crimes committed, and degrees of their trainability. Judges of juvenile courts are obliged to pay a visit to correctional and training centers within their jurisdiction once a month in order to assess the legal status of children and juveniles held there, as well as the process of their education and their moral progress (article 526 of the Criminal Procedure Code). According to the reports of managers of correctional centers, whenever the behavior of a child or juvenile causes moral corruption affecting others, and if this is ascertained by the judge of a juvenile court, they can order the child to be segregated from the other youngsters. The child can, at a later date, return to the previous center in the company of other children and juveniles if they correct their behavior (article 527). Furthermore, according to article 89 of the Penal Code, the juvenile court can review its decision once, based on the reports received regarding the behavior and condition of the juvenile in the correctional center; the court may reduce the detention term by up to one third or replace detention by an order to hand the juvenile over to their natural or legal guardians. The court's decision to review the original sentence shall be made if the child or youth has spent at least one fifth of the detention term in a correctional and rehabilitation center. The court's decision in these cases is deemed final; however, this does not prevent the juvenile from enjoying conditional release and other mitigations prescribed by the law if the respective requirements are met. According to the regulations regarding after-care centers of 2010, these centers were to be established to help prisoners in general, and children and juveniles deprived of liberty in particular, after release.

Despite the fact that no obstacles are standing in the way of executing international standards in these centers, there are no conflicts of interest with *Shari'a*, but due to restricted financial resources, some problems arise in practice. For example, most correctional centers established in the provincial capitals accept and register only male delinquents, because no special place is allocated for girls in these centers, so that they have to be sent to women prisons, which are not suitable for girls.

2.4.2.2 Informal institutions, proceedings and sanctions

According to a directive issued by the Head of the Judiciary of Iran in 2010, juvenile crimes should be addressed by informal authorities if possible. In Iran's legal system, informal addressing has been accepted according to the regulations of the Settlement Councils Regulation of 2002 and the Settlement Councils Act of 2015, although the rules determined in these legislations are different from the criteria of restorative justice in Western countries, and sometimes are even in contrast to them. Furthermore, in accordance with the Penal Code of 2013 and the Criminal Procedure Code of 2014, some informal institutions and proceedings have been introduced.

Under article 80 of the Criminal Procedure Code, if there is no complainant in criminal degrees seven and eight[476] – or if the complainant declares their consent and the accused has no former, more serious convictions[477] – and if after the charge, the social situation and the records of the accused have been explained (including the circumstances under which the crime has been committed) – and in case of necessity by obtaining a written recognition to observe the regulations –, the judge (including the prosecutor) can withdraw the prosecution once an order is kept on the file. Therefore, it seems that the legislator has granted the prosecutor the authority to ignore the prosecution. This refers to non-forgivable crimes, since the prosecution ends and thus the prosecutor is forced to stop it in such crimes, if the complainant declares his/her consent.

---

476 Crimes of degree 7 require 91 to 180 days imprisonment, a 10,000,000 to 20,000,000 Rials fine, whipping of 11 to 30 slashes, and deprivation of social rights of up to 6 months. Crimes of degree 8 require imprisonment of up to 3 months, an up to 10,000,000 Rials fine and whipping of up to 10 slashes. *Rahmdel* 2015, p. 851.

477 Stronger convictions include the death penalty, life imprisonment, cutting the member from the body, and physical retaliation if the amount of blood money is more than half of the total blood money, exile and imprisonment from 5 to 10 years, whipping as *Hadd* (some *Shari'a* crimes have fixed whipping like drinking alcohol, sexual intercourse without marriage, etc.), physical retaliation if the blood money equals half or less the total blood money, and imprisonment from 2 to 5 years.

According to article 81, in criminal degrees 6, 7, and 8, punishment could be suspended [478] if there is no complainant or if the complainant has declared their consent. This would also apply if the accused has compensated for the damages, or if both parties have agreed on a payment arrangement for the damage and the accused lacks a more severe former conviction. In the instance of consent of the accused, the prosecutor can suspend the prosecution for 6 to 24 months. They can order the following:

- providing services to the victim in an attempt to remove or reduce their damages, if agreed upon;
- abandonment of addiction by referring to a physician, clinic, or hospital within 6 months;
- refusal of following certain careers or professions for a maximum of one year;
- refusal to travel to specified destinations for a maximum of one year;
- introducing themselves to a specified authority or to the prosecutor at certain times for a maximum of one year;
- working on certain days or hours within public institutions or institutions of a public utility for a maximum of one year;
- taking part in educational, cultural, and professional courses during certain days or hours for a maximum of one year;
- surrendering driving privileges for certain means of transport and returning their driver's license for a maximum of one year;
- surrendering the right to carry or use a licensed weapon for a maximum of one year;
- relinquishing the right to have any contact with aiders and abettors or to the victim for a specified time;
- not leaving the country and surrendering their passport for a maximum of six months.

Furthermore, in accordance with article 82, in criminal degrees six, seven and eight, which can also be suspended, at the request of the accused and at the consent of the victim, as well as by obtaining proper security measures from the accused, the judge (including the prosecutor) can give the accused two months to obtain the consent of the victim and to compensate their damages. In order to bring about a compromise, they can also refer the case to the dispute resolution council or to a person or an institute for mediation between the accused and victim. The duration of mediation shall not exceed three months. In case of necessity, the judge can prolong this time, but only by three months. If the crime is a forgivable crime and the victim declares

---

478 Basically, punishment of the mentioned crimes (carrying a maximum of 2 years imprisonment), unless in cases that relate to the security of the country, of organized crime, sabotage in gas, oil, electricity, water, and communications. *Rahmdel* 2015, p. 852.

his/her consent, the prosecution ceases. Otherwise (if the crime is not forgivable), if the victim declares his/her consent, or if the accused provides compensation for damages or makes an agreement to do so (in case that he/she has no more serious former convictions), upon the accused'd agreement, the prosecutor can suspend his/her prosecution from 6 to 24 months on the basis of orders mentioned in article 81.[479]

2.4.3 Germany

In line with the criminological findings and international standards, Germany has developed an effective system of private and state welfare as well as justice institutions in the field of juvenile crime prevention and juvenile justice. The community youth welfare departments and the youth services in youth court proceedings are institutions with a double task. They fulfill purely welfare-oriented tasks and support the juvenile prosecutor and court by delivering personal and family background information for the trial on the one hand, and they are partly responsible for the execution of educational measures (mediation, social training, etc. based on the juvenile prosecutor's or judge's decision) on the other hand. The youth services in youth court proceedings are also responsible for avoiding unnecessary pre-trial detention. Therefore, they participate in the proceedings as early as possible and are immediately informed if a juvenile is placed in pre-trial detention. The personnel of the youth services are social workers or social pedagogues with at least three years of university education. The personnel of private welfare institutions in most cases have the same professional education. Sometimes they also have teachers, psychologists, and social workers with special training (e.g. as mediators) at their disposal. The Federal Probation Service also provides special courses for further professional specialization, for example mediation.[480]

Regarding juvenile rights in criminal proceedings, juvenile justice systems, particularly those following the welfare model, are often criticized for failing to guarantee human rights. The German juvenile criminal procedure preserves fundamental rights, and its disadvantages compared to adults are restricted to more exceptional cases. As far as the juvenile justice system is concerned, the German system does not share the shortcomings of welfare systems relying more on informal procedures than on formal legal rights.[481]

2.4.3.1 Formal institutions and proceedings

2.4.3.1.1 Pre-trial stage

It is notable that in Germany, criminal prosecution for adults in general, and for youth in particular, begins with an investigation of the crime by the police.[482] A specialized

479 *Rahmdel* 2015, p. 853.

480 *Dünkel* 2016, p. 7.

481 *Dünkel* 2016, p. 11.

482 *Albrecht* 2004, p. 470.

juvenile criminal police unit of officers qualified and trained to deal with children and juveniles exists in most major cities, although such specialized units for juveniles are not specifically mentioned in the Juvenile Judicature Act.[483] Police have no discretionary power to dismiss criminal cases (e.g. such as after cautioning the juvenile),[484] but must refer every suspect to the public prosecutor's office. The underlying reason for this is of a historical nature – more specifically, this is due to the possible abuse of police power as has occurred under the Nazi regime. Therefore, the responsibility for all forms of diversion is in the realm of the Juvenile Court prosecutor or the Juvenile Court judge. The police are strictly bound by the principle of legality. All criminal offenses have to be referred to the public prosecutor.[485] Without a formal indictment by their office, no juvenile case may be brought before the juvenile criminal court. The Youth Court Act regards the public prosecutor's domain as part of the juvenile judicature and to be under an obligation to specialize in juvenile issues just as juvenile judges. Juvenile public prosecutors are supposed to be pedagogically proficient and experienced in the education of juveniles (section 37). One or several juvenile public prosecutors are to be appointed depending on the caseload (section 36). By rule, they have to handle all cases placed within the jurisdiction of the juvenile courts.[486]

Once proceedings have been initiated, investigations should be conducted as soon as possible into the accused's life and family background, his/her development, previous conduct, and all other circumstances apt to assist in assessing his/her psychological, emotional, and character make-up. The parent or guardian and the legal representative, the school, and the person providing him/her with training should be heard insofar as possible. The school or person providing training shall not be heard if, as a result, the youth could fear suffering undesirable disadvantages, e.g. loss of his/her training place or job (section 43). If a youth penalty is to be expected, the public prosecutor or even (in the trial stage) the president of the youth court should question the accused before charges are brought forward (section 44).

As section 72 of the Youth Court Act stipulates that pre-trial detention is to be avoided whenever possible (last resort), family care can theoretically be a possibility to avoid pre-trial detention on the pre-trial level. Pre-trial detention is restricted in juvenile proceedings. For juvenile suspects aged 14 to 15, risk of escape in order to justify pre-trial detention may only be assumed if the juvenile has already made an attempt to escape or has no permanent place of residence (section 72). None of the reasons valid for adult offenders justify pre-trial detention for juvenile offenders. In

483 *Kaiser* 1994, pp. 324, 326.

484 However, the position of police in diversion procedures has been strengthened. Police in some jurisdictions are now empowered to initiate the conditions required to dismiss (or not prosecute) a case. Police thus influence who will be diverted and who will not. *Albrecht* 2004, p. 470.

485 *Dünkel* 2016, p. 8.

486 *Kaiser* 1994, pp. 324–326.

addition, the court, in deciding on pre-trial detention, must consider whether it is sufficient to place the juvenile offender in a foster home. If the latter is placed in pre-trial detention, the court aide must be informed (section 72a) so as to ensure that all relevant information is available, including possible alternatives such as foster care. Juvenile pre-trial detention must be organized in a way that favors education. As a consequence, a juvenile offender placed in a pre-trial detention center is obliged to work.[487] According to section 72(1) of the Act, pre-trial detention may be imposed and enforced only if its purpose cannot be achieved through a preliminary supervision order or other measures.[488] According to section 71 of the Youth Court Act, such a preliminary supervision order could end in (temporary) family care. It cannot only be imposed to avoid pre-trial detention but also to support the juvenile. (This possibility is not given to young adults if their development would otherwise be at risk and immediate action seems necessary.)

The juvenile and his/her parents have to agree to the measure. Placement in a residential home is only possible if other preliminary measures are not adequate. In practice, foster care and other preliminary supervision orders do not play any role.[489] Where prosecutorial diversion appears inappropriate and the likely sentencing outcome is a non-custodial sanction, the prosecutor submits a case file to the youth judge at the Local Court. In cases of more serious offending that could possibly result in a youth prison sentence, the prosecutor brings the accusation forward to the Youth Court of the Local Court, which is composed of one professional and two lay judges. Only in the most serious cases – usually homicide or manslaughter, but since 2006 also cases of sexual offenses against minors or others who should not be exposed to an appeal hearing –, the prosecutor submits the file to the Youth Chamber at the District Court (which consists of three professional and two lay judges).[490]

2.4.3.1.2 Trial stage

The trial stage is a very important stage. The youth court is required to open the trial if the accused is reasonably suspected of an offense – the elements of which include

---

487 *Albrecht* 2004, pp. 470–471.

488 According to section 72 II of the Youth Court Act, authorities may only use the risk of fleeing as a justification for pre-trial detention if an offender younger than 16 years of age has already made an attempt to escape or has no permanent place of residence. The practical procedure corresponds to the one applicable to adults. This means that young offenders may be interrogated only after they have been advised of their right to remain silent and to consult with an attorney of their choice. All phases of interrogation are strictly regulated by statute, and techniques such as exhausting the suspect, hypnosis, and deception are forbidden. The police may warn and release young offenders or detain them; but if young offenders are detained, they must be brought before a judge without delay and no later than one day after being taken into custody. The judge determines whether continued detention is warranted or if it can be avoided through other measures, such as bail or close outside supervision. Additionally, if the offender is a juvenile, the legal guardian and social services have to be informed immediately and have a right to be present during the interrogation. *Presson* 2017, p. 188.

489 *Pruin* 2012, p. 13.

490 *Dünkel* 2016, p. 9.

particularly the juvenile's responsibility within the meaning of section 3 of the Youth Court Act and if the procedural requirements have been met. Otherwise (i.e. in the absence of criminal responsibility because of immaturity), the judge will refuse to open the trial.[491] The possibility of a dismissal of proceedings following preferment of a charge will be discussed in the next section. The youth courts shall have the jurisdiction to hear cases involving youth misconduct. Youth courts shall mean the criminal court judge sitting as a youth court judge, the court of assessors (lay youth assessors' court), and the criminal panel (youth panel). The lay youth assessors' court shall be composed of the presiding youth court judge and two lay youth assessors. One man and one woman shall be present as lay youth assessors at each main hearing. However, it is notable that the lay youth assessors shall not participate in decisions taken outside the main hearing. The youth panel shall be composed of three judges, including the presiding judge and two lay youth assessors (grand youth panel); in appeal proceedings concerning the facts and law of judgments of the youth court judge, it shall be composed of the presiding judge and two lay youth assessors (small youth panel) (section 33a, b).[492] The youth court judge is charged with all tasks incumbent on a judge sitting in a local court in criminal proceedings. The supervisory functions incumbent on the family and guardianship judge for matters concerning youths should be transferred to the youth court judge. Deviation from the aforementioned is permissible for special reasons, e.g. if the youth court judge is appointed to sit in the district of several local courts. Supervisory functions incumbent on the family and guardianship judge shall mean: 1) supporting the parent, the guardian, or the carer by appropriate measures; and 2) those measures intended to ward off a danger to the youth (section 34).

As mentioned above, according to section 43 of the Youth Court Act, after criminal proceedings have been initiated, investigators gather information on personal and social circumstances relevant for evaluating the personality of the juvenile offender and the choice of sanction. Immediately after initiation, the juvenile court aide must

491 *Kaiser* 1994, p. 337.

492 The youth courts are departments of the district court or the regional court. Their respective jurisdictions depend on the expected legal consequences. As explained above, the first type is the single judge court. It has the competence to impose educational and corrective measures as well as juvenile imprisonment of up to one year, and it consists of one legally trained professional judge. The second type is the lay youth assessors' court (youth juror court). It possesses residual jurisdiction over all other cases concerning young offenders – in other words, the cases brought before the single magistrate court are the legislative exception. The hearings in the lay youth assessors' court are conducted by a magistrate presiding over two lay jurors or assessors. Each assessor has an equal vote with the magistrate in deciding both the guilt of the defendant and the legal consequence. The third type is the youth panel (youth chamber) of the district court, which has jurisdiction over special cases referred to it by the lay youth assessors' court and serious cases like murder. The youth panel is also an appellate court for decisions of the other two types of youth court. This panel is staffed by three judges and two lay assessors. All this means that jurisdiction in Germany is not only divided up horizontally (for example between family or criminal courts) but also vertically within the criminal court. *Presson* 2017, p. 216.

be notified. A social worker investigates the personal and social circumstances of the juvenile offender in order to provide the court with information on the appropriate sanction. The court aide has the right to be present during trial. The juvenile court has no discretion in admitting the social inquiry report. The court must hear the report in order to comply with the general procedural rule that any evidence relevant to the finding of guilt and the appropriate sentence must be heard.

Except for the parents of the juvenile offender, court hearings are not open to the public (including the media).[493] The victim may not only be present during the trial but may also act as a joint plaintiff. A series of victim's rights were added to the Code of Criminal Procedure in a major criminal law reform in 1987 (among them the rights of access to court files and to be informed about the outcomes of the criminal trial); these rights do not exist in juvenile court proceedings. Debate is under way on whether the goal of juvenile justice – that is education – justifies restricting the victim's rights. The right to appeal is restricted for juvenile offenders. Unlike adults, they must choose between two options: a full retrial or a review of their case on legal grounds. Adult offenders may have their cases reviewed twice.[494] According to section 55 of the Youth Court Act, the right to appeal is limited for the convicted young offender factually (subsection 1) as well as with regard to the different instances of the courts. To appeal judgments of the single judge, youth courts, and the lay youth assessors' courts, the appellant must choose between "full retrial" and a review of their case (subsection 2). Judgments which impose educational or corrective measures may be appealed upon the question of guilt only, i.e. not on the question of the severity of the sentence (section 55(1)). This means that some verdicts are not revisable in relation to the legal consequences.[495]

2.4.3.1.3 After-trial stage

According to section 82 of the Youth Court Act, the juvenile judge who issues the sentence is also responsible for its enforcement. This is a major difference to the rules applicable to adults. This rule constructs a much tighter net around the young offender and ensures that the same juvenile judge deals with the offender not fulfilling his/her

---

493 This is in contrast to the principle of public access to the trial that is applicable to adult criminal trials. The German juvenile criminal trial is always held behind closed doors in order to avoid any stigmatization of the young offender. "Public" in this sense means that everybody can attend a trial, regardless of who they are or what affiliations they might have with certain groups in society. However, as a deviation from the general rule, section 48 of the Youth Court Act is only applicable to juveniles and not to young adults. If a juvenile and a young adult are indicted together, the court has to decide whether the public should be excluded on the grounds of the educational interests of the juvenile offender. *Persson* 2017, pp. 200–201.

494 *Albrecht* 2004, pp. 470–471. The German Procedural Criminal Code grants an adult offender two ways of reviewing a judgment: "full retrial" (sections 312–323) or "review of their case" (sections 333–358). Both the defendant and the prosecution have the right to appeal. *Persson* 2017, p. 203.

495 *Persson* 2017, p. 204.

duties under the sentence as dealt with the trial itself. Furthermore, from a practical point of view, it provides for a faster procedural process since possible misconduct is always reported directly to the juvenile judge without having to go via the public prosecutor's office. The practical enforcement of the sentence is often in the hands of social services. These organize the enforcement of educational or corrective measures. If the young offender fails to fulfill the educational or corrective measure imposed on him/her, short-term detention may be imposed by the juvenile judge to enforce it. At the request of the public prosecutor, the juvenile judge first warns the convict to fulfill his/her duty. If the offender does not comply with the verdict at that point, short-term detention of up to four weeks can be imposed. It should be emphasized again that this sort of short-term detention does not replace the original sentence. The obligation to fulfill the originally imposed educational or corrective measures still stands after the short-term detention has been imposed. With regard to a suspended sentence or a conditional sentence/probation, a strong emphasis is placed on the intensive educative influence during the probation period. Even here, the Youth Court Act gives the responsibility to the juvenile judge, who "outsources" this duty to a probation officer for practical reasons. In contrast to adult criminal law, the appointment of a probation officer for a young convict is obligatory.

Another specific enforcement aspect to be mentioned with regard to young offenders is stipulated in section 56 the Youth Court Act: the partial enforcement of a verdict. This norm enables the juvenile court to declare part of a verdict to be enforceable for educational reasons even if the young offender has appealed another part.[496] Juveniles or young adults who have been convicted to deprivation of liberty should be placed in youth prisons. As already explained, in May 2006, the Federal Constitutional Court ruled that an adequate legal basis for running youth prisons had to be created until the end of 2007,[497] and the legal situation for young prisoners was to be changed at the beginning of 2008. Before 2008, only a few general legal provisions existed in the Youth Court Act and in the Prison Act for adult prisoners. No differentiated legal framework had previously covered the legal rights and duties of young prisoners. The Federal Constitutional Court outlawed this missing primary legislation as being unconstitutional, since in Germany, any restriction of fundamental human rights has to be based on regulations in law. Administrative rules are deemed an insufficient basis. The Federal Constitutional Court obliged the legislators of the Federal States to pass primary legislation before the end of 2007.

In September 2006, a general reform of the legislative competences came into force, transferring the competences for prison legislation to the Federal States. Their new laws vary to some extent and express different political orientations on what is to be seen as being the most promising concepts of rehabilitation. Nevertheless, there is a

---

496 *Persson* 2017, pp. 206–208.

497 *Boetticher et al.* 2004, p. 8.

strong consensus that the organization of youth prisons, even more than in adult prisons, must be oriented toward rehabilitation and education. Furthermore, the prevailing opinion is that youth prisoners shall be accommodated in small living groups and individual cells during the night. All youth prisons should also provide a variety of school and vocational training programs, special (social) therapeutic units, and a system of progressive preparation for release (including leaves of absence, early release schemes, and continuous care and aftercare). Although the competence of youth prison legislation has been transferred to the Federal States, legislation concerning prisoners' complaints, rights, and procedures are still Federal Law. The reform law of December 2008 brought major improvements, guaranteeing juvenile and young adult inmates an oral hearing as well as regular legal advice when complaining to the court.[498]

Furthermore, it is notable that an alternative to the youth prison sentence via the implementation of special institutions has been introduced with the Youth Prison Laws of the Federal States. Those institutions shall allow for the execution of the prison sentence in a quite open environment – comparable to a closed day- and night-time institution.

The oldest project in this context stems from Baden-Württemberg and has been copied in other Federal States. The "Project Chance" aims at avoiding imprisonment for a group of juvenile offenders who have been sentenced to youth imprisonment without probation by placing them in two young offender institutions. These are located in smaller villages, with space for 15 boys each, but with slightly different conceptions. Participation in the projects is voluntary. The youth prison is allowed to offer the possibility to offenders who are not drug-addicted, who have not been sentenced because of sexual offenses, and who have not shown rough violence in their offense. The director of the youth prison adds two informal reasons to that list. He offers the project to weak offenders who may normally be mistreated in the youth prison, based on his/her experience. The aim in both projects is to rehabilitate young offenders by offering new perspectives for their future lives. In both institutions, the youngsters have the possibility to receive a school degree or to finish vocational training. Unlike in most "normal" German youth prisons, the projects draw closer attention to peer groups and, in general, to contact with other people. Whereas in one of the projects, the boys live together in one house with single rooms, one kitchen, and one living room for the whole community, in the other project, the boys stay in two living communities that are organized like families: each community has two "house parents" or a "house family". It is important to mention the creation of a network set up for a certain amount of time after the offender's release from the institution. At least in one institution they try to find a place to live, a workplace, leisure time activities and, last but not least, a person the youngster is closely attached to for the time after release.[499]

---

498 *Dünkel* 2011, pp. 596–597; see also *Laubenthal* 2015.

499 *Pruin* 2012, p. 1; *Pichler* 2011, p. 7.

However, it should be noted that in accordance with the criminological findings on recidivism, the possibly positive effects of treatment or support offered in youth prisons are outweighed by negative influences of imprisonment and the prison subculture. Evaluations of dispositional alternatives in the juvenile justice system can take advantage of natural variations in decision-making between different juvenile court districts. Variations in the intensity of intervention are not associated with juvenile criminal behavior or juvenile crime. Other Western European countries rely on different policies vis-à-vis juvenile crime and juvenile offenders but face similar crime and delinquency problems.[500]

2.4.3.2 Informal institutions, proceedings, and sanctions

The interventions of the Youth Court Act are characterized by the principle of "subsidiarity" or minimum intervention under criminal law. Accordingly, penal action should only be taken if absolutely necessary on the one hand and must be limited by the principle of proportionality on the other. The legislative reform of the Youth Court Act in 1990, passed in the same year as that of the Youth Welfare Act, underlines the principle of youth court sanctions as a last resort. Therefore, priority is given to diversion, and where the youth courts do impose sanctions, primacy is given to educational or disciplinary measures instead of youth imprisonment. The most important response to petty offending is the dismissal of the case without any sanction being issued. In this context, police diversion, like cautioning or warnings, is not allowed in Germany. The 1990 reform of the Youth Court Act has extended the legal possibilities for diversion considerably. The legislature thus responded to the reforms that had been developed in practice since the end of the 1970s. The law now emphasizes the discharge of juvenile and young adult petty-crime offenders and those who have received other social and/or educational interventions (section 45(1) and (2)). Efforts to make reparation to the victim or on participation in victim-offender reconciliation (mediation) are explicitly put on a par with such educational measures. There is no restriction concerning the nature of offenses that are eligible; felony offenses can also be "diverted" under certain circumstances (e.g. a robbery) if the offender has repaired the damage or made another form of apology (restitution/reparation) to the victim. There are four different levels of diversion:[501]

- Diversion without any sanction (non-intervention) is given priority in cases of petty offenses. Section 45(1) of the Youth Court Act provides dismissal without any consequence in the case of minor offenses with minor guilt and no public interest in prosecution. It is applied when the public juvenile prosecutor does not see any further educative need apart from the discovery of the offense and the investigative procedure. The consent of the youth judge is not a necessary precondition.

---

500 *Albrecht* 2004, pp. 484–485; see also *Schäfer* 2004b.

501 *Persson* 2017, pp. 151–152; *Dünkel* 2011, pp. 5–7 and 11–15.

- Diversion with measures taken by other agencies (parents, the school) or in combination with mediation is the second level of diversion (diversion with education). Section 45(2) of the Youth Court Act provides dismissal with educative measures taken by other actors (for example the parents, the school) or measures initiated by the public prosecutor or in combination with mediation if the public juvenile prosecutor does not deem it necessary to involve the youth judge.
- Diversion with intervention. In these cases, the prosecutor proposes the youth court judge to impose a minor sanction, such as a warning, community service (usually between 10 and 40 hours), mediation, participation in a training course for traffic offenders, or certain obligations such as reparation/restitution, an apology to the victim, community service, or a fine (section 45(3) of the Youth Court Act). Once the young offender has fulfilled these obligations, the youth court prosecutor will dismiss the case in cooperation with the judge. Here the confession of the young offender is a precondition and the legal guardian has to consent. In this case, the formal criminal trial is replaced by an informal magisterial educative process, which often takes place in the chambers of the youth judge. Once the young offender has fulfilled the imposed obligations, the prosecutor will dismiss the case, in cooperation with the judge. Note here that the imposed legal consequences are the same as those in educational or correctional measures. The difference lies in that one measure is imposed through a formal sentence, while the other is imposed in the form of a divertive decision.
- The fourth level of diversion is the introduction of levels one to three in the youth court proceedings after a charge has been filed. In practice, the youth court judge will fairly often face a situation in which the young offender has in the meantime (after the prosecutor has filed the charge) undergone some form of educational measure such as mediation, which would deem a "formal" court sanction unnecessary (section 47 of the Youth Court Act).

Section 47 of the Youth Court Act enables the judge to dismiss the case in these instances. These broad options for diversion are that the reason why the vast majority (between 68 and 70 per cent) of cases against young offenders are closed without a verdict.[502]

Victim-offender mediation is considered to be the main form of restorative justice applied on young persons in Germany, which has gained in importance since the 1980s. The establishment of mediation projects was influenced by the international restorative justice movement as well as by criminal policy discussions in Germany aiming at promoting victims' rights and alternative measures for offenders. The first pilot projects on victim-offender mediation for juveniles and young adults were im-

502 *Dünkel* 2011, pp. 5–7 and 11–15; *Persson* 2017, pp. 151–152.

plemented in four German cities in the mid-1980s, and each project has been evaluated systematically. Until the mid-1990s, the number of mediation projects increased significantly across the country.[503]

A legal framework for victim-offender mediation is provided by the Juvenile Justice Act since 1990, as both a diversionary and educational/disciplinary measure. Hereby, the legislator offers a broad approach and underlines that a serious attempt by the offender to reconcile with the injured person is considered sufficiently. Under the Youth Court Act, within the diversion scheme, mediation can be applied by youth public prosecutors or judges (sections 45(2), 47(1) 2). Besides the possibility of applying mediation during diversion, mediation can also be ordered by the court as part of an educational measure (section 10(1) 7). The law provides that the judge may instruct the juvenile to attempt to achieve a settlement with the injured person (victim-offender mediation). Furthermore, mediation might be taken into account when applying the disciplinary measure of restitution (section 15(1) 1). During the probationary period or early release, the measures might also come into play in order to exercise a supervisory influence on the conduct of the young person (section 23(1), 88(6)). Both the educational and the disciplinary measures are considered critical, as the court order to achieve a mediation agreement is not in full accordance with the principle of voluntariness, which is one of the main bases of a mediation procedure. In principle, there is no restriction on eligible offenses. Usually, minor- or medium-severity offenses are considered applicable and petty offenses are excluded. In several Federal States in Germany, juvenile diversion guidelines underline categories of offenses that are eligible for victim-offender mediation. Generally, it is applicable at all levels of the criminal procedure. Even at the correctional level, victim-offender mediation and other restorative approaches have been applied over the last years.[504]

503 A study conducted in 1993/94 revealed that victim-offender mediation for juveniles and young adults was available in most parts of Germany. The number of mediation projects increased from 226 in 1992 to 368 in 1995. The majority of cases were registered with young offenders.

504 *Pruin* 2012, pp. 8–12.

Chapter 3

Attitudes of Elites to Juvenile Criminal Law: An Empirical Survey in Iran

3.1 Description of the Survey

3.1.1 Introduction

Juvenile criminal law/justice is one of the most important aspects of a criminal system because it is an area applicable to persons who are not old enough to incur full criminal responsibility. Thus, the most important step in establishing a juvenile justice system is to enact a special code which contains both substantive and procedural rules for juvenile delinquents. In order to enact or reform laws in this field, it is important to know the attitudes of the elites toward juvenile justice. Generally, the best ideas usually emerge from opinions and observations of scholars, practitioners, and others who work and reflect on these issues. Therefore, when enacting or reforming laws, the attitudes of elites should be taken into consideration by the legislature because these opinions are usually based on juveniles' needs.

As mentioned in the previous chapters, the foundation of the Iranian criminal justice system was changed after the 1979 Revolution and the abolition of the Juvenile Delinquency Act (1959) for Islamization purposes (Islamization of society). It should be noted that no special laws and regulations have to date replaced the 1959 Juvenile Delinquency Act, hence creating a legal vacuum. On the other hand, the Iranian legal system is a combination of Islamic jurisprudence, international conventions, and customary law, which has led to inconsistencies in the legal system in general – and specifically in laws and practices concerning children and juveniles. Therefore, there are some conflicts in the legislative, judicial, and executive criminal policies concerning juvenile delinquents in Iran, which is a very controversial subject matter.

Nevertheless, some attempts have been made under the Convention on the Rights of the Child as well as due to criminological, psychological, and educational findings to protect children's rights in the justice system. One of the most important efforts was to draft the Juvenile Justice Bill, which was approved in the Cabinet and sent to Parliament in March 2005. However, after several discussions over a long period of time, the bill was divided into two sections, that is, substantive and procedural criminal law. The articles relating to substantive criminal law (including the definition of criminal responsibility and juvenile sentencing systems) were merged and added to the new Islamic Penal Code, which was enacted in 2013, while the articles relating to procedural criminal law were merged and added to the Procedural Criminal Code, which was enacted in 2014. Consequently, it is apparent that the Iranian legal system will continue to be devoid of a special act for juvenile delinquents.

While analyzing the new Islamic Penal Code, the main object of this survey was to study the attitudes of elites toward juvenile justice, given their influence on law- and policy-making as well as on reforms. The results of the survey should form a basis for the Iranian legislature to review the rationale for legislation, because the most important factor for making a law are the views of elites and public attitudes. In this chapter, the results of the survey on the attitudes of elites toward juvenile justice in Iran will be discussed.

The questions asked in the empirical study were very technical; there design was based on the controversial nature of the subject matter and legal vacuum in Iran on the one hand – and the new interpretation of *Shari'a*, international standards, and the needs of juvenile delinquents on the other. In fact, the result of this study shows the potential in the Iranian system to revise/reform the laws and regulations relating to juvenile delinquency or to change the judicial criminal policy in the justice system. It is obvious that it was impossible to ask the participants about all the issues with vacuums and blind spots in the Iranian legal system in this regard, as the juvenile justice system has not been accepted by the legislature and there are a lot of vacuums and blind spots. Therefore, the most important issues were selected to be studied in this survey.

An examination of legal responses to juvenile delinquencies and related clashes of Islamic laws, judicial practice, international obligations, and social rules reveals the tragic anomaly of the Iranian legal system in its treatment of juvenile delinquency; however, these clashes are related to substantive rather than procedural criminal law. The age of criminal responsibility, the definitions of what is a child and juvenile delinquency, penal sanctions in the justice system, differential criminal proceedings, special courts, judges, and police are very controversial issues in Iran. It can be said that international standards have been ignored by the Iranian legislature in substantive criminal law due to various clashes between *Shari'a* and other resourses. Therefore, there is a wide gap between substantive and procedural criminal law, while the roots and obstacles of the administration of justice lie in substantive law.

In the Iranian Penal Code, the age of criminal responsibility is very low, and in most cases the type and nature of punishment is similar to the one of adults, which has created a problematic situation in practice. Therefore, some issues regarding substantive and procedural criminal law were asked about in this survey. Furthermore, so as to realize how the elites perceive the issues in the process and in practice, some faction cases were designed. In order to achieve clear views with the questionnaire, the only way was to draft multiple choices for each question; however, both the progressive and the conservative views/items were asked in the choices as the two main scales.

3.1.2 Hypotheses of the survey

The **first hypothesis** concerned the general views and attitudes of elites toward the juvenile justice system in Iran. It should be noted that most respondents were assumed to be supportive of children in conflict with the law.

As the **second hypothesis**, the survey assumed that there is potential to deal with juvenile delinquency according to the international obligations and the new interpretation of *Shari'a,* while the assumption which was tested was that Islamic lawyers are less supportive of the new interpretation of *Shari'a* and international standards.

The assumption of conservative views within the Islamic lawyers target group in substantive law (due to the obstacles) and its deep reflection in the new Iranian Penal Code (2013) was the **third hypothesis** in this survey. This is to be seen in light of no clashes between customary and *Shari'a* law occurring in procedural law.

Therefore, the assumption of progressive views of the Islamic lawyers target group in procedural criminal law was examined as the **forth hypothesis.**

The **fifth hypothesis** was based on the assumption that Iranian legislators ignored the elites' views when enacting the Iranian Penal Code, thus creating gaps between legislative and judicial criminal policy concerning juvenile delinquency on the one hand – and differences between the views of academic and Islamic lawyers on the other.

The **sixth hypothesis** assumed that practicing lawyers and staff who work in the judicial system, as well as faculty members and independent defense lawyers, agree with the progressive interpretation of *Shari'a* and customary law to a higher extent than conservative ones.

In this survey, the **seventh hypothesis** assumed that women are more receptive for customary law and the new interpretation of *Shari'a* (progressive view of *Shari'a*), more supportive of female delinquents in the justice system, and less supportive of severe punishments as well as the death penalty for (both male and female) delinquents.

The **eighth hypothesis** was that Islamic lawyers are less supportive than other groups of an elimination of severe punishments such as the death penalty, as well as of the equality between female and male delinquents in determining the age of criminal responsibility. Therefore, they were supposed to follow the conservative view of *Shari'a* more than others, while faculty members, defense lawyers, social workers, and master students were expected to be more supportive of these matters. Judges' and prosecutors' attitudes are closer to the latter than Islamic groups.

The **ninth hypothesis** assumed that all target groups support professionalism in procedural matters and the role of a special judge, prosecutor, and police officer in accordance with the Criminal Procedure Code of 2014.

The final and **tenth hypothesis** was that informal trials do not fit into the Iranian legal system, due to the lack of standards, even in juvenile's criminal cases which are not considered as serious cases and crimes; there is no consensus for diversion.

3.1.3 Research procedure

3.1.3.1 Sampling and administration of the questionnaire

In this survey, the attitude of elites toward juvenile justice in Iran has been studied. In order to evaluate how elites consider the issues concerning juvenile delinquents in Iran, two target groups were sampled. The main criteria for choosing these target groups was their potential impact on and influence in the revision and enactment of laws, as well as finally their recognition of juvenile justice as a differentiated and separate structure in the Iranian criminal justice system.

As shown in *Table 1*, the total number of respondents was 222, who were divided into two main groups: the first group comprised those who work in the judicial system, including criminal trial judges (of lower, appellate, and high courts), prosecutors and investigators, social workers, and defense lawyers (totaling 130 respondents). Due to their professional work, they are the most influential group in the formation of criminal policy and subsequently of juvenile justice in the legal system. The second group comprised those who do not work in the justice system but are influential in policy formulation and law-making processes, including law faculty members, Islamic lawyers, and Master's degree students in criminal law and criminology (totaling 92 respondents).

3.1.3.2 Target groups

3.1.3.2.1 Characteristics of group 1

In the first group, the questionnaires were distributed among different judicial jurisdictions across Tehran, Rasht, and other cities in the Gilan province, such as Bandar-Anzali and Lakan. Nevertheless, the majority of responding judges were lower cadre judges, although a few others were high court judges. Among the 41 trial judges who have been consulted for this survey, ten served in special branches dedicated to hearing cases pertaining to juveniles and children, five workd in Tehran, two in Rasht, one in Bandar-Anzali, and two other judges served in other cities in Gilan province. The most important point is that all judges who participated in this survey were criminal trial judges. It is noteworthy that according to the Iranian procedural criminal code, juvenile delinquents (girls above 9 years and boys above 15, but under 18 years) should be tried in the special branches of general courts mandated to hearing juvenile cases. These juvenile courts were created according to the administrative circular (issued by the head of judiciary). The mentioned persons should be tried in these separate branches of general courts, sentenced as adults and according to the general criminal code. The judges who serve in these branches are not the special judges. In other words, they are trial judges with general competence. However, due to the high population number in Tehran, a special building has been constructed for this reason. In other cities, one or two branches of the general courts have mandates to try juveniles, so there are no special buildings in other big, provincial cities.

In prosecutorial jurisdiction, most respondents are investigators and some others are general prosecutors. In this survey, in order to administer the questionnaire among social workers, not only those who served in correction centers were selected, but also some of those who worked in after-care organizations were asked, which are general organizations for both juveniles and adults. The last respondents in target group 1 were defense lawyers who were selected based on their experience in criminal trials, as well as with other work on human rights issues.

3.1.3.2.2 Characteristics of group 2

In group 2, the questionnaires were distributed among faculty members, master students, and Islamic lawyers across Tehran, Rasht, Bandar-Anzali, and Qum. All respondents who are categorized in the faculty member target group worked in law faculties, mainly in criminal law and criminology departments. Most of the master students in this survey were in their final year. The Islamic lawyers who were chosen to administer the questionnaire were from Tehran and Qum. Of the 23 respondents identified as Islamic lawyers, 52% had earned a PhD in criminal law and criminology, besides their education in the Islamic centers, and some of them were faculty members or lecturers at university. It should be noted that the Islamic lawyer target group has been categorized in this survey based on their Islamic education background. However, so as to know the progressive views of those Islamic lawyers who have been selected for this research, a second measurement was their academic background.

Table 1 Characteristics of the two target groups

Target groups	**N**	**%**
Group 1		
Criminal trial judges	41	19
Prosecutors and investigators	29	13
Social workers in the justice system	30	14
Defense lawyers	30	14
Total group 1	*130*	*60*
Group 2		
Law faculty members	25	11
Master students in criminal law and criminology	44	20
Islamic lawyers	23	11
Total group 2	*92*	*42*
Total groups 1 & 2	222	100

3.1.3.3 Provinces and Cities

The empirical study was conducted in three provincial capitals and other cities in Iran to evaluate the various attitudes of elites. The following provinces and cities were selected: Tehran province (City of Tehran, center of this province and capital of Iran), Gilan province (City of Rasht, center of this province), Bandar-Anzali, Lakan (and some other cities in these provinces), and Qum province (City of Qum, center of this province and the main center of Islamic studies).

Tehran and Gilan are the two provinces which were selected for administering the questionnaires. It is worth mentioning that the judicial system in these provinces is very standardized, and the best verdicts are said to be issued by the judges who work in these provinces, i.e. in the competition which was held in the Iranian judiciary to choose and introduce the best verdicts and judges, the above-mentioned provinces emerged as winners. Therefore, it can be said that these provinces are most influential in the revision and formulation of juvenile justice policies in Iran. In addition, Qum was selected in order to study the attitudes of Islamic lawyers because this city is one of the biggest and most important centers for Islamic studies in the *Shi'a* school in Iran. Therefore, some Islamic centers in Qum were requested to complete the questionnaires.

3.1.3.4 Organizations

The questionnaire was distributed in criminal courts, prosecutors' offices, correction centers, after-care organizations, universities, and Islamic centers. For conducting the questionnaire, I started visiting these organizations in the selected provinces and cities in Iran. However, in spite of repeated attempts to convince them, most of the organizations refused to cooperate in the first stage, and I had to explain in detail the importance of the survey and the aim of the empirical section of my dissertation in order to obtain their participation. Nevertheless, I had to eliminate some questions in the questionnaire in order to get permission to conduct the survey. Furthermore, I could not administer the questionnaire in some organizations such as the police and its units.

3.1.3.5 Total number of questionnaires

350 questionnaires were distributed among the target groups, out of which 222 were completed and returned. It should be noted that the questionnaires were administered in two stages after the pre-test. In the first stage in 2012, 200 questionnaires were distributed and about 120 completed. For completing the empirical research in the second stage, 150 further questionnaires were distributed in 2013, with 102 completed questionnaires having been collected.

3.1.3.6 Method of data analysis

The statistics SPSS software was used for the analysis of the survey data. Because the survey was qualitative, all data were analyzed at the first stage and were then explained according to the hypotheses and aims of the study in the next stage. At first, the questionnaires were analyzed according to the attitudes of the respondents in each target group and based on their occupations as the main objective in this survey. In the second stage, other variables and the nature of the various relationships between them were considered and analyzed. In addition, in this survey, the key questions were found to pinpoint the main aspects for analysis in all variables. In some questions, two important scales were considered, including the progressive and conservative views of *Shari'a*.

3.1.4 Description of variables

3.1.4.1 Occupation

As indicated in *Table 2*, 19 % of those who work in the justice system are criminal trial judges, 13% general prosecutors or investigators, 14% social workers, and 14% defense lawyers. The second group comprises those who do not work in the justice system: 11% are law faculty members, 10% Islamic lawyers and 20% master students in criminal law and criminology.

Table 2 Occupation of respondents

Occupation	N	%
Faculty members	25	11
Criminal trial judges	41	19
Social workers	30	14
Master students	44	20
Islamic lawyers	23	10
Defense lawyers	30	14
Prosecutors/investigators	29	13
Total	222	100

3.1.4.2 Education

As *Table 3* demonstrates, most respondents who participated in the survey hold a master's degree (M.A.) in criminal law and criminology, which constitutes 50% of respondents. 24% hold a Bachelor degree (B.A.) in judicial law, 23% a PhD in criminal law and criminology, 0.5% a diploma in human science, 1% academic Islamic law, and 1% Islamic studies. As a result, it can be said that most of the respondents are highly educated. In this survey, only one respondent, who works in an after-care organization as a trainee, holds a diploma. Furthermore, in the Islamic target group, only two respondents have higher education from Islamic schools (seminary schools) without any academic education; they work in the justice system. Nevertheless, when considering all target groups, it is noteworthy that only three respondents have law education exclusively. Thus, a majority of respondents in the survey are highly educated and qualified.

Table 3 Education of respondents

Education	N	%
PhD	50	23
M.A.	112	50
B.A.	54	24
Diploma	1	.5
Islamic law (equivalent to B.A.)	3	1
Seminary (Islamic studies)	2	1
Total	222	100

3.1.4.2.1 Distribution of respondents in relation to their education

As *Figure and Table 4* demonstrate, the highest percentage of distribution of the respondents who have a master's degree is university students (39%), criminal trial judges (21%), and defense lawyers (21%). The highest percentage of respondents who hold a bachelor's degree are social workers (47%), prosecutor-investigators (28%), and criminal trial judges (22%). And finally, the highest percentage of respondents who have a PhD are the faculty members (50%) and Islamic lawyers (24%), respectively.

Figure/Table 4 Distribution of respondents in relation to their education

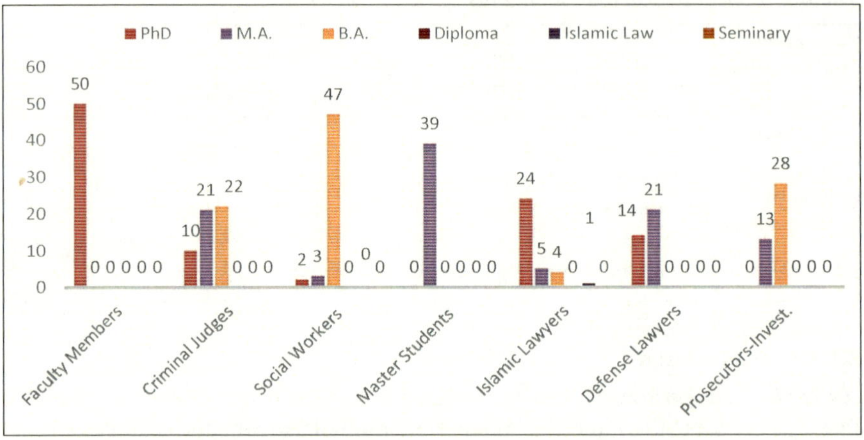

Education / Occupation	PhD		M.A.		B.A.		Diploma		Islamic law		Seminary	
	N	%	N	%	N	%	N	%	N	%	N	%
Faculty members	25	50	0	0	0	0	0	0	0	0	0	0
Criminal judges	5	10	23	21	12	22	0	0	0	0	0	0
Social workers	1	2	3	3	25	47	1	0.5	0	0	0	0
Master students	0	0	44	39	0	0	0	0	0	0	0	0
Islamic lawyers	12	24	5	5	2	4	0	0	3	1	2	1
Defense lawyers	7	14	23	21	0	0	0	0	0	0	0	0
Prosecutors/invest.	0	0	14	13	15	28	0	0	0	0	0	0
Total: 222	216						6					

3.1.4.3 Gender

As shown in *Table 5*, male respondents in this survey constitute 73%, while 27% are females.

Table 5 Gender of respondents

Gender	N	%
Male	162	73
Female	60	27
Total	222	100

3.1.4.3.1 Distribution of respondents in relation to their gender

It is noteworthy that in target group 1 (practical lawyers), all those who work as criminal trial judges are men (100%). This is due to restrictions that prevent women from working as trial judges (who can issue the final verdict in court) in the Iranian criminal system. On the other hand, there are no obstacles for women to work as prosecutors or investigators in the criminal justice system in Iran, so there are some women in this section. 79% of the respondents who are categorized in this group are women.

In target group 2, all of those identified as Islamic lawyers are men. While there are some female Islamic lawyers, a few questionnaires distributed amongst them were returned blank.

As demonstrated in *Figure and Table 6*, 39% of the female respondents serve as prosecutors or investigators, 22% are master students, 20% work as social workers, 13% are defense lawyers, and 7% serve as faculty members in the academic field.

Figure/Table 6 Distribution of respondents in relation to their gender

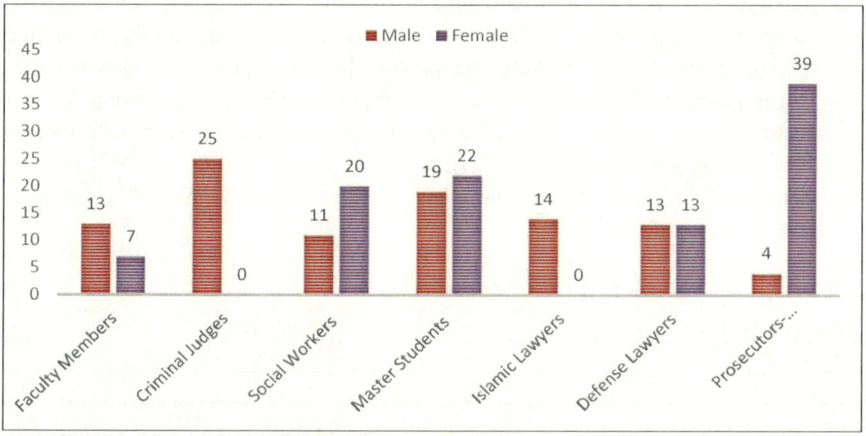

Occupation \ Gender	Male		Female	
	N	%	N	%
Faculty members	21	13	4	7
Criminal trial judges	41	25	0	0
Social workers	18	11	12	20
Master students	31	19	13	22
Islamic lawyers	23	14	0	0
Defense lawyers	22	13	8	13
Prosecutors/investigators	6	4	23	39
Total: 222	162		60	

3.1.4.4 Age

As shown in *Table 7*, there are six categories based on the age of respondents, i.e.: from 24 to 30 (21%), from 31 to 35 (26%), from 36 to 40 (23%), from 41 to 45 (16%), from 46 to 50 (10%), and above 51 (4%). As a result, most respondents are categorized as people who are between 31 and 40, which constitutes 50% of all respondents.

Table 7 Age of respondents

Age	N	%
24–30	47	21
31–35	57	26
36–40	52	24
41–45	36	16
46–50	21	10
51+	9	4
Total	222	100

3.1.4.5 Marital status

As shown in *Table 8*, 79% of respondents are married, while 21% are single. In the justice system, being married with children is considered essential for those working with juveniles and children because they presumably can more easily relate to them. It is worth noting that among all the respondents in the Islamic target group, 97% of prosecutors or investigators, 93% of social workers, 89% of criminal trial judges, and 76% of faculty members were married. When the marital status variable is taken into consideration, most single respondents were master students (48%), followed by defense lawyers (40%).

Table 8 Marital status of respondents

Marital status	N	%
Married	175	79
Single	47	21
Total	222	100

3.1.4.6 Place of residence

The respondents' places of residence are categorized into five groups, as shown in *Table 9*. As explained above, although the survey covered three provinces, the questionnaires were only distributed among the target groups in Tehran and Gilan provinces. In Qum province, questionnaires were only distributed in the Islamic target group. As shown in *Table 9*, 121 respondents in Tehran and 87 respondents in Gilan province participated in this survey. Nevertheless, as *Table 9* also demonstrates, more than half (55%) of respondents are living in Tehran. 22% of respondents are living in Rasht, 14% in Bandar-Anzali, 6% in Qum, and 4% in other cities in Gilan province.

Table 9 Places of residence of respondents

Place of residence	N	%
Tehran (Tehran province)	121	55
Qum (Qum province)	14	6
Rasht (Gilan province)	48	22
Bandar-Anzali (Gilan province)	30	14
Other cities in Gilan province	9	4
Total	222	100

3.1.4.6.1 Distribution of respondents in relation to their place of residence

As *Figure and Table 10* demonstrate, all faculty members (100%), 49% of criminal judges, 53% of social workers, 43% of master students, 39% of Islamic lawyers, 53% of defense lawyers, and 55% of prosecutors or investigators who participated in the survey are living in Tehran. The rest of respondents were Islamic lawyers who constituted 61% of this target group; they live in Qum, which is one of the main Islamic centers in Iran and also the center of Qum province. Also, 12% of criminal judges, 14% of social workers, 4% of master students, 12% of defense lawyers, and 6% of prosecutors or investigators live in Rasht, which is the center of Gilan province, while 15% of criminal judges, 48% of master students, 39% of Islamic lawyers, 7% of defense lawyers, and 3% of prosecutors or investigators are living in Bandar-Anzali, which is one of the most important cities in Gilan province. Finally, only 7% of criminal judges and 21% of prosecutors or investigators who participated in the survey are living in smaller cities of Gilan province.

Figure/Table 10 Distribution of respondents in relation to their place of residence

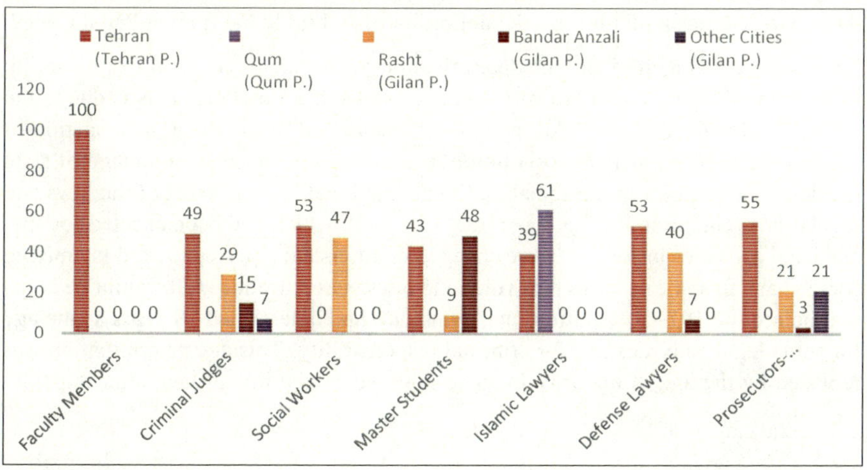

City / Occupation	Tehran (Tehran p.)		Qum (Qum p.)		Rasht (Gilan p.)		Bandar-Anzali (Gilan p.)		Other cities (Gilan p.)	
	N	%	N	%	N	%	N	%	N	%
Faculty member	25	100	0	0	0	0	0	0	0	0
Criminal judges	20	49	0	0	12	29	6	15	3	7
Social workers	16	53	0	0	14	47	0	0	0	0
Master students	19	43	0	0	4	9	21	48	0	0
Islamic lawyers	9	39	14	61	0	0	0	0	0	0
Defense lawyers	16	53	0	0	12	40	2	7	0	0
Prosecutors/investigators	16	55	0	0	6	21	1	3	6	21
Total: 222	183						39			

3.2 Analysis and Description of Survey Results and Findings

3.2.1 Attitudes to the definition of the child in criminal law

Children around the world are among the most vulnerable members of society and should be protected by the law until they are deemed capable of interacting as adults in society. Therefore, they should be treated differently from adults by a different criminal justice system when they commit a crime. There are many terms in use to identify someone within the scope of the criminal/juvenile justice system, such as child, minor, immature person, juvenile, youth offender, and young person. However, these terms should be defined by laws and regulations, and legislatures sometimes do not submit a very clear definition of them, which can cause problems in practice.

A child is defined as a young human being below the age of puberty (maturity) or below the legal age of majority.[505] Furthermore, the United Nations Convention on the Rights of the Child defines a child as "a human being below the age of 18 years unless under the law applicable to the child, majority is attained earlier". Which of these two criteria should inform the definition of a child in the Iranian Penal Code?

The definition of a child and the determination of the age of criminal responsibility in the Iranian legal system are very controversial subject matters. This is due to the adverse influence of *Shari'a* on policy. As mentioned in the theoretical section of this research, after the 1979 Revolution in Iran and the subsequent change of state ideology and policies in the social, political, and legal system, most of the laws and regulations relating to the legal age (majority age), which had been enacted for special legal activities in society, were changed or abolished as a step toward Islamizing society, and in some sections new rules and laws were introduced. It should be noted that before the 1979 Revolution, the legislature had determined 18 years as the age for some legal activities and for criminal responsibility. This age determination was replaced by the age of maturity in some laws, such as criminal law, after the 1979

505 *Ebadi* 1381/2002, p. 5.

Revolution. This is why the legislature has not clearly defined the age of maturity until the present day.

The question regarding the definition of the legal age of children in Iranian criminal law was one of the key questions in this survey. It was designed based on both the customary and the religious meaning of a child. In all target groups, the majority of respondents (about 70%) endorsed the customary meaning of a child, which is in conformity with the progressive the view of *Shari'a* (choices 1 and 3).

The first definition of a child in choice 1, which considered a child to be a person under 18 years of age, has been accepted by the Convention on the Rights of the Child. As discussed in the previous chapters, Iran has signed and ratified this Convention; therefore, the country is obliged to conform its national laws and regulations to international standards. Furthermore, according to the juvenile court's Act 1338, this definition had been accepted by the Iranian legislature but was abolished after the 1979 Revolution.

The second definition (choice 2) was designed based on discussions held during the drafting process of the new Penal Code in a bid to avoid two separate models in the juvenile justice system, which could have caused problems in practice. This definition does not encompass any minimum age for criminal responsibility, and the conception of maturity according to scientific findings has been replaced in order to stop discussions about obstacles in *Shari'a* concerning the exact age of criminal responsibility.

The third definition (choice 3) was based on the definition of a child in the previous Penal Code, which was very controversial and the main issue in the legal system. This choice defines a child as a person who has not reached the level of maturity according to *Shari'a*.

The forth definition (choice 4) was the model introduced by the new Penal Dode. It is noteworthy that due to the clashes between legal recourses in Iran, two separate models for determining the age of criminal responsibility and subsequently for dealing with juvenile delinquents were introduced based on the categories of crime and punishment in the Iranian legal system (flexible and fixed punishments), that is, 18 solar years (of age) for flexible punishment (*Ta'zir*) as well as 9 and 15 lunar years, respectively, for girls and boys in fixed punishment (*Qisas*, *Had*, *Diye*).

Table 11 Attitudes toward the definition of a child in criminal law

Child/juvenile should (in Iran Penal Code) be defined as a person who ...	**N**	**%**
is under the age of 18	86	39
has not reached the age of maturity according to scientific findings	66	30
has not reached the age of maturity according to *Shari'a*	29	13
is under the age of 18 in *Ta'zir* as well as under 9 and 15 lunar years, respectively, for girls and boys in *Qisas*, *Had*, *Diye*	41	19
Total	222	100

Contrary to the first choice and definition of child based on 18 years, which was endorsed by the majority (39%) of respondents (as shown in *Table 11*), only 13% of respondents chose the third choice, which is related to the definition of a child in the previous Penal Code. This choice defines a child as a person who has not reached the age of maturity according to *Shari'a* and in Islamic recourses.

Furthermore, 30% of respondents in all target groups believe that the age of a child should be determined based on their age of maturity and in accordance with scientific findings, which is one of the customary definitions of a child and the second-mostly chosen option by the respondents after choice 1. Only 19% of respondents agreed with the definition of a child in the new Iranian Penal Code, which embodies two models of the age of criminal responsibility. These findings show that the legislature has not followed the elites' view of the definition of a child.

3.2.1.1 Occupation and attitudes toward the definition of a child

Table 11-1 illustrates the attitudes of the various professional target groups toward different definitions of a child in the Penal Code. As shown in this table, the majority of respondents (39%) agree that a child is a person under 18 years of age (choice 1). As noted above, this age conforms to the definition of a child in international instruments on the one hand and is in accordance with their definition in the Juvenile Court's Act 1338, which, on the other hand, was abolished after the 1979 Revolution. In order to understand why the respondents favored this definition – besides the importance to fulfil international obligations –, it is important to state that the positive impact of the Juvenile Court's Act 1959 on the current judicial system should not be ignored. In fact, it can be said that suitable laws and regulations can create a commensurate and lasting judicial culture despite this abolishment. In other words, it should be mentioned that the positive message of this abolished act still lingers in the judicial system until the present day.

Figure/Table 11-1 Relationship between occupation and attitudes toward the definition of a child

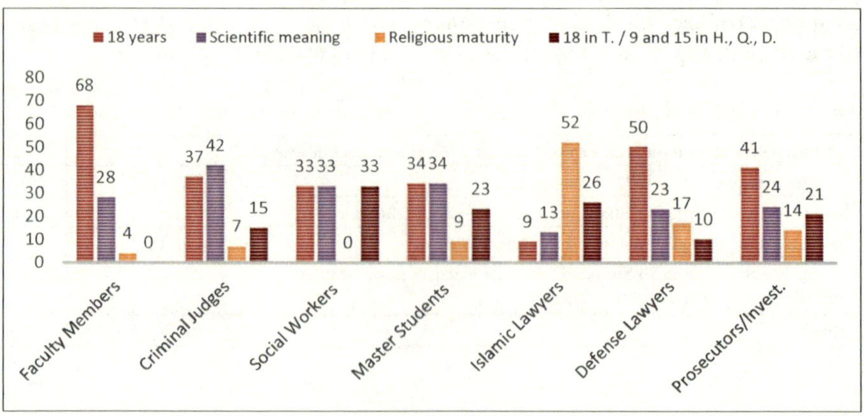

Occupation \ Choices	Progressive view of *Shari'a*				Conservative view of *Shari'a*			
	18 years		Scientific meaning		Religious maturity		18 in *T.* / 9 & 15 in *H.*, *Q.*, *D.*	
	N	%	N	%	N	%	N	%
Faculty members	17	68	7	28	1	4	0	0
Criminal judges	15	37	17	42	3	7	6	15
Social workers	10	33	10	33	0	0	10	33
Master students	15	34	15	34	4	9	10	23
Islamic lawyers	2	9	3	13	12	52	6	26
Defense lawyers	15	50	7	23	5	17	3	10
Prosecutors/investigators	12	41	7	24	4	14	6	21
Total: 222	152				70			

In order to assess the responses to this question in each target group as a most important choice, as shown in *Figure and Table 11-1*, the respondents in favor of this definition should be pointed out: faculty members with 68%, defense lawyers with 50%, prosecutors/investigators with 41%, judges with 37%, master students with 34%, and social workers with 33%. The number of Islamic lawyers in support of this customary and international definition of a child is lower than among other target groups. It should be noted that only 9% of them endorsed this choice. This means that Islamic lawyers are less supportive of the customary definition of a child (and thus a new interpretation and progressive view of *Shari'a*), while 52% of them endorsed choice 4, which defines a child as a person who has not reached the age of *religious* maturity.

In contrast, none of the social workers agreed with this definition (choice 4), the main reason being that they deal with juvenile delinquents personally nearly every day. In other words, they have realized that this kind of the age determination can cause problems in practice, which is contrary to the perspective of Islamic lawyers who have a more theoretical view due to being more out of touch to social realities. Once the choice endorsements were analyzed, it became evident that Islamic lawyers were more in favor of choices 3 (52%) and 4 (26%), which are both considered conservative definitions of a child and have been accepted by the legislature in the previous and current penal codes, respectively. It is thus apparent that Islamic lawyers exert more influence over law-making in Iran than any other group. Other factors include the *Shari'a*. It is more interesting that none of the faculty members agreed with the definition of a child as adopted by the Iranian legislature in the new Penal Code of 2013. Disregarding scientific findings and realities in society should be a cause for alarm in the Iranian justice system as this may result in problems that are unfavorable for juvenile delinquents.

It is worth mentioning that the most contrasting views are those categorized in target group 1, that is faculty members and Islamic lawyers. Furthermore, defense lawyers, who are independent practicing lawyers, leaned toward the view of faculty members. The endorsements of judges, prosecutors and investigators, master students, and social workers were similar: They support the progressive view of *Shari'a* to a larger extent than the conservative one. This means that even practicing lawyers who serve within the justice system agree with customary law despite their belief in Islam and their judicial obligations, which is a governmental section.

3.2.1.2 Place of residence and attitudes toward the definition of a child

Figure and Table 11-2 show the relationship between respondents' definition of a child and their place of residence. Respondents who were living in Qum were significantly less supportive of the progressive view of *Shari'a*. In fact, only 4% of Islamic lawyers believed that a child should be defined as a person who is under the age of 18, and just 14% of them believed that the child should be defined according to scientific criteria. It is worth mentioning that, as was shown in *Table 10*, most Islamic lawyers, who constituted 61% of this target group, live in Qum. In contrast, as shown in *Figure and Table 11-2*, respondents who were supportive of the progressive view of *Shari'a* tend to live in Tehran and Gilan provinces.

Figure/Table 11-2 Relationship between place of residence and attitudes toward the definition of a child

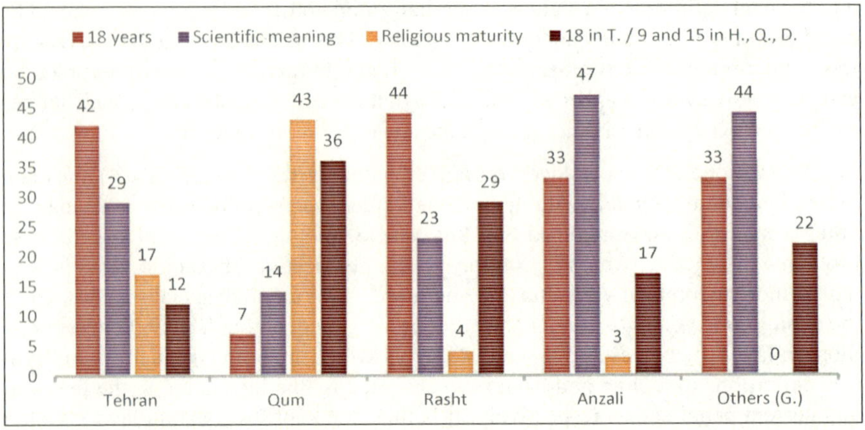

Choices / Residence	Progressive view of *Shari'a*				Conservative view of *Shari'a*			
	18 years		Scientific meaning		Religious maturity		18 in *T.* / 9 & 15 in *H., Q., D.*	
	N	%	N	%	N	%	N	%
Tehran (Tehran P.)	51	42	35	29	20	17	15	12
Qum (Qum P.)	1	7	2	14	6	43	5	36
Rasht (Gilan P.)	21	44	11	23	2	4	14	29
Anzali (Gilan P.)	10	33	14	47	1	3	5	17
Other cities (Gilan P.)	3	33	4	44	0	0	2	22
Total: 222	152				70			

3.2.1.3 Gender and attitudes toward the definition of a child

As seen in *Figure and Table 11-3*, there is no significant correlation between the gender of the respondents and their opinions on the definition of a child. Male and female respondents held a relativity homogenous view of the criteria for defining a child in criminal law. However, females were more supportive of the progressive view of *Shari'a* than males, who were more supportive of the conservative view in this regard.

Figure/Table 11-3 Relationship between gender and attitudes toward the definition of a child

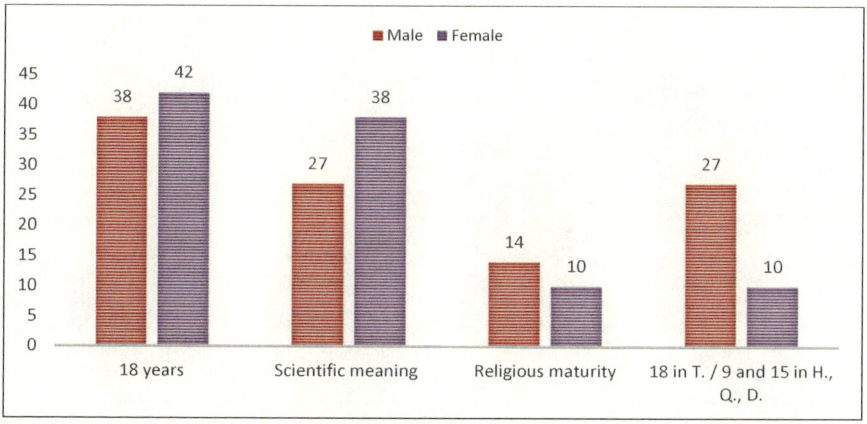

Choices / Gender	Progressive view of *Shari'a*				Conservative view of *Shari'a*			
	18 years		Scientific meaning		Religious maturity		18 in *T.* / 9 & 15 in *H.*, *Q.*, *D.*	
Gender	N	%	N	%	N	%	N	%
Male	61	38	43	27	23	14	35	27
Female	25	42	23	38	6	10	6	10
Total: 222	152				70			

3.2.1.4 Marital status and attitudes toward the definition of a child

As shown in *Figure and Table 11-4*, there is no significant correlation between the marital status of respondents and their opinion on the definition of a child. Both married persons and singles held a relatively homogenous view of the criteria for defining a child in criminal law. However, singles were more supportive of the progressive view of *Shari'a* than married respondents.

Figure/Table 11-4 Relationship between marital status and attitudes toward the definition of a child

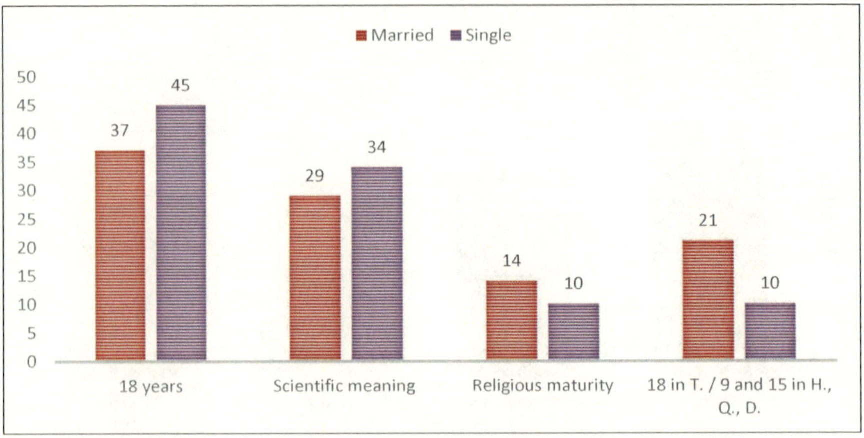

Choices / Marital status	Progressive view of *Shari'a*				Conservative view of *Shari'a*			
	18 years		**Scientific meaning**		**Religious maturity**		**18 in *T.* / 9 & 15 in *H., Q., D.***	
Marital status	N	%	N	%	N	%	N	%
Married	65	37	50	29	24	14	36	21
Single	21	45	16	34	5	10	6	10
Total: 222	152				70			

3.2.1.5 Age and attitudes toward the definition of a child

As seen in *Figure and Table 11-5*, which takes into consideration the relationship between the definition of a child and age differences between respondents, respondents who were between 31 and 45 years of age, respectively, rather believed in the progressive view of *Shari'a* more than others in this target group.

Figure/Table 11-5 Relationship between age and attitudes toward the definition of a child

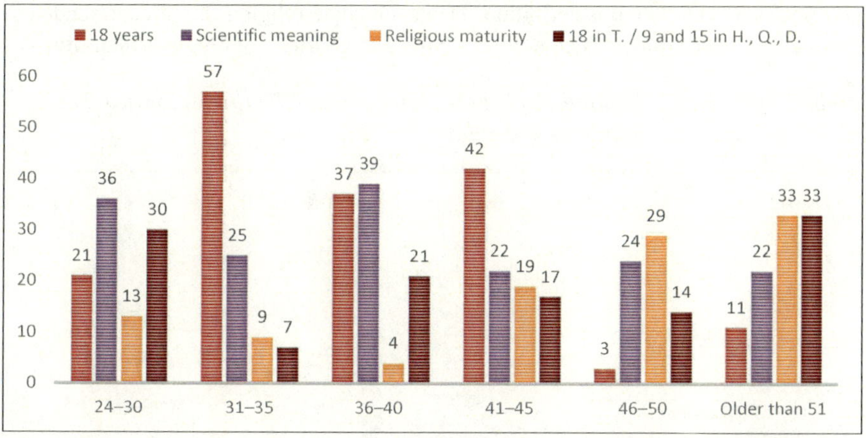

Choices / Age	Progressive view of *Shari'a*				Conservative view of *Shari'a*			
	18 years		Scientific meaning		Religious maturity		18 in *T.* / 9 & 15 in *H., Q., D.*	
	N	%	N	%	N	%	N	%
24–30	10	21	17	36	6	13	14	30
31–35	34	57	14	25	5	9	4	7
36–40	19	37	20	39	2	4	11	21
41–45	15	42	8	22	7	19	6	17
46–50	7	3	5	24	6	29	3	14
51+	1	11	2	22	3	33	3	33
Total: 222	152				70			

3.2.1.6 Education and attitudes toward the definition of a child

As can be seen in *Figure and Table 11-6*, which takes a look at the relationship between the definition of a child and education differences, respondents who held a PhD were more supportive of the progressive view of *Shari'a*. In fact, 52% of faculty members believed that the child should be defined as a person who is under the age of 18, which is in line with international obligations and standards. It is worth mentioning that most respondents who had an M.A. or B.A. endorsed the first choice that defines a child as someone who is under 18 years of age. In contrast, none of the respondents with other certificates agreed with this definition. Those who held religious certificates tended to endorse the third choice, which defines a child along criteria of religious maturity.

Figure/Table 11-6 Relationship between education and attitudes toward the definition of a child

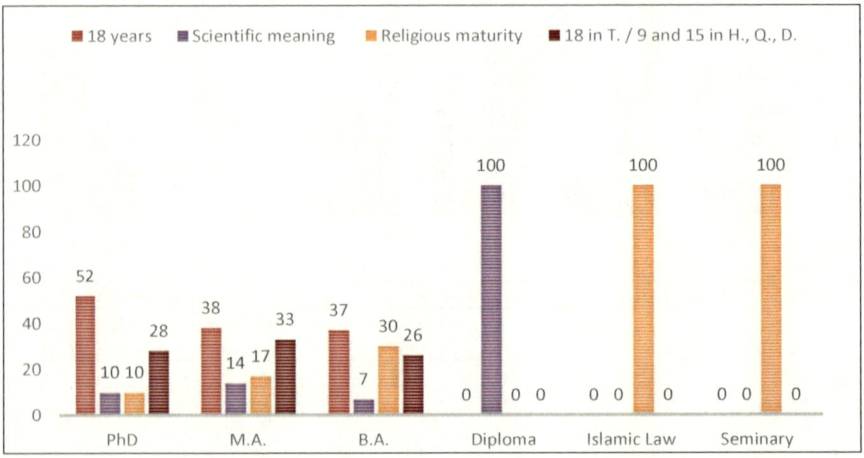

Education \ Choices	Progressive view of *Shari'a*				Conservative view of *Shari'a*			
	18 years		Scientific meaning		Religious maturity		18 in T. / 9 & 15 in H., Q., D.	
	N	%	N	%	N	%	N	%
PhD	26	52	5	10	5	10	14	28
M.A.	40	38	16	14	19	17	37	33
B.A.	20	37	4	7	16	30	14	26
Diploma	0	0	1	100	0	0	0	0
Islamic law	0	0	0	0	3	100	0	0
Seminary	0	0	0	0	2	100	0	0
Total: 222	112				110			

3.2.1.7 Results of a regression on the definition of a child

As could be seen, the perception concerning the age of criminal responsibility depends significantly on occupation, education, age, and place of residence. There are also interdependencies between these variables, and it may, e.g., be possible that if occupation is controlled, education loses its influence on this issue. In order to clarify the influence of occupation, education, gender, age, and marital status on these choices, a multivariate analysis was conducted. Since the answers could be ordered from "progressive" to "conservative", an ordered logit regression would

Table A Results of a regression on the definition of a child ($R^2 = 0.31$)

	Coefficient ± std. err.	p-value
Occupation		
Faculty members	Reference = 0	
Criminal trial judges	0.09 ± 0.32	0.78
Social workers	0.27 ± 0.38	0.48
Master students	0.22 ± 0.35	0.53
Islamic lawyers	1.64 ± 0.38	0.00
Defense lawyers	0.32 ± 0.32	0.33
Prosecutors/investigators	0.35 ± 0.37	0.34
Education		
PhD	Reference = 0	
M.A.	0.37 ± 0.25	0.14
B.A.	0.31 ± 0.30	0.30
Islamic law	-0.38 ± 0.50	0.72
Seminary school	1.20 ± 0.62	0.06
Diploma	1.21 ± 0.72	0.09
Age		
24–30	Reference = 0	
31–35	-0.87 ± 0.21	0.00
36–40	-0.39 ± 0.22	0.08
41–45	-0.41 ± 0.25	0.10
46–50	-0.21 ± 0.28	0.45
51+	0.18 ± 0.40	0.66
Gender		
Male	Reference = 0	
Female	-0.26 ± 0.17	0.13
Marital status		
Married	Reference = 0	
Single	-0.12 ± 0.18	0.52
Place of residence		
Tehran	Reference = 0	
Qum	-0.50 ± 0.41	0.23
Rasht	-0.09 ± 0.17	0.61
Anzali	-0.40 ± 0.22	0.08
Other Cities in Gilan province	-0.18 ± 0.35	0.61
Constant		2.02 ± 0.28	

have been the best choice. Nevertheless, a simple linear regression was finally preferred because it makes it much easier to interpret the results.[506] All independent variables were treated as being categorical.

The results concerning the definition of a child presented in *Table A* show only a few main (significant) effects even if the explained variance is moderate at .31. First of all, occupation has to be mentioned. Clearly and most importantly, Islamic lawyers have the most conservative view on the definition of a child. They stand in opposition to all other occupations, which do not differ (significantly) from each other. After having taken occupation into account, education had no effect. The small tendency toward a more conservative view in the case of Islamic studies and diplomas can be ignored because of their small numbers in the sample (n = 3). The age groups show a tendency, with 31–35-year-olds being the most progressive group. With increasing age, the respondents were more conservative. Interestingly, the youngest age group, which consisted by a good half of master students, was as conservative as the two oldest groups. Concerning gender, there was only a very small indication (not at all significant) that females were a bit more progressive.

3.2.2 Attitudes toward the criteria of criminal responsibility and maturity

Childhood is the first phase of human life, encompassing the cycle from birth to maturity. Thus, the childhood period includes a cyclical process and different phases. Generally, childhood is divided into two periods, that is, the period before the age of discretion and the discretion period. In the first period, a child is not aware of what is right and wrong. The second period starts after the emergence of discretion; in this period, the child's intellectual and apperception abilities are developed enough to recognize right and wrong; he/she is on the verge of maturity at this stage. In Islamic jurisprudence, the beginning of maturity marks the end of childhood. Therefore, maturity defines the boundary between childhood and adolescence. In a sense, the childhood age ranges from birth to maturity, with the child not having reached the age of religious maturity.

From the perspective of *Shari'a*, in order to determine the childhood age, we should examine the age of religious maturity. However, ascertaining the meaning and characteristics of maturity in *Shari'a* is problematic, since it is not clearly defined and therefore a controversial issue in the Iranian legal system. The age of maturity in *Shari'a* (*Shi'a* branch) is 9 lunar years for girls and 15 lunar years for boys, which is the main criterion for determining this age. Only physical/sexual (rather than intellectual/mental) maturity is relevant in this age determination, and it is noteworthy that in the fixed punishments (*Hadd*, *Qisas*, *Diye*), the new Penal Code defines the age of maturity according to the speeches of *Shi'a* leaders, i.e. 9 lunar years for girls and 15 lunar years for boys, which is based on physical and sexual maturity rather than mental and intellectual development. This is why the respondents were asked about the age criteria for

---

506 In order to verify the findings of this survey, an ordered logit regression was used, too. Its results were very similar to the linear regression presented here, so there was no need to make things more complicated.

determining criminal responsibility. As *Table 12* shows, 82% of respondents endorsed a combination of mental, physical, and sexual maturity (choice 4) as the most important criteria for determining a person's criminal responsibility.

Table 12 Attitudes toward the criteria of criminal responsibility and maturity

With regard to the criminal responsibility of juvenile delinquents, which characteristics should be considered?	N	%
Mental maturity	29	13
Physical maturity	5	2
Physical and sexual maturity	7	3
Mental, physical, and sexual maturity	181	82
Total	222	100

3.2.2.1 Occupation and attitudes toward the criteria of criminal responsibility and maturity

Figure and Table 12-1 show attitudes of different professionals toward the criteria of criminal responsibility and maturity. As can be seen, the respondents in the target group hold a relatively homogenous view of the criteria of criminal responsibility and maturity. A majority of respondents in each group endorsed choice 4, which considers the combination of mental, physical, and sexual maturity for determining criminal liability, and there is no variation in attitudes toward the characteristic of criminal responsibility between the different professions of the target groups. In fact, respondents' support of all aspects of maturity is considerably higher among those who support one or two aspects separately. However, prosecutors/investigators (96%) and criminal judges (83%) are more supportive of this choice, which shows to what extent the view of lawyers who work in the judiciary differs from the legislature's view. It should be noted that 13% of all respondents chose mental maturity (choice 1), while 2% endorsed physical maturity (choice 2), and 3% chose physical

Figure/Table 12-1 Relationship between occupation and attitudes toward the criteria of criminal responsibility and maturity

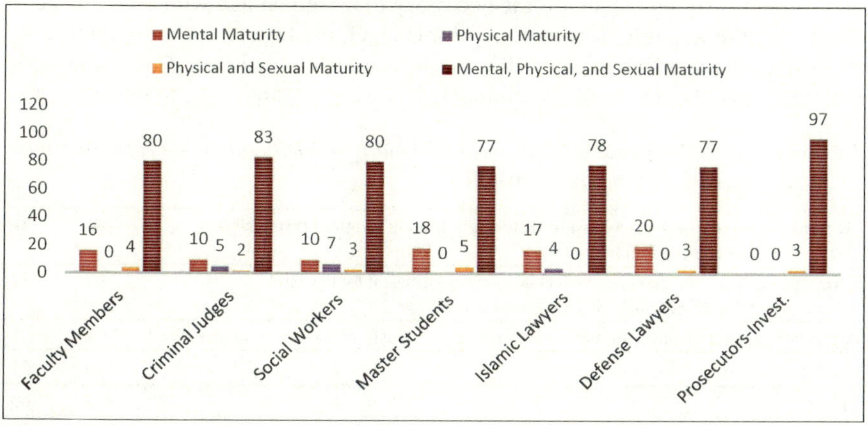

Choices / Occupation	Mental maturity		Physical maturity		Physical & sexual maturity		Mental, physical, and sexual maturity	
	N	%	N	%	N	%	N	%
Faculty members	4	16	0	0	1	4	20	80
Criminal judges	4	10	2	5	1	2	34	83
Social workers	3	10	2	7	1	3	24	80
Master students	8	18	0	0	2	5	34	77
Islamic lawyers	4	17	1	4	0	0	18	78
Defense lawyers	6	20	0	0	1	3	23	77
Prosecutors/investig.	0	0	0	0	1	3	28	97
Total: 222	34				188			

and sexual maturity (choice 3) as the most important characteristic. The most important point here is that, in spite of Islamic lawyers' conservative view of *Shari'a*, they largely believe that a combination of physical, sexual, and mental maturity is the most important criterion for criminal responsibility, as is believed in other professional groups. As discussed above, this is because they believe that girls and boys who have reached the ages of 9 and 15 years, respectively, are mentally, physically, and sexually mature according to Islamic rules, while other professional groups believe that criminal responsibility should be based on physical and sexual maturity, which is why they are not in support of these ages and prefer the international standard age of 18 years for criminal responsibility. As a result, all target groups agree with the importance of mental maturity, but it should be mentioned that the understanding of mental and intellectual maturity differs between professional groups.

3.2.3 Attitudes toward the criteria of mental maturity and their relation to the age of criminal responsibility

In order to consider mental maturity for the imposition of criminal responsibility and subsequently punishment, it is necessary to assign an age which is capable of perceiving the wrongfulness of acts. At this level, the legislature should introduce the age level according to scientific findings and social realities. Besides this age determination, the act should be conducted by a child who is able to understand its

Table 13 Attitudes toward the criteria of mental maturity and their relation to the age of criminal responsibility

If mental maturity should be considered for determining criminal responsibility, which criterion is more important?	**N**	**%**
Reaching the level of moral and intellectual maturity sufficient for the child to understand the wrongfulness of acts (generally)	83	37
Conducting in accordance with such understanding (specifically in the act committed)	3	1
Both above choices	136	61
Total	222	100

wrongfulness to impose a punishment, which can specifically be examined in the proceeding stage by a judge. In this survey, the criteria for mental maturity were asked. According to the findings shown in *Table 13*, only 37% of all respondents believed that reaching a level of moral and intellectual maturity that is sufficient for the child to (generally) understand the wrongfulness of acts is the most important criterion for determining mental maturity or intellectual development, while most respondents (61%) believed that in addition to the above criterion, conduct in accordance with such understanding (specifically regarding the act committed) should be considered to determine the mental maturity for imposing criminal liability.

3.2.4 Attitudes toward the differences between girls and boys in relation to their age of criminal responsibility

The crucial issue regarding the age of criminal responsibility in terms of fixed punishments is that there is a variance in determining this age for girls and boys, resulting in discrimination against girls. In essence, girls reach the age of criminal responsibility and consequently punishment earlier than boys if they commit crimes which are categorized in *Hadd*, *Qisas*, or *Diye*. Undoubtedly, men and women are of equal value in terms of human perfection, although some researchers focus on biological and physiological differences between females and males, which includes genetic (biological) and other differences with regard to sexual hormones, genitals and the reproduction system, body growth, cells, and brain weight. Other differences pertain to psychology, such as emotions, sensory and intellectual perceptions, as well as intelligence and memory. Some studies have focused on the impact of these differences on legal rights, obligations, as well as social and family inequalities.

However, what is very important concerning the situation of women is the sex and gender differences. The latter are of a more social and cultural nature. When determining the age of criminal responsibility, no differences should be made between girls and boys, as suggest to UN Member States by the international instruments. Nevertheless, in the new Iranian Penal Code, there is discrimination between girls and boys in the fixed punishments, which is a very controversial subject matter, and due to the obstacles in *Shari'a*, the legislature has persisted on this model of age determination.[507]

Table 14 Attitudes toward the differences between girls and boys in relation to their age of criminal responsibility

With regard to the criteria of criminal responsibility, there should be ...	N	%
no differences between girls and boys	111	50
differences between girls and boys	79	36
differences between girls and boys in *Hadd*, *Qisas*, and *Diye*	32	14
Total	222	100

507 *Assadi* 2007, p. 14.

As shown in *Table 14*, most respondents (50%) believe that no differences should be made between girls and boys in relation to their age of criminal responsibility. It follows that most respondents agree with the progressive view of *Shari'a* and international standards. This is a positive signal, since it reveals potential for legal reform in Iran. However, a very crucial point in this question is that the other 50% of respondents are opposed to this progressive view of *Shari'a*, which is rooted in the current discrimination against women in Iranian society in general. In fact, 36% of all respondents believed that there should be differences between girls and boys with respect to the age determination, with 14% supporting differences only in *Hadd*, *Qisas*, and *Diye* (fixed punishments). The differences between the sexes in fixed punishments are reflected in the new Penal Code of 2013, which indicates that the legislature has so far ignored the attitudes of elites in determining the minimum age of criminal responsibility.

3.2.4.1 Occupation and attitudes toward the differences between girls and boys in relation to their age of criminal responsibility

As *Figure and Table 14-1* show, most respondents support the first choice, i.e. there should be no differences between girls and boys. Faculty members show the highest level of support for this choice with 84%, followed by defense lawyers with 63%, social workers with 60%, master students with 46%, prosecutors/investigators with 45%, criminal judges with 39%, and Islamic lawyers with 17%. It is notable that 45% of the respondents in the target group of prosecutors/investigators believed that there should be differences between girls and boys. As explained above, most prosecutors and investigators (75%) are females. By implication, it can safely be concluded that despite their gender, most of these women support differences based on sex and discrimination against girls. To justify this attitude, it should be mentioned that women or girls committing crimes seem to be generally perceived as more reprehensible by a community. It seems that the (female) prosecutors and investigators who participated in the survey were under the influence of these perceptions of the norms, traditions, and culture in Iranian society. However, 39% of criminal judges who are governmental and serve in the justice system, and all of whom were males, believed that no differences should be made between girls and boys; 37% of them supported this differentiation, while 24% believed that the differences should only be applied in *Hadd*, *Qisas*, and *Diye*.

The pivotal point is that 82% of Islamic lawyers who participated in this survey followed the conservative view of *Shari'a* – 52% supported choice 2, 30% choice 3. An analysis of conservative views of *Shari'a* in terms of the differences between girls and boys reveals that in each target group, there were some respondents who supported these differences, which may cause discrimination in the justice system. This shows that these beliefs are rooted in the Iranian culture. In this light, the views of elites cannot easily be distinguished from those of laymen.

Figure/Table 14-1 Relationship between occupation and attitudes attitudes toward the differences between girls and boys

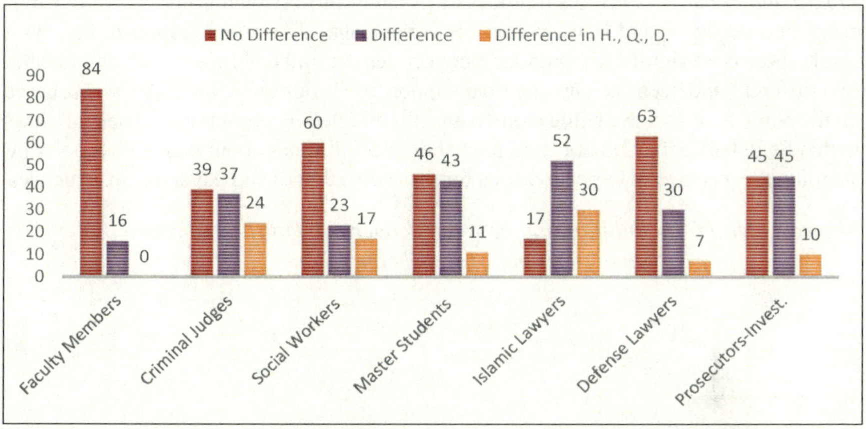

Choices / Occupation	Progressive view of *Shari'a*		Conservative view of *Shari'a*			
	No difference		**Difference**		**Difference in *H., Q., D.***	
	N	%	N	%	N	%
Faculty members	21	84	4	16	0	0
Criminal judges	16	39	15	37	10	24
Social workers	18	60	7	23	5	17
Master students	20	46	19	43	5	11
Islamic lawyers	4	17	12	52	7	30
Defense lawyers	19	63	9	30	2	7
Prosecutors/investig.	13	45	13	45	3	10
Total: 222	98		124			

3.2.4.2 Gender and differentiation between girls and boys with respect to the age of criminal responsibility

Figure and Table 14-2 reveals attitudes of different professional groups toward differences between girls and boys in relation to their age of criminal responsibility. As a result, there is no significant variation between female and male views, which indicates two different and separate signals: men support the progressive idea of *Shari'a* based on the realities in society, while women are still affected by conservative societal views in this field, hence they do not want to change their attitudes about this reality, notwithstanding the absence of logical reasons for these differences and adverse consequences.

Figure/Table 14-2 Relationship between genders and attitudes toward the differences between girls and boys

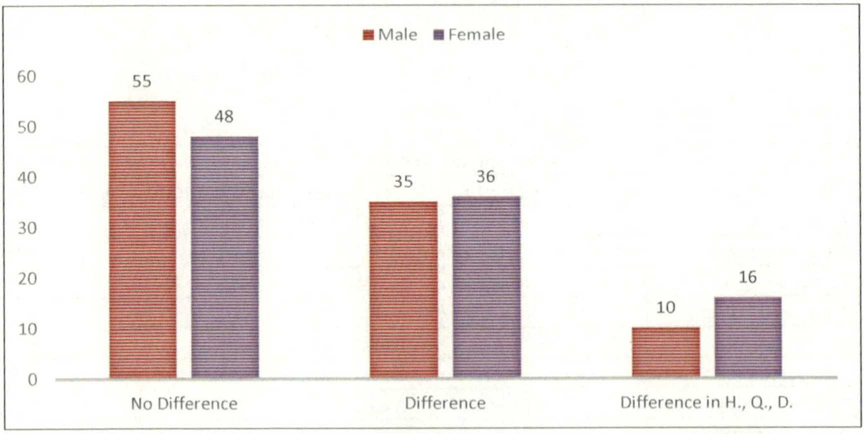

Gender \ Choices	Progressive view of *Shari'a*		Conservative view of *Shari'a*			
	No difference		Difference		Difference in *H., Q., D.*	
	N	%	N	%	N	%
Female	33	55	21	35	6	10
Male	78	48	58	36	26	16
Total: 222	111		111			

3.2.4.3 Education and differentiation between girls and boys in relation to their age of criminal responsibility

As can be seen in *Figure and Table 14-3*, regarding the relationship between the definition of a child and education differences, respondents who held a PhD were more supportive of the progressive view of *Shari'a*. In fact, 68% of faculty members believed that no differences should be made between girls and boys in relation to their age of criminal responsibility, which is in line with international obligations and standards. It is worth mentioning that most respondents who have a M.A. or B.A. endorsed the first choice, too. In contrast, none of the respondents with other certificates agreed with this standard. Those who have religious certificates endorsed the third choice, which holds that a differentiation should be made between girls and boys in *Hadd*, *Qisas*, and *Diye*. This shows to which extent the new legislation has been influenced by religious criteria in this field.

Figure/Table 14-3 Relationship between education and attitudes toward the differences between girls and boys

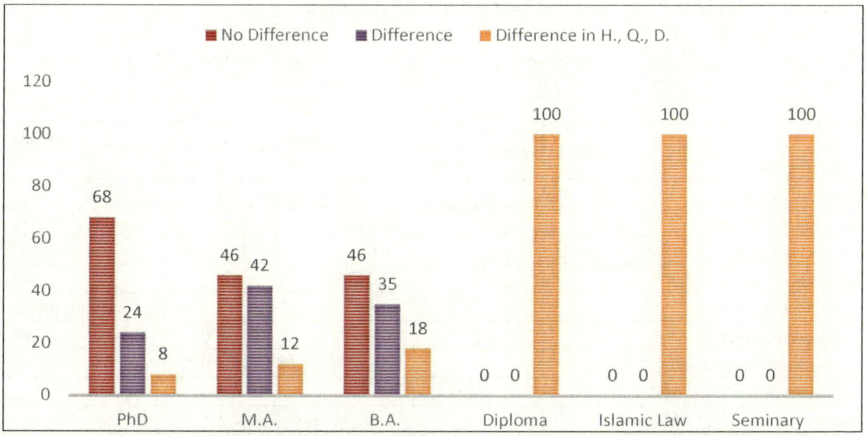

Choices / Education	**Progressive view of *Shari'a***		**Conservative view of *Shari'a***			
	No difference		**Difference**		**Difference in *H.*, *Q.*, *D.***	
	N	%	N	%	N	%
PhD	34	68	12	24	4	8
M.A.	52	46	47	42	13	12
B.A.	25	46	19	35	10	18
Diploma	0	0	0	0	1	100
Islamic law	0	0	0	0	3	100
Seminary	0	0	0	0	2	100
Total: 222	111		111			

3.2.4.4 Results of a regression on the differentiation between girls and boys

Concerning the differences between boys and girls before the law, the answers show minor effects of the independent variables concerning the definition of child, as shown in *Table B*. So at 0.18, R^2 is smaller than in the former case. Nevertheless, nearly the same structure shows up as before. The Islamic lawyers and the youngest age group prove to be the most conservative ones. As well, respondents from other cities in Gilan province seem to be rather conservative.

Table B *Results of a regression on the differentiation between girls and boys ($N = 222$; $R^2 = 0.18$)*

	Coefficient std. err.		p-value
Occupation			
Faculty members	Reference = 0		
Criminal trial judges	-0.22	0.30	0.470
Social workers	0.05	0.37	0.888
Master students	-0.01	0.34	0.976
Islamic lawyers	-0.88	0.37	0.018
Defense lawyers	0.01	0.31	0.966
Prosecutors/investigators	-0.11	0.36	0.756
Education			
PhD	Reference = 0		
M.A.	-0.29	0.24	0.234
B.A.	-0.26	0.28	0.362
Islamic law	-0.51	1.01	0.612
Islamic studies	0.03	0.59	0.956
Diploma	-0.40	0.69	0.564
Age			
24–30	Reference = 0		
31–35	0.55	0.20	0.007
36–40	0.45	0.21	0.031
41–45	0.58	0.24	0.015
46–50	0.29	0.27	0.280
51+	0.42	0.39	0.278
Gender			
Male	Reference = 0		
Female	0.08	0.17	0.627
Marital status			
Married	Reference = 0		
Single	0.16	0.17	0.350
Place of residence			
Tehran	Reference = 0		
Qum	0.10	0.40	0.803
Rasht	-0.09	0.17	0.586
Anzali	-0.22	0.21	0.303
Other cities in Gilan province	-0.79	0.34	0.020
Constant	2.13	0.27	0.000

3.2.5 Attitudes toward diminished criminal responsibility, minimum and maximum ages

One of the most difficult areas of criminal justice policy lies in establishing appropriate legal mechanisms to reflect the transition from the age of childhood innocence through to maturity and full responsibility under criminal law.[508] In every legal system, there are various ways in which age interacts with criminal responsibility. Legal systems tend to use two distinct directions of this age. One concerns the capacity of a child to commit a crime; the other relates to the exemption of a child from full criminal responsibility and the adult system.[509]

In criminal law, and in relation to the age of criminal responsibility, natural persons can be divided into three groups: a natural person who has no criminal responsibility (irresponsibility); a natural person who has diminished criminal liability; and a natural person who has full criminal responsibility. In fact, these categories indicate a gradual transition from lack of responsibility to full criminal liability. The Iranian legislature has ignored this need for a gradual transition and has subsequently diminished criminal responsibility for children for some crimes, which is a very controversial subject matter. Regarding this issue, the legislature had to consider two separate models: in fixed punishments, it has been influenced by *Shari'a*, and by the customary law in flexible punishments.

The new Penal Code has generally stipulated the age of maturity. According to article 145, immature persons do not have any criminal responsibility, and article 146 fixes the age of 9 lunar years for girls and 15 lunar years for boys as the age of maturity. This age is determined strictly without any minimum or maximum; therefore, there is no transitional period between the lack of responsibility and full responsibility for a child. Hence, a gradual change in criminal responsibility has not been accepted by the Iranian legislature. Basically, according to these articles applicable to fixed punishment (*Hadd*, *Qisas*, and *Diye*), either the subjects of criminal law are children (under the ages mentioned above) without any criminal responsibility, or they are adults (above the ages mentioned above) with full criminal responsibility. It should be noted that this age was determined based on the age of maturity under *Shari'a*, which is stipulated as the criterion of criminal responsibility. The age of evaluation in fixed punishments refers to the Islamic lunar calendar which is common in all Islamic countries. It is worth noting that the ages of 9 lunar years for girls and 15 lunar years for boys are equal to 8 and 14 solar years, respectively. The solar calendar is the official calendar in Iran.

It is important to mention that some *fatwas* (i.e. religious opinions) advocate for older ages for criminal responsibility and without any form of discrimination between girls and boys, but they are being ignored by the legislature. However, so as

---

508 *Urbas* 2000, p. 1.

509 *Maher* 2005, pp. 1–3.

to solve the practical problem, in addition to the age factor, the notions of mental maturity or intellectual development and the ability to reason were introduced in the Penal Code by the Iranian legislature for the first time (article 91), regarding the sentencing of juvenile delinquents in fixed punishments. In this respect, the judge is entitled to use forensic evidence or to seek the views of professionals in order to determine the intellectual development, maturity, and ability of the defendant to reason. Thus, this article gives judges the discretion to decide whether a child has understood the nature of the crime – and therefore whether he/she can be severely punished in *Hadd* and *Qisas*.

In contrast, in *Ta'zir* punishments, which are flexible, the age of criminal responsibility has not been defined based on the above-mentioned ages. According to the new Islamic Penal Code of 2013 (articles 87–94), the minimum age of criminal responsibility is 9 solar years and the maximum is 18 solar years, without any discrimination between girls and boys and taking into consideration the transitional period between irresponsibility and full responsibility. As can be seen, age assessment for *Ta'zir* punishments is based on the solar calendar. As mentioned above, with regard to *Ta'zir* punishments, the minimum and maximum ages were determined by the legislature, while in fixed punishments, the age of criminal responsibility has been strictly defined without any minimum or maximum; however, in the individual case, it is left to the judges to determine the age of criminal responsibility by using forensic evidence and the views of professionals if the judges are convinced that an intellectual assessment of the child is required.

As shown in *Table 15*, the majority of respondents (63%) believed that these minimum and maximum ages should be determined by the legislature for all crimes, while the third choice, which was designed according to the new Penal Code of 2013, was only endorsed by 16% of all respondents. In other words, overall, most respondents do not believe that the legislature should determine the minimum and maximum ages solely in *Ta'zir* and that judges should only do so by using forensic evidence and taking into account the views of professionals. This shows the gap between the new legislature and the views of the elites.

Table 15 Attitudes toward diminished criminal responsibility, minimum and maximum ages

According to the diminished criminal liability for juvenile delinquents,	**N**	**%**
the min. and max. ages should be determined by the legislature for all crimes	139	63
the min. and max. ages should be left to be considered by the judges (by using forensic evidence and the views of professionals) for all crimes	47	21
the min. and max. ages should be determined by the legislature for *Ta'zir* punishments, while it should be left to the judges in *Hadd*, *Qisas*, and *Diye* crimes by using forensic evidence and the views of professionals	36	16
Total	222	100

3.2.5.1 Occupation and attitudes toward diminished criminal responsibility

Figure and Table 15-1 show the attitudes of various professional groups toward diminished criminal responsibility. With regard to the first choice – designed according to international standards –, we should point out that the respondents support this definition as follows: 88% faculty members, 73% defense lawyers, 73% judges, 66% prosecutors/investigators, 57% Islamic lawyers, 34% master students, and 27% social workers. It is noteworthy that social workers are less supportive, which may lie in the duty of their job where they prefer solutions for society with regard to the situation at hand. The majority of faculty members (88%) believe that minimum and maximum ages should be determined for all crimes, while none commented on the statement that the minimum and maximum ages should be determined by the legislature in *Ta'zir* punishments, while they should be left to the judges with respect to *Hadd*, *Qisas*, and *Diye*, as well as by using forensic evidence and the views of experts (choice 3). The third choice was accepted by the legislature in the Penal Code of 2013, although it was endorsed by all target groups to a lesser extent than other choices. Again, this indicates the big gap between the legislature and the elite points of view in terms of revising the penal law.

Figure/Table 15-1 Relationship between occupation and attitudes toward diminished criminal responsibility

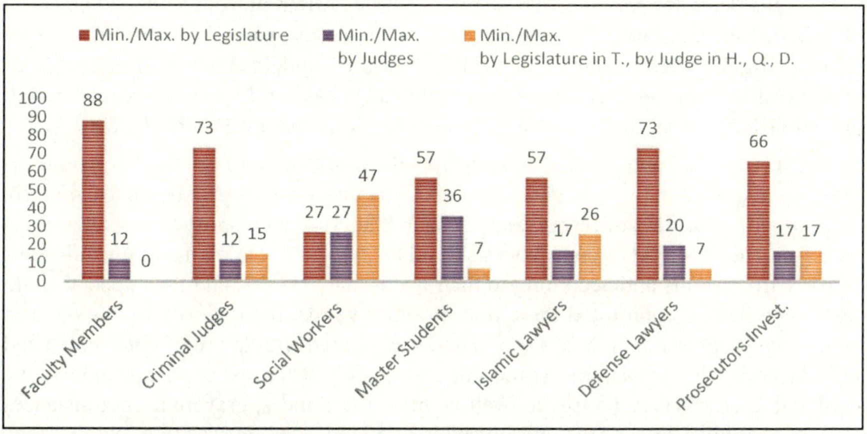

Choices / Occupation	Min./max. ages by legislature		Min./max. ages by judges		Min./max. ages by legislature in *T.*, by judge in *H.*, *Q.*, *D.*	
	N	%	N	%	N	%
Faculty members	22	88	3	12	0	0
Criminal judges	30	73	5	12	6	15
Social workers	8	27	8	27	14	47
Master students	25	57	16	36	3	7
Islamic lawyers	13	57	4	17	6	26
Defense lawyers	22	73	6	20	2	7
Prosecutors/investig.	19	66	5	17	5	17
Total: 222	139				83	

3.2.6 Attitudes toward criminal responses, mitigating and aggravating circumstances

According to the international standards, on the one hand, criminal responses as well as mitigating and aggravating factors should focus on the perpetrator's degree of criminal liability, the motives for committing the act, the intensity of danger or injury to the protected value, the circumstances in which the act was committed, the past conduct of the perpetrator, the entering of a guilty plea, the personal circumstances of the perpetrator, his/her behavior after committing the criminal offense, and the role of victims involved as well as their injury and harm. Considering these criteria for criminal responses as well as mitigating and aggravating circumstances are all the more crucial in juvenile justice.

In the Iranian legal system, due to the special category of crimes and subsequently the special sentencing system, there are two kinds of factors influencing criminal responses: mitigating and aggravating circumstances. In *Ta'zir* crimes, there is a separate and flexible sentencing system according to the differentiated criminal policies for juveniles. If it recognizes mitigating factors in this type of crimes, the court can reduce the sanction by up to half of the minimum allocated punishment and replace security as well as correctional measures for children and young people with other measures. In addition, in cases of reoffending, a penalty as severe as is applicable for adults should not be considered for juveniles. These provisions are different from adults. Thus, it can be said that the legislature has considered more supportive rules for juveniles in *Ta'zir* crimes in light of the Convention on the Rights of the Child. In contrast, in *Hadd, Qisas*, and *Diye*, due to the nature of these crimes, the judge has to refer to specific rules which are not different from those used with respect to adults and are affected by *Shari'a*.

In this survey, the central question addresses the implication of criminal responses as well as mitigating and aggravating circumstances for juveniles. As shown in *Table 16*, a vast majority of 91% of respondents believe that criminal responses as well as mitigating and aggravating circumstances should be applied to children and juveniles differently from adults and according to their needs, age, gender, and their special situation when having committed the crime. In other words, the majority of respondents follow the progressive view of *Shari'a* and don't agree with the rules determined and introduced by the legislature. At the same time, only 9% of respondents endorse the choice that criminal responses as well as mitigating and aggravating circumstances should be applied to children and juveniles in a similar manner as to adults in *Hadd* and *Qisas*, but differently in *Ta'zir*. However, only 0.5% of respondents believe that no differences at all should be made between adults and juveniles in this field.

Table 16 Attitudes toward criminal responses, mitigating and aggravating circumstances

Criminal responses, mitigating and aggravating circumstance for juvenile delinquency ...	**N**	**%**
should be applied differently from adults and according to children's needs (age, sex, and special situation)	201	91.0
should be applied just like for adults	1	0.5
should be applied like for adults in *Hadd* and *Qisas*, but differently in *Ta'zir*	20	9.0
Total	222	100

3.2.6.1 Occupation and attitudes toward criminal responses, mitigating and aggravating circumstances

As *Figure and Table 16-1* show, all faculty members believe that criminal responses, mitigating and aggravating circumstances should be applied differently to children and adults and according to children's needs, age, gender, and the crimes' particular (*Ta'zir*, *Hadd*, *Qisas*, and *Diye*) circumstances. But prosecutors and investigators, master students, criminal judges, social workers, defense lawyers, and Islamic lawyers support this progressive view of *Shari'a*, too. Only one respondent, a judge, believed that criminal responses, mitigating and aggravating circumstances should also be applied to adults. At the time of questioning, he was over 55 years of age and from Tehran. It should be mentioned that 30% of Islamic lawyers and 17% of social workers endorsed the third choice. Nevertheless, a more significant point here is that most respondents in these target groups (83% social workers and 70% Islamic lawyers) supported the first choice, which is a positive point in this survey. However, it is striking that the Islamic lawyers had their own idea about the definition of a child and the criteria for criminal responsibility. As a result, it can be said that in comparison with other target groups, the Islamic lawyers in this survey also believed in a differentiated criminal policy; but the problem is that according to them, a person who has reached the age of religious maturity cannot benefit from this differentiated policy, which is an unsolved issue in the Iranian legal system.

Figure/Table 16-1 Relationship between occupation and criminal responses, mitigating and aggravating circumstances

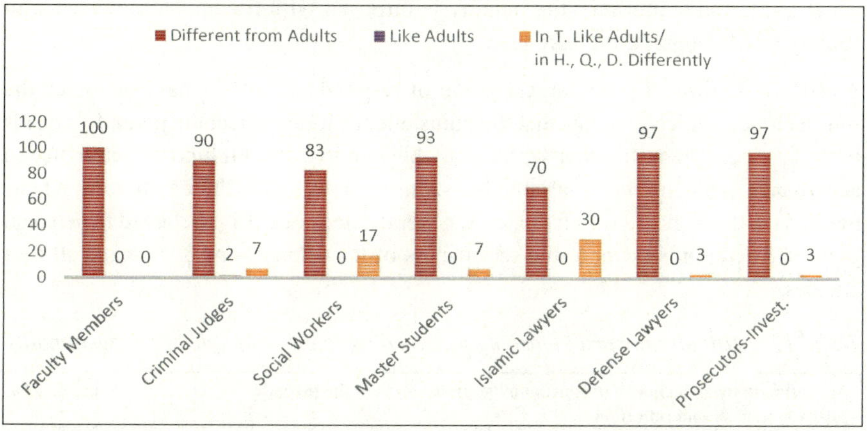

Occupation \ Choices	Progressive view of *Shari'a*		Conservative view of *Shari'a*			
	Different from adults		Like adults		*T.* as adults, *H.*, *Q.*, *D.* differently	
	N	%	N	%	N	%
Faculty members	25	100	0	0	0	0
Criminal judges	37	90	1	2	3	7
Social workers	25	83	0	0	5	17
Master students	41	93	0	0	3	7
Islamic lawyers	16	70	0	0	7	30
Defense lawyers	29	97	0	0	1	3
Prosecutors/investig.	28	97	0	0	1	3
Total: 222	201		21			

3.2.7 Attitudes toward punishment according to the principle of proportionality

The purpose of a punishment is one of the most important subjects in criminal law, especially when the punishment is imposed on children and juveniles. This is the reason why, according to the principle of proportionality, the purpose of punishment in juvenile justice was investigated in this survey. The choices in this question were selected based on the Iranian penal structure as well as on differences between fixed and flexible punishments. As has been discussed in the previous chapters, the purpose of a punishment in each crime, which has been categorized in the Penal Code, is completely different in each case. *Ta'zir* is considered a type of crime which disrupts the public order, and the state has the right to punish the lawbreaker in this section, while *Qisas* is a personal right and imposing a punishment in this section should be left to the victim or their family. *Hadd* is categorized as crimes against the right of God, and there are very unique proceedings for these types of crimes; under certain conditions, it is possible to efface such a punishment. *Diye* is a financial punishment which should be paid by the offender to the victim or their family for unintentional injury or killing. Therefore, the Iranian legal system provides for various purposes of punishment. Some of them focus on the seriousness of the offense, others are based on personal guilt, and sometimes the penalty is imposed with the aim of deterrence in society or of deterring the offender.

As shown in *Table 17*, the largest group of respondents (40%), have endorsed the fourth choice, which explains that the punishment should reflect the juvenile's needs in the justice system. In other words, they believe in a special kind of deterrence in accordance with international standards. In contrast, only 2% of all respondents marked the third choice which relates to general deterrence. 14% believed in personal guilt, 9% tied punishment to the seriousness of the offense, and 35% made all four choices.

Table 17 Attitudes toward punishment according to the principle of proportionality

According to the principle of proportionality, punishment in the juvenile justice system should reflect on ...	**N**	**%**
the seriousness of the offense	19	9
personal guilt	31	14
general deterrence	4	2
the offender and his/her needs, i.e. special deterrence	89	40
all of the above choices	73	35
Total	221	100

3.2.7.1 Occupation and attitudes toward punishment in the juvenile justice system

As shown in *Figure and Table 17-1*, respondents tended to support special deterrence, meaning that punishment should reflect the juvenile's needs for special deterrence in the justice system: 60% of faculty members, 50% of defense lawyers, 49% of criminal judges, 48% of Islamic lawyers, 46% of master students, 27% of social workers, and 17% of prosecutors/investigators preferred this choice. It is notable that 26% of Islamic lawyers believed that the seriousness of the offense should be considered when imposing a punishment, while other target groups tended not to endorse this choice: only 13% of social workers, 7% of criminal judges and defense lawyers, 5% of master students, and 4% of faculty members. This shows that practical lawyers' ideas are more like those of faculty members than of Islamic lawyers. So as to understand why Islamic lawyers believe in the seriousness of an offense, it should be noted that in *Shari'a*, some types of punishment are only imposed according to the seriousness of the offense and belong to the fixed crimes and punishments (*Qisas* and *Hadd*). It is impossible to change or remove them, so Islamic lawyers, in contrast to the other target groups, are more supportive of the proportionality between the seriousness of offenses and imposing penalties.

Figure/Table 17-1 Relationship between occupation and attitudes toward punishment in the juvenile justice system

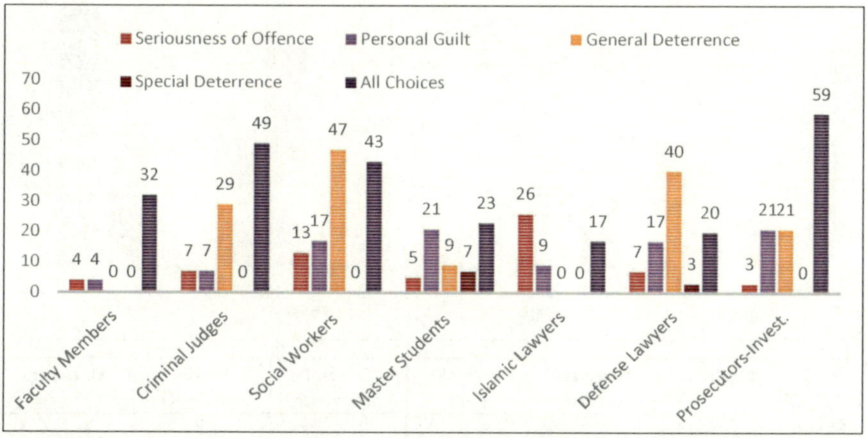

Choices / Occupation	Seriousness of offense		Personal guilt		General deterrence		Special deterrence		All choices	
	N	%	N	%	N	%	N	%	N	%
Faculty members	1	4	1	4	0	0	15	60	8	32
Criminal judges	3	7	3	7	0	0	15	37	20	49
Social workers	4	13	5	17	0	0	8	27	13	43
Master students	2	5	9	21	3	7	20	46	10	23
Islamic lawyers	6	26	2	9	0	0	11	48	4	17
Defense lawyers	2	7	5	17	1	3	15	50	6	20
Prosecutors/investigators	1	3	6	21	0	0	5	17	17	59
Total: 221	19		31		4		89		78	

3.2.7.2 Gender and attitudes toward punishment in the juvenile justice system

Figure and Table 17-2 consider aspects of different genders among the respondents as well as the differences between girls and boys in relation to their punishment in the juvenile justice system. As a result, there are no significant variations between female and male views in this regard. However, it should be noted that the third choice (regarding the proportionality between general deterrence and imposing penalties) was less endorsed by both males and females than the other choices, while the fourth choice (which covers the proportionality between special deterrence and imposing penalties designed according to international standards) was selected by more respondents of both genders than the other choices, which means that the respondents are aware of the aims of children's and juveniles' punishment in the juvenile justice system.

Figure/Table 17-2 Relationship between gender and attitudes toward punishment in the juvenile justice system

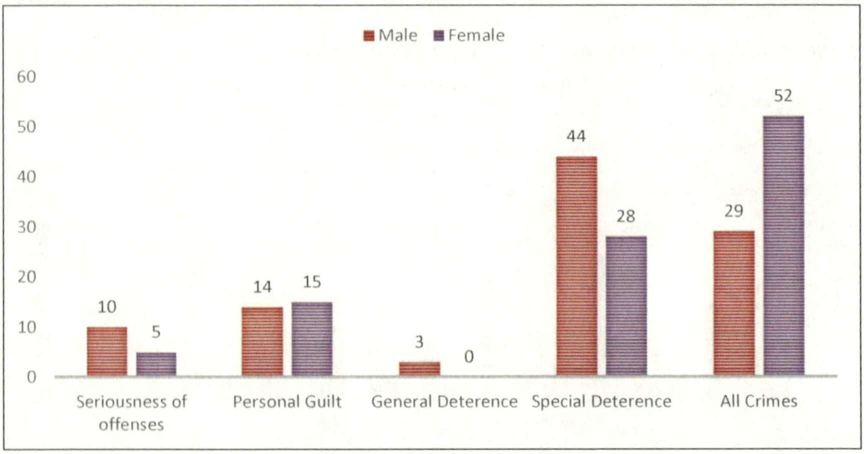

Choices / Gender	Seriousness of offense		Personal guilt		General deterrence		Special deterrence		All choices	
	N	%	N	%	N	%	N	%	N	%
Male	16	10	22	14	4	3	72	44	47	29
Female	3	5	9	15	0	0	17	28	31	52
Total: 221	19		31		4		89		78	

3.2.8 Attitudes toward educational and disciplinary measures

Educational and disciplinary measures cover a wide range of orders. Educational measures are designed to have a positive impact on the behavioral patterns of juvenile offenders in terms of securing and enhancing the conditions of socialization.[510] According to the new Islamic Penal Code, these vary, as has been explained in *Chapter 2*.

In this context, respondents were asked about the implications of educational and disciplinary measures for juveniles. As shown in *Table 18*, a majority of 81% believed that these measures should be employed in all crimes and according to the age, sex, as well as other physical and psychological characteristics of the respective juvenile delinquent. This indicates that most respondents support the progressive view of *Shari'a*. At the same time, only 19% of them endorsed the choice which elaborated that educational and disciplinary measures should be taken with regard to *Ta'zir* crimes as well as according to the age, sex, and other physical/psychological characteristics of juvenile delinquents, which is categorized in the conservative view of *Shari'a*.

Table 18 Attitudes toward educational and disciplinary measures

Educational and disciplinary measures ...	**N**	**%**
should be used in all crimes and according to the age, sex, and other physical/psychological characteristics of the juvenile delinquent/offender	179	81
should be used in *Ta'zir* crimes and according to the age, sex, and other physical/psychological characteristics of the juvenile delinquent/offender	43	19
Total	222	100

3.2.8.1 Occupation and attitudes toward educational and disciplinary measures

Figure and Table 18-1 illustrate the attitudes of the various professional target groups toward educational and disciplinary measures. As shown here, respondents agreed that educational and disciplinary measures should be used in all crimes and according to the age, sex, as well as other physical and psychological characteristics of juvenile offenders (choice 1). This choice conforms to the recognized rules in the international instruments and is in accordance with the content of the Juvenile Court Act of 1959, too, which was abolished after the 1979 Revolution. In order to assess the responses to this choice in each target group as considerably important with respect to this question the respondents in favor of this definition should be mentioned: 96% of faculty members, 91% of master students, 86% of prosecutors/investigators, 83% of defense lawyers, 81% of criminal judges, and 70% of social workers. At

510 *Winterdyk* 2002, p. 191.

48%, the number of Islamic lawyers who supported this choice was considerably lower than among the other target groups. This means that they are less supportive of this internationally recognized rule, while 52% of them endorsed choice 2, which relates to the educational and disciplinary measures that should be used in *Ta'zir* crimes as well as according to the age, sex, and other physical and psychological characteristics of juvenile delinquents.

Figure/Table 18-1 Relationship between occupation and attitudes toward educational and disciplinary measures

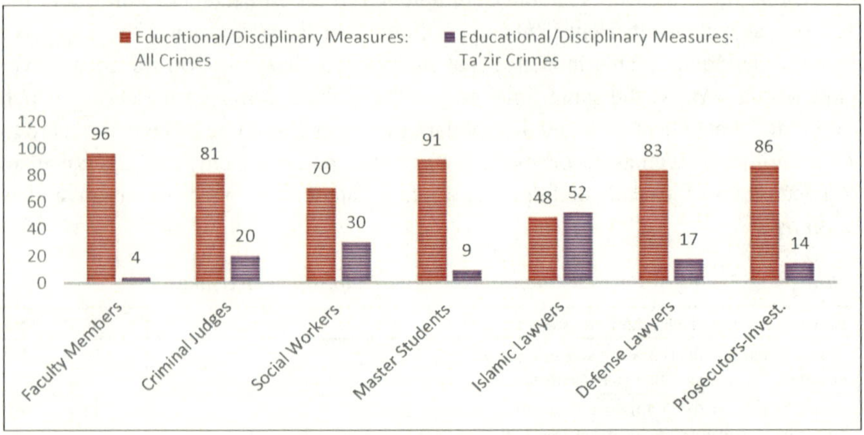

Occupation \ Choices	Progressive view of *Shari'a*: Educational/disciplinary measures: all crimes		Conservative view of *Shari'a*: Educational/disciplinary measures: *Ta'zir* crimes	
	N	%	N	%
Faculty members	24	96	1	4
Criminal judges	33	81	8	20
Social workers	21	70	9	30
Master students	40	91	4	9
Islamic lawyers	11	48	12	52
Defense lawyers	25	83	5	17
Prosecutors/investigators	25	86	4	14
Total: 222	179		43	

3.2.9 Attitudes toward deprivation of liberty

Deprivation of liberty of a juvenile "should be a disposition of last resort and for the shortest period of time and should be limited to exceptional cases. The length of the sanction should be determined by the judicial authority, without precluding the possibility of his or her early release".[511] Generally, in Iranian criminal law, deprivation of liberty is applicable for *Ta'zir* crimes as well as for other crimes in specific circumstances. In this section, deprivation of liberty is implied in line with international standards, as a last resort, and for a short period only. However, it is notable that in the international instruments, this punishment has been considered to be the most serious punishment, while in Iran's criminal law, other punishments applicable in *Hadd* and *Qisas* are perceived to be more serious, such as physical punishment and the death penalty. Therefore, the choices in this question have been designed based on international instruments. Accordingly, deprivation of liberty should be imposed only if educational or disciplinary measures are insufficient to educate the juvenile offender, or if the seriousness of the crime requires criminal punishment; and it should be imposed as a last resort as well as for the shortest possible period of time for serious crimes and in correctional institutions.

As demonstrated in *Table 19*, a slight majority of respondents (51%) believed that all circumstances should be considered when imposing deprivation of liberty. However, among the individual measures, the third choice (deprivation of liberty that should be imposed as a last resort as well as for the shortest possible period of time for serious crimes and in correctional institutions) was endorsed slightly more often (19%) than the other ones, with the second choice (emphasizing the seriousness of the crime) was endorsed by the fewest respondents (14%).

Table 19 Attitudes toward the deprivation of liberty

Deprivation of liberty (i.e. imprisonment) should be imposed on a child/juvenile ...	N	%
only if educational or disciplinary measures are insufficient to educate the juvenile offender	35	16
only if the seriousness of the crime requires criminal punishment	31	14
as a last resort as well as for the shortest possible period of time for serious crimes and in correctional institutions	43	19
all the above choices	113	51
Total	222	100

3.2.9.1 Occupation and attitudes toward deprivation of liberty

As indicated in *Figure and Table 19-1*, what appears to be important for this aspect is that by considering and analyzing the first three choices, it became clear that the faculty members generated the same values for all of them. The endorsement of defense lawyers and social workers was also close to this group of respondents. However, the insufficiency of disciplinary and educational criteria was taken into account

511 *Emerson* 1992, p. 25.

by more Islamic lawyers (*Tulab*), 39% of whom selected this choice, while only 9% of them believed that deprivation of liberty should be applied if a serious crime is committed. Among the practical lawyers, a small percentage of prosecutors/investigators (3%) and judges (10%) selected the first choice; however, these groups, like the Islamic lawyers, paid less attention to the seriousness of the crime, and only 7% of participants in these two groups selected the second option (that the seriousness of the crime should be a criterion for deprivation of liberty). From one perspective, this indicated that Islamic lawyers believed in deterrence in the context of *Ta'zir* crimes, so the intensity of the crime was simply not taken into account by them. This is a very effective criterion in order to adjust children's laws in *Ta'zir* crimes in accordance with international standards. Such a view was also accepted by the legislature in the new Iranian Penal Code in 2013.

Figure/Table 19-1 Relationship between occupation and attitudes toward deprivation of liberty

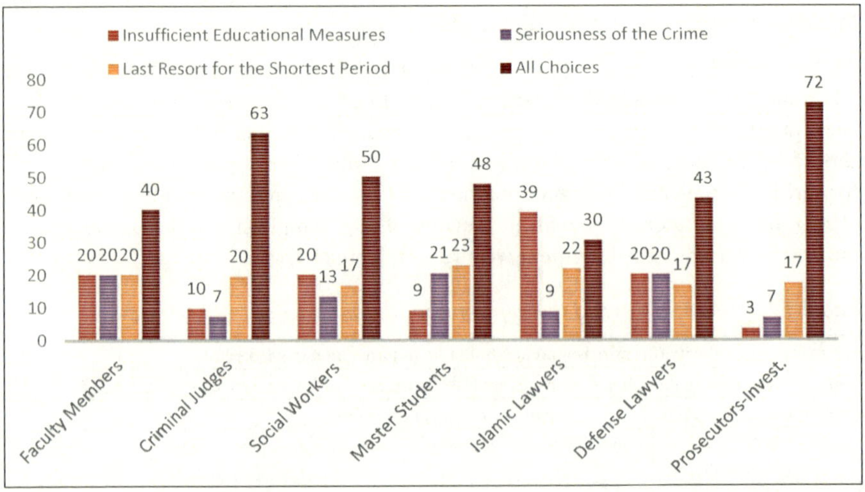

Choices / Occupation	Insufficient educational measures		Seriousness of the crime		As a last resort for the shortest possible period		All choices	
	N	%	N	%	N	%	N	%
Faculty members	5	20	5	20	5	20	10	40
Criminal judges	4	10	3	7	8	20	26	63
Social workers	6	20	4	13	5	17	15	50
Master students	4	9	9	21	10	23	21	48
Islamic lawyers	9	39	2	8	5	22	7	30
Defense lawyers	6	20	6	20	5	17	13	43
Prosecutors/investig.	1	3	2	7	5	17	21	72
Total: 222	35		31		43		113	

3.2.10 Attitudes toward physical punishment

Sentencing child offenders to capital or corporal punishment has been consistently denounced by the United Nations Treaty Bodies and Special Procedures as well as by regional human rights mechanisms as a violation of human rights. Furthermore, in this survey, the respondents were asked about physical punishment such as the death penalty, dismemberment of limbs, and whipping, which all exist in the Iranian penal system. The first choice in this question – which states that physical punishment should not be applied to any person under the age of 18 at all, in line with international standards – was designed based on international standards, while the other two choices were asked according to the legislative policies since the Islamic Revolution of 1979. Choice 2 explains about the legislative policies in the previous Penal Code. Accordingly, physical punishment could be applied to any person under the age of 18 and for all crimes. The third choice represents the new legislative policy in the new Penal Code of 2013. According to this choice, physical punishment can be applied to any child who commits a *Hadd*, *Qisas*, or *Diye* crime, but cannot be applied for the *Ta'zir* crimes. As seen in *Table 20*, most respondents (70%) endorsed the first choice.

Table 20 Attitudes toward physical punishment

Physical punishment (death penalty, defect of limb, and whipping) ...	N	%
should not at all be applied to any person under the age of 18	155	70
can be applied to any person under the age of 18	14	6
can be applied to any child who commits *Hadd*, *Qisas*, or *Diye* crimes, but should not be applied for *Ta'zir* crimes	53	24
Total	222	100

3.2.10.1 Occupation and attitudes toward physical punishment

Figure and Table 20-1 illustrates the attitudes of the various professional target groups toward physical punishment. As a result, the majority of respondents agree that physical punishment should not be applied to any person under the age of 18 at all (choice 1). This choice conforms to the recognized rules in the international instruments on the one hand and is in accordance with the Juvenile Court Act of 1959 on the other hand, which was abolished after the 1979 Revolution. This choice is categorized in the progressive view of *Shari'a* as well. In order to assess the responses to this most important choice in each target group, the responses in favor of this definition were as follows: 92% of faculty members, 84% of master students, 83% of defense lawyers, 70% of social workers, 69% of prosecutors/investigators, and 59% of criminal judges. At 22%, the share of Islamic lawyers in support of this choice was lower than in the other target groups. This means that they are less supportive of this rule, while 74% of them endorsed choice 3 (where physical punishment can be applied to any child who commits *Hadd*, *Qisas*, or *Diye* crimes, but not in cases of *Ta'zir* crimes.

As shown in *Figure and Table 20-1*, all target groups largely resembled each other when selecting the options, except for the Islamic lawyers. However, criminal judges, prosecutors, and social workers were more similar to each other, and there was another similarity between faculty members, master students, and defense lawyers. As mentioned before, the first choice was selected by many more respondents than the other ones combined. In contrast to all other groups, only a small share of Islamic lawyers selected this option, whereas the majority of them (74%) selected the third choice (representing the conservative view but being inconsistent with international documents). This choice was accepted by the legislator in 2013, indicating the extent to which they were influenced by the views of this group. The legislator obviously did not consider the beliefs and notions of other groups when codifying this law. Although the third choice also tended to be selected by a significant share of judges, prosecutors, and social workers, the notable point here is that the majority of respondents in these groups, who were classified as the practicing lawyers, selected the first choice.

As we know, a good legislator is one who modifies political practice in accordance with criminal, judicial, and executive policies. This important rule was ignored by the legislator in the Iranian Penal Code of 2013, and it is obvious to what extent this decision was estranged from social realities. This issue will certainly have reverberations

Figure/Table 20-1 Relationship between occupation and attitudes toward physical punishment

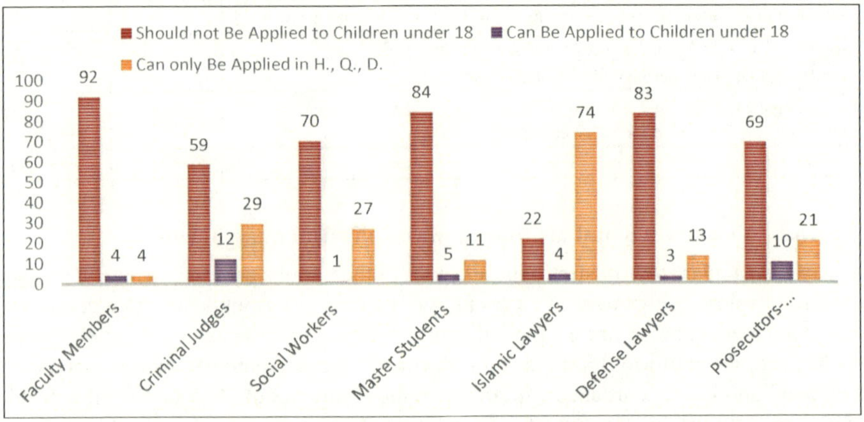

Choices / Occupation	Progressive view of *Shari'a*: Should not be applied to children under 18		Conservative view of *Shari'a*: Can be applied to children under 18		Conservative view of *Shari'a*: Can only be applied in *H., Q., D.*	
	N	%	N	%	N	%
Faculty members	23	92	1	4	1	4
Criminal judges	24	59	5	12	12	29
Social workers	21	70	1	1	8	27
Master students	37	84	2	5	5	11
Islamic lawyers	5	22	1	4	17	74
Defense lawyers	25	83	1	3	4	13
Prosecutors/investig.	20	69	3	10	6	21
Total: 222	155		67			

in the future; moreover, it violates international obligations. However, what is striking in this question is that a large majority of respondents (even among the Islamic lawyers) voted against the second option, which can be interpreted as an absolute rejection of physical punishment as a general measure. The legislator accepted this matter for *Ta'zir* crimes, in which all types of physical punishment were abolished for children and juvenile delinquents. However, it should be mentioned in this context that other variables did not indicate significant differences in the respondents' selection of choices, except for the variable of occupation illustrated here.

3.2.10.2 Education and attitudes toward physical punishment

As can be seen in *Figure and Table 20-2*, regarding the relationship between physical punishment and educational differences, respondents who held a PhD were more supportive of the progressive view of *Shari'a*. In fact, 76% of faculty members believed that physical punishment should not be applied to children under 18, which is in line with international obligations and standards.

Figure/Table 20-2 Relationship between education and attitudes toward physical punishment

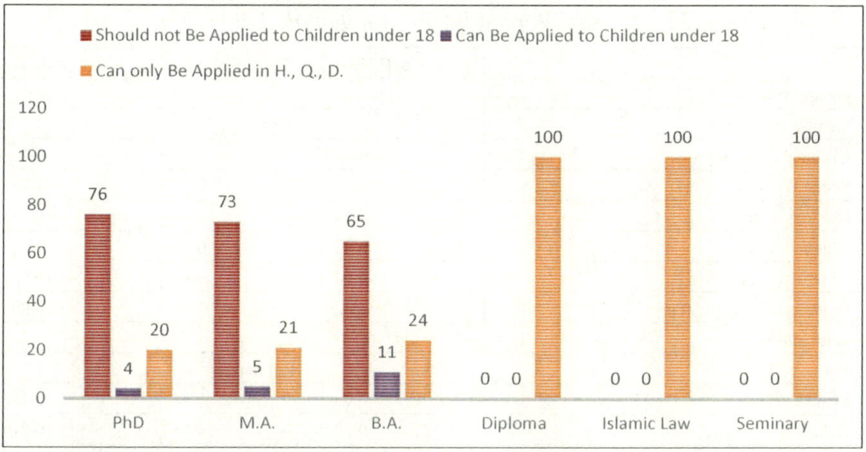

Choices / Education	Progressive view of *Shari'a*		Conservative view of *Shari'a*			
	Should not be applied to children under 18		Can be applied to children under 18		Can only be applied in *H.*, *Q.*, *D.*	
	N	%	N	%	N	%
PhD	38	76	2	4	10	20
M.A.	82	73	6	5	24	21
B.A.	35	65	6	11	13	24
Diploma	0	0	0	0	1	100
Islamic law	0	0	0	0	3	100
Seminary	0	0	0	0	2	100
Total: 222	155		67			

It is worth mentioning that most respondents who held a M.A. or B.A. degree endorsed the first choice, too. In stark contrast, none of the respondents with other certificates of education agreed with this view. Respondents who had religious certificates endorsed the third choice, which rules that physical punishment should only be applied to children under 18 in *Hadd, Qisas*, and *Diye* cases. These results show to what extent the new legislation has been influenced by religious criteria in this respect.

3.2.10.3 Results of a regression on attitudes toward physical punishment

In order to achieve mean full results in a regression on the attitudes toward physical punishment, both answer categories can be applied to any person under the age of 18 who commits *Hadd, Qisas*, or *Diye* crimes, but should not be applied in cases of *Ta'zir* crimes. Thus, a R^2 of .29 could be achieved. Again, as can be seen in *Table C*, the same structure shows up as before: Islamic lawyers and both the youngest and oldest age groups proved to be the most conservative ones. This is not so apparent for the age groups since only the 41–45-year-olds differ significantly from the youngest group, while the middle age groups (31–50) all show the same tendency. Furthermore, respondents from other cities in Gilan province seem to be more conservative.

Table C Results of a regression on attitudes toward physical punishment ($N = 222$; $R^2 = 0.29$; recoded 1 = 1 not at all; 2/3 = 2)[512]

	Coefficient	Std. err.	p-value
Occupation			
Faculty members		Reference = 0	
Criminal trial judges	0.18	0.14	0.22
Social workers	0.05	0.17	0.76
Master students	-0.18	0.16	0.26
Islamic lawyers	0.43	0.17	0.01
Defense lawyers	-0.07	0.14	0.63
Prosecutors/investigators	-0.02	0.17	0.90
Education			
PhD		Reference = 0	
M.A.	0.08	0.11	0.46
B.A.	0.09	0.13	0.51
Islamic law	0.53	0.47	0.26
Islamic studies	0.26	0.28	0.34
Diploma	0.15	0.32	0.64
Age			
24–30		Reference = 0	
31–35	-0.16	0.09	0.09
36–40	-0.18	0.10	0.06
41–45	-0.28	0.11	0.01

512 Can be applied to all but *Ta'zir* crimes.

	Coefficient	Std. err.	p-value
46–50	-0.19	0.13	0.14
51+	0.10	0.18	0.57
Gender			
Male	Reference = 0		
Female	0.03	0.08	0.71
Marital status			
Married	Reference = 0		
Single	-0.02	0.08	0.84
Place of residence			
Tehran	Reference = 0		
Qum	0.31	0.18	0.10
Rasht	0.04	0.08	0.63
Anzali	0.07	0.10	0.46
Other cities in Gilan province	0.46	0.16	0.00
Constant	1.28	0.12	0.00

3.2.11 Attitudes toward the role and duties of judges in the juvenile justice system

The most important person in the juvenile court is the juvenile court judge. The descriptions of the different systems reveal the unique role of the juvenile court judge, a role that includes many non-traditional functions. In general, this role combines judicial, administrative, collaborative, and advocacy components. The most traditional role of the juvenile court judge is to decide the legal issues in each of the described categories of cases. Besides this duty, he/she must determine issues such as whether certain facts are true, whether a child should be removed from a parent, which types of services should be offered, or whether the child should be returned to the family and the community or placed permanently in another setting.[513]

In this survey, respondents were asked about the role and duty of the judge in the juvenile justice system. The choices were designed based on international standards. As shown in *Table 21*, 70% of respondents endorsed the fourth choice, which specifies that all duties in this context should be considered by the juvenile court judge.

Table 21 Attitudes toward the role and duties of the judge in the juvenile justice system

What should be the role and duties of the judge in the juvenile justice system?	N	%
To have a comprehensive knowledge of how cases should be resolved	29	13
To know what types of alternative dispute resolutions can be employed	27	12
To know how he/she can use his/her discretion	10	5
All of the above choices	156	70
Total:	222	100

513 *Edwards* 1992, p. 27.

3.2.11.1 Occupation and attitudes toward the role and duties of the judge in the juvenile justice system

Figure and Table 21-1 show that the majority of respondents agree that all given duties should be considered by a juvenile court judge. It is apparent that the answers from all groups are similar. In fact, even the Islamic lawyers are largely in compliance with other groups, which indicates their awareness of standards and customary law. However, it is also notable that the first choice (regarding the judge's comprehensive knowledge of how cases should be resolved) and the second choice (that the judge should know the types of alternative dispute resolutions) were endorsed more often than the third one, which is related to using discretion. Since the question about the duties of juvenile court judges relates to procedural laws, there is no difference in opinions. This also indicates that procedural criminal law is influenced by customary law, and recognized standards can change as well as support the rights of children as far as there is no religious prohibition. In order to assess the responses to the very important choice 4 in each target group, the respondents in favor of this definition should be pointed out: 80% of defense lawyers, 71% of master students, 71% of criminal judges, 70% of Islamic lawyers, 68% of faculty members, 67% of social workers, and 66% of prosecutors/investigators.

Figure/Table 21-1 Relationship between occupation and attitudes toward the role and duties of the judge in the juvenile justice system

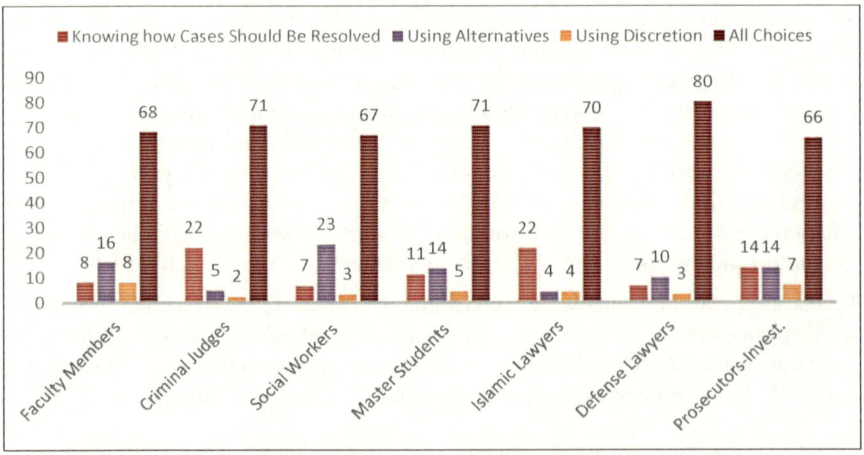

Occupation \ Choices	Knowing how cases should be resolved		Using alternatives		Using discretion		All choices	
	N	%	N	%	N	%	N	%
Faculty members	2	8	4	16	2	8	17	68
Criminal judges	9	22	2	5	1	2	29	71
Social workers	2	7	7	23	1	3	20	67
Master students	5	11	6	14	2	5	31	71
Islamic lawyers	5	22	1	4	1	4	16	70
Defense lawyers	2	7	3	10	1	3	24	80
Prosecutors/investig.	4	14	4	14	2	7	19	66
Total: 222	29		27		10		156	

3.2.12 Attitudes toward the role and duties of the prosecutor in the juvenile justice system

While modifying the laws concerning juvenile delinquents and considering the special prosecutor's office for juveniles, can the prosecutor stand as an official who protects the rights of children in order to uphold the rights of society against the accused? This discussion was brought to an end by explaining that the responsibilities of a children's prosecutor were different from those of the prosecutor in an adult court. Finally, this establishment was accepted by the legislator. *Table 22* illustrates that most respondents in this survey (87%) selected the third choice: to seek justice in light of the social interests and needs of the juveniles as well as the safety and welfare of the community. This choice was designed based on recognized standards.

Table 22 Attitudes toward the role and duties of the prosecutor in the juvenile justice system

What should be the role and duties of the prosecutor in the juvenile justice system?	N	%
To seek justice in light of the social and personal interests of the juveniles	7	3
To seek justice in light of the safety of the community	21	10
To seek justice in light of the social interests and needs of both the juveniles and the safety and welfare of the community	194	87
Total	222	100

3.2.12.1 Occupation and attitudes toward the role and duties of the prosecutor in the juvenile justice system

Figure and Table 22-1 illustrate the attitudes of the various professional target groups toward the role and duties of the prosecutor in the Iranian juvenile justice system. As a result, the majority of respondents accepted the multiple roles of the prosecutor (choice 3). As stated before, since this question relates to procedural laws, there are no major differences between opinions. Since the responses between the different target groups to choice 3 are significant, the respondents in favor of this definition are as follows: 97% of social workers, 97% of prosecutors/investigators, 89% of master students, 88% of criminal judges, 87% of Islamic lawyers, and 68% of faculty members. However, it is important to note that the general role of a prosecutor to ensure the security and protect the rights of society was selected by more respondents than role of the juvenile's prosecutor to make provisions for the child with respect to their personal and social interests – apart from providing the two other duties.

On the other hand, the only exception to the aforementioned consideration concerns the prosecutors and investigators: no representative of this group confirmed the duty of simply making provisions for whatever is in the interest of society. It is a very important point that the employees of the criminal justice system consider their duties with respect to the best interests of children. This also indicates that these employees are trained well. Moreover, this kind of awareness was also observed among the social workers. Choice 1, which seeks justice in light of the social and personal interests of the juvenile, was not endorsed by the social workers employed in the criminal justice system. Although the nature of their job requires this confirmation and endorsement, the awareness of people who work in the criminal justice system regarding how to design supportive criteria for children has led to a situation where each of these groups considered this problem from a comprehensive perspective.

Figure/Table 22-1 Relationship between occupation and attitudes toward the role and duties of the prosecutor in the juvenile justice system

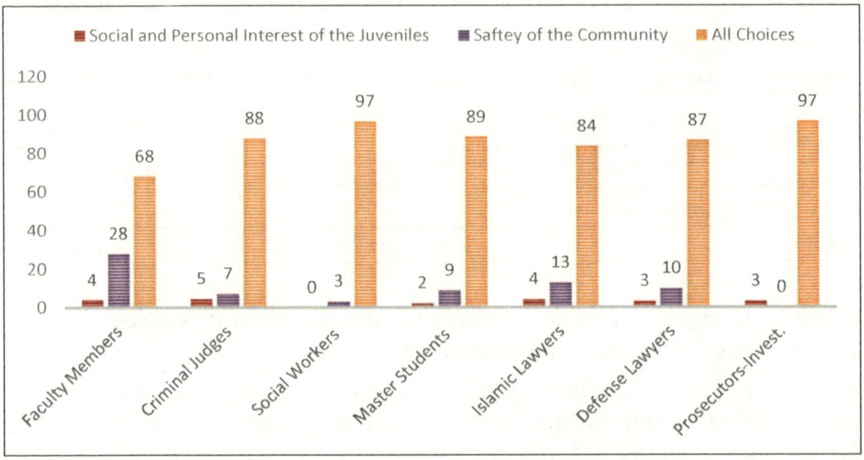

Choices / Occupation	Social and personal interests of the juveniles		Safety of the community		All choices	
	N	%	N	%	N	%
Faculty members	1	4	7	28	17	68
Criminal judges	2	5	3	7	36	88
Social workers	0	0	1	3	29	97
Master students	1	2	4	9	39	89
Islamic lawyers	1	4	3	13	19	84
Defense lawyers	1	3	3	10	26	87
Prosecutors/investig.	1	3	0	0	28	97
Total: 222	7		21		194	

3.2.13 Attitudes toward the role and duties of the police in the juvenile justice system

The first contact a young juvenile offender has with the juvenile justice system usually takes place with a police officer. The nature and circumstances of this police contact are likely to be significant and have a lasting impression on a young person. The traditional law enforcement role of police is to detect and investigate crimes as well as to apprehend those responsible for committing them. This traditional law enforcement role gives police visibility in the public, as they "serve and protect". To finalize their law enforcement role and to ensure that suspects are brought to trial, the police engage in the interrogation of suspects, the collection of physical evidence at a crime scene, and the presentation of the evidence in court.[514] Nowadays, besides their traditional role, police officers have the authority to resolve the problem alternatively in minor cases.

In this survey, the question regarding the role of the police was designed according to the very important circumstance of them having different roles and duties. The first choice illustrates the traditional role and duty of the police, which is to arrest juvenile delinquents and take them into custody as well as to the judicial section, while the second and third choices describe more recent roles and duties within the juvenile justice system, which takes a different direction by guiding juvenile delinquents toward their parents, guardians, or county social service agencies. As can be seen in *Table 23*, nearly half of respondents (48%) believed that all duties and roles combined should be taken into account by the police in the juvenile justice system (choice 4).

Table 23 Attitudes toward the roles and duties of the police in the juvenile justice system

What should be the roles and duties of the police in the juvenile justice system?	**N**	**%**
To arrest and take the juvenile delinquent into custody	53	24
To take no formal action against the juvenile delinquent (discretion, diversion)	24	11
To refer the juvenile delinquent to a parent, guardian, or county social service agency	39	18
All of the above choices	106	48
Total	222	100

3.2.13.1 Occupation and attitudes toward the roles and duties of the police in the juvenile justice system

Figure and Table 23-1 illustrate the attitudes of the various professional target groups toward the role and duties of the police in the juvenile justice system. The majority of respondents supported simultaneous duties of the police. It should be

514 *Bates et al.* 2021, p. 10; see also *Cox et al.* 2017, p. 103.

noted that only 10% of social workers believed that the traditional duty (first choice) should be fulfilled by the juvenile police. This also indicates that social workers, who are in closer contact to juvenile delinquents in society, can feel the negative consequences of police interventions and arrests by taking juvenile delinquents into custody.

Figure/Table 23-1 Relationship between occupation and attitudes toward the role and duties of the police in the juvenile justice system

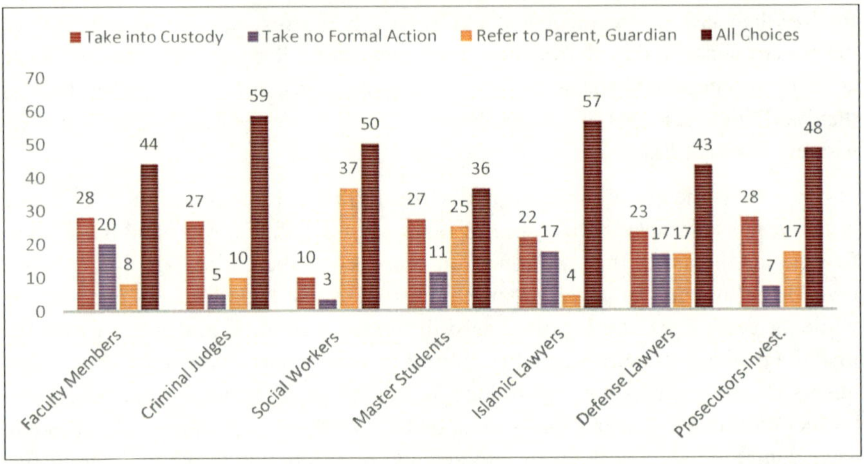

Choices / Occupation	Take into custody		Take no formal action		Refer to parent, guardian		All choices	
	N	%	N	%	N	%	N	%
Faculty members	7	28	5	20	2	8	11	44
Criminal judges	11	27	2	5	4	10	24	59
Social workers	3	10	1	3	11	37	15	50
Master students	12	27	5	11	11	25	16	37
Islamic lawyers	5	22	4	17	1	4	13	57
Defense lawyers	7	23	5	17	5	17	13	43
Prosecutors/investigators	8	28	2	7	5	17	14	48
Total: 222	53		24		39		106	

3.2.14 Attitudes toward the competent authority for diversion

Diversion describes an attempt to divert (or channel out) youthful offenders from the juvenile justice system. The primary objective of diversion programs is to redirect youths away from formal processing in the juvenile justice system while still holding them accountable for their actions. Diversion practices vary depending on the juvenile justice contact point at which the youth is diverted, as well as the types of services provided. Juveniles may be diverted by law enforcement before arrest, during court intake, or even after adjudication but before disposition. Depending on the point at which youths are diverted, a diversion program may involve outright release with minimal services, referral to a community agency, or the direct provision of services.[515] In this survey, respondents were also asked about the diversion authority that can be implemented by the legislature, police, prosecutor, or juvenile court judge. As *Table 24* illustrates, 33% of respondents believed that all mentioned authorities should be allowed to divert the case.

Table 24 Attitudes toward the competent authority for diversion

Diversion measure should be considered as a temporary or permanent procedure by ...	N	%
legislature	70	32
police	14	6
prosecutor	11	19
juvenile court judge	53	24
all of the above choices	74	33
Total	222	100

3.2.14.1 Occupation and attitudes toward the competent authority for diversion

Figure and Table 24-1 illustrate the attitudes of the various professional target groups toward the competent authority for diversion in the juvenile justice system. What is striking in this table more than anything else is that (when leaving out the fifth choice) it considers all authorities – including the legislator, the police, the prosecution, and the juvenile court – to be qualified for diversion. A majority of faculty members, criminal judges, and prosecutors selected this option. The legislator and then the juvenile court judge were endorsed more than the others. At any rate, this indicates that Iran's legal system and lawyers allocate a high importance to the law and the solutions suggested by it. It is quite obvious to the respondents that, after the legislator, the juvenile court is most highly qualified to determine if the crime committed by the child should be dealt with officially. On the other hand, the choice that the police and prosecution are qualified to intervene and change the path of proceedings was selected only by a small number of respondents. However, what is more important in this question is that none of the prosecutors considered the police, i.e. the first organization to deal with the child offender, to be qualified for diversion. Therefore, none of the judges considered the prosecutor, who deals with the child before the case is referred to the court, to be qualified for diversion. This indicates that these authorities rather consider themselves to be qualified for diversion. However, as mentioned before, the police refused to participate in this study; it would have been interesting to evaluate their opinions on which authority is most qualified for diversion. However, it is notable that the Islamic lawyers (*Tulab*) did not consider

515 *Kary* 2015, p. 2.

the prosecutor to be qualified for diversion. When investigating the problem, it was understood that since the prosecutor is an authority in modern legal systems not known by the *Shari'a*, the Islamic lawyers seemed to believe that the prosecutor should not hold the duty of diversion.

Figure/Table 24-1 Relationship between occupation and attitudes toward the competent authority for diversion

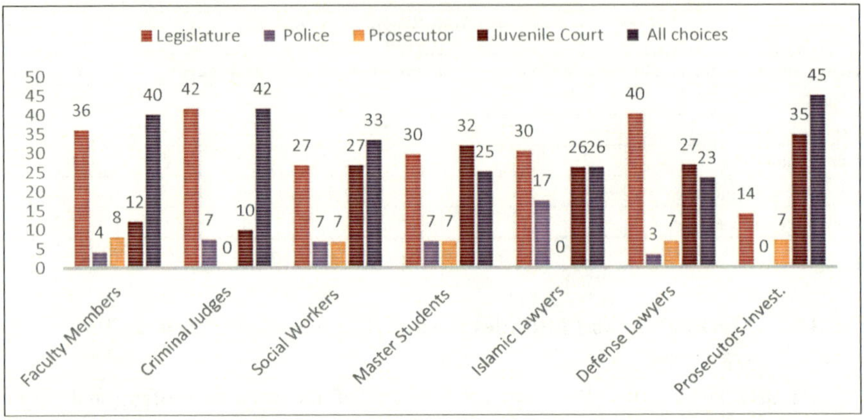

Occupation \ Choices	Legislature		Police		Prosecutors		Juvenile court		All choices	
	N	%	N	%	N	%	N	%	N	%
Faculty members	9	36	1	4	2	8	3	12	10	40
Criminal judges	17	42	3	7	0	0	4	10	17	42
Social workers	8	27	2	7	2	7	8	27	10	33
Master students	13	30	3	7	3	7	14	32	11	26
Islamic lawyers	7	30	4	17	0	0	6	26	6	26
Defense lawyers	12	40	1	3	2	7	8	27	7	23
Prosecutors/investig.	4	14	0	0	2	7	10	35	13	45
Total: 222	70		14		11		53		74	

3.2.15 Attitudes toward sexual intercourse and drug trafficking under pressure of others

Due to the relatively low age of criminal responsibility in Iran on the one hand and the inflation of criminalizing special behaviors on the other, we are faced with a range of juvenile offenders who are not actually criminals or delinquents. In fact, they are victims who, on the whole, forcefully commit crimes under the given social and individual circumstances. In Iran, many children and adolescents are sentenced for sexual crimes or drug dealing, but actually, they are forced to do so by their families or older criminals who dominate the child in some way. Sometimes, such behavior becomes a profession and a way for them to earn a living.

The main question in this context is: how should we deal with these individuals? Do they deserve to be punished? Or should they be supported and protected through certain measures instead?

Considering a fictitious case in this question, the respondents were asked how a 13-year-old girl should be treated by the law for committing an illegal sexual act or for being involved in drug trafficking, having been forced to do so by her father in both cases. The first and second choices in this question were drafted according to international standards. In this regard, the first choice explains that this child should be considered a victim and protected within the criminal justice system. As demonstrated in *Table 25*, 40% of respondents selected this choice. The second choice states that this child should be considered a victim-delinquent; however, she should be treated like a child in the juvenile justice system. 48% of respondents endorsed this choice. Nevertheless, the first and second choices found most supporters, which indicates an awareness of the participating target groups in this study. The third and fourth choices, which reflect a rather conservative viewpoint, consider the given girl to be a child delinquent who should be punished in the juvenile justice system (third choice) – or to be an adult delinquent who should be punished in accordance with adult rules (fourth choice). Only a small number of respondents chose these latter options, with only 13% endorsing the third and fourth choices each.

Table 25 Case 1: Attitudes toward sexual intercourse and drug trafficking under pressure of others

A 13 year-old girl is accused of illegal sexual intercourse and drug trafficking. She committed these crimes under pressure and guidance of her father. In your opinion she should be considered a ...	**N**	**%**
victim and protected by the criminal justice system	89	40
victim-delinquent and treated as a minor in the juvenile justice system	107	48
delinquent and treated and punished as a minor in the juvenile justice system	13	6
delinquent and punished as an adult	13	6
Total	222	100

3.2.15.1 Occupation and attitudes toward sexual intercourse and drug trafficking under duress by others

Figure and Table 25-1 illustrate the attitudes of the various professional target groups toward sexual intercourse and drug trafficking due to duress by others. As shown in this table, the majority of respondents endorsed the first and second choices, which were designed in compliance with social realities. The first choice, which considers the child to have been victimized and to be in need of support was confirmed by 56% of faculty members, 47% of lawyers, 46% of master students, 44% of judges, 33% of social workers, and 31% of prosecutors/investigators. However, it is apparent that only

17% of Islamic lawyers endorsed this choice, which indicates that victimization was not considered for the child by this target group. Concludingly, the Islamic lawyers' viewpoint was in strong opposition to the group of faculty members. The second choice, which considers the given child as a victim-delinquent, was chosen by 66% of prosecutors, 53% of social workers, 52% of Islamic lawyers, 47% of defense lawyers, 46% of master students, 44% of faculty members, and 37% of criminal judges. What is significant is that although a small number of Islamic lawyers believed in the victimization of the child (choice 1), 52% of them still believed that the child should be considered as victim-delinquent in this case.

Regarding the conservative opinions, which explain that the delinquent girl should be considered either as a child (choice 3) or as an adult (choice 4), were not endorsed by faculty members, social workers, and prosecutors/investigators at all, which reveals a stark difference between the opinions of these target groups and that of the legislator. The third choice (delinquent to be considered as a child) was chosen by 13% of social workers, 12% of criminal judges, 5% of master students, and 3% of both defense lawyers and prosecutors/investigators. It should be pointed out that this choice was not selected by any of the Islamic lawyers. Nevertheless, a strong 30% of the Islamic lawyers opted for the even more conservative fourth choice, which equals a completely non-supportive viewpoint. However, it should also be noted that a large majority of 70% of Islamic lawyers believed in the progressive view of *Shari'a*. On the other hand, choice 4 had by far the smallest number of supporters among all groups.

Figure/Table 25-1 Case 1: Relationship between occupation and attitudes toward sexual intercourse and drug trafficking under pressure of others

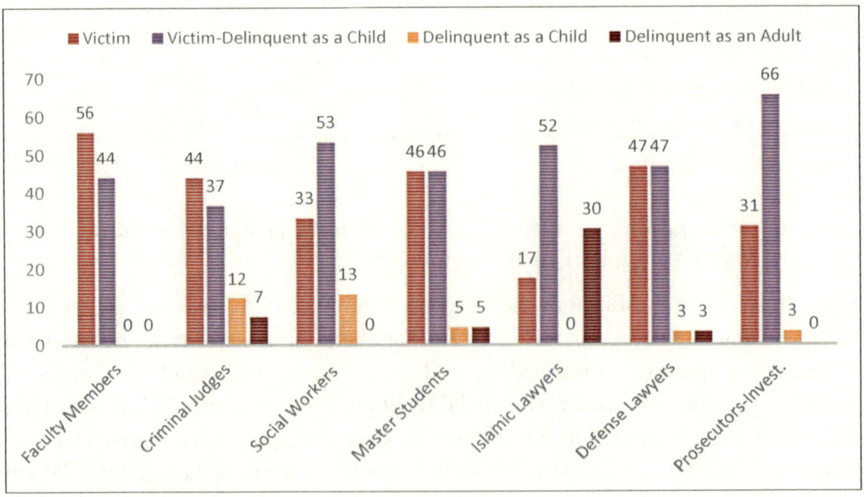

Occupation \ Choices	Progressive view of *Shari'a*				Conservative view of *Shari'a*			
	Victim		Victim-delinquent as a child		Delinquent as a child		Delinquent as an adult	
	N	%	N	%	N	%	N	%
Faculty members	14	56	11	44	0	0	0	0
Criminal judges	18	44	15	37	5	12	3	7
Social workers	10	33	16	53	4	13	0	0
Master students	20	46	20	46	2	5	2	5
Islamic lawyers	4	17	12	52	0	0	7	30
Defense lawyers	14	47	14	47	1	3	1	3
Prosecutors/investig.	9	31	19	66	1	3	0	0
Total: 222	196				26			

3.2.15.2 Place of residence and attitudes toward sexual intercourse and drug trafficking under pressure of others

Figure and Table 25-2 reveal the relationship between place of residence and attitudes toward sexual intercourse and drug trafficking under pressure by others. The first choice, which is in compliance with the progressive view of *Shari'a* and in accordance with social realities, was endorsed by 60% of respondents who live in Anzali, 54% in Rasht, 34% in Tehran, 22% in other cities in Gilan province, and 14% of respondents living in Qum. A closer look reveals that Qum houses by far the highest percentage of respondents with a conservative view of *Shari'a*. It is obvious that the respondents who live in Qum were less supportive of the progressive view of *Shari'a* (and thus of international standards), which corresponds with the majority (67%) of participating Islamic lawyers living in Qum.

Figure/Table 25-2 Case 1: Relationship between place of residence and attitudes toward sexual intercourse and drug trafficking under pressure of others

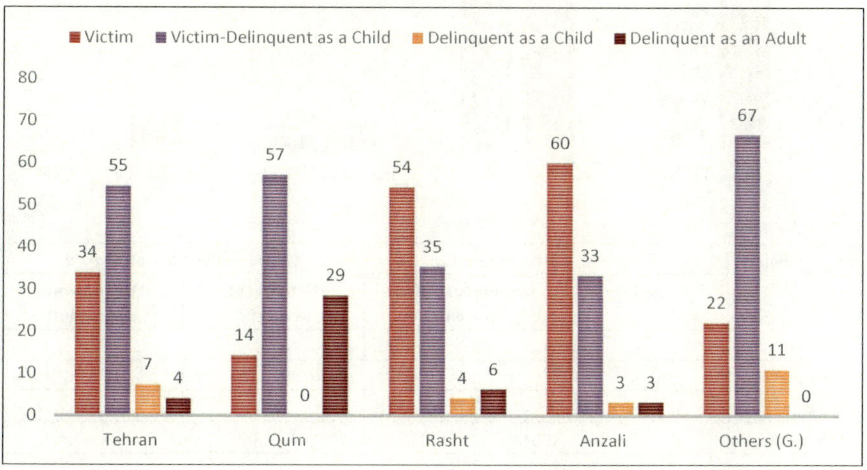

Choices / Residence	Progressive view of *Shari'a*				Conservative view of *Shari'a*			
	Victim		Delinquent-victim as a child		Delinquent as a child		Delinquent as an adult	
	N	%	N	%	N	%	N	%
Tehran (Tehran P.)	41	34	66	55	9	7	5	4
Qum (Qum P.)	2	14	8	57	0	0	4	29
Rasht (Gilan P.)	26	54	17	35	2	4	3	6
Anzali (Gilan P.)	18	60	10	33	1	3	1	3
Other cities (G. P.)	2	22	6	67	1	11	0	0
Total: 222	196				26			

3.2.15.3 Gender and attitudes toward sexual intercourse and drug trafficking under pressure of others

Figure and Table 25-3 show the relationship between gender and attitudes toward sexual intercourse and drug trafficking under pressure by others. Significant differences were detected in this variable between choices 2 and 4, with women selecting the choices within the progressive view of *Shari'a* more often than men.

Figure/Table 25-3 Case 1: Relationship between gender and attitudes toward sexual intercourse and drug trafficking under pressure of others

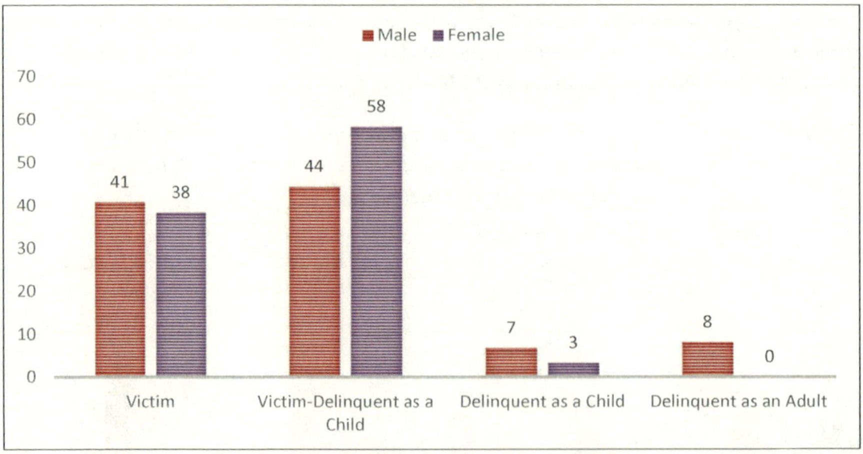

Choices / Gender	Progressive view of *Shari'a*				Conservative view of *Shari'a*			
	Victim		Delinquent-victim as a child		Delinquent as a child		Delinquent as an adult	
	N	%	N	%	N	%	N	%
Male	66	41	72	44	11	7	13	8
Female	23	38	35	58	2	3	0	0
Total: 222	196				26			

3.2.15.4 Marital status and attitudes toward sexual intercourse and drug trafficking under pressure of others

As can be seen in *Figure and Table 25-4*, in the variable of marital status, those who were single nearly exclusively endorsed the first and second choices, i.e. the progressive view of *Shari'a*.

Figure/Table 25-4 Case 1: Relationship between marital status and attitudes toward sexual intercourse and drug trafficking under pressure of others

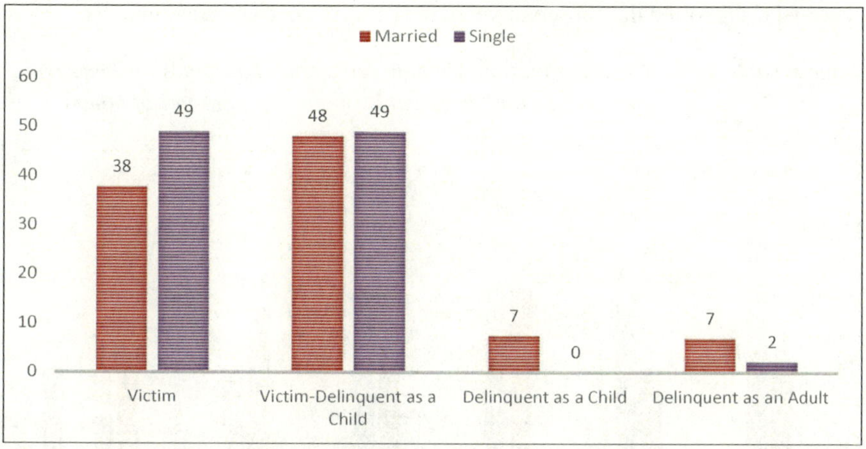

Choices / Marital status	Progressive view of *Shari'a*				Conservative view of *Shari'a*			
	Victim		Delinquent-victim as a child		Delinquent as a child		Delinquent as an adult	
	N	%	N	%	N	%	N	%
Married	66	38	84	48	13	7	12	7
Single	23	49	23	49	0	0	1	2
Total: 222	196				26			

3.2.15.5 Age and attitudes toward sexual intercourse and drug trafficking under pressure of others

Regarding the variable of age, as presented in *Figure and Table 25-5*, attention should be paid to the age group of 24–30 years, which turned out to be the largest group supporting the choices within the progressive views of *Shari'a* and international standards, whereas the age group above 45 endorsed the more conservative view with a slightly higher percentage than the younger groups. However, the number of those supporting the progressive view of *Shari'a* was still much higher than those who supported the conservative view of *Shari'a* in each age group.

Figure/Table 25-5 Case 1: Relationship between age and attitudes toward sexual intercourse and drug trafficking under pressure of others

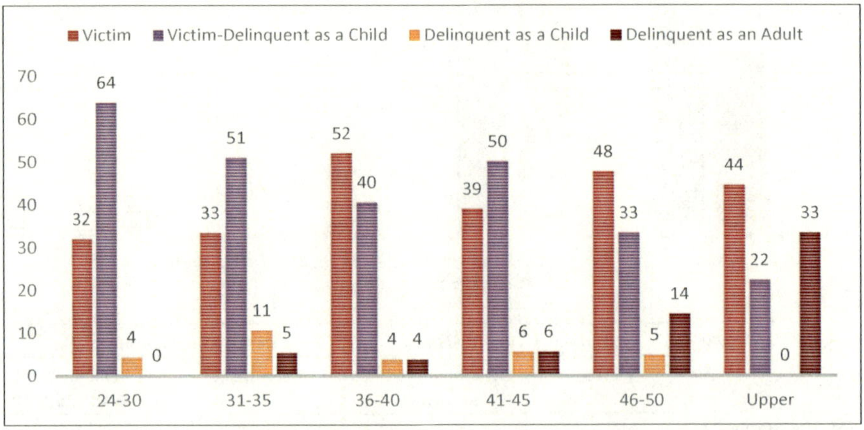

Age \ Choices	Progressive view of *Shari'a*				Conservative view of *Shari'a*			
	Victim		Delinquent-victim as a child		Delinquent as a child		Delinquent as an adult	
	N	%	N	%	N	%	N	%
24–30	15	32	30	64	2	4	0	0
31–35	19	33	29	50	6	11	3	5
36–40	27	52	21	40	2	4	2	4
41–45	14	39	18	50	2	6	2	6
46–50	10	48	7	33	1	5	3	14
51+	4	44	2	22	0	0	3	33
Total: 222	196				26			

3.2.15.6 Education and attitudes toward sexual intercourse and drug trafficking under pressure of others

As seen in *Figure and Table 25-6*, most respondents who hold a PhD, M.A. or B.A. degree, or a diploma endorsed the first and second choices, i.e. they were supportive of the progressive view of *Shari'a* – although it has to be considered that only one respondent held a diploma at all. In contrast, none of the respondents with other certificates of education agreed with the progressive view of *Shari'a*.

Figure/Table 25-6 Case 1: Relationship between education and attitudes toward sexual intercourse and drug trafficking under pressure of others

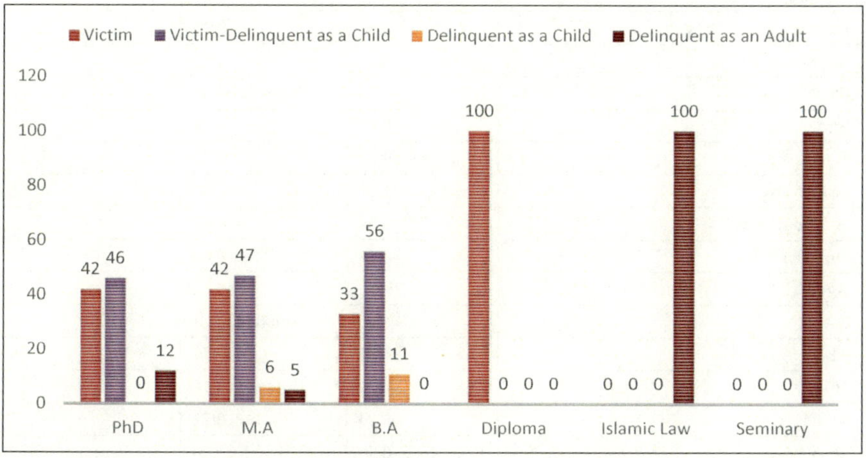

Choices / Age	Progressive view of *Shari'a*				Conservative view of *Shari'a*			
	Victim		Delinquent-victim as a child		Delinquent as a child		Delinquent as an adult	
	N	%	N	%	N	%	N	%
PhD	21	42	23	46	0	0	6	12
M.A.	47	42	53	47	7	6	5	5
B.A.	18	33	30	56	6	11	0	0
Diploma	1	100	0	0	0	0	0	0
Islamic law	0	0	0	0	0	0	3	100
Seminary	0	0	0	0	0	0	2	100
Total: 222	161				61			

3.2.15.7 Results of a regression on attitudes toward sexual intercourse and drug trafficking under pressure of others

Differences in the results of the regression for attitudes toward sexual intercourse and drug trafficking under pressure of others can mainly be found between occupations (see *Table D*). While faculty members were very progressive, Islamic lawyers were most conservative, with criminal trial judges and social workers also tending toward the conservative view. The other variables showed no significant differences.

Table D Results of a regression on attitudes toward sexual intercourse and drug trafficking under pressure of others (N = 222; $R^2 = 0.19$)

	Coefficient	Std. err.	p-value
Occupation			
Faculty members		Reference = 0	
Criminal trial judges	0.59	0.26	0.03
Social workers	0.73	0.32	0.02
Master students	0.50	0.29	0.09
Islamic lawyers	0.92	0.32	0.00
Defense lawyers	0.35	0.27	0.19
Prosecutors/investigators	0.50	0.31	0.11
Education			
PhD		Reference = 0	
M.A.	-0.11	0.21	0.61
B.A.	-0.26	0.25	0.29
Islamic law	-1.38	0.87	0.12
Islamic studies	0.34	0.52	0.51
Diploma	-1.18	0.60	0.05
Age			
24–30		Reference = 0	
31–35	0.01	0.17	0.96
36–40	-0.27	0.18	0.13
41–45	-0.24	0.21	0.25
46–50	-0.16	0.24	0.50
51+	0.24	0.34	0.48
Gender			
Male		Reference = 0	
Female	-0.08	0.14	0.59
Marital status			
Married		Reference = 0	
Single	-0.21	0.15	0.16
Place of residence			
Tehran		Reference = 0	
Qum	0.10	0.34	0.76
Rasht	-0.24	0.14	0.10
Anzali	-0.37	0.19	0.05
Other cities in Gilan province	0.08	0.29	0.77
Constant	1.65	0.23	0.00

3.2.16 Attitudes toward criminal responsibility if a person under 18 years kills another one without malice aforethought

Murder is the most serious of all homicide charges; it is categorized as intentional killing. In different legal systems, the responses to this crime vary. However, it is quite obvious that in every legal system, the most severe punishments are implied for murder. In most legal systems in which physical punishment is still applicable, the punishment for murder is often the death penalty. However, according to international standards, no severe punishment should be applied to children under 18 years of age. Therefore, it is very important to understand the criteria for age-related criminal responsibility in every legal system.

In the Penal Code of 2013, the Iranian legislator has introduced four different types of murder, all of which are punishable by *Qisas*, which is a very controversial subject matter. In the first type of murder, the murderer acts with malice aforethought. The second type is given when the perpetrator acts without malice aforethought but their act is fatal nevertheless. The third category is related to the victim's situation and vulnerability: if the perpetrator acts without malice aforethought, nor is their behavior intended to be fatal, butthe act results in the victim's death due to their vulnerability, and if the perpetrator has known about the victim's vulnerability, they will be accused of murder. The fourth form of murder is given when the perpetrator acts with malice aforethought but has no intention of killing any person(s), such as in a bomb attack.

It is quite obvious that some types of murder have more severe consequences for society, and therefore, the perpetrator deserves a more severe punishment in comparison to others; this, however, has not been considered by the Iranian legislator. Regarding persons under 18 years of age, it should be noted that most juvenile cases correspond to the second form of murder; on the other hand, juveniles should be assumed as being immature, although (as has been mentioned before) the Iranian legislature has not yet introduced clear criteria for determining the age of criminal responsibility.

In this survey, attitudes were asked toward the criteria of criminal responsibility for killing another person without malice aforethought when the perpetrator is under 18 years of age. As can be seen in *Table 26*, the majority of respondents endorsed the first and second choices, with a total of 70%. These choices are categorized within the progressive view of *Shari'a* and are thus in line with international instruments. Furthermore, 24% of respondents selected choice 4, which has been accepted by the legislator in the new Penal Code of 2013.

However, the most noteworthy result of this question – as can be seen in *Table 26* – is that only 7% of all respondents endorsed choice 3, which treats the 16-year-old perpetrator like an adult to be tried and punished according to the general criminal code. This categorization had been accepted by the legislature in the previous Penal Code. Choice 3 and 4 are considered to be conservative views of *Shari'a*. However, the new Penal Code has improved compared to the previous one in this respect but is still not in accordance with international standards and more progressive views of *Shari'a*.

Table 26 Case 2, part 1: Attitudes toward killing another person without malice aforethought

A 16-year-old boy kills another boy in a street fight by knife, but without malice aforethought. He is accused of murder. **Part 1: In your opinion, the accused ...**	N	%
should be considered a minor and should be tried and punished according to his diminished criminal responsibility in the juvenile justice system	100	45
should be examined in order to determine if he is mentally mature by using forensic evidence and consulting the views of professionals	55	25
should be considered an adult and should be tried and punished like an adult according to the general criminal code	15	7
can be examined in order to determine if he is mentally mature by using forensic evidence and consulting the views of professionals at the discretion of the judge	52	24
Total	222	100

3.2.16.1 Occupation and attitudes toward criminal responsibility if a person under 18 years kills another person without malice aforethought

Figure and Table 26-1 illustrate the attitudes of the various professional target groups toward the criteria of responsibility when a person under 18 years of age kills another person without malice aforethought. The majority of respondents endorsed choice 1: 76% of faculty members, 53% of social workers, 50% of defense lawyers, 44% of criminal judges, 41% of prosecutors/investigators, and 39% of master students. Less Islamic lawyers were in support of this customary and international standard than respondents of other target groups. It should be noted that only 13% of Islamic lawyers endorsed this choice, which indicates that they are less supportive of the customary law, a new interpretation and progressive view of *Shari'a*, while 26% of them endorsed choice 3, which states that that the delinquent should be considered an adult nevertheless. In contrast, none of the faculty members or social workers agreed with this assumption.

Figure/Table 26-1 Case 2, part 1: Relationship between occupation and attitudes toward criminal responsibility if a person under 18 years kills another person without malice aforethought

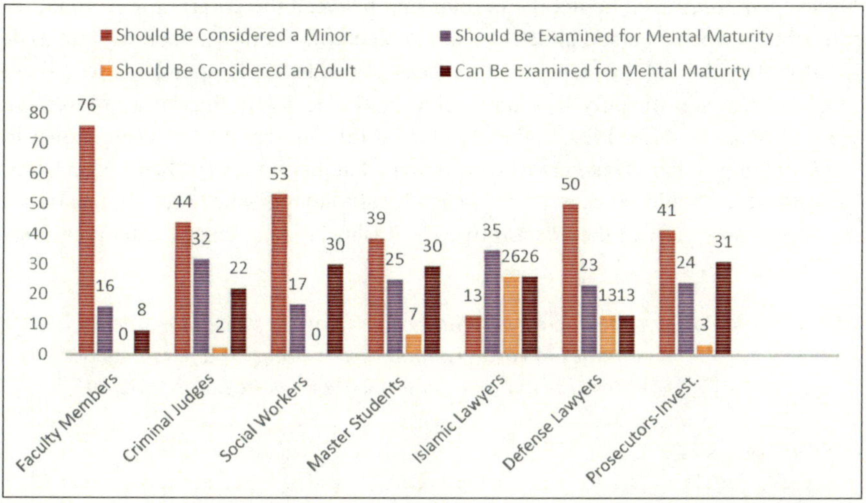

Choices / Occupation	Progressive view of *Shari'a*				Conservative view of *Shari'a*			
	Should be considered a minor		Should be examined for mental maturity		Should be considered an adult		Can be examined for mental maturity	
	N	%	N	%	N	%	N	%
Faculty members	19	76	4	16	0	0	2	8
Criminal judges	18	44	13	32	1	2	9	22
Social workers	16	53	5	17	0	0	9	30
Master students	17	39	11	25	3	7	13	30
Islamic lawyers	3	13	8	35	6	26	6	26
Defense lawyers	15	50	7	23	4	13	4	13
Prosecutors/investig.	12	41	7	24	1	3	9	31
Total: 222	155				67			

It is worth mentioning that the most contrasting views occurred between faculty members and Islamic lawyers.Social workers and defense lawyers leaned toward faculty members, while the endorsements of criminal judges, prosecutors/ investigators, and master students resemble each other in supporting the progressive view of *Shari'a* rather than the conservative view. In other words, even practicing lawyers who serve in the justice system agree with customary law, despite their loyalty to Islamic thought and obligations with respect to the *Shari'a*-dominated justice system.

3.2.16.2 Place of residence and attitudes toward criminal responsibility if a person under 18 years kills another person without malice aforethought

Figure and Table 26-2 depict the relationship between the criteria for criminal responsibility when a person under 18 years of age kills another person without malice aforethought and the respondents' places of residence. Respondents who lived in Qum were significantly less supportive of choice 1 and thus of a progressive view of *Shari'a*. Only 14% of them endorsed this choice. As has been shown in *Table 10*, most Islamic lawyers who participated in this survey (61%) lived in Qum. In contrast, as could be seen in *Table 11-2*, respondents who were supportive of the progressive view of the *Shari'a* lived in Rasht, Tehran, Anzali, and other cities in Gilan province.

Figure/Table 26-2 *Case 2, part 1: Relationship between place of residence and attitudes toward criminal responsibility if a person under 18 years kills another person without malice aforethought*

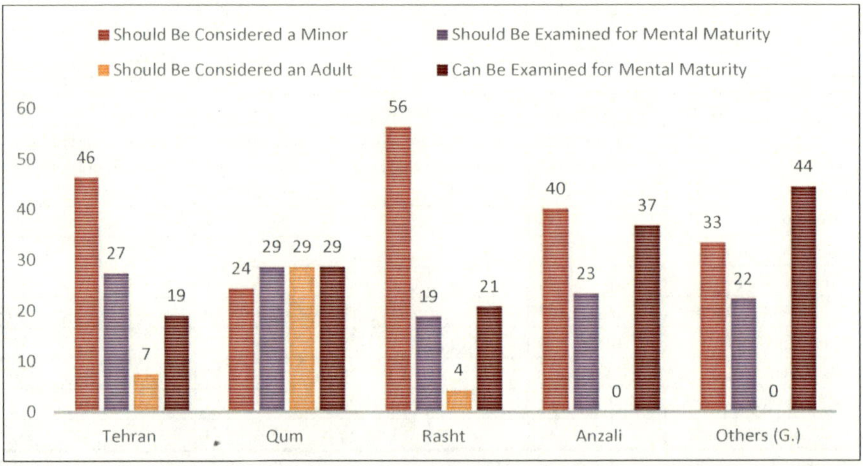

Choices / Residence	Progressive view of *Shari'a*				Conservative view of *Shari'a*			
	Should be considered a minor		Should be examined for mental maturity		Should be considered an adult		Can be examined for mental maturity	
	N	%	N	%	N	%	N	%
Tehran (Tehran P.)	59	46	33	27	9	7	23	19
Qum (Qum P.)	2	14	4	29	4	29	4	29
Rasht (Gilan P.)	27	56	9	19	2	4	10	21
Anzali (Gilan P.)	12	40	7	23	0	0	11	37
Other cities (G. P.)	3	33	2	22	0	0	4	44
Total: 222	155				67			

3.2.16.3 Gender and attitudes toward criminal responsibility if a person under 18 years kills another person without malice aforethought

As can be seen in *Figure and Table 26-3*, there is no significant correlation between the gender of the respondents and their opinions on the criteria of responsibility when a person under 18 years of age kills another person without malice aforethought. Male and female respondents held a relativity homogenous view in this regard.

Figure/Table 26-3 Case 2, part 1: Relationship between gender and attitudes toward criminal responsibility if a person under 18 years kills another person without malice aforethought

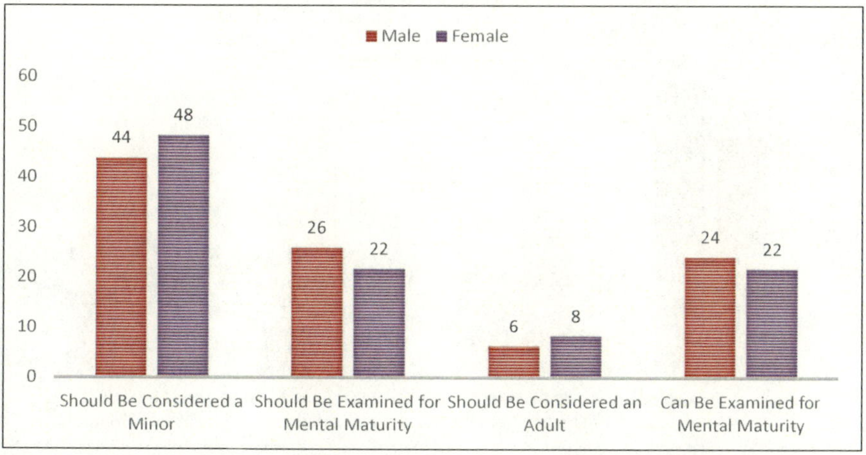

Choices / Gender	Progressive view of *Shari'a*				Conservative view of *Shari'a*			
	Should be considered a minor		Should be examined for mental maturity		Should be considered an adult		Can be examined for mental maturity	
	N	%	N	%	N	%	N	%
Male	71	44	42	26	10	6	39	24
Female	29	48	13	22	5	8	13	22
Total: 222	155				67			

3.2.16.4 Marital status and attitudes toward criminal responsibility if a person under 18 years kills another person without malice aforethought

As shown in *Figure and Table 26-4*, there is no significant correlation between the marital status of the respondents and their opinions on the criteria of responsibility when a person under 18 years of age kills another person without malice aforethought. Married and single respondents alike held a relativity homogenous view.

Figure/Table 26-4 Case 2, part 1: Relationship between marital status and attitudes toward criminal responsibility if a person under 18 years kills another person without malice aforethought

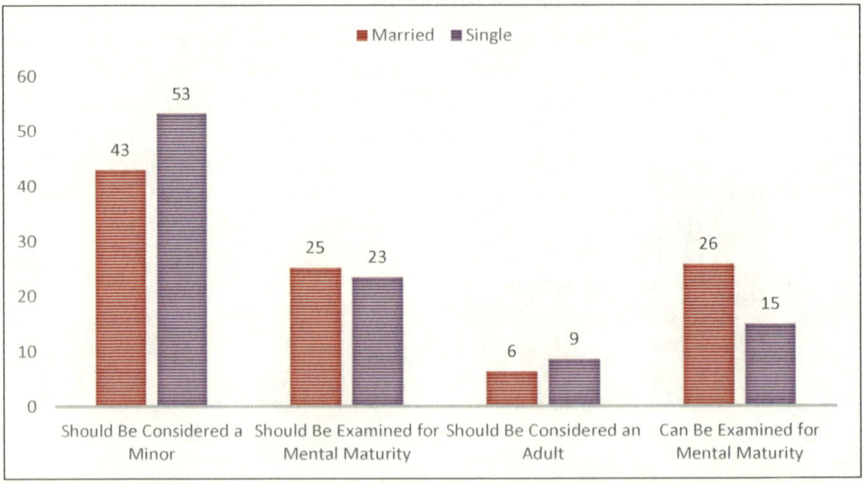

Choices / Marital Status	Progressive view of *Shari'a*				Conservative view of *Shari'a*			
	Should be considered a minor		Should be examined for mental maturity		Should be considered an adult		Can be examined for mental maturity	
	N	%	N	%	N	%	N	%
Married	75	43	44	25	11	6	45	26
Single	25	53	11	23	4	9	7	15
Total: 222	155				67			

3.2.16.5 Age and attitudes toward criminal responsibility if a person under 18 years kills another person without malice aforethought

As can be seen in *Figure and Table 26-5*, which takes respondents' age differences into account, respondents who were between 31 and 45 years of age tended to believe in the progressive view of *Shari'a* more strongly than others. The conservative view of *Shari'a* only found a majority among resondents aged 51 or older.

Figure/Table 26-5 Case 2, part 1: Relationship between age and attitudes toward criminal responsibility if a person under 18 years kills another person without malice aforethought

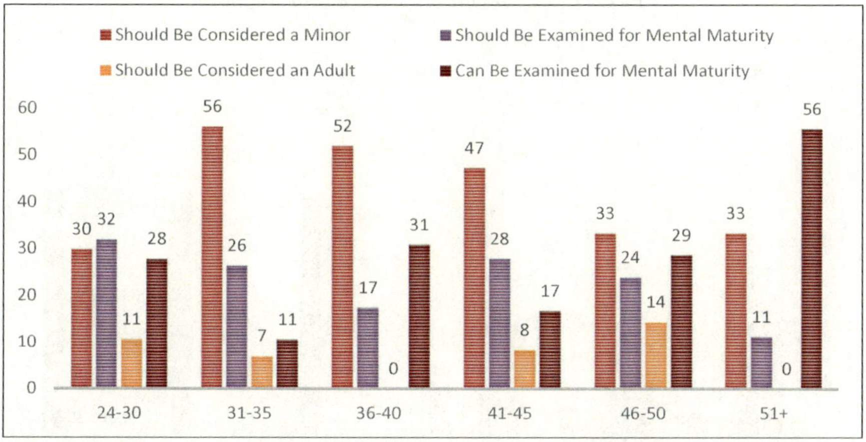

Choices / Age	Progressive view of *Shari'a*				Conservative view of *Shari'a*			
	Should be considered a minor		Should be examined for mental maturity		Should be considered an adult		Can be examined for mental maturity	
	N	%	N	%	N	%	N	%
24–30	14	30	15	32	5	11	13	28
31–35	32	56	15	26	4	7	6	11
36–40	27	52	9	17	0	0	16	31
41–45	17	47	10	28	3	8	6	17
46–50	7	33	5	24	3	14	6	29
51+	3	33	1	11	0	0	5	56
Total: 222	155				77			

3.2.16.6 Education and attitudes toward criminal responsibility if a person under 18 years kills another person without malice aforethought

Figure and Table 26-6 consider education differences when assessing criminal responsibility by a person under 18 years of age who has killed another person without malice aforethought. Here, respondents who held a PhD were most supportive of the progressive view of *Shari'a*. Overall, 82% of faculty members endorsed the first two choices. The same was found for respondents who held a M.A. or B.A. degree.

In contrast, none of the respondents with other certificates agreed with these choices. All respondents holding religious certificates endorsed the fourth choice, which reflects a conservative view of *Shari'a*. These results indicate to which extent the new legislation has been influenced by religious criteria for determining the criminal responsibility of persons under 18 years of age.

Figure/Table 26-6 Case 2, part 1: Relationship between education and attitudes toward criminal responsibility if a person under 18 years kills another person without malice aforethought

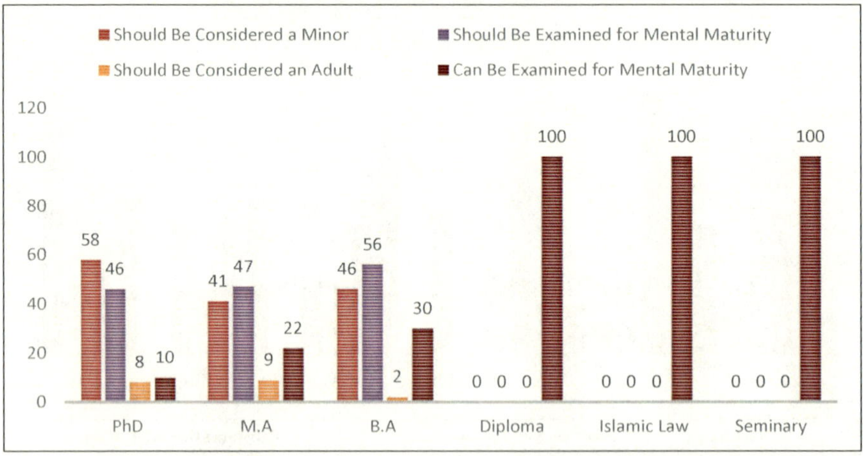

Choices / Education	Progressive view of *Shari'a*				Conservative view of *Shari'a*			
	Should be considered a minor		Should be examined for mental maturity		Should be considered an adult		Can be examined for mental maturity	
	N	%	N	%	N	%	N	%
PhD	29	58	12	24	4	8	5	10
M.A.	46	41	31	28	10	9	25	22
B.A.	25	46	12	22	1	2	16	30
Diploma	0	0	0	0	0	0	1	100
Islamic law	0	0	0	0	0	0	3	100
Seminary	0	0	0	0	0	0	2	100
Total: 222	155				67			

3.2.16.7 Results of a regression on criminal responsibility for killing another person without malice aforethought

Concerning the liability for killing another person – as presented in *Table E* –, the data has only a weak structure. The Islamic lawyers show a conservative tendency, as does the youngest age group (especially in contrast to the group of 31-35-year-olds). Here, a significant effect can be seen in those respondents educated in Islamic studies and those with diplomas, since either target group consisted of only three people; this suggests that the effect has no significant meaning among all target groups.

Table E Results of a regression on criminal responsibility for killing another person without malice aforethought ($N = 222$; $R^2 = 0.18$)

	Coefficient	**Std. err.**	**p-value**
Occupation			
Faculty members	Reference = 0		
Criminal trial judges	0.34	0.40	0.39
Social workers	0.48	0.48	0.32
Master students	0.56	0.44	0.21
Islamic lawyers	0.87	0.48	0.07
Defense lawyers	0.49	0.40	0.23
Prosecutors/investigators	0.56	0.47	0.23
Education			
PhD	Reference = 0		
M.A.	0.09	0.31	0.78
B.A.	0.09	0.37	0.82
Islamic law	2.06	1.31	0.12
Islamic studies	1.89	0.78	0.02
Diploma	2.08	0.90	0.02
Age			
24–30	Reference = 0		
31–35	-0.70	0.26	0.01
36–40	-0.25	0.27	0.36
41–45	-0.49	0.31	0.12
46–50	-0.04	0.35	0.90
51+	-0.33	0.50	0.51
Gender			
Male	Reference = 0		
Female	-0.12	0.22	0.59
Marital status			
Married	Reference = 0		
Single	-0.28	0.23	0.21
Place of residence			
Tehran	Reference = 0		
Qum	-0.18	0.52	0.73
Rasht	-0.14	0.22	0.52
Anzali	0.11	0.28	0.70
Other cities in Gilan province	0.55	0.44	0.22
Constant	1.93	035	0.00

3.2.17 Attitudes toward punishment of a person under 18 years of age for killing another person without malice aforethought

In most legal systems in which physical punishment is still applicable, a frequent punishment for murder is the death penalty. *Qisas*, which means retaliation in retributive justice, is applicable for murder in *Shari'a* and thus in Iran. In *Qisas*, applying a mediatory mechanism can also change retaliation to *Diye*, with consent of the victim's family. In Islam, murder is categorized as a crime against a person's rights, which is why in all cases of murder, *Qisas* can only be demanded by the victim or their heirs. However, if *Qisas* is not executed for any reason, imprisonment of 3 to 10 years will be replaced by *Qisas* for the purpose of ensuring public order. It should be noted that according to international standards, severe punishments should not be meted out at all to persons under 18 years of age who commit serious crimes. Therefore, in Iran, the prescribed punishment for this group for murder violates international obligations.

In this survey, attitudes were asked toward punishment of a person under 18 years of age for killing another person without malice aforethought. The first choice in this regard stated that punishment in the given fictitious case should be deprivation of liberty and *Diye*, applied according to the progressive view of *Shari'a*. As shown in *Table 27*, the majority of respondents (69%) endorsed this view. In contrast, choice 3 – which catered to the conservative view of *Shari'a* and allows *Qisas* for a child under 18 years of age – was endorsed only by 14% of respondents.

Table 27 Case 2, part 2: Attitudes toward punishment of a person under 18 years for killing another person without malice aforethought

A 16-year-old boy kills another boy during a street fight by knife but without malice aforethought. He is accused of murder. Part 2: Which of the following responses should be applied?	**N**	**%**
Deprivation of liberty and *Diye* (blood money)	152	69
Diye	16	7
Qisas	31	14
Deprivation of liberty for a long time	23	10
Total	222	100

3.2.17.1 Occupation and attitudes toward punishment of a person under 18 years for killing another person without malice aforethought

Figure and Table 27-1 illustrate the attitudes of the various professional target groups toward punishment of a person under 18 years of age for killing someone else without malice aforethought. As can be seen in the table, the vast majority of respondents endorsed choice 1, which states that in such a case, deprivation of liberty and *Diye* should be enforced. With 48%, only the Islamic lawyers supported this international standard to a lesser extent than the other target groups. Another 48% of Islamic lawyers endorsed choice 3, which states that *Qisas* should be applied in such a case. In contrast, none of the faculty members agreed with this punishment.

Figure/Table 27-1 Case 2, part 2: Relationship between occupation and attitudes toward punishment of a person under 18 years for killing another person without malice aforethought

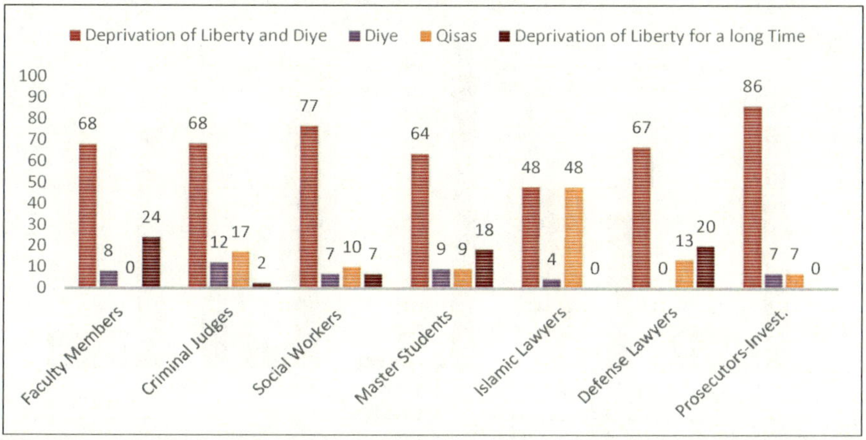

Choices / Occupation	Progressive view of *Shari'a*				Conservative view of *Shari'a*			
	Deprivation of liberty and *Diye*		*Diye*		*Qisas*		Deprivation of liberty for a long time	
	N	%	N	%	N	%	N	%
Faculty members	17	68	2	8	0	0	6	24
Criminal judges	28	68	5	12	7	17	1	2
Social workers	23	77	2	7	3	10	2	7
Master students	28	64	4	9	4	9	8	18
Islamic lawyers	11	48	1	4	11	48	0	0
Defense lawyers	20	67	0	0	4	13	6	20
Prosecutors/investig.	25	86	2	7	2	7	0	0
Total: 222	168				54			

3.2.17.2 Place of residence and attitudes toward punishment of a person under 18 years for killing another person without malice aforethought

Figure and Table 27-2 depict the correlation between punishment of a person under 18 years of age for killing another without malice aforethought and respondents' place of residence. Those who lived in Rasht were significantly more supportive of choices 1 or 2, i.e. the progressive view of *Shari'a*. 67% of respondents who live in Tehran, endorsed choice 1. It is notable that half of the respondents who lived in Qum endorsed choice 3 (*Qisas*), while only 10% of those living in Anzali made the same choice.

Figure/Table 27-2 Case 2, part 2: Relationship between place of residence and attitudes toward punishment of a person under 18 years for killing another person without malice aforethought

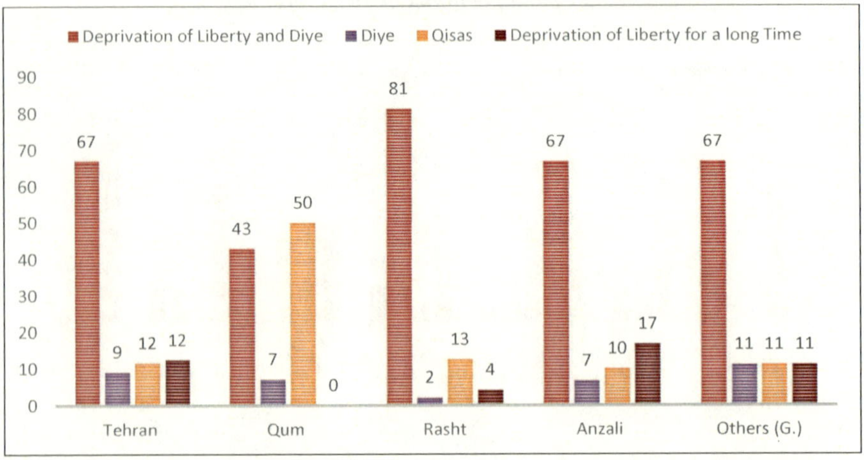

Residence \ Choices	Progressive view of *Shari'a*				Conservative view of *Shari'a*			
	Deprivation of liberty and *Diye*		*Diye*		*Qisas*		Deprivation of liberty/long time	
	N	%	N	%	N	%	N	%
Tehran (Tehran P.)	81	67	11	9	14	12	15	12
Qum (Qum P.)	6	43	1	7	7	50	0	0
Rasht (Gilan P.)	39	81	1	2	6	13	2	4
Anzali (Gilan P.)	20	67	2	7	3	10	5	17
Other Cities (G. P.)	6	67	1	11	1	11	1	11
Total: 222	168				54			

3.2.17.3 Gender and attitudes toward punishment of a person under 18 years for killing another person without malice aforethought

Figure and Table 27-3 show the relationship between gender and attitudes to the given circumstances. No significant differences were found for this variable, although women favored choices in the progressive view of *Shari'a* more than males.

Figure/Table 27-3 Case 2, part 2: Relationship between gender and attitudes toward punishment of a person under 18 years for killing another person without malice aforethought

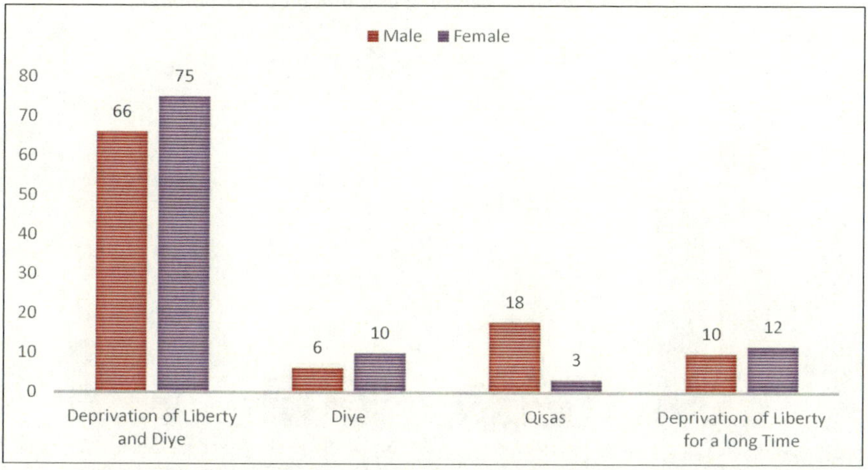

Choices / Gender	**Progressive view of *Shari'a***				**Conservative view of *Shari'a***			
	Deprivation of liberty and *Diye*		***Diye***		***Qisas***		**Deprivation of liberty for a long time**	
Gender	N	%	N	%	N	%	N	%
Male	107	66	10	6	29	18	16	10
Female	45	75	6	10	2	3	7	12
Total: 222	168				54			

3.2.17.4 Marital status and attitudes toward punishment of a person under 18 years for killing another person without malice aforethought

As can be seen in *Figure and Table 27-4*, no significant differences were found for the variable of marital status, although those who were married endorsed choice 1 to a higher degree, while singles had a much higher tendency toward deprivation of liberty for a long time.

Figure/Table 27-4 Case 2, part 2: Relationship between marital status and attitudes toward punishment of a person under 18 years for killing another person without malice aforethought

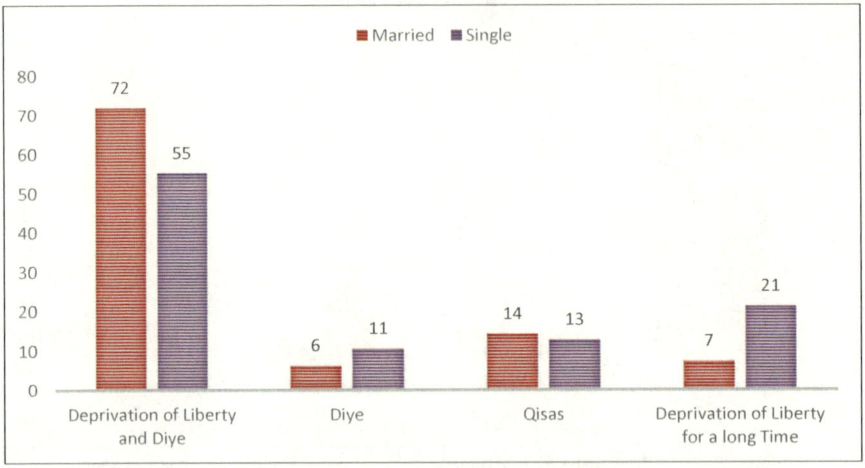

Choices / Marital status	Progressive view of *Shari'a*				Conservative view of *Shari'a*			
	Deprivation of liberty and *Diye*		*Diye*		*Qisas*		Deprivation of liberty for a long time	
	N	%	N	%	N	%	N	%
Married	126	72	11	6	25	14	13	7
Single	26	55	5	11	6	13	10	21
Total: 222	168				54			

3.2.17.5 Age and attitudes toward punishment of a person under 18 years for killing another person without malice aforethought

Figure and Table 27-5 show that most respondents in all target groups endorsed the first choice, which is categorized as the progressive view of *Shari'a*. Nevertheless, 33% of those who were between 46 and 50 years of age endorsed the third choice, making it the age group most supportive of *Qisas*.

Figure/Table 27-5 Case 2, part 2: Relationship between age and attitudes toward punishment of a person under 18 years for killing another person without malice aforethought

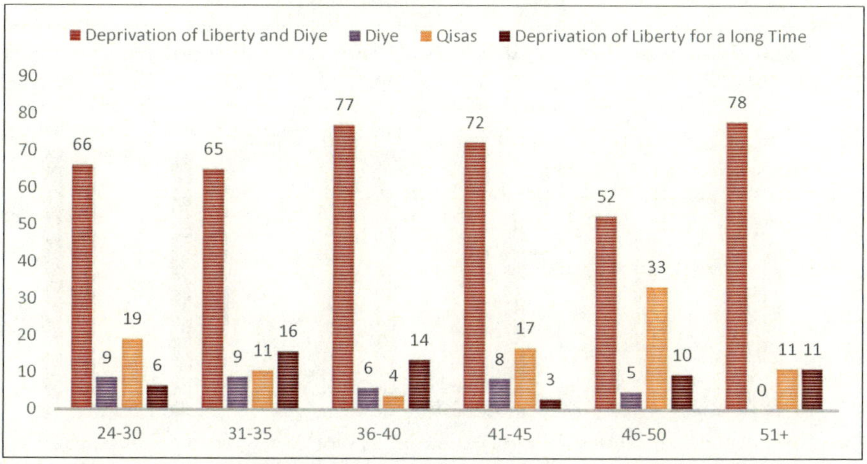

Choices / Age	Progressive view of *Shari'a*				Conservative view of *Shari'a*			
	Deprivation of liberty and *Diye*		*Diye*		*Qisas*		Deprivation of liberty for a long time	
	N	%	N	%	N	%	N	%
24–30	31	66	4	9	9	19	3	6
31–35	37	65	5	9	6	11	9	16
36–40	40	77	3	6	2	4	7	14
41–45	26	72	3	9	6	17	1	3
46–50	11	52	1	5	7	33	2	10
51+	7	78	0	0	1	11	1	11
Total: 222	168				54			

3.2.17.6 Education and attitudes toward punishment of a person under 18 years for killing another person without malice aforethought

Figure and *Table 27-6* show that respondents who were in possession of a diploma, were employed as social workers, or held religious certificates did not endorse the third choice (*Qisas*). What is important to know about this target group in this context is that according to *Shari'a*, they believe in taking forgiveness into consideration – as opposed to *Qisas*, due to lack of intention.

Figure/Table 27-6 Case 2, part 2: Relationship between education and attitudes toward punishment of a person under 18 years for killing another person without malice aforethought

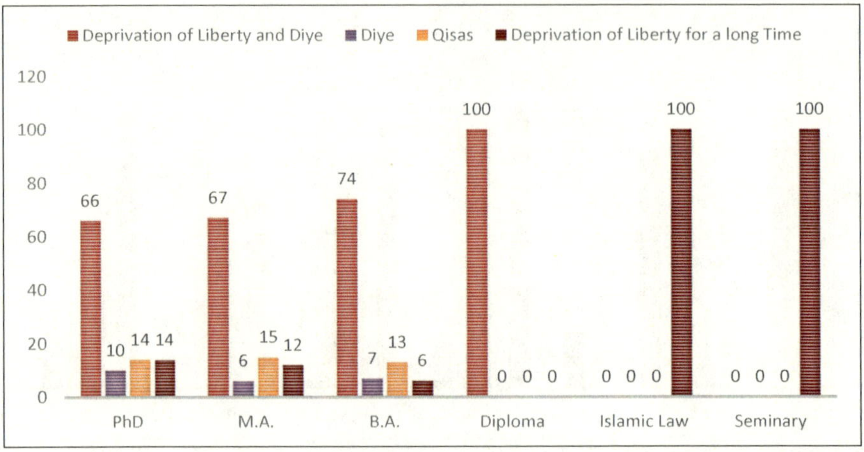

Choices / Education	Progressive view of *Shari'a*				Conservative view of *Shari'a*			
	Deprivation of liberty and *Diye*		*Diye*		*Qisas*		Deprivation of liberty for a long time	
	N	%	N	%	N	%	N	%
PhD	31	66	5	10	7	14	7	14
M.A.	75	67	7	6	17	15	13	12
B.A.	40	74	4	7	7	13	3	6
Diploma	1	100	0	0	0	0	0	0
Islamic law	0	0	0	0	0	0	3	100
Seminary	0	0	0	0	0	0	2	100
Total: 222	168				54			

3.2.17.7 Results of a regression on punishment for killing another person without malice aforethought

Concerning punishment for killing another person, only a weak structure can be found in the survey data. As expected, the Islamic lawyers show a conservative tendency. But interestingly, as *Table F* shows, the criminal trial judges, master students, and prosecutors/investigators seem to be slightly more progressive in this respect than the faculty members. A minor effect of being more conservative can be seen for those holding a M.A. At this point, no effect of age is seen, but a gender effect was identified due to the women being more progressive.

Table F Results of a regression on punishment for killing another person without malice aforethought ($N = 222$; $R^2 = 0.16$)

	Coefficient	Std. err.	p-value
Occupation			
Faculty members	Reference = 0		
Criminal trial judges	0.08	0.11	0.49
Social workers	0.11	0.13	0.42
Master students	-0.07	0.12	0.57
Islamic lawyers	0.40	0.13	0.00
Defense lawyers	0.03	0.11	0.77
Prosecutors/investigators	0.05	0.13	0.70
Education			
PhD	Reference = 0		
M.A.	0.12	0.09	0.17
B.A.	0.06	0.10	0.59
Islamic law	-0.07	0.37	0.85
Islamic studies	-0.32	0.22	0.14
Diploma	-0.53	0.25	0.04
Age			
24–30	Reference = 0		
31–35	-0.13	0.07	0.08
36–40	-0.18	0.08	0.02
41–45	-0.14	0.09	0.10
46–50	0.01	0.10	0.95
51+	-0.17	0.14	0.24
Gender			
Male	Reference = 0		
Female	-0.14	0.06	0.02
Marital status			
Married	Reference = 0		
Single	0.03	0.06	0.65
Place of residence			
Tehran	Reference = 0		
Qum	0.11	0.14	0.46
Rasht	-0.02	0.06	0.72
Anzali	-0.03	0.08	0.71
Other cities in Gilan province	0.00	0.12	0.97
Constant	0.15	0.10	0.13

3.2.18 Attitudes toward informal responses in the juvenile justice system

In terms of diversion for juvenile delinquents, the international rules state that social services and programs should be implemented for the prevention of juvenile delinquency, especially where no special institute or entity has been established for children. Official institutions responsible for social control should take action only as a last resort. Even where there are special institutions for children, the competent authorities (courts, prosecutors, etc.) should not resort to a formal trial in investigations of juvenile delinquents, as far as possible. Even the police and other criminal authorities involved in investigations of statute offenses should be authorized to overlook the official proceedings; according to the criteria established for this purpose in the legal systems as well as at the discretion of the relevant authority, they should refuse to refer the child to such authorities and rather resort to non-judicial methods.

As initially explained, diversion means non-referral of juvenile delinquents to formal institutions such as courts or the prosecutor's office because of the possible negative effects of proceedings (such as labeling and stigmatization) in these entities. In many cases, no intervention of such organizations is considered to be the best reaction. Hence, diversion could be a good reaction at the beginning, particularly when the crime is not too serious and if the family, school, or other informal social control institutions and settlement councils take action appropriately and constructively.

In this survey, the respondents were furthermore asked to consider the following case: a 14-year-old girl and her 17-year-old brother coming from precarious family circumstances steal a car tape recorder. What kind of trial should be provided? As *Table 28* illustrates, 55% of respondents in all target groups believed that informal responses are more suitable in this case, while 45% believed that a formal trial should be provided.

Table 28 Case 3: Attitudes toward informal responses

A 14-year-old girl and her 17-year-old brother are accused of having stolen a car tape recorder. Their parents don't take care of or support them, and therefore, they have financial and emotional problems. In your opinion, what kind of trial and response should be applied in this case?	N	%
Informal/unofficial responses according to the diversion measures (settlements councils)	121	55
Formal/official responses by official and judicial authorities	101	46
Total	222	100

3.2.18.1 Occupation and attitudes toward informal responses in the juvenile justice system

Figure and Table 28-1 illustrate the attitudes of the various professional target groups toward informal responses in the juvenile justice system: 73% of defense lawyers, 68% of faculty members, 64% of master students, 57% of social workers,

44% of Islamic lawyers, 32% of criminal judges, and 14% of prosecutors/investigators were in favor of international standards. The pivotal point in this question is that mainly the criminal judges and prosecutors/investigators who served in the justice system were less supportive of applying diversion measures. It is worth mentioning that the most contrasting views can be found between faculty members and prosecutors/investigators. Furthermore, defense lawyers, faculty members, master students, and social workers were similar in their support of informal responses.

Figure/Table 28-1 Case 3: Relationship between occupation and attitudes toward informal responses

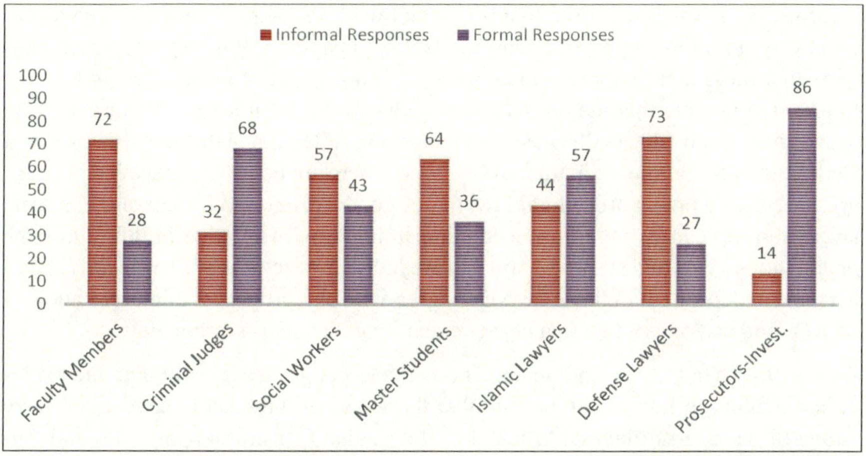

Choices / Occupation	Informal/unofficial responses		Formal/official responses	
	N	%	N	%
Faculty members	18	72	7	28
Criminal judges	13	32	28	68
Social workers	17	57	13	43
Master students	28	64	16	36
Islamic lawyers	10	44	13	57
Defense lawyers	22	73	8	27
Prosecutors/investigators	4	14	25	86
Total: 222	121		101	

3.3 Discussion

After considering all target groups that contributed to this survey, as well as investigated attitudes, the key finding is that the progressive views are generally stronger than the conservative ones. It is notable that among the given choices, most respondents endorsed the ones that were designed according to a modern interpretation of *Shari'a*, international standards, and customary measures. In other words, the attitudes of elites toward the juvenile justice system in Iran can be evaluated as being supportive.

Among the investigated professional groups, the most striking difference could be observed between faculty members and Islamic lawyers in that former are more supportive of the progressive views, while Islamic lawyers are less supportive. However, it should be noted that all other target groups' views were closer to those of the faculty members than to those of the Islamic lawyers. In fact, if we want to consider the respondents, it can be said that these law faculty members, defense lawyers, master students, prosecutors and investigators, social workers, and criminal trial judges respectively are more supportive than the Islamic lawyers in this regard, which suggests that there is a majority which supports changing laws in this direction in the future. It is notable that the conservative views of the Islamic lawyers target group in substantive law (due to obstacles) and its strong reflection in the new Iranian Penal Code could also be shown in this survey, while there are no clashes between customary and *Shari'a* law in procedural law; therefore, the progressive view of the Islamic lawyers group can be seen in procedural criminal law. As shown in this study, the practicing lawyers and staff who work in the judicial system, as well as faculty members and independent defense lawyers, agree with the progressive interpretation of *Shari'a* and customary law to a higher extent than the conservative view.

On the other hand, if we want to categorize the target groups according to their roles in law-making in Iran, it can be said that the views of the Islamic lawyers are very strong because, according to Article 4 of the Iranian Constitution, all rules and regulations must conform to *Shari'a*, which is why their view is very important for this survey. Findings illustrate that this target group can be considered to lean to a more conservative view of *Shari'a* than other target groups. As already explained, there are some new interpretations of *Shari'a* which can be used by law- or policymakers, but this new interpretation, which tends toward an international harmonization and is in line with the needs of juveniles, has not been chosen in some sections of this survey by the Islamic lawyers.

The Islamic lawyers are less supportive than other groups of an elimination of severe punishments, such as the death penalty, or of establishing equality between female and male delinquents with regard to the minimum age of criminal responsibility. They rather endorsed the conservative view of *Shari'a*, while faculty members, defense lawyers, social workers, and master students all were more supportive of these changes. Judges and prosecutors also tended to lean in this direction.

A very important point here is that the Iranian legislator did not share these elites' views when enacting the Iranian Penal Code of 2013, which is now regarded as substantive law. Thus, gaps broke open between legislative and judicial criminal policy concerning juvenile delinquency, and differences can be seen in this survey between the views of academic and Islamic lawyers in terms of substantive law, while the legislature has in fact been influenced by the views of elites from all target groups, which supports professionalism in procedural matters as well as the role of the special judge, the prosecutor, and police officers when enacting the Criminal Procedure Code 2014.

Another important point illustrated in the survey is that the informal trials do not fit into the Iranian legal system even in criminal cases involving juveniles which are not considered as serious cases or crimes. This is due to the lack of standards in these informal trials. It should be noted that among the participants of this survey, there was no consensus for diversion.

This study shows that most respondents share relatively similar views of the surveyed items, in spite of their different age and gender, places of residence, and marital status. Nevertheless, it should be noted that in some respects, women were more supportive than men. For example, they were more respectful of customary law and the new interpretation (progressive view) of *Shari'a*, more supportive of female delinquents in the justice system, and less supportive of severe punishments and the death penalty for (both male and female) juvenile delinquents.

Conclusion

People are considered as the human resources of the society they live in. In addition to their financial potential, social development is closely associated with the quality of human resources. Children and juveniles, the main and most important component for developing these resources, play a key role in this regard. The right to enjoy human resources burdens society with mutual duties; if everybody's rights and duties are applied in line with each other and according to the common societal goals, the true concept of justice will be realized. It is noteworthy that (as well as addressing the health and nutrition of children as part of the process of fostering human resources) taking into account the mental health of children ensures their behavior in accordance with social requirements in the future. Whenever this attention is paid to economic, political, cultural, and social issues, it will be acknowledged that society has undergone some sort of relative and comprehensive progress. Otherwise, it might have to face harmful consequences that are sometimes irreversible or at best controllable only at a high price. A society that does not value and respect its human resources will suffer the losses and damages resulting from this negligence. The emergence and spread of phenomena such as delinquency, maladjustment, and deviance among juveniles is one of the most negative consequences a society may face if it neglects its youth.

Delinquency is a psychological, social, and biological phenomenon that needs to be addressed by society through adopting comprehensive measures and strategies in the framework of social and situational (i.e. non-criminal) prevention, especially for children and juveniles who, in addition to being granted differentiated preventive measures, should be treated with more reserve than adults in these issues because of their special physical and mental condition. Particular attention to children and juveniles should be extended when they get into the criminal realm and where their behaviors (in the form of acts or omissions) violate the provisions of the Penal Code. In this case, they are not just viewed as any ordinary children and juveniles, but they are treated as a delinquents or deviants. Therefore, society should take measures to help such children and juveniles to be released from the vicious cycle they have got into and to start a more purposeful life that is more beneficial to society. The presence of large numbers of delinquents in a society corroborates the negligence of that society with respect to them. Hence, it is very important that society, and especially the government, pay special and additional attention to this issue in order to prevent delinquency. Such an approach will ensure the long-term prevention of an increase in delinquency among children and juveniles. In this regard, the identification of delinquents coincides with their entry into the criminal realm and into contact with the judicial system. According to research in this field, they should be treated differently from adults through a special section in the criminal justice system called the 'juvenile justice system'. The most important step in establishing such a juvenile justice system is to enact a special code that contains both substantive and procedural rules for juvenile delinquents, in line with the best interests of the child and in accordance with the requirements of due process.

At the international level, the United Nations have developed rules and regulations regarding juvenile delinquents and children in conflict with the law. In fact, the since 1959, the UN have issued many binding and non-binding documents concerning the standards and norms on crime prevention and criminal justice – and to juvenile justice in particular. Accordingly, juvenile justice shall be conceived as an integral part of the national development process of each country, within a comprehensive framework of social justice for all juveniles, and, at the same time, by contributing to the protection of the young as well as the maintenance of a peaceful order in society. States parties shall seek to promote the establishment of laws, procedures, authorities, and institutions specifically applicable to children alleged as, accused of, or recognized as having infringed the penal law. In terms of the promotion and application of juvenile criminal law, an attempt should be made to develop a set of rules and regulations in each national jurisdiction unit in order to address the crimes of juvenile offenders. The United Nations have differentiated the criminal policy toward children and juveniles and have considered a desirable criminal pattern for the treatment of juvenile delinquents; it goes without saying that this can play an essential role in harmonizing policies and spreading juvenile justice throughout the world. The international community has to play an important role in setting out widely accepted guidelines, and once general principles are set out in international human rights agreements, the primary locus of human rights exists within nations; therefore, the national governments are the primary agents for the realization of children's rights.

It is important to ensure that human rights are clearly articulated in national law. When states undertake obligations to honor particular human rights by becoming parties to international human rights agreements, they are expected to elaborate their understanding of those obligations by spelling them out in their own national laws. Therefore, national governments and legislatures should take responsibility for the realization of child rights and juvenile justice standards by providing special provisions, procedures, and institutions. Enacting special laws as the first and most important step clarifies the individual's rights, the government's obligations that correspond to those rights, the government's response mechanisms (including both identification of the agencies responsible for fulfilling the obligations and the procedures through which they are to operate), and also the accountability mechanisms ensuring that these obligations are honored.

After ratifying the UN Convention on the Rights of the Child as a binding instrument, State Parties take on obligations under international law to implement its contents. In this context, implementation is the process whereby State Parties take action to ensure the realization of rights in the Convention for all children within a jurisdiction. The core obligation to be implemented is set out in Article 4, which requires State Parties to undertake all appropriate legislative, administrative, and other measures for the implementation of the rights recognised in the present Convention on the Rights of the Child. With regard to economic, social, and cultural rights, States parties shall undertake such measures to the maximum extent of their available resources. In general, international human rights treaties do not specify how States parties are to give effect to their obligations at the domestic level, but they do require them to take all appropriate measures. As such, it is up to the states to determine how best to implement their international treaty obligations, subject to the satisfaction of those obligations in practice. According to the United Nations Committee on Economic, Social and Cultural Rights, each State Party must decide for itself which means are the most appropriate under the circumstances with respect to each of the rights. Despite ratification of the Convention on the Rights of the Child, not all State parties have adopted the standards for establishing juvenile justice systems which are specialized, comprehensive, and separate from their adult criminal justice systems and should lay the primary aims on rehabilitation and social reintegration. This is due to the power of state sovereignty.

Criminal law has been closely associated with sovereignty and can be seen from the state's virtually unlimited discretion to discipline members of society. In an autocracy – contrary to a democracy, because the state and society, as basic entities, are fortified –, they form the foundation of the criminal law and its culture within such a regime; human rights are delegated to a more or less insignificant status, and sociological as well as criminological findings are often also ignored. Iran is one of the countries that have not paid full attention to the conception of child rights or to sociological and criminological findings in the formulation of its juvenile criminal law. However, over the past few years, there have been attempts to re-establish a juvenile justice system based on a more differentiated criminal policy. After examination of juvenile criminal law in Iran and the related clashes between Islamic laws, judicial practices, international obligations, and social rules, it can be seen that the Iranian legal system has not yet found a definite integrated strategy for the treatment of juvenile delinquency. As has been explained, collisions occur more often in substantive criminal law than in procedural law. The minimum age of criminal responsibility, the definition of a child or juvenile delinquent, as well as penal sanctions in the juvenile justice system remain controversial issues. In general, it can be said that due to several conflicts of interest between *Shari'a* and other legal sources, certain international standards have been ignored by the Iranian legislature in substantive criminal law.

In contrast, in Germany, the legislature is in line with international instruments and standards, as there is an integrated system for dealing with juvenile delinquents in the German justice system. The German Youth Court Act, as a specific act, comprises substantive and procedural criminal law applicable to juvenile delinquents. It can be said that a separate system with special institutions and procedures has been established for juvenile delinquents here, and the juveniles who are considered as persons in need are handled by the Youth Welfare Act. This can be taken as a good example for reforming juvenile criminal law in Iran, too.

References

Abachi, M. (ed.) (2017): Child's Rights. Tehran.

Abachi, M. (2015): The Modern System of Juvenile Justice in Iran. Zeitschrift für die gesamte Strafrechtswissenschaft 127/3, pp. 822–841.

Abachi, M. (2009): Juvenile Justice System in United Nations Instruments. 2nd ed. Tehran.

Abachi, M. (2008): Crime Prevention and National Program. Crime Prevention Journal 9, pp. 33–72.

Abachi, M. (2005): Prevention of Juvenile Delinquency and Juvenile Victimization. The Judiciary Law Journal 46, pp. 49–86.

Abbasi, B. (2012): Iranian Constitution. Tehran.

Abbasi, B. (2011): Exploring Constitutional Law Sources. The Law Journal 3, pp. 247–260.

Abbott, K. & Snidal, D. (2000): Hard and Soft Law in International Governance. International Organization Journal 54/3, pp. 421–456.

Akhondi, M. (1990): Criminal Procedural Law. Tehran.

Albrecht, H.-J. (2013), Sentencing in Germany: Explaining Long-Term Stability in the Structure of Criminal Sanctions And Sentencing. Law and Contemporary Problems 76, pp. 211–236.

Albrecht, H.-J. (2004): Youth Justice in Germany. Crime and Justice 31, pp. 443–493.

Albrecht, H.-J. (2002a): Juvenile Crime and Juvenile Law in the Federal Republic of Germany, in: J.A. Winterdyk (ed.), Juvenile Justice Systems: International Perspectives. 2nd ed. Toronto, pp. 171–205.

Albrecht, H.-J. (2002b): Post-adjudication Dispositions in Comparative Perspective, in: M. Tonry & R.S. Frase (eds.), Sentencing and Sanctions in Western Countries. Oxford, pp. 293–330.

Amin, S.H. (2007): History of the Iranian law. Tehran.

Aramesh, R. (2001): The Necessities of Cognition of the History of Law. Journal of Lawyer 6, pp. 70–72.

Ardebili, M.A. (2014): General Criminal Law II. Tehran.

Ardebili, M.A. (1995): Crime and Punishment. Tehran.

Asadi, L.S. (2003): Children's Criminal Proceedings in Iran's Law. The Judiciary Law Journal 45, pp. 103–123.

Ashori, M. (2002): Criminal Justice. Tehran.

Ashori, M. (1998): Criminal Procedural Law. Tehran.

Bahadori Jahromi, A. & Hosseini, S.L. (2012): Fair Trial in Islamic Thoughts and National Law. Islamic Human Rights Journal 7, pp. 9–32.

Bates, K.A. & Swan, R.S. (2021): Juvenile Delinquency in a Diverse Society, 3rd ed. San Marcos.

Bernard, T.J. & Kurlychek, M.C. (2010): The Cycle of Juvenile Justice. Oxford.

Biesanz, J. (1941): Nazi Influence on German Youth Hostels. Social Forces 19/4, pp. 554–559.

Bigdeli, A. (1996): International Law. Tehran.

Birnbaum, L., Cetinkaya, H. & Harper, E. (2014): Legal Status of the Child: Iran's International Human Rights Obligations. Essex.

Boes, M. (2013): Crime and Punishment in Early Modern Germany. Surry.

Boetticher, A. & Feest, J. (2004): German Criminal and Prison Policy, in: P. Tak & M. Jendley (eds.): Prison Policy and Prisoners' Rights, pp. 361–381.

Bohlander, M. (2009): Principles of German Criminal Law. Oxford.

Boyne, S.M. (2014): The German Prosecution Service. Heidelberg/New York/Dordrecht.

Bueren, V.G. (1998): The International Law on the Rights of the Child. Leiden.

Buse, D.K. (2005): The Regions of Germany: A Reference Guide to History and Culture. Westport.

Cappelaere, G., Grandjean, A. & Naqvi, Y. (2005): Children Deprived of Liberty: Rights and Realities. Leiden.

Carvalho, E. (2008): Measuring Children's Rights: An Alternative Approach. International Journal of Children's Rights 16, pp. 545–563.

Cashmore, J. & Parkinson, P. (2007): What Responsibility Do Courts Have to Hear Children's Voices? International Journal of Children's Rights 15, pp. 43–60.

Chamelin, N.E. & Thomas, A.C. (2012): Essentials of Criminal Law. New Jersey.

Child Rights Information Network (2009): Guide to non-discrimination and the CRC.

Cipriani, D. (2009): Children's Rights at the Minimum Age of Criminal Responsibility. Farnham.

Cohen, C.P. & Naimark, H. (1991): United Nations Convention on the Rights of the Child: Individual Rights Concepts and their Significance for Social Scientists. American Psychologist 46/1, pp. 60–65.

Cox, St.M., Allen, J.M. & Hanser, R.D. (2017): Juvenile Justice: A Guide to Theory, Policy, and Practice, 9th Edition. Los Angeles.

Coy, J.P. (2011): A Brief History of Germany. New York.

Crisan, L. (2010): The Principles of Legality "nullum crimen, nulla poena sine lege" and their Role. Effectius Newslettern 5, pp. 1–3.

Crofts, T. (2008): The Criminal Responsibility of Children, in: G. Monahan & L. Young (eds.), Children and the Law in Australia. Sydney, pp. 167–186.

Crofts, T. (2004): The Rise of the Principle of Education in the German Juvenile Justice System. International Journal of Children's Rights 12/4, pp. 401–417.

Crofts, T. (2002): The Criminal Responsibility of Children and Young Persons: A Comparison of English and German Law. Aldershot.

Danesh, T.Z. (1995): Children and Juvenile Delinquents. Tehran.

Dausab, Y. (2009): The Best Interest of the Child, in: O.C. Ruppel (ed.), Children's Rights in Namibia. Windhoek, pp. 145–157.

Davis, W.S. (1912): Readings in Ancient History. Boston.

Detrick, S. (1995): Convention on the Rights of the Child: 170 States Parties. The International Journal of Children's Rights 3/3–4, pp. 446–447.

Diakonoff, I.M. (1968): the Cambridge History of Law. Cambridge.

Doob, A. & Torny, M. (2004): Varieties of Youth Justice. Crime and Justice Journal 31, pp. 1–20.

Dubber, M. (2005): Comparative Criminal Law. Buffalo.

Dünkel, F. (2016): Youth Justice in Germany. Oxford.

Dünkel, F. (2014): Juvenile Justice Systems in Europe: Reform Developments between Justice, Welfare and New Punitiveness. Kriminologijos Studijos 1, pp. 31–76.

Dünkel, F. (2011): Studies on Prisons, Juvenile Justice and Criminology. Greifswald.

Dünkel, F. (2005): Juvenile Justice in Germany. Greifswald.

Dussich, J. (1989): Juvenile Delinquency: Definitions, Character, and Theory, in: Wegener, H., Lösel, F., Haisch, J. (eds.), Criminal Behavior and the Justice System. Berlin/Heidelberg, pp. 70–80.

Ebadi, Sh. & Zamirian, M. (2002): Tradition and Modernity. Tehran.

Entessar, N. (1988): Criminal Law and the Legal System in Revolutionary Iran. Boston College Third World Law Journal 8, pp. 91–102.

Feeny, F. (1998): German and American Prosecutions: An Approach to Statistical Comparison. California.

Fisher, H. (2013): The German Legal System and Legal Language, 5th ed. Dublin.

Flowers, F. (2016): A Short History of Human Rights. Minnesota.

Foster, N. & Sule, S. (2010): German Legal System and Laws. Oxford.

Fox, R.L. (1994): Alexander the Great. London.

Frase, R.S. (2005): Punishment Purpose. Stanford Law Review 58/1, pp. 67–83.

Fulbrook, M. (2015): A History of Germany 1918–2014: The Divided Nation, 4th ed. Madlen.

Gharavi, M. (1980): A Research on Juvenile Delinquency in Iran. Tehran.

Ghassemi, Gh. (2013): Criminal Policy in Iran Following the Revolution of 1979: A Comparative Analysis of Criminal Punishment and Sentencing in Iran and Germany. Berlin.

Ghassemi, Gh. (2009): Criminal Punishment in Islamic Societies: Empirical Study of Attitudes to Criminal Sentencing in Iran. European Journal on Criminal Policy and Research 15, pp. 158–180.

Ghassemi, M. (2015): Comparative Criminal Law: An Economic Perspective. Asian Journal of Law and Economics 7/2, pp. 249–278.

Goldoost, R. & Seyed Esfehani, H. (2003): Pre-trial Stage in Juvenile Proceedings. Legal Research Journal 39, pp. 259–398.

Grahn Farley, M. (2002): A Child Perspective on the Juvenile Justice System. The Journal of Gender, Race & Justice 6, pp. 297–335.

Guzman, A. (2010): Internaitonal Soft Law. Berkeley.

Habibzade, T. & Attar, M.S. (2012): Conceptual Evaluation of Soft Law in the International Law Sources System. Comparative Law Journal 2, pp. 1–28.

Hamilton, C. (2011): Guidelines for Legislative Reform on Juvenile Justice. New York.

Hanjani, F. (2006): International Law. Tehran.

Hashemi, S.H. (2005): Girls, Age of Intellectual Development and Criminal Responsibility. Journal of the Cultural and Social Council for Women 23, pp. 247–300.

Heinrich, B. & Kherkheulidze, I. (2014): Perfection of the Trends in Juvenile Justice Systems on the Basis of Comparison of German, Georgian and the U.S. Juvenile Law. Zeitschrift für Internationale Strafrechtsdogmatik 12, pp. 610–621.

Herz, D. (2012): Addressing the Needs of Multi-System Youth: Strengthening the Connection between Child Welfare and Juvenile Justice. Georgetown.

Hodgson, D. (1994): The Child's Right to Life, Survival and Development. The International Journal of Children's Rights 2, pp. 369–394.

Holman, B. & Ziedenberg, J. (2013): The Dangers of Detention: The Impact of Incarcerating Youth in Detention and Other Secure Facilities. Washington.

Holthusen, B. (2004): Ultima ratio? Deprivation of Liberty Measures for Children and Young People – Findings and Need for Research, in: Centre for the Prevention of Youth Crime (ed.), Prevention of Youth Crime in Germany: Educational Strategies. München, pp. 133–157.

Holthusen, B. & Schäfer, H. (2004): Strategies and Organization of Youth Crime Prevention in Germany: The Importance of the Educational Approach, in: Centre for the Prevention of Youth Crime (ed.), Prevention of Youth Crime in Germany: Educational Strategies. München, pp. 11–23.

Interagency Panel on Juvenile Justice (IPJJ) (n.d.): Juvenile Justice; www.ipjj.org/en/juvenile-justice/overview/ [20/05/2021].

Islamic Council (1981): Universal Islamic Declaration of Human Rights. The Islamic Council in Paris on 19 September 1981.

Jackson, W. (1899): Zoroaster the Prophet of Ancient Iran. London.

Jamshidi, A. (2003): Criminal Policy Discourse in the Judicial Bill of Establishment of Juvenile Courts. The Judiciary Law Journal 45, pp. 61–102.

János, J. (2007): Criminal Justice in Sasanian Persia. Iranica Antiqua 42, pp. 347–386.

Jareborg, N. (2005): Criminalization. Ohio State Journal of Criminal Law 2, pp. 521–534.

Jehle, J. M. (2015): Criminal Justice in Germany. Berlin.

Junger Tas, J. & Dünkel, F. (eds.) (2009): Reforming Juvenile Justice. New York.

Kaiser, G. (1994): The Juvenile Justice System: The Case of Germany. Legal Studies Forum 3, pp. 319–348.

Karami, M.B. (2001): Introduction to the Criminal Justice System in Ancient Persia. Tehran.

Katebi, H. (1967): Advocacy in the Old Time. People Rights Journal 4, pp. 57–58.

Katoziyan, M. (1393/2014): Philosophy of Law. Tehran.

Kent, G. (1987): Realizing International Children's Rights through the Implementation of National Law. The International Journal of Children's Rights 5/4, pp. 439–456.

Khajeh Noori, Y. (2017): The Globalization of Juvenile Penal Law, Tehran.

Khaleghi, A. (2009): Criminal Procedural Law. Tehran.

Khazani, M. (1998): Criminal Process, Tehran.

Khazani, M. (1993): Presumption of Innocence. Law Research Journal 13/14, pp. 125–144.

Krehl, Ch. (2003): Reforms of the German Criminal Code – Stock-taking and Perspectives – also from a Constitutional Point of View. German Law Journal 4/5, pp. 421–431.

Laubenthal, K., Baier, H. & Nestler, N. (2015): Jugendstrafrecht. Berlin/Heidelberg.

League of Nations (1924): Geneva Declaration of the Rights of the Child, 1924.

Leblanc, L.J. (1996): Reservations to the Convention on the Rights of the Child: A Macroscopic View of State Practice. The International Journal of Children's Rights 4/3, pp. 357–381.

Ledogar, R.J. (1993): Practice and Implementation. Implementing the Convention on the Rights of the Child through National Programs of Action for Children. The International Journal of Children's Rights 1, pp. 377–391.

Loeber, R., Farrington, D.P. & Petechuk, D. (2003): Child Delinquency: Early Intervention and Prevention. U.S. Department of Justice.

Lotfi, A. (2005): Principles of Islamic Jurisprudence. Tehran.

Lundy, L., Kilkelly, U. & Byrne, B. (2013): Incorporation of the United Nations Convention on the Rights of the Child in Law: A Comparative Review. The International Journal of Children's Rights 21/3, pp. 442–463.

Maher, G. (2005): Age and Criminal Responsibility. Ohio State Journal of Criminal Law 2/2, pp. 493–512.

Malekian, F. & Nordlöf, K. (2012): The Sovereignty of Children in Law. Cambridge/Newcastle.

Malekian, F. & Nordlöf, K. (2013): International Legal Status of Children, in: Jay S. Albanese (ed.), Encyclopedia of Criminology and Criminal Justice VI. Hoboken.

Markham C.R. (1874): A General Sketch of the History of Persia. Whitefish/Montana.

Matthews, S., Schiraldi, V. & Chester, L. (2018): Youth Justice in Europe: Experience of Germany, the Netherlands, and Croatia in Providing Developmentally Appropriate Responses to Emerging Adults in the Criminal Justice System. Justice Evaluation Journal 1/1, pp. 59–81.

Maxim, P. & Whitehead, P. (2008): Youth in Conflict With The Law, UK.

McCrudden, C. (2008): Human Dignity and Judicial Interpretation of Human Rights. The European Journal of International Law 19, pp. 655–724.

McPherson, T. (1967): Punishment: Definition and Justification. Analysis 28/1, pp. 21–27.

Mehra, N. (2006): Iran's Criminal Laws and Regulations concerning Children and Juvenile Delinquents: Present and Future. Journal of Religious Research 62/63, pp. 41–60.

Meyers, J.A. (1999): The Determinant of Juvenile Justice Policy in France and Germany. Richmond.

Mir Hosseini, Z. (2010): Sharia and National Law in Iran, in: J.M. Otto, (ed.): Sharia Incorporated- A Comparative Overview of the Legal Systems of Twelve Muslim Countries in past and Present. Leiden.

Moazzenzadegan, H.A. (2010): Special Criminal Proceedings of Juveniles in Iran's Laws. Jurisprudence and Family Law Journal 32, pp. 139–161.

Mohammadi, M.H. & Ghaeni, Z. (2004): The History of Children's Literature in Iran: Oral Literature and the Ancient Era, 4th ed. Tehran.

Morag, T. (2014): The Principle of UNCRC and their Influence on Israeli Law. Michigan State International Law Review 22/2, pp. 531–556.

Morgan, D. (1994): Persian Perceptions of Mongols and Europeans, in: S.C. Schwartz (ed.), Implicit Understandings: Observing, Reporting, and Reflecting on the Encounters between Europeans and Other Peoples in the Early Modern Era. Cambridge.

Mosavi Bojnordi, M. (2017): Investigating of the Concept and Criteria of Childhood in Islamic Thought and Iranian Laws, in: M. Abachi (ed.), Child's Rights. Tehran, pp. 17–54.

Nadjafi Abrandabadi, A.H. (2009): Preface, in: M. Abachi, Juvenile Justice System in United Nations Instruments, 2nd ed. Tehran.

Nadjafi Abrandabadi, A.H. & Hashem Beigi, H. (2018): Encyclopedia of Criminology. Tehran.

Nadjafi Abrandabadi, A.H. & Jamshidi, A. (2009): Criminal Justice in the Iranian Legal System and UN Instruments. Tehran.

Nobahar, R. (2012): Age and Criminal Responsibility: Re-reading the Famous Theory of Imami Jurists. Criminal Law Journal 3/2, pp. 161–185.

Noorbaha, R. (2005): General Criminal Law. Tehran.

Oshidari, J. (1992): Mazdista Encyclopedia. Tehran.

Overland, M.E. & Newhouse, J. (1968): Juvenile Criminal Law in the Federal Republic of Germany and in England. California Western Law Review 4, pp. 35–64.

Parvaresh, H. (1957): Administrative Law before Islam. Tehran.

Pasha Saleh, A. (1969): Legal History of Iran. Tehran.

Persson, M. (2017): Caught in the Middle? Young Offenders in the Swedish and German Criminal Justice Systems. Lund.

Pichler, S. & Römer, D. (2011): Juvenile Law and Recidivism in Germany: New Evidence from the Old Continent. Heidelberg.

Pirnia, H. (1983): History of Ancient Iran. Tehran.

Pirnia, H. & Ashtiyani, E. (2008): History of Iran. Tehran.

Pourshariati, P. (2008): Decline and Fall of the Sasanian Empire: The Sasanian-Parthian Confederacy and the Arab Conquest of Iran. London/New York.

Pruin, I. (2012): Recalling Conditionally Released Prisoners in Germany. European Journal of Probation 4/1, pp. 63–72.

Rahami, M. (2002): Intellectual Development. Law and Political Science Journal 58, pp. 187–198.

Rahmdel, M. (2015): New Developments in the Iranian Code of Criminal Procedure. Zeitschrift für die gesamte Strafrechtswissenschaft 127/3, pp. 788–821.

Rasekh, M. (2002): History of Beynonahrain. Tehran.

Robbers, G. (2012): An Introduction to Criminal Law. Baden-Baden.

Saneie, P. (2000): General Criminal Law. Tehran.

Schäfer, H. (2004a): Combating Criminal Behaviour among Children under 14 Years of Age: The Responsibility of the Educational System, in: Centre for the Prevention of Youth Crime (ed.), Prevention of Youth Crime in Germany: Educational Strategies. München, pp. 45–61.

Schäfer, H. (2004b): How to remand juveniles in young offender institutions is avoided. Everything is better than prison – educational alternatives in Germany, in: Centre for the Prevention of Youth Crime (ed.), Prevention of Youth Crime in Germany: Educational Strategies. München, pp. 113–132.

Schaffstein, F., Beulke, W. & Swoboda, S. (2014): Jugendstrafrécht. Stuttgart.

Shaffer, G.C. & Pollack, M.A. (2010): Hard vs. Soft Law: Alternatives, Complements, and Antagonists in International Governance. Minnosota Law Review 94, pp. 706–799.

Shambayati, H. (2010): Juvenile Delinquency. Tehran.

Shamloo, B. (2011): Juvenile Justice. Tehran.

Shaygan, M. (2010): The Model of Criminal Justice concerning Children and Juveniles in Conflict with the Law, in: The National Conference Proceeding on Criminal Justice of Children and Juveniles in the Iranian Legal System. Tehran, pp. 350–321.

Siegel, L.G. & Senna, J.J. (1997): Juvenile Delinquency: Theory, Practice, and Law. Eagan/Minnesota.

Slobogin, C., Fondacaro M. R. & Cross, T. (2006): Reconceptualizing Due Process in Juvenile Justice: Contributions from Law and Social Science. Hastings Law Journal 57, pp. 955–990.

Smith, D. (2005): The Effectiveness of the Juvenile Justice System. Edinburgh.

Sokhansanj, A. (2001): Sources and Periods of Islamic Jurisprudence. Religion Moon Book 49/50, pp. 21–29.

Soler, M., Shoenberg, D. & Schindler, M. (2009): Symposium Issue Juvenile Justice: Lessons for a New Era. Georgetown Journal on Poverty Law & Policy XVI, pp. 483–541.

Spencer, W. (1975): From the Arab Invasion to the Saljuqs. The Cambridge History of Iran 4, pp. 1–56.

Steans, W. (1912): Readings in Ancient History: Greece and the East. Boston.

Steinerte, E. & Wallce, R.M.M. (2009): United Nations Protection of Human Rights Section: A Mechanism for Human Rights Protection by United Nations Bodies. London.

Subramanian, R. & Shames, A. (2013): Sentencing and Prison Practices in Germany and the Netherlands: Implications for the United States. Brooklyn.

Tobin, J.W. (2002): Time to Remove the Shackles: The Legality of Restraints on Children Deprived of their Liberty under International Law. The International Journal of Children's Rights 9, pp. 213–239.

UN Committee on the Rights of the Child (2019): Children's Rights in the Child Justice System. General Comments No. 24, 2019.

UN Committee on the Rights of the Child (2016): The Implementation of the Rights of the Child during Adolescence. General Comments No. 20, 2016.

UN Committee on the Rights of the Child (2013a): Protecting Children's Rights in Criminal Justice Systems. Penal Reform International, 2013.

UN Committee on the Rights of the Child (2013b): The Right of the Child to Have his or her Best Interests Taken as a Primary Consideration. General Comments No. 14, 2013.

UN Committee on the Rights of the Child (2011): The Right of the Child to Freedom from all Forms of Violence. General Comments No. 13, 2011.

UN Committee on the Rights of the Child (2009): The Right of the Child to Be Heard. General Comments No. 12, 2009.

UN Committee on the Rights of the Child (2007): Children's Rights in Juvenile Justice. General Comments No. 10, 2007.

UN Committee on the Rights of the Child (2006): The Right of the Child to Protection from Corporal Punishment and other Cruel or Degrading Forms of Punishment. General Comments No. 8, 2006.

UN Committee on the Rights of the Child (2003): General Measures of Implementation for the Convention on the Rights of the Child. General Comments No. 5, 2003.

UN Committee on the Rights of the Child (2002): The Role of Independent Human Rights Institutions. General Comments No. 2, 2002.

UN Economic and Social Council (2005): Guidelines on Justice Matters involving Child Victims and Witnesses of Crime. UN Economic and Social Council resolution 2005/20 of 22 July 2005

UN Economic and Social Council (1957): Standard Minimum Rules for the Treatment of Prisoners. UN Economic and Social Council resolution 663 C (XXIV) of 31 July 1957.

UN General Assembly (2000a): Optional Protocol to the Convention on the Rights of the Child on the Involvement of Children in Armed Conflicts. UN General Assembly resolution A/RES/54/263 of 25 May 2000.

UN General Assembly (2000b): Optional Protocol to the Convention on the Rights of the Child on the Sale of Children, Child Prostitution and Child Pornography. UN General Assembly resolution A/RES/54/263 of 25 May 2000.

UN General Assembly (1990a): United Nations Guidelines for the Prevention of Juvenile Delinquency. UN General Assembly resolution 45/112 of 14 December 1990.

UN General Assembly (1990b): United Nations Rules for the Protection of Juveniles Deprived of their Liberty. UN General Assembly resolution 45/113 of 14 December 1990.

UN General Assembly (1990c): United Nations Standard Minimum Rules for Noncustodial Measures. UN General Assembly resolution 45/110 of 14 December 1990.

UN General Assembly (1989): Convention on the Rights of the Child. UN General Assembly resolution 44/25 of 20 November 1989.

UN General Assembly (1985): United Nations Standard Minimum Rules for the Administration of Juvenile Justice. UN General Assembly resolution A/RES/40/33 of 29 November 1985.

UN General Assembly (1984): Convention against Torture and Other Cruel, Inhuman or Degrading Treatment or Punishment. UN General Assembly resolution 39/46 of 10 December 1984.

UN General Assembly (1979a): Convention on the Elimination of All Forms of Discrimination against Women. UN General Assembly resolution 34/180 of 18 December 1979.

UN General Assembly (1979b): United Nations Code of Conduct for Law Enforcement Officials. UN General Assembly resolution 34/169 of 17 December 1979.

UN General Assembly (1966a): International Covenant on Civil and Political Rights. UN General Assembly resolution 2200A (XXI) of 16 December 1966.

UN General Assembly (1966b): International Covenant on Economic, Social and Cultural Rights. UN General Assembly resolution 2200A (XXI) of 16 December 1966.

UN General Assembly (1959): Declaration of the Rights of the Child. UN General Assembly resolution 1386 (XIV), A/RES/14/1386 of 20 November 1959.

UN General Assembly (1948): Universal Declaration of Human Rights. UN General Assembly resolution 217 A (III) of 10 December 1948.

UN Office of the High Commissioner for Human Rights and International Bar Association (2003): A Manual on Human Rights for Judges, Prosecutors and Lawyers No. 9.

UNOCD (2010): Criteria for the Design and Evaluation of Juvenile Justice.

Vahedi Navaei, A. (2001): The Iranian Criminal Justice System in the Sassanid Era. Tehran.

Verhellen, E. (ed.) (1996): Monitoring Children's Rights. Leiden.

Viehmann, H. (2009): Strategies of Violence Prevention within the German Framework of Juvenile Criminal Law. Cologne.

Vogel, J. (2009): How to Determine Individual Criminal Responsibility. Stuttgart, pp. 151–162.

Vormbaum, T. (2014): A Modern History of German Criminal Law. Berlin/Heidelberg.

Wall, J. (2008): Human Rights in Light of Childhood. International Journal of Children's Rights 16, pp. 523–543.

Wiese, K.K. (2009): Juvenile Justice: A Comparison between Laws of Germany and the Netherlands. Canterbury.

Woll, L. (2000): Reporting to the UN Committee on the Rights of the Child: A Catalyst for Domestic Debate and Policy Change? The International Journal of Children's Rights 8, pp. 71–81.

Yaghmaeei, E. (1993): Judiciary in Iran. Tehran.

Yasami, R. (1982): Iran in the Sassanid Era. Tehran.

Yazdgerdi, A. (1975): The Administrative and Judicial Structure in the Sassanid Era. Tehran.

Zarrinkub, A.H. (1975): The Arab Conquest of Iran and its Aftermath, in: R.N. Frye (ed.), The Cambridge History of Iran 4, The Period from the Arab Invasion to the Saljuqs. Cambridge, pp. 1–56.

Zermaten, J. (2010): The Best Interests of the Child: Literal Analysis, Function and Implementation. The International Journal of Children's Rights 18/4, pp. 483–499.

Max-Planck-Institut für ausländisches und internationales Strafrecht

Schriftenreihe des Max-Planck-Instituts für ausländisches und internationales Strafrecht

Die zentralen Veröffentlichungen des Max-Planck-Instituts für ausländisches und internationales Strafrecht werden in Zusammenarbeit mit dem Verlag Duncker & Humblot in den folgenden fünf Unterreihen der „Schriftenreihe des Max-Planck-Instituts für ausländisches und internationales Strafrecht" vertrieben:

- „Strafrechtliche Forschungsberichte",
- „Kriminologische Forschungsberichte",
- „Interdisziplinäre Forschungen aus Strafrecht und Kriminologie",
- „Publications of the Max Planck Partner Group for Balkan Criminology" sowie
- „Sammlung ausländischer Strafgesetzbücher in deutscher Übersetzung".

Diese Publikationen können direkt über das Max-Planck-Institut unter www.csl.mpg.de oder über den Verlag Duncker & Humblot unter www.duncker-humblot.de erworben werden.

Darüber hinaus erscheinen im Hausverlag des Max-Planck-Instituts in der Unterreihe „research in brief" zusammenfassende Kurzbeschreibungen von Forschungsergebnissen und in der Unterreihe „Arbeitsberichte" Veröffentlichungen vorläufiger Forschungsergebnisse. Diese Veröffentlichungen können über das Max-Planck-Institut bezogen werden.

Detaillierte Informationen zu den einzelnen Publikationen sind unter www.csl.mpg.de abrufbar.

Research Series of the Max Planck Institute for Foreign and International Criminal Law

The main research activities of the Max Planck Institute for Foreign and International Criminal Law are published in the following five subseries of the "Schriftenreihe des Max-Planck-Instituts für ausländisches und internationales Strafrecht/Research Series of the Max Planck Institute for Foreign and International Criminal Law", which are distributed in cooperation with the publisher Duncker & Humblot:

- "Strafrechtliche Forschungsberichte" (Reports on Research in Criminal Law),
- "Kriminologische Forschungsberichte" (Reports on Research in Criminology),
- "Interdisziplinäre Forschungen aus Strafrecht und Kriminologie" (Reports on Interdisciplinary Research in Criminal Law and Criminology),
- "Publications of the Max Planck Partner Group for Balkan Criminology", and
- "Sammlung ausländischer Strafgesetzbücher in deutscher Übersetzung" (Collection of Foreign Criminal Laws in German Translation).

These publications can be ordered from the Max Planck Institute at www.csl.mpg.de or from Duncker & Humblot at www.duncker-humblot.de.

Two additional subseries are published directly by the Max Planck Institute for Foreign and International Criminal Law: "research in brief" contains short reports on results of research activities, and "Arbeitsberichte" (working materials) present preliminary results of research projects. These publications are available at the Max Planck Institute.

Detailed information on all publications can be found at www.csl.mpg.de.

Max-Planck-Institut für ausländisches und internationales Strafrecht

Auswahl weiterer noch lieferbarer Publikationen aus der kriminologischen Veröffentlichungsreihe K:

K 169 *Andreas Schwedler, Gunda Wössner*
Elektronische Aufsicht bei vollzugsöffnenden Maßnahmen
Implementation, Akzeptanz und psychosoziale Effekte des baden-württembergischen Modellprojekts
Berlin 2015 • 126 Seiten • ISBN 978-3-86113-252-3 € 25,-

K 168 *Daniela Hunold*
Polizei im Revier
Polizeiliche Handlungspraxis gegenüber Jugendlichen in der multiethnischen Stadt
Berlin 2015 • 288 Seiten • ISBN 978-3-86113-251-6 € 35,-

K 167 *Christopher Murphy*
"Come in Spinner" – Money Laundering in the Australian Casino Industry
Berlin 2014 • 152 Seiten • ISBN 978-3-86113-250-9 € 29,-

K 166 *Ramin Tehrani*
Die „Smart Sanctions" im Kampf gegen den Terrorismus und als Vorbild einer präventiven Vermögensabschöpfung
Berlin 2014 • 256 Seiten • ISBN 978-3-86113-247-9 € 35,-

K 165 *Daniela Cernko*
Die Umsetzung der CPT-Empfehlungen im deutschen Strafvollzug
Eine Untersuchung über den Einfluss des Europäischen Komitees zur Verhütung von Folter und unmenschlicher oder erniedrigender Behandlung oder Strafe auf die deutsche Strafvollzugsverwaltung
Berlin 2014 • 455 Seiten • ISBN 978-3-86113-246-2 € 39,-

K 164 *Franziska Kunz*
Kriminalität älterer Menschen
Beschreibung und Erklärung auf der Basis von Selbstberichtsdaten
Berlin 2014 • 387 Seiten • ISBN 978-3-86113-244-8 € 35,-

K 163 *David Jensen*
Maras
A study of their origin, international impact, and the measures taken to fight them
Berlin 2013 • 245 Seiten • ISBN 978-3-86113-243-1 € 35,-

K 161 *Gunda Wössner, Roland Hefendehl, Hans-Jörg Albrecht* (Hrsg.)
Sexuelle Gewalt und Sozialtherapie
Bisherige Daten und Analysen zur Längsschnittstudie „Sexualstraftäter in den sozialtherapeutischen Abteilungen des Freistaates Sachsen"
Berlin 2013 • 274 Seiten • ISBN 978-3-86113-241-7 € 35,-

K 159 *Andreas Armborst*
Jihadi Violence
A study of al-Qaeda's media
Berlin 2013 • 266 Seiten • ISBN 978-3-86113-119-9 € 35,-

Max-Planck-Institut für ausländisches
und internationales Strafrecht

Auswahl weiterer noch lieferbarer Publikationen aus der kriminologischen Veröffentlichungsreihe K:

K 158 *Martin Brandenstein*
Auswirkungen von Hafterfahrungen auf Selbstbild und Identität rechtsextremer jugendlicher Gewalttäter
Berlin 2012 • 335 Seiten • ISBN 978-3-86113-118-2 € 35,-

K 157 *Ghassem Ghassemi*
Criminal Policy in Iran Following the Revolution of 1979
A Comparative Analysis of Criminal Punishment and Sentencing in Iran and Germany
Berlin 2013 • 265 Seiten • ISBN 978-3-86113-116-8 € 35,-

K 156 *Gunther Olt*
Pressefreiheit im Kontext strafrechtlicher Ermittlungsmaßnahmen
Berlin 2013 • 265 Seiten • ISBN 978-3-86113-114-4 € 35,-

K 155 *Anna-Maria Getoš*
Politische Gewalt auf dem Balkan
Schwerpunkt Terrorismus und Hasskriminalität: Konzepte, Entwicklungen und Analysen
Berlin 2012 • 330 Seiten • ISBN 978-3-86113-113-7 € 35,-

K 154 *Alke Glet*
Sozialkonstruktion und strafrechtliche Verfolgung von Hasskriminalität in Deutschland
Berlin 2011 • 387 Seiten • ISBN 978-3-86113-112-0 € 35,-

K 153 *Mustafa Abdelbaqi*
The Administration of Criminal Justice in Palestine
Berlin 2011 • 387 Seiten • ISBN 978-3-86113-105-2 € 35,-

K 152 *Tim Lukas*
Kriminalprävention in Großsiedlungen
Berlin 2010 • 315 Seiten • ISBN 978-3-86113-104-5 € 35,-

K 151 *Carolin Quenzer*
Jugendliche und heranwachsende Sexualstraftäter
Berlin 2010 • 292 Seiten • ISBN 978-3-86113-103-8 € 35,-

K 150 *Ying Wang*
Der strafrechtliche Schutz des Urheberrechts
Eine vergleichende Untersuchung zum deutschen und chinesischen Strafrecht
Berlin 2011 • 262 Seiten • ISBN 978-3-86113-102-1 € 35,-

K 149 *Anne Wildfang*
Terrorismus
Berlin 2010 • 304 Seiten • ISBN 978-3-86113-101-4 € 31,-

K 148 *Figen Özsöz*
Rechtsextremistische Gewalttäter im Jugendstrafvollzug
Berlin 2009 • 284 Seiten • ISBN 978-3-86113-100-7 € 35,-

Max-Planck-Institut für ausländisches und internationales Strafrecht

Auswahl aktueller Publikationen aus der kriminologischen Reihe BC und der interdisziplinären Reihe I:

BC 6 *Reana Bezić*
Juvenile Delinquency in the Balkans
A Regional Comparative Analysis based on the ISRD3-Study Findings
Berlin 2020 • 201 Seiten • ISBN 978-3-86113-281-3 € 32,-

BC 5 *Filip Vojta*
Imprisonment for International Crimes
An Interdisciplinary Analysis of the ICTY Sentence Enforcement
Berlin 2020 • 375 Seiten • ISBN 978-3-86113-280-6 € 40,-
Ausgezeichnet mit der Otto-Hahn-Medaille der Max-Planck-Gesellschaft

BC 4 *Gorazd Meško, Eszter Sárik, Anna-Maria Getoš Kalac* (eds.)
Mapping the Victimological Landscape of the Balkans
A Regional Study on Victimology and Victim Protection with a Critical Anylysis of Current Victim Policies
Berlin 2020 • 581 Seiten • ISBN 978-3-86113-279-0 € 46,-

BC 3 *Lucija Sokanović*
Fraud in Criminal Law
A Normative and Criminological Analysis of Fraudulent Crime in Croatia and the Regional Context
Berlin 2019 • 280 Seiten • ISBN 978-3-86113-273-8 € 35,-

BC 2 *Sunčana Roksandić Vidlička*
Prosecuting Serious Economic Crimes as International Crimes
A New Mandate for the ICC?
Berlin 2017 • 530 Seiten • ISBN 978-3-86113-264-6 € 44,-

BC 1 *Anna-Maria Getoš Kalac, Hans-Jörg Albrecht, Michael Kilchling* (eds.)
Mapping the Criminological Landscape of the Balkans
A Survey on Criminology and Crime with an Expedition into the Criminal Landscape of the Balkans
Berlin 2014 • 540 Seiten • ISBN 978-3-86113-248-6 € 44,-

I 27 *Zishi Zhou*
Das Sexualstrafrecht in Deutschland und China
Eine vergleichende Darstelung von Geschichte, Stand und Entwicklung
Berlin 2020 • 270 Seiten • ISBN 978-3-86113-278-3 € 35,-

I 25 *Chenguang Zhao*
The ICC and China
The Principle of Complementarity and National Implementation of International Criminal Law
Berlin 2017 • 245 Seiten • ISBN 978-3-86113-266-0 € 35,-

I 23 *Hans-Jörg Albrecht* (Hrsg.)
Kriminalität, Kriminalitätskontrolle, Strafvollzug und Menschenrechte
Internationales Kolloquium zum Gedenken an Professor Dr. Günther Kaiser am 23. Januar 2009
Berlin 2016 • 176 Seiten • ISBN 978-3-86113-258-5 € 30,-